4th International Conference on Civil Engineering Research (ICCER 2020)

AA007710

IOP Conference Series: Materials Science and Engineering
Volume 930

Surabaya, Indonesia
22 - 23 July 2020

ISBN: 978-1-7138-4767-0
ISSN: 1757-8981

Printed from e-media with permission by:

Curran Associates, Inc.
57 Morehouse Lane
Red Hook, NY 12571

proceedings.com

Some format issues inherent in the e-media version may also appear in this print version.

This work is licensed under a Creative Commons Attribution 3.0 International Licence.
Licence details: http://creativecommons.org/licenses/by/3.0/.

No changes have been made to the content of these proceedings. There may be changes to pagination and minor adjustments for aesthetics.

Printed with permission by Curran Associates, Inc. (2022)

For permission requests, please contact the Institute of Physics at the address below.

Institute of Physics
Dirac House, Temple Back
Bristol BS1 6BE UK

Phone: 44 1 17 929 7481
Fax: 44 1 17 920 0979

techtracking@iop.org

Additional copies of this publication are available from:

Curran Associates, Inc.
57 Morehouse Lane
Red Hook, NY 12571 USA
Phone: 845-758-0400
Fax: 845-758-2633
Email: curran@proceedings.com
Web: www.proceedings.com

TABLE OF CONTENTS

IOP Conf. Series: Materials Science and Engineering **930** (2020) 011001 doi:10.1088/1757-899X/930/1/011001

PREFACE

The Fourth International Conference on Civil Engineering Research (ICCER) is a biannual international conference hosted by Department of Civil Engineering of Sepuluh Nopember Institute of Technology with the theme of *Towards a Safe and Sustainable Construction using Advanced Materials and Technology for Structural Failures Prevention, Rehabilitation, and Retrofitting*. Departing from the theme, variety of topics related are presented by five keynote speakers to address the common challenges in building and infrastructure development as well as critical assessments which are interrelated with technological advancement both in materials and structures.

With the world facing unprecedented global CoVid-19 pandemic which is still undergoing and ramping up considerably, however, it is not possible– if not too risky– to hold an international conference in a conventional way. Gathering keynote speakers, presenters, and participants from around the world are considered harmful as it would trigger a higher possibility of people getting imposed by the Coronavirus. The situation is also exacerbated by the current travel restrictions applied by the majority of the countries which hence prevent the participants, either local or international, to travel from their origin country or region. Given that the situation of CoVid-19 in Indonesia and beyond is still uncertain with the cases growing drastically and not showing any sign of decreasing shortly, the possibility of the face-to-face conference is extremely low even when being postponed at later dates. Also, pressing the conference on hold by an uncertain amount of time would have knock-on effects on the upcoming ICCER conferences in the future. Therefore, it is of sensible (right) call that we organise this year Fourth ICCER in a virtual format.

The Fourth ICCER was held in two days on exact dates of 22 to 23 July 2020. It was run virtually from the server of the Department of Civil Engineering at Sepuluh Nopember Institute of Technology in Surabaya, Indonesia, a similar location as of the organising committee. The technology used to deliver the conference was license 'Zoom Video Conferencing' software which was personally preferred due to comprehensive tools and user-friendly interface. Accordingly, Zoom is also known as the leader in modern enterprise video communications, with an easy and reliable cloud-based platform for video and audio conferencing, thereby making it suitable for the optimised virtual conference that the organising committee envisaged.

In the Fourth ICCER, a total of 174 participants were sighted on Zoom with 94 identified as authors, whilst 80 recognised as the audience. The location of participants varied with the majority was from Indonesia, while others were from the United Kingdom, Thailand, Japan, Malaysia, Australia, and Prague. The conference model was generally similar to that of the conventional (face-to-face) method where there was an opening ceremony, sponsorship sessions, keynote sessions, plenary sessions, and a closing ceremony. In keynote sessions, the typical time allocated for keynote speaker was 60 minutes for presentation followed by 30 minutes for question and answer (Q&A) session. In plenary (parallel) sessions, 15 virtual

Content from this work may be used under the terms of the Creative Commons Attribution 3.0 licence. Any further distribution of this work must maintain attribution to the author(s) and the title of the work, journal citation and DOI.
Published under licence by IOP Publishing Ltd

rooms (links) were provided; each of which was detailed with information of scheduled presentations which typically consisted of 9 authors/presenters. Each presenter was given 10 minutes to disseminate their research with the video and screen share on as a means of drawing people attention, thereby creating a more active virtual environment. After the presentation, a 5-minute Q&A session was led by the moderator of the room in which all the questions from the audience were read from the chatbox. For more information on the technical quality and success of delivery of the virtual conference, please visit our website (iccer.ce.its.ac.id).

Last but not least, please allow us to thank you all for participating in the ICCER conference. Despite organising in a virtual format, we hope that the essence of this conference is much as similar in many respects as the conventional conference.

With kind regards
Chair of Organising Committee
Prof Priyo Suprobo
Sepuluh Nopember Institute of Technology, Indonesia

Peer review declaration

All papers published in this volume of *IOP Conference Series: Materials Science and Engineering* have been peer reviewed through processes administered by the Editors. Reviews were conducted by expert referees to the professional and scientific standards expected of a proceedings journal published by IOP Publishing.

- **Type of peer review: Single-blind**
  The Fourth International Conference on Civil Engineering Research (ICCER) adopts the single-blind peer review in which the reviewers are aware of the identity of the author(s), while the latter is not informed about the identity of reviewers. This is preferred to keeping our reviewers anonymous which therefore allow them to critique work without being influenced by the authors, providing them with more space to judge research more objectively.

- **Conference submission management system:**
  WP (WordPress) Abstract. The website (iccer.ce.its.ac.is) is specifically developed and designed for the first time by means of interactive tools to provide convenient access to committees to organise the filing system as well as reviewers and authors to review and submit the papers respectively.

- **Number of submissions received:** 111 papers

- **Number of submissions sent for review:** 111 papers

- **Number of submissions accepted:** 94 papers

- **Acceptance Rate (Number of Submissions Accepted / Number of Submissions Received X 100):** 84.68%

- **Average number of reviews per paper:** 2

- **Total number of reviewers involved:** 36

- **Any additional info on review process:**
  In addition to single-blind review, similarity and plagiarism detection checks by means of commercially available iThenticate software are also taken into account to verify the originality of written work. Papers will be reverted to authors as if the similarity index is beyond the limit specified. As if this encountered, authors are obliged to provide further amendments and send the revised version of their work to the editor.

- **Contact person for queries:**
Dr Trihanyndio Rendy Satrya, E-mail: rendy_star@yahoo.com
Dr Asdam Tambusay, E-mail: asdam@its.ac.id

Content from this work may be used under the terms of the Creative Commons Attribution 3.0 licence. Any further distribution of this work must maintain attribution to the author(s) and the title of the work, journal citation and DOI.
Published under licence by IOP Publishing Ltd

Developing activities of green design, green purchasing, and green transportation as the part of green supply chain management in construction sector

L B Setyaning[1*], I P A Wiguna[1], and F Rachmawati[1]

1 Department of Civil Engineering, Faculty of Civil, Planning and Geo Engineering, Sepuluh Nopember Institute of Technology, Surabaya, East Java, Indonesia.

*Corresponding authors: btari.18031@mhs.its.ac.id

Abstract. Green Supply Chain Management (GSCM) is a concept that integrates environmental factors into the traditional SCM started from the process of design, material procurement, production, distribution and end-of life. The benefits of implementing GSCM does not only improve the company's environmental performance, but also improve economic performance such as reducing operational costs and organizational performance such as improving the company's image. In manufacture and automotive industry, GSCM is able to improve company performance, therefore construction industry can consider of adopting this approach. Some GSCM practices in construction sector include green design, green purchasing, green transportation, green construction, green warehousing, green recycling and facilitating green practices. The aim of this study is to identify relevant green design, green purchasing and green transportation practices as part of GSCM if applied in the construction sector. The research variables were obtained through a literature study on GSCM both in construction sector and other sectors, then the variables were validated by respondents, practitioners in construction sector. Based on three respondents, there are ten green design practices, six green purchasing practices and five green transportation practices that are relevant to the construction sector, especially building projects.

1. Introduction

Green Supply Chain Management (GSCM) is a concept that integrates environmental factors into traditional SCM started from the process of design, material procurement, production, distribution and management of products after their useful life [1] while in the construction sector the series of activities are planning, design and production on site [2]. An environmentally friendly construction project is the result of environmentally friendly practices carried out at each stage of the supply chain by each stakeholder started from the design stage to the demolish stage, so stakeholders must be able to work together and align their respective interests [3]. Green practices in the construction sector include green design, green purchasing, green transportation, green warehousing, green construction, green recycling and green practices [4], [5], [6].

GSCM implementation provides several benefits related to the environment, namely reducing waste production, saving energy consumption, reducing the use of hazardous materials, increasing the use of reuse and recycle materials and reducing the level of carbon emissions. Other benefits that are

Content from this work may be used under the terms of the Creative Commons Attribution 3.0 licence. Any further distribution of this work must maintain attribution to the author(s) and the title of the work, journal citation and DOI.
Published under licence by IOP Publishing Ltd

not related to environment are increased customer satisfaction, improved corporate image, expanding market share, and increased customer social responsibility. Beside benefits to environment, cost reduction is the most important and the main benefit at once for the organization. If the implementation of green innovation is made broadly and effectively, there will be a decrease in costs of operational activities, both in the manufacturing and construction sectors [6].

The implementation of GSCM in construction sector that can be proved from limited research with GSCM in construction sector topic is still rare. According to Balasubramanian and Shukla [4], research about GSCM in construction sector is still fragmented and they only use single view from supply chain stakeholder for example contractor or developer perspective and just explore one of stage from supply chain stage, for example green transportation or green purchasing. The study about GSCM in construction sector is important because this study provides a whole understanding about steps that must be done to make this sector more. From explanation above about limited research and implementation of GSCM in construction sector especially in Indonesia, the aim of this study is to get relevant green practices of green design, green purchasing and green transportation as part of GSCM so that the practices can be well implemented in the construction sector.

2. Literature review

2.1. Green supply chain management (GSCM)

Supply Chain is described as the activity to meet consumer needs containing flow and goods transformation started from raw material until accepted by end consumer, and there are flow and money transformation too [7]. Activities in supply chain are looking for raw materials, manufacturing and assembling, storage, list the order and tracking, distribution until accepted by consumer [8], while in construction sector the activities are planning, design and on site production [2]. Supply chain in construction sector is more complicated, diverse and fragmented if compared to another sector because it involves many stakeholders [9]. Green Supply Chain Management (GSCM) is a concept that involves environmental consideration into traditional SCM [1] both in upstream and downstream [10], while Jung [11] described GSCM as a primary effort to integrate environmental requirement with SCM. GSCM is an important organization philosophy which plays a significant role in supporting efficiency and synergy between partner, facilitating environmental performance, minimize waste, cost saving for corporate profits and market share through risk reduction and environmental impact [12].

2.2. GSCM Practices

GSCM practices are practices carried out by companies to minimize adverse environmental impacts associated with the activities they undertake [4]. The environment is affected at every stage of the production process [13] and it is very important for companies to ensure the production process and supply chain implement green practices to minimize environmental impacts [14]. In several previous studies, each GSCM researcher in the construction sector grouped GSCM practices into several sections (see Table 1). In this study, the practices of GSCM that will be discussed are green design, green purchasing and green transportation.

2.2.1. Green design. Green design or better known as eco-design consists of activities designed to minimize the adverse impact on the environment of products and services during their lifetime [14]. The design consists of planning in which the material will be purchased, how to make products from the material and how products that have expired will be discarded [6]. The selection of appropriate materials and components can minimize negative impacts on the environment, such as the use of prefabricated components which can increase the ease of construction as well as reduce waste. In addition, the selection of the right material also influences the demolis stage, namely the amount of material and components that can be recycled increases [15], so that energy and resource consumption, air emissions, water pollution and waste generation can be minimized [16].

2.2.2. Green purchasing. Green purchasing aims to ensure that purchased materials and components have environmentally friendly characteristics such as the ability to be reused, the ability to be recycled and the absence of hazardous materials [14]. Green purchasing requires companies to consider aspects of sustainability in the material purchasing process in addition to aspects in traditional purchasing such as cost, quality and reliability [17].

Table 1. GSCM practices in the construction sector.

GSCM Practices	Balasubramanian and Shukla [4]	Shurab et. al. [5]	Ali et al [6]
Green Design	√	√	√
Green Purchasing	√	√	√
Green Transportation	√		√
Green Construction	√	√	√
Green Recycling	√		√
Green Warehousing			√
Facilitating Green Practices	√	√	

2.2.3. Green transportation. Green transportation is an activity related to transportation and carried out by the construction sector with the aim of minimizing negative impacts on the environment [18]. According to Ng et. al. [15] 6-8% of carbon produced during the construction process comes from transportation material, therefore it is necessary to apply green practices related to transportation such as transporting material to full trucks and using fuel-efficient vehicles. In addition, the use of technology such as video calls during meetings, the use of public transportation for employees, and choosing the location of employee residence close to the project location can also reduce the adverse effects of transportation activities [18], [19]. Green transportation can save expenses and contribute a lot to the development and economic sustainability.

3. Research methodology

The research process started from background explanation, problem formulation and research objectives, then continued to a literature study on GSCM both in construction sector and other sectors to obtain research variables. The research questionnaire was based on research variables to validate activities in green design, green purchasing and green transportation practices that were relevant for construction sector, especially building projects in Indonesia. Respondents in this study were practitioners in construction sector involved in building projects with green construction concept namely Grand Sungkono Lagoon and Grand Shamaya. Questionnaires that had been filled in by respondents were analyzed. Activities in green design, green purchasing and green transportation practices were relevant to the construction sector especially building projects if:

a. Three respondents stated it was relevant.
b. Two out of three respondents stated it was relevant.

4. Result and discussion

There were three respondents in this study who were practitioners in building projects with green construction concept. The respondent profiles can be seen in table 2.

Table 2. Respondent profiles.

Respondent	Company	Working Experience
Site Engineer Manager	Contractor	5-10 years
Site Operational Manager	Contractor	5-10 years
Project Director	Developer	> 15 years

Research data were obtained by meeting the respondent one by one to fill out the questionnaire. The questionnaire was filled out immediately so that respondents could ask questions if there were variables in the questionnaire that they did not understand. If the respondent could not fill out the questionnaire at that time, then an appointment would be made. Analysis of the results of the questionnaire was carried out after all the questionnaires had been filled out with the method described in the research methodology chapter.

4.1. Green design

Out of 14 green design practices *obtained from literature study*, there are only 10 relevant practices to be applied in building projects which can be seen in table 3. The four irrelevant practices are designing a building with less material and energy consumption, consideration of the ease of the demolish stage, consideration of building life cycle and incorporating consumer feedback in the eco-design phase.

Table 3. Green design questionnaire result.

Green Design Practices	Respondents			Conclusion
	1	2	3	
Provision for natural ventilation [4]	√	√	√	Relevant
Provision for natural lighting [4]	√	√	√	Relevant
Integration of photovoltaic panels [4], [5]	√	√	√	Relevant
Consideration for energy efficient lighting system [4], [5]	√	√	√	Relevant
Consideration for energy efficient heating and air conditioning (HVAC) systems [4], [5]	√	√	√	Relevant
Consideration of materials with high recycled content [4], [6], [16]	√	√	√	Relevant
Consideration of materials with low embodied energy [4], [16]	√	√	√	Relevant
Design a building with less material and energy consumption [6], [16]	√	-	-	Not relevant
Provision for the use of prefabricated components [4]	√	√	√	Relevant
Consideration to reduce the use of hazardous materials [4], [6]	√	√	√	Relevant
Consideration of the ease of the demolish stage [16]	-	-	√	Not relevant
Consideration of building life cycle [16]	-	-	√	Not relevant
Provision for waste water recycling [4], [5]	√	√	√	Relevant
Incorporate consumer feedback in the eco-design phase [6]	-	-	√	Not relevant

In Gharzeldeenand Beheiry's [20] research on the application of green design in building projects in UEA, there are 15 green design practices in which 5 practices are similar to green design practices in this study namely provision for natural ventilation, integration of photovoltaic panels, consideration for energy efficient heating and air conditioning (HVAC) systems, consideration of materials with high recycled content and provision for waste water recycling. The highest level of implementation is the consideration for energy efficient heating and air conditioning (HVAC) systems by 63.6% and the lowest is the integration of photovoltaic panels by 47.5%. In the study of Zhu et. al. [21] in nine manufacturing industries, the practice of designing products with harmless materials is part of eco design and has the highest level of implementation. Whereas in the Balasubramanian and Shukla [4] research specifically for the construction sector, the highest implementation is provision for the use of prefabricated components. Inhibiting factors in the application of green design include lack of knowledge about green parameters, lack of trust in recycled materials, high costs, lack of interest from both government and private parties and lack of green material inventory [20].

4.2. Green purchasing

From ten green purchasing practices obtained from the literature study, there are three practices that are not relevant to be applied in the construction sector especially building projects, namely providing design specifications to suppliers that include environmental requirements for purchased items, environmental audit for suppliers' internal management, and second-tier supplier environmentally friendly practice evaluation. It can be seen in table 4.

Table 4. Green purchasing questionnaire result.

Green Purchasing Practices	Respondents			Conclusion
	1	2	3	
Purchasing recycled material [5]	√	√	-	Relevant
Purchasing non-toxic material [5]	√	-	√	Relevant
Purchasing eco label material [21]	√	√	√	Relevant
Requires the adoption of an Environmental Management System (EMS) by suppliers to take part in tenders [5], [6]	√	√	√	Relevant
Requires suppliers to have ISO 14001 certification to be able to participate in tenders [5], [21]	√	-	√	Relevant
Requires the supplier to have prior experience in providing green material to be able to participate in the tender [5]	√	√	√	Relevant
Providing design specification to suppliers that include environmental requirements for purchased item [21]	√	-	-	Not relevant
Cooperation with suppliers for environmental objectives [21]	√	√	√	Relevant
Environmental audit for suppliers' internal management [6], [21]	√	-	-	Not relevant
Second-tier supplier environmentally friendly practice evaluation [21]	√	-	-	Not relevant

There are differences between the results of this study and the results of the study by Zhu et. al. [21], which providing design specifications to suppliers that include environmental requirements for purchased items obtain the highest level of implementation in nine manufacturing industries, whereas in this study the practice is considered irrelevant by respondents. Then for second-tier suppliers the environmentally friendly practice evaluation has the lowest level of implementation where nine manufacturing industries just consider the implementation, whereas in this study the practice is not relevant. The second highest level of implementation is requiring suppliers to have ISO 14001 certification to be able to participate in tenders. According to Murphy and Bendell [22] improving the environmental performance of an organization through the application of green purchasing is more complex compared to other green practices. Several steps were suggested by Murphy and Bendell [22] to overcome these difficulties including the application of EMS by suppliers and working with suppliers to increase awareness of the environment which both practices are presented in this study and are considered relevant by respondents. Lack of data on the environment in general is a major obstacle in implementing green purchasing and procurement.

4.3. Green transportation

Out of five green transportation practices taken from literature study, all of them are relevant if applied in the construction sector. It can be seen in table 5 below. In Balasubramanian and Shukla [4] research, the highest implementation was provision of accommodation to employees near project sites. The provision of mess for employees, especially daily workers near the project site, has many benefits for the company, namely workers are not tired in travelling from home to the project so that they can be more productive, workers rarely come late and the company more easily controls the cleanliness and health of workers' mess. The lowest implementation in the study was the use of video conferencing. This is a bit odd, because in an era with extremely rapid technological developments like today, the use of video conferencing is common and very easy to do, and its implementation does not require a large cost. According to Ali et. al. [6] green transportation can save expenses and contribute a lot to the development and economic sustainability. More importantly, green transportation forces its users to live a healthy lifestyle. With the reduction of pollution to the environment, especially the work environment, the immune system of workers will be better.

Table 5. Green transportation questionnaire result.

Green Transportation Practices	Respondents			Conclusion
	1	2	3	
Provision of accommodation to employees near project sites [4]	√	√	√	Relevant
Use of video conferencing [1]	√	-	√	Relevant
Employees are encouraged to use shared transport and public transport [4]	√	-	√	Relevant
Materials are transported in full truckload quantities [4]	√	√	√	Relevant
Materials are transported in fuel efficient vehicles [4]	√	√	√	Relevant

5. Conclusion

This study concludes that there are ten green design practices, six green purchasing practices and five green transportation practices that are relevant to be applied in the construction sector in Indonesia. These practices do not only come from the literature on GSCM in the construction sector, but also from the literature on GSCM in the manufacturing sector. Based on the research by Malviya and Kant [23] the manufacturing sector has benefited greatly from the adoption of GSCM, so the construction sector needs to consider participating in implementing GSCM. Further research could develop the detailed measurement scale to complete the measurement tools and then conduct empirical studies in construction projects to validate the practices.

Acknowledgement

The data related to this paper have been supported by PT. PP Persero (Tbk) and PT. PP Properti (Tbk) especially Grand Sungkono Lagoon and Grand Shamaya Project.

References

[1] Zhu Q, Sarkis J, and Lai KH 2007 Green Supply Management: Pressures, Practices And Performance Within The Chinese Automobile Industry *J. Clean. Prod* **15** pp 1041-52

[2] Winch G M 2010 *Managing Construction Projects :An Information Processing Approach2nd edition* (West Sussex: John Wiley & Sons)

[3] UN Global Compact Office and Business for Social Responsibility 2010 *Supply Chain Sustainability:A Practical Guide For Continuous Improvement*
[4] Balasubramanian S and Shukla V 2017 Green Supply Chain Management An Empirical Investigation On The Construction Sector *Supply Chain Manag.* **22** 58-81
[5] Shurrab, J Hussain M and Khan M 2018 Green and Sustainable Practices in The Construction Industry : A Confirmatory factor Analysis Approach *Eng. Constr. Archit. Manag.* **26** 1063-86
[6] Ali Y, Saad T B, Sabir M, Muhammad N, Salman A, and Zeb K 2019 Integration of Green Supply Chain Management Practices in Construction Supply Chain of CPEC *Manag. Enviro.l Qual.*
[7] Li S C M L 2007 *Concepts, Techniques And Practices Enhancing Value Through Collaboration* (Singapore: World Scientific Publishing Co. Pte. Ltd)
[8] Hervani A A, Helms M M, and Sarkis J 2005 Performance Measurement For Green Supply Chain Management *BM Int. J.* **12** 330-53
[9] Rezgui, Yand, Miles J 2009 Exploring The potential Of SME Alliances In The Construction Sector *J. Constr. Eng. Manag.* **136** 558-67
[10] Shipeng L 2011 A Study On Green Supply Chain Management Of Enterprises Based On Self-Locking Theory *Proceedings of International Conference on E-Business and E-Government (ICEE)* Shanghai pp 1–4
[11] Jung 2011 A Bibliometric Analysis On Green Supply Chain Management : A Pre-liminary Result *Proceedings Of The IEEE 13th Conference On Commerce and Enterprise Computing* Luxembourg pp 418–20.
[12] Rao P and Holt D 2005 Do Green Supply Chains Lead To Competitiveness And Economic Performance? *Int. J. Oper. Prod. Manag.* **25** 898-916.
[13] Walker H, Di S L, and Mc Bain D 2008 Drivers And Barriers To Environmental Supply Chain Management Practices: Lessons From The Public And Private Sectors *J. Purch. Supply Manag.* **14** 69-85.
[14] Rostamzadeh R, Govindan K, Esmaeili A, and Sabaghi M 2015 Application of Fuzzy VIKOR For Evaluation of Green Supply Chain Management Practices *Ecol. Indic.* **49** 188-203
[15] Ng S T, Wong J M W, Skitmore S, and Veronika A 2012 Carbon Dioxide Reduction in The Building Life Cycle: A Critical Review *Eng. Sustain.* **165** 281-92
[16] Mitra S 2014 A Framework For Research On Green Supply Chain Management *Supply Chain Forum : An International Journal* **15** 34-51
[17] Eltayeb T K, Zailani S, and Ramayah T 2011 Green Supply Chain Initiatives Among Certified Companies in Malaysia and Environmental Sustainability: Investigating The Outcomes *Resour. Conserv. Recy.* **55** 495-506
[18] BRE Department of Trade and Industry United Kingdom 2003 *Construction Site Transport*
[19] Tema Nord Nordic Council of Ministres 2010 *Innovative Green Public Procurement OfConstruction, IT and Transport Services in Nordic Countries*
[20] Gharzeldeen M N and Beheiry S M 2014 Investigating the use of green design parameters in UAE construction projects *Int. J. Sustain. Eng.*
[21] Zhu Q, Sarkis J, and Geng Y 2005 Green Supply Chain Management In China: Pressures, Practices And Performance *Int. J. Oper. Prod. Manag.* **25** 449-68
[22] Murphy D F and Bendell J 1998 Do It Yourself Or Do It Together? The Implementation Of Sustainable Timber Purchasing Policies by DIY Retailers In The UK *Greener Purchasing. Opportunities And Innovations* 118–34
[23] Malviya R K and Kant R 2015 Green Supply Chain Management (GSCM): A Structured Literature Review And Research Implications *BM. Int. J.* **22** 1360-94

Risks of Public Procurement for Construction Works

A O F Dita[1*], M A Rohman[1], and C B Nurcahyo[1]

[1] Department of Civil Engineering, Faculty of Civil, Planning and Geo Engineering, Institut Teknologi Sepuluh Nopember, Surabaya, East Java, Indonesia.

*Corresponding author's e-mail: arif@ce.its.ac.id

Abstract. Conducting a public procurement process is prone to conflict if it deviates from procurement principles. Based on Presidential Regulation number 16 of 2018, the public procurement principles are efficient, effective, transparent, open, competitive, fair, and accountable. Problems in public procurement process may occur at every stage, starting from identifying procurement needs to delivering the results of the completed work. Many problems are likely to occur if risk identification is not implemented at public procurement process. If the risk is not identified early, it will be a potentially significant budget loss of government's construction projects. This paper is an initial research to discover the risk variables in the public procurement process. The purpose of this study is to identify risks that may arise in the process of public procurement. Research variables in this study were obtained from related literature studies. The data in this study were acquired from questionnaires and interviews with several parties related with the public procurement. Mean and Standard Deviation (SD) were employed to assess the relevant variables from questionnaires. According to the experts' opinion, there were 13 variables in the Planning and Preparation Stages and 9 variables in the Implementation Stage.

1. Introduction

A procurement is the client's effort to obtain goods or services from prospective suppliers using certain methods and processes for determined budget, time, quality and others. Public procurement process involves several related parties who have to comply with ethics, norms and principles to regulate the procurement. In contrast with the past, procurement process is currently conducted as an internet-based activity named as e-procurement. E-procurement refers to procurement processes carried out online and through information and networking systems, which can provide several benefits such as reducing cycle time, levelling uneven expenses, increasing visibility, and lowering costs as outputs of efficient, integrated and automated processes [1]. E-procurement provides a sense of security and comfort in the procurement process because it has to comply with the rules regulated electronically by prioritizing the procurement principles. E-procurement system is expected to resolve manual procurement problems which are severe condition of corruption, collusion, and nepotism, because e-procurement system restricts physical contact between client and prospective suppliers in Procurement Service Unit. All activities are recorded by ICT system which makes it easier to be checked and supervised.

Inadequate governance in procurement process can interfere the implementation of construction work, cause detrimental effects to stakeholders, and inflict impact on the owner budget. Close interaction between client and prospective suppliers makes public procurement process have several risks. Many cases of government's construction works were behind schedule, over budget, low quality, and inefficient because the public procurement process was deviated from the procurement principles. Public

Content from this work may be used under the terms of the Creative Commons Attribution 3.0 licence. Any further distribution of this work must maintain attribution to the author(s) and the title of the work, journal citation and DOI.
Published under licence by IOP Publishing Ltd

procurement process has three stages, in which there are several elements of risk inherent in every stage. Therefore, client needs to deliver an exhaustive risk management to handle the impact to optimize the chances of procurement process to succeed.

Important aspects that must be considered in analysing risk are finding and differentiating the problems that generate risks [2]. The first step of risk management is identifying risks. Risk identification is used to find risk variables that may arise in public procurement process. Risks are explained as events that can arise throughout the project and affect critical factors of the project [35]. Furthermore, managing the risks associated with public procurement process is very important for the success of the upcoming construction work [36]. However, the rules and regulations set by the government are not easy to implement as the intended purpose.

Each construction work cannot avoid risk [3] because the project has dynamic and complex nature with a lot of uncertainties [4]. Public procurement process becomes more complex because of the involvement of various parties, including the government as the client and the public sectors as the prospective suppliers. The risk management approach is considered able to provide a concept that is appropriate enough to see the various problems in the public procurement process. Risk management is a proactive approach so it is expected to minimize negative impacts and maximize positive impacts. The risk management approach can also provide a way to find out the risk response and risk allocation to the eligible party. An appropriate risk response can be applied to maintain or transfer the negative impacts [5]. By using a risk concept approach, problems that might occur in the public procurement process can be seen from a broader view and make it easier to determine appropriate handle of the risk priorities.

Based on the description above, this research takes a risk management approach to solve the problems occurring in public procurement process. Risk variables were obtained from various literature studies. The data in this study were collected by questionnaires and interviews with several parties involved in public procurement process. Identifying risks was conducted to observe risk variables that may arise in public procurement process and to determine the next steps for further analysis. In addition, risk identification can contribute to related parties before the implementation of public procurement process.

2. Literature review

2.1. Public procurement

Indonesia as a developing country continues to carrying out its national development efforts, especially in infrastructure sector. Infrastructure development programs are implemented by the national and local governments. Continuous economic growth in Indonesia has resulted in the construction industry rise. According to data released by the Ministry of Finance, there are about 5.9% year-on-year growth of infrastructure development budget provided by the national government, amounted to 423.3 trillion in 2020.

Every construction project undertaken by the government needs a good public procurement process. Procurement is an activity to acquire goods or services in a transparent, effective, and efficient manner as well as compliance with the needs and desires of the client [6]. Public procurement process needs involvement from several parties who have to comply with ethics, norms, and principles in its implementation.

Procurement is client's endeavour to obtain or fulfil the desired goods or services using certain methods and processes to reach an agreement on price, time, and other arrangements [2]. Traditional procurement process was implemented manually by face-to-face interactions, but nowadays procurement process is conducted by electronic systems named as e-procurement [19]. E-procurement is the process of procurement by utilizing information and communication technology facilities as the statutory regulations [33]. The parties involved in the procurement process can be classified into three main participants, namely: Client, Supplier, and Procurement Service Unit [7].

Each client uses different methods of obtaining goods or services, depending on the type and quantity of goods or services to be purchased, the location of the purchase, budget availability, and other considerations [8]. The methods of selecting suppliers are e-purchasing, direct procurement, direct appointment, quick tender, and tender [9]. E-purchasing is a procedure for purchasing goods and services that have been listed at an official electronic catalogue system. Direct Procurement is a method of direct elections to the goods or services prospective suppliers without any auction process or selection. Direct Appointment is a method of selection by directly appointing a goods or services prospective supplier in certain circumstances. Quick Tender is a selection process of qualified suppliers listed in Client's Suppliers Performance Information Systems. Moreover, tender is an open selection of suppliers of goods or construction works and other services that meet the predefined requirements.

Tender is a series of bidding activities carried out to select a prospective supplier that is suitable and feasible to do a job [10]. The tenders examined in this study are a series of selection, evaluation, election, and determination of prospective suppliers to perform construction work funded by government and related on applicable government regulations in its throughout process. This study takes the process of selecting a tender for the construction work as the object because it can represent all the stages of the process and other methods, and tender can be followed by all potential suppliers of goods or services. The stages of public procurement process are divided into three stages, namely preparation, planning, and implementation [9]. The stages are arranged to the public procurement process, but there are still various problems and violations even though it has been carried out electronically [11]. Collusion carried out on public procurement process can create fake competition among bidders [12].

Violation caused by weak governance can obstruct competition in procurement markets and raise the prices paid by the administration for acquiring goods or services, and as the result, it will have a direct impact on higher public expenditure [13]. Referring to several cases of procurement, violations that often occur are inaccuracy in technical specifications and owner estimate (OE) price documents [14]. The common procedure for selecting suppliers for construction works in public procurement is traditionally decided based on the lowest bid price [15], [17], [19]. This method can lead to eliminating more qualified contractors, selecting inappropriate contractors, and accepting parties who cannot carry out the project successfully. As a result, it has impacts on poor project quality, schedule delays, and additional costs due to rework, leading to claims and disputes [18], [19], [20], [24]. Considering lowest bid prices as the sole selection criterion is not the best approach [21], [24]. Therefore [22], [24] the selection criteria have been classified into three aspects: 1. Technology: related to quality control capabilities, design, technical capabilities, and engineering solutions. 2. Bid prices: in the public sector, bid prices remain as a main criterion for selecting a prospective supplier due to public accountability of the client [23], [24]. 3. Health, safety and environment (HSE): this aspect includes the supplier's HSE policies, plans, and performances. In this case, this aspect focuses on supplier's ability to be socially responsible for its employees and the wider community affected by the construction work.

2.2. Review of public procurement process from the risk concept

Procurement is a risk-prone activity. Risk is an uncertain event or condition that can occur because of the current process or further process in the future [25]. Risk in procurement process is a condition faced by a person or institution when there is a possibility of negative consequence [26]. Risk Source is a fundamental driver that causes risks in a project or organization. In carrying out the business process, a client will encounter various risks that have impacts on its business process [27]. Therefore, Client's Procurement Work Unit must be responsible for its expenditure to manage public procurement process. Therefore, [28] the framework for procurement risk is divided based on the following principles: (i) based on a risk management process, (ii) applies to different organizational/entity levels, (iii) applies to all stages of procurement, (iv) sustainable and applied throughout all stages of procurement, (v) identified risks must be managed.

There are three important elements in public procurement risk: events or conditions, probability, and impact on objectives. Events or conditions indicate what may or will happen. Probability indicates the likelihood of an event or situation to occur. Impact on objectives signifies the consequences of the risk event. Risk analysis is used to determine the probability and impact of the identified risks. The approach to managing risk is known as risk management. However, risk management is becoming more complex not only because of the increase in various risks, but also because the relationship between risk variables and risk sources is rising [27]. Therefore, the importance of analysing risk is to separate small risks that can be accepted from large risks that have to be avoided. It also can be used as data for prioritizing risks. Many studies indicate that the highest risk level occurs during the bidding evaluation process [24]. The risk factors related to public procurement process potentially generate destruction, damage, or loss [24].

Even qualified suppliers still have to manage their own risk factors such as financial risk, inadequate tender management, and internal miscommunication among their employees [24].

Based on previous research, there are several discussions of risk identification in public procurement process. A research by [12] conducted to map the fraud risks in auditing process of the public procurement. The background of this research is the rampant corruption cases that occur in public procurement process. A qualitative descriptive method is used to analyse problems and subsequently provide problem solving. As research findings, there are 15 stages of cheating mode in public procurement process. These stages include procurement planning, appointment of procurement committee, selection of the procurement system, determination of schedules, preparation of Owner Estimate price, preparation of procurement documents, announcement of registration, determination of qualification stages for suppliers, aanwijzing (information session), bidding document opening, bidding evaluation, verification of suppliers' qualification, minutes of bidding result documents, award notification, objection and appeal of bidding result, contract signing, and handover of the completed work. One of the frauds committed in procurement planning is a conduction of a procurement which does not comply with the Client's real needs. Therefore, it is very important to understand the critical stages in the procurement process to find potential fraud at every stage. A study is conducted by [30] on the risk management analysis of procurement process in Klaten district. The background of the problems in this study is that the public procurement in Klaten has not fully paid attention to the procurement principles, resulting in 30 identified risk variables in the public procurement. The most dominant risk category that occurs is the legislative and regulatory frameworks. Meanwhile, other categories are between moderate and high risks.

Table 1. Risk variables of public procurement in the planning and preparation stages

No.	Variable	Source
A1	Public procurement does not comply with client's needs	[6], [12], [26]
A2	Political intervention/leadership or controlled by other parties	[2], [36]
A3	Error in determining procurement system (Method, Evaluation Criteria, and Document Submission)	[26]
A4	Client Project Manager does not understand the work package/activity items in the budget planning	[12], [26]
A5	Intervention from other parties in the preparation of bidding documents	[29]
A6	Lack of coordination between Client Project Manager and Procurement Service Unit resulting in the different perceptions at scope of work	[29], [34], [36]
A7	Incorrect goods or services technical specification determination by Client Project Manager	[2]
A8	Client Project Manager does not understand the construction method of the work being tendered	[26], [29]
A9	Different perception between Client Project Manager and Client for the tendered work	[29], [36]
A10	Misinterpreting the technical aspects of goods or services needed	[29]
A11	Lack of competence of officers and staffs in the preparation of Owner Estimate price	[26], [36]
A12	Biased specifications	[26]
A13	Procurement plan advantages certain parties	[2], [6]
A14	Inflated budget in procurement plan document	[2], [6], [12], [26]
A15	Procurement documents are leaked before the bidding process starts officially	[6], [29]
A16	Incorrect in splitting/merging the bidding scope package	[2], [12]
A17	Procurement plan is incomplete	[2], [6]
A18	Evaluation criteria are unfair and violate the regulations	[26]

A research by [29] focused on identification and analysis of the issues of public procurement for construction works. The purpose of this study is to identify any problems that occur in any procurement process of construction works, from preparation stage to implementation stage. As a result, there are 135 identified problems in procurement process. The main problems are the lack of technology in e-procurement, lack of details on rules and policies related to procurement, and limited capable human resources. Several risk variables for public procurement summarized from previous studies are presented in Table 1 and Table 2.

Table 2. Risk variables of public procurement in the implementation stage

No.	Variable	Source
B1	Technical problems of internet access for downloading and uploading documents	[29]
B2	Passive bidder	[29]
B3	Lack of communication among Client Project Manager, Procurement Service Unit, and the Group of Expert at the bidding stage	[2], [36]
B4	Procurement Service Unit does not understand the technical substance and important information that will be described in *aanwijzing* (information session)	[29]
B5	The ICT system is being hacked to intervene bidding	[12], [26]
B6	Client's Audit Officer (PPHP) doesn't understand the contents of the contract	[29]
B7	Status in the ICT system notifies a successfully uploading process, but the contents of document is empty when downloaded	[29]
B8	There is no client's periodic evaluation of the public procurement process	[2]
B9	Limiting bandwidth server to intervene the participants to submit the bidding documents	[30]
B10	Bidding file is corrupted and cannot be opened	[29]
B11	Procurement Service Unit delivers less accurate result due to excessive amount of bidding evaluation processes	[29]
B12	Leakage of Owner Estimate price including the detailed unit price analysis	[2], [36]
B13	No clear standards and criteria to determine fair-price evaluation	[12], [29]
B14	*Aanwijzing* (information session) duration is too short for complex construction work	[2]
B15	Acceptance criteria of goods or services are biased	[2]
B16	Not all client's officer understands the technical construction method	[2]
B17	Horizontal conspiracy among prospective suppliers to generate unfair competition	[2], [11]
B18	The content of the award notification does not meet the minimum standard of announcement	[12]

3. Research methodology

Methods used in this paper are risk identification, preliminary survey, and data analysis. After identifying the risk variables, a preliminary survey was conducted by involving four experts directly involved in the public procurement. The number of experts participated in survey is three or more experts, which will be considered as proof of content validity [37]. The experts involved in this preliminary survey are those who have experience at least 5 years in public procurement subject.

Preliminary survey was conducted to verify the relevance of identified variables from the literature study to this research. The rating scale used in this study was a scale of 1-5. This scale was used to calculate relevance attributes based on experts' opinion. Scale 1 represents a highly irrelevant variable and scale 5 represents a highly relevant variable. The next step of the preliminary survey was determining the sample, designing the questionnaire, conducting a trial survey, and distributing the questionnaire. Data obtained from the preliminary survey through a questionnaire were then analysed

The data analysis was carried out with validity and reliability tests. Validity test was conducted to determine the accuracy of the questionnaire. An instrument can be said valid if it successfully measures what must be measured and the results obtained can meet all the requirements of scientific research [31]. Meanwhile, a reliability test was needed to ensure that the measurement instrument has consistency to assess variables repeatedly. Reliability test shows how far the instrument can be used without bias (error free) [31]. Reliability was usually measured using the Cronbach Alpha coefficient and an instrument is said reliable if this coefficient is greater than 0.60 [32].

4. Result and analysis

This paper presents the results of preliminary survey to determine risk variables that may occur in the public procurement process of construction works. The result of preliminary survey was risk variable rank based on its relevance to public procurement process.

Data analysis from the experts' opinion was presented as mean and standard deviation (SD) which can be seen in Table 3 and Table 4. Because this study used a scale of 1-5, therefore, scale 3 was taken to determine the relevance because its value is the middle score between 1 to 5. Then, the variable was considered relevant if the mean of the variable is greater than three (>3). Based on the preliminary survey analysis, there were several risk variables omitted because they were considered irrelevant with values of less than three.

Table 3. Variable risk of public procurement in the planning and preparation stage results

No.	Variable	Rank	Mean	STD
A6	Lack of coordination between Client Project Manager and Procurement Service Unit results in the different perceptions at scope of work	1	4,00	0,00
A12	Biased specifications	2	4,00	0,00
A16	Incorrect in splitting/merging the bidding scope package	3	4,00	0,00
A17	Procurement plan is incomplete	4	4,00	0,00
A3	Error in determining procurement system (Method, Evaluation Criteria, and Document Submission)	5	3,75	0,50
A5	Intervention from other parties in the preparation of bidding documents	6	3,50	0,58
A18	Evaluation criteria are unfair and violate the regulations	7	3,50	1,00
A7	Incorrect goods or services technical specifications determination by Client Project Manager	8	3,25	0,96
A13	Procurement plan advantages certain parties	9	3,25	1,50
A15	Procurement documents are leaked before the bidding process starts officially	10	3,25	0,96
A2	Political intervention/leadership or controlled by other parties	11	3,00	1,41
A9	Different perceptions between Client Project Manager and Client for the tendered work	12	3,00	1,41
A11	Lack of competence of officers and staffs in the preparation of Owner Estimate price	13	3,00	1,83
A14	Inflated budget in procurement plan document	14	2,75	1,50
A8	Client Project Manager does not understand the construction method of the work being tendered	15	2,50	1,29
A10	Misinterpreting the technical aspects of goods or services needed	16	2,50	1,29
A4	Client Project Manager does not understand the work package/activity items in the budget planning	17	2,25	1,26
A1	Public procurement does not comply with client's need	18	2,00	1,41

There were five irrelevant risk variables at the preparation and planning variables. The irrelevant variables were Public procurement does not comply with client's need (A1), Client Project Manager does not understand the work package/activity items in the budget planning (A4), Misinterpreting the technical aspects of goods or services needed (A10), Client Project Manager does not understand the construction method of the work being tendered (A8), and Inflated budget in procurement plan document (A14).

In experts' opinion, the variable of public procurement does not comply with client's need (A1) is not relevant with the process because it is based on [9] paragraph 18 which had been regulated in the procurement process. In addition, [16] it has been regulated in Chapter 3 that the procurement planning consists of the task and authority, planning activities, and identification of the needs of goods or services. In the planning process, procurement has to comply with the national government in accordance with Short-Term, Medium-Term, and Long-Term Development Plans.

Then, Client Project Manager does not understand the work package/activity items in the budget planning (A4), Misinterpreting the technical aspects of goods or services needed (A10), Client Project Manager does not understand the construction method of the work being tendered (A8), Inflated budget in procurement plan document (A14) were considered irrelevant because all arrangements for planning and procurement preparations had been determined in the regulation [9] [16]. The preparation of a public procurement plan is the first step to increase the efficiency and effectiveness of a clear and profitable procurement, therefore, the implementation needs to be monitored and controlled.

In addition, there were 9 irrelevant risk variables at the implementation stage according to experts' opinion as the results of preliminary survey. The irrelevant variables were: Technical problems of internet access for downloading and uploading documents (B1), Procurement Service Unit does not understand the technical substance and important information that will be described in aanwijzing (information session) (B4), Bidding file is corrupted and cannot be opened (B10), Not all client's officer understands the technical construction method (B16), Passive bidder (B2), Limiting bandwidth server to intervene the participants to submit the bidding documents (B9), The ICT system is being hacked to intervene bidding (B5), Status in the ICT system notifies a successfully uploading process, but the contents of document are empty when downloaded (B7), The content of the award notification does not meet the minimum standard of announcement (B18)

Table 4. Variable risk of public procurement in the implementation stage results

No.	Variable	Rank	Mean	STD
B11	Procurement Service Unit delivers less accurate result due to excessive amount of bidding evaluation processes	1	4,00	0,00
B12	Leakage of Owner Estimate price including the detailed unit price analysis	2	4,00	0,00
B14	*Aanwijzing* (information session) duration is too short for complex construction work	3	4,00	0,00
B17	Horizontal conspiracy among prospective suppliers to generate unfair competition	4	3,75	0,50
B3	Lack of communication among Client Project Manager, Procurement Service Unit, and the Group of Expert at the bidding stage	5	3,50	0,58
B6	Client's Audit Officer (PPHP) doesn't understand the contents of the contract	6	3,50	1,73
B13	No clear standards and criteria to determine fair-price evaluation	7	3,50	1,00
B8	There is no client's periodic evaluation of the public procurement process	8	3,25	1,50
B15	Acceptance criteria of goods or services are biased	9	3,25	1,50
B1	Technical problems of internet access for downloading and uploading documents	10	2,75	1,50
B4	Procurement Service Unit does not understand the technical substance and important information that will be described in *aanwijzing* (information session)	11	2,75	1,50
B10	Bidding file is corrupted and cannot be opened	12	2,75	1,26
B16	Not all client's officer understands the technical construction method	13	2,75	1,50
B2	Passive bidder	14	2,50	1,29
B9	Limiting bandwidth server to intervene the participants to submit the bidding documents	15	2,50	1,29
B5	The ICT system is being hacked to intervene bidding	16	2,25	1,26
B7	Status in the ICT system notifies a successfully uploading process, but the contents of document are empty when downloaded	17	2,25	1,26
B18	The content of the award notification does not meet the minimum standard of announcement	18	1,75	0,96

Based on experts' opinion, variable of Bidding file is corrupted and cannot be opened (B10) is not the responsibility of the Procurement Service Unit. Prospective suppliers must follow document submission procedure generated by ICT system to minimize failures in uploading bidding documents.

Technical problems of internet access for downloading and uploading documents (B1), Limiting bandwidth server to intervene the participants to submit the bidding documents (B9), The ICT system is being hacked to intervene bidding (B5), Status in the ICT system notifies a successfully uploading process, but the contents of document are empty when downloaded (B7), and The content of the award notification does not meet the minimum standard of announcement (B18) variables considered irrelevant because they have been already regulated [9] [16] and afterward implemented using nationwide e-procurement system developed and managed by national government.

Other irrelevant variables are considered having a low influence on the purpose of public procurement, so it can be omitted from risk register.

In addition to the discussion of irrelevant variables, this study focuses on relevant variables based on experts' opinion. Experts declared that at the planning and preparation stages, there were 13 relevant variables in the public procurement process. The complete list of these variables can be seen in table 3. Four out of thirteen variables are claimed as the most relevant variables.

The first variable is Lack of coordination between Client Project Manager and Procurement Service Unit results in the different perceptions at scope of work (A6), which is caused by lack of maturity of tender preparation. Moreover, inconsistency in using many terms/names for certain items leads to different perceptions that inhibit procurement process and give significant impact on procurement objectives. Biased specifications (A12) is the second most relevant variable, caused by non-standardized process, choosing goods or services which are not suitable with current conditions, tendency to determine specification based on its availability (not based on actual need), and having less functional contribution.

The third variable is Incorrect in splitting/merging the bidding scope package (A16) to avoid open tender [29]. This variable might occur because there was an unprofessional planning process that leads to causing collusion and nepotism [29], causing unfair competition among prospective suppliers and giving benefit for certain parties. The last most relevant variable is Procurement plan is incomplete (A17) causing procurement process cannot be executed as soon as possible, so that eventually it is often carried out at the end of the fiscal year [29].

There were 9 variables considered relevant at the implementation stage according to experts' opinion. Three out of nine variables are determined as the most relevant variables. The first variable was Procurement Service Unit delivers less accurate result due to excessive amount of bidding evaluation processes (B11). This variable existed because a lot of bidding processes were conducted at the same time, in contrast to limited number of Procurement Service Unit staffs.

Leakage of Owner Estimate price including the detailed unit price analysis was the second variable considered the most relevant. This variable might occur because the existence of hidden interests, and no client's financial statements, and collusion [29] that leads to unhealthy procurement process. Aanwijzing (information session) duration is too short for complex construction work was the third most relevant variable. It was raised because Aanwijzing (information session) usually focuses on discussing the fulfilment of administrative requirements rather than that of the technical, afterward result in construction work delays due to lack of understanding of construction method [29].

This result may be explained by the fact that these variables are relevant in the public procurement for construction projects. These relevant variables can be further analysed to determine the level of risk in the public procurement process. Identifying variables is very important to do at an early stage because these variables become the objects of observation to be studied.

5. Conclusion and recommendation

In accordance with the research background and literature review conducted, there were 18 risk variables in the Planning and Preparation Stages and 18 risk variables in Implementation Stage of Public Procurement. Preliminary survey determined 13 variables at the Planning and Preparation Stages as relevant, and another 5 variables as irrelevant based on experts' opinion. On the other hand, for the Implementation Stage, 9 variables were found relevant to the Public Procurement process and the other 9 variables were claimed irrelevant based on experts' opinion.

Acknowledgements

The authors appreciate the experts involved in the preliminary survey of this research in 2020.

References

[1] Ramkumar M Tobias S and Mamata J 2016 Risk assessment of outsourcing e-procurement services: integrating swot analysis with a modified ANP-based fuzzy inference system *J. Prod. Plan. Control* **27** 1-20

[2] Sutedi A 2012 Aspek Hukum dalam Pengadaan Barang & Jasa and Berbagai Permasalahannya 2 Tarmizi (Jakarta: Sinar Grafika)

[3] Subramanyan H, Sawant P H and Bhatt V 2012 Construction project risk assessment: development of model based on investigation of opinion of construction project experts from india *J. Const. Eng. Manag.* 409–21

[4] Al-Kharashi A and Skitmore M 2009 Causes of delays in Saudi Arabian public sector construction projects *Const. Manag. Eco.* **27** 3–23

[5] Wang S Q, Dulaimi F and Aguria M Y 2004 Risk management framework for construction projects in developing countries *Const. Manag. Eco.* 22 237–52

[6] Arsana I (2016) Manajemen pengadaan barang dan jasa pemerintah *Deepublish* 315

[7] Rolas S, Toyfur MF and Juliantina I 2019 An analysis of the effect of job implementation methods in the failure of the selection procurement process of electronic construction service providers (e-procurement) on the district of musi banyuasin *Int. J. Civ. Eng. Tech.* **10** 8

[8] Turban E Lee J K King D and Liang T P 2010 Electronic commerce 2010: a managerial perspective 8 Springer International Publishing Switzerland

[9] Peraturan Presiden No 16 2018 Pengadaan barang dan jasa pemerintah 16

[10] Malik Alfian 2010 Pengantar bisnis jasa pelaksana konstruksi *ANDI*

[11] Kautsariyah S and Hardjomuljadi S 2016 Analisis penyimpangan pada proses pemilihan penyedia jasa konstruksi elektronik di pemerintah daerah *J. Konstr* 75-86

[12] Alfian 2015 Pemetaan jenis dan risiko kecurangan dalam audit pengadaan barang dan jasa **4** *ISSN* 2089-2861

[13] Organisation for Economic Co-operation and Development 2009 Principles for integrity in public procurement (Paris)

[14] Adlina B A and Simatupang D P N 2015 Risiko hukum dalam pengadaan barang/jasa pemerintah *Univ. Indonesia J. Pub.*

[15] Russell J and Skibniewski M Decision criteria in contractor prequalification *J. Manag. Eng.* **4** 2 148–64

[16] Peraturan LKPP No 7 2018 Pedoman perencanaan pengadaan barang dan jasa pemerintah 7

[17] Wong CH, Holt GD and Cooper P A Lowest price or value? Investigation of UK construction clients' tender selection process *Constr. Manag. Eco.***18** 67–74

[18] Mbachu J Conceptual framework for the assessment of subcontractors eligibility and performance in the construction industry *Constr. Manag. Eco.* **26** 5 471-484

[19] Holt G D, Olomolaiye P O and Harris F C Factors influencing UK construction clients choice of contractor *Build Environ.* **139** 241–8.

[20] Horta I M, Camanho A S and Lima A F 2013 Design of Performance Assessment System for Selection of Contractors in Construction Industry E-marketplaces *J. Const. Eng. Manag.*

[21] Semaan N and Salem M 2017 A deterministic contractor selection decision support system for competitive bidding *Eng. Constr. Archictect Manag.* **24** 61–77

[22] Liu B, Yang X, Huo T, Shen G Q and Wang X 2017 A linguistic group decision making framework for bid evaluation in mega public projects considering carbon dioxide emissions reduction **148** 811–25

[23] Hatush Z Skitmore M 1998 Contractor selection using multicriteria utility theory: an additive model *Build Environ.* **33** 2–3 105–15

[24] Cheaitou A, Larbi R and Housani B A 2018 Decision making framework for tender evaluation and contractor selection in public organizations with risk considerations (UAE: Elsevier) **68** 1-12

[25] Hanafi M M 2009 Manajemen risiko (Yogyakarta: Penerbit dan Percetakan)

[26] Meirsha I M 2019 Modul analisis risiko pengadaan barang dan jasa pemerintah (Inspektorat Jendral Ristekdikti)
[27] Oktavia W C 2013 Analisis dan mitigasi risiko pada proses pengadaan barang dan jasa dengan pendekatan metode interpretive structural modelling (ISM), analytic network process (ANP) dan house of risk (HOR) (Surabaya: National Proceeding Seminar on Technology XIX) A19-1 – A19-8
[28] Asian Development Bank 2018 Procurement risk framework (Philippines: ADB)
[29] Kusumarukmi E I 2018 Identifikasi dan analisis permasalahan pelelangan umum proyek konstruksi (Surabaya: Tesis)
[30] Andriyani Y I 2015 Analisis manajemen risiko pengadaan barang dan jasa (studi pada DPPKAD kabupaten klaten) (Yogyakarta: Tesis)
[31] Mohajan and Haradhan 2017 Two criteria for good measurements in research: validity and reliability *MPRA Paper* 83458 **17** 56-82
[32] Ghozali I 2005 Aplikasi analisis multivariate dengan program SPSS *Semarang Handbook*
[33] Christopher and Schooner 2007 Incrementalism: eroding the impediments to a global public procurement market *J. Int. Law* 529
[34] Santoso B, Handajani L and Arifiantie R 2015 Perspektif triangle fraud theory dalam pengadaan barang/jasa di pemerintah provinsi NTB *J. Invest.* **11** 195-213
[35] Khumpaksal S 2007 Risks in the construction project procurement process and the mitigation methods *J. Arch. Plan. Res. Stud.* 135-45
[36] Oyewobi L O, Ibrahim A D and Ganiyu B O 2011 Evaluating the impact of risk on contractor's tender figure in public buildings projects in Northern Nigeria *EPPM* 2(1) 2-13
[37] Polit D F, Beck C T and Owen S V 2007 Focus on research methods is the CVI an acceptable indicator of content validity appraisal and recommendations *Wiley InterScience* **30** 459–467

IOP Conf. Series: Materials Science and Engineering **930** (2020) 012003 doi:10.1088/1757-899X/930/1/012003

A framework for mapping stakeholders interests related social sustainability in residential building

F R Rivai[1], M A Rohman[1*] and Supani[1]

[1] Department of Civil Engineering, Institut Teknologi Sepuluh Nopember
Kampus ITS Sukolilo, Surabaya, Indonesia

*Corresponding author's e-mail: arif@ce.its.ac.id

Abstract. Dwelling is the primary necessity of every human being. Population growth and urbanization movements without being matched by land availability in urban areas make the need for residential buildings higher. Land use from horizontal to vertical patterns gives not only advantages but also disadvantages. Buildings need to adopt sustainable construction concepts to provide maximum benefits in short as well as long term throughout their life cycles. Social sustainability is one aspect of sustainable development that has a relationship with community needs. The different roles of each stakeholder make them have a different interest too and very important to fulfil it, so it is necessary to identify each stakeholder's interest. Commonly used method only indicating which stakeholders or criteria are more significant separately does not explain how stakeholders can influence it. To bridge the research gap, this paper aims to propose a framework that can analyze stakeholders' influence to achieve social sustainability. The result of paper is a model using two-mode social network analysis to map stakeholders and their claims relating to social sustainability in residential buildings. The model can show the importance of the social sustainability criteria and stakeholders who have a significant role and should be prioritized more comprehensively.

1. Introduction

Housing is a basic need for everyone and often defined as places or buildings where people live or abode. More than that, housing is among basic social conditions that defines the quality of life and well-being of people and places. It is a place where part of relationships between society and environment supports various human activities[1]. Over time, population growth and urbanization movements continue to rise, and those cause increased density in a city. To meet the basic human need for dwelling with limited land availability, the pattern of urban development has changed from horizontal spread to vertical growth. The use of less area to be built residential buildings can provide more rooms for green spaces and parks. The existence of residential buildings has many benefits as well as negative impacts on residents or people who live nearby. Space is now separated by different levels of the floor affecting the quality of social interaction among residents, and it can reduce social bonds such as lack of caring and sharing. In connection with that, it is crucial to ensure buildings can provide maximum benefit and reduce negative impact during their life cycles.

The concept of sustainable development (SD) is quite relevant to adopt in construction projects to achieve success. There are three main aspects known as the triple bottom lines of SD, including: economic, environmental, and social aspects [2]. So far, social aspects have not been developed much in construction project management, in contrast to the two other aspects that have received more

Content from this work may be used under the terms of the Creative Commons Attribution 3.0 licence. Any further distribution of this work must maintain attribution to the author(s) and the title of the work, journal citation and DOI.
Published under licence by IOP Publishing Ltd

attention. The aspect of social sustainability is essential to be realized in project implementation so that it can provide long-term benefits for the wider community [3].

Stakeholders can be defined as a group or individuals who can affect or be affected by the decision to achieve organizational goals [4]. Added by [5], stakeholders in the project have interests and expectations on project performance. Therefore, different expectations on each of their interests need to be managed properly so that the actual results can be appropriate and provide stakeholders satisfaction [6]. The satisfaction of stakeholders is one of the project success criteria [7]. Failure to respond and meet community expectations can cause stakeholder opposition which can lead to project failure [8], [9]. According to [10], achieving the product's customer satisfaction is more important than meeting the specifications of the project. The Sydney Opera House is an example of a project that considered success. The project took three times longer than the planned time and almost five higher than the budgeted, but it becomes the Australia's most famous landmark [11]. However, before satisfying the stakeholders, the most fundamental thing to do is to identify the interest of each stakeholder.

Most previous studies analyze the importance of stakeholders and social sustainability criteria that should be prioritized separately and do not see the relationship between them. Based on these conditions, this paper proposes a framework model by applying social network analysis (SNA) to visualize stakeholders' interests by connecting them with appropriate social sustainability criteria in residential buildings in the operational phase, which is the longest phase in project life cycle and where the project has become a product. It is hoped it can provide complete information that helps in formulating solutions to evaluate and improve the social sustainability performance in residential buildings.

2. Literature review

2.1. Sustainability concept

SD is defined as meeting the needs of the current generation without sacrificing the ability of future generations to meet theirs [12]. In addition, [13] defines SD as the improvement of the quality of human life while living within the carrying supporting ecosystem capacities. Extensive research and discussion on the concept of SD produce three fundamental aspects to be balanced, namely economic, environmental, and social aspects.

For decades, construction industry practitioners worldwide have been aware that their activities can negatively affect communities and environment [14]. Sustainable construction (SC) should be applied to minimize the negative impacts. [15] has identified several key features of SC as follows: (1) Considering the life cycle of project; (2) Integration of all three aspects namely: environmental, social and economic; (3) Considering non-technological as well as technological solutions, which means achieving SC does not fully use technology; (4) Considering the needs of present and future stakeholders.

Sustainable housing is not simply as a green building viewed in resource-saving but socially-enhancing and environment friendly which is integrated into the comprehensive urban/settlement system [1]. [16] stated that aesthetic and structural aspects are not the only basic considerations when providing a house but also the need to pay attention to aspects related to resident, cultural, social, and economic values of a society.

2.2. Social sustainability

Social sustainability is about to satisfy human needs and being maintained over long periods of time regarding social justice, human dignity, and participation that are fulfilled [17]. In the context of construction, [18] describes the principles of social sustainability are improving the quality of human life, making provision for social self-determination and cultural diversity, protecting and promoting human health, implementing skills training and capacity enhancement of disadvantaged people, seeking equitable of social cost and benefit, and seeking intergenerational equity. Social sustainability has a different perspective with social impact assessment (SIA). SIA has the tendency to eliminate the negative impacts [19]. Meanwhile, social sustainability aims to deliver maximum benefits and to avoid negative impacts in achieving satisfaction and promoting the success of the project [20].

Several studies have been conducted to consider the social context in residential buildings. [16] stated that there are six categories such as cultural, heritage, indoor environment quality, health and well-being, safety and service quality, accessibility that are considered in providing buildings. Other studies [21] develop several criteria including: safety, security, sense of belonging to a place, comfort, and aesthetics. [22] develop the indicator that is divided by two, namely: functional, aesthetic & innovative design and user comfort & safety. The other perceptives say that accessibility to public facilities and emergency, crime prevention and safety, and traffic calming are required to be considered in providing buildings [23].

2.3. Previous methods
[24], [25], [26] and [27] use mean analysis to rank the most important attributes to achieve social sustainability in construction industry. Other methods used by [16] and [22] are pair wise comparison to determine the weight of the attributes. Furthermore, [28] and [29] develop fuzzy logic to define linguistic scale to get important weight of social sustainability criteria. [30] combines pair wise comparison and fuzzy logic to get more objectivities of importance indicators. [24] suggests that social sustainability should fully reflect the perspective of the stakeholders in a project. Moreover, [3], [19] and [31] have started the concept of social sustainability using SNA that can be reflected by managing the needs and interests of different stakeholders but not yet specific to residential buildings.

2.4. Social network analysis (SNA)
SNA is a methodology to analyze the structure of social relations in the group by interaction and interrelationship (ties) among the actors (nodes) [32]. SNA has been widely used in construction project management because it can provide a relationship structure and a lot of information that are illustrated in the graph [33]. There are two kind networks of SNA, i.e.: one-mode and two-mode networks. Unlike one-mode network which only links the actor to actor, the two-mode network can analyze the relationship between actors and their associated attributes.
The steps of two-mode SNA mainly involve [33]:
1. Identifying nodes and boundaries of the network
2. Linking the relationships
3. Visualizing and projecting the network
4. Analyzing the network
5. Finding the results

3. Proposed framework
The proposed framework in this paper is using social network analysis (SNA) to map stakeholders and their interests relating to social sustainability in residential buildings. The step process of the network model is presented in figure 1. The term usage of 'criteria', 'indicators', and 'attributes' is used interchangeably.

Step	Nodes	Links	Structure Network	Projecting	Network analysis	Discussion
Process	Identifying stakeholder groups and identifying social sustainability criteria	Connecting the relationship between stakeholders and criteria	Visualization of network	Converting two-mode into one-mode network	Analyzing of Projected one-mode network	Finding the result
Outcome	List of Stakeholders group and social sustainability criteria	Matrix of Two-mode	Illustration of Model	Projected one-mode network	Topology Measurement	Result interpretations

Figure 1. Process of the network model

3.1. Identifying the nodes

The first is to identify the nodes to form boundaries on the network. In the two-mode network, there are two types of nodes. In this paper, it consists of stakeholder groups and the attributes of social sustainability in residential building context. The stakeholder groups will adopt stakeholder classification by [19] categorizing into three communities, namely industry, user, and neighbourhood community. Industry communities are people who provide or operate the buildings, such as owners, facility managers, and operational staff/employees. Users are people who use the buildings, such as residents and visitors, while the neighbourhood is the people who live nearby and/or are affected by the buildings, such as residential neighbours, commercial neighbours and who use the path and road in the surrounding areas.

Based on the literature review, 22 indicators were proposed for the social sustainable criteria that might be appropriate for residential buildings as shown in table 1. It is important to perform a preliminary survey to verify the research variables by interviewing experts. Likert scale from 1-5 can be used to calculate the attribute relevancy based on experts' opinion in which scale 1 represents very irrelevant variables and scale 5 represents very relevant ones. The attributes were considered relevant if the mean score results of opinions from experts are larger than or equal to three (≥ 3) because this value is the middle score between 1 to 5 [34].

Table 1. Social sustainability criteria based on literature

No.	Criteria	Source
1	The building gives a sense of security	[16], [21]–[23]
2	The building gives a sense of safety	[16], [21]–[23]
3	The building pays attention to health from pollution/environmental problems	[16], [19], [22], [35]
4	The building provides a physical comfort	[19], [21], [25]
5	The building provides a local job opportunity	[25], [36], [37]
6	The building provides economic benefits for the surrounding community	[20], [25], [35]
7	Stakeholders are involved in the decision-making process	[20], [19], [25], [35]
8	Stakeholders' conflicts can be resolved	[37]
9	Easy communication and information exchange among stakeholders	[19]
10	The surrounding community does not feel disturbed by the buildings	[35], [36]
11	The building respects the cultural values of the surrounding community	[16], [19]

12	The building provides parking area	[19]
13	The building provides proper traffic management	[19], [22]
14	The building provides an easy access to public facilities	[16], [22], [23]
15	The building provides an easy access to emergency facilities	[22], [23]
16	The building provides an easy access to the disabilities	[22]
17	The building creates a sense of place	[21]
18	The building can be a place for social interaction	[21]
19	The building pays attention to aesthetic aspect and functionality	[22], [25]
20	The building provides a communal open space area	[22]
21	The building provides a private open space area	[16], [22]
22	The building provides a visual privacy area	[16]

3.2. Linking the nodes

The structure of SNA consists of nodes and links. Links are the representation of interaction and interrelationship among the nodes [32]. The questionnaire will be distributed to stakeholders who have been identified at the previous stage. Stakeholders will choose attributes that are interests that must be met in residential buildings related to social sustainability that has been assessed by experts. If the stakeholders give a value of 1 to the attribute, it means that the attribute is their interest; otherwise if they give a value of 0 then the attribute is not their interest. The answers from the stakeholders are collected and made into a two-mode matrix. Equation 1 is shown as the matrix formulation.

$$X = [UA]_{U_iA_n, U_jA_m} = \begin{cases} U_{i,j}A_{n,m} = 1, \; if \; related \\ U_{i,j}A_{n,m} = 0, \; if \; not \; related \end{cases} \quad (1)$$

where:

X = Two-mode matrix between stakeholders (A) and social sustainability criteria (U) in which $U_{i,j}A_{n,m}$ =1, if stakeholders i,j have interest criteria n,m and $U_{i,j}A_{n,m} = 0$ if otherwise.

3.3. Network visualization

This matrix X is an input of the visualization of the network. Network visualization can be illustrated using applications such as UCINET, NetMiner, NetDraw, Pajek, etc [38].

3.4. Projecting two-mode network

In addition to analyzing relationships in the network, SNA can be used to find out how big an actor's role is or the level of importance of attribute in the network is. The two-mode matrix needs to be projected into one-mode matrix before the analysis. There are 2 types of one-mode matrix i.e., the matrix of relationships between stakeholders and other stakeholders, and the matrix of relationships between criteria and other criteria. The two-mode will be converted into one-mode network by using equations 2 and 3 [39].

$$[AA]_{nm} = X^T.X \quad (2)$$

$$[UU]_{ij} = X.X^T \quad (3)$$

where:

X^T = The transpose of matrix X.

$[AA]_{nm}$ = One-mode matrix that represents the relations between stakeholders.

$[UU]_{ij}$ = One-mode matrix that represents the relations between criteria.

3.5. Network analysis
The topology measurement of node relation used is centrality measurement. Several kinds of centrality commonly used are degree, betweenness, and eigenvector centrality. Degree centrality is defined as the importance of node which refers to how many nodes are directly connected [40]. Betweenness centrality is defined as the number of time nodes passed by the shortest path between 2 nodes that are not directly linked [40]. Eigenvector centrality is defined as the importance of node which has higher score when the neighbours node also has high connection [33]. The highest value can be interpreted as an important node of the network.

3.6. Result and discussion
After the results are obtained in the previous step, the result is discussed to get the findings of the relations of networks, the importance of the criteria and stakeholder prioritization, so it can help the decision-makers in formulating solutions to evaluate and improve the building performance and management of stakeholders to ensure social sustainability. The illustration of the flowchart network is presented in figure 2.

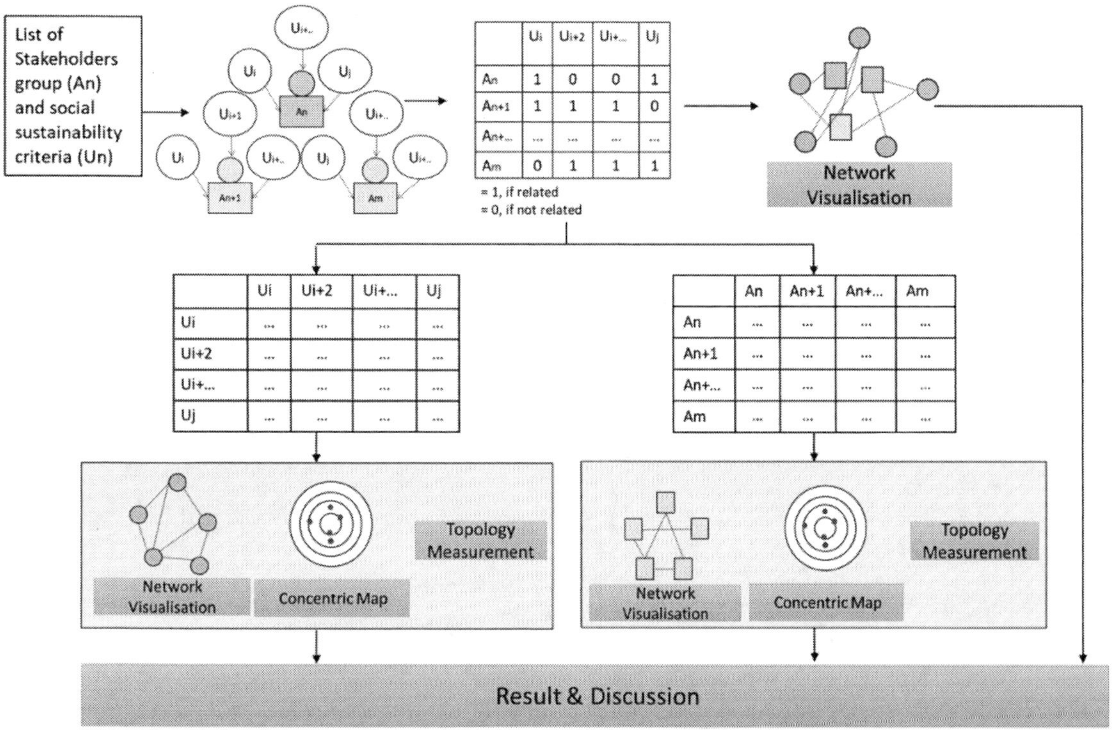

Figure 2. Illustration of network analysis model flowchart

4. Discussion
Implementing social sustainability in project construction can deliver the project's success. Satisfying the social need of stakeholders indicates that the project has been implemented well. Based on literature review that was conducted, there are 22 criteria of social sustainability in residential buildings. Each criterion cannot satisfy all stakeholders because of differences in needs and interests. Hence, this research proposes a framework model to achieve social sustainability more comprehensively. The model creates a network that connects stakeholders with social attributes and shows how stakeholders can influence social sustainability.

IOP Conf. Series: Materials Science and Engineering **930** (2020) 012003 doi:10.1088/1757-899X/930/1/012003

5. Conclusion and recommendation

Based on the background of the problem and the literature review conducted, this study discusses about social sustainability that considers the needs and interests of different stakeholders in residential buildings which are limited. Therefore, this study proposes the framework model to map stakeholders' interests relating to social sustainability criteria. The paper also shows the steps to use the model to determine the importance of 22 criteria based on literature review and the stakeholders that have major roles in residential buildings in the operational phase. The hope is that this research can contribute to evaluate and improve the residential building performance in terms of sustainability, especially in the social aspect to provide maximum benefit.

In further research, the proposed model needs to be tried on the actual buildings as a case study and should pay attention to each phase of the project from initiation to demolition. The weight of importance that can be generated in the analysis of the model can be continued to assess the performance and choose alternatives to decision making.

Acknowledgment

The authors would like to thank the Ministry of Research and Technology/National Agency for Research and Innovation who provided a grant for this research in 2020.

References

[1] Golubchikov O and Badyina A 2012 Sustainable housing for sustainable cities: a policy framework for developing countries *Nairobi* (Kenya: UN Habitat)

[2] Elkington J 1998 Partnerships from cannibals with forks: the triple bottom line of 21st-century business *Environ. Qual. Manag.* **8** 1 37–51

[3] Doloi H 2012 Assessing stakeholders' influence on social performance of infrastructure projects Hemanta *Facilities* **30** 11/12 531-50

[4] Freeman E and McVea J 2001 A stakeholder approach to strategic management *Darden Bus. Sch. Work. Pap.* 01-02

[5] Newcombe R 2003 From client to project stakeholders: A stakeholder mapping approach *Constr. Manag. Econ.* **21** 8 841–48

[6] Li T H Y, Ng S T and Skitmore M 2013 Evaluating stakeholder satisfaction during public participation in major infrastructure and construction projects: A fuzzy approach *Autom. Constr.* **29** 123–35

[7] Silvius G and Schipper R 2016 Exploring the relationship between sustainability and project success *Int. J. Inf. Syst. Proj. Manag.* **4** 3 5–22

[8] El-Gohary N M, Osman H and El-Diraby T E 2006 Stakeholder management for public private partnerships *Int. J. Proj. Manag.* **24** 7 595–604

[9] Majamaa W, Junnila S, Doloi H and Niemistö E 2008 End-user oriented public-private partnerships in real estate industry *Int. J. Strateg. Prop. Manag.* **12** 1 1–17

[10] Baccarini D 1999 The logical framework method for defining project success *Proj. Manag. J.* **30** 4 25–32

[11] Shenhar A J, Dvir D, Levy O and Maltz A C 2001 Project success: a multidimensional strategic concept *Long Range Plann.* **34** 6 699–725

[12] WCED 1987 *Our common future* New York

[13] IUCN/UNEP/WWF 1991 *Caring for the Earth: A strategy for sustainable living* Gland Switzerland

[14] Wirahadikusumah R D and Ario D 2015 A readiness assessment model for Indonesian contractors in implementing sustainability principles *Int. J. Constr. Manag.* **15** 2 126–36

[15] Gunatilake S 2013 The Uptake and Implementation of Sustainable Construction: Transforming Policy into Practice University of Central Lancashire

[16] Ardda N, Mateus R and Bragança L 2018 Methodology to identify and prioritise the social aspects to be considered in the design of more sustainable residential buildings-Application to a developing country *Buildings* **8** 10

[17] Littig B and Grießler E 2005 Social sustainability: A catchword between political pragmatism

and social theory *Int. J. Sustain. Dev.* **8** 1/2 65–79
[18] Hill R C and Bowen P A 1997 Sustainable construction: Principles and a framework for attainment *Constr. Manag. Econ.* **15** 3 223–39
[19] Almahmoud E and Doloi H K 2015 Assessment of social sustainability in construction projects using social network analysis *Facilities* **33** 3/4 152–76
[20] Rohman M A, Doloi H and Heywood C A 2017 Success criteria of toll road projects from a community societal perspective *Built Environ. Proj. Asset Manag.* **7** 1 32–44
[21] Maleki B, Casanovas Rubio M D M, Hosseini S M A and De La Fuente Antequera A 2019 Multi-criteria decision making in the social sustainability assessment of high-rise residential buildings *IOP Conf. Ser. Earth Environ. Sci.* **290** 1
[22] Ahmad T and Thaheem M J 2017 Developing a residential building-related social sustainability assessment framework and its implications for BIM *Sustain. Cities Soc.* **28** 1–15
[23] Yigitcanlar T, Kamruzzaman M and Teriman S 2015 Neighborhood sustainability assessment: Evaluating residential development sustainability in a developing country context *Sustain.* **7** 3 2570–602
[24] Valdes-Vasquez R and Klotz L E 2013 Social sustainability considerations during planning and design: Framework of processes for construction projects *J. Constr. Eng. Manag.* **139** 1 80–9
[25] Li H, Zhang X, Ng S T, Skitmore M and Dong Y H 2018 Social sustainability indicators of public construction megaprojects in China *J. Urban Plan. Dev.* **144** 4 04018034
[26] Sodangi M 2018 Social sustainability efficacy of construction projects in the pre-construction phase in *Proc. Inst. Civ. Eng. - Eng. Sustain.* **172** 2 57–67
[27] Rohman M A and Hidayat Y A 2019 Understanding Social sustainability criteria for public school building *IOP Conf. Ser. Mater. Sci. Eng.* **636** 1
[28] Rajak S and Vinodh S 2015 Application of fuzzy logic for social sustainability performance evaluation: A case study of an Indian automotive component manufacturing organization *J. Clean. Prod.* **108** 1184–92
[29] Hendiani S and Bagherpour M 2019 Developing an integrated index to assess social sustainability in construction industry using fuzzy logic *J. Clean. Prod.* **230** 647–662
[30] Xiahou X, Tang Y, Yuan J, Chang T, Liu P and Li Q 2018 Evaluating social performance of construction projects: An empirical study *Sustainability* **10** 7 2329
[31] Wang H, Zhang X and Lu W 2018 Improving social sustainability in construction: conceptual framework based on social network analysis *J. Manag. Eng.* **34** 6 1–9
[32] Wasserman S and Faust K 1994 Social Network Analysis: Methods and Applications : *Camb. Univ. Press*
[33] Liang X, Yu T and Guo L 2017 Understanding stakeholders' influence on project success with a new sna method: A case study of the green retrofit in china *Sustainability* **9** 10 1927
[34] Rohman M A and Wiguna I P A 2019 Evaluation of road design performance in delivering community project social benefits in Indonesian PPP *Int. J. Constr. Manag.* 1–13
[35] Sierra L A, Yepes V and Pellicer E 2018 A review of multi-criteria assessment of the social sustainability of infrastructures *J. Clean. Prod.* **187**. 496–513
[36] Nasirzadeh F, Ghayoumian M, Khanzadi M and Cherati M R 2019 Modelling the social dimension of sustainable development using fuzzy cognitive maps *Int. J. Constr. Manag.* 1–14
[37] Zuo J, Jin X H and Flynn L 2012 Social sustainability in construction - An explorative study *Int. J. Constr. Manag.* **12** 2 51–63
[38] Yang R J 2014 An investigation of stakeholder analysis in urban development projects: Empirical or rationalistic perspectives *Int. J. Proj. Manag.* **32** 5 838–49
[39] Everett M G and Borgatti S P 2013 The dual-projection approach for two-mode networks *Soc. Networks* **35** 2 204–10
[40] Prell C, Hubacek K and Reed M 2009 Stakeholder analysis and social network analysis in natural resource management *Soc. Nat. Resour.* **22** 6 501–18

A framework to assess success criteria performance of public private partnership (PPP) toll road projects in Indonesia

F Kristiawan[1*], M A Rohman[1], and Machsus[2]

[1]Department of Civil Engineering, Faculty of Civil, Planning, and Geo Engineering, Institut Teknologi Sepuluh Nopember, Surabaya, Indonesia.
[2]Department of Civil Infrastructure Engineering, Faculty of Vocational, Institut Teknologi Sepuluh Nopember, Surabaya, Indonesia.

*Corresponding author's e-mail: freedykristiawan.18031@mhs.its.ac.id

Abstract. Public Private Partnership (PPP) is widely used to bridge infrastructure projects in the world. PPP arises as a result of the ongoing budget constraints faced by various governments and part of the limited private sector opportunities in infrastructure development. In Indonesia, the implementation of PPP has not been successfully implemented. In 2017, 22 PPP projects were offered, but only 5 projects were signed. According to previous studies, there are some problems related to PPP implementation in Indonesia. Referring to this, PPP project success criteria are very important to be defined to assess the success of this type of project. Data collection procedure consists of preliminary survey involving experts in the field to verify the research attributes, followed by main survey to collect respondents' perception regarding the attributes using Likert Scale. Importance Performance Analysis (IPA) is proposed to measure the performance of each success criterion in each respondent sector and followed by gap analysis to investigate critical gaps between them. The results of this paper are a methodological framework model that presents a gap implementation and a performance of the successful PPP criteria for toll road projects in Indonesia. These results can be used to assess the performance of PPPs in Indonesia.

1. Introduction

Economic growth in Indonesia gradually began to rise since the economic crisis that struck in 1990. Recent data show that from 2015 to 2018 Indonesia's economic growth experienced a consistent increase and was still maintained at the 5% level [1]. To help the pace of economic growth, it needs to be balanced with an increase in adequate infrastructure. Understanding that this is not just an easy problem, the government is beginning to realize the important role of the private sector to meet the needs for this infrastructure. Public Private Partnership (PPP) is widely used to bridge infrastructure projects in the world [2]. PPP arises as a result of budget constraints that are constantly faced by various governments and part of the limited opportunities for the private sector in infrastructure development [3].

In the early 1990s, PPP was an innovation in private sector financing for infrastructure and at the same time, PPP provided an overview of the provision of public assets and services through the participation of the public, private, and consumer sectors [4]. In its development, the government began to explore any alternatives, and the PPP can facilitate various infrastructure development needs, such as transportation infrastructure, sports facilities, water treatment facilities, and waste management. In recent decades, the concept of PPP has spread rapidly, especially among developing countries. Governments in many developing countries are now considering the concept of PPP as an innovative

Content from this work may be used under the terms of the Creative Commons Attribution 3.0 licence. Any further distribution of this work must maintain attribution to the author(s) and the title of the work, journal citation and DOI.
Published under licence by IOP Publishing Ltd

approach in bridging their large infrastructure gaps [5]. In Indonesia, PPP implementation has been tried since 1997-1998, but there are still many that have not been optimal. In Indonesia, one of the most dominant sectors is the toll road.

The Indonesian toll road industry has a positive outlook. The potential for strong demand is due to the limited construction of new public roads and the slow growth of public transportation infrastructure. Demand in the toll road industry is still promising, especially for inter-city toll roads. Toll road construction has been carried out in Indonesia since 1978 and currently around 1500 km of toll roads have been in operations [6]. The main purpose of developing toll roads is to reduce traffic congestion in urban areas and to support regional growth. Newly data for PPP projects offered in 2017 stated that 22 projects with PPP scheme were offered but only 8 projects in 2018 were tendered, 5 projects signed and worked on toll road projects and one of these projects was located in East Java [7].

Apart from the various success stories regarding PPP implementation in toll road construction in Indonesia, there are still a number of problems and issues that need to be addressed. Related to this, the Indonesian government has also worked hard to make PPP successful in this sector. There are many issues related to PPP problems that cause failures including: high costs in the tender process, complex negotiations, cost limitations in innovation, different or conflicting goals among project stakeholders [8]. Even though Indonesia is one of the countries that offers the biggest opportunity in Asia to invest in the national infrastructure sector [9], the lack of PPP implementation in Indonesia is because the private sector is aware that conducting PPP business in Indonesia is still very difficult [10]. If it is related to the market maturity curve [11], Indonesia can be considered having a high PPP activity but a low institutional and legal framework. This is confirmed by the statement that one of the reasons for the slow progress of the PPP is due to the inadequate policy framework, which includes the legal framework and institutional framework [12]. Moreover, the willingness of the private sector to take part in a project is based on where the project environment takes place [13]. Many issues related to the success of the PPP project give a question about the meaning given to how the PPP is considered successful and what constitutes success for the PPP itself.

Basically, the success of the PPP project has different meanings for various stakeholder groups [14]. It is important to see different understandings of 'success' that have emerged in some literature about PPP and how success can be related to various forms of PPP. The term Critical Success Factor (CSF) is often used as an additional success factor in the literature. In PPP, this factor will affect the private sector that brings superior resources and financial capital and knowledge. In line with the importance of success factors that has been widely studied, doubts often arise about what and who actually determines the success of the project. Masuring project success is difficult to understand, despite of that, to measure success in certain dimensions there is a need to identify adequate success criteria that will enable the assessment of success [15].

Success criteria are often considered similar as success factors in the literature. However, these two terms must be distinguished because these two terms are actually different [10]. A criterion is a set of principles or standards by which an assessment is made; whereas factors are a set of circumstances, facts, or influences that contributes to the outcome [16]. Success criteria can be described as a standard on which the success of the project is assessed. Success/failure of a project is determined by the outcome measure of the selected success criteria [15]. Although there are a lot of literature studies on the success factors of general construction projects and project perspectives related to PPP, many previous studies have only raised the success factors for PPP projects in Indonesia; however, there is still very little research that discusses PPP performance in Indonesia based on standards that are described from the project's success criteria. Later research is not limited to measuring performance alone, but the use of the PPP success criteria for the project should be expanded by using it to evaluate performance and by prioritizing efforts for improvement. The general management technique that fits this purpose is Importance Performance Analysis (IPA).

This paper tries to evaluate the current conditions in the PPP implementation process for toll road projects in Indonesia that will be able to assist the government and the private sector in identifying and measuring their performance based on success criteria.

IOP Conf. Series: Materials Science and Engineering **930** (2020) 012004 doi:10.1088/1757-899X/930/1/012004

2. Literature review

The theory that underlies this research includes Public Private Partnership (PPP) and concept of success criteria as an alternative assessment for measuring success in Indonesia's toll road PPP.

2.1. Public private partnership (PPP)

Public Private Partnership (PPP) is described as a long-term arrangement between public institutions and private sector organizations where both parties carry out their additional skills and assume various levels of responsibility and risk for the purpose of developing public facilities [8], [17]. The PPP concept consists of broad relationships between the public and private sectors to develop projects or services. This contractual arrangement is built as a joint commitment, where cooperative work ideally takes place together with shared goals and authority, utilizing the strengths of each party such as creativity, efficiency, joint investment, and the allocation of risks and rewards [18]. The government can also utilize the flexibility of the private sector to innovate, obtain the latest expertise or knowledge and promote efficiency in government sector personnel by incorporating private best practices and the sophistication of the technology used [19]. As an operational definition, PPP in government is a contractual mechanism between private and public institutions which aims to provide certain assets to electronic facilities or services. They not only share responsibilities and risks, but they also invest together in organizational resources [20]. In return for this effort, both parties are valued based on agreed joint revenue and a performance-based cost collection and payment model [21]. Currently, governments in many developing countries are considering PPP as an innovative/alternative in covering the need for large infrastructure [5].

From a theoretical point of view, initially, public and local authorities assumed full responsibility for their services, provision, and territorial development. Meanwhile, in the case of PPP, the government sector delegates the provision of public services which require large investments in infrastructure and certain risks associated with the private sector in the long term, for the purpose of improving the situation [22]. Thus, the State is more likely to buy services from the private sector rather than has to involve itself in property development and public service delivery organizations related to the property. In other situations, PPP aims to accommodate situations where and when business activities are forced to reduce their level of activity as a result of reduced demand. PPP will allow to maintain and even increase the level of activity of some businesses, because the public sector does not invest in developing public services during the initial period, but this task is left to the private business sector instead and engaged in the process at a later stage. A significant advantage of this model is that the managerial experience of the private sector allows it to implement projects faster and with higher quality. The fact is that private companies have more specialization in their fields, because they are more experienced in investing and have more opportunities to obtain capital loans from banks. Partnership conditions like that make it possible to do what they know, private businesses must develop infrastructure and provide services, and the public sector will create favorable conditions.[23], [35].

2.1.1. Current situation on toll road infrastructure in Indonesia. Infrastructure development is an important and vital aspect for a country's economic development. A country's economic growth cannot be separated from the availability of infrastructure such as transportation, telecommunication, sanitation and energy. The development of transportation sector is the main dimension to encourage and contribute more to the country's economic progress [24]. Toll road development in Indonesia has started since 1978 with a State-Owned Enterprise (SOE) PT Jasa Marga as the operator and regulator of toll roads [12]. Initially, the PPP concept was not introduced to toll road development. The involvement of the private sector in the provision of toll roads began in 1983. The Government of Indonesia invited the private sector to take part in providing toll roads through the BOT (Build, Operate and Transfer) scheme and in 2004. Since 2004, the management of the implementation of Government Cooperation with Business Entities in toll road development can be done by the Government and/or Private [12]. Since then, every year, the development of toll infrastructure in Indonesia has always been included in the priority projects offered in Indonesia's PPP book [25]. For the East Java region at present, about 8 toll road projects are included in the national strategic projects. Apart from the various success stories regarding PPP implementation in toll road construction in Indonesia, there are still a number of problems and issues that still need to be addressed.

2.2. Success criteria

The project success criteria are a set of principles or standards by which project success can be assessed. This is a condition under which judgments can be made. On the other hand, the factors for project success are a series of circumstances, facts, or influences that contributes to the project's results. This is an influential force that facilitates or impedes project success. They contribute to the success or failure of a project, but do not form the basis of assessment [16]. In general, project success is described as achieving a number of criteria that are felt externally [26]. It was explained that there were 5 types of success in the collaborative process including: (1) achieving results, (2) getting the process to work, (3) achieving the existing milestone, (4) getting recognition from others, and (5) acknowledging personal pride in fighting for partnerships [27]. In the first point, what is meant by achieving results is achieving a predetermined mission. In the second part, it is explained that if the process runs smoothly and helps create new ideas and satisfaction among participants, then the process can be a criterion for success. The third point is about an innovation content that can be produced by a good process. Meanwhile, the remaining two factors relate to personal investment and perspective of the actors to make changes. Specifically, project success is the main objective of the project manager, and to achieve this goal, a series of main objectives must be achieved [28].

Success criteria for construction projects are grouped into two broad frameworks, namely the macro and micro viewpoints. The macro criteria include time, utility, and operation, while the micro criteria include time, cost, quality, performance, and safety [16]. Meanwhile, [33] presented two groups of criteria for the success of a construction project as an objective and subjective measure. The criteria for objective success are more easily measured and objective compared to measures of subjective success. Objective criteria include time, cost, accident rate, and net present value, while subjective measures include quality, functionality, end-user satisfaction, client satisfaction, and design and construction team satisfaction. Still related to the criteria for success, [29] divides project success into three dimensions: project management success, product success, and market success. Project management success focuses on achieving management targets in terms of time, quality, and cost. The second dimension relates to final product targets in terms of customer satisfaction, technical specifications, and functionality, while the third category focuses on project potentials that contribute to the company's long-term success in terms of reputation, revenue, market share, and competitive advantage. Not much different from the general success criteria, [30] identified success criteria in developing countries and concluded that there were four groupings that underlie success measures, including costs, time, quality, customer satisfaction, and environmental impact. Meanwhile, [31] assess the success of large-scale public projects based on safety, efficient use of resources, reduced conflicts and disputes, stakeholder satisfaction, and effectiveness. This is very different from the general definition, where [31] argue that traditional measures of time, quality, and cost no longer apply in evaluating the success of large-scale public development projects. Each project has a set of objectives to achieve, and they serve as a standard for measuring performance. The criteria are needed to compare the level of objectives with the level of performance, and project success is to achieve project goals and stakeholder satisfaction. This resulted in 25 success criteria which can be seen in table 1.

Table 1. Variable success criteria from several literatures

No	Success Criteria	Definition	Source
1	Local economic development	The project contributes to the economic development of the surrounding community	[32]
2	Reduced litigations and disputes	Contract litigations and disputes are minimized throughout the project lifecycle	[32], [33]
3	Effective risk management	Risks are properly identified. The risk sharing and transfer mechanism are agreed upon and implemented effectively by the government & business entities	[34], [35]

No	Success Criteria	Definition	Source
4	Budget solution	Solve the problem of restraining the public sector budget	[36]
5	Reduced public administration costs	Lower costs are incurred by the public sector in project administration because the risk of large projects is allocated to the business entity sector	[34], [37], [38]
6	Reduced public and political protest	Reduction in agitation (resentment due to provocation) and protests, which often arises due to increased tariffs, lack of transparency, corruption, etc.	[38]
7	Project on time	Projects in construction / before the time schedule for commissioning (implementation)	[5], [29]
8	Accordance with the budget	Projects are built according to estimated costs and without excess of operational costs	[29]
9	Additional financial resources for another priority projects	This relates to the allocation of public money for other important projects.	[38]
10	Environmental performance	The project does not affect the health and safety of residents or the environment	[28], [38]
11	Attract towards toll road users	Should attract sufficient traffic so project benefit will exceed project cost	[10], [39]
12	Vehicle operational optimization	Reduce vehicle operating costs and reduce travel time	[10], [40]
13	Development of alternative public transportation	Should contribute to transportation costs and time savings that support the economy, alleviate poverty, increase community employment and increase government revenue	[10], [40]
14	Reducing project life cycle costs	Lower lifecycle costs are realized, which increase the value of the project money	[37]
15	Reliable and quality service operations	Continuous and uninterrupted project services are provided and in accordance with user satisfaction	[5], [41]
16	Satisfy the needs of public facilities / services	The PPP project carried out fully meets the needs for public facilities / services	[37]
17	Effective technology transfer and innovation	Technology and technical innovation are shared effectively among stakeholders, especially with local practitioners	[5], [14]
18	Long term relationship and partnership	Good relationships and established coordination are formed among the stakeholders	[5], [36]
19	Long project life span	Longer life span means the availability of products or services in a longer period. For PPP projects with certain concession periods, a longer range means the remaining service period is longer after the project transfer to the client	[38]
20	Increased construction and maintenance capabilities	Constructability and maintainability are two important issues that must be considered in the design process. PPP projects encourage adequate attention at these points.	[5], [38]
21	Low tariffs/tolls	The tariff / toll level measures the cost of using project facilities. Determine the level of profit of the concession holder. Increased efficiency allows low tariff / toll level	[32]
22	Profitability	Earnings / profits are continuously received by the collaborating agents during the project operation	[14]
23	Modular and repeatable design / construction	Facilitate public clients to develop similar projects in the future	[5], [38]

No	Success Criteria	Definition	Source
24	Addition of facilities	The public sector can obtain additional facilities / services outside the requirements of the business sector	[5], [32]
25	Public sector guarantee	The private sector can get government sponsorships, guarantees and tax reductions	[5], [37]

3. Research methodology

This study uses the Gap analysis (GA) method which is a representative variant of the IPA method to measure the level of importance and to prioritize criteria that require improved performance. GA is a methodological framework at which gaps are weighted according to the importance of factors [9]. Importance-performance analysis (IPA) is a method that provides a two-dimensional grid of performance interests where the values of importance and performance among the various attributes are plotted with each other and the resulting interest-performance space is generally divided into four quadrants [32]. IPA provides a measure of how well a service is to meet the needs of consumers. Before plotting with IPA, a GAP analysis is performed to show the critical gap between each variable, then a Paired T-test is conducted to show that there may be a significant gap between the level and importance of the criteria in understanding the respondents. The results of the identification of critical gaps obtained from paired T-test and gap analysis will be plotted to the IPA to provide an indication of which criteria get an assessment of their performance and to be included in the recommendations for performance improvement efforts. Based on this, this study aims to identify the criteria for success of the Indonesian PPP project, to measure the level of importance, and to prioritize criteria that require performance improvement.

3.1. Data collection

The initial step in obtaining data is to compile a list of experts and practitioners affiliated with the main organizations in developing PPP, from which the list is then carried out by distributing preliminary questionnaires to validate the variables obtained from literature studies in accordance with field implementation. Then, a pilot test is carried out to ensure that the questionnaires can be understood by respondents and there are no problems with the questions asked [10]. After the preliminary questionnaires and pilot test, the main survey was conducted with a questionnaire filled out by key stakeholders from public and private sector organizations, who have adequate knowledge of PPP projects in Indonesia including the Ministry of Finance, coordinating institutions and contractors, the Indonesian Infrastructure Guarantee Fund, BAPPENAS, The Indonesian Investment Agency, PT Jasa Marga, and other related organizations. To accommodate the limitations of questionnaire and respondent distribution, questionnaire surveys will be used through online questionnaire platforms. The Snowball method is also used to identify potential respondents who can be contacted for the main survey. The main questionnaire survey was conducted to obtain the value of approval and satisfaction with the practice of toll road PPP in Indonesia.

3.2. Data analysis

Data received from the results of questionnaires will be tested for validity and reliability. Data validity test is used to measure valid or invalid questionnaires. In testing the validity using SPSS aids, decision making is based on a significance value of less than 0.05 (5%), then the question items are declared valid. Then, the reliability test is a tool to measure the questionnaire which is an indicator of the variable or construct. The construct or variable is said to be reliable if it gives an Cronbach Alpha value> 0.60 [33]. Testing the validity and reliability will be assisted by the SPSS for Windows program. After going through the two tests, data analysis will be continued with GAP analysis which is a representative variant of the IPA method.

3.3. Gap analysis

The Gap Analysis consists of four main steps, consisting of Paired Sample T-test, Weighted Gap Analysis, Importance Performance Analysis, and Critical Gap Identification. First, the next step in data analysis involves identifying the trust criteria which show a critical gap between the level of expected performance and the level of perceived performance. The identification of these criteria can give an indication of which criteria should be prioritized in relation to actions to be taken in an effort to improve the improvement of PPP performance in Indonesia.

3.3.1. Paired T-test. Paired sample T-tests aim to reveal significant differences between the level and importance of trust for each PPP success criterion. The results of Paired sample T-tests indicate that there may be a significant gap between the level and importance of trust in each success criterion.

3.3.2. Weighted gap analysis. Weighted gap analysis is used as a methodological framework in which the gap is weighed according to interests [9]. Weighted gaps for each confidence criterion are achieved by multiplying the average interest and the average gap together [34].

3.3.3. Importance performance analysis (IPA). Importance Performance Analysis is used to map the relationship between interests and the performance attributes of each offered and the gap between performance and attribute expectations. Importance Performance Analysis consists of two components: quadrant analysis and gap analysis. With quadrant analysis, it can be seen the response of respondents to the attributes plotted based on the order of importance (perception) and performance (expectations) attributes. Meanwhile, the gap analysis is used to see the gap between the attribute interests and the expectations of consumers / respondents for this attribute [33]. In this IPA analysis, SPSS for Windows program will be used as the software tool.

3.3.4. Critical gaps identification. The selection criteria for determining criteria with critical gaps are based first on the criteria that have the strongest indicative results from paired sample T-tests. The second criterion concerns existing criteria rated 1-6 results from weighted gap analysis. The third criterion is based on criteria that are included in the 'critical' or 'significant' areas resulting from the IPA analysis [34].

4. Result and discussions

Applying a methodological framework can provide an overview of the current situation related to the performance conditions of the PPP success criteria in Indonesia. Based on literature review that was conducted, there are 25 success criteria. This research proposed a framework model to identify and measure the success criteria and their performance as the basis for the PPP project implementation improvement. This model creates a network that identifies and measures PPP success criteria performance in Indonesia.

5. Conclusion and further works

Based on the background of the problem and the literature review, this study discusses about evaluating the current conditions in the PPP implementation process for toll road projects in Indonesia that will be able to assist the government and the private sector in identifying and measuring their performance based on success criteria. This study proposes the framework model to identify and measure the success criteria and their performance as the basis for the PPP project implementation improvement. This paper can be replicated with a strong theoretical basis to assist governments and the private sector in identifying and evaluating the success criteria for PPP under their control, and in measuring their performance based on these success criteria.

In further research, the framework model needs to be tried on the actual project toll road as a case study. The weight of the success criteria generated in the analysis of the model can be taken into consideration for the prioritization of proposed improvements to the criteria that have low performance.

References

[1] Bisnis 2019 available at: https://ekonomi.bisnis.com/read/20191010/9/1157640/bank-dunia-pertumbuhan-ekonomi-indonesia-2019-terjaga-pada-level-5-persen
[2] Cui C et al 2018 Review of studies on the public–private partnerships (PPP) for infrastructure

projects *Int. J. Proj. Manag.*

[3] Solomon O B, Akintayo O and Olusegun E A 2012 Critical success factors in public-private partnership (PPP) on infrastructure delivery in Nigeria *J. Facil. Manag.* **3** 212 – 22

[4] Grimsey D and Lewis M K 2005 The economics of public private partnerships (Cheltenham: Edward Elgar in press)

[5] Osei-Kyei R and Chan A P C 2015 Developing transport infrastructure in Sub-Saharan Africa through public–private partnerships: Policy practice and implications *Transp. Rev* **36** 2 170–86

[6] Jawa Pos 2019 available at: https://www.jawapos.com/infrastruktur/07/11/2019/jokowi-targetkan-2024-indonesia-miliki-5-000-kilometer-jalan-tol/

[7] Badan Perencanaan Pembangunan Nasional (BAPPENAS) 2018 Public-private partnerships: infrastructure projects in Indonesia *Ministry of National Development Planning*

[8] Akintoye A, Beck M, Hardcastle C, Chinyio E and Asenova D 2001 Framework for risk assessment and management of private finance initiative projects *Glasgow Caledonian Univ. Scot.*

[9] Wibowo A and Alfen H W 2015 Government-led critical success factors in PPP infrastructure development *Built Env. Proj. Asset Manag.* **5** 1 121-34

[10] Rohman M A, Doloi H and Heywood C 2017 Stakeholders' perspective on achieving project social benefit from Indonesian toll roads (The University Of Melbourne: Australia)

[11] Siemiatycki M 2009 Delivering transportation infrastructure through public-private partnerships: planning concerns *J. Am. Plann. Assoc.* **76** 1 43-58

[12] Rostiyanti S F and Tamin R Z 2010 Identification of challanges in public private partnerships implementation for indonesian toll road *First Makassar Int. Conf. Civ. Eng. MICCE 2010* 1131-35

[13] Zhang J, Zou W and Kumaraswamy M 2015 Developing public private people partnership (4P) for post disaster infrastructure procurement *Int. J. Disaster Resilience Built Env.* **6** 4 468 – 84

[14] Mladenovic G, Vajdic N, Wundsch B and Temeljotov-Salaj A 2013 Use of key performance indicators for PPP transport projects to meet stakeholders' performance objectives *Built Environ. Proj. Asset Manag.* **3** 2 228–49

[15] Marenjak S and Kušljić D 2012 Evaluating the success of PFI projects in Croatia applying success criterion *Starting Date of Operation*

[16] Lim C S and Mohamed M Z 1999 Criteria of project success: an exploratory re-examination *Int. J. Proj. Manag.* **17** 4 243-48

[17] Jefferies M, Gameson R and Rowlinson S 2002 Critical success factors of the BOOT procurement system: reflection from the stadium Australia case study, engineering *Constr. Arch. Manag.* **9** 4 352-61

[18] Kwak Y H, Chih Y and Ibbs C W 2009 Towards a comprehensive understanding of public private partnerships for infrastructure development *California Manag. Review* **51** 51-78

[19] Taher M, Yang Z and Kankanhalli A 2012 Public-private partnerships in e-government: insights from singapore cases *PACIS Publishing* 116

[20] Palaco I, Park M J, Kim S K and Rho J J 2018 Public–private partnerships for e-government in developing countries: an early stage assessment framework

[21] Basílio M 2017 The degree of private participation in PPPs: evidence from developing and emerging economies, the emerald handbook of public–private partnerships in developing and emerging economies: perspectives on public policy *Entrepreneurship and Poverty Publishing* 81-111

[22] Meitude L and Paliulis N R 2011 Feasibility study of public- private partnership *Int. J. Strateg. Prop. Manag.* **15** 3 257-74

[23] Abdel A A M 2007 Successful delivery of public–privatepartnerships for infrastructure development *J. Constr. Eng. Manag.* **133** 12 918–13

[24] Carnis L and Yaliawati E 2013 Nusantara: between sky and earth could the ppp be the solution for Indonesian airport infrastructures? *J. Case Studies Transport Policy* **1** 18–26

[25] Badan Perencanaan Pembangunan Nasional (BAPPENAS) 2019 Public-private partnerships: infrastructure projects in Indonesia *Ministry of National Development Planning*

[26] Rosenau M D 1984 Project management for engineers *Univ. Michigan Ann Arbor MI*
[27] Huxman C and Hubbert P 2009 Hit or myth? stories of collaborative success' in O'Flynn, Janine & Wanna, John.eds. collaborative governance (Canberra: ANU Press)
[28] Chan A P C, Scoot D and Lam E W M 2002 Framework of success criteria for design/build projects *J. Proj. Manag.* **18** 3 120–8
[29] Al-Tmeemy S M H M, Abdul-Rahman H and Harun Z 2011 Future criteria for success of building projects in Malaysia *Int. J. Proj. Manag.* **29** 3 337–48
[30] Ahadzie D K, Proverbs D G and Olomolaiye P O 2008 Critical success criteria for mass house building projects in developing countries *Int. J. Proj. Manag.* **266** 75–687
[31] Toor R S and Ogunlana S O 2010 Beyond the 'iron triangle': stakeholder perception of key performance indicators (KPIs) for large-scale public sector development projects *Int. J. Proj. Manag.* **28** 3 228-36
[32] Zhang X 2006 Factor analysis of public clients' bestvalue objective in public–privately partnered infrastructure projects *J. Constr. Eng. Manag.* **132** 9 956–65
[33] Chan A P C and Chan A P 2004 Key performance indicators for measuring construction success *Benchmarking Int. J.* **11** 2 203–21
[34] Li B, Akintoye A, Edwards P and Hardcastle C 2005 Perceptions of positive and negative factors influencing the attractiveness of PPP/PFI procurement for construction projects in the UK *Eng. Constr. Arch. Manag.* **12** 2 125–48
[35] Liu J, Love P, Davis P, Smith J and Regan M 2015 Conceptual framework for the performance measurement of public-private partnerships *J. Infrastr. Sys.* **21** 1
[36] Yuan J, Wang C, Skibniewski M and Li Q 2012 Developing key performance indicators for public-private partnership projects: questionnaire survey and analysis *J. Manag. Eng.* **28** 3 252–64
[37] Yuan J, Zeng A Y, Skibniewski M and J Li Q 2009 Selection of performance objectives and key performance Indicators in public-private partnerships projects to achieve value for money *Constr. Manag. Econ.* **27** 3 253–70
[38] Zhang X 2006a Factor analysis of public clients' bestvalue objective in public–privately partnered infrastructure projects *J. Constr. Eng. Manag.* **132** 9 956–65
[39] Liyanage C and Romero F V 2015 Measuring success of PPP transport projects: a cross-case analysis of toll roads *Transport Reviews* **35** 2 140–61
[40] Duncan T and Liman R 2006 Learning from successful road projects. Operations evaluation department (Asian Development Bank)
[41] Ng S T, Wong Y M and Wong J M 2012 Factors influencing the success of PPP at feasibility stage–a tripartite comparison study in Hong Kong *Habitat Int.* **36** 4 423–32
[42] Feng M, Mangan J, Wong C, Xu M and Chandra L 2014 Investigating the different approaches to importance–performance analysis *Service Industries J.* **34** 12 1021-41
[43] Tileng M Y, Utomo W H and Latuperissa R 2013 Analysis of service quality using servqual method and importance performance analysis (IPA) in popu-lation department, Tomohon City *Int. J. Com. App.* (0975-8887) **70** 19 24-30
[44] Paterson I, Maguire H and Al-Hakim L 2008 Analysing trust as a means of improving the effectiveness of the virtual supply chain *Int. J. Netw. Virtual Organ.* **5** 3/4 325–48.
[45] Wang S Q, Tiong R L K, Ting S K and Ashley D 2000 Evaluation and management of political risks in China's BOT projects *J. Constr. Eng. Manag.* **126** 3 242–50

A concept to evaluate procurement principles implementation of public construction project in Surabaya

P V Sandi[1], M A Rohman[1*] and C Utomo[1]

[1]Department of Civil Engineering, Institut Teknologi Sepuluh Nopember
Kampus ITS Sukolilo, Surabaya, Indonesia.

*Corresponding author's E-mail: arif@ce.its.ac.id

Abstract. Public construction procurement is a Government program to meet infrastructure needs. As a government program that uses the public budget as a source of funding, public procurement tends to be vulnerable to various problems and criticisms, especially related to public trust as users. Therefore, the government must prove that the implementation of procurement including the procurement of construction can be relied upon, so it must be carried out professionally and ethically. To guarantee that, construction procurement must adhere to the principles set out in the regulation. Presidential Regulation of the Republic of Indonesia Number 16/2018 regulates Public Procurement of Goods and/or Services and establishes 7 principles of procurement implementation, including: efficient, effective, transparent, open, competitive, fair, and accountable. This paper is an initial stage of research to evaluate the implementation of the principles of construction public procurement. The purpose of this study is to identify the indicators that must be met by the government in the implementation of these principles. These principles will become the research variables, while the indicators are obtained from related literature studies. The data in this study were obtained by distributing questionnaires and interviews with several parties concerned with the public procurement in Surabaya. According to the experts' opinion, there are found that 48 indicators of the principles which could be used to evaluate the principles of the implementation of public construction project procurement in Surabaya.

1. Introduction

The construction sector as an inseparable part of the procurement system is also strengthened by the general definition of procurement as the acquisition of goods, services, or construction, from a third party, at the best possible price, in an appropriate quantity, at the right time and place [1]. Specifically, in the construction projects, procurements are the acquisition of new buildings, space in the buildings, either by directly buying, renting, or leasing from the open market or by designing and building the facilities to meet a specific need [2].

Procurement and construction, especially those related to government projects are complex things. Procurement is very complex because it involves cultural, leadership, management, economic, environmental and political issues [3] while construction is very complex because it must integrate the interests of a large number of stakeholders towards the ultimate goal to realize the facilities being built. Furthermore, government procurement is very vulnerable to public trust because it uses the

Content from this work may be used under the terms of the Creative Commons Attribution 3.0 licence. Any further distribution of this work must maintain attribution to the author(s) and the title of the work, journal citation and DOI.
Published under licence by IOP Publishing Ltd

State/Regional Government Budget, which is a public fund as a source of funding [1]. The use of public funds has an impact on the vulnerability of public trust in the procurement system.

To address the vulnerability of public trust and various problems in a procurement, the government as well as other countries use information technology (IT)/ electronic procurement (e-procurement) as a medium for the procurement of goods and services, including construction sector. E-procurement is a system that can be relied upon by an organization in making contracts for product acquisition and service purchase by utilizing IT systems. [4]. E-procurement positively contributes to helping developing countries that tend to experience problems with transparency, accountability, and public participation to access information [5]. Specifically, e-procurement has good benefits in reducing labor costs and errors, reducing material costs with strategic purchases, enabling companies/government to more efficiently and accurately capture and aggregate how much they spend on the company for the various products purchased, and has an overall impact on the company/government's asset base [6].

Beside the problems that have been explained before, there are some barriers in the implementation of e-procurement including security of transaction issues, high technology costs, lack of legal framework, lack of technical expertise, lack of knowledge related to e-procurement, and lack of working relationships with business partners [7]. Therefore, the implementation of this system must be carried out carefully and professionally, and develop continuous strategies and innovations.

One of the cities in Indonesia that first implemented e-procurement and was considered successful is Surabaya. E-procurement has been started and developed in Surabaya since 2004. The procurement system in Surabaya was even adopted as a national pilot system in Indonesia. Surabaya is considered as the most successful city in implementing e-procurement in Indonesia and is often used as a model for other regions in Indonesia. This is very interesting to review as well as to find out how Surabaya implemented the system when there were many problems and obstacles that occurred globally as previously explained.

One thing that can be evaluated is how Surabaya applies the policies and procedures in the form of procurement principles. This is because procurement principles are the cornerstones of public procurement implementation which in detail should regulate public procurement management, including the framework and code of conduct of public procurement for all relevant parties [8]. The principle is clearly stated in Presidential Regulation of the Republic of Indonesia Number 16/2018 which stipulates seven principles of procurement implementation, including: efficient, effective, transparent, open, competitive, fair, and accountable.

Based on the description of these conditions, this research wants to find out the indicators of procurement principles that can be used to evaluate the implementation of procurement principles, specifically in the construction project sector in Surabaya. The 7 principles will become the research variables while the indicators are obtained from related literature studies. The data in this study were obtained by distributing questionnaires and interviews with the experts. After identifying the indicators, it can be known what indicators are relevant to be used in evaluating the implementation of the principles of public procurement in Surabaya, and which can be used as a benchmark for improvement for further analysis.

2. Literature review

2.1. Public procurement

Public procurement is a product of public policy issued and endeavored by the government. It is defined as a very important government system for spending public money with the aim of obtaining goods, jobs and services needed for public programs and projects [9]. The main purpose of public procurement is a guarantee of the effectiveness of the contract to related parties, both to contractors or suppliers who meet the requirements to provide goods and services required to support public services in accordance with established principles and regulations [8].

2.2. Electronic procurement (E-Procurement)

E-procurement is a procurement system that is applied to an integrated data base system and wide area internet-based communication system network in part or all of the purchasing processes [10]. Generally, e-procurement is defined as a comprehensive process in which organizations use IT systems to establish agreements for the acquisition of products or services (contracting) or to purchase products or services in exchange for payment (purchasing) which includes electronic ordering, internet bidding, card purchases, auctions, and automatic procurement system integration [11].

E-procurement offers promising opportunities as regards the efficiency, transparency, and opening up of public procurement, ensuring that public contracts are awarded to the bidder offering the best value for money [12].

E-procurement is often exposed to issues of quality of resources or the performance of procurement and provider employees, procedural errors, lack of transparency, and so forth. One research found that the problem of professional resistance is one of the very serious problems in e-procurement. This is indicated by the lack of understanding related to bureaucracy and finance which leads to the less competitive preferences so that e-procurement also leads to certain bidders [13].

2.3. Construction procurement

Construction helps capital formation and employment [14]. Construction procurement is the purchase of construction-related services with the ultimate goal of creating new buildings or new structures, including all related work or location changes, repairs, maintenance, expansion, or demolition of new buildings or existing structures [1].

The procurement activities are distributed among multiple parties in each project phase, according to the selected procurement method. The activities of construction procurement are described into 4 phases: (i) the pre-contractual phase : define requirements, plan the procurement process, pre-tender survey, obtain the necessary approvals, bid solicitation; (ii) the contracting phase: bidder selection, pre-bid conference, tender evaluation, purchase order; (iii) the contract administration phase: issue contract amendments, monitor progress, follow up on delivery, administer progress payments; and (iv) the post-contractual phase: file final action contractor agreement to final claim, issue final contract amendment, complete of financial audits, check for proof of delivery, return of performance bonds and close-out, and ensure completeness and accuracy of file documentation [1].

2.4. Principles of public procurement in Indonesia

The principle of public procurement regulates the management of public procurement, including the establishment of a framework for implementing the code of conduct for all parties directly or indirectly involved in the process of implementing public procurement [8].

Each country basically has its own principle of public procurement based on economic, political, and other circumstances and conditions in the country. In Indonesia, these principles are emphasized in Presidential Regulation of the Republic of Indonesia Number 16/2018 which stipulates 7 principles of procurement implementation, including: efficient, effective, transparent, open, competitive, fair, and accountable.

Efficiency can be seen as a situation where procurement can reduce losses or waste of resources, but with these limitations, it is able to produce benefits and better output in terms of time, cost, and energy [17] [4]. This principle requires that the government procurement process be proportional to the values and risks of the project that underlie the activity, which is generally related to time sensitivity and delay avoidance, meaning that with a minimum budget the quality of public procurement results can be obtained [9].

The effective principle can be interpreted productively, namely the available resources obtaining goods and services that have the highest benefits [18]. This principle is a condition where the use of government procurement can achieve the desired results [17] [19].

Transparent is the provision of complete information to all prospective participants. The transparent is a principle that requires the disclosure of relevant information related to procurement to the public through sources that are easily accessible and widely available at reasonable prices or at no cost, the existence of reporting related to the implementation of activities, and the use of confidentiality in contracts [20]. A research in Scandinavia found that one of the causes of the low corruption rate in Scandinavian countries was the application of information transparency, namely providing maximum freedom for the public to access documents owned [21].

Open principle provides opportunities for all competent providers of goods and services to be involved [20]. This principle also emphasizes the involvement of the community in carrying out social control to meet the demands of civil society for information needs [22].

Competitive principle refers to the involvement of two or more bidders, preventing competition restrictions [23]. That competition is based on ethics and norms including avoiding the practices of Corruption, Collusion, and Nepotism [20].

Fair principle is related to the equal and equitable treatment of each individual or organization that offers, which means that the supervisor of the applicant is in legal and factual circumstances [8],[9],[21]. This principle also needs a code of conduct as which reflects the policies, practices and procedures, and the same perceptions for all parties [24].

The principle of accountable refers to the existence of an effective control mechanism with a clear chain of responsibility based on the authority of the parties involved at each stage and complementary internal and external controls to ensure the principles of procurement benefits, the handling procurement complaints and disputes in accordance with the mechanism apply, and the involvement of the civil society organizations, the media and the wider community to participate in evaluating the procurement implementation process [25].

The reviews above are just a few of the many opinions and reviews from various literatures. Therefore, table 1 containing summary of opinions is provided as an indicator of procurement principles and the result of the survey.

3. Research methodology

This research method is to identify the indicators of variables, preliminary surveys involving experts, and survey analysis. The preliminary survey was conducted by involving four experts who were directly involved in public procurement in Surabaya. The experts in this survey are those who have experienced at least 5 years in this field.

The preliminary survey is to verify the indicators of the principle of public procurement that have been identified from literatures. This stage is also to get opinions from experts if there are indicators of a principle that overlaps with other principle indicators. To verify the indicators, the preliminary survey used a 1-5 scale representing highly irrelevant to highly relevant. Then the indicators are considered relevant if the mean of the variable is greater than three (>3).

4. Result and analysis

Based on the preliminary survey analysis, there are some indicators that are omitted because they are considered irrelevant with a value of less than 3, or have similarities with other indicators in the principles. There are total 48 relevant indicators and 14 irrelevant indicators in this survey. The results of the survey can be seen in table 1.

Table 1. Indicators of public procurement principles and the result of survey

NO	Indicators	Source	Rank	Mean	STD
A.	EFFICIENT				
1.	Utilization of electronic system	[26]	1	4.50	0.58
2	Less time process	[4], [9], [17], [18], [27]	2	4.25	0.96

3.	Decreasing number of claims	[28]	3	4.25	0.50
4.	Single point access	[26]	4	3.75	1.50
5.	Less cost process	[4], [17], [27], [28], [29]	5	3.75	0.96
6.	Reduction of human error	[28]	6	3.75	1.26
7.	A classification system for goods, works, and services	[26]	7	3.50	1.73
8.	Decreasing number of administrative staff	[28]	8	2.50	1.73
9.	Increased number of participation	[31]	9	2.25	0.96
10.	A classification of procurement value	[26]	10	1.00	0.58
B.	EFFECTIVE				
1.	Achievement of procurement goals	[17], [30], [32]	1	5.00	0.00
2.	Completion on time	[19], [60]	2	4.50	0.58
3.	In accordance with people needs	[61], [62]	3	4.25	0.96
4.	Program integration	[30], [33]	4	4.25	0.50
5.	Control mechanism that regulates procurement	[26]	5	4.00	0.82
6.	System quality	[32]	6	3.75	1.89
7.	Minimize face to face communication	[34]	7	3.75	1.89
8.	There is a management that regulates the communication process, strategy, innovation, and program adaptation	[15]	8	2.75	1.50
C.	TRANSPARENT				
1.	Rules and procedures are simple and easy to apply	[35], [36],[37], [38]	1	4.75	0.50
2.	Access to information	[21],[26], [34], [37], [38], [40], [39], [42], [40], [41], [42]	2	4.75	0.50
3.	Information availability	[8], [9], [20], [34], [39], [43], [23]	3	4.50	0.58
4.	Ease of accessing the system	[20]	4	4.50	0.58
5.	Budgeting visibility	[38]	5	3.75	1.89
6.	Monitoring and sanctions mechanism	[35]	6	3.75	1.89
7.	Archiving documents	[38]	7	3.50	1.91
8.	Mass media participation	[37], [38]	8	3.00	1.62
9.	Society participation	[44]	9	2.50	1.91
10.	Information justification	[8]	10	2.25	1.50
D.	OPEN				
1.	Admission based on qualifications	[8]	1	4.75	0.50
2.	Blacklist/suspend list publication	[16]	2	4.50	0.58
3.	Dissemination of procurement information	[45]	3	3.50	0.58
4.	There are no restrictions on participants based on origin, location, religion, ethnicity, or political orientation	[20]	4	3.25	0.58
5.	Utilization of local resources	[26]	5	2.25	1.50
6.	Privacy protection policy	[8]	6	2.25	1.73
E.	COMPETITIVE				
1.	There was no intervention from any party	[8]	1	4.75	0.50
2.	Clear procedures and requirements related to procurement competition	[42], [23], [49]	2	4.75	0.50
3.	The same opportunity to compete	[42]	3	4.25	1.50
4.	Procurement Agency is not affiliated	[48]	4	4.25	0.96
5.	The completeness of document	[42]	5	4.25	0.96

6.	Involving three or more bidders	[42], [23], [46]	6	4.00	1.41
7.	Obey the procurement norms and ethics	[20], [50]	7	3.50	0.58
8.	The proportion of the number and value of direct appointment contracts is smaller	[47]	8	2.50	0.58
9.	Participation of small and medium enterprise (SME)	[58]	9	2.25	1.00
F.	FAIR				
1.	Determination of winners based on qualifications	[8]	1	4.75	0.50
2.	Avoiding conflicts of interest	[22]	2	4.75	0.50
3.	Equal distribution of rights and obligations between candidates	[9]	3	4.50	0.58
4.	Code of Conduct	[24]	4	4.50	0.58
5.	Equal opportunity and treatment for participants	[8], [9], [21], [39], [42], [51], [52]	5	3.75	1.89
6.	The technical specifications required do not refer to specific brands	[26][59]	6	3.75	1.89
7.	The low number of complaints received	[54]	7	3.75	1.89
8.	Bidding considerations are based on compliance	[21]	8	3.25	1.71
9.	Credible mechanism to solving complaints	[53]	9	2.75	1.53
G.	ACCOUNTABLE				
1.	There is a control or supervision mechanism	[8], [38], [41]	1	4.75	0.50
2.	Internal and external audit	[34], [38], [41]	2	4.75	0.50
3.	Open reporting system	[26]	3	4.50	0.58
4.	Professionalism standard	[56]	4	4.25	0.96
5.	Action due to performance in the form of rewards and sanctions	[8], [38], [42], [57]	5	4.00	1.41
6.	Organizational framework	[38], [55]	6	3.75	1.26
7.	Conflict/complaint resolution	[25], [38]	7	3.25	1.71
8.	Evaluation by society	[25], [42]	8	2.75	1.26
9.	Supervisory agency	[42]	9	2.25	1.73
10.	Publication of uncompetitive procedures	[42]	10	2.00	1.53

Fourteen (14) irrelevant indicators (<3) of each principle are as follows: 3 out of 10 Efficient principles include irrelevant, they are decreasing numbers of administrative staffs, increased numbers of participations, and a classification of the procurement value, while the most relevant indicator is the Utilization of Electronic System.

One (1) of 8 indicators of Effective principle is irrelevant, namely the management that regulates the communication process, strategy, innovation, and program adaptation, while the most relevant indicator is the Achievement of procurement goals. For the principle of transparency, 2 of 10 indicators are irrelevant, they are society participation and information justification. For relevant indicators, there are 2 most relevant namely Rules and procedures are simple and easy to apply and Access to information. For the open principle, there are 2 of 6 indicators are irrelevant. They are utilization of local resources and privacy protection policy while the most relevant indicator is admission based on qualifications.

There are 2 of 9 indicators of competitive are irrelevant. They are the proportion of the number and value of direct appointment contracts is smaller and participation of small and medium enterprise (SME). On the other hand, the most significant relevant indicators are there was no intervention from any party and clear procedures and requirements related to procurement competition. One (1) of 9 fair principle indicators are irrelevant namely credible mechanism to solve complaints while the most relevant are determination of winners based on qualifications and avoiding conflicts of interest.

For accountability principle, there are 2 of 10 indicators are irrelevant. They are publications of uncompetitive procedures and supervisory agency while the most relevant indicators are: There is a control or supervision mechanism and internal and external audit.

5. Discussion

There are total 14 indicators that are categorized as irrelevant, so it is necessary to explain a number of things as the main reasons for which they were given a low score and then not relevant.

Indicator decreasing number of administrative staff is basically related to the use of electronic systems, which then results in the number of staff that should not be as many as if using a traditional system. However, experts believe that the indicator is not very big if used to measure procurement efficiency because the needs of resources (staff) are very dependent on the internal needs of the organization. Indicator increased number of participations, according to experts, is irrelevant because the number of bids is basically not limited, this is due to every incoming bid regardless of the amount to be selected in accordance with the provisions of the terms and conditions set for each type of work. Indicator a classification of procurement value is irrelevant because detailed procurement rules have classified each type of work into the procurement category with their respective methods with the amount of the value of their respective projects.

Indicator the management that regulates the communication process, strategy, innovation, and program adaptation, according to the expert, is irrelevant because this relates to the procurement rules themselves. This indicator is related to upper management because it is related to the character of each organization. Indicator society participation is irrelevance in transparent principles. Transparency does prioritize access to information, but the right to information for the community in a system can be accessed through the website system. In addition, the role of the community can be directly categorized more or less the same as mass media participation, in which the results of the analysis include relevant indicators.

Indicator information justification is basically related to the provisions of the categorization of types of information whether the information is public or confidential information so that it can be a reference, whether it can be accessed publicly or not. According to experts, this indicator is less significant because basically all types of information must be displayed to the public, except if it has entered the realm of procurement which indeed at this stage must be kept secret. Indicator utilization of local resources is related to the alignment of local resources. According to experts, this can be categorized as irrelevant because in the rule of non-local resources, including certain brands may be used on condition that the resources or specifications cannot be met domestically.

Privacy protection policies, is not related to open principles. According to experts, this is basically the same as indicator information justification, which is related to information, but there should be nothing to keep secret if it follows the participant data. Indicator the proportion of the number and value of direct appointment contracts is smaller, is similar to indicator a classification of procurement value, that the experts argue that procurement rules explicitly regulate and categorize the types of work in the procurement system based on the amount of each value with their own method. Therefore, the size of the amount and the value of procurement indeed affect each type of work.

Indicator Participation of Small and Medium Enterprise (SME), according to experts, cannot be used as a measurement tool for competitive principles. This indicator is similar to indicator the proportion of the number and value of direct appointment contracts is smaller, that the SME has its own portion in the type of work in the procurement system, including construction. The participation requirements are basically explained in each type of work, including the level of business of the participant.

On the principle of fair, there is one irrelevant indicator which is indicator credible mechanism to solve complaints. According to experts, this indicator is more inclined to the principle of transparency for the indicator of monitoring and sanctions mechanism. This is because basically it has the same intention related to handling complaints and giving sanctions that must be known and understood as rules for the participants.

In the principle of accountable, there are 3 irrelevant indicators. Indicator evaluation by society is the same as indicator society participation, that this role can already be achieved through the participation of the media depicted in the mass media participation indicator. Indicator publication of uncompetitive procedures, according to experts, is every action that leads to fraud should be processed internally and the final result in the form of a new blacklist action can be disclosed to the public, as confirmed in an open principle indicator: Blacklist publication. Finally, indicator supervisory agency in which the experts stated that the indicator is the same as the internal and external audit, indicator of the principle of accountable.

6. Conclusion and recommendation

In accordance with the background of this study, this study aims to obtain indicators that can be used to evaluate the implementation of the principles of public procurement. Based on the literature studies, 62 indicators were obtained which were grouped into 7 Indonesian public procurement principle variables. After the preliminary survey analysis with the experts, 48 indicators were obtained which could be used to evaluate the implementation of the principles of public procurement and 14 other indicators were declared irrelevant. This research is expected to help further research in evaluating and designing strategies for improving public procurement for government.

Acknowledgements

The authors thank the Indonesian Ministry of Research and Technology/National Agency for Research and Innovation for providing grants to perform this research in 2020. We also appreciate the experts involved in the preliminary survey in this research.

References

[1] Ruparathna R and Hewage K 2015 Review of contemporary construction procurement practices *J. Manag. Eng.* **31** 3
[2] Mohsini R and Davidson C H 1991 Building procurement- key to improve performance *Build. Res. Inf.* **19** 2 106-13
[3] Noor M A, Khalfan M M A, and Maqsood T 2013 The role of procurement practices in effective implementation of infrastructure projects in Pakistan *Int. J. Manag. Proj. Business* **6** 4 802-26
[4] Gunasekaran 2009 Adoption of E-Procurement and participation E-Marketplace on firm procurement tools *J. Phys. Distrib. Logis. Manag.* **33** 8 720-34
[5] Olken B 2007 Monitoring corruption: evidence from a field experiment in Indonesia *J. Political Econ.* **115** 2 200-49
[6] Presutti Jr W D 2003 Supply management and e-procurement: Creating value added in the supply chain *Ind. Mark. Manag.* **32** 219–26
[7] Hawking P, Stein A and Forsters S 2004 E-Procurement: is the ugly duckling actually a swan down under *Asia Pacific J. Mark. Logist.* **16** 1 1-26
[8] Lynch J 2013 Public procurement: principle, categories and methods (Leanpub)
[9] Khan N 2018 What Is Public Procurement: Introduction, Public Procurement Fundamentals *Emerald Pub. Limited* 1-20
[10] Croom S and Brandon-Jones A 2005 Key issuesin E-Procurement: Procurement implementation and operation in the public sector *J. Public Proc.* **5** 3
[11] Moon M 2005 E-Procurement Management in state governmnets: A diffusion of e-procurement practices and its determinants *J. Public Proc.*) **5** 1 54-72
[12] Oliveira L M S and Amorim P P 2001 Public e-procurement *Int. Fin. Law Review* **20** 43-47
[13] Sepehri M and Qadiri M 2018 Professional resistance against competitive public procurement in Iran *J. Public Proc.* **18** 2 165-81

[14] Hillebrandt P M 2000 Economic theory and the construction industry (London: Macmillan)
[15] Steers R M 1986 *Efektivitas Organisasi* (Jakarta: Erlanga)
[16] Thai Ki V 2017 Global public procurement theories and practices (Switzerland: Springer International Publishing)
[17] Teo T and Lai K H 2009 Usage and performance impact of electronic procurement *J. Bussiness Logist.* **30** 125-39
[18] Vaidya K and Campbell J 2016 Multidisclipinary approach to defining public e-procurement and evaluating its impact on procurement efficiency *Inf. Syst. Frontiers* **18** 2 333-48
[19] Panayiotou N A, Gayialis S P, and Tatsiopoulos I P 2004 An e-Procurement system for governmental purchasing *Int. J. Prod. Econ.* 79-102
[20] Hardjowijono B and Muhammad H 2009 Prinsip dasar dan kerangka hukum pengadaan barang dan jasa pemerintah (Jakarta: Indonesian Procurement Watch)
[21] Krivinsh A and Vilks A *2013* Prevention of corruption in public procurement: Importance of general legal principles *Jurisprudence* **20** 1 235-47
[22] Organisation for Economic Co-operation and Development 2016 Preventing corruption in public procurement (Paris)
[23] Lembaga Kebijakan Pengadaan Barang/Jasa Pemerintah and Millenium Challenge Corporation United States of America 2016 *Modul 1: Menerapkan Prinsip-Prinsip Dasar Pengadaan, Pelatihan Pengadaan – Tingkat Dasar, Proyek Modernisasi Pengadaan LKPP/MCA-I* (Bogor: LKPP/MCA-I)
[24] Byung G S, Chang H L, Byoung C H, and Hyunjeonget N 2019 Investigating the fair treatment of suppliers and its trust fostering role and performance benefits *Int. J. Prod. Econ.* **216** 54-66
[25] Organisation for Economic Co-operation and Development 2009 OECD principles for integrity *Public Procure.* (Paris)
[26] Institute for Development of Freedom of Information 2018 Reccomendations for the public procurement system of the Republic of Indonesia
[27] Asian Development Bank, Inter-American Development Bank, and World Bank 2004 Electronic governmnet procurement roadmap
[28] Vaidya K and Campbell J 2014 Multidisciplinary approach to defining public e-procurement and evaluating its impact on procurement efficiency *Inf. Sys. Frontiers* 1-16
[29] Zunk M B, Marchener M J, Uitz I, Lerch C and Schiele H 2014 The role of e-procurement in the austrian construction industry: Adoption rate, benefits, and barriers *Int. J. Ind. Eng. Manag.* **5** 1 13-21
[30] Siagian S P 1982 Manajemen modern (Jakarta: Gunung Agung)
[31] Daniel I 2014 Principal of public procurement: the basis for the good performance *OECD/SIGMA*
[32] Joesbury P 2016 Improving the effectiveness of procurement identification and improvement of key determinant factors – The pepps project (Birmingham: Doctorate Thesis Aston University)
[33] Indrawijaya A L 1983 Perilaku organisasi (Bandung: Sinar Baru)
[34] Neupane A 2014 The Potential of public e-procurement technology to reduce corruption in public procurement (Australia: Doctorate Dissertation University of Southern Queensland)
[35] Asian Development Bank 1999 International survey of government procurement system *Int. Gov. Solution Ltd*
[36] Vicente R S 2008 Evaluation of efficiency and effectiveness of the regional procurement service depots in the Philipines: A case of region 1 (New Zealand: Master Thesis Massey University)
[37] Jones D S 2012 Reforms to promote competition in government procurement in Southeast Asis: achievements and constraints *Asia Pacific J. Public Adm.* **34** 2 113-36
[38] Organisation for Economic Co-operation and Development 2016 High-level principle for integrity, transparency, and effective control of major events and related infrastructures (Paris)

[39] Tabish Z S H and Jha N K 2011 Analyses and evaluation of irregularities in public procurement in India *Constr. Manag. Eco.* **29** 3 261-74
[40] Da Cruz N F, Tavarez A F, Marques R C and De Sousa L 2016 Measuring local government tranparency *Pub. Manag. Review* **18** 6 866-93
[41] Hasan M S M 2016 Public procurement in roads and highway department: A tussle of transparencyand accountability (Bangladesh: Dissertation North South University)
[42] Komakech R A 2016 Public procurement in developing countries: Objectives, principles and required professional skills *Public Policy Adm. Res.* **6** 8
[43] Ochrana F and Pavel J 2013 Analysis of the impact of transparency, corruption, openness in competition and tender procedures on public procurement in the Czech Republic *Central Euro. J. Pub. Policy* **7** 2 116-36
[44] Thai K V 2001 Public procurement re-examined *J. Public Procure.* **1** 1 9-50
[45] Eady R, Parera S, Heany G and Carlisle J 2007 *Construction Contractor's Perspective on Web-based Advertising of Public Sector Construction Contracts* ed Boyd D (Belfast: Proc. 23rd Annual ARCOM/Association of Researchers in Construction Management) 767-76
[46] Kang H and Kim B 2018 A study on reforming the Korean public procurement system (Korea: Korea Institute of Public Finance)
[47] The World Bank 2016 Public procurement indicators *World Bank Group*
[48] Racca G M, Perrin R C, and Albano G L 2016 Competition in the execution phase of public procurement *Public Contract Law J.* **41** 1 89-108
[49] Comission Nacional De La Competencia 2011 Guide on Procurement and Competition (Spain: CNC)
[50] McRudden C 2012 Procurement and fairness in the workplace. Making employment right effective (Hart Publishing: Oxford and Portland, Oregon)
[51] Lindskog H, Brege S and Brehmer P O 2010 Corruption in public procurement and private sector purchasing *J. Organ. Transformation Soc. Change* **7** 2 167-88
[52] Kim D I 2011 The principle of fairness: Theory, defence, and application (United Kingdom: Doctorate Thesis University of Warwick)
[53] Chimia L A and Trepte P 2019 Public procurement and aid effectiveness: A roadmap under construction (United Kingdom: Hart Publishing)
[54] Jourdain R and Balgobin N 2003 Controlling corruption in asia and pacific (Kuala Lumpur: 4th Regional Anti-Corruption Conference of the ADB/OECD Anti Corruption Innitiative for Asia and the Pacific)
[55] Armstrong E 2005 Integrity, transparency and accountability in public administration: recent trends, regional and international developments and emerging issues *United Nations*
[56] Diggs S N and Roman A V 2012 Understanding and tracing accountability in the public procurement process *Public Perform. Manag. Review* **36** 2 290-315
[57] Sohail M and Cavill S 2008 Accountability to prevent corruption in construction projects *J. Constr. Eng. Manag.* **134** 729-38
[58] Organisation for Economic Co-operation and Development 2018 SMEs in public procurement:Practices and strategies for shared benefits: OECD government reviews (Paris)
[59] Jacobs K 2018 Corporate and global standardization initiatives in contemporary society (USA: IGI global)
[60] Purwosusilo 2014 Aspek hukum pengadaan barang dan jasa (Jakarta: Penerbit Kencana)
[61] Tanesia R K 2015 Studi efektivitas pengadaan barang dan jasa pemerintah secara tradisional dan elektronik *J. Tek. Sipil* **13** 2 136-45
[62] Andriani S 2012 Analisa efektivitas hukum dalam penerapan pengadaan barang dan jasa secara elektronik (e-procurement) serta peranan lembaga pengawas terhadap pengadaan barang dan jasa pemerintah (Jakarta: Tesis Universitas Indonesia)
[63] Peraturan Presiden 2018 Pengadaan barang/jasa pemerintah 16

Work breakdown structure (WBS) dictionary and checklist development of stadium architectural and interior works for safety planning

J T Amin[1*], L Sagita[1], and Y Latief [1]

[1] Civil Engineering Department, Faculty of Engineering, Universitas Indonesia, Kampus Baru UI, Depok, Jawa Barat, 16424, Indonesia.

*Corresponding author's e-mail: Junita.taskia@alumni.ui.ac.id

Abstract. Nowadays, architectural and interior works in stadium construction have become the most critical elements. The stadium is required not only to have excellent sports facilities but also to have a high-level architecture and interior. Therefore, the works need to be more managed and planned so that the stadium project can run better and more effectively. The project planning phase has a crucial role which is when the project plans are documented, the project deliverables and requirements are defined, and subsequently, the project schedule is created, which means the development of Work Breakdown Structure (WBS) and its associated WBS dictionary with the checklist is undoubtedly imperative. The development is defining and breaking down the work into smaller components, which finally become such a project implementation reference in order to minimize the errors and ensure the project's implementation can be in accordance with the plan. The results of this study describe the details of the development of WBS dictionary and checklist of architectural and interior work in stadium construction for safety planning that includes six (6) levels such as project name, division, sub of work, work package, activity, and resource and other information related.

1. Introduction

One of the crucial infrastructure developments quite highly concerned in Indonesia recently is a stadium that provides a multitude of purposes, such as sports activities, particularly football, as the most popular sport in Indonesia, even across the world. The popularity in Indonesia has increased quite significantly in each year. As a result, the stadium most often becomes landmarks on the scale of the city [1].

The establishment of a high-quality and sophisticated stadium that complies with the standards required has become mandatory and should be supported by proper facilities. Not only that, but it is also necessary to have an attractive and unique architectural design as it for football stadiums has developed extremely in recent years. Formerly, stadiums were characterized mainly as feats of engineering with less emphasis on architectural finesse. Progressively, a football stadium is regarded as an architectural icon within the urban landscape that have a huge impact on infrastructure. However, the crucial consideration for every detail of the architectural design is safety [2].

Content from this work may be used under the terms of the Creative Commons Attribution 3.0 licence. Any further distribution of this work must maintain attribution to the author(s) and the title of the work, journal citation and DOI.
Published under licence by IOP Publishing Ltd

Besides the importance of the architectural side, interior work also becomes a critical element of building a stadium-related to furniture arrangement inside the building. Stadiums have improved significantly in the level of comfort that they provide for spectators. It has become mandatory to have appropriate facilities, such as seats and sanitary facilities, which are convenient for not only players but also spectators [3].

Implementation of a stadium project is a complex works that require an appropriate approach and structured planning such as Work Breakdown Structure (WBS) development to manifest how project deliverables are broken down into smaller elements which are work packages and provide a way of showing high-level areas of responsibility. Moreover, WBS is known as a framework for project implementation which is used to describe project scope to improve project estimates adequately, control project implementation better and verify the completion of a project, since it allows the work to be planned and controlled [4].

In the execution of the project, there will be found a variety of problems that will lead to disadvantageous impacts on the project target. Many factors lead to several project stumbling blocks, such as over budget, delayed projects, poor quality, and rework. These problems affect the project objectives of cost, quality, and time of the project. As a result of that, it is completely necessary to do something in order to overcome various issues that may occur. The leading cause of these issues is strongly related to lack of adequate work completion, so developing WBS to become a WBS Dictionary is necessary [5].

WBS Dictionary is a tool used to segregate the work into manageable components and define each WBS element, including its scope, activities, and resources. Furthermore, developing a detailed WBS dictionary as the supplementary document would lead to a successful project as it can minimalize the issues. However, there is another crucial thing required to support the tool, which is a WBS checklist that is aimed to review each component in the WBS Dictionary. The Checklist is for consistency in the plan which means it is sincerely critical to avoid any problems [6]. As a result of that, this research would develop the WBS dictionary and checklist of stadium architectural and interior work for safety planning.

2. Literature review

2.1. Stadium architectural and interior WBS

2.1.1. Stadium architectural and interior. Stadium is a venue for sports, concerts, and other events which consists of a field and is surrounded by a tiered structure designed to let spectators stand, sit, and view the event. It is a place that brings thousands of people together and creates a very incredible architectural atmosphere. Stadiums should be accessible for spectators and it is necessary to know spectators fully [7]. Moreover, it should seat thousands of people to create a festive atmosphere and it should have a significant architecture to remind this atmosphere after the festivity. Thus, architectural and interior works have become associated with stadium development. The architecture itself means an art implementation in building construction or a sophisticated structure with high accuracy. Moreover, the work itself includes visual appearance works of the buildings before structural design while not affecting the strength of the building structure. Meanwhile, Interior works are associated with the arrangement of what inside the stadium [8].

2.1.2. Work breakdown structure (WBS). WBS is essential for defining tasks and managing scope which naturally leads to planning and scheduling so that it is also crucial for the practical application of the technique. It provides the basis for identifying work as it relates to project objectives and also develops the framework for managing the work to its completion. This structure plays a primary role in project and program management in each of the process groups which ensures a consistent definition of the scope of the work to be undertaken. It comprises multi-level hierarchy describing the entire scope to be accomplished by the performing organization; however, the specific number of levels should be appropriate for effectively managing the project in question. In overall, the WBS is used in projects as follows [9]:

1. To define the project's scope of work in terms of deliverables and to further break down these deliverables into components.
2. To provide the project management team with a framework on which to base project status and progress reports.
3. To communicate information regarding the project scope
4. As a key input to other project management processes and deliverables

2.2 Safety performance
Safety performance is usually defined as an accident that occurred to a person resulting in various degrees of injury. In construction, it is implied the most significant factor in the cause of the site accidents is dangerous or unsafe behaviour and therefore it could be an evidence of a poor safety culture. Excellent safety performance is also related to the management style and that applying excessive pressure by any means to the workmen resulted in increased injuries [10].

2.3. WBS dictionary
Work Breakdown Structure (WBS) dictionary is used as a document that provides detailed deliverables, activity, and scheduling information about each component in the WBS which could supports the WBS. Most of the information included in the WBS dictionary is developed by other processes and added to this document at a later stage. Information in the WBS dictionary may include but is not limited to [4]:

1. Code of account identifier,
2. Work description,
3. Assumptions and constraints,
4. Responsible organization,
5. Schedule milestones,
6. Associated schedule activities
7. Resources required,
8. Cost Estimates,
9. Quality requirements,
10. Acceptance criteria,
11. Technical references, and
12. Agreement Information

2.4. WBS checklist
WBS checklist is aimed to evaluate throughout tasks defined previously in WBS as it consists of questions related to the functions. The form of checklist itself varies depends on the project requirement related. Furthermore, it is the result of the questions. This research utilizes a checklist with four WBS levels: project name, division, sub of work, and work package. A checklist is a structured tool, usually component-specific, that aims to verify that a set of required steps has been performed successfully. Checklists range from simple to complex based on project requirements and practices. Many organizations have standardized checklists available to ensure consistency performed tasks infrequently. When checklists are used to support to project completion, the completed checklists become part of the project's records [4].

2.5. Correlation between safety performance and WBS
By considering safety performance in developing WBS which is vital as its primary function is to avoid risks, the planning can result in a variety of detailed planning recommendations for a project. Without any safety consideration, a project can lead to various problems that have not defined previously so that risk identification is necessary. Identifying and analyzing potential events that may negatively impact on the project can reduce and minimalize risks and any problems unexpected [4]. It is proven by the previous research that said safety as a key performance indicator of a project. The critical intersection of project and

IOP Conf. Series: Materials Science and Engineering **930** (2020) 012006 doi:10.1088/1757-899X/930/1/012006

safety management involves the management and reduction of income erosion, while enhancing quality and creating cultural enrichment so that safety should be one of key components of the planning process [11].

3. Methodology

This research methodology, used to develop the WBS dictionary and checklist for architectural and interior works of a stadium in this study, includes:

1. Identifying the WBS standard for a stadium construction project, the author developed it through archive analysis compiled and several project data of stadium used as the literature for benchmarking.
2. Developing the WBS dictionary for a stadium, the benchmarked WBS standard was subsequently validated for its format and contents by five experts with a minimum of ten years of experience in managing stadium projects in Indonesia and other projects related to the stadium, and positioned as the Project Manager.
3. After validated, the WBS dictionary and checklist were adjusted based on the final WBS standard and were then validated by three experts with the same qualifications as the previous one

4. Result and discussion

The WBS level in this research is defined and benchmarked from BoQ (Bill of Quantity) on five previous stadium projects to determine the activities and resources of each level. A six-level of the WBS then formed based on those data which consists of:

1. WBS Level 1: Project Name,
2. WBS Level 2: Working Group / Division,
3. WBS Level 3: Sub of work,
4. WBS Level 4: Work Package,
5. WBS Level 5: Activities, and
6. WBS Level 6: Resources

Furthermore, the six-level WBS validated by the five experts with a background of WBS-based project planning knowledge and at least ten years experiences in the field of stadium construction projects to gain feedback, suggestion, and comments regarding the WBS standard of the stadium project. The Stadium WBS standard is illustrated in Table 1.

Table 1. Validated stadium WBS standard

WBS LEVEL 1		WBS LEVEL 2		WBS LEVEL 3		WBS LEVEL 4		WBS LEVEL 5		WBS LEVEL 6		
KODE	**Project Name**	**KODE**	**Division**	**KODE**	**Sub of Work**	**KODE**	**Work Package**	**KODE**	**Activity**	**KODE**	**Type of Resource**	**Resource**
0	**Stadium**	3	Architectural work	3.2	Wall Installation Work	3.2.2	Wall Panel Installation Work	3.2.2.1	Hollow Aluminium Frame Installation work		Workman	Foreman, Worker, Operator

The validated stadium WBS standard shows that there is no change compared to the previous one after validated. Then, After validated, the WBS standard is converted into WBS dictionary and checklist which is then validated by three experts with the same qualifications as the previous one to get feedback and comments regarding the elements such as work components from level 1-6, person in charge, deliverable, reference, and other aspects required in WBS dictionary. Based on the validation and interview results, here is the stadium WBS dictionary and checklist illustrated in Figure 1 and Figure 2.

WBS DICTIONARY OF STADIUM
WALL WORK
Wall Panel Work

3	Work Family / Division	:	Architectural work
3.2	Sub of Work / Section	:	Wall work
3.2.2	Work Package	:	Wall panel work
	Person in Charge	:	
	Description	:	The wall panel work consists of Hollow aluminium frame work, wall panel work, finishing work of hpl and upholstery, hardware work/ additional work for accessories according to qualification and specification determined
	Deliverable	:	Wall panel
	Reference	:	1. Contract Document (BoQ) 2. RKS/ Technical Specification Document 3. Material Brochure 4. Expert Validation

Code	Activities	Resources		
		Man	Material	Equipment
3.2.2.1	Hollow aluminium frame work installation work	1. Foreman 2. Worker 3. Operator	1. Hollow Aluminium frame 2. Frame binder: fishcer, screw, nail	1. Lightweight Tools
3.2.2.2	Wall panel installation work	1. Foreman 2. Worker 3. Operator	1. plywood panel 2. frame binder: anchor, nail, screw	1. Lightweight Tools
3.2.2.3	Finishing work of hpl and upholstery	1. Foreman 2. Worker 3. Operator	1. Hpl (High Pressure Laminate) 2. Upholstery 3. Adhesive	1. Lightweight Tools
3.2.2.4	Hardware work/ additional work for accessories	1. Foreman 2. Worker 3. Operator	1. Handle 2. Lockset 3. Accessories binder: bolt and nut, screw	1. Lightweight Tools

Figure 1. Validated of stadium WBS dictionary for each work package

WBS Level 1	:	Project Name
WBS Level 2	:	Work Family / Division
WBS Level 3	:	Work Package
WBS Level 4	:	Person in Charge

WBS CHECKLIST OF STADIUM

ARCHITECTURAL WORK

WBS LEVEL	**CODE**	**DETAILS**	
1		Stadium	Project name
2	3	Architectural Work	The architectural work consists of flooring, wall installation work, plafond installation work, wall and door installation work, hardware work, sanitary work.
3	3.1	Wall Work	The wall installation work consists of wall, wall panel installation work, toilet cubicle installation work with specification required.
4	3.1.1	Wall	This work consists of mortar plastering work, tail water installation work, bonding plastering work, basic painting work, final painting work, brick wall installation work, fire brick installation work, glass installation work, frame installation work, gypsum board installation work, textile tape and compound, grouting.

Figure 2. Validated stadium WBS checklist for each division.

WBS Dictionary is mainly used to describe each element of project activity (WBS level 5) and resources (WBS level 6). The validated format provided the descriptions of the primary elements such as the code, person in charge, resource, and deliverable of every defined activity. Hence, the stadium WBS dictionary is necessary to ease potential problems as it is more readable and understandable. During the construction process, it can be used very effectively and particularly for monitoring and controlling every defined work package. Furthermore, the validation process results also found that the development of the WBS Dictionary is not just about a project management tool. Still, it is the primary document that is crucial for construction project planning. It fully supports the sustainability of the project itself, which supports its time, cost, and quality performance within the project's lifecycle. For instance, to provide a sustainable product, in this case, a stadium. The stadium's material quality can be thoroughly defined using the WBS dictionary as it is the basis for determining the activities, resources, and quality requirements. Issues which may occur due to environmental or sustainability considerations can also be facilitated through the details of its WBS dictionary [4].

Regarding the WBS checklist, the experts recommended no significant change but simplify some descriptions. It resulted from the Stadium WBS checklist; architectural work has more detailed information and description due to its more complicated task than others in other divisions. Moreover, it is supported as well by previous research that architectural work has detailed works, whereas the interior work has much simpler work so that the description of each work is not as detail as the architectural work's [12].

The researcher also conducted a direct interview with the experts regarding the Stadium WBS Dictionary and Checklist which concluded that WBS dictionary and checklist could become a guide because it can help the person in charge control the whole works in a project with detailed information and description for each work. As a result, it showed that the WBS dictionary and checklist are essential as necessary planning of constructions, particularly in stadium construction with a high level of difficulty. Moreover, since the WBS dictionary and checklist are used to identify all works up to resources with a detailed description, it has brought tremendous beneficial impacts on a project. However, the WBS dictionary and checklist have often been found not easy to understand and read because of the ambiguous of each work description so that words adjustment and modification, especially by using familiar words, are needed in developing the WBS dictionary and checklist. Finally, risk identification could be identified based on the validated WBS dictionary. This risk identification development is aimed to thoroughly succeed each work by minimalizing the risks and determining how to control them. An example of the development can be seen in Table 2.

Table 2. HSE risk identification.

No	WBS Level 4	WBS Level 5	Hazard Identification	Risk Assessment			Priority Scale	Risk Management	Person in Charge
	Work Package	Activity		Frequency	Severity	Risk Rating			
1	Wall work	Hollow aluminium frame work	the aluminium frame falls and injures the workman	1	2	2	3	Use glasses and helmet	Foreman, Safety Team
			the point of frame can injure the workman	1	2	2	3	Use glasses and helmet	Foreman, Safety Team

The table above shows that the hollow aluminium work has two risks with each risk assessment, priority scale, management, and person in charge. Risk assessment consists of three categories: level of frequency, severity, and probability produced by the hazard. The risk rating is obtained by multiplying the frequency value and severity value. Priority scale is a value of the fact or condition of being regarded or treated as more important. From that, identifying the priority scale, the alternative solution can be described, for example, using glasses and helmet [13].

5. Conclusion

To conclude, the validated stadium WBS dictionary of architectural and interior works has changed slightly to be more detailed, based on the WBS standard. Some descriptions and requirements of the workman have been modified so that it becomes more comprehensive. The WBS dictionary can be used effectively as a basis to identify the project's scope, all of the the job activities along with the sequence of work on the project, the person in charge of a job within the project, the resources of each activity, and the quality requirements of each material. This tool, finally, can be used to control and monitor all of the works in order to keep the project from performing any work outside the scope, which may lead to aftermaths such as overbudget, rework, delayed project, poor quality, and the failure of project implementation. Regarding the WBS checklist, it has several description adjustments to simplify the meaning and the words. It has also been found that it is required to support the WBS Dictionary by evaluating every element and defining each work. After obtaining the WBS dictionary and checklist, the possible hazards were identified to avoid work accidents in the first place. Finally, further studies of the WBS dictionary can be conducted with the selection of implementation methods in stadium construction projects within depth risk management analysis for each defined activity.

Acknowledgement

The authors would like to thank you for the financial support provided from Universitas Indonesia through PUTI (*Publikasi Terindeks Internasional*) 2020 funding scheme NKB-1132/UN2.RST/HKP.05.00/2020, managed by Directorate for Research and Public Services (DPRM) Universitas Indonesia.

References

[1] Andrew R and Suryawan I N 2015 Studi literasi pengembangan manajemen klub sepakbola di Indonesia *Modus J.* **27** 175-82

[2] UEFA 2010 UEFA guide to qualify stadiums (Switzerland: UEFA) 2–7

[3] FIFA 2011 Football stadiums: Technical recommendations and requirements, 5th edition (Switzerland: FIFA) 108–12

[4] Anonymous 2017 A guide to the project management body of knowledge 6th ed. *Pennsylvania Proj. Manag. Inst.* 162-329

[5] Latief Y et al 2018 IOP *Conf. Ser. Earth Environ. Sci.* 195 012003

[6] Kerzner H 2009 *Project Management: A System Approach to Planning, Scheduling, and Controlling*, 10th ed. (Hoboken: John Wiley & Sons, Inc) 418

[7] Kazemian F and Pour S A 2014 The architectural formation of stadiums in different periods of time *Eur. J. Nat. Soc. Sci.*, ISSN 1805-3602

[8] Erten S and Ozfiliz S 2006 Stadium construction and sustainibility: The review of mega-evant stadiums (1990-2012) *1 st International CIB Endorsed METU Postgraduate Conference* 525-41

[9] Anonymous 2006 *Practice standard for work breakdown structures 2nd ed.* (Pennsylvania: Project Management Institute) 1–7

[10] Sawacha E, Naoum S, and Fong D 1999 Factors affecting safety performance on construction sites *Int. J. Proj. Manag.* **17** 309–15

[11] Bulger R A 2013 Integrating project management and safety management (Los Angeles: Project Management Institute)

[12] Siahaan F 2015 Tinjauan tentang pekerjaan arsitektur dalam proyek konstruksi dengan pendekatan pada bangunan gedung bertingkat *Scale J. Arch.* 344–59

[13] Nurdin L 2018 Identifikasi bahaya penilaian risiko dan pendendalian (IBPRP) dalam RK3K *Sibima Konstruksi*

IOP Conf. Series: Materials Science and Engineering **930** (2020) 012007 doi:10.1088/1757-899X/930/1/012007

Development of dictionary and checklist based on Work Breakdown Structure (WBS) at seaport project construction for cost estimation planning

A A Ilmi[1], L S R Supriadi[1], Y Latief[1], and F Muslim[1]

[1] Civil Engineering Department, Faculty of Engineering, Universitas Indonesia, Kampus Baru UI, Depok, Jawa Barat, 16424, Indonesia

*Corresponding author's e-mail: alifa.amalia71@ui.ac.id

Abstract. A construction project consists of many activities. To be more easily managed, activities are divided into smaller called work packages. The elements of construction project work are relatively similar, and these can be standardized and used as a basis for a universal program for construction project. The WBS standards can be developed into a dictionary and checklist that will help the project planning process and is expected to minimize the occurrence of cost management errors on the project. The purpose of this research is to develop a dictionary and checklist standard based on Work Breakdown Structure (WBS) for planning cost estimates for seaport project construction, identifying sources that can potentially have an impact on a project's cost overrun. This research method uses a survey method to seaport expert's validation and interviews using questionnaires to find the WBS-based dictionary and checklist standards. This research is expected to be a standard reference dictionary and checklist based on Work Breakdown Structure (WBS) at seaport construction projects for cost estimation planning. The importance of this research to knowledge is to know the steps of construction work especially at seaport project and to know WBS dictionary and checklist as a format that can be used as a reference base and easy-to-understand guide for seaport project, which provide work methods, activities, and resources.

1. Introduction

The port is a sea transportation building facility which has an important role in supporting economic activities. Knowing the important role of the port, making the port needs to have a development plan to keep up with the times, the port development plan is realized with well-planned planning before project implementation. Therefore, Work Breakdown Structure (WBS) is needed to determine the work packages, activities, and project resources [1].

WBS is one of the main tools for planning, monitoring, and controlling the work package of a project. WBS is the process of dividing work into smaller parts and making it easy to effectively describe the scope of the project. The aim is to improve project estimates, to better control project implementation and to verify project completion more accurately [2].

In implementing the port project, various problems have been found that will have a negative impact on the project target. Many factors cause problems in port projects, one of them is the cost estimation problem. There are several phenomena related to the problem of cost estimation in port projects due to

Content from this work may be used under the terms of the Creative Commons Attribution 3.0 licence. Any further distribution of this work must maintain attribution to the author(s) and the title of the work, journal citation and DOI.
Published under licence by IOP Publishing Ltd

inaccuracies when planning WBS. One of them is the case of the Tanah Ampo Cruise Port project, which was carried out in 2007 to 2012. According to Balitribune (2017), Tanah Ampo Cruise Port has cost up to Rp 86 billion, but at this time it cannot be used because the length of the pier is still not appropriate and the current waves that cause the yacht cannot lean on pier. To overcome this problem, a budget of Rp 50 billion is allocated from the state budget for water break construction so that yachts can pier at this port by lowering anchors at sea, then being picked up by lifeboats to get to the port.

Inaccurate cost estimates can have a negative impact on the entire construction process and all parties involved. Cost overruns that occur at the project implementation stage are very dependent on planning. The causes of problems arising from estimated port project costs are generally caused by several factors such as unclear identification of all the resources needed in project activities, including material, manpower, and equipment. Furthermore, another cause is the complexity of the design and size of the project [3]. These causes must be considered to minimize the risks that arise in the project cost estimation process.

For this reason, a standard or framework for port construction projects planning is needed to assist in the preparation of accurate and detailed WBS, where it also needs the description and requirements of each element of work in the form of WBS dictionary. The WBS dictionary is a work definition tool by explaining the scope of each work element, so that all resources needed in project activities can be identified. The WBS dictionary is expected to be a dynamic tool for terms of reference that can manage the agreed project scope and review estimated costs for each activity.

Besides WBS Dictionary, Checklist is also needed. The checklist is a more detailed work breakdown that covers all work activities of a project. With the identification of risks that arise in each activity in the grouped structure work is expected to be well registered and produce a cost estimation plan that can be used as input for contractors.

This background becomes the author's foundation for conducting research on the topic "Development of Dictionary and Checklist based on Work Breakdown Structure (WBS) in Seaport Project Construction for Cost Estimation Planning". The purpose of this research is to develop a dictionary and checklist standard based on Work Breakdown Structure (WBS) for planning cost estimates for seaport project construction, identifying sources that can potentially have an impact on a project's cost overrun.

2. Literature review

2.1. Seaport

According to Government Regulation No.69 of 2001 Article 1 paragraph 1 concerning Ports, a port is a place that consists of land and waters around it with a certain limit as a place of government activity and economic activity used as a place for ships for lean, passenger up and down to load and reduce objects with safety facilities and intra and inter modes of transportation. The port has the following functions:

1. Interface, the port as a meeting place for two modes of land and sea transportation so that the port must be able to provide various facilities and services needed.
2. Link, the port is a connecting transportation system, so the port is very important for all transportation activities.
3. Gateway, the port functions as a gateway for a country or region, so that it can play an important role for the economy of a country or region.
4. Industrial entities, ports as export-oriented industrial development of a country or region.

2.2. Work breakdown structure

According to the (Project Management Institute, 2006), WBS is a practical tool that can help the project planning team overcome large project uncertainties by making the project a lower level of uncertainty [4]. The function of WBS itself is as follows:

1. To determine project targets into small components so that they are easily managed.
2. To provide project management and progress frameworks for project status.

3. To facilitate communication between project managers and stakeholders about the life of a project.
4. As input for other project management processes.

The best method for preparing WBS is to use a hierarchical decomposition technique for the entire scope planned according to the project goals.

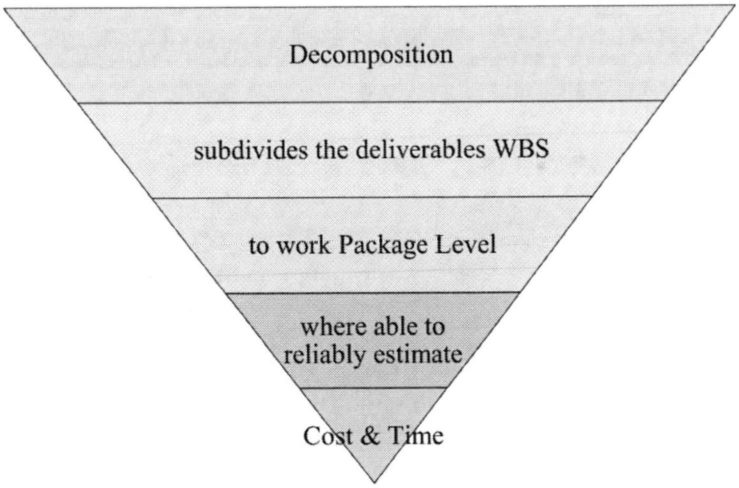

Figure 1. Decomposition technique

The decomposition technique, as shown in figure 1, is a technique that is used to divide the project scope and deliverables into smaller, and easier to manage. The results of the decomposition of each level in WBS are described as follows, and illustrated in figure 2:

1. Level 1: the full scope of work needed to produce the overall product. The top level is represented by a different project name for each WBS.
2. Level 2: the first level of decomposition and as the main field in the scope of work called division.
3. Level 3: describes each of the main areas of level 2 into sections consisting of various sub of work.
4. Level 4: every work in level 3 will be further broken down into a work package and checks are always made on the components formed whether they meet the requirements as a work package, one of which is considered to be sufficient to be managed or planned in terms of cost and time.
5. Level 5: activities that explain the details of each work package
6. Level 6: resources, such as items needed in work package. Resources here are in the form of materials, manpower and equipment.

The 5th and 6th levels will be explained in detail in a format called the WBS Dictionary. The WBS dictionary is a supporting document for the WBS that provides data in the form of detailed deliverables, activities, and resources information for each component in WBS.

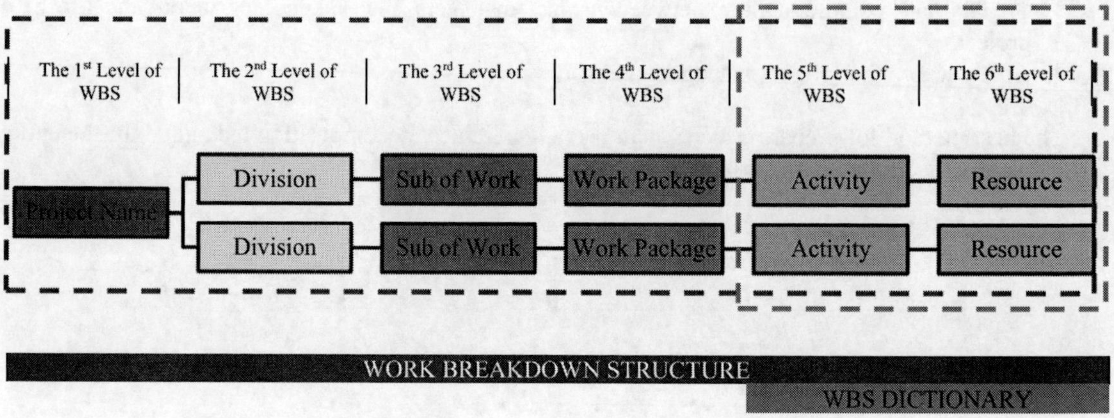

Figure 2. Work breakdown structure and WBS dictionary

2.3. WBS dictionary

Based on (Project Management Institute, 2013), the WBS Dictionary is a document that provides data detailed deliverables, work activities, and resources information about each component in Work Breakdown Structure [5]. The WBS dictionary helps identify and explain each work package (the lowest level) available on WBS. Work packages at WBS are represented as names and are not detailed. However, project stakeholders need to communicate in the project with sufficient and complete information about each deliverable. Such information can be found in the WBS dictionary and will improve the communication process between project stakeholders.

The WBS dictionary format varies depending on project needs. Some of the information contained in the contents of the WBS Dictionary is about identification codes, job descriptions, assumptions and limits, responsible organizations, activity schedules, required resources, estimated costs, quality requirements, acceptance criteria, technical references, approval information and still many more are not limited to the ones mentioned before.

2.4. WBS checklist

WBS Checklist is a more detailed job breakdown that covers all work activities of a project. Checklists are useful for the project team to achieve the desired level of work detail following from WBS. Checklists can be developed and used in the project management process such as a list of work to be done in the project, documents needed in the procurement process, project meeting agenda, identification of risks to the project, resume of project problems and status of completion and quality that is not according to the works. The development of the checklist depends on each project. But in principle, the checklist will help complete the work related to WBS [5].

2.5. Cost estimation

The cost estimation process (Project Management Institute, 2017), based on the figure 3, consists of inputs, which are:

1. The project management plan, which consists of a cost management plan, a quality management plan, and a basic scope
2. Project documents, which consist of lessons learned about project information, project schedule, resource requirements and risk list
3. Company or organization, which consists of corporate environmental factors and organizational process assets. [2]

While the output of the process is cost estimation, basics of estimation and updated project documents.

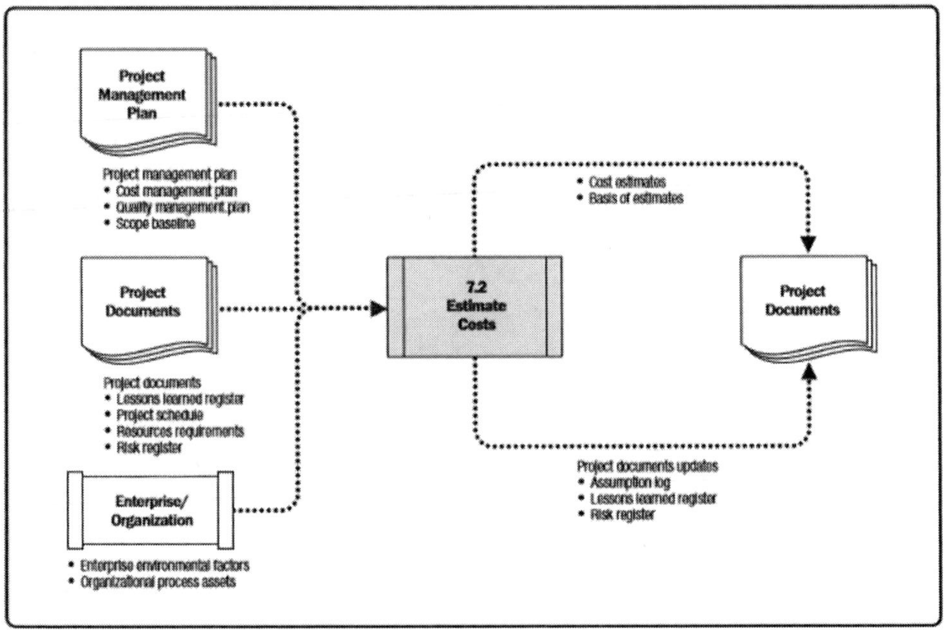

Figure 3. Cost estimation process

3. Methodology

The development of dictionary and checklist based on Work Breakdown Structure (WBS) at Seaport project construction use qualitative and descriptive approach, where each identified components are compiled and synthesized using the validated format for WBS dictionary and checklist, with the following data sources and sampling:

1. The validated WBS standards of seaport projects construction are used to form the initial format and content of the WBS-based dictionary and checklist, based on archive analysis.
2. WBS-based dictionary and checklist is made using primary data obtained from questionnaires and also interviews with experts in seaports with a minimum experience of 20 years. The analysis used for this research is descriptive statistical using a mode or value which often arises from respondents' answers on questionnaires related to the WBS-based dictionary and checklist.
3. In addition to primary data, there is also secondary data used in the form of project document files of Project BOQ and RAB data and analysis of literature studies according to Minister of Transportation Regulations Number PM 51 of 2015 concerning Seaport Operations.

4. Result and discussion

Figure 4 shows the four levels of WBS that have been validated for seaport construction projects, from this data finally developed into a WBS dictionary that containing the various activities and resources required for each work package described in details at level 5 and 6 of the WBS. This phase begins with the analysis from archive and related data literature on the WBS dictionary as a reference for determining its format. The WBS dictionary format also includes details of additional important information that is needed in a construction project such as person in charge, description, deliverables, and references for each different work package. These components will relate to WBS, but with more specific content. This is useful to guide project planning. These components can be seen in figure 5.

IOP Conf. Series: Materials Science and Engineering **930** (2020) 012007 doi:10.1088/1757-899X/930/1/012007

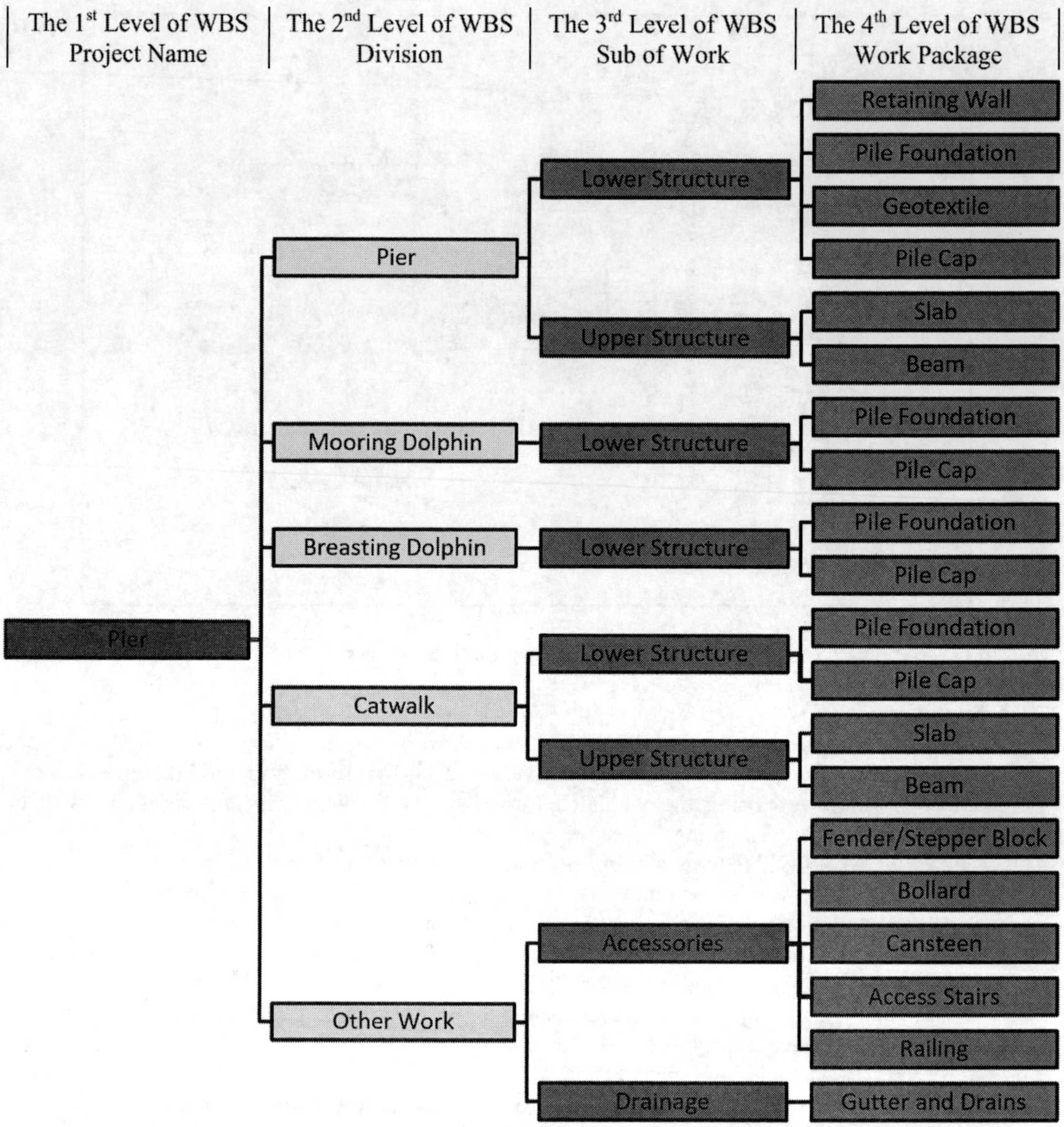

Figure 4. Validated work breakdown structure

IOP Conf. Series: Materials Science and Engineering **930** (2020) 012007 doi:10.1088/1757-899X/930/1/012007

DICTIONARY BASED ON WBS AT SEA PORT PROJECT CONSTRUCTION FOR COST ESTIMATION PLANNING

Pile Foundation

B	Project Name	:	Pier
B.3	Division	:	Mooring Dolphin
B.3.1	Sub of Work	:	Lower Structure
B.3.1.1	Work Package	:	Pile Foundation
	Person in Changes	:	
	Description	:	Pile foundation are the part of the structure used to carry and transfer the load of the structure to thc bearing ground located at some depth below ground surface.
	Deliverable	:	Pile Foundation
	Reference	:	1. Bill of Quantity 2. Technical Specification Documents 3. Expert Validation

Code	**Activity**	**Resources**		
		Manpower	**Material**	**Equipment**
B.3.1.1.1	Material procurement		1. Precast concrete piles	
B.3.1.1.2	Pile erection	1. Workers 2. Foreman	1. Precast concrete piles	1. Jack in pile 2. Drop hammer 3. Diesel hammer 4. Hydraulic hammer 5. Vibratory pile driver 6. Crane
B.3.1.1.3	Joining piles	1. Workers 2. Foreman	1. Precast concrete piles	1. Welding tools

Figure 5. Dictionary based on WBS

To avoid ambiguities that can occur in the project and clarify descriptions in project activities, checklist based on WBS was also developed. The checklist validation produces a format that consists of breaking down each level of WBS, starting from the name of the project, division, sub of work, to the work package, as shown in figure 6. This is used to adjust WBS by including inspection guidelines that refer to details at each level of WBS and used to provide clarity and understanding as a reference for planning [6]. Then the checklist is also still equipped with a WBS code, this WBS code will function to find out the sequence of activities.

IOP Conf. Series: Materials Science and Engineering **930** (2020) 012007 doi:10.1088/1757-899X/930/1/012007

CHECKLIST BASED ON WBS AT SEA PORT PROJECT CONSTRUCTION FOR COST ESTIMATION PLANNING

PIER

WBS Level 1	:	Project Name
WBS Level 2	:	Division
WBS Level 3	:	Sub of Work/ Section
WBS Level 4	:	Work Package

WBS LEVEL	CODE	DESCRIPTION	
1	B	Pier	Name of sub-project being worked on.
2	B.3	Mooring Dolphin	Mooring dolphin is planned to be able to stand up to the horizontal force caused by the impact of the ship, wind gusts, currents that hit the ship when moored.
3	B.3.1	Lower Structure	Lower structure is the work of foundations and building structures that are below the surface of the land.
4	B.3.1.1	Pile Foundation	Pile foundation are the part of the structure used to carry and transfer the load of the structure to the bearing ground located at some depth below ground surface.
4	B.3.1.2	Pile Cap	Pile cap is a structural member placed on, and usually fastened to, the top of a pile and used to transmit loads into the pile or group of pile, to connect them into a bent.

Figure 6. Checklist based on WBS

After validation of WBS, dictionary and checklist based on WBS, it can be related to the dictionary and checklist based on WBS with cost estimation planning. In planning project cost, project documents in such as resource requirements and risk registers are used as input to plan costs, because all risks that arise will ultimately affect the cost estimation planning, as shown in figure 7 and figure 8. The risk cost calculation is then performed (H. W. Kang and Y. S. Kim, 2018). The total rate is the direct ratio of risk costs to the total value of the work, which is summed from RC.E (Risk Cost of Engineering), RC.P (Risk Cost of Procurement), RC.C. (Risk Cost of Engineering) [7].

Table 15. Model B-Total Risk Cost of a New Project (Unit: 1,000 KRW)

Item		Risk Factors	RC.E		RC.P		RC.C		Total	
			Cost	Rate	Cost	Rate	Cost	Rate	Cost	Rate
RF . CP	RF-01	Error in basic design	20,526	10.5%	444,571	4.2%	347,718	4.7%	812,814	4.5%
	RF-02	Error in detail design	27,185	13.8%	767,895	7.2%	423,631	5.7%	1,218,710	6.7%
	RF-03	Error in mechanical design	16,747	8.5%	717,209	6.7%	296,642	4.0%	1,030,597	5.7%
	RF-04	Change in construction method	22,813	11.6%	841,822	7.9%	474,092	6.4%	1,338,727	7.3%
	RF-05	Variation in bill of quantity	32,436	16.5%	996,219	9.4%	546,574	7.4%	1,575,228	8.6%
	RF-06	Delay in procurement schedule	-	-	236,217	2.2%	229,301	3.1%	465,518	2.6%
	RF-07	Error in fabrication	-	-	326,557	3.1%	182,472	2.5%	509,029	2.8%
	RF-08	Error in PID design	7,853	4.0%	313,624	2.9%	147,019	2.0%	468,497	2.6%
	RF-09	Lack of local manpower	-	-	442,536	4.2%	369,951	5.0%	812,486	4.5%
	RF-10	Strike by workers	-	-	-	-	143,734	1.9%	143,734	0.8%
	RF-11	Error in connection (pipe to pipe)	8,422	4.3%	583,277	5.5%	535,597	7.3%	1,127,296	6.2%
	RF-12	Contract change in subcontractor	18,852	9.6%	1,008,201	9.5%	776,836	10.5%	1,803,889	9.9%
	RF-13	Delay in construction	-	-	-	-	363,421	4.9%	363,421	2.0%
	RF-14	Change of design(Contractor)	27,802	14.2%	779,211	7.3%	478,793	6.5%	1,285,806	7.1%
	RF-15	Construction accident	-	-	-	-	203,903	2.8%	203,903	1.1%
	RF-16	Lack of local material	-	-	247,343	2.3%	133,799	1.8%	381,142	2.1%
	RF-17	Variation in material cost	-	-	515,379	4.8%	318,674	4.3%	834,053	4.6%
	RF-18	Lack of temporary facilities	-	-	102,247	1.0%	97,906	1.3%	200,152	1.1%
	RF-19	Transportation environment	-	-	444,571	4.2%	220,729	3.0%	665,299	3.7%
	RF-20	Transportation equipment	-	-	355,657	3.3%	131,996	1.8%	487,653	2.7%
RF . NP	RF-21	Constructional defect	-	-	304,305	2.9%	384,572	5.2%	688,877	3.8%
	RF-22	Lack of local manpower	13,730	7.0%	456,458	4.3%	256,382	3.5%	726,569	4.0%
	RF-23	Error in equipment manufacturing	-	-	304,305	7.1%	320,477	4.3%	1,081,240	5.9%
Sum			196,366	100%	10,644,059	100%	7,384,217	100%	18,224,642	100%

Figure 7. Model risk cost new project

So that for each risk arising from WBS, the WBS dictionary and checklist that has been made can be identified and known for the actual cost of each risk making it easier to estimate project costs.

Risk		Model Risk Cost New Project (Kang, 2018)			Actual Cost
Risk ID	Risk Description	Item	Risk Factor	Rate (%)	
R01	High wave	RF13	Delay in Construction	2	363.421 x Value of Total Work

Figure 8. Actual cost calculation

Moreover, in this study, there are specific works that has been identified and compiled structurally in the WBS dictionary and checklist. In the port construction project, there are special works that characterize the port project differently from other construction projects, which are in preparatory work namely bathymetry surveys and pier construction work, namely dolphin breasting work and dolphin mooring. The layout of these works can be seen in figure 9.

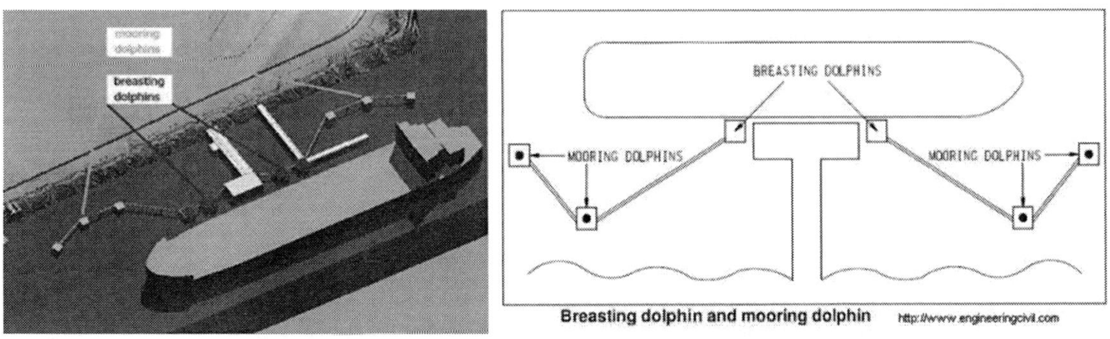

Figure 9. Breasting dolphin and mooring dolphin

Dolphin is a construction used to tether a ship to use a pier to shorten the length of the building. This mooring is planned to be able to withstand the horizontal force caused by the impact of the ship, the blowing of the wind, the current that hits the ship when it is moored.

Breasting dolphin has a larger size, because it is planned to withstand the waves and wind shocks when the ship is anchored, breasting dolphins are also equipped with fenders to withstand boat impacts and boilers to place the ship's rope, in order to move the ship along the pier and withstand the pull of the ship. Mooring dolphin is not used to withstand collisions, but only as a fastening.

5. Conclusion

The Work Breakdown Structure (WBS) dictionary is developed after WBS is created and validated according to WBS standard. The WBS dictionary is more specific than WBS because it explains the person in charge, description, deliverables, and references for each different work package. The difference between the WBS dictionary and checklist is on the content. The contents of the WBS dictionary contain information about the work in detail for each work package, which contains activities from the work package to the resources needed (material, manpower, and equipment). While the contents of the checklist contain a list of descriptions starting from the name of the construction project to the work package on the construction project to reduce the ambiguity that can occur in the project.

The result of the validated dictionary and checklist based on WBS is a format that can be used as a reference and guide for managing seaport construction project, to provide information about work methods, activities, and resources. Also, the detailed contents from the WBS dictionary and checklist can become essential documents for identifying risks in estimating the seaport construction costs. The details in the WBS dictionary and checklist also highlights the specializations of the seaport work packages, which differs from other type of construction projects.

Acknowledgement

The author would like to thank the financial support provided by the Universitas Indonesia through the PUTI (*Publikasi Terindeks Internasional*) 2020 funding scheme NKB 1132/UN2.RST/HKP.05.00/2020, managed by the Directorate for Research and Public Services (DRPM) Universitas Indonesia.

References

[1] Riantini L S, Latief Y, Soepandji B S and Rajasa M 2017 Development of Risk-based Standardized WBS (work breakdown structure) for cost estimation of apartment's project *Int. J. Civ. Eng. Tech.*
[2] Project Management Institute 2017 A guide to the project management of body of knowledge 6th edition *Proj. Manag. Inst. Inc.*
[3] Fischer M and Akinci B 1998 Factors affecting contractors' risk of cost overburden, *J. Manag. Eng.* 67-77
[4] Practice standard for work breakdown structures 2nd edition 2006 *Proj. Manag. Inst. Inc.*
[5] Project Management Institute 2013 A guide to the project management body of knowledge 5th edition *Proj. Manag. Inst. Inc.*
[6] Haugan G T 2002 Effective work breakdown structure (United States of America)
[7] Kang H W and Kim Y S 2018 A model for risk cost and bidding price prediction based on risk information in plant construction projects *KSCE J. Civ. Eng.*

Critical factors that influence the success of construction projects procurement in Surabaya

F A Safitri[1*], M A Rohman[1], and R Indryani[1]

[1]Department of Civil Engineering, Institut Teknologi Sepuluh Nopember
Campus of ITS Sukolilo, Surabaya, Indonesia

*Corresponding author's e -mail: arif@ce.its.ac.id

Abstract. Construction procurement for public infrastructure projects is important with regard to the national development to fulfil infrastructure needs. As public construction works uses the public budget, its procurement process should be conducted appropriately. Proper procurement process will result the best contractors and consultants. This paper aims to analyze the factors that influence the procurement success of the public construction project in Surabaya. The right procurement process will produce good service providers. There has been a lot of research related to procurement, but only a few have focused on the determinants of construction procurement. Therefore, research is needed related to the critical success factors of construction procurement by reviewing presidential regulation number 16 of 2018. This study attempts to identify factors determining the success of construction procurement based on a literature review and applicable presidential regulations. The preliminary survey was carried out by involving several experts who assessed the variables obtained from the results of related literature studies.

1. Introduction

Construction procurement for public infrastructure projects is a very important thing because the process of public procurement in order to meet development needs is determined through the procurement process. The procurement process is carried out competitively and taking into account the principles of fair business competition can also improve the welfare of the community because most government projects aim to spur government activities and economic growth.

One type of procurement that has been the government's priority for some time is the construction sector. The construction sector includes road projects, school buildings, drainage projects, and many others. Construction services have an important role in producing facilities and infrastructure that support growth and development in all fields, especially in national development [1].

Based on the description above, it is necessary to develop a proactive policy to develop Indonesian construction so that it can play a positive role in reducing the risk of failure of construction projects [2]. One of the things that can be done is to implement good public procurement [3]. Information technology systems that are often referred to as electronic procurement (E-procurement) in recent years have been used in the procurement process in Indonesia. This system is expected to be a means to increase cost efficiency and productivity in product revenue and customer satisfaction [4]

However, even though E-procurement has been implemented, there is no guarantee that the existing procurement process and system has no shortcomings at all. The Corruption Eradication

Content from this work may be used under the terms of the Creative Commons Attribution 3.0 licence. Any further distribution of this work must maintain attribution to the author(s) and the title of the work, journal citation and DOI.
Published under licence by IOP Publishing Ltd

Commission (KPK) noted that in 2019, corruption in Indonesia was dominated by procurement by 80% [5]. Public procurement systems are considered to be the main source of the budget, where the combination of conventional and collusive intervention systems that use a major decline in public services [6]. Considering the importance of the construction sector, it is very important that procurement implementation in this field can be carried out cleanly and efficiently for the success of national development.

Surabaya City has been one of the pioneers in implementing E-procurement in Indonesia since 2004. In general, the Surabaya City Government is considered successful in implementing e-procurement systems in the public sector and has made the implementation of e-procurement systems in preventing Corruption Collusion and Nepotism (KKN) [7]. The implementation of this policy has a positive impact on the achievement of the City of Surabaya so that it has managed to get several awards related to procurement, one of which is the 2013 E-procurement award. Some of the advantages of E-procurement are that it is more transparent so that it can be monitored by the public, more efficiently because it facilitates the document research process, and in terms of bidding competition, it can save 20-30% of the budget from auction efficiency [8].

However, the achievement of the procurement success of the Surabaya City government does not automatically guarantee the success of all its construction projects. However, achieving the success of the Surabaya City government procurement does not automatically guarantee the success of all construction projects. The impact of less successful construction projects will be detrimental to the parties involved in the project, the community and the surrounding environment [7]. Several previous studies related to procurement have been conducted. However, not much research has focused on finding the critical success factors for procurement and its development strategies [9].

Based on the description above, it is necessary to conduct the research to find out what factors determine the success of construction procurement for public infrastructure projects in the City Government of Surabaya. By identifying the critical success factors, efforts or strategies can be formulated to increase procurement success in the city of Surabaya. The purpose of this research is to find out what factors influence the success of construction procurement for public infrastructure projects in the city of Surabaya through a preliminary survey.

2. Literature review

2.1. Public procurement

Public procurement is an activity to obtain goods or services by ministries, institutions, regional apparatuses, or other institutions whose processes start from the planning of needs until the completion of all activities whose budgets are financed by the state/ region [10]. The purpose of procurement is to produce the right goods / services measured in terms of quality, time, cost, location of the right service provider of every money spent [10]. Public procurement is the essence of the duties of public sector organizations. The main proportion in public expenditure at every level of public sector organizations is the public procurement and construction activities.

2.2. E-procurement

Use of integrated communication systems that are usually web-based to carry out part or all of the purchasing process. The process combines the activities of early identification of needs by users, search, sourcing, negotiation, ordering, acceptance, and review after purchase [11]. In the process e-procurement reduces things that are not needed in a business process. In addition, in practice, e-procurement reduces paper usage, saves time and reduces labor usage[12].

The purpose of e-procurement is to realize the effective, efficient, transparent, fair, non-discriminatory and accountable public procurement in accordance with the principles of procurement [13]. Therefore, the Corruption Eradication Commission (KPK) encourages all government agencies to implement e-procurement in public procurement [14]. The implementation of e-procurement is considered to be able to save budget, time and increase transparency in the implementation of public procurement[15].

2.3. Success factors
The success factor is a matter or condition that can lead to the achievement of a desired goal. The success factor is used as a planning tool to reach an ideal condition or success [16]. The success factor is the term for an element that is needed for an organization or project to achieve its mission. With the approval of a previous study [9] what was agreed in this case was the best procurement, the efficient, fair and transparent procurement process.

2.4. Review of public procurement from the concept of critical success factors
The success factor as a limited area of all the results obtained, if satisfactory, can certainly lead to success towards the goal[17]. In this context it is assumed that the existence of critical factors related to the public procurement ensures that their implementation will be successful. There are several factors that influence the success or failure of a public procurement. Success factors related to the construction procurement for public infrastructure projects must consider all aspects involved [9].
Some researchers [9,18,19,20,21,22] found that the influence of e-procurement implementation at the local level, particularly leadership, human resources, planning and management, policies and regulations, system integration, infrastructure and standardization, and the dependent variable of the efficiency and effectiveness of e-procurment.

Of the eleven factors that were modeled [19], the most influential factors were top management support and performance measurement. The six factors formulated [9] produce that human resources most influence the success of the occupation. In addition to this, training and education in the new procurement process and technology are very important to ensure the successful adoption of e-procurement. This is evidenced by the research produced [20] in identifying the determinants of procurement success. Based on the results of the identification of literature found 10 variables that can be seen in table 1.

3. Research methodology
The methodology used in this study started from the identification of variables obtained from several previous studies. Then a preliminary survey involving several experts was appropriate in their field [23]. The number of experts involved in this matter was 4 experts and it had been considered sufficient to measure the relevance of the variable i.e. at least three to five experts [24]. Another special requirement was that the experts involved in the preliminary survey were those who had at least 5 years experiences in this field.

The purpose of the preliminary survey was to validate the variables obtained from studies of research literature related to actual conditions in the field [25]. At this stage the practitioner could judge whether the chosen variable actually happened or not by assessing the relevance level of one to five using a Likert scale[26]. The variables to be measured were translated into indicator variables. Then the indicator was used as a starting point for compiling instrument items which could be statements or questions[27, 28, 29, 30]. Scale 1 representing variables was highly irrelevant and scale 5 representing variables that were highly relevant[26]. In addition, if there were other variables not listed in the preliminary questionnaire, practitioners could add new variables.

Then a preliminary survey was conducted. Resource persons from this survey were experts to assess the relevance of variables obtained from literature studies. Variables were developed with several measurement indicators which would then be given a rating scale. After the data were obtained, the next step was data analysis. Check the completeness of the answers carried out to ensure no data were lost in distributing the questionnaire. In addition, researchers might also ensure that the answers given by experts were complete and clear. This was done in order to get accurate results.

Then the data obtained would be transferred to Excel for analysis. The average test was carried out on the answers obtained from the questionnaire. In order to make it easier to see the relevance of the most relevant assessment indicators, the average test results were sorted by rank.

4. Result and analysis

This paper presents preliminary results of research to investigate the critical success factors for the procurement of goods and services in Indonesia. It covers the preliminary survey result based on the experts' survey which is presented with regard to the attributes relevancy and their rankings.

Table 1. Variables of Critical Success Factors Procurement

Variable ranking	Variable	Indicator	Source	Mean	SD	Rank	Average	Conclusion
1	Procurement Success (PS)	Efficient procurement process	[9], [19], [20]	5.00	0.00	1	4.57	Relevant
		Punctuality		5.00	0.00	2		Relevant
		Fair procurement process		4.75	0.71	3		Relevant
		Transparent procurement process		5.00	0.00	4		Relevant
		Guaranteeing the selection of the best providers		4.25	0.71	5		Relevant
		Quality accuracy		4.00	0.71	6		Relevant
		Cost accuracy		4.00	0.71	7		Relevant
2	Security and Authentication (SA)	The system guarantees the privacy of data privacy	[9], [18], [19]	4.50	0.71	1	4.50	Relevant
		The system is equipped with multi-factor authentication to secure access		4.50	0.71	2		Relevant
		The system is equipped with cryptography to secure network communications		4.50	0.71	3		Relevant
3	Technology Standards (TS)	Software availability	[9], [18], [19], [20], [21]	4.75	0.71	1	4.45	Relevant
		Availability of adequate internet access		4.75	0.71	2		Relevant
		Relevant content or information		4.75	0.71	3		Relevant
		Accurate content or information		4.75	0.71	4		Relevant
		Up to date (updated) content or information		4.75	0.71	5		Relevant
		Fast system response time		4.75	0.71	6		Relevant
		Guidelines and regulations can be accessed online		4.50	0.71	7		Relevant

		There are a number of features to the procurement implementation that can be accessed online		4.50	0.71	8		Relevant
		There are features in the form of standard operational procurement procedures that can be accessed online		4.50	0.71	9		Relevant
		Hardware availability		4.25	0.96	10		Relevant
		Hardware reliability		4.25	0.71	11		Relevant
		Software reliability		4.25	2.12	12		Relevant
		Providing bidding room services		3.75	0.71	13		Relevant
		Large bandwidth capacity		3.75	1.41	14		Relevant
4	Planning and Managemen t (PM)	Evaluate achievement targets	[9], [19]	4.75	0.71	1		Relevant
		Own a clear SOP		4.75	0.71	2		Relevant
		Setting success targets related to procurement		4.75	0.71	3		Relevant
		Follow international e-procurement standards		4.75	0.71	4	4.46	Relevant
		Appropriate procurement implementation schedule		4.50	0.71	5		Relevant
		Procurement officers carry out daily monitoring		3.25	1.41	6		Relevant
5	Policies and Regulations (PR)	Own a coherent policy	[9], [19], [20], [21]	4.75	0.71	1		Relevant
		Own a policy related to accountability		4.75	0.71	2		Relevant
		There are strict legal sanctions related to fraud		4.75	071	3		Relevant
		The procurement process is carried out in accordance with applicable regulations		4.75	0.71	4	4.42	Relevant
		Conformity of laws / regulations regarding policies with technological		3.75	2.12	5		Relevant

		changes						
		Conformity of laws / regulations regarding policies with environmental changes		3.75	2.12	6		Relevant
6	Reengineering Process (RP)	There is no conspiracy between providers of goods and services	[19]	4.75	0.71	1		Relevant
		It is not possible for cooperation between procurement officers and providers		4.75	0.71	2		Relevant
		Compliance with Standard Operating Procedures (SOP)		4.75	0.71	3	4.40	Relevant
		The procurement master plan is clearly and systematically structured		4.50	0.71	4		Relevant
		Evaluation of things that can be reengineered		3.25	1.41	5		Relevant
7	System Integration (SI)	System stability	[9], [18], [19]	4.50	1.41	1		Relevant
		Ease of electronic transactions		4.25	0.71	2		Relevant
		Accessibility of service users / suppliers		4.00	1.41	3	4.19	Relevant
		Easy data exchange between providers and service users		4.00	1.41	4		Relevant
8	User Willingness (UW)	Share clear experiences of goals and objectives	[9], [18], [20], [21], [22]	4.50	0.71	1		Relevant
		Establishment of an appropriate work culture		4.00	1.41	2		Relevant
		Having expertise / competency certification according to their fields		3.75	2.12	3	3.70	Relevant
		Identify measurable performance indicators		3.75	0.71	4		Relevant
		Conduct education and training related to public procurement		2.50	2.12	5		Irrelevant

9	Provider Involvement (PI)	Having expertise / competency certification according to their fields	[9], [18], [20], [21], [22]	4.75	0.71	1		Relevant
		Having the ability to use technology		4.00	0.71	2		Relevant
		Actively participating in education and training related to public procurement		3.00	2.12	3	3.56	Relevant
		Develop independent training		2.50	0.71	4		Irrelevant
10	Leadership (L)	Establishment of Goods and Services Procurement Work Unit	[9], [19], [20], [21], [22]	4.75	0.71	1		Relevant
		Own a good vision and procurement goals		4.50	0.71	2		Relevant
		Top management support		4.25	1.41	3		Relevant
		Appropriate allocation of resources		4.00	2.12	4		Relevant
		Publish the vision and objectives of procurement		3.75	0.71	5		Relevant
		Integration between top management and procurement audits		3.50	1.41	6	3.48	Relevant
		Involvement of top management		2.75	1.41	7		Irrelevant
		There is a mechanism for vision consultation with stakeholders		2.75	1.41	8		Irrelevant
		Involving the private sector related to public procurement		2.25	2.12	9		Irrelevant
		Involving the community related to pubic procurement		2.25	2.12	10		Irrelevant

Data analysis from the experts' opinion can be represented in the attributes mean and standard deviation (SD) as can be seen in table 2 In this study, 3.00 (three) score was used as the cut-off to determine the attribute relevance as this value is the middle score between 1 to 5. It means that the attributes are considered relevant if their means are larger than three (≥ 3) as the middle value. As the means of the overall attributes were greater than three, all attributes were considered relevant to be used for the main survey through the questionnaire distribution. Based on the results of this preliminary survey, there are several variables that are omitted because they are considered irrelevant by showing mean values smaller than 3 (three).

There are 6 critical success factors that are not relevant according to procurement experts in Surabaya, so 58 others are considered relevant. Following some opinions from experts in Indonesia during the preliminary survey show that top management involvement, consultation mechanism, involving other parties, evaluations in the reengineering process, education and training for users are considered insignificant as determining factors for successful procurement in Surabaya.

While experts agree that the benchmark for the success of procurement of goods and services is to produce the right goods and services related to quality, quantity, time, and cost, and to ensure the selection of the best providers, and an efficient, fair and transparent procurement process.

5. Discusision

Data analysis from the experts' opinion can be represented in the attributes mean and standard deviation (SD) as can be seen in table 1 In addition to the attributes relevancy, it is also important to find the attributes ranking based on experts' opinion. The highest ranking is the success of the procurement itself (average 4.57). Experts agree that the success of procurement is the most important thing. All indicators in this variable are relevant with the highest average value being 5.00 and the lowest being 4.00. Efficient, timely and transparent procurement processes are indicators that have very high relevance values. This is in line with previous research [9]. The most relevant measurement indicators of this attribute are the efficient procurement process, timeliness, and transparent procurement processes

While security and authentication (average 4.50) is the second most dominant critical success factor in procurement according to experts. Electronic procurement requires good safety standards to achieve successful procurement. Access management and identity protection and required to avoid system threats from outside interference. All indicators in this attribute are relevant to the average value of all the indicators being 4.50.

The next attribute of success determination is the technology standard (average 4.45). Experts agree that this element of technical standards is a basic need that encourages successful procurement. By increasing the role of technology standards, the higher the chances of successful procurement. This is in line with previous research that good technology standards will influence procurement success. All indicators in the technology standard are relevant with the highest average being 4.75 and the lowest being 3.75.

The fourth variable is planning and management (average 4.46). All indicators in this measurement are relevant to the highest mean of 4.75 and 3.32. The experts agree that before the procurement process is carried out, procurement actors must carry out good planning and management of all activities carefully. Evaluation of activities and disciplined SOPs are important points.

The fifth variable is policy and regulation (average 4.42). All indicators in this measurement are relevant to the highest mean of 4.75 and lowest of 3.75. Experts agree that the provision of good construction is also determined by the policies and regulations adopted by the government. Procurement must be in line with regulations that become the legal umbrella and regulate all procurement activities. So that procurement can be carried out according to procurement objectives.

The rating attribute that determines the success of procurement is the reengineering process (average 4.40). All measurement indicators in this variable are considered relevant to the highest mean of 4.75 and lowest of 3.25. Experts agree that reviewing failed bidding or improper planning is a success for procurement.

The seventh variable is system integration (average 4.19) All measurement indicators in this variable are relevant to the highest mean 4.5 and the lowest mean 4.00. Experts agree that system integration is a success. In this case, reviewing the ease of transaction and the convenience of access are important points.

While the attribute with the lowest average according to experts in determining the success of procurement is leadership (an average of 3.48). Within this attribute, there are four indicators deemed irrelevant, including top management involvement (average 2.75), there is a vision consultation mechanism with stakeholders (average 2.75), involving the private sector and other parties (community) in procurement (average 2.25).

The involvement of top management is not necessary because in this case top management is only authorized to delegate part of the budget execution tasks. Top management is expected to support the procurement process well, without having to be directly involved in each process. The vision consultation mechanism with stakeholders is also not considered relevant by experts in determining procurement success. This is because the procurement vision and mission must be clearly defined in the procurement plan so that it can run efficiently. In addition, the involvement of the private sector and the community is also considered less relevant by experts. Procurement will be held transparently and openly so that all parties involved are able to know the planning to the results.
Another irrelevant indicator is developing your own training at home (average 2.50). This indicator is contained in the provider involvement attribute. This also relates to other indicators deemed relevant by experts to provide education and training related to procurement (average 2.50). This indicator is contained in the user's willingness attribute. Experts agree that providers must be professional in using technology systems and keep abreast of the latest developments in information technology. In addition this may also be possible if there are additional human resources or new policies.

6. Conclusion and recommendation

According to the preliminary survey, 58 out of 64 relevant factors were found to be important for the successful procurement of construction for public infrastructure projects. The most dominant determinants of procurement success are security and authentication, technology standards, and planning and management. 6 other factors are considered irrelevant in influencing the success of construction procurement. The next research step will be carried out to conduct the main survey and analysis using multiple regression to study the effect of these factors on the performance of construction procurement.

Acknowledgments

The authors appreciate the experts involved in the preliminary survey of this research in 2020.

References

[1] Taufik and Saputra 2012 Pengaruh implementasi sistem pengadaan secara elektronik (e-procurement) terhadap fraud pengadaan barang/jasa pemerintah, studi pada satuan kerja perangkat daerah kabupaten Magetan *J. Admin. Publik* **3** 1905-11
[2] Oktaviani L 2017 Kajian persepsi pelaku jasa konstruksi tentang kegiatan pengadaan jasa konstruksi secara e-procurement di kota Kupang *J. Tek. Sipil Univ. Neg.* Padang **2**
[3] Purwadi A 2019 Praktik persekongkolan tender pengadaan barang dan jasa pemerintah *J. Hukum Magnum Opus* **2** 2
[4] Peleg et al 2002 Understanding the incremental nature of eprocurement implementation at the state and local levels *J. Public Procure.* **2** 1 5-28.
[5] Hakim L and Mula A 2019 80% Kasus korupsi terjadi di sektor pengadaan: Koran sindo (Surabaya)
[6] Nugroho et al 2015 Pengaruh implementasi sistem pengadaan secara elektronik (e-procurement) terhadap fraud pengadaan barang/jasa pemerintah, studi pada satuan kerja perangkat daerah kabupaten Magetan *J. Admin. Publik* **3** 11 1905-11
[7] Saputra R et al 2016 Analisis kegagalan konstruksi dari perspektif socio- engineering system *J. Rekayasa Sipil* **15** 1
[8] Jasin et al 2007 Mencegah korupsi melalui e-procurement meninjau keberhasilan pelaksanaan e-procurement di pemerintah kota Surabaya (Komisi Pemberantas Korupsi: Jakarta)
[9] Achmad N and Sunhyuk K 2015 Making e-procurement work in a decentralized

procurement system: *A* comparison of three Indonesian cities *Int. J. Public Sector Manag.* **28** 198-220

[10] Peraturan Presiden 2018 Pengadaan barang dan jasa pemerintah 16

[11] Croom and Brandon J A 2007 Impact of e-procurement: Experinces from implementation in the uk public sector *J. of Purch. Supply Manag.* **13** 4 294-303

[12] Salim M A et al 2015 Pengadaan barang dan jasa konstruksi dengan e-procurement *J. Tek. Sipil Univ. 17 Agustus*

[13] Aristami I G and Hendrawati T 2020 Evaluasi pengaruh teknologi , pengguna, dan organisasi terhadap manfaat dari penerapan e-procurement *J. Sistem Informasi* **9** 201-14

[14] Endianingsih D 2015 Peran e-catalogue dalam proses pengadaan elektronik *J. Kalibrasi* **13** 1

[15] Heriawati L 2018 Pengaruh penerapan electronic procurement dan good governance terhadap kinerja pengadaan barang/jasa pada bagian pengadaan bbarang dan jasa sekretariat daerah kabupaten Pandeglang (Institut Ilmu Sosial dan Manajemen STIAMI : Jakarta)

[16] Grunert K G and Ellegaard C 1992 The concept of key success factors *Theory and Methods MAPP Working Paper* **4**

[17] Maqbool et al 2018 Critical success factors for renewable energy projects: Empirical evidence from Pakistan *J. Clean. Prod.*

[18] DeLone W H and McLean E R 2003 The delone and mclean model of information systems success: a ten-year update *J. Manag. Inf. Syst.* **19** 4 9-30

[19] Vaidya K et al 2009 Critical factors that influence e- procurement implementation success in the public sector *J. Public Procure.* **6** Nos 1/3 70-99

[20] Gunasekaran A and Ngai E W T 2008 Adoption of e-procurement in Hong Kong: an empirical research *Int. J. Prod. Econ.* **113** 1 159-75

[21] Panda P and Sahu G P 2010 Combating corruption in public procurements through e-procurement: challenges and opportunities in shri bhagwan dahiya, kavita chakravarty, nandita rathee and seema v lamba (eds.) governance issues and corruption (Haryana, India: Intellectual Foundation) 173-96

[22] Kaliannan et al 2009 Government purchasing: A review of e-procurement system in Malaysia *J. Knowledge Eco. Know. Manag.* **4**

[23] Siregar A Z and Harahap N 2019 *Strategi dan teknik penulisan karya tulis ilmiah dan publikasi* (Sleman: Depublish) 27

[24] Sugiyono 2013 *Metode penelitian kuantitatif, kualitatif dan R&D* (Bandung: Alfabeta)

[25] Mufarrikoh Z 2019 *Statistika Pendidikan: Konsep sampling dan uji hipotesis* (Surabaya: Jakad Media Publishing)

[26] Duli N 2019 *Metodologi penelitian kuantitatif: Beberapa konsep dasar untuk penulisan skripsi dan analisis data dengan SPSS* (Yogyakarta: Penerbit Deepublish) 78

[27] Djaali H and Pudji M 2008 *Pengukuran dalam bidang pendidikan* (Jakarta: Grasindo) 60

[28] Muchson 2017 *Statistik deskriptif* (Online Publisher : Guepedia) 878 602 443 085 6 16

[29] Sanjaya W 2015 *Penelitian Pendidikan: Jenis,metode, dan prosedur* (Jakarta: Kencana) 249

[30] Kristanto V H 2018 *Metodologi penelitian: Pedoman penulisan karya tulis ilmiah (KTI)* (Yogyakarta: Deepublish) 72

IOP Conf. Series: Materials Science and Engineering **930** (2020) 012009 doi:10.1088/1757-899X/930/1/012009

Identification of fire safety indicators for shopping centre buildings in Surabaya

G F Marantika[1], M A Rohman[1*], F Rachmawati[1] and I P A Wiguna[1]

[1] Department of Civil Engineering, Faculty of Civil, Planning and Geo Engineering, Sepuluh Nopember Institute of Technology, Surabaya, East Java, Indonesia.

[*]Corresponding author's e-mail: arif@ce.its.ac.id

Abstract. The convenience of shopping and getting entertainment at the same time is an advantage which is offered by the shopping centres. But the different functions of building and the crowds of visitors at the shopping centre inflict a greater risk of fire. Therefore, the safety against fire needs to be considered by the shopping centres in order to make users and the building structures more secure. Current fire safety regulation is still dominated by the standards and the codes from NFPA. The lack of conducting researches and knowledge related to fire safety affects the suitability and the application of regulations in Indonesia. This research aimed to identify the fire's safety criteria according to the current local conditions of shopping centres. A survey was conducted by distributed questionnaires. The variables used in the questionnaire were obtained based on the literature studies and then it was assessed by several experts. According to the experts' judgment, all variables are relevant with the current local conditions and there were three new variables added in this research. The result showed that the construction practitioners needed to improve their knowledge related to the fire safety in shopping centre starting from the initial stages of design.

1. Introduction

The movement of people from rural areas to urban areas or urbanization is a common phenomenon that is found around us. It triggers the economic growth of a region and the level of community wages positively, and then it will affect the pattern of public consumption. Besides, the industry's development always presents a variety of goods and services, and also advances science and technology encourage the modern lifestyles to be more consumptive. These phenomena make the retail businesses such as Shopping Centres develop rapidly.

The shopping centre is not only as a place for economic transactions to meet their needs, but it has another purpose such as to get entertainment and to interact socially. Visiting the shopping centre has become a culture that spreads in society regardless of age, economic status, and social status. Based on this phenomenon, the construction practitioner makes a project to build a Shopping Centre to support the market demand. Similar with the other buildings, Shopping Centre needs to suit with the quality of building, this is one of the main keys that need to be considered by the construction practitioner in planning and constructing a building.

The quality of a building is assessed from the structure's readiness, the functions' fulfilment, and the reliability's fulfilment and the feasibility of buildings. These criteria determine that the building has

Content from this work may be used under the terms of the Creative Commons Attribution 3.0 licence. Any further distribution of this work must maintain attribution to the author(s) and the title of the work, journal citation and DOI.
Published under licence by IOP Publishing Ltd

been suited with the technical standards and the administration of quality buildings. This is confirmed in *Undang-Undang* Number 28 in 2002 about The Building Construction and also The Minister of Public Works and Public Housing's Regulation Number 29/PRT/M/2006 in 2006 about The Guidelines on Technical Requirements for Buildings. One of the things which is regulated in these regulations regarding the building structure standards in supporting loads and the ability to dealing with hazards, including fire.

The danger of fire can be defined as any situation that trigger a fire or explosion, and then it can threaten life or property [1]. The building and its system, human activities, and fire establish a complex system that creates the dynamic changes in fire behaviour. In the shopping centre buildings, the different functions of buildings create a greater fire risk to the building itself. The different building functions affect the design of fire response system [2].

The safety guarantees in shopping centre need to be a major concern from the planning stage until the operational stage. Indonesia has regulations that are stated in Minister of Public Works and Public Housing's Regulation Number 26/PRT/M/2008 about The Technical Requirements for Fire Protection Systems in Buildings and Environments and equipped with normative references such as SNI, that regulates the building planning procedures, fire prevention and fire protection that focus on the main safety controls, namely the physical control of buildings, management control, and human control. Until nowadays, the safety standards for fire still based on National Fire Protection Association (NFPA) 101 Life Safety Code [3]. A few researches had been found according to the local conditions in Indonesia. These researches should become the basis of current regulatory standards.

There are also deficiencies and negligence of the current standards application in Indonesia. According to [4], the gap between regulatory standards and fire protection systems in buildings is based on the lack of knowledge about building management and fire protection standards. Some components are not implemented according to the standards and the regulations. Besides, the shopping centre buildings design that tends to be complex with different functions also allows the building was not designed effectively according to the applicable standards. Based on the statement above, it is necessary to review whether the applicable standards are following the current local conditions and whether there are still deficiencies that need to be added to the existing standards.

2. Literature review

2.1. The concept of fire safety

The fire safety can be defined as a set of practices to prevent fires, manage fire growth, and manage the fire effects either intentionally or unintentionally while keeping the resulting losses at an acceptable level [5]. There are four main goals of fire safety, namely the life safety, the protection of building contents, the protection of building structures, and the minimization of threats to the environment [6]. The fire safety can be achieved through preventive and protective measures [7].

2.2. Fire safety criteria

Safety The safety against fire can be achieved by controlling how the combustible materials such as solids, liquids, and gasses that is available in an environment. When there are cases where a fire ignites, the amount of fire growth can be controlled because flammable materials have a limited amount. The structural stability also has an important role because it can help localize the fire, enable safe rescue operations by the fire fighters, and prevent property loss due to the collapse of entire buildings [5].

The carrying capacity of construction loads and structural elements is very important to be considered by the construction practitioner. One of the burdens that must be able to be ensured by a structure is the accidental events such as fire. If the functions change of a building modifies the life span of the building and it produces the enhancement of incombustible material, then to keep up the structure's stability, it is necessary to repair the structural system or the amount of fire exposure needs to be reduced into the original structure stage of building which is able to withstand loading [6].

The fire detectors functions are to detect fire growth that can be a threat to the building users. The detector is a device that will produce a signal that can be captured, used, and forward it to other devices in order to give a warning. The detectors can be categorized according to the type of these devices detection [8], [6]. Furthermore, the signal that is sent will be forwarded to the system control unit, and then it will process the alarm to sound and inform the building user that there is fire [8]. The purpose of fire alarm system is to begin the evacuation process by giving as early notice as possible that fire has been identified [9].

When fire has been detected, then the further efforts are needed to control the fire including controlling the fire, suppressing the fire, and extinguishing the fire. Controlling the fire means limiting the growth of the fire, suppressing the fire means reducing the size of the fire and extinguishing the fire means stopping the fire completely [10]. The sprinkler is an automatic fire suppression tool. While it can use a system of standpipes and hoses manually which allows the water supply to reach outlets on each floor of the building [5].

A smoke management system must be installed to provide the fire safety in large spaces. The smoke exhaust systems can work in keeping smoke at a certain level effectively. The height of smoke exhaust system follows the fire size and the buoyancy generated by fire [6], [11]. Besides the good smoke management, the existence of escape route signs from the room and information about what needs to be done in the event of fire is very important to make sure the safety of building users' lives. Lane markers make it easier for the building users to find the escape routes [6]. Fire safety management is the last thing to be a concern. Fire safety plans must be prepared before, not only by the building manager but also including the designer team at the earliest design stage. There must be a prevention and maintenance plan until the staff training plan [10].

Based on the results of literature identification, it found that there were 18 variables that could be seen in Table 1 below.

Table 1. The Variable of Fire Safety Criteria in the Shopping Centre

No.	Variable	Source
1.	Available fuel such as flammable liquid, flammable solids, and flammable gases inside building can be controlled.	[5], [6], [12]
2.	Building structure is able to withstand fire.	[5], [6]
3.	Fire compartment in building is able to inhibit the spread of fire according to planning time.	[5], [12]
4.	Fire detectors installed in building are able to detect fire.	[6], [8], [12]
5.	Fire alarm systems in building is working properly.	[8], [12]
6.	Water source in building are sufficient to extinguish fire.	[5]
7.	Water tank to extinguish fire in building are separated from daily water tank.	[5]
8.	Automatic fire suppression devices in building such as sprinklers can reduce fire growth.	[6], [8], [10], [13]
9.	Manual fire extinguishing systems such as standpipe and fire hose are able to supply water to extinguish fire each floor.	[5]
10.	Smoke control system can drain smoke outside the building.	[6], [11], [14], [15]
11.	Exit signs inside the building are easily found.	[5], [6]

12.	Fire escape routes including emergency stairs inside the building are safe from fire.	[5], [12]
13.	Exit path from the front of the building to the evacuation point free of obstacles.	[5], [6], [12]
14.	Emergency lighting inside building works well.	[12]
15.	Special area for parking of fire trucks in building are always available.	[5]
16.	Fire fighters are involved in fire control in building.	[5]
17.	Fire protection inspection as an effort to prevent fire in building.	[16], [17], [18]
18.	Information regarding the procedures for evacuation and emergency contact during fires can be accessed by building users.	[6], [8], [16], [19]

2.3. The concept of shopping centre

A Shopping Centre is a group of other retail and commercial businesses that are planned, are developed, and are managed as a single property, that consist of multi-brand commercial rental units and public areas [20]. The rapid population growth in urban areas and the number of people who live alone or independently, for example those who work at home or live in isolated suburbs, it encourages the rise of business places with a variety of products, services, and facilities, such as the building of Shopping Centre [2]. Shopping Centre customers not only buy for certain products but also do another things including getting entertainment [21].

3. Research methodology

After having variables identification from literature review, it was followed by the process of preparing a survey questionnaire. The preparation was done by adjusting the variables that had been known. Then, the variables were modelled in the form of statements to be measured. The chosen method for compiling the questionnaire was Likert scale. Likert scale had four or more questions which were combined in the questionnaire to create a score that represented the individual traits [22]. The measurement scale that was applied to measure the respondents' perceptions of the building safety criteria in shopping centre towards the fire was 1 to 5 scales. Scale 1 reflected an intensely irrelevant assessment while on the contrary, scale 5 was for a highly relevant.

Furthermore, a preliminary survey was conducted by the researchers. This survey was a survey which was conducted by several resource persons (namely: the experts) in order to verify the suitability of the research variables which were obtained from the literature study process according to the current local conditions. From this activity, it was able to be found out whether the existing variables were well-suited or it might need a reduction or addition of a new variable. This survey had been conducted by sending questionnaire and interviews with the experts, which was produced the relevant variables. The experts had been requested to give a variable relevance scale according to the range of the available scale.

The data was obtained from the survey results then were analysed by the researchers. The initial step taken by the researcher was checking the respondents' answers completeness in questionnaires. This activity purposed to check if there were any incomplete or unclear answers. The data was sorted based on the results of checks and was prepared for the data cleaning process. Hereinafter the mean test on the answers of each respondent was collected by the researcher to know the mean score of the test. After that, the mean test results were sorted by rank.

4. Results and analysis

In this research, the preliminary survey results were described to find out the relevance level of several fire safety criteria in shopping centre building. The data analysis was conducted according to the research method.

The results of data analysis from the preliminary survey based on four experts could be seen in Table 2. According to [24], in general, the experts needed to conduct an assessment on a questionnaire tended to be small. The research conducted by [25], [26], and [27] used two to four experts to participate in the research. Thus, there were four experts in this research were considered able to provide an assessment based on their knowledge and experiences.

The data analysis was begun by calculating the mean and the standard deviation of variables that had been assessed by an expert based on Likert Scale. Scales 1 to 2 were considered irrelevant, while scales 4 to 5 were considered relevant. For scale 3, the assessment weight was quite relevant, so scale 3 was also considered relevant.

Table 2. The Variable of Fire Safety Criteria in the Shopping Centre Result

No.	Variable	Rank	Mean	SD	Conclusion
4.	Fire detectors installed in the building are able to detect fire.	1	4.75	0.50	Relevant
5.	Fire alarm systems in the building is working properly.	2	4.75	0.50	Relevant
6.	Water source in the building are sufficient to extinguish fire.	3	4.75	0.50	Relevant
11.	Exit signs inside the building are easily found.	4	4.75	0.50	Relevant
12.	Fire escape routes including emergency stairs inside the building are safe from fire.	5	4.75	0.50	Relevant
13.	Exit path from the front of the building to the evacuation point free of obstacles.	6	4.75	0.50	Relevant
17.	Fire protection inspection as an effort to prevent fire in building.	7	4.75	0.50	Relevant
2.	Building structure is able to withstand fire.	8	4.50	0.58	Relevant
3.	Fire compartment in building is able to inhibit the spread of fire according to planning time.	9	4.50	0.58	Relevant
8.	Automatic fire suppression devices in building such as sprinklers can reduce fire growth.	10	4.50	0.58	Relevant
10.	Smoke control system can drain smoke outside the building.	11	4.50	0.58	Relevant
14.	Emergency lighting inside building works well.	12	4.50	0.58	Relevant
18.	Information regarding the procedures for evacuation and emergency contact during fires can be accessed by building users.	13	4.50	0.58	Relevant
1.	Available fuel such as flammable liquid, flammable solids, and flammable gases inside building can be controlled.	14	4.25	0.96	Relevant
16.	Fire fighters are involved in fire control in building.	15	4.25	0.96	Relevant
7.	Water tank to extinguish fire in building are separated from daily water tank.	16	4.00	0.00	Relevant
15.	Fire fighters are involved in fire control in building.	17	3.75	0.96	Relevant
9.	Manual fire extinguishing systems such as standpipe and fire hose are able to supply water to extinguish fire each floor.	18	3.50	1.00	Relevant

Based on the results of preliminary survey analysis which was conducted by the researchers, all variables were relevant. The highest mean of variable was fire detector with a value of 4.75 and the lowest mean of variable was a manual fire extinguishing system with a value of 3.5. There were 16 variables that were most relevant based on the experts with mean score were 4.00 to 4.75. The first variable was the fire detector. This variable was considered to provide safety in the Shopping Centre against fire, but good care was needed in order to make the devices function properly. The same thing applied to the second variable, it was the fire alarm system. The third variable was the water source in the Shopping Centre might always be available with a discharge according to applicable regulations. The fourth to sixth variables were the exit sign, fire escape route, and exit path, these needed to be considered in order to ensure the building user safety. The seventh variable, which was the fire protection inspection, it was needed to be checked routinely so that it could detect more easily whether the protection device was functioning properly or not. The eighth and ninth variables were the structure of buildings and compartments in Shopping Centre should be planned well from the beginning of development. The tenth variable was automatic fire suppression, it would function very well if the placement was adjusted to a combustible material and certain areas were considered to be very vulnerable to fire. The eleventh variable was the smoke control system, it was deemed necessary to be well designed and to pay attention with the air conditioning in the room. The twelfth variable was emergency lighting inside the shopping centre, it was very necessary during the evacuation process; therefore the equipment for emergency lighting must always be ready and functioning properly. The thirteenth variable was the information about how to escape during a fire, it might be available in a place that was visible easily and it might be understood easily, so that this information could be used during the evacuation process. The fourteenth variable was available fuel. It should be considered in quantity and then controlled for their uses so it could be adapted to the fire extinguish equipment. The fifteenth variable was the involvement of fire fighter. It was needed in the process of fire control in according to the main tasks and the purpose of fire fighter. The sixteenth variable was a separate water tank for extinguish the fire. It was needed with the capacity adjusted with applicable regulations.

The experts also provided three new variables that were needed to be added in the existing variables. The first variable was the fire simulation, such as how to use the fire protection equipment, how is the evacuation process, how is the process of extinguishing fire as a preventive measure against fire as well as education for Shopping Centre users. The second variable was the analysis of shopping centre users including the level of knowledge, number of users, users' experience, users' gender, diffability, etc. The analysis of users' behaviour was needed to understand how the behaviour of each individual in dealing with fire. This analysis was considered to be able to support the safety facilities that were available and allowed a fast and independent response from users so it could minimize the loss of lives due to the fire danger. [23] stated that there were three general aspects that affected the fire response model, such as the human aspect, the building aspect, and the fire aspect. The human aspect included individual, social, and situational aspect. The existence of knowledge about human behaviour in facing the fire was very important so the policy be measured could be effective. Furthermore, the last variable was the wide range of entrance and exit access also the wide range of the building perimeter for fire trucks used a ladder. The good design was needed from the start of planning so that the fire trucks could move without obstacles through the entrance and the exit access or around the building.

5. Conclusion and recommendation

Based on the background study and literature review, the researchers obtained that there were 18 variables of shopping centre buildings safety towards the fire danger. In the preliminary survey that was conducted by the researcher, it was found that all variables were relevant to the current local condition and 3 variables were added to complement the existing variables. It is hoped that this research has a contribution to increase the knowledge and it can be a reference research to help the building practitioners and shopping centre managers, also it can help the future research regarding the fire safety.

Acknowledgments
The authors appreciate the experts who involved in the preliminary survey in this research that is conducted in 2020.

References

[1] National Fire Protection Association (NFPA) 1 Fire Code 2018
[2] Endangsih T 2008 *Keselamatan Bangunan Pusat Perbelanjaan Terhadap Bahaya Kebakaran, Studi Kasus Senayan City* (Depok: Universitas Indonesia)
[3] Sujatmiko W 2016 Penerapan standar keselamatan evakuasi kebakaran pada bangunan gedung di Indonesia *Jurnal Pemukiman* **11** 2 p 116–27
[4] Djunaidi Z, Tuah N A A and Rafifa G 2018 Analysis of the active and passive fire protection systems in the government building, Depok City, Indonesia *KnE Life Sci.* **4** 5 p 384
[5] Kodur V, Kumar P and Rafi M M 2019 Fire hazard in buildings : review , assessment and strategies for improving fire safety buildings *PSU Research Review* **4** 1 p 1-23
[6] Marchant E W 2000 Fire safety systems – interaction and integration *Facilities* **18** p 444–55
[7] Gupta A K, Kumar R, Yadav P K and Naveen M 2001 Fire safety through mathematical modelling *Curr Sci* **80**
[8] Fischer R J, Halibozek E P and Walters D C 2019 Fire prevention and protection, and occupational safety *Introduction to Security* (Oxford/Cambridge: Elsevier) chapter 12 p 269–92
[9] Lyman D 2018 Ambulatory surgery center safety guidebook *Managing Code Requirements for Fire and Life Safety* (Oxford/Cambridge: Elsevier) p 27–31
[10] Chow C L, Chow W K and Yuan H Y 2004 A preliminary discussion on selecting active fire protection systems for atria in green or sustainable buildings *Archit. Sci. Rev.* **47** 3 p 229–36
[11] Chow C L and Li J 2010 An analytical model on static smoke exhaust in atria *J. Civ. Eng. Manag.* **16** 3 p 372–81
[12] Della-Giustina D E 2014 Fire Safety Management Handbook *Fire Safety Management Handbook* (London: Taylor & Francis)
[13] Chow W K and Chan L Y 2011 Possibility of using water mist fire suppression system in hong kong *J. Eng. Des. Tech.* **9** 2 p 157–63
[14] Short CA Whittle G E and Owarish M 2006 Fire and smoke control in naturally ventilated buildings *Build. Res. Inf.* **34** 1 p 23–54
[15] Woods A A T 2007 Improving energy efficiency improves smoke safety and occupant comfort *Strateg. Plan Energy Environ.* **27** 2 p 42–50
[16] Lo S M 1998 A building safety inspection system for fire safety issues in *Struct. Survey* **16** 4 p 209–17
[17] Kikwasi G J A Study on the awareness of fire safety measures for users and staff of shopping malls: the case of Mlimani city and quality centre in dar es salaam *J. Civ. Eng. Archit.* **9** 12 p 1415–22
[18] Taylor M, Appleton D, Keen G and Fielding J 2019 Assessing the effectiveness of fire prevention strategies *Public Money. Manag.* **39** 6 p 418–27
[19] Tsui S and Chow W 2004 Legislation Aspects of Fire Safety Management in Hong Kong *Facilities* **22** 5/6 p 149-64
[20] International Council of Shopping Centre 2015 Asia-Pacific Shopping Centre Classification (New York: New York)
[21] Makgopa S 2016 Determining shopping mall visitors perceptions on mall attributes (*Problems Perspective Manag* (open-access)) **14** 3-2 p 522-27
[22] Budiaji W 2013 Skala pengukuran dan jumlah respon skala likert *J Ilmu Pertan Perikan* **2** 2 p 127–33
[23] Kobes M, Helsloot I, de Vries B and Post J G 2010 Building safety and human behaviour in fire: a literature review *Fire Safety J* **45** 1 p 1–11
[24] Olson K 2010 An examination of questionnaire evaluation by expert reviewers *Field Methods* **22** 4 p 295–318

[25] Demaio T and Landreth A 2004 Examining expert reviews as a pretest method *QUEST 2003: Proc. of the 4th Conf. on Quest. Eval. Std.* p 60–73

[26] Jansen H 2005 The productivity of the three-step test-interview (TSTI) compared to an expert review of a self-administered questionnaire on alcohol consumption *J. Off. Stat.* **21** 1 p 103

[27] Ikart E M 2019 Survey questionnaire survey pretesting method: an evaluation of survey questionnaire via expert reviews technique *J. Soc. Sci. Stud.* **4** 2 p 1

A method to develop performance indicators based on performance criteria of Indonesian National Occupational Competency Standards (SKKNI) for construction safety technician competency

A Okviana[1*] and Y Latief[2]

[1] Post Graduate Student, Department of Civil Engineering, Faculty of Engineering, University of Indonesia, Depok, West Java 16424, Indonesia
[2] Department of Civil Engineering, Faculty of Engineering, University of Indonesia, Depok, West Java 16424, Indonesia

*Corresponding author's e-mail: anna.okviana@ui.ac.id

Abstract. In the Indonesian National Occupational Competency Standards (SKKNI), performance indicators are not described in detail, where performance indicators are important measures for assessing performance achievements. Due to this gap, the researchers intend to propose a method to develop performance indicators based on performance criteria of the Indonesian National Occupational Competency Standards (SKKNI) for the Construction Safety Technician positions. The competency unit to be developed for its performance indicators was the Competency Unit "Hazard Identification and Risk Control". The researchers propose a method to develop the performance indicators based on the performance criteria of Indonesian National Occupational Competency Standards (SKKNI) for Construction Safety Technician competency. The method was based on competency levels in Bloom's taxonomy theory which was combined with a functional analysis theory. This method resulted in performance indicators for each performance criterion in three competency domains. By testing the performance indicators Construction Safety Professional and Human Resource Management Professional and also conforming to expert judgments, the results of compiled performance indicators are able to assess any performance achievements of each performance criterion.

1. Introduction

A reference for construction OHS competency standards in Indonesia consists of the Indonesian National Occupational Competency Standards (SKKNI) for construction safety human resource based on the Manpower and Transmigration Ministerial Decree number 350 of 2014. SKKNI for construction safety human resource is a basis for issuing OHS construction skill certificates through competency tests for workers, which can guarantee that the construction OHS human resources have work competency qualifications required to carry out related tasks/jobs properly and correctly, complying the mandate of regulations. In the mandate of Government Regulation No. 31 of 2006 concerning the National Occupational Training System, competency standards are the reference in developing competency-based training programs and the extent to where graduates have met the required competency standards [9].

Content from this work may be used under the terms of the Creative Commons Attribution 3.0 licence. Any further distribution of this work must maintain attribution to the author(s) and the title of the work, journal citation and DOI.
Published under licence by IOP Publishing Ltd

SKKNI for construction safety human resource is according to the Regional Model Competency Standard (RMCS). The RMCS is a model for developing competency standards introduced by the International Labor Organization (ILO), whose developments use a functional approach from the work process of a similar business/industrial activity [9].

In the RMCS model of each competency element, there is a measure of performance criteria. Performance criteria are the performance standards or tasks involved in each relevant job function. In the SKKNI, performance criteria are set for each element of competency. Further, there are also variable limits and assessment guidelines for each competency unit. The assessment guidelines consist of the assessment contexts, competency requirements, required knowledge and skills, work attitudes, and critical aspects.

Unit Code :	
Unit Title :	
Unit Description :	
Competency Element	**Performance Criteria**
1.	1.1
	1.2
	1.3
2.	2.1
	2.2
	2.3
3.	3.1
	3.2
	3.3
Variable Limit	
• **Variable Context** • **Tools and Equipment** • **Regulations Requirement** • **Norm and Standard**	
Assessment Guidance	
• **Assessment Context** • **Competency Requirement** • **Knowledge & Skill Required** • **Attitude Required** • **Critical Aspect**	

Figure 1. SKKNI structure for each competency unit.

Performance standards are the minimum benchmarks of performance which must be achieved by any individual or groups on all performance indicators. Each performance indicator meets particular criteria. In measuring performance, thus, there must be any criteria or sizes which have to be taken into concern [4].

Referring to the SKKNI structure in figure 1, the performance indicators are not described in detail for each performance criterion, where performance indicators are the important measures for assessing performance achievements. Due to this gap, the researchers intend to propose a method for developing performance indicators based on the performance criteria of the SKKNI for the Construction Safety Technician positions.

2. Literature Review

A functional analysis is a method used to examine the proficiencies of the laborers which are imprescriptible in a productive function. The function could be set on the level of an occupational sector, enterprise, group of enterprises, or the whole sector of production/services. The implementation of a functional analysis may be established in different initial stages: an occupation (PC repairman); mainstream occupations at diverse fields (occupational safety and health); or occupational sector (hotel). This thing evidently shows the functional analysis' flexibility. Being designed as a large-scale analysis tool, it does not mean that the functional analysis cannot be used to examine professions in particular subsectors or organizations [10].

A functional analysis cannot be defined as a precise method. By applying a deductive strategy, this analysis may be defined as a working approach to the expected competencies. The analysis starts by determining the primary goals of the productive function or service understudy. After that, the analysis is continued by raising inquiries to investigate the functions which are needed to be executed for achieving the prior function. The analysis is ideally conducted towards a group of laborers who already recognize the functional object of the analysis. This approach is worth to be used as a tool since it is derived directly from its representative qualities. To maintain uniform criteria, particular instructions are adhered to during the preparation. The structure below represents how the main purpose or key function is usually explicated [10] :

Figure 2. Function Analysis Structure [10].

The analysis of functions aims to identify any crucial functions to meet the major goal, which is, recognizing the value-added of the functions. The analysis results are then presented in a functional tree or functional map [10].

The functional tree or functional map is a graphic representation of functional analysis results. It is designed like a tree which is displayed horizontally. The functional tree reflects the methodology which is followed by its drawing-up. The map presents how the key function disassembles in succession – when it has been set- into its component functions. The branches of the tree are "causes" which are graphically connected to the left (or downwards, according to how the branches are presented) to their corresponding "consequences". Reading the tree branches from left to right or bottom to top, we will discover the answers to "how" the main functions are implemented by executing the basic functions which constitute them. By reading the tree branches from right to left, conversely, we will figure out the answers to "what for" of each function, which are embodied in the function of the immediate following level. Below is a graphic representation depicting an example of a functional map [10].

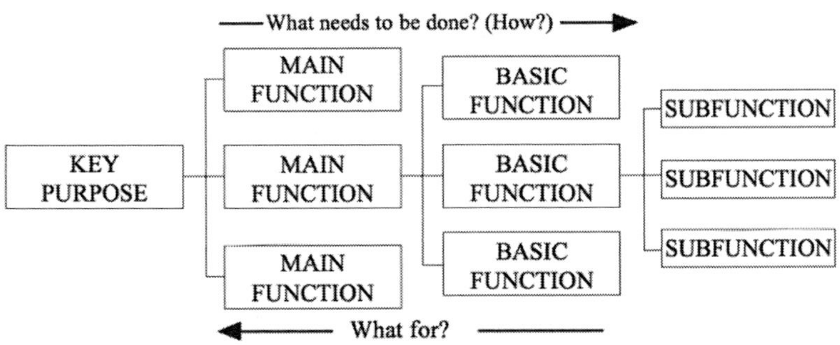

Figure 3. General Outlines of a Functional Map [10].

IOP Conf. Series: Materials Science and Engineering **930** (2020) 012010 doi:10.1088/1757-899X/930/1/012010

Referring to the figure 3, in the SKKNI, a basic function is represented as a competency unit/unit title, whereas sub function is represented as a competency element. Questions of "How?" or "What needs to be done?" can reflect the sub function from the basic function. The researchers propose to implement this method for developing performance indicators for each performance criterion.

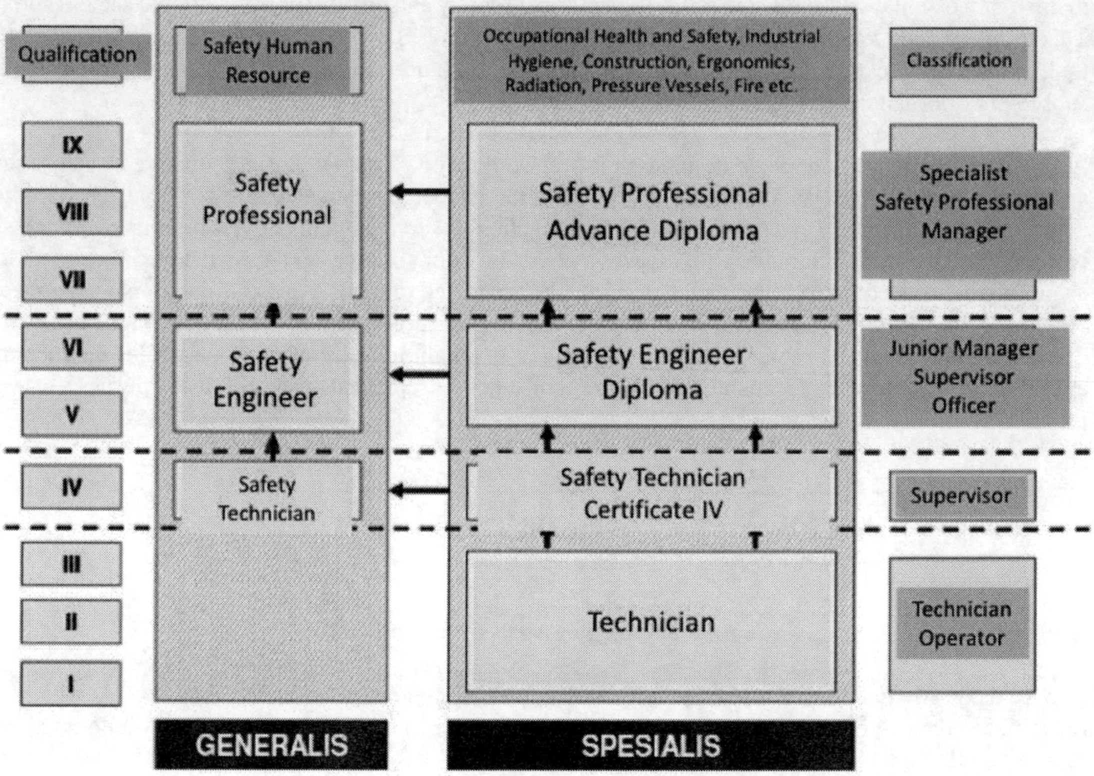

Figure 4. OHS profession certification pattern [6].

The figure 4 presents the pattern of OHS profession certification nationally. According to the certification pattern above, the Construction Safety Technician is categorized in level 4 of qualification. Each qualification level has its own knowledge parameter which can be observed in figure 5.

Moving on to the theory of taxonomy, taxonomy in education is used to classify instructional objectives; commonly called learning goals, performance goals, or learning goals. There are three general classifications or domains of the instructional objectives, namely: (1) cognitive domains (related to learning goals and oriented to thinking abilities); (2) the affective domains (related to feelings, emotions, value systems, and heart attitudes); and (3) psychomotor domains (oriented to motor skills or the use of skeletal muscle). Currently, there are various types of instructional objective taxonomies which are named after their inventors, one of which is: Bloom's taxonomy [2]. The relationship between the level of categories in Bloom's taxonomy and the level of qualification for Construction Safety Technician positions is illustrated in figure 5.

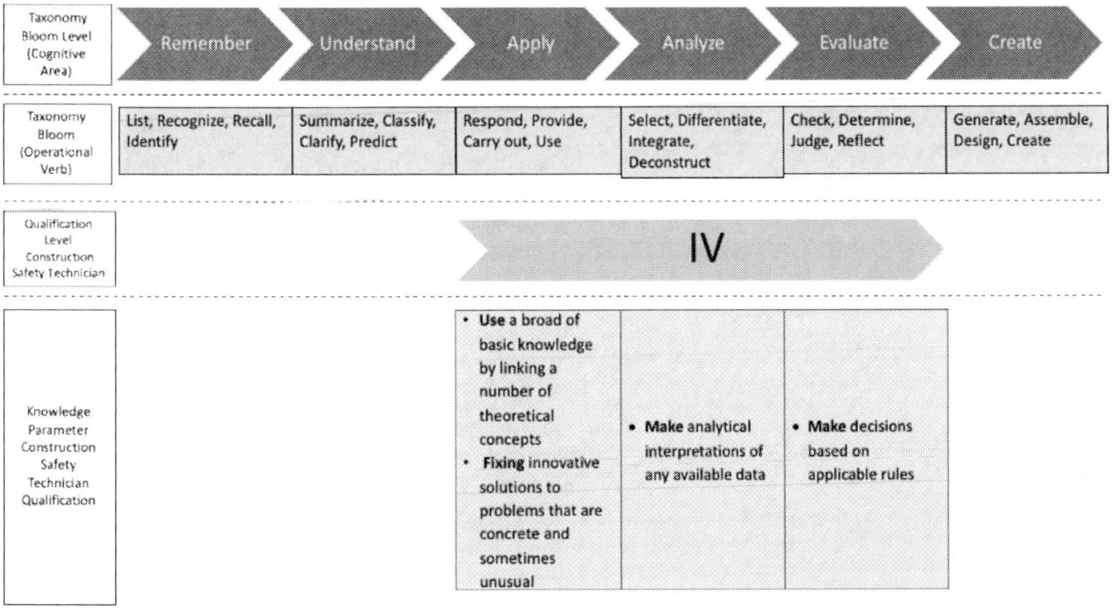

Figure 5. Relationship of cognitive domains and competency levels in Bloom's taxonomy and Construction Safety Technician competencies along with their knowledge parameters.

As presented in the figure 5, the knowledge parameter of Construction Safety Technician qualification is at the levels "Apply" until "Evaluate". It means that all tasks becoming Construction Safety Technician responsibilities are located between the levels "Apply" until "Evaluate". Researchers suggest combining Bloom's taxonomy operational verbs with functional analysis theory to develop performance indicators.

3. Methodology

In this study, the competency unit to be developed for its performance indicators was the Competency Unit "Hazard Identification and Risk Control". In developing the performance indicators, the researchers combined Bloom's taxonomy operational verbs with functional analysis theory.

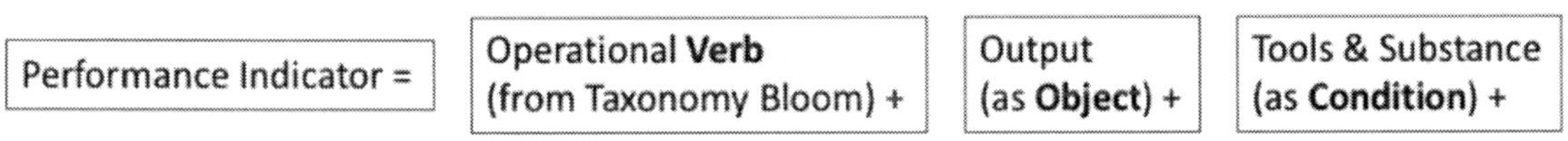

Figure 6. Diagram to develop performance indicators.

The first stage was to determine any output from each performance criterion. The output must be a noun and used as an object. The next stage was to determine any condition and aspect which influenced the performance indicators, including substances and tools. Substances and tools must be selected from variable limits and assessment guidelines for each competency unit in SKKNI. The last stage was to determine any operational verb which could accurately describe the level of each performance. In a tabular form, each step was arranged in an easy to understand form. Below is an example of the performance criteria from the Competency Unit "Hazard Identification and Risk Control".

IOP Conf. Series: Materials Science and Engineering **930** (2020) 012010 doi:10.1088/1757-899X/930/1/012010

Table 1. Example of performance indicator development process.

Performance Criterion [a]	Output	Substance [b]	Tool [b]	Operational Verb (Cognitive / Affective / Psychomotor) [c]
1.2 Construction documents and method statement, assessing the potential hazards	List of potential hazards	• Construction document • Method Statement	• Manpower and Transmigration Ministerial Regulation No. 13 of 2011 concerning Threshold Value of Physical Factors and Chemical Factors at Workplace • Manpower and Transmigration Ministerial Regulation No. PER-01/MEN/1980 on Occupational Health and Safety in Building Construction • Public Work Ministerial Regulation No. 05/PRT/M/2014 • Environmental information and data and demographic conditions of the construction workplace community • Results of workshop with related parties	• Cognitive : Deconstruct • Affective : Adjust • Psychomotor : Make

[a] Performance Criterion in "Hazard Identification and Risk Control" Competency Unit in SKKNI.
[b] Based on variable limits and assessment guidelines for "Hazard Identification and Risk Control" Competency Unit in SKKNI.
[c] Based on operational verbs for each knowledge level and domain in Bloom's taxonomy.

4. Result

The performance indicator resulted from the example of performance criteria in the table 1 above is then described in the table 2 below.

Table 2. Example of performance indicator result.

Performance Criterion [a]	Performance Indicator		
	Cognitive Domain	Psychomotor Domain	Affective Domain
1.2 Construction documents and method statement, assessing the potential hazards	**Deconstruct** potential hazards in construction documents and method statement based on the of Manpower and Transmigration Ministerial Regulation No. 13 of 2011 concerning Threshold Value of Physical Factors and Chemical Factors at Workplace	**Make a list** of potential hazards in construction documents and method statement based on the Manpower and Transmigration Ministerial Regulation No. 13 of 2011 concerning Threshold Value of Physical Factors and Chemical Factors at Workplace	**Adjust the list** of potential hazards in construction documents and method statement based on the Manpower and Transmigration Ministerial Regulation No. 13 of 2011 concerning Threshold Value of Physical Factors and Chemical Factors at Workplace

	Deconstruct potential hazards in construction documents and method statement based on the Manpower and Transmigration Ministerial Regulation No. PER-01/MEN/1980 on Occupational Health and Safety in Building Construction	**Make a list** of potential hazards in construction documents and method statement based on the Manpower and Transmigration Ministerial Regulation No. PER-01/MEN/1980 on Occupational Health and Safety in Building Construction	**Adjust the list** of potential hazards in construction documents and method statement based on the Manpower and Transmigration Ministerial Regulation No. PER-01/MEN/1980 on Occupational Health and Safety in Building Construction
	Deconstruct potential hazards in construction documents and method statement based on environmental information data and demographic conditions of the construction workplace community	**Make a list** of potential hazards in construction documents and method statement based on environmental information data and demographic conditions of the construction workplace community	**Adjust the list** of potential hazards in construction documents and method statement based on environmental information data and demographic conditions of the construction workplace community
	Deconstruct potential hazards in construction documents and method statement based on the results of workshop with related parties	**Make a list** of potential hazards in construction documents and method statement based on the results of workshop with related parties	**Adjust the list** of potential hazards in construction documents and method statement based on the results of workshop with related parties

[a] Performance Criterion in "Hazard Identification and Risk Control" Competency Unit in SKKNI.

The performance indicator result of all performance criteria in Competency Unit "Hazard Identification and Risk Control" is attached at the end of this paper. Performance indicators were also tested to Construction Safety Professional and Human Resource Management Professional, and also based on expert judgments. The results of compiled performance indicators can be used to assess any performance achievement in each performance criterion.

5. Conclusion, Limitation and Recommendation

Researchers propose a method of approach to develop performance indicators based on performance criteria in the Indonesian National Occupational Competency Standards (SKKNI) for Construction Safety Technician competency. The method was based on the competency level theory of Bloom's taxonomy which was combined with the functional analysis theory. By implementing the approach, this research resulted in the development of performance indicators for each performance criterion on three competency domains. Further, by testing the performance indicator to Construction Safety Professional and Human Resource Management Professional and conforming to expert judgments, the results of compiled performance indicators are able to assess performance achievement in each performance criterion.

However, this research still lacks enough sufficient expert validation to test the validity of the performance indicators. This research still can be developed until becoming a case study to evaluate the performance of Construction Safety Technician by using performance indicators which will have been validated. The evaluation, then, can be used to discover any competency gaps which need to be improved.

IOP Conf. Series: Materials Science and Engineering **930** (2020) 012010 doi:10.1088/1757-899X/930/1/012010

6. Appendix

Table 3. Performance indicator result of "hazard identification and risk control" unit competency.

Unit Code	: S.941200.011.01			
Competency Unit Title	**: Hazard Identification and Risk Control**			
Competency Element	**Performance Criterion**	**Performance Indicator**		
		Knowledge	**Skill**	**Attitude**
1. Identify potential hazards in the construction work environment	1.1 Various types of construction work to be executed are inventoried.	**Classify** various types of construction work in construction work contract documents	**Make a list** of construction work types	**Adjust the list** of construction work types based on construction work contract documents
	1.2 Construction documents and a method statement, assessing the potential hazards	**Deconstruct** potential hazards in the construction documents and method statement based on the Manpower and Transmigration Ministerial Regulation No. 13 of 2011 concerning Threshold Value of Physical Factors and Chemical Factors at Workplace	**Make a list** of potential hazards in the construction documents and method statement based on the Manpower and Transmigration Ministerial Regulation No. 13 of 2011 concerning Threshold Value of Physical Factors and Chemical Factors at Workplace	**Adjust the list** of potential hazards in the construction documents and method statement based on the Manpower and Transmigration Ministerial Regulation No. 13 of 2011 concerning Threshold Value of Physical Factors and Chemical Factors at Workplace
		Deconstruct potential hazards in the construction documents and method statement based on the Manpower and Transmigration Ministerial Regulation No. PER-01/MEN/1980 on Occupational Health and Safety in Building Construction	**Make a list** of potential hazards in the construction documents and method statement based on the Manpower and Transmigration Ministerial Regulation No. PER-01/MEN/1980 on Occupational Health and Safety in Building Construction	**Adjust the list** of potential hazards in the construction documents and method statement based on the Manpower and Transmigration Ministerial Regulation No. PER-01/MEN/1980 on Occupational Health and Safety in Building Construction
		Deconstruct potential hazards in the construction documents and method statement based on the Public Work Ministerial Regulation No. 05/PRT/M/2014 on Occupational Health and Safety Management	**Make a list** of potential hazards in the construction documents and method statement based on the Public Work Ministerial Regulation No. 05/PRT/M/2014 on Occupational Health and Safety Management System	**Adjust the list** of potential hazards in the construction documents and method statement based on the Public Work Ministerial Regulation No. 05/PRT/M/2014 on Occupational Health and Safety Management System

		System (SMK3) of Public Work Sector	(SMK3) of Public Work Sector	(SMK3) of Public Work Sector
		Deconstruct potential hazards in the construction documents and method statement based on environmental information data and demographic conditions of the construction workplace community	**Make a list** of potential hazards in the construction documents and method statement based on environmental information data and demographic conditions of the construction workplace community	**Adjust the list** of potential hazards in the construction documents and method statement based on environmental information data and demographic conditions of the construction workplace community
		Deconstruct potential hazards in the construction documents and method statement based on workshop results with related parties	**Make a list** of potential hazards in the construction documents and method statement based on workshop results with related parties	**Adjust the list** of potential hazards in the construction documents and method statement based on workshop results with related parties
	1.3 The list of potential hazards is arranged in a working order.	**Classify** potential hazards according to the working order.	**Make a list** of potential hazards in the working order	**Adjust the list** of potential hazards based on any changes in contract documents, environmental conditions, and workshop results
2. Conduct an assessment of hazard risk in the construction work environment	2.1 Facilities, techniques, processes and methods of hazard risk assessment are established.	**Choose** the tools, techniques, processes, and risk assessment methods to be used, in accordance with the Public Work Ministerial Regulation No. 05/PRT/M/2014 on Occupational Health and Safety Management System (SMK3) of Public Work Sector	**Make a list of** tools, techniques, processes, and risk assessment methods to be used, in accordance with the Public Work Ministerial Regulation No. 05/PRT/M/2014 on Occupational Health and Safety Management System (SMK3) of Public Work Sector	**Adjust the list** of tools, techniques, processes, and risk assessment methods to be used, in accordance with the Public Work Ministerial Regulation No. 05/PRT/M/2014 on Occupational Health and Safety Management System (SMK3) of Public Work Sector
	2.2 The hazard risk assessment is conducted.	**Carry out** a hazard risk assessment on the list of potential hazards according to the established tools, techniques, processes, and methods	**Make a list of** hazard risk assessment on the list of potential hazards according to the established tools, techniques, processes, and methods	**Adjust the list** of hazard risk assessment on the list of potential hazards according to the established tools, techniques, processes, and methods
	2.3 The level of risk is determined.	**Choose** the classification level of hazard risk based	**Match** the level of hazard risk in the risk assessment table	**Adjust** the level of hazard risk based on any changes in the

		on the Public Work Ministerial Regulation No. 05/PRT/M/2014 on Occupational Health and Safety Management System (SMK3) of Public Work Sector	based on the classification level of hazard risk	list of potential hazards
3. Conduct any hazard risk control in the construction work environment	3.1 A hazard risk control plan is arranged.	**Make** any target and hazard risk control plan based on the work contract and the Public Work Ministerial Regulation No. 05/PRT/M/2014 on Occupational Health and Safety Management System (SMK3) of Public Work Sector	**Make** a table for establishing the targets and hazard risk control programs based on the work contracts and the Public Work Ministerial Regulation No. 05/PRT/M/2014 on Occupational Health and Safety Management System (SMK3) of Public Work Sector	**Adjust** the table of established targets and hazard risk control programs based on the work contracts and the Public Work Ministerial Regulation No. 05/PRT/M/2014 on Occupational Health and Safety Management System (SMK3) of Public Work Sector
	3.2 The danger risk control is conducted.	**Implement** any risk control based on the target and OHS program table which has been prepared	**Practice** the hazard risk control monitoring in the OHS target and program table	**Adjust** the monitoring of risk control based on any changes in targets and the OHS program table
	3.3 The risk control is evaluated.	**Check** the risk control of hazards based on the results of monitoring the risk control of hazards, and **arrange** changes/improvements, to the discrepancy of findings in the field, with plans and targets that have been made, at each step of the activity/work	**Make** a summary of hazard control, based on the results of monitoring of risk control, and **make** a list of changes/improvements, to the discrepancy of findings in the field, with plans and targets that have been made, at each step of the activity/work	**Adjust** the summary of hazard control, based on changes in the results of monitoring of hazard risk control, and **adjust** the list of changes/improvements, based on changes in nonconformity findings in the field, with plans and targets that have been made, at each step of the activity/work

Acknowledgements
The author would like to thank University of Indonesia for providing financial support through PUTI Proceedings of the Fiscal Year 2020 with the contract number: NKB-1139/UN2.RST/HKP.05.00/2020 managed by the Directorate of Research and Community Engagement (DRPM) University of Indonesia.

References

[1] Elsye V, Latief Y and Sagita L 2018 Development of work breakdown structure (WBS) standard for producing the risk based structural work safety plan of building *MATEC Web. Conf.* **147** 06003 (*Preprint* gr-qc/06003)

[2] Gunawan I and Palupi A R 2012 Taksonomi bloom – revisi ranah kognitif: kerangka landasan untuk pembelajaran, pengajaran dan penilaian [Bloom taxonomy – cognitive domain revision : platform framework for learning, teaching and assesment] *Jurnal Pendidikan Dasar dan Pembelajaran* **2** 02

[3] Latief Y, Machfudiyanto R A, Arifuddin R, Setiawan R M F and Yogiswara Y 2017 Study of evaluation OSH management system policy based on safety culture dimensions in construction project *IOP J. Phys. : Conf. Series* **877** 012028 (*Preprint* gr-qc/012028)

[4] Rahadi D R 2010 *Manajemen Kinerja Sumber Daya Manusia* (Malang: Tunggal Publishing) p 20-28

[5] RoI 1980 *Peraturan Menteri Tenaga Kerja dan Transmigrasi No. PER-01/MEN/1980 tentang Keselamatan dan Kesehatan Kerja pada Konstruksi Bangunan [Manpower and Transmigration Ministerial Regulation No. PER-01/MEN/1980 concerning Occupational Health and Safety in Building Construction]* (Jakarta: Republic of Indonesia)

[6] RoI 2008 *Keputusan Menteri Tenaga Kerja dan Transmigrasi Republik Indonesia No. KEP.42/MEN/III/2008 tentang Penetapan Standar Kompetensi Kerja Nasional Indonesia Sektor Ketenagakerjaan Bidang Keselamatan dan Kesehatan Kerja [Manpower and Transmigration Ministerial Decree No. KEP/42/MEN/III/2008 concerning the Establishment of Indonesian National Work Competency Standards in Occupational Health and Safety Employment Sector]* (Jakarta: Republic of Indonesia)

[7] RoI 2011 *Peraturan Menteri Tenaga Kerja dan Transmigrasi Republik Indonesia No. PER-13/MEN/X/2011 tentang Nilai Ambang Batas Faktor Fisika dan Faktor Kimia di Tempat Kerja [Manpower and Transmigration Ministerial Regulation No. 13 of 2011 concerning Threshold Value of Physical Factors and Chemical Factors at Workplace]* (Jakarta: Republic of Indonesia)

[8] RoI 2014 *Peraturan Menteri Pekerjaan Umum No. 05/PRT/M/2014 tentang Pedoman Sistem Manajemen Keselamatan dan Kesehatan Kerja (SMK3) Konstruksi Bidang Pekerjaan Umum [Public Work Ministerial Regulation No. 05/PRT/M/2014 concerning Occupational Health and Safety Management System of Public Work Sector]* (Jakarta: Republic of Indonesia)

[9] RoI 2014 *Keputusan Menteri Tenaga Kerja dan Transmigrasi Republik Indonesia No. 350 tahun 2014 tentang Penetapan Standard Kompetensi Kerja Nasional Indonesia Kategori Jasa Profesional, Ilmiah dan Teknis Golongan Pokok Jasa Arsitektur dan Teknik Sipil; Analisis dan Uji Teknis Bidang Keselamatan dan Kesehatan Kerja (K3) Konstruksi [Manpower and Transmigration Ministerial Decree No. 350 of 2014 on the Establishment of Indonesian National Work Competency Standards for the Professional, Scientific, and Technical Services of the Basic Group in Architectural and Civil Engineering Services; Analysis and Technical Test in the Field of Construction Occupational Health and Safety]* (Jakarta: Republic of Indonesia)

[10] Vargas F 2004 *40 Question on Labour Competency* (Montevideo: IOT/Cinterfor) p 39-45

IOP Conf. Series: Materials Science and Engineering **930** (2020) 012011 doi:10.1088/1757-899X/930/1/012011

4D BIM implementation to improve EPC project performance from contractor's perspective. A case study

M Piselia[1*]

[1] Department of Civil Engineering, Faculty of Engineering, Uniersity of Indonesia

[*]Corresponding author's e-mail: maretha.piselia@tripatra.com

Abstract. The most significant benefits of 4D BIM are better visualization of construction work, better communication between project teams and improving project time and cost performance. 4D BIM allows stakeholders to visualize construction activities during project duration and identify potential spatial conflicts and schedule conflicts. Although researches related to 4D BIM has proven the benefits of 4D modeling in construction projects, these benefits are still questionable for EPC projects, whether it can compensate the initial costs and efforts to create the model. The implementation strategies across EPC project lifecycle also become fundamental issue. Through literature studies and structured interviews, this research identifies resource requirements, implementation strategies and the benefits of 4D BIM in improving EPC project performance.

1. Introduction

EPC projects in mining and oil & gas sector face major challenges in completing projects on time and on budget, as many as 98% of EPC mega projects experience 30% over budget and 77% of projects experience schedule delay of at least 40%. Despite that, manufacturing sector is experiencing faster growth in productivity due implementation of new technologies, the rate of improvement of productivity in EPC Projects remain the same in the last decades [1].

In terms of performance and progress control, managers traditionally base their decisions on S-Curves, Gantt Charts, Cost Reports and other visual aids to determine if the performance of a project is satisfactory or not, and based on this, make the necessary adjustments. However, traditional construction planning and control tools such as Gantt Charts, S-Curves and Time Location Charts do not provide a clear visualization of information, since the sequence of activities is not linked to the physical objects represented by those activities. Project progress is difficult to be tracked and visualized in Gantt chart method, whereas the works are done simultaneously in large-scale projects [2].

In S curve method, the cumulative amount of activity can be shown based on the costs incurred, the number of Man Hour spent and the amount of material used by the project, however this only shows the progress of the project over time and the deviation from plan progress in numerical value [3]. Furthermore, the form of the report does not provide information about specific activities that need to be completed next, interface information, spatial boundaries and critical activity information does not show visual references of deviations or delays that occur [4]. In brief the problems that arise are visualization and limited communication between the project planner and the execution manager in the field.

Content from this work may be used under the terms of the Creative Commons Attribution 3.0 licence. Any further distribution of this work must maintain attribution to the author(s) and the title of the work, journal citation and DOI.
Published under licence by IOP Publishing Ltd

Building Information Modeling (BIM) technology has been developed by adding a time dimension known as 4D BIM which can be one of the solutions to visualize the progress of construction work into three-dimensional images. The benefits of using 4D BIM as a project planning method include increasing accuracy and enabling more detailed scheduling, construction work sequences can be better simulated, improve safety performance and better communication among project stakeholders [5]. Although researches related to 4D BIM has proven the benefits of 4D modeling in construction projects, these benefits are still questionable whether it can compensate the initial costs and efforts to create the model.

This research aims to identify the 4D BIM implementation strategies to optimize the benefits, identify resources requirements and analyze the impact of 4D BIM use in improving EPC project performance. A case study carried on a Power Plant 25 MW EPC project in North Sumatra, Indonesia from EPC contractor's perspective.

2. Literature review

2.1. 4D BIM definition

The 4D BIM model can be defined as three dimensional (3D) combined with time (schedule) information-makes it easier for the project team to visualize the sequential construction process such that the appropriateness of construction plan can be thoroughly reviewed. 4 D model is created by linking construction schedule to 3D model. This process of linking a schedule to a 3D model occurs in a 4D simulating program [6] Many specialized software appear to produce 4D models by connecting 3D models and schedules from separate platforms. Typically, this software requires 3D data models imported from CAD or BIM applications and Schedule data imported from planning/scheduling applications such as MS Project and Primavera. The scheduler then connects the components of the 3D model to the construction activities of the schedule and forms a 4D model based on time.

2.2. LOD and implementation strategies

The Level of Development is breakdown of the proportions of what is included in the model element. The level at which the component specifications, geometry, and information attached have been thought to what extent project team members can depend on information when using the model. In essence, Level of Development is to specify the information that the model must contain according to its use at the different stages of a project lifecycle.

The American Institute of Architects (AIA) defines five levels of developments: LOD 100 to LOD 500. At LOD 100, which is the pre-design stage, the model consists of 2D symbols to indicate the presence of elements. In LOD 200, elements are partially defined by describing estimates of quantity, size, shape, and location. With LOD 300, elements are defined with their exact dimensions and relative position to support precision. LOD 350 explains information about an element precisely and outlines the relationship and connection of elements with other components. The LOD 400 level outlines basic information about building various elements. With the LOD 500, the model begins to represent the functions of real life elements in buildings. It was shown that it is required to manage multiple graphical LOD corresponding to the different usages of the model either expected or unexpected. The graphical LOD necessary for construction phase is higher than the graphical LOD at pre-construction phase [7]. This research has study at which LOD, 4D BIM implementation shall be implemented across EPC project lifecycle.

2.3. Project key performance indicators

To measure a project's performance and to apply a benchmarking approach, it must first determine the Key Performance Indicators (KPIs) that are most important in determining the overall success of the project. A KPI is a compilation of data sizes used to assess the performance of construction operations [8].

Research conducted by David D. John (2018) [9] identified results and process-oriented Key Performance Indicators (KPI). The results-oriented KPI consists of cost, time, quality, safety and client satisfaction levels. Whereas KPI which is process oriented consists of communication management, resource management, cost management, effectiveness of schedule control, effectiveness of quality management, effectiveness of safety management, management and coordination of Human Resources. To achieve the objectives of this study two principal methodologies have been considered: 1. an extensive literature review on the performance measurement to identify the initial set of KPIs. 2. Exploration of the practitioner view to formulate the KPI measures from a 4D BIM implementation view point. The analysis of semi-structured interviews resulted in the development of following 4Dbased KPIs as represented in table 1.

Table 1. 4D based Key Performance Indicators.

Factor	Performance Indicator	Attributes
KPI (Result oriented)	Financial	1. Gross Profit
		2. Total CO Approved
		3. Actual Billing vs. Plan
	Schedule Goal	1. Schedule Performance Index (SPI)
		2. Key Milestone Achieved
	Cost Goal	1. Cost Performance Index (CPI)
	Quality Goal	1. Document Review Cycles
		2. Number of Non-Conformances
	Safety Goal	1. Fatalities
		2. Total Recordable Injury Rate (TRIR)
	Customer Satisfaction	1. Customer Satisfaction Survey
	Productivity Performance	1. Engineering Man Hours Productivity
		2. Equipment Utilization Index
		3. Overall Construction Productivity
KPI (Process oriented)	Communication Management	1. Communication Frequency
		2. Communication Effectiveness
		3. Communication Method
	Resources Management	1. Material management
		2. Labor management
		3. Subcontractor management
	Schedule Management	1. Schedule control
		2. Planning efficiency
		3. Smooth work process
	Quality Management	1. Effective quality management
		2. Earlier detection of problems
		3. Design efficiency
		4. Rework efficiency
	Safety Management	1. Effective Safety Management
	Productivity Performance	1. Man Hours spent to complete construction

3. Methodology

A Power Plant 25 MW EPC Project is used for the case study which was located in North Sumatra, Indonesia. 4D BIM model has been developed at the engineering (pre-construction phase) using Autodesk Navisworks Manage 2016 as project review software that supports intelligent 3D model-based design with scheduling, visualization, and collaboration tools. Primavera P6 Professional R8.1 is used for scheduling the project plan and imported into Autodesk Navisworks Manage software. The resources

Man Hour to develop 4D BIM model at pre-construction phase was defined based on project data records; meanwhile the resources Man Hour for construction phase was defined based on contractor's database and expert estimation.

In the following parts, the author characterizes the Level of Development of 4D BIM across phases of a project, and on the basis of the above mentioned LOD definitions. The next aim is to analyses the benefits of 4D BIM technology to improve EPC project Key Performance Indicator based on survey result, qualitative feedbacks and structured interviews with EPC contractor practitioner.

4. Findings and Discussions

4.1. Resource requirements

Autodesk Navisworks Manage 2016 and Primavera P6 Professional R8.1 are required for 4D BIM implementation as explained above. Additionally 3D survey also required at the beginning of engineering phase to included project site condition into 3D model.

Figure 1. Interfaces of Autodesk Naviswork Manage of 4D BIM

In this study the additional Man–hours was spent for 4D BIM implementation in Power Plant 25 MW Project having the following elements such as pile, foundation, steel structure, mechanical equipment, gas engine equipment, pipes, electrical and instrumentation. The hours were broken into the following phases:

1. Personnel software study (initiation phase), two months durations
2. 4D BIM - EPC schedule development (pre-construction phase), four months duration
3. 4D BIM - EPC schedule monitoring and controlling (construction phase), eleven months duration

Man Hour was calculated for additional activities to implement 4D BIM at initiation phase, pre-construction phase and construction phase as represented in table 2. About 19 project members were involved and additional 3,392 Man Hours were required for 4D BIM implementation across 21 months project lifecycle. In percentage, the total of additional Man Hour required was 1.11% of total project indirect Man Hours.

Table 2. Man Hours spent across project phases.

No	Project Phase	Additional Activities	Number of Manpower Involved	Total Man Hours
1	Study & learning of software (initiation phase)	1. Personnel Study	19	456
2	4D BIM development (pre-construction phase)	1. Conduct survey and field investigation - 3D Survey and photogrammetry	6	24
		2. Modelling the structure of the project site into BIM	4	40
		3. Determine the LOD (level of Detail) that will be used in integrating the Schedule and 3D models	4	40
		4. Analyzes additional project activities needed based on project conditions	2	64
		5. Sequence of project activities simulation into a 4D BIM	2	20
		6. Construction activities schedule integration into 3D model elements	15	48
		7. Validation of Schedules and 3D models integration results by all disciplines	4	124
		8. 3D model optimization and design based on simulation and integration results	3	64
		9. Schedule optimization based on simulation and integration results	19	40
		Subtotal		**464**
3	4D BIM schedule monitoring and controlling (construction phase)	1. Input actual weekly construction progress into 4DBIM	2	260
		2. Analyse project activities based on actual conditions at construction site	9	636
		3. Analyze the remaining work in the field with the team with 4D BIM media	9	558
		4. Present work plans for the next two and six weeks to the project team with 4D BIM media	8	146
		5. To coordinate the process of checking on schedule conflicts, space-time conflicts, design conflicts, design conflicts in the field, and conflicts between stakeholders	9	162
		6. To coordinate the process of checking the layout of the field for construction activities virtually to analyze the access road, the need for heavy construction equipment space and the area to place temporary material	9	162
		7. To coordinate the process of checking to avoid conflicts with Construction activities	11	266
		8. Shows delay activities virtually using 4D BIM media	9	162
		9. Shows construction performance virtually using 4D BIM media	9	162
		10. Shows project critical path activities virtually using 4D BIM media	9	162
		11. Schedule re-baseline based on the results of 4D BIM evaluation and analysis	3	56
		Subtotal		**2,472**
		Total		**3,392**

4.2. Implementation strategies

In the initiation phase case study, 4D BIM implementation started since contract award and the first step is to define Work Breakdown Structure and discuss with engineering team who is responsible to develop 3D model. This step is very important to prevent major adjustment in engineering (pre-construction) phase. The schedule breakdown structure shall be in line with 3D level of development, for example LOD 300 represents project master schedule level 3. The other important activity is to study and learn 4D BIM supporting software by all project members involved in 4D BIM implementation.

In the pre-construction phase, it is recommended to link 3D model with project master schedule when 3D model has achieved LOD 300, thus in line with project master schedule level 3. In the case study project LOD 300 also defined as 30% 3D model progress and design review meeting was conducted between all project stakeholders at the end of month# 2. LOD 100 and LOD 200 are not recommended to start 4D BIM implementation due to the maturity of 3D model is not sufficient and the 4DBIM benefits will not optimal.

In the construction phase, the 4D BIM implementation was started at the end of month# 6 or when 3D model has achieved LOD 500 or also defined as 60% 3D model progress. This arrangement is consistent with rolling wave planning scheduling method. In the construction phase it is important to implement 4D BIM up to daily activity schedule or schedule level 5. LOD for 4D BIM implementation across EPC project lifecycle is shown in figure 2.

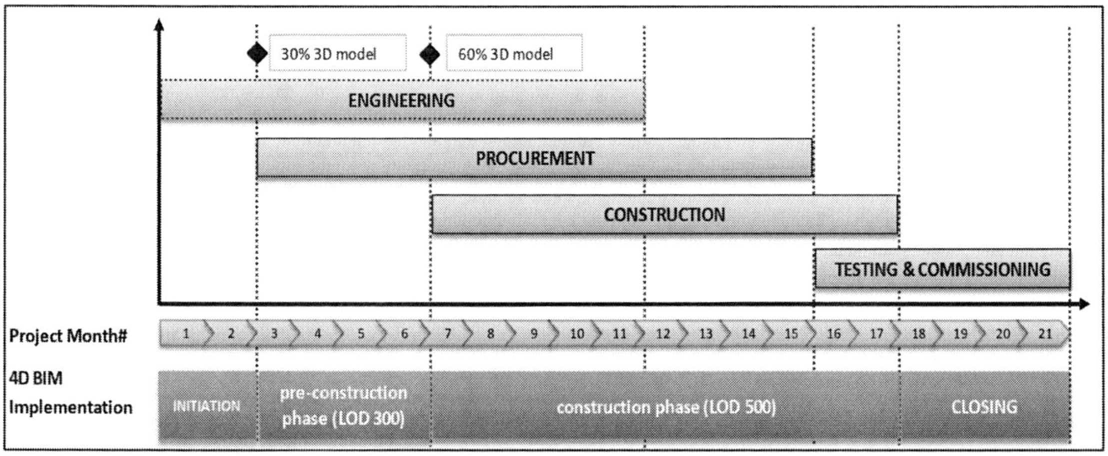

Figure 2. 4D BIM implementation across EPC project lifecycle

4.3. The impact of 4D BIM implementation to improve key performance indicators

Interviewees were asked to rank the impact of 4D BIM implementation to improve identified KPIs. The ranking of the impact was done by using a five (5) point Likert Scale (where 1= has no impact, 2 = has very low impact, 3 = has fairly impact, 4 = has impact, and 5 = has high impact). The impact will be further classified in qualitative terms (rating on a scale).

Figure 3 and figure 4 represents the weighted (%) ranking of the KPI on the basis of the views of the respondents. As shown in figure 3, the implementation of 4D BIM has the highest impact to improve project productivity (83%) for result oriented KPI and the highest impact to improve schedule management (97%) for process oriented KPI as shown in figure 4.

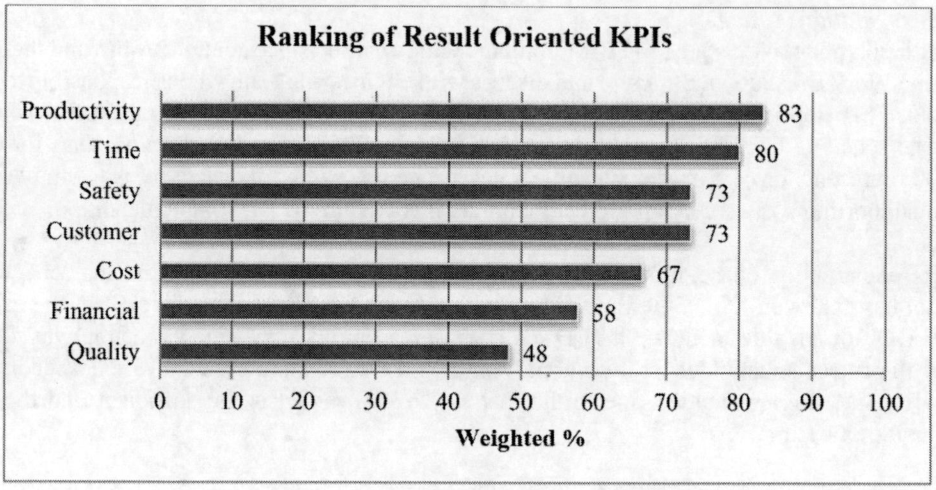

Figure 3. Ranking of result oriented KPIs

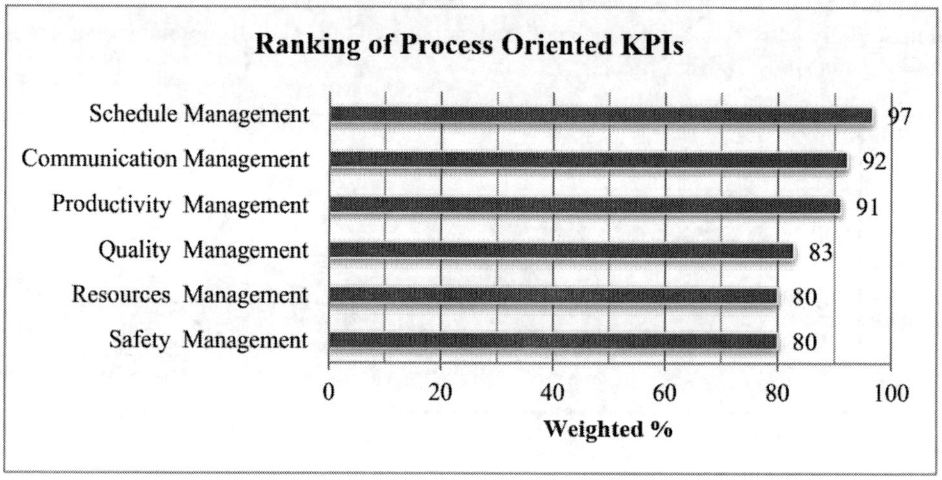

Figure 4. Ranking of process oriented KPIs

5. Conclusions

This paper reported a 4D BIM implementation case study on Power Plant 25 MW Project in North Sumatra, Indonesia. The first part of the research defined resources requirements for 3D BIM implementation across project lifecycle. About 19 project members were involved and 3,392 Man Hours were utilized for initiation phase, develop 4D BIM with LOD 300, develop 4D BIM with LOD 500 and to implement 4D BIM as construction activity controlling and monitoring tools. 4D BIM implementation was started since project award for initiation phase.

In the pre-construction phase, 4D BIM was developed by connected 3D model LOD 300 with project master schedule level 3. Furthermore, in the construction phase, 4D BIM was implemented to control and monitor construction activity by connected 3D model LOD 500 with project schedule level 5 which shows construction daily activities.

It is shown 4D BIM implementation has potential benefits to improve project productivity and time performance. Moreover the benefit of 4D BIM implementation has higher impact to improve schedule and communication management process during construction phase.

References

[1] IHS Herold Global Project Database 19 November 2013
[2] Mubarak S 2015 *Construction Project Scheduling and Control* (New Jersey: Wiley)
[3] Nicholas J and Steyn H 2012 *Project Management for Engineering, Business and Technology* (Oxon: Routledge)
[4] Koo B and Fischer M 2000 Feasibility study of 4D CAD in commercial construction *J. Constr. Eng. Manag.* **126** 4 p 251-60
[5] Swapnesh P R and Valunjkar S S 2017 Improve the productivity of building project using building information modelling (bim) based 4d simulation model *IJRASET* **5** p 45-98
[6] Pitake S A and Patil D S 2013 Visualization of construction progress by 4D modeling application *IJETT* **4** p 3000
[7] Boton C, Kubicki S and Halin G 2015 The challenge of level of development in 4D/BIM simulation across AEC project lifecysle. A case study. *CCC 2015* **123** p 59-67
[8] Elshakour A H A, Al-Sulaihi I A, Al-Gahtani K S 2012 Indicators for measuring performance of building construction companies in Kingdom of Saudi Arabia *JKSUES* **25** p 125–34
[9] John D D 2018 *Building Information Modeling (BIM) Impact on Construction Performance* (Georgia: Georgia Southern University)

IOP Conf. Series: Materials Science and Engineering **930** (2020) 012012 doi:10.1088/1757-899X/930/1/012012

Implementation strategy of total quality management and quality culture to increase the competitiveness of contractor companies in Indonesia

G Pinandhita[1*] and Y Latief[1]

[1] Department of Civil Engineering, Faculty of Engineering, Universitas Indonesia, Depok, West Java 16424, Indonesia.

*Corresponding authors: gpinandhita@yahoo.com

Abstract. When compared to the manufacture industry, the construction industry is being a concern because of their poor quality produced. For this reason, many contractor companies adopted and implemented management strategies to improve their product quality. One of management strategy approaches used is Total Quality Management (TQM). Implementation of TQM requires a fundamental change by changing the culture, processes, strategies, and beliefs in a company. This quality orientation is the key to fulfill customer satisfaction and business practices. Quality Culture is a part of organizational culture related to habits, beliefs, values, morals, and behavior to improve quality of the product. In Indonesia, contractor companies are generally weak in many ways. This will cause contractor company in Indonesia have great difficulties to be competitive with foreign contractors. This increasing global competition makes contractor companies in Indonesia have to adopt TQM and Quality Culture as one of the strategies to always make improvements continuously so as to provide competitive advantage and increase company performance. For Indonesian contractor companies to compete with foreign contractors, the alternative strategy is needed. Framework of the alternative strategy is the result of the implementation of TQM and Quality Culture, because they can be seen as a strategy that improves economic conditions and helps companies achieve competitive advantage.

1. Introduction

When compared to the manufacture industry, the construction industry is being a concern because of their poor quality produced. This is not only seen from the final product produced but also in terms of process, workers and material. For this reason, many contractor companies adopted and implemented management strategies to improve their quality. One of management strategy approaches used is Total Quality Management (TQM). [10].

TQM is believed to be able to increase customer satisfaction, increase the involvement of company employee, strengthen suppliers relationships, and create an continuous quality improvement [10]. TQM has been widely adopted by many contractor companies to fix their quality problems, and to meet the customer's wants and needs. Also, TQM has other goal in the construction industry that is to reduce production costs, and increase competitiveness value of the company [7].

Content from this work may be used under the terms of the Creative Commons Attribution 3.0 licence. Any further distribution of this work must maintain attribution to the author(s) and the title of the work, journal citation and DOI.
Published under licence by IOP Publishing Ltd

Quality Culture is a part of organizational culture related to habits, beliefs, values, morals, and behavior to improve quality of the product. Quality Culture also can be defined as an organizational culture with great quality oriented in every activity undertaken.

TQM and Quality Culture have a concepts that is a holistic approach to project developing that involves continuous improvement, customer satisfaction, top level management support, and employee training to create products without defects and without doing additional work [10]. Data from the Ministry of Public Works and Housing (PUPR) informs that the total number of national contractors in Indonesia up to 2016 based on data from Bina Konstruksi is 108,852 that consisting of individual 29 contractors, 95.014 small contractors, 13.225 medium contractors, and 584 large contractor. Data in 2015 shows that 85% of Indonesia's construction market value is controlled by large contractors while the remaining 15% of the construction market value is contested by medium and small contractors. This situation causes business competition become very tight between the middle and small qualification construction market. Also, this has become a matter of how to increase the class of medium and small construction service business contractor companies to become large qualified business entities in order to be competitive and survive in this highly competitive era.

Increasing global competition makes contracting companies in Indonesia should adopt TQM and Quality Culture as one of the strategies in meeting customer satisfaction [6]. An alternative strategy is needed for Indonesian contractor companies to compete with foreign contractors [11]. Framework of the alternative strategy is the result of the implementation of TQM and Quality Culture, because they can be seen as a strategy that improves economic conditions and helps companies achieve competitive advantage [13].

2. Literature review

2.1 Total quality management

TQM is an approach method to increase the competitive advantage, effectiveness, and flexibility of an company by utilize all aspects such as plan, organize, and deepen all of the activities. Also, TQM can be interpreted as a process that improving aspects of the quality in terms of processes, human resources, products, and services in a company, with the main goal is to increase customer satisfaction through continuous improvement in processes and systems [2]. Aspects that included in TQM [4] [1] [12] are:

1. Leadership and Top Level Management's Commitment
   Commitment from top level management is very important. It is very important to change the company environment to achieve changing staff behavior and improve the quality. The company's efforts for TQM will fail without this fundamental cultural change. This fundamental change cannot be achieved unless top level management has an obsession to focus on quality and continuous improvement.
2. Training
   Each training program in the company have to include the orientation to the basic concepts and procedures of TQM. TQM requires participatory, disciplined, and organized approaches, so team training is also very important. The training program have to cover many topics such as analysis, problem solving, interpersonal communication and interaction, statistical methods, cost of quality measurement, and collection-evaluation of quantitative information.
3. Communication
   Good communication and a good system are important so that employees convey ideas to top level management and to provide the necessary changes. The effective strategy is an open communication line that allows direct access for every employee, at any level, to connect top level management to ideas for specific improvements or problems. Top level management reaction of employees' concerns and ideas is very important.
4. Teamwork

In achieving company's goals, teams are very important. Work together in teams is generally more effective to achieve common goals than work alone. The team approach is not only to internal organizational teams, but also include external vendors and customers.

5. Customer Satisfaction
   The main goal of TQM is to achieve customer satisfaction whether the customer is internal (such as: department in the company) or external (such as: the consumer). Defining customer needs and desires is the first step to achieve customer satisfaction, then translating these needs and desires into a standard. Customer satisfaction have to be limited to meet customer expectations, but still have to try to exceed through continuous improvement. To meet customer expectations, company should conduct the information gathering programs that can measure customer satisfaction level. The program will help the company to identify the customer dissatisfaction, so the company can directly take the corrective action.
6. Continuous Improvement
   Continuous improvement requires a focus on processes so they can be changed to be more efficient. By comparing progress against certain criteria, the level of success can be determined. The process of measuring and comparing success rates against predetermined criteria is known as "benchmarking." Benchmarking is a systematic search for best practices that lead to superior performance.
7. Process Management
   Process improvement has a relationship with continuous improvement. Process improvement is a statistical method and process control, because data measurement and analysis are very important for process improvement. Accurate data is very important for employees and management to take better decisions. A quality improvement team can be formed in any company to check the process.
8. Focus on Employees
   TQM views employee satisfaction as an important factor in increasing their contribution in company. Employee is the company's internal customers to exchange information and services. Top level management should encourage advice and open the work environment, so that employees can share the honest comments without fear of punishment. Recognition and appreciation should be extended to valuable suggestions for the company.
9. Supplier Management
   The quality of each stage in a process depends on the quality of the previous stage. There should be careful identification of each other's needs. TQM pays attention to suppliers or vendors of labor and materials. Maintaining close and long-term relationships with suppliers gives an economic achievement and the best quality. One of the strategies to improve relations between suppliers and companies is by conducting regular visits and regular communication. Also, it helps the supplier to have a good understanding and feel the needs of the company.

2.2 Quality culture

Quality Culture also can be defined as an organizational culture with great quality oriented in every activity undertaken. Company need the focus of each individual to be able to utilize all resources to improve quality [3]. Here are the forming aspects of Quality Culture [3] [8] [5]:

1. Leadership
   One of the main determinants of the successful implementation of Quality Culture is Leadership. Most of the quality problems are related to top level management. This shows that successful quality management is highly dependent on top level management. Top level management should have the philosophy that is the quality will receive priority higher than cost or schedule, so in the long run, consistent and superior quality will lead to increased performance.
2. Customer Focus
   Emphasis on customer satisfaction or quality driven by customers is considered by many experts and proposed as the great success of the quality management effort. Strategic concepts of customer satisfaction with withdrawals such as customer retention and market penetration.

IOP Conf. Series: Materials Science and Engineering **930** (2020) 012012 doi:10.1088/1757-899X/930/1/012012

3. Teamwork
   Teamwork allows many various parts of the company to meet customer needs in ways that cannot be done by individual work. The teamwork provides more mixed of skills in thought and process in company management and those who hold employee supervisory roles.
4. Employees Involvement
   Employee involvement as a process of transforming company culture to gathers the creativeness of all employees for problem solving and to make improvements. Top level management should adopt the Taoist teachings of people-based management. They recognize the importance of daily interactions with all employees and solve the problems together.
5. Employees Empowerment
   The impact of human resources in an company depends on the type of empowerment given to them. The employee empowerment program can be seen as an opportunity for company in the current competitive environment. Some experts consider employee empowerment and commitment as a condition for successful implementation.
6. Motivation
   It takes huge determination and desire for continuous improvement in the very long term in forming Quality Culture. So that the motivation is needed in making continuous improvements.
7. Communication
   Effective communication is important in the process of empowering employees. The complicated management structures cause the communication problem. Effective internal communication in the company is important for achieving quality success because it can provide ongoing customer satisfaction.
8. Relationship with Supplier
   Establishing relationships with suppliers is an important aspect of establishing Quality Culture because materials and parts purchased are often the main source of quality problems. Poor supplier product quality results in additional costs for the buyer. Therefore, most quality problems will be caused by suppliers. For both parties to succeed and their business to grow, a good relationship between the two parties is needed. Companies should treat their suppliers as long-term business partners.
9. Organization's Structure
   The organizational structure is also an important part in the process of improving the process. This is closely related to the lines of communication, leadership, teamwork, and employee involvement. An efficient organizational structure can produce good communication, good teamwork, and good employee involvement, so as to improve the quality of work.
10. Recognition and Reward
    Another important aspect of quality improvement program is recogniting every individual, section, and department in the company as an increase in performance. Recognition is one of the most important steps in the quality improvement process, because it is the basis for increasing the involvement of all employees in business operations. Recognition and also reward should stimulate employee commitment to quality improvement effectively.
11. Strategy and Policy for Quality
    The integration of quality problems and customer satisfaction into operational plans is needed to conduct strategic quality plan. Also, the important factor for success in quality management is the policy development because the best company use the policy development process to ensure employee understanding of company goals and how to contribute.
12. Power Distance
    Power Distance is a culture where employee in a company do not have power unless there is an injustice in the distribution of power to the company. The inequality power depends on social, educational, or occupational levels. Countries with high power distance have superiors who can use their power and create high gaps between superiors and subordinates. So, subordinates are

reluctant to question or disagree with their superiors. In the other hand, countries with low power distance show an equal relationship between superiors and subordinates.

13. Uncertainty Avoidance
    Uncertainty Avoidance relates to employees who feel uncomfortable dealing with an unknown or uncertain future. The essence of this factor is how a society reacts to the fact that time only goes one direction and the future is unknown and whether it tries to control the future or leave it alone. Countries with a culture of high uncertainty avoidance tend to be more emotional. They try to minimize the occurrence of unknown or unusual circumstances. When changes occur they undergo them carefully, step by step with planning and implementing applicable laws and regulations. Whereas a country with a culture of low uncertainty avoidance accepts and feels comfortable in unstructured situations or environments that are often subject to change. They try to have some rules in their activities. Employees in this culture tend to be more pragmatic, they are far more tolerant of change.
14. Individualism
    Individualism is the company 's tendency to pay attention to themselves and those closest to them (spouse, children, parents). Culture of individualism shows more selfish tendencies and emphasizes individual goals. The culture of individualism emphasizes work success and achievement or wealth and career advancement. In contrast, collectivism is the company 's tendency to join groups and then look after one another with the consequences of exchanging loyalties between employees. Collective culture places greater emphasis on groups.
15. Masculinity – Feminimity
    Masculinity culture values the value of work performance and firmness. Feminity culture has a submissive value and supports social life, which is more respectful of others and sympathy for the needy. Masculinity culture is considered more in line with the male character who is assertive, more ambitious, and brave to compete. Feminity culture is very balanced between the sexes and accept parenting between women and men and is more focused on the quality of life.
16. Short Term – Long Term Orientation
    This is related to the choice of focus for human effort: the future, present, or past. Long-term orientation and short-term orientation describe the focus and culture of values that concern the mindset of the company. A long term orientation is more concerned with the future. They encourage pragmatic values oriented towards respect, status, saving attitude, including perseverance, savings, and adaptation capacity. A company with a dimension of short-term relationship orientation related to the past and present, including stability, respect for tradition, always maintaining the public appearance, and fulfilling social obligations.
17. Indulgence – Restraint
    Indulgence reflects a company which in its social order is very tolerant of expressing desires and feelings, especially those related to the use of free time, seeking entertainment with employees, purchasing goods, consumption, and things that smell sexual. While the restraint reflects on the company that holds back the pleasures mentioned earlier on indulgence. Employees who fall into this condition tend to be less able to enjoy life.

2.3 Contractor competitiveness
In this research, the competitiveness of contractors will be considered as the company's ability to meet products, services, values, experience demands and abilities [2]. Contractor competitiveness has six parameters [9]:

1. Company's Image
2. Technology and Inovation
3. Financial Ability
4. Project Management Ability
5. Marketing Ability
6. Human Resource and Company

3. Research methodology
This research flowchart can be seen in the following figure.

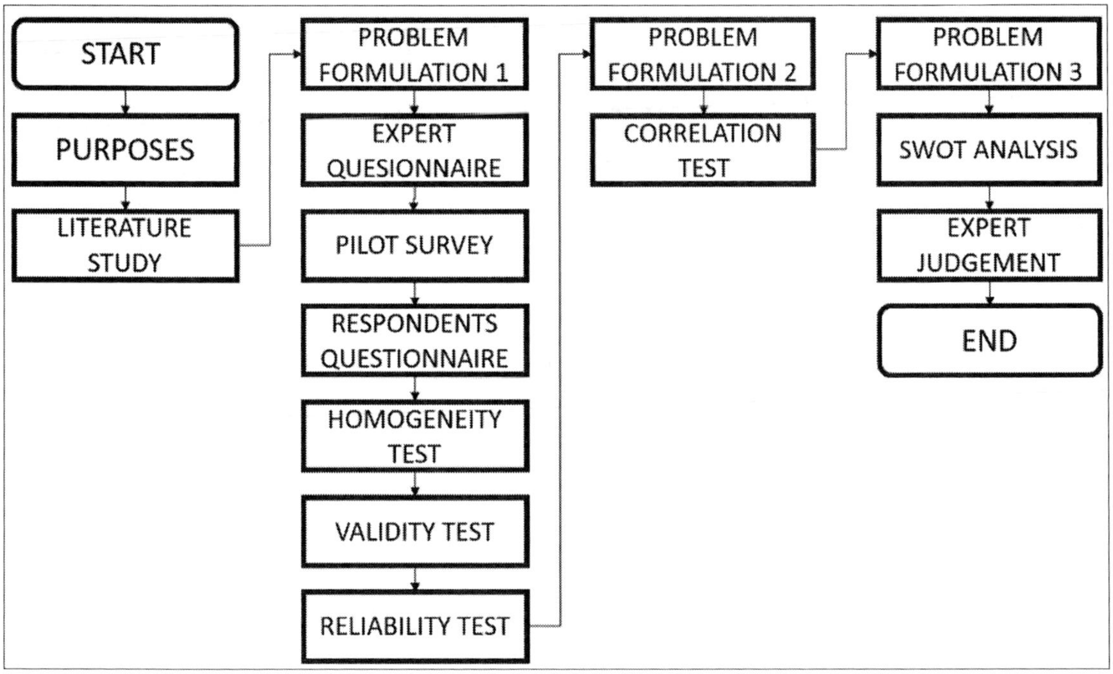

Figure 1. Research flowchart.

4. Analysis and results

4.1 Pilot Survey
In this research, a pilot survey was conducted on 10 respondents with each background. From the pilot survey obtained several suggestions in writing sentences on the research questionnaire.

4.2. Respondents questionnaire
In this research, there were 49 respondents. The questionnaire was distributed to various contractor companies in Indonesia, both with BUMN or private ownership. The respondents were also obtained from various ages, educational backgrounds, and experiences. It is hoped that this respondent will be sufficient to represent many contracting companies in Indonesia.

4.3 Homogeneity test
Homogeneity test is used to determine the variance of data populations whether between two or more groups of data have the same or different variants. This homogeneity test will use the Kolmogorov-Smirnov Z Test and the Kruskal-Wallis H Test. In the Kolmogorov-Smirnov Z Test, the data group is said to be homogeneous if the significance value (Asymp.Sig.)> 0.05, while in the Kruskal-Wallis H Test, the data group is said to be homogeneous if the significance value (Asymp.Sig.)> 0.05 and the chi- value square <chi square table value. This test is carried out with the help of SPSS software.

4.4 Validity test
Validity test is used to find out how accurate a data is in measuring a study. Data is said to be valid if there is a correlation to the total score. In this study the validity test uses the Pearson Correlation method, which uses a correlation of data scores with total data scores. Then, the significance test was carried out with the criteria using r table at the 0.05 significance level with a two-tailed test. If the value

is positive and r count> r table (with N = 49, then r table = 0.282), the data is declared valid, otherwise, the data is declared invalid.

Hasil Uji Validitas											
Var.	R Hitung	Hasil	Var.	R Hitung	Hasil	Var.	R Hitung	Hasil	Var.	R Hitung	Hasil
X1	.407**	Valid	X25	.682**	Valid	X46	.768**	Valid	Y9	.780**	Valid
X2	.728**	Valid	X26	.778**	Valid	X47	.708**	Valid	Y10	.751**	Valid
X3	.685**	Valid	X27	.614**	Valid	X48	.738**	Valid	Y12	.733**	Valid
X4	.654**	Valid	X28	.845**	Valid	X50	.724**	Valid	Y13	.642**	Valid
X5	.589**	Valid	X29	.799**	Valid	X51	.792**	Valid	Y15	.726**	Valid
X6	.607**	Valid	X30	.789**	Valid	X52	.727**	Valid	Y16	.659**	Valid
X8	.663**	Valid	X31	.703**	Valid	X54	.792**	Valid	Y17	.656**	Valid
X9	.668**	Valid	X32	.786**	Valid	X56	.783**	Valid	Y18	.868**	Valid
X11	.739**	Valid	X33	.734**	Valid	X57	.658**	Valid	Y19	.774**	Valid
X12	.731**	Valid	X34	.751**	Valid	X58	.660**	Valid	Y20	.838**	Valid
X14	.640**	Valid	X35	.664**	Valid	X59	.606**	Valid	Y22	.770**	Valid
X16	.706**	Valid	X36	.592**	Valid	X60	.497**	Valid	Y23	.770**	Valid
X17	.653**	Valid	X37	.657**	Valid	Y1	.659**	Valid	Y24	.637**	Valid
X18	.694**	Valid	X39	.829**	Valid	Y2	.632**	Valid	Y25	.714**	Valid
X19	.720**	Valid	X40	.717**	Valid	Y3	.604**	Valid	Y28	.599**	Valid
X20	.717**	Valid	X41	.766**	Valid	Y4	.759**	Valid	Y29	.720**	Valid
X21	.582**	Valid	X42	.732**	Valid	Y5	.631**	Valid	Y30	.699**	Valid
X22	.686**	Valid	X43	.634**	Valid	Y6	.820**	Valid			
X23	.712**	Valid	X44	.739**	Valid	Y7	.819**	Valid			
X24	.661**	Valid	X45	.639**	Valid	Y8	.762**	Valid			

Figure 2. Table results of validity test.

4.5 Reliability test

Reliability Test is used to find out the consistency or consistency of measuring instruments that usually use questionnaires. The method often used in research to measure the scale of stretches is Cronbach Alpha. Following are the results of the reliability test conducted in this study.

Table 1. Result of reliability test.

Reliability Statistics	
Cronbach's Alpha	N of Items
.986	77

Based on the table above, the Cronbach's Alpha value = 0.986 is obtained. So it can be concluded that the instrument used in this study has a very high reliability value

4.6 Correlation test

In this research, correlation testing was performed using Pearson Correlation. In the calculation of the Pearson Correlation coefficient will be obtained which shows the closeness of the relationship between variables. The correlation coefficient values range from 0 to 1 or 0 to -1. If it approaches 0, then the relationship between these variables gets weaker, and vice versa. Here is a table of Pearson correlation test results.

Table 2. Result of correlation test.

	Correlation													
		X1	X2	X3	X4	X5	X6	X8	X9	X11	X12	X14	X16	X17
Y	**Pearson Correlation**	.292*	.657**	.583**	.503**	.550**	.601**	.619**	.606**	.644**	.661**	.519**	.704**	.598**
		X18	**X19**	**X20**	**X21**	**X22**	**X23**	**X24**	**X25**	**X26**	**X27**	**X28**	**X29**	**X30**
Y	**Pearson Correlation**	.624**	.656**	.576**	.456**	.618**	.637**	.585**	.640**	.772**	.607**	.822**	.798**	.771**
		X31	**X32**	**X33**	**X34**	**X35**	**X36**	**X37**	**X39**	**X40**	**X41**	**X42**	**X43**	**X44**
Y	**Pearson Correlation**	.654**	.751**	.719**	.642**	.553**	.547**	.585**	.842**	.667**	.653**	.630**	.516**	.668**
		X45	**X46**	**X47**	**X48**	**X50**	**X51**	**X52**	**X54**	**X56**	**X57**	**X58**	**X59**	**X60**
Y	**Pearson Correlation**	.525**	.725**	.642**	.668**	.692**	.784**	.642**	.755**	.712**	.675**	.657**	.557**	.412**

4.7 Relationship model

The following is a model of the relationship between TQM and Quality Culture to increase the competitiveness of contractors in Indonesia.

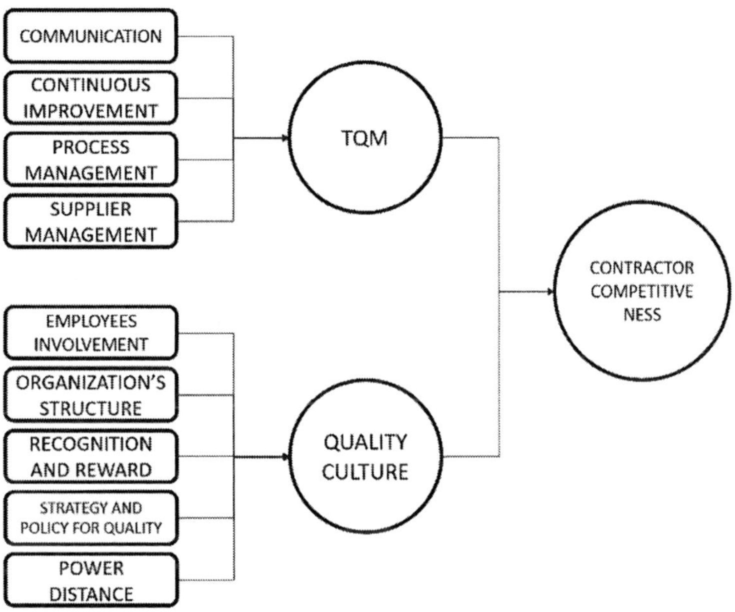

Figure 3. Relationship model.

4.8 SWOT analysis

SWOT analysis is the systematic identification of various factors to formulate a company's strategy. This analysis is based on logic that utilizes Strength, Weakness, Opportunity, and Threats. SWOT analysis compares between external factors, Opportunity and Threats, and internal factors, Strength and Weakness.

The research variables are then grouped so that they become input from the SWOT matrix in this study. The grouping process is done based on the correlation coefficient value that was done before. The SWOT variables that are input in the SWOT Matrix below are the dominant variables (correlation coefficient value> 0.7).

Table 3.. SWOT matrix.

	Strength	Weakness
SWOT analysis	1. top level management is committed to upholding quality. 2. the team works well together 3. the company makes continuous improvements 4. the company has a clear and efficient organizational structure	1. communication is not going well 2. employees are not given the opportunity to develop themselves 3. companies choose suppliers based on cheaper prices
Opportunity - the company has a policy of maintaining quality - the company maintains good relations with stakeholders	1. always focus on high quality both in the quality of work and the quality of employees 2. improve good relations with stake holders 3. manage lesson learn appropriately	1. not only focus on the customer, but also enhance employee self-development 2. pay attention to the quality of the suppliers so that they always produce the highest quality products
Threats - lack of training for employee development - customers do not meet expectations for quality, cost, and schedule	1. perform employee self-improvement so that they can always improve the quality of themselves and their work 2. with high human resources, customers' expectations of quality, cost, and schedule will be met	1. perform value engineering. it aims to reduce costs with better results. 2. certainly increase the ability of employees. improve employee quality, don't focus on quantity.

5. Conclusion and recommendations

5.1 Conclusion

The following conclusions obtained in this research, in accordance with the research objectives that have been submitted previously.

1. Dominant factors influencing TQM and Quality Culture to improve the competitiveness of contractors in Indonesia are determined by correlation coefficient values > 0.7. So it can be concluded that the dominant factors in TQM variables are Communication, Continuous Improvement, Process Management, and Supplier Management. While the dominant factors in the Quality Culture variable are Employee Engagement, Organizational Structure, Recognition and Rewards, Quality Strategy and Policy, and the Use of Power.
2. The relationship model between TQM and Quality Culture to increase the competitiveness of contractor companies is obtained by knowing the correlation coefficient value> 0.7. The Relationship Model can be seen in figure 3.

3. The strategy for implementing TQM and Quality Culture to improve the competitiveness of contractors in Indonesia is as follows.
   a. Always focus on high quality both in the quality of work and quality of employees
   b. Improve good relations with stake holders
   c. Manage lesson learn appropriately
   d. Not only focus on customers, but also enhance employee self-development
   e. Pay attention to the quality of the suppliers so that they always produce the highest quality products
   f. Perform employee self-improvement so that they can always improve the quality of themselves and their work
   g. With high human resources, customers' expectations of quality, costs and schedules will be met
   h. Perform value engineering. It aims to reduce costs with better results.
   i. Certainly, improve the ability of employees. Improve employee quality, don't focus on quantity.

5.2 Recommendation

The following are suggestions for this research so that they can be used as input in further developing this research.

1. The TQM variable is considered too often used. In subsequent studies, the authors suggest that one of the independent variables used is Six Sigma.
2. The limitations of the writer cause the respondents found not in accordance with expectations. The questionnaire should be made concise and attractive, and be prepared for more time to obtain the optimal number of respondents.
3. The authors' limitations also lead to the formulation of strategies using literature study and SWOT analysis. The author hopes for further research the formulation of strategies can be done by means of case studies.

References

[1] Al-Otaibi F M, Alharbi M F and Abdullah 2015 *Effect of TQM Practices Faktors on Competitiveness, Evidence from Saudi Arabia*
[2] Alotaibi F, Yusoff R Z and Islam R 2013 *Assessing The Impact of TQM Practices and Quality Culture with Competitiveness of Saudi Contractors*
[3] Campos A C, Mendes J d, Silva J A and Valle P O 2014 *Critical Success Faktor for Quality Culture*
[4] Harrington J and Voehl F 2012 *Applying TQM to The Construction Industry*
[5] Hofstede G 2011 *Dimensionalizing Cultures: The Hofstede Model in Context*
[6] Maddeppungeng A, Abdullah R and Kartika D D 2016 *Pengaruh TQM dan SCM terhadap Daya Saing pada Industri Konstruksi*
[7] Mazher U, Gharleghi B and Fah B C 2015 *A Study on The Faktors Affecting TQM in Saudi Arabian Construction Industry*
[8] Samawi G A, Abu-Tayeh B A, Yosef F, Mdanat M and Al-Qatawneh M I 2018 *Relation Between TQM Practices and Business Excellence*
[9] Tan Y, Shen L Y, and Langston C 2010 *A Fuzzy Approach for Assessing Contractors Competitiveness*
[10] Wanderi E N, Mberia H and Oduor J 2015 *Evaluation of Faktors Influencing TQM Implementation in Rwadan Construction*
[11] Wibowo K, Adhy D S and Utomo S B 2013 *Model Strategi Pengembangan Daya Saing Konsultan Jasa Konstruksi*
[12] Yeng S K, Jusoh M S and Ishak N A 2018 *The Impact of TQM on Competitive Advantage*
[13] Yunis M and Yung J 2012 *TQM, strategy, and performance a firm-level analysis.*

IOP Conf. Series: Materials Science and Engineering **930** (2020) 012013 doi:10.1088/1757-899X/930/1/012013

The role funding based non cash loan in the form of supply chain financing on elevated road project performance (case study join operation project PT.X)

S A Stani[1*]

[1] Civil Engineering Department, Engineering Faculty, Universitas Indonesia, Depok, 16424, Indonesia.

[*]Corresponding author's e-mail: sesde@adhi.co.id

Abstract. Indonesia infrastructure road development in recent years has experienced a very significant increase, but in its implementation the issue of land acquisition and limited funding are the main problem in realizing this development. Elevated Road Project is one of the way to minimize the effect land acquisition, and also Joint operation (JO) and applying Non Cash Loan (NCL) at the form Supply Chain Financing (SCF) in project can have good performance effect that can be see positif Project Cash Flow (PCF). Supply Chain Financing (SCF) is a financial instrument issued by financial institutions with various terms and conditions applicable to a supply chain that occurs between buyers and sellers, which aims to optimize the management of working capital and company liquidity. However, the selection of the SCF financing scheme also has a weakness where there is a cost of money that must be spent in the process. The purpose of this study was to determine the NCF SCF Cost Structure and evaluate the implementation of the use of SCF facilities on the elevated road project performance with a JO project case study at PT.X. Research methodology using survey and project cases study, while for the conducted to determine cost structure using statistical descriptive analysis. This study produce cost structure table which is majority for material, equipment, and sub contractor can be implemented using NCL SCF, also implementation procurement procedure using SCF until invoicing.

1. Introduction

Indonesia has a tremendous number of islands and population. The government has to build many infrastructures such as many harbour, not only sea harbours in every island but also airports to connect the regions and economics centre of the country. Indonesia has to increase and extend roadway, toll road, elevated-road which are connecting between cities and villages to the border area of Indonesia (source : kppip,2017). According to Directorate General of State Assets, Finance Ministry of the Republic Indonesia's data, there are problems to build the infrastructures. The main problem of infrastructure development is land problem or land acquisition at 44% which is affect next phase of development (table 1). The others problems are inadequate and long-grawn planning and preparation at 25 %, followed by funding limitations problem at 17 % and licensing problem at 12%. Therefore, According to data from Toll Road Regulatory Agency (BPJT) and Diractorate General of Highways in the Ministry of Public Works and Public Housing that one of infrastructure development which growing rapidly is elevated-road infrastructure development. Because, It can be done on a trace Right of Way (ROW) of an existing road. It will minimize land acquisition for the infrastructure development (BPJT, 2016).

Content from this work may be used under the terms of the Creative Commons Attribution 3.0 licence. Any further distribution of this work must maintain attribution to the author(s) and the title of the work, journal citation and DOI.
Published under licence by IOP Publishing Ltd

In funding limitation problem of infrastructure development, According to Committee for the Acceleration of Priority Infrastructure Provision (KPPIP), Coordinating Ministry for Economic Affairs' data that Government of Indonesia has 222 National Strategic Project (PSN) and 3 additional program whose investment needs are estimated to reach Rp. 4.092 trillion. It is budgeted from APBN at Rp 423 trillion, Rp 1.255 trillion from BUMN and BUMD,the rest is budgeted from private sector investment at Rp 2.414 trillion. The role of BUMN, BUMD and Private sector are very dominant for infrastructure development. Therefore, the government of Indonesia through KPPIP encourages private sector to support infrastructure development in Indonesia by providing some funding schemes such as funding schemes for Government Cooperation with Business Entities (KPBU), Public Private Partnership (PPP) and Non APBN Infrastructure Financing (PINA). KPBU as one of reliable scheme to persue infrastructure development in Indonesia is a provision and funding scheme infrastructure development based on a contract between Government represented by Minister/Head of Institution/Regional Government as Person in Charge of Cooperation Projects (PJPK) and private sector or business entity by taking into account the principle of risk sharing between the parties.

However, KPBU scheme will emerge a new problem for business entity, namely working capital will increased significantly. It couse the debt obligation increase and generate higher Debt to Equity Ratio (DER). According to Global Rating Agency's data, Moody's Investor Service (moody's) which was reported on CNBC Indonesia mentions growth in debt of business entities increased significantly and at high risk in last five years since infrastructure project become mainstay project for Government (table 1). Therefore, Working Capital Management (WCM) and Financial Flow become the critical working strategy for state owned construction company (BUMN Karya) to maintain growth and liquidity of company with apply Non Cash Loan system based on Supply Chain Financing. Moreover, Joint Operation (KSO) form believed to be able to help sharing working capital that occurs in the implementation of infrastructure project, especially elevated-road.

Table 1. Debt Equity Ratio (DER) BUMN Karya 2014 – 2019 [1].

No.	Emiten Code	Debt Amount 2014*	Debt Amount 2019*	Debt Deviation (%)	DER Ratio 2014	DER Ratio 2019
1.	ADHI	8,70	25,00	187,30	4,97	3,92
2.	JSMR	20,80	68,30	234,3	1,79	3,20
3.	PTPP	12,20	37,10	203,6	5,11	2,26
4.	WIKA	10,90	41,80	282,5	2,20	2,35
5.	WSKT	9,70	103,70	970,00	3,40	3,59

*Rp. In Triliun

PT. X, one of state owned construction company (BUMN Karya) who play an active role supporting government programs in advancing national infrastructure. PT. X implementing various national strategic project and other projects using diference funding scheme in accordance with the payment scheme offered by the owner. It has an impact on working capital needs and company liquidity. PT. X appliying a policy, assign payment methods to third parties (sub-contractor and vendor) by using a *Non Cash Loan* (NCL) atau *Non* Tunai methods based on *Supply Chain Financing* (SCF) as one of action to mantain working capital balance and extend the maturity period of the debt. Figure 1 shows comparison of the use of Plafon of Cash Loan and Non Cash Loan in PT. X, Non Cash Loan (NCL) payment become dominant or majority corporately. In this research, we investigate the implementation strategies *Cash Loan* (NCL) SCF which impact Project Cash Flow (PCF) in elevated-road infrastructure project. In order to find the implementation strategies, we first identify identifying the cost component of the construction project that can be implemented in the Non Cash Loan process based on Supply Chain Financing (SCF).

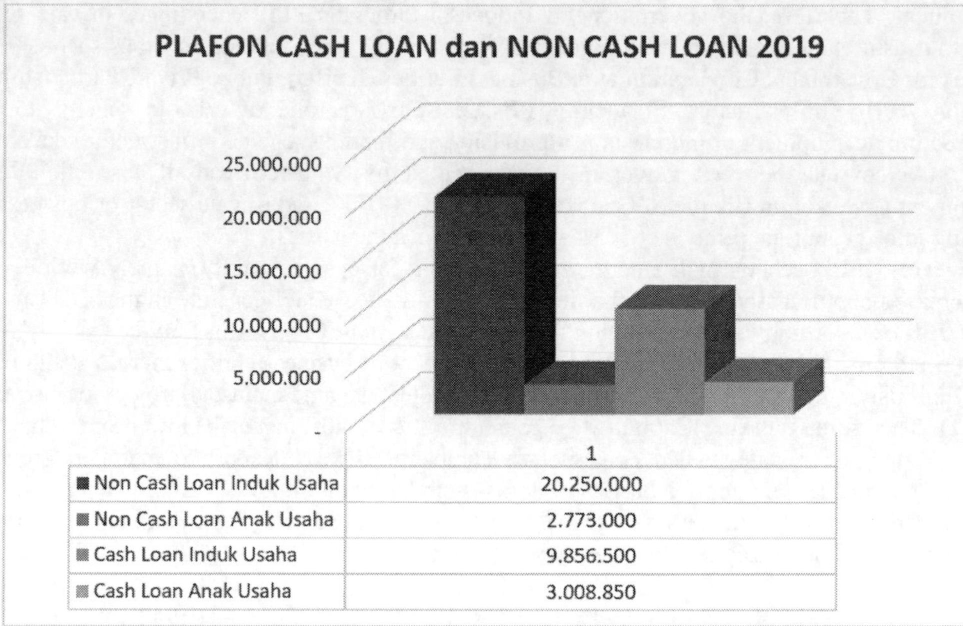

*Rp. In Miliar

Figure 1. Data plafon cash loan dan non cash loan pada PT.X [2]

2. Literature Review

2.1 Supply Chain Management (SCM)

Some literature say Supply Chain Financing as part of Supply Chain Management because it is one of supply chain part. According to Chopra dan Meindl (2007), Supply Chain Management is supply chain including whole parts such as produsen, distributor dan customer in directly and/or indirectly in order to fulfill customer demand. Supply chain encompasses not only producer and suppliers but also distributors, warehouses, retailers and the customers them self.

2.2 Working Capital Management (WCM)

From accounting viewpoint, Sanders (2006) says working capital as "discrepancy between short term asset and short term obligation. Whereas, Cronie (2008) says working capital as "cash amount which is required by company to fund discrepancy between payment and colection". Cronie was criticized by Van Thienen (2011), He argued that working capital is business liquidity which required for operational days.

2.3 Supply Chain Financing (SCF)

Supply Chain Financing definition from some literature are:

- An approach for two or more organizations in a supply chain, including external service providers, to jointly create value through means of planning, steering, and controlling the flow of financial resources on an inter-organizational level (Hofmann, 2005).
- A combination of trade financing provided by a financial institution, a third party vendor, or a corporation itself, and a technology platform that unites trading partner and financial institutions

electronically and provides the financing triggers based on the occurrence of one or several supply chain events (Aberdeen Group, 2006).

- Optimizing the financial structure and the cash-flow within the supply chain can be named SCF (Supply Chain Finance) (Moritz Leon Gomm, 2010).
- Supply Chain Finance (SCF) can be defined as the use of financial instruments, practices and technologies for optimizing the management of the working capital and liquidity tied up in supply chain processes for collaborating business partners – the buyer, the supplier and the financing institution (EBA, 2014).

However, we define Supply Chain Financing (SCF) as financial instruments issued by financial institurions with various terms and conditions apply in a supply chain between buyer and seller. It purposed to optimizing working capital management and company liquidity. From definition above, we can figure the parties involved SCF process such as financial institutions (Bank or other financial institutions), buyer, seller and suplier in a pyramid diagram below.

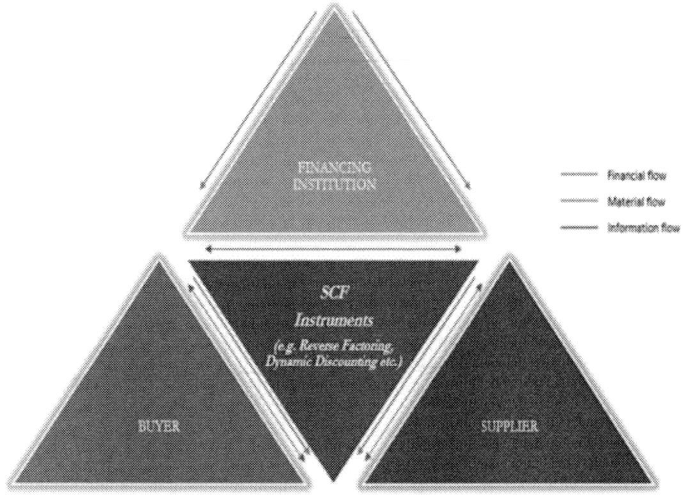

Figure 2. SCF Pyramid [3].

The explanations of this SCF pyramid as a simple picture of SCF proces are showed below:

<u>Financial Institutions</u>

This financial institutions is Banking parties. The Banking parties provide plafon of credit facilities to buyer with terms and conditions apply. It provided with non cash form for buyer because the banking parties will transfer it to supplier as a payment of working invoice of supplier to buyer directly. Non-cash credit will be paid by the buyer to bank in a certain time periode in accordance the maturity agreement.

<u>Buyer</u>

Buyer is a contractor who gain SCF facilities or credit recepient from financial institution parties or Banking parties. Whereas, contractor as credit recipient is not gain the credit on cash directly because banking will distribute the approved credit to supplier as payment form from contractor to supplier. The Bank will provide cost of money accordance to the maturity period agreement between the contractor and the Bank because of its credit transaction. The contractor will pay the credit on the maturity period to The Bank.

<u>Supplier</u>

Supplier is sub-contractor or vendor or foreman, the parties who make agreement with a contractor to execute a certain work. Sub-contractor will implement the obligation accordance to the contract between sub-contractor and contractor. After completing the works, the sub-contractor can collect the payment of the works from contractor. The billing will be verified and approved by contractor and distribute to the Bank. If the billing approved by the contractor, the Bank will receive the billing from the sub-contractor to make payment from the Bank to sub-contractor account.

SCF Instrument

SCF instrument is factors which part of terms or conditions in SCF pyramid cycle as follows:

- Working capital company. It will determine how big the credit plafon for SCF process.
- Discount rate per year. It is part of cost of money which happen because of SCF process. Discount rate value is various depend on agreement between the Bank and contractor. Discount rate range is 7% - 10 %.
- Outstanding duration. Generaly outstanding duration offer is 30 days/60 days/90 days/180 days.
- Administrative document. It is requirement that must be met by all third parties (The Bank, Contractor and Sub-contractor). These administrative requirements are various depend on company policy. Basically, the requirement that should be completed are eligible, contract between Contractor and Sub-contractor, completion work, employment report and SCF approval letter.

Key of perform indicator which include for SCF financial flow are (Hausman, 2003) :

✓ *Days of working capital* (DWC)
A time to collect company working capital. SCF lengthen company DWC indirectly. So that, it can be used to gain other investment oportunities.

✓ *Days of sales outstanding* (DSO)
A duration between sales of goods and invoice to customers and actual payment date from customers.

✓ *Days of inventory* (DIO)
A duration between delivery of goods with invoice from Sub-contractor and goods sales with invoice to customer. It shows average number of items amount per day in supply before sale by company. It is the focus of whole activity in classic supply chain management.

✓ *Days payables outstanding* (DPO)
A duration between goods sales, invoice to customer and expectation payment date. Indicator of Days payables outstanding is similar with Days of sales outstanding. It shows the average time of credit outstanding in units of days. The best DSO is duration of long due because the company will collect all credit before due date.

3. Research Metodology

This research is a qualitative study using literature studies, expert judgment, surveys, descriptive analysis and case studies used to answer each research question. The case study in this research was carried out on the 9 km KSO elevated road project which is located across Jakarta. There are 6 variable to be studied cash in, cash out, progress, procurement procedure, invoicing procedure and cost of money invoicing procedure.

Table 2. Variable, sub variable and indicator.

Variable	Sub Variable	Indicator
1. Cash In	Cash In	1. Advanced Payment 2. Term Payment 3. Paid up capital 4. Asset Sales
2. Cash Out	Direct costs for wages	1. Foreman's wage costs 2. Daily Labor Wage Costs 3. Wage Mobilization Costs
	Direct Cost of Material	1. Ready mix material 2. Steel Rebar material

Variable	Sub Variable	Indicator
		3. Etc
	Direct Cost of Equipment	1. Service Crane 2.Manlift 3.Etc
	Direct Cost of Subcontractor	1. Bore Pile Subcontractor 2. Erection Subcontractor 3. Stressing Subcontractor 4. Etc
	Indirect Cost	1. Overhead 2. Insurance 3. Cost of Money SCF 4.Etc
	Project Profit	1. Profit
3. Progress	Owner	1. Term of payment 2. Collection Period Owner 3. Outstanding Payment Owner
	Third Party	1. Collection Period Third Party 2. Outstanding Payment Third Party
4. Procurement Procedure		1. Review Contract Document 2. Technical Specification Document 3. Etc
5. Invoicing Procedure		1. Progress Submission 2. Approval Progress Submission 3. Etc
6. Cost of money invoice procedure		1. Approval payment

4. Finding and Discussions

4.1. Cost Structure

Survey was conducted by expert and also the respondent was choose the person who involve and have experience related with Non Cash Loan based on Supply Chain Financing (SCF). The results of survey for cost structure component of the construction project that can be implemented in the Non Cash Loan process based on Supply Chain Financing (SCF) are:

Table 3. Summary cost structure.

Variable	Sub Variable	Results
1. Cash In	Cash In	Can't be implemented in SCF
2. Cash Out	Direct costs for wages	Cost structure in SCF except daily worker.
	Direct Cost of Material	Cost structure in SCF
	Direct Cost of Equipment	Cost structure in SCF
	Direct Cost of Subcontractor	Cost structure in SCF
	Indirect Cost	Can't be implemented in SCF
	Project Profit	Can't be implemented in SCF
3. Progress	Owner	Cost structure in SCF
	Third Party	Cost structure in SCF

IOP Conf. Series: Materials Science and Engineering **930** (2020) 012013 doi:10.1088/1757-899X/930/1/012013

4.2. Implementation strategies SCF

1. Procurement SCF Procedure

   Based on the survey result variable X4. Procurement SCF Procedure are:

Table 4. Procurement SCF Procedure.

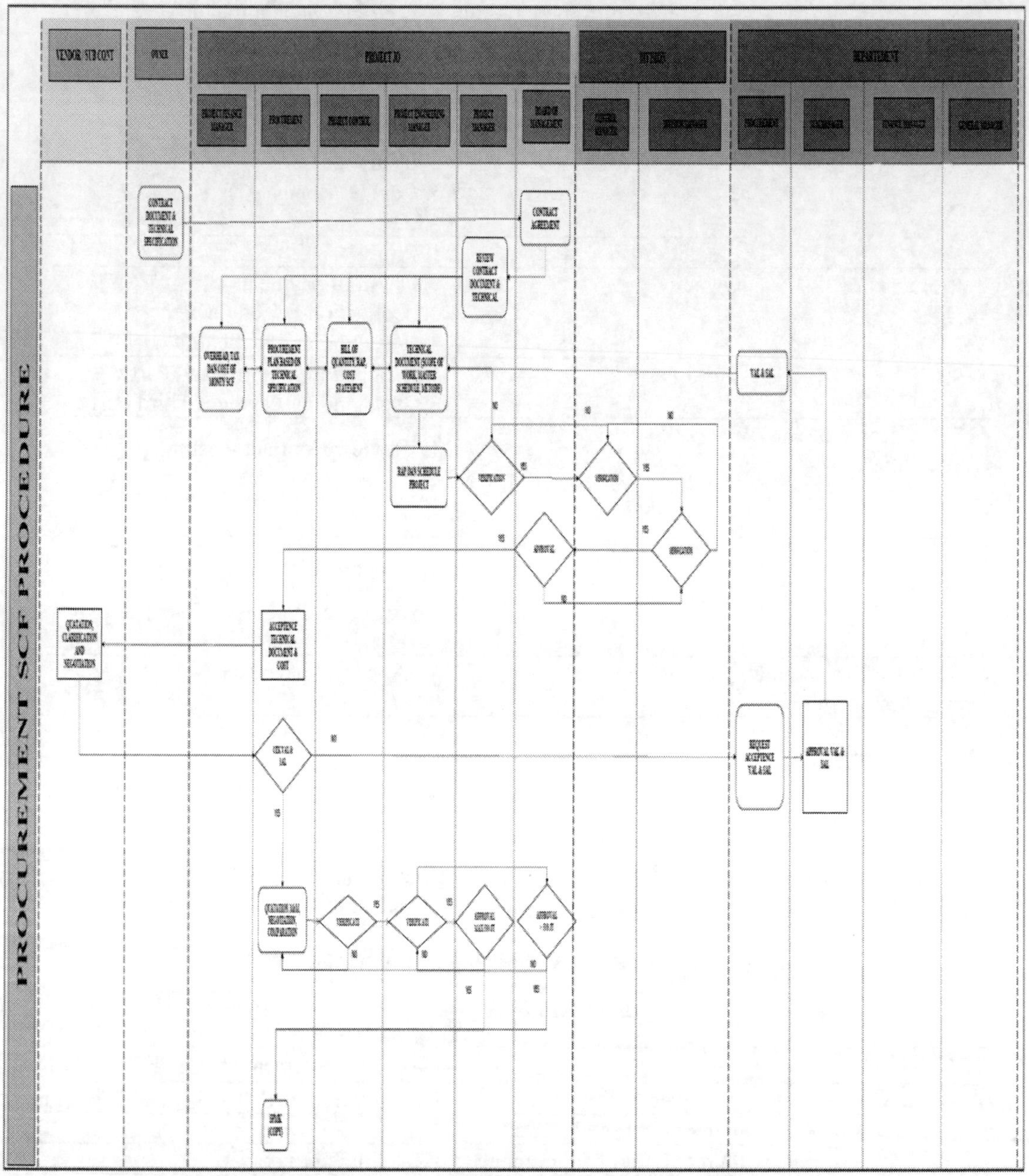

2. Invoicing SCF Procedure

   Based on the survey result variable X5. Invoicing SCF Procedure are:

Table 5. Invoicing SCF Procedure.

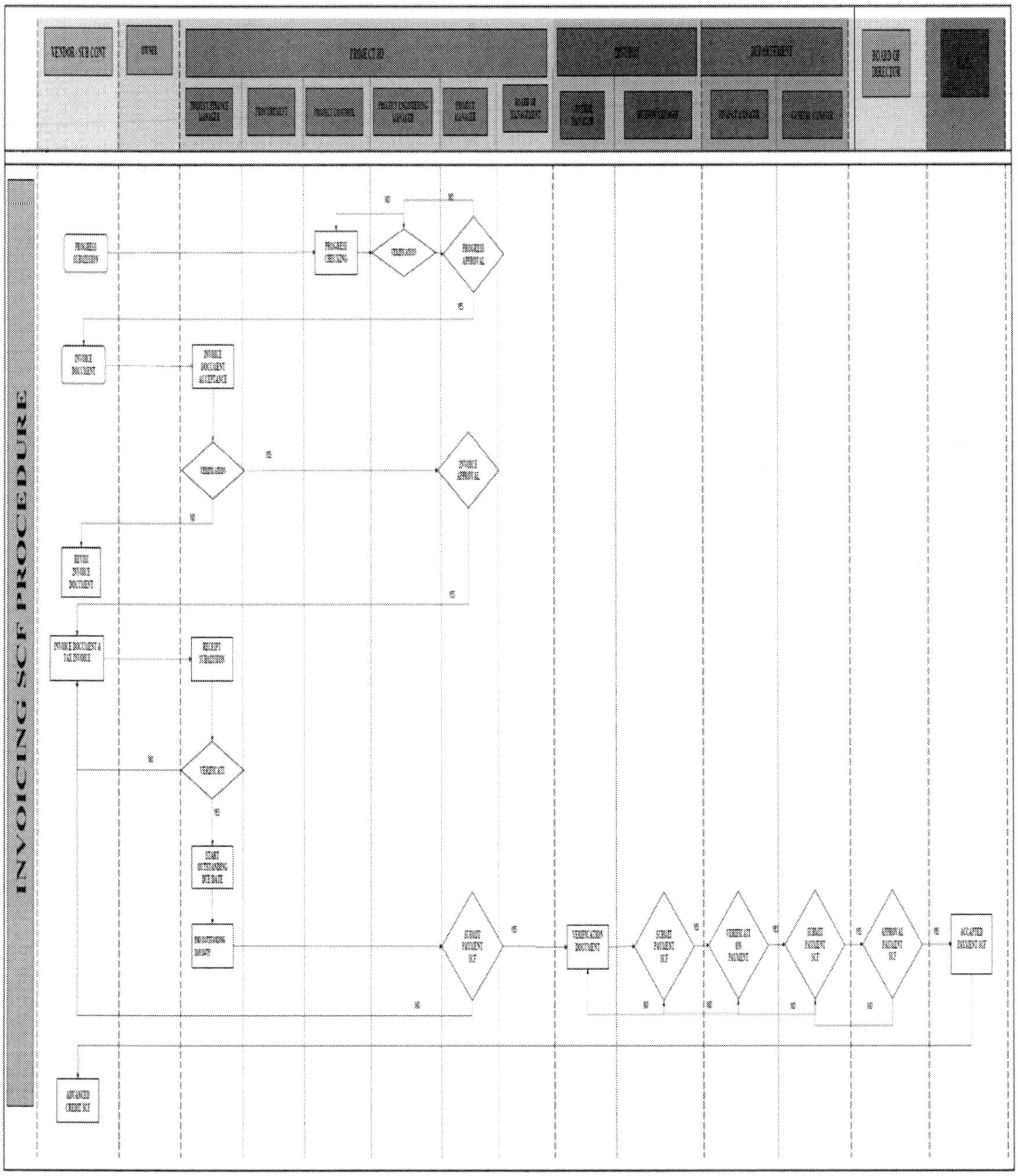

3. Invoicing Interest SCF Procedure
   Based on the survey result variable X6. Invoicing Interest SCF Procedure are:

Table 6. Invoicing Interest SCF Procedure.

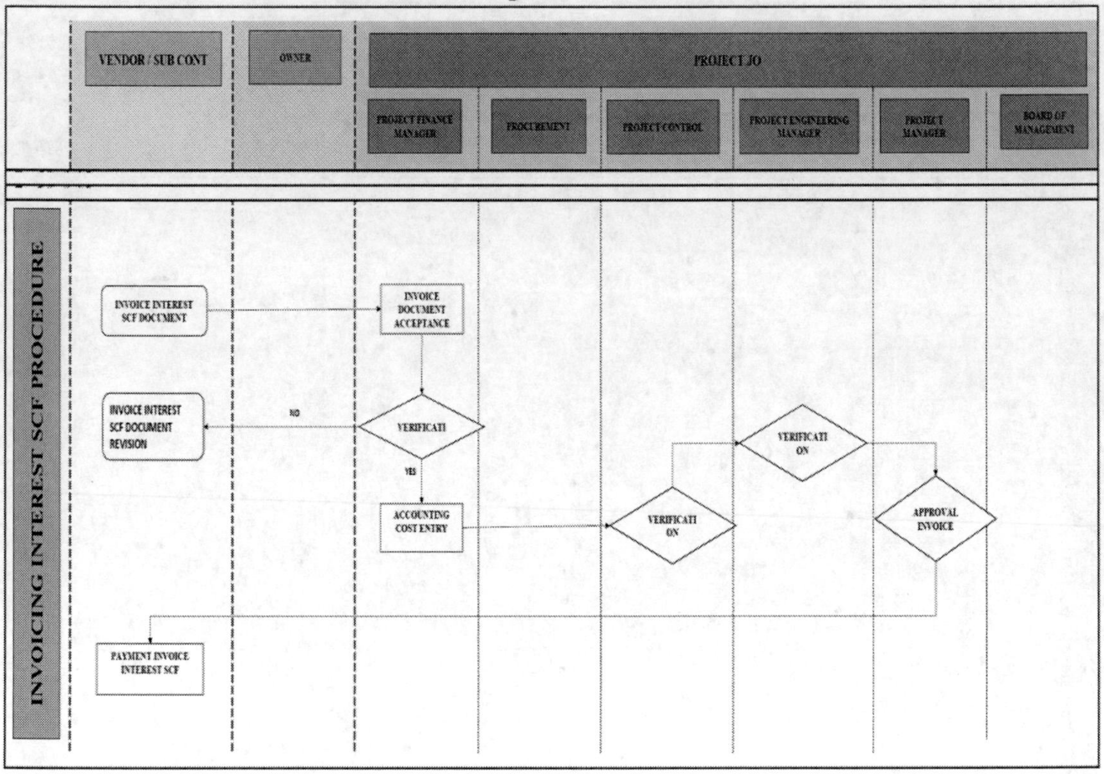

4.3. Interview Result

There are 4 question in interview form for respondent to be answer and summary the result are:

Table 7. Interview Result.

No.	Question	Days	Results
Q1	How many days ideal for Days Sales Outstanding (DSO) from owner to contractor	14 days	16,00%
		30 days	52,00%
		45 days	12,00%
		60 days	4,00%
		90 days	12,00%
		180 days	4,00%
Q2	How many days ideal for Days Payable Outstanding (DPO) from contractor to third party.	14 days	4,00%
		30 days	20,00%
		45 days	32,00%
		60 days	24,00%
		75 days	4,00%
		90 days	8,00%
		180 days	8,00%
Q3	Can the Supply Chain Financing (NCF) Non Cash Loan (NCL) financing method increase the working capital of the construction project	Yes it can	92,59%
Q4	Can the Supply Chain Financing (SCF) Non Cash Loan (NCL) financing method improve the performance of the project's financial cash flow or Project Cash Flow (PCF) into Positive Cash Flow Nett	Yes it can	96,30%

From table Q1 for Days Sales Outstanding (DSO) majority respondent choose 30 days, Q2 for Days Payable Outsanding (DPO) majority respondent choose 45 days. For Q3 majority respondent agree that NCL can increase working capital of construction project. And for Q4 majority respondent also agree that NCL SCF can improve the performance of the financial Project Cash Flow (PCF) into positif PCF.

4.4. Case Study Project Cash Flow (PCF)

Case Study for the project Joint Operation was elevated toll road along 9,3 KM in Jakarta. Total amount for this project around IDR 2,8 Trillion, one of the project charter that must be done in this project was investment construction equipment for a long term purpose. This table was cost breakdown and estimating for this project, simulation overdraft cashflow with or without SCF.

Table 8. Cost Breakdwon and Estimating.

No.	DESCRIPTION	TOTAL COST		PERCEN TAGE
		COST WITHOUT SCF	COST WITH SCF	
1	**CASH IN**			
a	**Cash In from Progress**			
	- Advance Payment	236.138.084.534,46	236.138.084.534,46	10,00%
	- Monthly Payment	2.609.120.160.000,00	2.609.120.160.000,00	100,00%
	- Advance Money Return	(236.138.084.534,46)	(236.138.084.534,46)	15,00%
	- Cut Retention	(45.659.602.800,00)	(45.659.602.800,00)	5,00%
	- Payment Retention	45.659.602.800,00	45.659.602.800,00	5,00%
	- Tax PPN	260.912.016.000,00	260.912.016.000,00	10,00%
	- Tax PPh 3%	(78.273.604.800,00)	(78.273.604.800,00)	3,00%
b	**Cash In from SCF**			
	PT. Y		943.790.546.201,16	
	PT. X		542.137.564.836,74	
	Asset Sales	304.695.748.018,00	304.695.748.018,00	
	COMULATIVE TOTAL CASH IN	**2.791.758.571.200,00**	**4.277.686.682.237,90**	
2	**CASH OUT**			
a	**Cash Out**			
	- Wages	102.751.496.873,51	102.751.496.873,51	3,94%
	- Material	831.190.023.205,70	831.190.023.205,70	31,86%
	- Equipment & Heavy Duty	259.282.517.270,28	259.282.517.270,28	9,94%
	- Subcontractor	859.432.565.387,03	859.432.565.387,03	32,94%
	- Operational Cost	75.639.970.629,00	75.639.970.629,00	2,90%
	- Over Head	129.454.860.790,33	129.454.860.790,33	4,96%
	- Insurance and Guarantee	11.164.360.580,00	11.164.360.580,00	0,43%
	- Design Cost	50.045.619.392,48	50.045.619.392,48	1,92%
	- Equipment Investment (Owning Cost)	133.345.060.324,31	133.345.060.324,31	5,11%
	- Cost of Money	-	66.625.391.207,89	2,55%
	- PPN	260.912.016.000,00	260.912.016.000,00	10,00%
	- Equipment Investment (Residual Value)	304.695.748.018,00	304.695.748.018,00	11,68%
b	**SCF Return**			
	PT. Y	-	943.790.546.201,16	
	PT. X	-	542.137.564.836,74	

IOP Conf. Series: Materials Science and Engineering **930** (2020) 012013 doi:10.1088/1757-899X/930/1/012013

No.	DESCRIPTION	TOTAL COST		PERCENTAGE
		COST WITHOUT SCF	COST WITH SCF	
	TOTAL COMULATIVE CASH OUT	2.713.218.490.452,65	4.265.771.992.698,44	
3	TOTAL CASH IN	2.791.758.571.200,00	4.277.686.682.237,90	
4	TOTAL CASH OUT	2.713.218.490.452,65	4.265.771.992.698,44	
5	TOTAL COMULATIVE CASH FLOW	78.540.080.747,35	11.914.689.539,46	
6	CASH FLOW CONDITION	NEGATIVE CASH FLOW CURRENT MONTH	POSITIF CASH FLOW CURRENT MONTH	

The result show without NCL SCF profit clould be increase significantly better than with NCL SCF because cost of money for SCF. The weakness without NCL SCF Project Cash Flow (PCF) had negative during project, PCF had better performance if with NCL SCF.

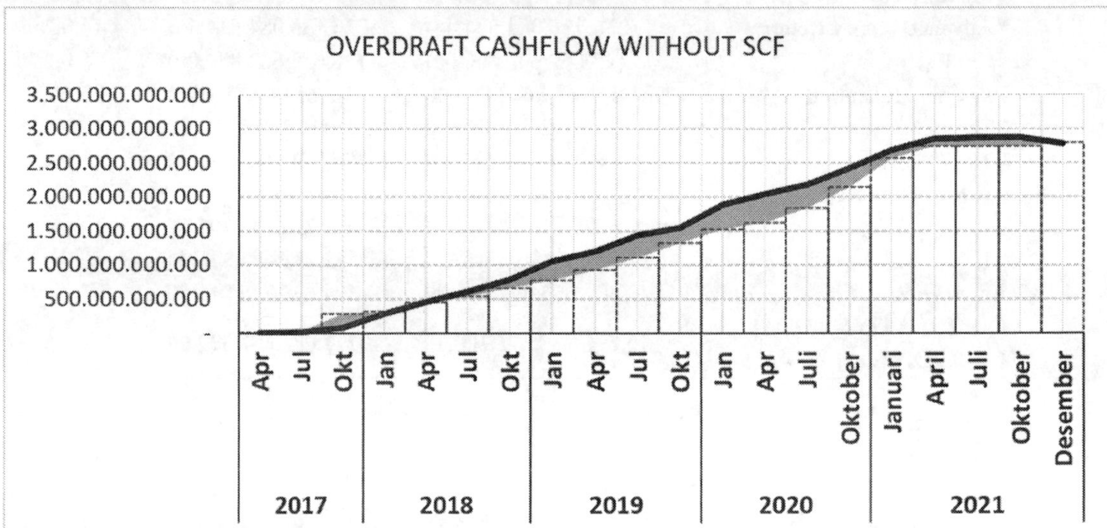

Figure 3. Shows overdraft cashflow without SCF.

The figure 3 bove explain negative cash flow in the current month. Red curve show cash out position during project and blue chart shows income or cash in position during project. Overdraft cashflow show without SCF red curve shows greater than blue chart. It mean cash out bigger than cash in during project. The figure 4 explain overdraft cashflow with SCF have positif cash flow during project. Blue chart or cash in are bigger than red curve it mean cash in bigger than cash out.

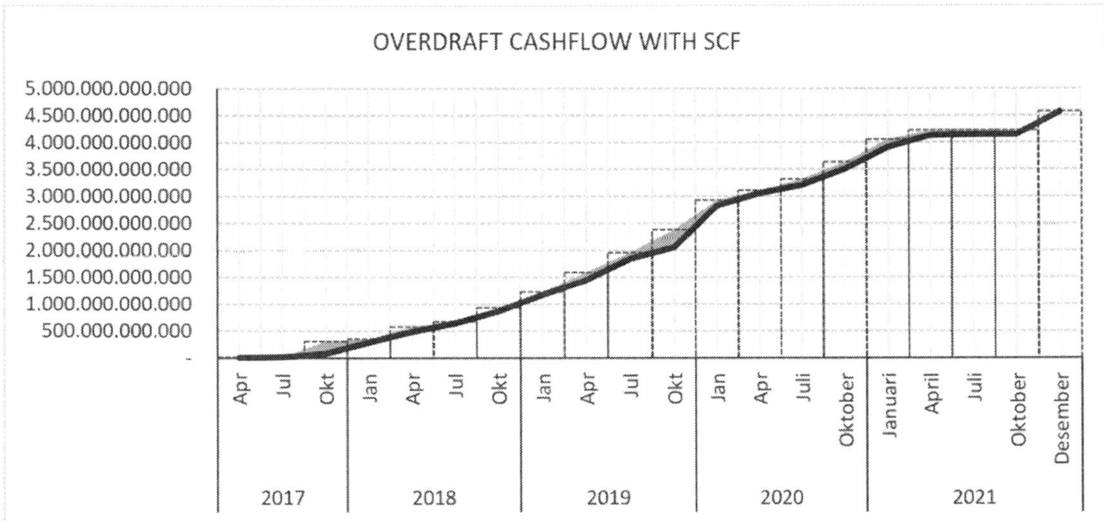

Figure 4. Shows overdraft cashflow without SCF.

5. Conclusion

This paper discuss an effort to identify the cost structure NCL SCF and the strategy implementation NCL SCF that would be influenced Project Cash Flow on joint operation elevated toll road project. There are some point that could be conclude form this paper:

1. Mostly all variable cost in elevated toll road project could become cost factor for NCL SCF except Cash in, Direct cost and Profit.
2. Strategy implementation for NCL SCF could be develop in procurement procedure, invoice procedure and invoice interest procedure.
3. Ideal Days Sales Outstanding (DSO) are 30 days and ideal for Days Payable Outsanding (DPO) are 45 days. NCL can increase working capital of construction project and also NCL SCF can improve the performance of the financial Project Cash Flow (PCF) into positif cash flow.
4. NCL SCF could improve Project Cash Flow or Overdraft Cash Flow into positive.

References

[1] Chart RTI Analisys – CNBC Indonesia 2019

[2] Financial and Cash flow report PT.X, 2019

[3] Georgios L. Vousinas, *Supply Chain Finance: Definition, Modern Aspects and Research Challenges Ahead.* Researchgate 2019

[4] Ayers and James B 2009 *Supply Chain Project Management A Strucutred Collaborative and Measureable Approach* Second Edition (New York : CRC Press Taylor) pp 11 – 21

[5] Hartmann A and Caerteling J 2010 Subcontractor procurement in construction the interplay of price and trust *Supply Chain Manag. Int. J.* **15** 5 354-62

[6] Anonim 2007 *Peraturan Menteri Nomor 43 Tahun 2007 tentang Pedoman Kualifikasi Pelelangan Nasional Pekerjaan Jasa Pelaksanaan Infrastruktur (Pemborongan)* (Jakarta)

[7] Baar J E and Jacobson S M 2004 The keys to forecasting #2 Work Breakdown Sructure *Cost Eng.* 12-4

[8] Badan Pusat Statistik 2018a *Statistik Indonesia, Statistical Year Book of Indonesia 2018* (Jakarta: Badan Pusat Statistik)

[9] Sadeghpour F and Isaac S 2015 A comparative study of owner, contractor, and consultant perspective on the selection criteria for subcontractors *Org. Tech. Manag. Constr. Int. J.* **7** 2

[10] Project Management Institute 2013 *A Guide to Project Management Body of Knowledge* fifth Edition, pp 47-61

IOP Conf. Series: Materials Science and Engineering **930** (2020) 012014 doi:10.1088/1757-899X/930/1/012014

A Literature Review of Methods in Research on Green Building Cost Analysis

R B Yasinta[1*], C Utomo[1] and Y Rahmawati[2]

[1] Department of Civil Engineering, Institut Teknologi Sepuluh Nopember, Indonesia.
[2] Department of Civil and Environmental Engineering, Universiti Teknologi PETRONAS, Malaysia.

[*]Corresponding author's e-mail: christiono@ce.its.ac.id

Abstract. Development greatly influences environmental sustainability and quality. The concept of green building is one solution to minimize environmental damage due to construction. The concept of green building requires greater investment than conventional building. Life Cycle Cost (LCC) impacts the cost of planning, building's operational and maintenance cost and residual cost. Recently, research of green building on cost analysis is developing rapidly. There are so many studies of green building on cost analysis with different methods. This research addressed several relevant methods and is expected to set clear which methods to use for the future research under green building topics. This journal is meant to find out research method used in determining green building on cost analysis. The method to be carried out is a literature review of 14 relevant reputable journals from 2013-2020. The most widely used method is the Life Cycle Cost Analysis (LCCA).

1 Introduction

The construction industry in developing countries account for around 30% of total greenhouse gas emissions [1]. CO_2 gas emissions generated by the building sector accounted for 34.8% of the total energy use in the world [2]. The concept of green building is one solution to minimize environmental damage due to construction [1]. Not only maximizes savings in energy, green building also makes an effective use of space. It harmonizes with nature the way it preserves the environment, result in less pollution and a maintained health. The characteristics of green building are environmental improvement and human health, efficiency of natural resources and building materials, and water and energy efficiency [3]. Green building has attracted widespread attention since the early 1990s. Until now, research on green building has been carried out in many different disciplines [4]. Until now there is still debate about the differences used for green building. However, there are some studies mentioning that the application of the concept of green building requires greater investment than conventional building. The high initial cost of construction is a major obstacle in applying the concept of green building.

There are several methods to measure and analyze the effect of applying green building on cost analysis, namely LCC analysis, payback period, Net Present Value (NPV, sensitivity analysis, regression analysis and fuzzy analysis. Some studies applied LCC calculation methods such as in studies [19], [22], [23], [27], [28], [29] and [32] while studies that applied regression analysis are those such as research [20] and [21]. In addition, studies [19], [22] and [25] applied the sensitivity analysis method. This journal is meant to find out research method used in determining green building on cost analysis.

Content from this work may be used under the terms of the Creative Commons Attribution 3.0 licence. Any further distribution of this work must maintain attribution to the author(s) and the title of the work, journal citation and DOI.
Published under licence by IOP Publishing Ltd

2 Conceptual Background

2.1 Cost Analysis of Green Building
Cost of planning, operational and maintenance cost and residual cost are part of the Life Cycle Cost (LCC). LCC is the total cost of a building over its lifetime and includes the cost of planning, design, operation, maintenance, minus the residual value [5]. At present, most building are only concerned with initial capital cost such as land, projects and construction cost without regard to the operational and maintenance cost of the building. This does not happen if you apply the concept of green building. Emphasis on initial cost by using green building materials has a small operational cost. While conventional building that do not apply the concept of green building result in higher energy use cost and also lower sustainable level (such as higher carbon emissions) [6]. Cost analysis have been carried out in previous studies, such as the Weerasinghe and Ramachandra study [22] who analyzed the cost differences in applying green building and conventional building using survey methods, LCC analysis and sensitivity analysis.

2.2 Methods of Previous Research

2.2.1 *Types and Data of Research.* In general, the type of research is divided into 2 namely quantitative and qualitative research. Quantitative method is a method that explains the data obtained into statistics, while the qualitative method is a method that explains the data obtained to be descriptive to produce meaning and understanding of a thing [7]. Primary data is original data collected by researchers directly. While secondary data is data that has been collected and analyzed by others [8].

2.2.2 *Research Techniques.* The following are the methods used by previous researchers regarding the effect of applying green building on cost analysis.

a. Life Cycle Cost Analysis (LCCA) is a method that analyze the total cost in the form of construction, operational cost and the residual value of building [9]. LCC considers all cost including initial cost, such as capital investment cost and construction cost; future cost, such as energy cost, operational and maintenance cost; and depreciation cost. The advantage of LCCA is calculate all cost over the age of the building thereby reducing the risk the failure of investment. Meanwhile, the disadvantage of this method is can't calculate budget allocation [10].
b. Sensitivity analysis is a method that measures the uncertainty impacts one or more variables. This method is useful to improve the model prediction that will be made. The results of various parameters can be used to evaluate resilience, alternative different outcomes and analyze values with results that can change significantly. Disadvantage of sensitivity analysis is this method based on assumptions [11].
c. Payback period is the time needed to recover investment. This method is to calculate the breakeven point of investment. Payback period only counts how quickly a company will recover cash investment [12]. Disadvantage of payback period is ignored time value of money and too much emphasis on short-term benefits [13].
d. Regression analysis is a quantitative method that measures the effect of dependent variables with an independent variable [14]. In the regression graph there is a sufficiently low-dimensional summary plot. These plots, which do not require a model for their construction, all contain information about the responses available from the predictors. This method is used to visualize dependencies, find unexpected relationships and examine the right model. Disadvantage of regression analysis is easily affected by outlier [15].
e. Fuzzy analysis is a method used to analyzc the uncertainty, such as decisions making, planning and production problems [16]. Disadvantage of fuzzy analysis is required a lot of data and expertise to develop this method [17].

f. Net Present Value (NPV) is present value of cash flow at the project rate of return compared to the initial investment. This method is used to calculate the Return on Investment (ROI) for projects or expenses. NPV analyzes cash flow and time value of money [18]. The disadvantage of this method is the inaccurate calculation between projects with different economies [13].

3 Results and Discussion

3.1 Methodology

This research is to analyze the method used in influencing the application of green building in stages of cost analysis. Various sources from reputable journals in this study to be analyzed. There are 14 publication journals published from 2013 - 2020. The following is a list of journals with rank and SJR of each journal listed in table 1.

Table 1. Selected journal's repute

Journal Publication	SJR	Quartiles Category	Number of Papers
Renewable Sustainable Energy Reviews	3,632	Q1 in Renewable Energy, Sustainability and the Environment	3
Journal of Environmental Economics and Management	2,393	Q1 in Economics and Econometrics	3
Energy Policy	1,988	Q1 in Energy (miscellaneous)	1
Journal of Cleaner Production	1,886	Q1 in Renewable Energy, Sustainability and Environment	1
Building and Environment	1,871	Q1 in Building and Environment	2
Cities	1,606	Q1 in Urban Studies	1
Regional Science and Urban Economics	1,419	Q1 in Urban Studies	1
Sustainable Cities and Society	1,047	Q1 in Civil and Structural Engineering	1
Built Environment Project and Asset Management	0,515	Q2 in Civil and Structural Engineering	1

Based on table 1, information about 14 reputable journals was obtained to measure the effect of the application of green building on cost analysis.

3.2 Result

Based on the literature review, a description of a number of methods used in analyzing the effect of applying green building on cost analysis was obtained. The following is an explanation of the types and research data used in 14 journal publications. There is an explanation Y (Yes) is used to state the use of research types and data that are appropriate and N (No) is used to express the use of research types and data that are not appropriate. The following is an explanation of the theoretical mapping from previous research presented in table 2.

Table 2. Theoretical mapping of previous research.

Cost Analysis of Green Building	Type and Data of Research				Methods	Objectives
	Quantitative	Qualitative	Primary Data	Secondary Data		
Li et al [19]	Y	N	N	Y	Literature Review, LCC Analysis and Sensitivity Analysis	Calculate and analyze LCC and sensitivity analysis of green building
Chegut et al [20]	Y	N	Y	N	Survey and Regression Analysis	Analyze the difference in marginal cost used between green building and conventional building
Devine and McCollum [21]	Y	N	N	Y	Literature Review and Regression Analysis	Analyze the relation between development of finance and green building
Weerasinghe and Ramachandra [22]	Y	N	Y	N	Survey, LCC Analysis and Sensitivity Analysis	Compare LCC of green building and conventional building
Vyas and Jha [23]	Y	N	Y	N	Survey, LCC Analysis, Payback Period and NPV	Develop the concept of green building in accordance with the rules of the Government
Khoshbakht [24]	Y	N	Y	N	Interview, Survey and Fuzzy Analysis	Developing research systematically with the criteria of green building materials that consider the 3 pillars of sustainability
Deng and Wu [25]	Y	N	N	Y	Literature Review and Sensitivity Analysis	Evaluate energy efficiency in green building from a developer's perspective
Fan et al [26]	N	Y	Y	N	Survey and Descriptive Analysis	Improve the efficiency of green building with transaction cost analysis
Illankoon and Lu [27]	Y	N	N	Y	Literature Review	Analyze literature review of journals from 2005 to 2018

IOP Conf. Series: Materials Science and Engineering **930** (2020) 012014 doi:10.1088/1757-899X/930/1/012014

Cost Analysis of Green Building	Type and Data of Research				Methods	Objectives
	Quantitative	Qualitative	Primary Data	Secondary Data		
					and LCC Analysis	on critical success factors for green building
Dwaikata and Ali [28]	Y	N	Y	N	Survey and LCC Analysis	Analyze and develop LCC of green building that reach 60 years the age of building
Tam et al [29]	Y	N	Y	N	Literature Review, Interview and LCC Analysis	Analyze LCC of green building wood materials to fulfill the requirements of the Green Start rating tools
Kim et al [30]	Y	N	N	Y	Literature Review and LCC Analysis	Analyze the LCC net-zero energy building design with energy efficient heating, ventilating and air conditioning (HVAC) system and solar power generation system
Zuo et al [31]	Y	N	N	Y	Literature Review and LCC Analysis	Evaluate concept of green building using LCC analysis
Annibaldi et al [32]	Y	N	Y	N	Survey, Interview, and LCC Analysis	Analyze the optimal condition between building energy performance and investment cost using LCC analysis

The use of questionnaire distribution and expert opinion as data collection tools has been carried out by [20], [22], [23], [24], [26], [28], [29] and [32]. The dominant analysis used is LCC analysis as in the studies of [19], [22], [23], [27], [28], [29], [30], [31] and [32]. Meanwhile sensitivity analysis is used by researchers [19], [22] and [25]. Regression analysis is used by researches [20] and [21]. The only one Fuzzy analysis is used by researcher [24].

3.3 Discussion

Judging from previous research, a mapping diagram is formed. Mapping diagram is a group of previous research based on data sources and types of research. Data sources are on the X axis in the form of primary and secondary data, while the Y-axis in the form of research types is qualitative and quantitative. The mapping diagram can be seen in figure 1.

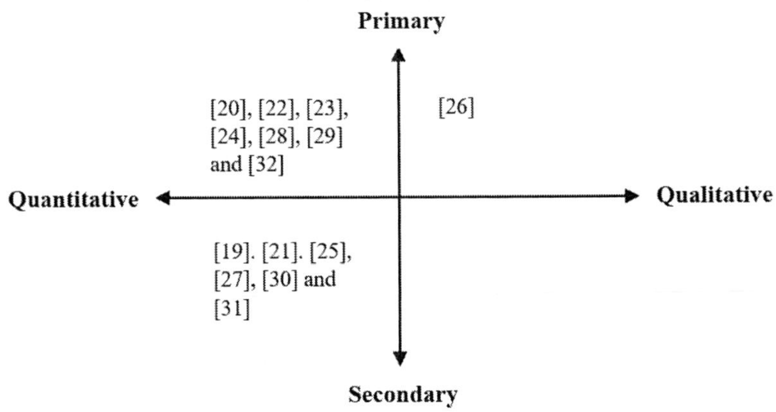

Figure 1. Mapping diagram of type and data research

Based on figure 1, it is found that the most dominant method used is research that uses quantitative types with primary data.

4 Conclusion

The most widely used method is research with quantitative methods with the types of primary data described in journals [20], [22], [23], [24], [28], [29] and [32] with differences in analysis namely regression analysis, LCC analysis, sensitivity analysis, payback period, NPV and fuzzy method. From the explanation above, it is also obtained that the most widely used analysis is the LCC analysis. This paper can help future research on the application of green building to cost analysis by using appropriate methods and analysis.

Acknowledgement

The authors gave their consent and awards in the form of research grants and fellowships from "Penelitian Tesis Magister ITS 2020 based on contract number 917/PKS/ITS/2020".

References

[1] Geng Y, Ji W, Wang Z, Lin B and Zhu Y 2019 *Energy Build* **183** 500 – 14
[2] International Energy Agency 2018
[3] Darko A and Chan A 2016 *Hab. Inter.* **57** 53 – 63
[4] Li Q, Long R, Chen H, Chen F and Wang J 2020 *J. Clean. Prod.* **245** 118775
[5] Marenjak A 2012 *Croat. Assoc. Civ. Eng.* **64** 293 – 303
[6] Sesana M M and Salvalai G 2013 *Build. Env.* **67** 211 – 6
[7] Creswell J W 2003 *Research Design: Qualitative, Quantitative and Mixed Method Approaches* (London: Sage Publications) p 152
[8] Mesly O 2015 *Creating Models in Psychological Research* p 51
[9] Fuller S 2016 *Life-Cycle Cost Analysis (*LCCA*)* p 10
[10] Sam K 2010 Green Construction Project Management and Cost Oversight pp 304 – 42
[11] Pichery 2014 Encyclopedia of Toxicology pp 236 – 7

[12] Reniers G, Talarico L and Paltrinieri 2016 Dynamic Risk Analysis in the Chemical and Petroleum Industry p 103
[13] Femi A, Olawale A O and Ogundele 2008 The Importance of The Payback Method in Capital Budgeting Decision p 26
[14] Rendall R, Pereira A and Reis M 2016 *Comp. Aid. Che. Eng.* **38** 1629 – 34
[15] Cook R D 2001 *Inter. Ency. Soc. Behav. Sci.* 8888 – 93
[16] Coroiu A M 2015 *Mat. Sci. Eng.* **95** 012154
[17] Godil R R, Shamim M S, Enam S A and Qidwai U 2011 *Surg. Neur. Intr.* **2** 985 – 90
[18] Gallo A 2014 *A Refresher on Net Present Value* p 31
[19] Li S, Lu Y, Kua H W and Chang R 2020 *J. Clean. Prod.* **252** 119771
[20] Chegut A, Eichholtz P and Kok N 2019 *J. Environ. Eco. Manag.* **98** 102248
[21] Devine A and McCollum M 2019 *Cities* **92** 303 – 17
[22] Weerasinghe A S and Ramachandra T 2018 *Built. Environ. Proj. Asset Manag.* **8** 528 – 43
[23] Vyas G S and Jha K N 2017 *Sustain. Cities Soc.* **36** 107 – 15
[24] Khoshbakht S M, Rostami R, Valipour A, Ismail M and Rahmat A R 2018 *J. Clean. Prod.* **173** 82 – 99
[25] Deng Y and Wu J 2013 *Reg. Scien. Urban. Eco.* **47** 82 – 99
[26] Fan K, Chan E H and Qian Q K 2018 *Energy Policy* **119** 563 – 73
[27] Illankoon I M and Lu W 2019 *Build. Environ.* **158** 182 – 91
[28] Dwaikata L N and Ali N A 2018 *J. Build. Eng.* **18** 303 – 11
[29] Tam V W Y, Senaratne S, Le K N, Shen L, Perica J and Illankoon I M 2017 **147** 458 – 69
[30] Kim D, Cho H, Koh J and Im P 2020 *Renew. Sust. Energy Rev.* **118** 109508
[31] Zuo J, Pullen S, Rameezdeen R, Bennetts H, Wang Y, Mao G, Zhou Z, Du H and Duan H 2017 *Renew. Sust. Energy Rev.* **70** 358 – 68
[32] Annibaldi F, Cucchiella F, Berardinis P D, Rotilio M and Stornelli V 2019 **116** 109441

Research on real estate investment trust (REIT) as real estate financing for developers: a methodology review of previous study

A F Sari[1], C Utomo[1*] and Y Rahmawati[2]

[1] Departement of Civil Engineering, Sepuluh Nopember Institute of Technology, Indonesia
[2] Department of Civil & Environmental Engineering, Universiti Teknologi PETRONAS, Perak Darul Ridzuan, Malaysia.

*Corresponding author's e-mail: christiono@ce.its.ac.id

Abstract. Project financing commonly found in Indonesia is conventional project financing, namely financing with own capital and long-term loans with a certain composition. The use of conventional financing often produces an un-optimal and unprofitable cost of capital. Real Estate Investment Trust (REIT) is one of alternative financing that can be applied in global property projects, but the use of this strategy is needed to be further analyzed to be used as a project financing option in Surabaya. The developer companies tend to agree with the development of REIT in Surabaya and also the benefits obtained from the implementation of REIT. However, they are still having doubts about implementing it. The purpose of this study is to examine using REIT as a method of real estate financing for developers. This research is focused on finding the consequences of using REIT from the developer side. Therefore, the literature review method will be carried out.

1. Introduction

The economic development of each country is dependent on the development of a good real estate market. Development is one important factor in the country's economy. Surabaya is the second-largest city in Indonesia with 3,158,943 populations according to 2019 statistical bureau. Population growth and community migration are expected in Surabaya since it is also a city with a strategic position that has economic development potential. Thus, the need for property development will keep rising. For successful property development, one needs a successful project. A successful project for property development is reflected in a good financing organization [26].

One characteristic of a construction project is that it always requires financing. Project financing is providing funds for a project from various sources whether it is from equity or loan [26]. Financing in the most common way, is done with own capital or with a loan with a certain composition. Financing must produce the most appropriate financing composition to produce the optimum cost of capital or the lightest and profitable to the company. However, this method often results in an un-optimal cost of capital and is detrimental to the company [21].

The development of the Real Estate Investment Trust (REIT) the first time in the United States of America in 1960 under the REIT Act policy was spread to various parts of the world [17]. REIT (Real Estate Investment Trusts) is a real estate investment fund, which is an important means of real estate

Content from this work may be used under the terms of the Creative Commons Attribution 3.0 licence. Any further distribution of this work must maintain attribution to the author(s) and the title of the work, journal citation and DOI.
Published under licence by IOP Publishing Ltd

securitization. Specifically, REIT is a trust fund collected from many investors by issuing shares or certificates of income, real estate investment management by specialized investment institutions, and distributing income as regulated to investors. As of October 2018, 37 countries have adopted laws and issued REIT [14].

The REIT system is designed like a mutual fund, to provide opportunities for small and medium investors to invest in real estate. REIT also reduces the debt rate, because it can access public capital, besides, REITs have an average debt level of 40% of their market value, whereas the real estate industry usually has a debt level in the range of 80% [5]. Even so, REIT can still access debt through banks, to achieve certain objectives such as acquiring property and then issuing public securities to reconstruct capital structures [9].

The establishment of REIT can also be done to avoid some property development problems in Indonesia such as price fluctuations and capital outflows in the national property industry which is quite large. REIT can reduce these two. Also besides, the fund market media is also beneficial for increasing the frequency and volume of property asset trading in the form of secondary market management [22].

The only few companies that issue REITs in Indonesia show that REITs are underdeveloped in Indonesia. The underdevelopment of REIT is unfortunate because companies can benefit from it, such as 1) Recycle capital, which is the funds obtained from REIT can be used to expand or build property further, 2) REIT also strengthens the solvency ratio, which is REIT funds can be used to pay off debt-debt related to property payments [23]. The impact of cutting taxes into a single tax can also attract investors so that it can improve the performance and value of the company [24].

The developer company tends to agree with the development of REIT in Surabaya and also the benefits obtained from the implementation of REIT. However, they are still doubting implementing it [25]. This paper will focus on mapping the previous research literature that will be used to obtain perceptions about financing using Real Estate Investment Trust (REIT) for developers in Surabaya, Indonesia.

2. Literature Review of the Previous Study

2.1. Methodology

The papers used in theoretical mapping are papers obtained from various reputable journal sources to make sure the academic standards of the literature analyzed in this paper. In this review literature, there are 20 publications from 2010-2020. Table 1 will list the journals used along with the ranking based on the SJR.

Table 1. Quality levels of selected journals and number of papers.

Journals Publication	SJR	Q Category	Number of Papers
Economic Modelling	0.97	Q2 in Economics and Econometrics	1
Journal of Property Investment & Finance	0.48	Q1 in Economics, Econometrics, and Finance	5
Journal of Real Estate Finance and Economics	0.63	Q1 in Urban Studies	6
Corporate Governance	1.47	Q1 in Business, Management, and Accounting	1
International Review of Economics and Finance	0.81	Q1 in Finance	1
North American Journal of Economics and Finance	0.54	Q2 in Economics and Econometrics	1
Real Estate Economics	1.15	Q1 in Accounting	1
Physica A: Statistical Mechanics and its Applications	0.71	Q2 in Condensed Matter Physics	1
Journal of Corporate Finance	1.57	Q1 in Business and International Management	1
Journal of Behavioral and Experimental Economics	0.7	Q1 in Social Sciences (miscellaneous)	1
European Journal of Management and Business Economics	0.64	Q1 in Business and International Management	1

2.2. Result

From the literature review, it is known that from previous studies, the scope of research on REIT revolves around the topic of financing, the risk of REIT, the benefits of REIT, as well as the concept and character of REIT itself. In table 2, is a theoretical mapping from previous research related to REIT and financing. The scope of the research paper is written for each paper, as well as the methodology used in the study, data types, and notes regarding what kind of methods were used in the researches.

REIT (Real Estate Investment Trusts) is a real estate investment fund, which is an important means of real estate securitization. Specifically, REIT is a trust fund collected from investors by issuing shares or certificates of income, establishing real estate investment management by specialized investment institutions, and distributing after-tax income as regulated to investors [14].

Tax concessions received by REIT companies make REIT must distribute a minimum after-tax profit of 90% to shareholders. However, this large profit-sharing also causes a decrease in retained earnings growth which is a component of shareholder funds [7].

Because of the limitations of REIT companies to maintain their capital, access to public capital markets is essential. However, accessing the capital market is not easy, so one alternative is to acquire and manage existing real estate assets [5].

As for acquiring these assets, relying on a mortgage from REIT alone is not enough. Many REIT companies are starting to turn to bank loans for acquisition funding while issuing public securities to reconstruct capital structures [9]. The acquisition of property by REIT makes REIT able to maintain high returns by reducing costs [5].

This asymmetrical information can affect the liquidation of a management company and can disrupt REIT credit lines from banks to acquire or investing [10]. It was also found that REIT might be sensitive to speculative bubbles. This speculative bubble affects the average return. Research result states that the highest average returns are in the REIT type equity, then the lowest average return is in type Mortgage REIT [1].

2.3. Discussion

The position diagram of each study in the literature is shown in figure 1. Based on figure 1, it can be depicted that the most commonly found research about REIT is about its characteristics, which means research about REIT is relatively new hence a lot of researchers are researching on REIT characteristics.

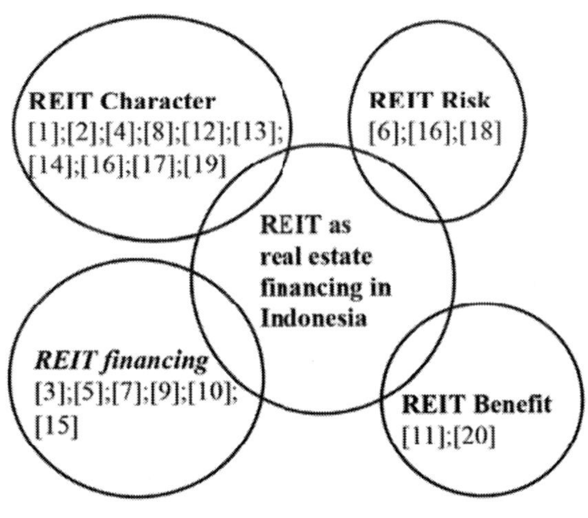

Figure 1. Mapping positioning diagram

IOP Conf. Series: Materials Science and Engineering **930** (2020) 012015 doi:10.1088/1757-899X/930/1/012015

Table 2. Literature review and theoretical mapping of the previous study

NO.	AUTHOR/ YEAR	SCOPE				METHODS		DATA TYPE		Notes
		Financing	Risk	Benefit	Character	Quantitative	Qualitative	Primary	Secondary	
1	Lee et al, 2014				√	√			√	Quantile Auto Regression (QAR)
2	Cardona, 2016				√	√	√		√	3 liquidity measurement and GMM regression
3	Ghosh et al, 2011	√				√	√		√	Mathematic equation model
4	Morri and Baccarin 2016				√	√	√		√	Regression analysis
5	Scherrer, 2016	√					√		√	Literature review
6	Gerlach et al, 2015		√			√	√		√	EGARCH model
7	Seok et al, 2020	√					√		√	Mathematic equation model
8	Marzuki et al, 2019				√	√			√	Correlation analysis and mean-variance framework
9	Hardin III and Wu, 2010	√				√			√	Descriptive and regression analysis
10	An et al, 2012	√				√			√	Mathematic equation and regression analysis

Table 2. Literature review and theoretical mapping of the previous study (cont'd)

NO.	AUTHOR/ YEAR	SCOPE				METHODS		DATA TYPE		Notes
		Financing	Risk	Benefit	Character	Quantitative	Qualitative	Primary	Secondary	
11	French dan Price, 2018			√			√		√	Mathematic calculation
12	Lu-Andrews, 2017				√	√			√	Mathematic calculation
13	Loo dkk, 2015				√	√			√	Descriptive and mathematic calculation
14	Liu dkk, 2019				√	√			√	Hurst Exponent
15	Doan dan Nguyen, 2018	√				√			√	Descriptive and multivariate analysis
16	Beracha dkk, 2019		√		√	√			√	Multivariate regression
17	Abdullah dkk, 2016				√	√			√	3 performance measurement index
18	An dkk, 2016		√			√			√	Regression analysis
19	Bao dan Gong, 2017				√	√			√	Mathematic equation and descriptive
20	Grybauskas dan Pilinkiene, 2019			√		√			√	Data Envelope Analysis (DEA)

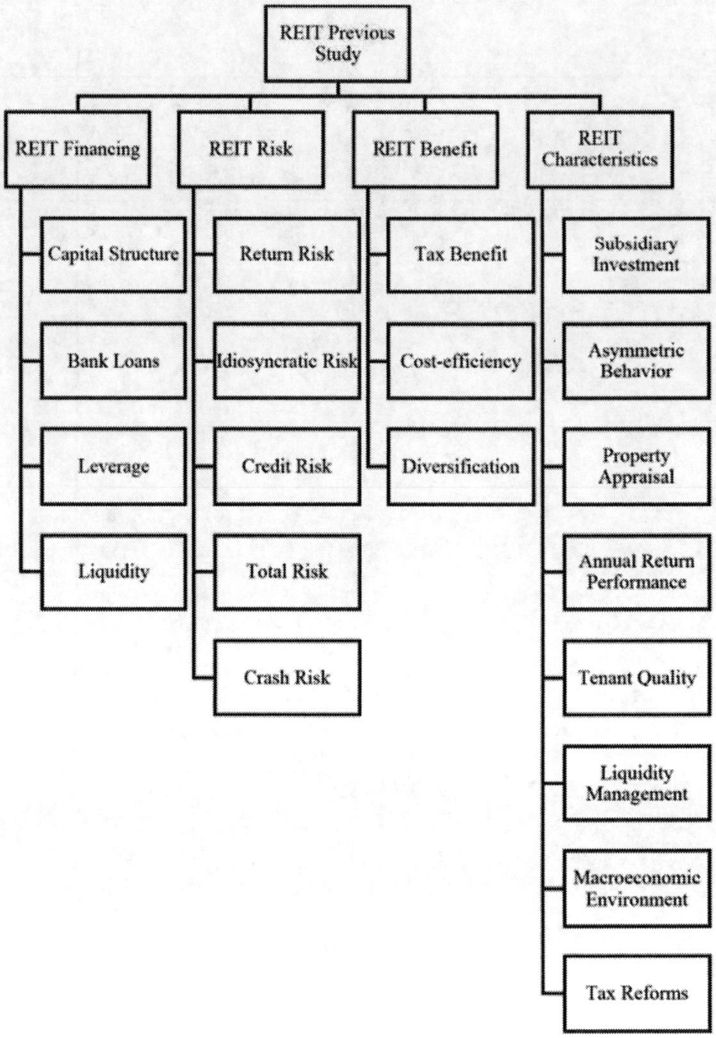

Figure 2. Scope of the previous study

Literature review in previous studies for theoretical mapping can be seen in table 2. Based on table 2, there are four scopes for REIT related paper used to be researched, namely REIT financing, REIT risk, REIT benefit, and REIT character. REIT financing scope is talking about the way REIT does its financing such as its capital structure, debt ratio, leverage, liquidity. REIT risk scope is about the risk that can be happened while using REIT. REIT benefit scope is about the benefit that could be obtained while using REIT. The last REIT character is talking about REIT's basic concept or character when REIT met some occasions or conditions. The scope of the previous study and its content are described in figure 1.

Since it was mentioned in the literature mapping, there are literatures about REIT benefit and its characters to help the developers utilize it to its full advantage, and there are literatures that explained about its risks to be learned and be prepared for risks mitigation. Thus, the implementation of REIT as real estate financing for developers in Surabaya, Indonesia is encouraged.

3. Conclusion

This paper includes review of literature on real estate financing using Real Estate Investment Trust (REIT) for developers in Indonesia. The conclusions that can be drawn are as follows:

a. Using REIT as financing has advantages and risks for developers who will use it.
b. From the literature review, a lot of previous studies revolve around topics such as REIT characteristics, REIT financing, REIT benefit, and also REIT risk.
c. In this review, found that the most commonly found research about REIT is about its characteristics, which means research about REIT is relatively new hence a lot of researchers are researching on REIT characteristics.

As a continuation of the literature review and formulation of variable concepts to obtain a perception of using REIT as real estate financing for developers in Indonesia, a review needs to be conducted to analyze the methods to be used as research tools.

Acknowledgment

The authors appreciate the recognition and awards in forms of research grant and fellowship from "Penelitian RistekBrin 2020".

Reference

[1] Lee C C, Lee C F and Lee C C 2014 *Eco. Model.* **42** 29-37
[2] Cardona J C 2016 *J. Prop. Inv. Fin.* **34** 387-406
[3] Ghosh C and Giambona E 2011 *J. Real Estate Fin. Eco.* **43** 39–72
[4] Morri G and Baccarin A 2016 *J. Prop. Inv. Fin.* **34** 347-74
[5] Scherrer 2016 *Corp. Govern.* **4** 78-82
[6] Gerlach R, Obaydin I and Zurbruegg R 2015 *Int. Rev. Eco. Fin.* **38** 207-19
[7] Seok S I, Cho H and Ryu D 2020 *North American J. Eco. Fin.* **51** 101063
[8] Marzuki M J, Newell G and McGreal S 2019 *J. Prop. Inv. Fin.* **38** 56-70
[9] Hardin III W G and Wu Z 2010 *Real Estate Eco.* **38** 257-84
[10] An H, Hardin III W and Wu Z 2012 *J. Real Estate Fin. Eco.* **45** 678–704
[11] French D W and Price S M 2018 *J. Real Estate Fin. Eco.* **57** 43-63
[12] Lu-Andrews R 2017 *J Real Estate Fin. Eco.* **54** 272-96
[13] Loo W K, Anuar M A and Ramakhrisnan S 2015 *J. Prop. Inv. Fin.* **34** 68-82
[14] Liu J, Cheng C, Yang X, Yan L and Lai Y 2019 *Phys. A: Stat. Mech. App.* **534** 122035
[15] Doan T and Nguyen N Q 2018 *J. Corp. Fin.* **51** 109-24
[16] Beracha E, Feng Z and Hardin III W G 2019 *J. Real Estate Fin. Eco.* **58** 408–37
[17] Abdullah N A, Mohd K N T and Wong W C 2016 *J. Prop. Inv. Fin.* **35** 184-90
[18] An H, Wu Q and Wu Z 2016 *J. Real Estate Fin. Eco.* **53** 527–58
[19] Bao H X H and Gong C M 2017 *J. Behav. Exp. Eco.* **69** 38-49
[20] Grybauskas A and Pilinkiene V 2019 *Europ. J. Manage. Business Eco.* 2444-8494
[21] Putra M A A, Utomo C and Nurcahyo C B 2013 *J. Tek. POMITS* **2** 1
[22] Lim A 2011 *Peran Notaris Dalam Pelaksanaan REITs (Real Estate Investment Trust)/DIRE-KIK (Dana Investasi Real Estat-Kontrak Investasi Kolektif) Di Indonesia* (Depok: Universitas Indonesia) p 1-9
[23] Wibowo M T 2019 *J. Manaj. Bis. Kewir.* **3** 61-7
[24] Wibowo F I 2018 *Pengaruh Good Corporate Governance dan Tax Amnesty Terhadap Kinerja dan Nilai Perusahaan: Studi Kasus Pada Perusahaan Sektor Properti dan Real Estate yang terdaftar di Bursa Efek Indonesia Tahun 2014-2016* (Yogyakarta: Universitas Islam Indonesia) p 1-9
[25] Christianto S 2014 *Analisa Faktor-Faktor yang Mempengaruhi Penggunaan Real Estate Investment Trust (REIT) untuk Pembiayaan Properti di Surabaya* (Surabaya: Institut Teknologi Sepuluh Nopember) pp 59-60
[26] Gorshkov R and Epifanov V 2016 *Proc. Eng.* **165** 1211-15

IOP Conf. Series: Materials Science and Engineering **930** (2020) 012016 doi:10.1088/1757-899X/930/1/012016

Evaluation of the maturity level and critical success factors of the knowledge management implemented in state-owned construction company in Indonesia

A W Septari[1*], Y Latief[1]

[1] Civil Engineering Departement, Faculty of Engineering, Universitas Indonesia, Depok, Indonesia

[*]Corresponding author's e-mail: ade.widya@ui.ac.id

Abstract. The implementation of Knowledge Management (KM) is considered a sustainable approach to overcome the challenges in the increasingly competitive construction field. This means there is a need to recognize the importance of this concept as a strategic asset to improve organizational performance in the industry. The purpose of this paper is to present an overview of KM implementation in State-Owned Construction Company (SOCC) in Indonesia. The research was conducted in three stages with the first focusing on the KM maturity level (KMML) while the second was on KM critical success factor (KM CSF). Both stages made use of literature review, expert validation, and questionnaires but KMML was developed using a spreadsheet to identify strengths and future development of KM factors while KM CSF was formulated with the use of descriptive analysis and Relative Importance Index to rank KM factors. Furthermore, the third stage involved recommended strategies to improve KM using literature review, gap analysis, and expert validation. The results showed the KMML of SOCC is at level 3 of 5, 7 CSFs ranking based on the level of importance and the KM culture factor was recommended to be prioritized to improve KM implementation in SOCC in Indonesia.

1. Introduction

Infrastructural development is one of the focuses of the Indonesian government from 2014 to the present [1] and this is reflected in the efforts being made to accelerate the realization of projects considered strategic and of high urgency in a short period. The country currently has a total of 223 Projects and 3 National Strategic Programs (PSN) with a total investment value of 4,183 Trillion [2].

In the Indonesia Infrastructure Report Q4 2018 prepared by Fitch Solutions, the State-Owned Construction Company (SOCC) was observed to be dominating the construction of infrastructural projects in Indonesia with Wijaya Karya, Waskita Karya, and Pembangunan Perumahan holding the largest share despite the increasingly competitive tendering process. This is believed to be an opportunity for SOCC to stay afloat in the construction industry of Indonesia and also to become a market leader.

Several problems have been associated with the projects in the construction industry due to different reasons and an example is the project management methodology being used which involves planning, executing, and monitoring observed not to be contributing enough to the achievement of the desired goals [3]. Moreover, challenges from competitors are required to be overcome using different

Content from this work may be used under the terms of the Creative Commons Attribution 3.0 licence. Any further distribution of this work must maintain attribution to the author(s) and the title of the work, journal citation and DOI.
Published under licence by IOP Publishing Ltd

approaches due to the increasingly complex and competitive construction industry caused by globalization and others [4] [5]. This was believed by several researchers to have led organizations to seek a sustainable approach which includes the implementation of KM [6,7]. The Project Management Institute has also added KM discussion to Chapter 4 of Project Integration Management in the sixth edition of the Project Management Body of Knowledge (PMBOK) [8]. This further supports KM as an approach to find solutions to the problems in the construction field. There are, however, limited discussions studies on KM CFS and the maturity level in the construction industry of Indonesia, therefore, there is a need for further research on the concept.

2. Literature Review

2.1. Knowledge Management
Several definitions have been associated with Knowledge Management (KM) in line with the development and diversity involved in understanding the concept. KM has been reported to be a management tool justifying the belief that knowledge is an asset to increase organizational capacity and the ability to work more effectively [9]. It is the process of identifying or creating, assimilating, and applying organizational knowledge to exploit new opportunities and improve performance [10]. The concept is about ensuring the skills, experience, and expertise of the project team and other stakeholders are used before, during, and after the project [8].

2.2. Knowledge Management Maturity Model (KMMM)
Maturity Model is a phased approach to improve business processes over a long period and has been observed to have several types such as InfoSys KMMM (2000), Siemens KMMM (2002), General KMMM, APQC, IAEA, etc. The International Atomic Energy Agency (IAEA) also uses a scale of 1-5 measurements [11] and 7 main KM elements have been identified to include policies and strategies, human resource planning and processes, training and improvement of human performance, document management, technical solutions (IT), approaches to obtain tacit knowledge, and KM Culture.

2.3. Knowledge Management Critical Success Factor (KM CSF)
KM CSF in the construction industry is recognized as an internal problem managed by companies in relation to the mobilization of limited resources to achieve the desired project management performance results [12]. Davenport et al. have identified eight KM CSFs which are industrial value, organizational infrastructure, standard flexible knowledge structures, culture, motivational practices, different channels for sharing knowledge, support from senior management, and language clarity and purpose [13]. Meanwhile, Wang et al. [14] and Zang [15] found 11 CSFs for public-private infrastructure organizations including leadership and support, information technology (IT), strategy and planning, culture, organizational infrastructure, KM resources, conferences, training and education, motivation tools, processes and activities, measurement, and human resource management.

Valmohammadi [16] also identified 12 KM CSF from different studies to include management leadership and support, organizational culture, information technology, KM strategy, performance measurement, organizational infrastructure, processes and activities, rewards and motivation, removal of limitation, training and education, resource management human power, and comparative analysis.

2.4. *Relationship between KMMM & KM CSF*
The maturity model helps organizations to develop, implement, and compare the successful implementation of KM [17]. It is used as a framework for organizations to measure the level of KM maturity and help them focus and prioritize areas to increase [18]. Meanwhile, CSF is related to more significant matters required by each organization or industry to concentrate based on its limited resources to succeed at a certain time or the elements deemed necessary for an organization to achieve its mission [19]. Several studies have shown KM maturity and KM CSF to be interrelated to achieve the desired level of KM implementation. This, therefore, means KMMM can be used to measure the

maturity level of KM currently being utilized while CSF determines the required elements to achieve the desired maturity level.

3. Research Methods

This study is divided into 3 research questions (RQ) and these include what is the KM maturity level in SOCCs in Indonesia (RQ1)? What are the CSFs needed to improve KM implementation (RQ2)? and What is the strategy to improve KM performance and consequently the organizational performance of SOCCs in Indonesia (RQ3)?

3.1. Data collection

The questionnaire was first validated by 5 experts with at least 15 years of experience to provide suggestions and improvements. This was followed by a preliminary survey respondents to provide input and correct questions before the main survey. The questionnaire was distributed to 9 SOCCs and 69 represented by 76.67% of the 90 distributed were returned but 2 were rejected due to the non-fulfillment of the requirements by the respondents. This means only 67 were used. The male respondents were 67% while women were 33%.The job positions were divided into 3 groups with the top manager discovered to be 61%, middle management was 24% while the staff was 15%. Moreover, the work experience was categorized into three, specifically <5 years, 5-10 years and 10-20 years with 1%, 81%, and 18%, respectively. Most of the respondents as represented by 84% have a Bachelor's degree as their highest educational qualification while 16% have a Master's degree.

3.2. Methods and measures

KM maturity assessment tools with a total of 39 indicators from 7 elements adapted from IAEA were used to answer RQ1 [11]. This involved determining the current and desired conditions. The current condition is the KM maturity level applied in the company while the desired level expected to be achieved to ensure optimal application and improvement in company performance. The indicators were scored on a five-point Likert scale with 1 indicating "not used at all", 2 "little extent", 3 "some extent", 4 "great extent", and 5 "very great extent". The data obtained were tested for validity and reliability using Pearson Correlation and Cronbach's Alpha respectively and later calculated, grouped based on each rating scale, and the average value for each variable determined using Microsoft Excel. The results are presented using a radar diagram to make the differences of each variable in the current and desired conditions category visible.

The measuring tools with 71 indicators from 12 KM CSF adapted from Valmohammadi [16] were used to answer RQ2. The indicators were scored on a five-point Likert scale 1-5 based on the importance of applying KM. For the purposes of data processing, 12 factors were combined into 7 main factors to ensure the criteria used were the same with those in the Maturity Level RQ1 using the methods used in Valmohammadi [16] and Kozjek and Ovsenik [20]. The factors obtained were also validated by experts and were used as the KM CSF indicators after the data has been tested for homogeneity, validity, and reliability using the same method in RQ 1. This was followed by descriptive analysis to determine the distribution of variable data in the form of mean, median, and standard deviation after which each variable was calculated using the Relative Importance Index (RII) method to determine the rank of the KM factors.

$$RII = \frac{\Sigma W}{A.N} = \frac{5.n5+4.n4+3.n3+2.n2+n1}{5N} \quad (1)$$

Where, W is the weighting associated with each factor by the respondent, ranging from 1 to 5. For example, n1 = number of respondents for "not important at all", n2 for "little important", n3 for "quite important", n4 for "important", and n5 for "very important". A is the highest weight which is 5 for this study while N is the total number of respondents.

The strategies to answer RQ3 include literature review and expert validation. This involved processing the results from RQ1 using Gap Analysis and linked to RQ2 results after which they were both analyzed and used to develop a strategy to increase the KM maturity level and subsequently organizational performance.

4. Findings

The validity was tested by correlating each item score to the total score using the Pearson Correlation technique and the questionnaire items were declared valid or confirmed to have the ability to measure the variables when the correlation coefficient (r) $\geq$ r table. The results showed the correlation coefficient value for each item was greater than r table (0.2369), therefore, they were all declared valid.

The reliability was tested using Cronbach's Alpha and the decision-making criteria require the Cronbach's Alpha coefficient value to be $\geq$ 0.6. The value for the items was observed to be more than 0.6 and they were all declared reliable.

4.1. RQ1 : KM Maturity Level (KMML) of SOCC in Indonesia

Table 1 and figure 1 show the KMML for each element of knowledge and its indicators. The "KM Culture" category was observed to have the lowest value with the current level being 3,14, while the desired level was 4,54. Meanwhile, the highest element was "Training and human performance improvement" with the current level found to be 3,46 while the desired level was 4,66. The KM Maturity Level in SOCC was discovered to be 3,35 and this means it is being currently used "to some extent" while the desired condition was found to be 4,58 and this means it is expected to be used "to a great extent" (IAEA).

Table 1. KM Maturity Level Analysis of SOCCs in Indonesia.

Indicators	Extent currently utilized	Extent desired
Policy and Strategy for KM	**3,33**	**4,52**
Human Resource Planning and Proccesses	**3,37**	**4,57**
Training and human performance improvement	**3,46**	**4,66**
Document Management	**3,44**	**4,68**
Technical solutions (IT)	**3,44**	**4,56**
Tacit knowledge capture	**3,30**	**4,54**
Knowledge management culture	**3,14**	**4,54**
Average Maturity Level	**3,35**	**4,58**

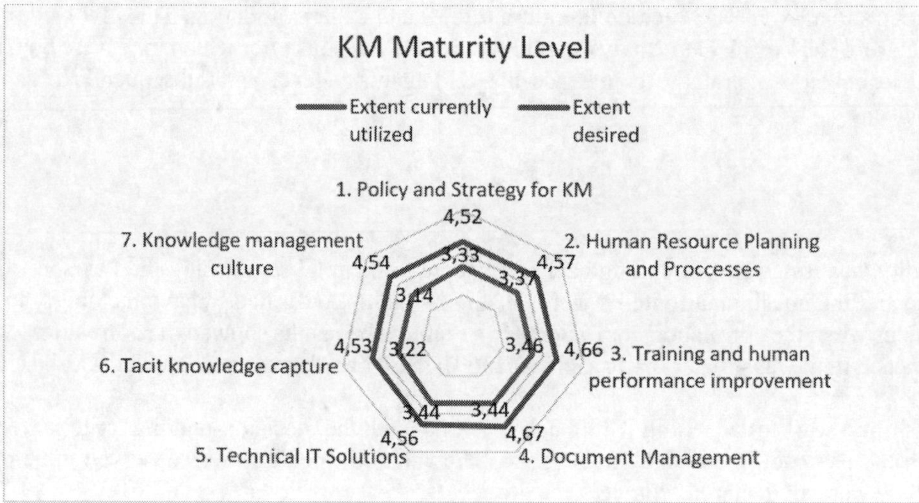

Figure 1. Radar Diagram of KM Maturity Level.

4.2. RQ2 : Critical Success Factors needed to improve implementation of KM

Table 2 shows the KM CSF to improve organizational performance in SOCC in Indonesia in general and the category with the highest level of importance was found to be "training and human performance improvement" with an RII value of 0.879 while "Document management" has the lowest with 0.816.

Table 2. KM CSF Analysis of SOCC.

Indicators	Mean	SD	Med	RII	*Rank*
Training and human performance improvement	4,396	0,608	4,0	0,879	1
Knowledge management culture	4,391	0,526	4,0	0,878	2
Policy and Strategy for KM	4,384	0,613	4,0	0,877	3
Technical solutions (IT)	4,261	0,684	4,0	0,852	4
Tacit knowledge capture	4,184	0,688	4,0	0,837	5
Human Resource Planning and Proccesses	4,109	0,737	4,0	0,822	6
Document Management	4,080	0,710	4,0	0,816	7

4.3. RQ3 : Strategies to improve KM implementation, so that improve organizational performance in SOCC in Indonesia

The strategy to develop KM implementation was obtained by analyzing KMML and KM CSF using the analysis to determine the biggest difference/gap between the current and desired conditions of KMML and this was confirmed to be "KM Culture" category with a difference of 1.40 as shown in table 3. This, therefore, means this category needs to be prioritized in order to improvise or raise the KM maturity level.

Table 3. Gap Analysis of KM Maturity Level.

Indicators	Extent currently utilized	Extent desired	Gap
Policy and Strategy for KM	3,33	4,52	1,19
Human Resource Planning and Proccesses	3,37	4,57	1,20

Training and human performance improvement	3,46	4,66	1,20
Document Management	3,44	4,68	1,24
Technical solutions (IT)	3,44	4,56	1,12
Tacit knowledge capture	3,30	4,54	1,24
Knowledge management culture	**3,14**	**4,54**	**1,40**

The recommended strategies were to increase the KM maturity or minimize the gap to achieve the desired maturity level using KM CSF with the result from the gap analysis of KM ML (RQ1) and CSF in each category (RQ2) used as input after the results have been validated by experts. This study only discussed the categories to be prioritized for improvement and "KM Culture" was found. The recommendations validated to improve this category are, therefore, presented in the following table.

Table 4. Recommended Strategies for KM Improvement on SOCC in Indonesia

Category need improvement	Maturity Level Gap	Rank	Recommended Strategies	RII
KM Culture	1,40			
		1	Leaders encourage the creation, sharing and use of knowledge	0,910
		2	High level of trust among employees in sharing knowledge	0,901
		3	Management acts as a role model to show desirable behavior	0,899
		4	Employee motivation to seek knowledge	0,899
		5	Tying a motivational approach as a job performance appraisal system	0,899
		6	Collaboration between employees is important	0,890
		7	Reward employees with an emphasis on group performance	0,890
		8	Management of organizational change towards the adoption of KM through employees	0,887
		9	Empower employees to explore new possibilities	0,887
		10	Individual impulse to ask questions	0,884

5. Discussion

The KM maturity level in SOCC in Indonesia is currently estimated at level 3 out of 5 and this means it is being used "to some extent" [11]. Meanwhile, Zain [21] showed the national construction service companies in Indonesia to be at level 2 and this means they use it "to a little extent" [11]. This is related to the strategies, processes, and discussions of KM identified in SOCCs while those in national private contractors are being developed [22]. The similarity between the two studies is observed from the IT Technical Solutions category which was recorded to have the highest current maturity level for both state-owned and private national contractors.

The category was observed to have the smallest value difference of 1.119 based on gap analysis and this means the respondents were quite satisfied with the IT system provided by the company in the application of KM, even though it was not optimal nor in accordance with the expected conditions. Moreover, the interviews with several SOCCs showed some systems have been designed such as integrated IT system applications for KM, Human Capital, etc. which are accessible from both laptops and mobile phones, and the employees were observed to be interested in using them to manage knowledge. This means information and communication technology has the ability to support the KM process by making it possible to quickly capture and transfer the knowledge obtained to others through different channels such as email, databases and other means such as Intranets, video conferencing, etc. [22,37], therefore, allowing project team members to learn, solve problems, and make decisions faster [23].

"KM Culture" with a difference of 1,400 has the biggest gap rank and "Example of a leader" variable was found to have the highest gap in the element. Meanwhile, despite the importance of leadership example in implementing KM [13], the process is expected to fail as long as the management does not understand its value as a promising alternative to existing techniques [24] required to achieve the expected project performance. This means a leader needs to show an appropriate leadership style while applying KM [25]. A strategy is, therefore, required to improve the KM maturity, especially in the Culture variable.

The element ranked first in KM CSF for both SOCC and National Private contractors in Indonesia is the same which is the "Training and human performance improvement but those in 2nd to 7th were different. This shows the element is considered very important by respondents and in line with a previous study which identified employee training as one of the important CSFs supporting the application of KM to achieve expected project performance [26]. This means the training budget is a strategic investment rather than a cost and its effect should be seen as a long-term action [27]. Moreover, learning increases interaction between employees and this further creates new knowledge for the organization [28].

The strategies considered very important and recommended for immediate improvement include "Leaders encourage the creation, sharing, and use of knowledge" followed by "High levels of trust among employees in sharing knowledge" and "Management acts as a role model to show desirable behavior". They are ranked based on their level of importance in the KM CSF analysis and are expected to be applied in KM implementation to be able to achieve the project results desired [29]. This is important because a clear identification of these internal CSFs allows construction managers to take proactive actions to avoid the negative impacts caused by project delays, poor quality, and high costs [30].

The "Leaders encourage the creation, sharing, and use of knowledge" strategy is very important due to the significant roles played by leaders of construction organizations in recognizing the value of KM and devising appropriate strategies for its implementation. This means strong leadership is needed to provide strategic direction, plans, and authority to achieve the goals and objectives of each project [31]. It is also required to apply the right KM strategy which allows a suitable environment to create, store, share, and use knowledge to ensure the effective functioning of project planning and successful implementation [32].

"A high level of trust among employees in sharing knowledge" strategy is very important due to the significance of the KM culture founded on trust, collaboration, and learning to enable project team members to think, behave consistently, and encourage the use of knowledge in project activities [33] without underestimating their skills.

"Management acts as a role model to show desirable behavior" strategy is also very important due to the need for the leaders to demonstrate the leadership style appropriate to achieve successful KM implementation [34]. The role of the management is to empower and delegate responsibility to project managers towards maintaining the level of motivation required by the team members to set collective goals and share knowledge [35]. This, therefore, means management needs to continually search for methods to motivate the team members rather than concentrating on individuals and competitive goals [6,36].

6. Conclusion

The first objective of this research was to determine the current level of KM maturity achieved by SOCC in Indonesia and, after series of analysis using seven categories, it was found to be 3 out of 5 and this means it is being used "to a certain extent" (IAEA).

Relative Importance Index (RII) analysis was used to analyze 7 categories to obtain KM CSF. The highest level of importance was found with "Training and human performance improvement", followed by "Knowledge management culture", "Policy and Strategy for KM", "Technical solutions (IT)", "Tacit knowledge capture", "Human Resource Planning and Processes" and the lowest was "Document Management".

The strategies recommended to improve the application of KM were determined by combining the results obtained from analyzing the largest gap between the current and expected maturity level and those from the KM CFS analysis. The maturity level needs to be improved and, in this case, the "KM Culture" is prioritized with the three most important strategies recommended found to be "Leaders encourage the creation, sharing, and use of knowledge", "Management acts as a role model to show desirable behavior", and "Leaders act as catalysts for KM". They are recommended to achieve KM maturity according to the expected conditions and to enable organizations to react to environmental changes by considering the strengths and weaknesses of their environment to achieve the desired project performance.

Further research is expected to improve the recommended strategies of KM implementation from each knowledge factor instead of focusing on only KM culture. Moreover, a follow-up survey is needed after CSF KM is applied to the company to determine the level of increase in KM maturity based on the strategies recommended in this study. It is also possible for construction service companies in Indonesia to evaluate the maturity level of the KM application implemented and determine the key success factors to improve organizational performance according to the conditions and needs of each company.

Acknowledgements

The author appreciates the University of Indonesia for providing financial support through PUTI Proceedings of the Fiscal Year 2020 with the contract number: NKB-1139/UN2.RST/HKP.05.00/2020 managed by the Directorate of Research and Community Engagement (DRPM) University of Indonesia.

References

[1] Gewati M 2019 *Alasan Pemerintah Jokowi-JK Fokus Bangun Infrastruktur* available at https://money.kompas.com/read/2019/03/26/084500826/alasan-pemerintah-jokowi-jk-fokus-bangun-infrastruktur?page=all.
[2] Komite Percepatan Penyediaan Infrastruktur Prioritas. (n.d). Proyek Strategis Nasional. Retrieved September 01,2019, from https://kppip.go.id/proyek-strategis-nasional/
[3] Flyvbjerg B, Holm M K S and Buhl S L 2003 How common and how large are cost overruns in transport infrastructure projects *Transport Review* **23** 1 71-88.
[4] Hari S, Egbu C and Kumar B 2005 A knowledge capture awareness tool: an empirical study on small and medium enterprises in the construction industry *Engineering, Construction and Architectural Management* **12** 6 533-67.
[5] Rangasamy S and Ghosch S 2002 Critical success factors for SPS implementation in UK small and medium enterprises: some key findings from a survey *The TQM Magazine* **14** 4 217-24.

[6] Enshassi A, Falouji I and Al-Kilani S 2016 Knowledge management critical success factors in construction projects *International Journal of Sustainable Construction Engineering and Technology* **7** 1 69-84.

[7] Nilashi M, Zakaria R, Ibrahim O, Majid M Z, Zin R M and Farahmand M 2015 MCPCM: a DEMATEL-ANP-based multi-criteria decision-making approach to evaluate the critical success factors in construction projects *Arabian Journal for Science and Engineering* **40** 2 343-61.

[8] PMI Book Service Center 2017 *A guide to the project management body of knowledge (PMBOK guide)* Project Management Institute, Inc.

[9] Nonaka I and Takeuchi H 1995 *The Knowledge-Creating Company: How Japanese Companies Create the Dynamics of Innovation* Oxford University Press (Boston, MA).

[10] Yang C W, Fang S C and Lin J L 2010 Organisational knowledge creation strategies: a conceptual framework *International Journal of Information Management* **30** 3 231-8.

[11] Ajmal M, Helo P and Kekale T 2010 Critical factors for knowledge management in project business *Journal of Knowledge Management* **14** 1 156-68.

[12] Davenport T H, De Long D W and Beers M C 1997 Building successful knowledge management projects *Centre for Business Innovation Working Paper* (Ernst&Young LLP:Cambridge, MA).

[13] Wang J, Li Z and Tam V W 2014 Critical factors in effective construction waste minimisation at the design stage: a Shenzhen case study, China *Resources, Conservation and Recycling* **82** 1-7.

[14] Zang X 2005 Critical success factors for public–private partnerships in infrastructure development *Journal of Construction Engineering and Management* **131** 1 3-14.

[16] Valmohammadi C 2010 Identification and prioritization of critical success factors of knowledge management in Iranian SMEs: An experts' view *African Journal of Business Management* **4**(6) 915-24.

[17] Robinson H S, Carrillo P M, Anumba C J and Al-Ghassani A M 2005 Knowledge management practices in large construction organisations *Engineering, Construction and Architectural Management* **12** No 5 431-45.

[18] Kochikar V P 2000 *The knowledge management maturity model - A staged framework for leveraging knowledge* Infosys Technologies Ltd.

[19] Rockart J F 1982 The changing role of the information systems executive: a critical success factors perspective *Sloan Management Review* **6** No 1 3-13.

[20] Kozjek D and Ovsenik M 2017 Model of knowledge management factors and their impact on the organization's success *De Gruyter Open Access Jounal* **50** 112-31.

[21] Zain A H 2019 Evaluation of the maturity level and critical success factors in the implementation of knowledge management in the national private construction service company in Indonesia *Thesis* (University of Indonesia)

[22] Berraies S, Chaher M and Yahia K B 2014 Knowledge management enablers, knowledge creation process and innovation performance: an empirical study in Tunisian information and communication technologies sector *Business Management and Strategy* **5** No 1 1-26.

[23] Jeng D J and Dunk D 2013 Knowledge management enablers and knowledge creation in ERP system success *International Journal of Electronic Business Management* **11** No 1 49-59.

[24] Idris K M and Kolawole A R 2016 Influence of knowledge management critical success factors on organisational performance in Nigeria construction industry *Ethiopian Journal of Environmental Studies and Management* **9** No 3 315-25.

[25] Singh S K 2008 Role of leadership in knowledge management *Journal of Knowledge Management* **12** No 4 3-15.

[26] Chong S C, Wong K Y and Lin B 2006 Criteria for measuring KM outcomes in organisations *Industrial Management& Data Systems* **106** No 7 917-36.

[27] Wong K Y 2005 Critical success factors for implementing knowledge management in small and mediumenterprises *Industrial Management& Data Systems* **105** No 3 261-79.

[28] Lee H and Choi B 2003 Knowledge management enablers, processes, and organisational performance: an integrative view and empirical examination *Journal of Computer Information Systems* **20** No 1 179-228.

[29] Gunasekera V S and Chong S C 2018 Knowledge management critical success factors and project management performance outcomes in major construction organisations in Sri Lanka: A case study *VINE Journal of Information and Knowledge Management systems* **48** Issue: 4 537-58.
[30] Sambasivan M and Soon Y W 2006 Causes and effects of delays in Malaysian construction industry *International Journal of Project Management* **25** No 5 517-26.
[31] Bolisani E and Scarso E 2015 Strategic planning approaches to knowledge management: a taxonomy *VINE: The Journal of Information and Knowledge Management Systems* **45** No 4 495-508.
[32] Hansen M, Nohria N and Tierney T 1999 What's your strategy for managing knowledge? *Harvard Business Review* **77** No 2 106-16.
[33] Scarborough H, Swann J and Preston J 1999 Knowledge Management: A Literature Review *Issues in People Management* Institute of Personnel Development (London).
[34] Singh S K 2008 Role of leadership in knowledge management *Journal of Knowledge Management* **12** No 4 3-15.
[35] Geraint J 1998 Share strength: developing a culture of knowledge sharing *People Management* **4** No 16 44-7.
[36] Luca J and Tarricone P 2001 Does emotional intelligence affect successful teamwork? *Proceedings of the 18th Annual Conference of the Australasian Society for Computers in Learning in Tertiary Education at the ASCILITE* (Melbourne) 367-76.
[37] Riggins F J and Rhee H 1999 Developing the learning network using extranets *International Journal of Electronic Commerce* **4** No 1 65-83.

The ratio of changes in construction costs and development index of the standard analysis of construction costs for state buildings

Yusroniya Eka Putri Rachman Waliulu [1*]

[1] Department of Civil Engineering Institut Teknologi Sepuluh Nopember, Indonesia

[*]Corresponding author's e-mail: iput@ce.its.ac.id

Abstract. Wages rate and material price are an important part of a construction cost. The Construction Cost Index (CCI) that has been developed which applies to a certain area or period a basis for determining the bid price. This index is also very useful to forecasting construction cost which should reflect the comparison of wages rate and material price changes over time. Standard Construction Cost Analysis per m2 is one of the performance budget instruments to assess the budget of a construction project. Surabaya City Government applies that standard in set up government budget. SCC is composed of several Unit Price Analysis with a certain volume that design has been standardized. The composition of the labour and material price index, as a cost driver for Standar Construction Cost Analysis based on time units are expected to be used in the preparative of cost estimates for Building Construction and further research related to the effect of price increases on construction costs. The results obtained from this study are the construction cost change ratio in Surabaya for 2013- 2016 increased average 17.3% per year; start from 2016 - 2020 increased not over 6,5% per year. Where as material cost change ratio tends to increase every year, except for 2014 - 2015 which has increase dramatically. In the other hand the labour price index indicate a constant increase because it was related to the Surabaya City Minimum Wage Rates, but in the early year in data (2013-2014) the Minimum Wage Rates significantly increased. In 2017-2018 labour cost change ratio increased more higher and material cost change ratio turn to decreased because the analysis used new guidelines for Analysis of Work Unit Prices for Public Works (Regulation of the Minister of Public Works and Public Housing of the Republic of Indonesia number 28 / PRT / M / 2016). The average component weight of the CCI according to the development of price are labour and foreman (41,3%), stone / brick (10,7%), roof truss and cover (10,4%), cement (8%), floor cover and paint (7,4%), wood (6,8%), iron bar (5,6%), accessories and complement (5,6%) and sand (4,2%). Material cost index tends to increase every year, except for 2018 and 2019 which has decreased. Whereas labour cost index indicated a constant increase because it was related to the UMK (City Minimum Wage) Surabaya. So that the CCI from 2013 to 2020 has increased from 1,00 to 1,93.

1. Introduction

The demand for material goods in an area is high if the economy of a region grows rapidly. Improper pricing policies can give the program negative affect [1]. The cost of a building can be very high if material and wages are imported from other regions. On the other hand when construction activities are low, competition is intense and construction companies are begining to bid on other regional markets [2].

Content from this work may be used under the terms of the Creative Commons Attribution 3.0 licence. Any further distribution of this work must maintain attribution to the author(s) and the title of the work, journal citation and DOI.
Published under licence by IOP Publishing Ltd

Construction costs are one of the main criteria for project success which is needed to supports the Good Government Programs. To avoid high costs, information about price from suppliers in Surabaya needs to be updated every year. Surabaya City Government has conducted a survey which is stated in the Basic Unit Price of the Surabaya City Government every year. It is necessary for the contractor to get material prices and local wages in short term period [3].

Pricing can be based on the consumer price index that issued by BPS (Badan Pusat Statistik) every month. However, this index does not reflect actual construction costs because it is an aggregation of consumer goods groups that are largely unrelated to the construction process. Therefore a more specific cost index is needed. Fundamentally, Construction Cost Index (CCI) should reflect the comparison of price changes over time a fixed product [4]. This index is very useful to forecasting the cost of construction activities [5]. Construction costs are a crucial part of a construction project, the Construction Cost Index (CCI) has been developed which applies to a certain area or period. That index provides a base for determining bid prices.

Standard Construction Cost (SCC) is one of the performance-based budget instruments that have been publish since 2000 in Surabaya. SSC is one of the performance budget instruments to assess the budget amount of an activity. Most local governments in Indonesia have not implemented SCC in preparing their budgets. SCC is the standard cost of a program or activity so that the budget allocation becomes more rational. SCC is composed of several Unit Price Analysis with a certain volume that has been standardized by units per square meter of building.

Based on this background, it is necessary to conduct research on the development of wages rate and material prices as a support to compile a price index. Price index of building materials, wages rates, and building construction costs based on the time index. The time index is a comparison of the price of building materials, wages rates, and building construction costs from year to year. This preliminary research is carried out by collecting data on the prices of building materials and labour in Surabaya from 2013 to 2020 which form Standard Design Building State Type C. The material price and wages rate index and Building Cost Analysis based on time units are expected to be used in the estimate cost for building construction and further research related to the effect of price change on construction costs.

2. Methods

The data of this study are secondary data of Basic Unit Price material and wages of Surabaya City (e-budgeting) in the form of unit price analysis of building materials from 2013 to 2020 and wages rates from 2013 to 2020. Determination of material and wage rate requirements by processing data on work unit prices based on the Indonesian National Standard concerning procedures for calculating unit prices for relevant buildings and housing issued by the Indonesian National Standards Agency such as SNI 2835: 2008, SNI 2836: 2008, and SNI 2837: 2008 and Regulation of the Minister of Public Works and Public Housing of the Republic of Indonesia number 28 / PRT / M / 2016 concerning Guidelines for Analysis of Work Unit Prices for Public Works. In this study refers to the analysis that still refers to technical standards. Data were analyzed with a price index formula based on time index. The cost of building construction is taken from the Standard Construction Cost of Building which are issued by the Surabaya City Government from 2013 to 2020.

Estimated cost of building construction is carried out by taking one Construction Building Standard (Type C Building), then estimated prices in 2013 to 2020 and from the construction costs can be seen changes in prices for building construction projects each year.

This research began by Analysis of the Surabaya City Construction Cost Unit in e-budgeting program for state building construction work which was divided into 3 types of design standard, there are Type A, Type B and Type C. Then proceed to determine the material needs and wages of workers. There are 72 types of materials and 4 types of wages rate used to arrange Standard Construction Cost Type C

Building.The next step is to collect data in the form of material prices and wages in the period of 2013 to 2018 using secondary data from the Development Administration Division of the Surabaya City Government. Then the materials and wages are carried out by processing data using the Indonesian National Standard concerning the procedure for calculating unit prices for buildings and housing.

2.1. Previous Research
The Construction Cost Index using Pratt Method [6],[7]. The research results index is limited to the total value of the building and has not reviewed the prices for each material and wages. Limit sensitive cost items in the city of Jakarta which is a component of the construction cost index. [8] Determination of the Construction Cost Index using building total cost more fully reflects the price index of the building. Furthermore, sensitive cost components are different between building types [6].

Many countries in Europe use Laspeyres Formula in forming the Cost Index [9]. The calculation of the Construction Cost Index by using the total value of the building is more representative of the index of a building, because all the material and wages are involved. So using the method automatically can also calculate the cost index of each material and wage.

Studied about the Construction Cost Index (CCI) in several countries in Europe [10]. This index is limited to 2 constituent elements, namely material costs and wages. The Construction Cost Index is calculated from the weights of material and wage in the base city. Material index and wages between regions are obtained using the Laspeyres formula which is the cost ratio between regions. From each of the material indexes and wages, a construction cost index can be calculated. The resulting index applies with the assumption that implementation and organizational techniques do not change, worker productivity tend to be the same between regions and there is no material or equipment change.

2.2. Price Index
The price index is defined as the price comparison rate at a time (a particular year) of the price at the time (year) used as a basis. Determining the price index based on time requires three variable, which is the price in a given year, the price in the base year, and the base year price index [11].

$$Price\ Index\ year\ n: \frac{price\ year\ n}{price\ of\ base\ year}\ x\ price\ index\ of\ base\ year \tag{1}$$

The price index of the base year is given a value of 100% or 1 (one) in the calculation of the price index, there are two factors that need to be considered in choosing the base year, there is the base year should be chosen in normal conditions and stable economic conditions or from the earliest data, because the base is 100% or 1 (one), the base year which is to be used as a basis for comparison is not too far from the year that is compared because if it is too far away it will cause some consequences that make the results doubtful [12].

2.3. Cost Weight
The total value of the building produced is then used to calculate the cost weight. The formula for calculating material and wage rate weight each year is as follows:

$$wMT = (V \times UMJ) / TJ \times 100\,\% \tag{2}$$

$$wLT = (V \times ULJ) / TJ \times 100\,\% \tag{3}$$

Where wMT is material cost weights to the total cost of the building; wLT is labour cost weights to the total cost of the building; V is volume; UMJ ismaterial price per unit; ULJ is labour price per unit; and TJ is total building costs.

The formula to calculate the cost weight of each material against the total cost of the material in the base city and the cost weight of each wage to the total cost in the following basic year is the formula:

$$wMM = (V \times UMJ) / TMJ \times 100 \% \quad (4)$$

$$wLL = (V \times ULJ) / TLJ \times 100 \% \quad (5)$$

Where wMM is material cost weights to total material costs; wLL is labour cost weights to total labour costs; V is volume; UMJ is material price per unit; TMJ is total material costs; ULJ is labour price per unit; TLJ is total labour costs.

2.4. Cost Index

Material prices and wage rates are used to calculate an index of material costs and wage rates using the Laspeyres Formula [10]. Following is the formula of Material Cost Index and Labour Cost Index in year n:

$$IMA = UMA / UMJ \times 100 \quad (6)$$

$$ILA = ULA / ULJ \times 100 \quad (7)$$

Where IMA is Material Cost Index per unit; UMA is Material Price per unit; UMJ is Material price per unit in the base year; ILA is Labour Cost Index per unit; ULA is Labour Price per unit; ULJ is Labour price per unit in the base year.

The weight of material items and wages in the base year, the material cost index and the wage cost index in the second year are used to construct the Construction Cost Index, namely the index of the total value of buildings in year n.

$$CCI = \Sigma(wMTxIM) + \Sigma(wLTxIL) \quad (8)$$

Where CCI is *Construction Cost Index;* IM is Material Cost Index; IL is Labour Cost Index; wMT is the cost weight of each material to the total cost in the base year; wLT is the cost weight of each wage for the total cost in the base year.

The following is the formulation of the total material cost index and the index of total wage costs in year n:

$$MCI = \Sigma (wMM \times IM) \quad (9)$$

$$LCI = \Sigma (wLL \times IL) \quad (10)$$

Where MCI is *Material Cost Index;* LCI is *Labor Cost Index*; wMM is the cost weight of each material to the total material cost in the base year; wLL is the cost weight of each wage for the total labour cost in the base year.

The total material cost index and the total wage cost index are used to see the difference in prices in the city of Surabaya each year.

3. Results

The total construction cost per m2 for Type C Buildings in Surabaya City in 2020 is Rp6,053,612 - which includes the total material cost of Rp 3,100,303,- the total labour cost of Rp 2,953,309. Based on type C design standard, the recapitulation results of calculation and analysis of others construction cost/m2 and construction cost change ratio can be seen in table 1.

Table 1. Construction cost/m2 and construction cost change ratio (year 2013-2020)

year	2020	2019	2018	2017	2016	2015	2014	2013
Labour Cost (Rp) / m2	2,953,309	2,739,249	2,574,718	2,081,597	1,995,980	1,747,556	1,665,194	1,241,229
Material Cost (Rp) / m2	3,100,303	3,251,110	3,355,271	3,486,958	3,358,108	2,939,583	2,192,319	2,080,071
Construction Cost (Rp) / m2	6,053,612	5,990,359	5,929,989	5,568,555	5,354,088	4,687,139	3,857,513	3,321,300
Construction Cost Change Ratio	1.1%	1.0%	6.5%	4.0%	14.2%	21.5%	16.1%	
Labour Cost Change Ratio	7.8%	6.4%	23.7%	4.3%	14.2%	4.9%	34.2%	
Material Cost Change Ratio	-4.6%	-3.1%	-3.8%	3.8%	14.2%	34.1%	5.4%	

The recapitulation results of the standard analysis of building construction costs per m2 obtained the construction cost change ratio every year in Surabaya for 2013- 2014 increased 16.1%; in 2014 – 2015 raised to 21.5%; in 2015 - 2016 amounted 14.2%; 2016 - 2017 by 4,0%; 2017- 2018 increase 6,5%; 2018- 2019 increased 1%; in 2019 – 2020 raised to 1.1%.
The calculation of the volume of work, analysis of material requirements and labour cost, and analysis of the unit price of the material group data and wages by recapitulation can be seen in table 2 and figure 1.

Table 2. Average component weight of the CCI according to the development of wage rates and material price

year	**2020**	**2019**	**2018**	**2017**	**2016**	**2015**	**2014**	**2013**	**Average**
Labours Cost	**48.8%**	**45.7%**	**43.4%**	**37.4%**	**37.3%**	**37.3%**	**43.2%**	**37.4%**	**41.3%**
Foreman	3.7%	3.6%	3.5%	3.4%	3.5%	3.0%	3.5%	3.0%	3.4%
Labour	45.0%	42.1%	39.9%	34.0%	33.8%	34.3%	39.7%	34.4%	37.9%
Material Cost	**51.2%**	**54.3%**	**56.6%**	**62.6%**	**62.7%**	**62.7%**	**56.8%**	**62.6%**	**58.7%**
Stone Brick	9.4%	9.9%	10.1%	12.2%	12.7%	13.9%	8.3%	9.2%	10.7%
Roof Truss and Cover	12.7%	10.9%	10.8%	12.1%	12.4%	7.8%	7.8%	8.5%	10.4%
Cement	5.6%	7.1%	7.3%	7.9%	7.8%	8.7%	9.2%	10.2%	8.0%
Wood	5.7%	6.3%	6.1%	6.9%	6.8%	7.3%	7.1%	8.5%	6.8%
Paint	3.6%	3.6%	4.2%	4.4%	4.5%	6.6%	5.7%	6.0%	4.8%
Iron Bar	3.4%	5.4%	5.5%	5.6%	5.6%	6.2%	6.1%	6.6%	5.6%
Sand	3.7%	4.1%	4.4%	4.6%	4.1%	3.9%	4.2%	4.6%	4.2%
Floor Cover	1.8%	2.0%	2.6%	2.8%	2.8%	2.7%	3.0%	3.1%	2.6%
Accessories	2.5%	2.4%	2.5%	2.7%	2.6%	2.2%	2.2%	2.3%	2.4%
Complement	2.8%	2.8%	3.0%	3.3%	3.4%	3.5%	3.2%	3.6%	3.2%

Based on the results, the average component weight of the CCI according to the development of price are labour and foreman (41,3%), stone / brick (10,7%), roof truss and cover (10,4%), cement (8%), floor cover and paint (7,4%), wood (6,8%), iron bar (5,6%), accessories and complement (5,6%) and sand (4,2%).

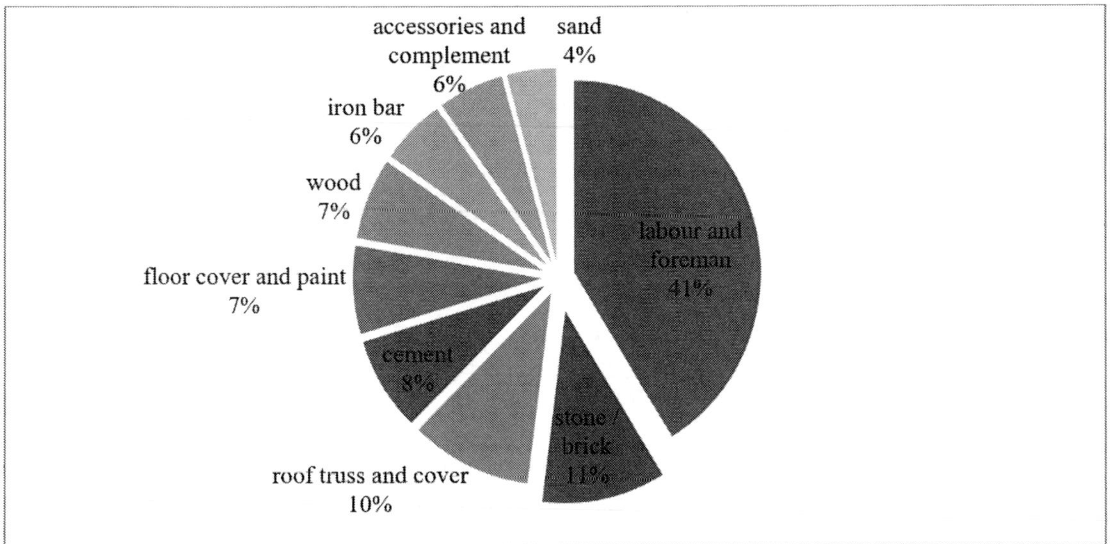

Figure 1. Component weight of the Construction Cost Index (CCI) according to the development of wage rate and material price

Weight based on the total material value (wMM) or the total wage value (wLL) are useful to form a Construction Cost Index (CCI), Material Cost Index (MCI) and Labour Cost Index (LCI), recapitulation results can be shown in table 3 and table 4.

Table 3. Calculation of Material Price Index and Labour Price Index per component per year (2013-2020).

Year	**2020**	**2019**	**2018**	**2017**	**2016**	**2015**	**2014**	**2013**
Labours Cost	**IL**	**IL**	**IL**	**IL**	**IL**	**IL**	**IL**	**IL**
Foreman	2.28	2.18	2.07	1.92	1.85	1.39	1.36	1.00
Labour	2.39	2.21	2.07	1.66	1.59	1.41	1.34	1.00
Material Cost	**IM**	**IM**	**IM**	**IM**	**IM**	**IM**	**IM**	**IM**
Cement	1.00	1.25	.28	1.31	1.23	1.20	1.05	1.00
Wood	1.21	1.34	1.29	1.37	1.29	1.21	0.97	1.00
Floor Cover	1.07	1.14	1.49	1.48	1.42	1.23	1.13	1.00
Paint	1.10	1.07	1.26	1.24	1.21	1.56	1.10	1.00
Accessories	1.96	1.86	1.94	1.96	1.83	1.37	1.12	1.00
Iron Bar	0.95	1.47	1.47	1.42	1.36	1.31	1.06	1.00
Stone Brick	1.85	1.93	1.95	2.22	2.22	2.12	1.04	1.00
Sand	1.49	1.60	1.73	1.68	1.45	1.19	1.05	1.00
Roof Truss	2.73	2.28	2.24	2.31	2.28	1.25	1.05	1.00
Roof Cover	2.73	2.50	2.50	2.91	2.81	1.58	1.18	1.00
Miscellaneous	1.40	1.39	1.51	1.56	1.53	1.36	1.04	1.00

Table 4. Calculation of Construction Cost Index(CCI), Material Cost Index (MCI) and Labour Cost Index per year (2013-2020).

Year	Construction Cost Index (CCI)	Labour Cost Index (LCI)	Material Cost Index (MCI)
2013	1.00	1.00	1.00
2014	1.18	1.34	1.06
2015	1.41	1.41	1.48
2016	1.61	1.61	1.74
2017	1.68	1.68	1.79
2018	1.81	2.07	1.70
2019	1.86	2.21	1.67
2020	1.93	2.38	1.72

Based on the table 4, by using 2013 as the base year with index = 1,00, it can be seen that the material cost index tends to increase every year, except for 2018 and 2019 which has decreased. Whereas labour cost index indicated a constant increase because it was related to the UMK (City Minimum Wage) Surabaya. So that the CCI from 2013 to 2020 has increased from 1,00 to 1,93.

4. Conclusion

Based on the result mentioned above, then obtained some conclusion namely:

1. The average component weight of the CCI according to the development of price are labour and foreman (41,3%), stone / brick (10,7%), roof truss and cover (10,4%), cement (8%), floor cover and paint (7,4%), wood (6,8%), iron bar (5,6%), accessories and complement (5,6%) and sand (4,2%).
2. The calculation of the standard analysis of building construction costs per m2 obtained the construction cost change ratio every year in Surabaya for 2013- 2016 increased average 17.3% per year; start from 2016 - 2020 increased not over 6,5% per year.
3. Material cost index tends to increase every year, except for 2018 and 2019 which has decreased. Whereas labour cost index indicated a constant increase because it was related to the UMK (City Minimum Wage) Surabaya. So that the CCI from 2013 to 2020 has increased from 1,00 to 1,93.

References

[1] Deng *et al.* 2010 *Economic Growth and the Expansion of Urban Land in China* (Urban Studies) **47** 4 813 – 34.

[2] Skitmore *et al.* 2006 Construction price formation: full-cost pricing or neoclassical microeconomic theory? *Journal of Construction Management and Economics* **24** 7 773-83.

[3] Memon *et al.* 2010 Factor Affecting Construction Cost in Mara Large Construction Project : Perspective of Project Management Consultant *International Journal of Sustainable Construction Engineering and Technology* **1** 2 41-54.

[4] Williams T P 1994 Predicting Changes in Construction Cost Indexes using Neural Networks *Journal of Construction Engineering and Management* **120**(2) 306-20.

[5] Hwang S 2009 Dynamic Regression Models for Prediction of Construction Cost *Journal of Construction Engineering and Management* **135** 360-75.

[6] Wibowo *et al.* 2013 Menuju Indeks Biaya Konstruksi Rumah Sejahtera Murah (IBK-RSM) *Proceeding of Seminar Nasional IX-2013* Civil Engineering ITS, Surabaya.

[7] Pratt J W 1987 Dividing the Indivisible: Using Simple Symmetry to Partition Variance Explained *Proceeding of the 2nd International Tampere Conference* 245-60.

[8] Libianto *et al.* 2016 The Development of the Construction Cost Indices of Low-Cost Housing in Several Regions in Indonesia *Media Teknik Sipil, ISSN 1693-3095* **14** 2 215-26 http://ejournal.umm.ac.id/index.php/jmts/article/view/3712

[9] Yu M K W and Ive G 2008 The Compilation Methods of Building Prices in Britain : a Critical Review *Journal of Construction and Economics* **26** 693-705.
[10] Eurostat 2006 *Methodology of Short-term Business Statistics : Interpretation and Guidelines.*
[11] Soeharto I 1997 *Manajemen Proyek Dari Konseptual Sampai Operasional* Erlangga, Jakarta.
[12] Santosa P *et al.* 2007 *Statistika Deskriptif dalam Bidang Ekonomi dan Niaga* Erlangga, Jakarta.

Weighting variables for building performance evaluation

L N S Simbolon[1*], I P A Wiguna[1], T J W Adi[1]

[1]Faculty of Civil, Planning, and Geo Engineering, Institut Teknologi Sepuluh Nopember, Keputih, Sukolilo, Surabaya, Indonesia

[*]Corresponding author's e-mail: tri_joko@ce.its.ac.id

Abstract. Building performance appraisal needs to be done as a form of evaluation, so that the level of usability of a building can be seen at the operation and maintenance stages. The purpose of this study is to give weight to the variables used in evaluating the performance of a building. In this study, the identification of building performance variables is based on literature studies and semi-structured interviews from the experts of the field. The determined variables are then filtered using the Relative Important Index (RII) method, and weighted using the Pairwise Comparison method. All this process is conducted based on the opinion of experts who are competent in the field of building, and all stages are conducted by questionnaire. The results of the filtering and weighting of variables produce 5 (five) variables and 24 (twenty four) sub-variables used in building performance evaluation, with the results of weighting for safety variable with a weight of 59.2%, health variable with a weight of 24.6%, comfort variable with a weight of 10.2%, ease variable with a weight of 4.3%, and utility variable with a weight of 1.8%.
Keywords: infrastructure management, variable weighting, building performance evaluation

1. Introduction

The construction of high rise buildings in Indonesia is increasing from year to year. Increased construction of multi-storey buildings occurs due to increasing population conditions, economy, and reduction of land area. Building construction was also done to support government programs such as the master plan for accelerating economic development according to Regulation of the Minister of Public Works and Public Housing No. 22 / PRT.M / 2018 [1].

The role of buildings is very important in the development of a country's progress because according to Regulation of the Minister of Public Works No. 24/PRT/M/2008 [2] regarding Guidelines for Building Maintenance and Repairs, the function of the building is as a place for humans to carry out their activities, as a place to both live, religious activities, business activities, social activities, culture, as well as special activities and general activities.

Based on Law of the Republic of Indonesia No. 28 of 2002 regarding Buildings [3], the definition of a building is a physical form of the results of construction work that is integrated with the place of domicile, partly or wholly on top and/or in land and/or water, which functions as a place humans do their activities, both for residential, business activities, social activities, culture, and for special activities.

Content from this work may be used under the terms of the Creative Commons Attribution 3.0 licence. Any further distribution of this work must maintain attribution to the author(s) and the title of the work, journal citation and DOI.
Published under licence by IOP Publishing Ltd

To maintain the function of the building as an asset, proper infrastructure asset management is needed. Infrastructure asset management is a program or knowledge to manage an infrastructure so that it can continue to function properly, continuously as long as it is still needed, economically, efficiently, effectively, and meet the principles of green or sustainability [4]. Infrastructure asset management must be based on good knowledge of the characteristics of the infrastructure being managed or discussed because infrastructure characteristics can be very different from one to another.

One of the stages in asset management is the evaluation of the performance of assets. In buildings, performance can be measured from 4 (four) requirements, namely: functional requirements (related to building functions), performance requirements (related to the physical performance of buildings and complete installation of facilities and infrastructure), legality requirements and compliance with regulations, and user requirements (related to the convenience and costs that users must spend [5].

The government provides rules regarding the construction of buildings to control land use, spatial planning, and impact on the environment and humans. Law of the Republic of Indonesia No. 28/2002 is a government regulation regarding building construction, the implementation of which is regulated in Government Regulation of the Republic of Indonesia No. 36 of 2005 [6]. The law states that construction of buildings must meet administrative and technical requirements. One of the administrative requirements of a building is a building permit and the building technical requirements in the form of building reliability which includes aspects of safety, health, comfort and ease.

The Ministry of Public Works and Public Housing is one of the Ministries that has the task of managing state assets under the supervision of the Ministry of Finance as the manager of state assets.

In accordance with Presidential Regulation No. 15 of 2015, the authority for building infrastructure construction is the Ministry of Public Works and Public Housing, assigned to an Echelon I (one) level organizational unit within the Ministry of Public Works And Public Housing, namely the Directorate General of Human Settlements (DG of Human Settlements). One of the assets that is the responsibility of the Ministry of Public Works and Public Housing, through DG of Human Settlements, to manage properly is buildings [7].

The number of buildings in the Ministry of Public Works and Public Housing office is approximately 10 (ten) units, one of which is the Directorate General of Human Settlements building and the Directorate General of Highways building. The two buildings are two of the oldest buildings in the Ministry of Public Works and Public Housing. From 2010 to 2015, many new and higher buildings were built around the buildings with the latest and best standards, and the main building (Ministerial Building) is one of the newest examples of buildings, with green building standards and utility standards as well as supporting facilities and infrastructure. Therefore, it is better and more complete.

In Indonesia, one of the basic regulations used as a reference to determine the performance evaluation variable of buildings is the Regulation of the Minister of Public Works and Public Housing No. 27/PRT/2018 regarding Certificate of Worthiness of Building Function [8]. In this Ministerial Regulation, it is explained that there are several variables that can be used as a requirement for assessing building performance, in terms of building administrative requirements, technical requirements, building layout requirements, and building reliability.

Having been established for a long period of time, the Directorate General of Human Settlements building and the Directorate General of Highways building will naturally experience a decline in performance. This decline in performance can have an impact on the operation and maintenance of the buildings. In terms of operations, it will hamper the main tasks and functions of the buildings as an office building. While in terms of maintenance, it will need to be done in the form of routine maintenance (daily), monthly or yearly to maintain building performance in operational terms [9]

Therefore, it is important to identify the building performance variables, specifically the Directorate General of Human Settlements Building and the Directorate General of Highways Building. This study aims to weight each variable in evaluating a building by using performance evaluation variables in terms of building reliability.

2. Literature Review

Research on building performance has been carried out by many researchers. Antonius & Poedjiastoeti [10] conducted research on evaluating environmental conditions and structures in public buildings (with a case study: Performance of Two Buildings in Semarang). The research method used is descriptive analysis on As Built Drawings. The research aims to evaluate environmental conditions and structures in buildings.

Selowati & Indryani [11] conducted a study with the topic of the feasibility analysis of *The Via and The Vue Surabaya* apartments in terms of architecture. This descriptive analysis method research aims to determine the performance of buildings related to the demands of comfort and health of the buildings as well as matters that are at high risk to human safety based on applicable standards and regulations.

Research that evaluates building damage was carried out by Ery [12]. This research on the topic of determining the performance variables of educational building used the analytical hierarchy process method. The purpose of the study was to assess the damage to the Politeknik Pos, Politeknik Negeri Bandung, and Politeknik Manufaktur buildings; both through visual inspection and testing, as well as non-destructive testing for latent damage. From this study, the weight of each building component can be calculated which is reflected in the performance requirements of building reliability, such as security, safety, health, comfort and accessibility of building requirements.

This study has similarities with the previous studies which were evaluating the performance of buildings. However, in this study, the scope of building performance evaluation only addresses the technical requirements of buildings in terms of reliability.

In Indonesia, the provisions governing buildings are regulated in several statutory regulations, including the Law of the Republic of Indonesia, and also the Minister of Public Works and Public Housing Regulation. Starting from the Law of the Republic of Indonesia Number 28 of 2002 regulating Buildings, Regulation of the Minister of Public Works Number 24 / PRT / M / 2008 on Guidelines for Building Maintenance and Repair, Regulation of the Minister of Public Works Number 45 / PRT / M / 2007 on Technical Guidelines for Building Construction of State Building [13], Regulation of the Minister of Public Works Number 05 / PRT / M / 2007 on the Technical Guidelines for High-Rise Simple Flats [14], Regulation of the Minister of Public Works No. 30 / PRT / M / 2006 on the Technical Guidelines for Facilities and Accessibility in Buildings and Environments [15].

In measuring the performance of building infrastructure, there are many variables that can be used from various literature studies. However, this research only uses the Minister of Public Works and Public Housing Regulation No.27 / PRT / M / 2018 because this regulation is the latest regulation issued by the Ministry of Public Works and Public Housing as a guide in regulating building performance. Among the sub-variables that will be used, if there are almost similar sub-variables, they will be summarized into one sub-variable that is considered representative. Considering previous research, the variables and sub-variables that are relevant to be applied in Indonesia as shown in table 1.

Table 1. Building Performance Variables in Indonesia

No	Variables	Sub-Variables	Source
1	Allotment of buildings	Conformity of building function with designation in regional spatial planning or building and environmental plannings	Regulation of the Minister of Public Works and Public Housing of the Republic of Indonesia Number 27 / PRT / M / 2018
2	The intensity of the building	Building density Building height Clearance of buildings	Regulation of the Minister of Public Works and Public Housing of the Republic of Indonesia Number 27 / PRT / M / 2018
3	Building architecture	Appearance of buildings Inner space layout Balance Harmony The harmony of the building with its environment	Regulation of the Minister of Public Works and Public Housing of the Republic of Indonesia Number 27 / PRT / M / 2018
4	Control of environmental impacts	Environmental permits for buildings in accordance with statutory provisions	Regulation of the Minister of Public Works and Public Housing of the Republic of Indonesia Number 27 / PRT / M / 2018
5	Safety	The structure of the building Fire hazard protection Lightning rod Safety and reliability of electrical installations for buildings that are equipped with electrical installations Disaster safety of public interest buildings	Regulation of the Minister of Public Works and Public Housing of the Republic of Indonesia Number 27 / PRT / M / 2018
6	Health	Ventilation system Lighting system Clean water system	Regulation of the Minister of Public Works and Public

No	Variables	Sub-Variables	Source
		Sewerage system and/or wastewater Sewage and garbage disposal system Rainwater distribution system Use of building materials	Housing Republic of Indonesia Number 27 / PRT / M / 2018
7	Comfort	Comfort of wiggle room Comfort of air conditions in the room Comfort of the view Comfort from vibration and noise	Regulation of the Minister of Public Works and Public Housing of the Republic of Indonesia Number 27 / PRT / M / 2018
8	Ease	Ease of connection to, from, and inside buildings (means of horizontal relationships between spaces / between buildings vertical connection between floors in buildings and evacuation facilities) Completeness of infrastructure and facilities for building utilization	Regulation of the Minister of Public Works and Public Housing of the Republic of Indonesia Number 27 / PRT / M / 2018

In determining building performance variables and sub-variables, several countries use certain approaches and perspectives in accordance with the objectives to be achieved by building operators in the country or region. According to Daniel et al [16], Hong Kong uses 5 variables in building performance measurements consisting of structural systems, closure systems, environmental modification systems, protection systems, and utility systems which are measured through several sub-variables on each variable.

In the Minister of Public Works and Public Housing Regulation No.27/2018, there are no variables that assess the building's performance in terms of the utility system, even though the building cannot be separated from the utility system itself. On the other hand, for other systems such as structural systems, it has been regulated in the Minister of Public Works Regulation No.45 of 2007 concerning Technical Guidelines for the Construction of State Buildings in the section on building structure requirements. The closure system is regulated in the Minister of Public Works Regulation No.24/2008 concerning Guidelines for Building Maintenance and Repairs in the house keeping requirement section. Environmental modification system and protection system have been regulated in the Minister of Public Works and Public Housing Regulation No.27/2018 concerning Building Worthy Certificate of Building in the comfort and safety requirement sections.

In this research, variable synthesis is carried out by simplifying and grouping similar variables and sub-variables. In general, this study uses 5 (five) variables, where 4 (four) variables, namely safety, health, comfort, and ease are taken from the Minister of Public Works and Public Housing Regulation No.27 of 2018, and the 5th variable is utility system which is taken from Daniel et al. [16]. This is done because this research only evaluates building performance in terms of reliability.

3. Research Method

To identify building performance variables within the Ministry of Public Works and Public Housing, variables that affect the buildings' performance evaluation appropriately need to be determined. In Indonesia, one of the basic regulations used as a reference for determining building performance

evaluation variables is the Regulation of the Minister of Public Works and Public Housing No. 27 / PRT / 2018 regarding Certificate of Worthiness of Building Function. It is explained that there are several variables that can be used as a requirement to evaluate building performance, from both in terms of building administrative requirements, technical requirements, building layout requirements, and in terms of building reliability.
The evaluation related to building performance, either private or government buildings, has been reviewed by several researchers. The assessment variables used are also diverse, some of which examine all aspects of structure, architecture, accessibility, and building, and environmental governance. However, there are also those who only evaluate from one aspect.

The method used in the previous research included a descriptive analysis that was considered less accurate in determining influential variables or sub-variables. In other studies, multidimensional scaling and interpretive structural modelling which are quite difficult to implement were used. As for this research, the method includes identifying variables which is done by literature studies and preliminary survey. Literature studies are taken from the rules and previous research results, in which there are variables and sub-variables needed in research. Meanwhile, the preliminary survey is carried out by discussion and consultation of experts who are considered competent in the field of building to ensure the suitability of the variables and sub-variables used in accordance with the research objectives. The next stage is distributing questionnaires, to filter variables that are used. This is done to ensure that the sub-variables used are really important in the opinion of respondents who are experts in the field of building, using the relative important index (RII). This method was chosen because from the results of the questionnaire collected, it can be analysed using statistical tests and RII at the same time, making it more practical in data collection and more comprehensive in the determination of sub-variables. The weighting method uses a paired comparison questionnaire which is one of the stages in the Analytical Hierarchy Process (AHP).

The pairwise comparison method is used because this research does not determine alternatives such as the AHP, but only determines the importance weight. On the other hand, not everyone understands and understands the selection of variables and sub-variables used in this study, so that the respondents chosen are organizers in the field of building construction, in this case the Directorate of Building Management as an institution that has the task of carrying out the formulation and implementation of policies, preparation of technical plans, and implementation of building the buildings. Based on this consideration, the pairwise comparison method is considered more appropriate in determining the weights. This is also in accordance with the opinion of Sarachaga [17] which states that AHP is used because of its simplicity and flexibility when combined with other methods.

Population in this study is officials who are still active at the structural level of Directorate of Building and Neighbourhood Development, Directorate General of Human Settlements, officers at the General Affairs Division of Secretariat of Directorate General of Human Settlements, and officers at the General Affairs Division of Secretariat of Directorate General of Highways) in the Ministry of Public Works and Public Housing.

This study uses non-probability sampling, which is a technique that does not provide equal opportunity for each element or member of the population to be selected as a sample. The method used is purposive sampling which is a sampling technique with certain considerations and cannot be given to everyone randomly. This is because not everyone understands the selection of variables and sub-variables used in this study.

4. Research Analysis

4.1. Identification of Building Performance Evaluation Variables

This study uses research variables in the form of variables and sub-variables taken based on literature studies particularly the Regulation of the Minister of Public Works and Public Housing No. 27 / PRT / 2018 regarding Certificate of Worthiness of Building Function, previous research by adding 1 (one) building performance variable, namely utility system, and preliminary surveys that have been carried out in the previous stages. All variables and sub-variables that are considered relevant for building performance measurements are identified and refined to be used as variables and sub-variables based on table 2.

Table 2. Variables and Sub-Variables

Variable	Code	Sub-Variable
Safety Aspects (X1)	X1.1	Strong building structure
	X1.2	Fire hazard protection
	X1.3	Lightning rod
	X1.4	Safety and reliability of electrical installations for buildings that are equipped with electrical installations
	X1.5	Building disaster security
Health Aspects (X2)	X2.1	Ventilation system
	X2.2	Lighting system
	X2.3	Clean water system
	X2.4	Dirty and/or waste water disposal system
	X2.5	Sewage and sewage system
	X2.6	Rainwater delivery system
Comfort Aspects (X3)	X3.1	Comfort of space
	X3.2	Comfort of air condition in the room
	X3.3	View comfort
	X3.4	Vibration and noise comfort
Ease Aspect (X4)	X4.1	Ease of relationship to, from, and inside buildings (means of horizontal relations between spaces / between buildings and facilities of vertical connections between floors in buildings)
	X4.2	Evacuation facilities
	X4.3	Completeness of infrastructure and facilities for building utilization
Utility Aspects (X5)	X5.1	Underground and aboveground piping system
	X5.2	Underground and aboveground drainage system
	X5.3	Low and high volt electrical installations
	X5.4	Elevator
	X5.5	Gas installation
	X5.6	Other large installations (Waste Water Treatment Plant/WWTP)

4.1.1. Variables Filtering

Variables filtering are done to eliminate variables in the form of sub-variables that are considered not important for the purpose of the assessment by the Relative Important Index (RII) method. With variable filtering, it is expected that the sub-variables that will be used can measure the performance of buildings properly. Decision making in RII is based on a minimum value of ≥ 0.750 so that a variable and sub-variable are considered important and influential. The summary of the questionnaire variable filtering based on survey can be seen in table 3 as follows:

IOP Conf. Series: Materials Science and Engineering **930** (2020) 012018 doi:10.1088/1757-899X/930/1/012018

Table 3. Variables Filtering

Variables	Code	Total	Resp.	Score	Note
Safety	X1	57	13	0.8769	Used
Health	X2	53	13	0.8154	Used
Comfort	X3	55	13	0.8462	Used
Ease	X4	52	13	0.8000	Used
Utility	X5	55	13	0.8462	Used

Based on the table above, it can be seen that all variables can be used in the next analysis, namely weighting variables. The results of sub-variable filtering based on the questionnaire can be seen in table 4 as below:

Table 4. Sub-Variables Filtering

Sub-Variables	Code	Total	Resp.	RII	Decision
Safety Aspects (X1)					
Strong building structure	X1.1	56	13	0.86	Used
Fire hazard protection	X1.2	54	13	0.83	Used
Lightning rod	X1.3	50	13	0.76	Used
Safety and reliability of electrical installations for buildings that are equipped with electrical installations	X1.4	55	13	0.84	Used
Building disaster security	X1.5	52	13	0.80	Used
Health Aspects (X2)					
Ventilation system	X2.1	56	13	0.86	Used
Lighting system	X2.2	55	13	0.84	Used
Clean water system	X2.3	61	13	0.93	Used
Dirty and / or waste water disposal system	X2.4	60	13	0.92	Used
Sewage system	X2.5	60	13	0.92	Used
Rainwater delivery system	X2.6	56	13	0.86	Used
Comfort Aspects (X3)					
Comfort of space	X3.1	55	13	0.84	Used
Comfort of air condition in the room	X3.2	58	13	0.89	Used
View comfort	X3.3	50	13	0.76	Used
Vibration and noise comfort	X3.4	53	13	0.82	Used
Ease Aspect (X4)					
Ease of relationship to, from, and inside buildings (means of horizontal relations between spaces / between buildings and facilities of vertical connections between floors in buildings)	X4.1	56	13	0.86	Used
Evacuation facilities	X4.2	56	13	0.86	Used
Completeness of infrastructure and facilities for building utilization	X4.3	54	13	0.83	Used
Utility Aspects (X5)					
Underground and aboveground piping system	X5.1	54	13	0.83	Used
Underground and aboveground drainage system	X5.2	55	13	0.84	Used
Low and high volt electrical installations	X5.3	56	13	0.86	Used
Elevator	X5.4	54	13	0.83	Used

Gas installation	X5.5	53	13	0.81	Used

4.1.2. Weighting Variable Results

After 24 sub-variables are selected, a paired comparison questionnaire will be re-arranged to determine the importance of the variables and sub-variables. Taking into account the opinions of experts through a weighting questionnaire given to 13 (thirteen) respondents at the Directorate General of Human Settlements and Directorate General of Highways, the research obtains weights of 5 (five) variables, the output of which can be seen in the following table:

Table 5. Weighting Variables

Variables	**Weight (%)**
Safety	59.2%
Health	24.6%
Comfort	10.2%
Ease	4.3%
Utility	1.8%
Total	**100%**

After obtaining the results of weighting variables as the table above, to make it easier to visually see the results of the weighting of variables, the histogram has been made as follows:

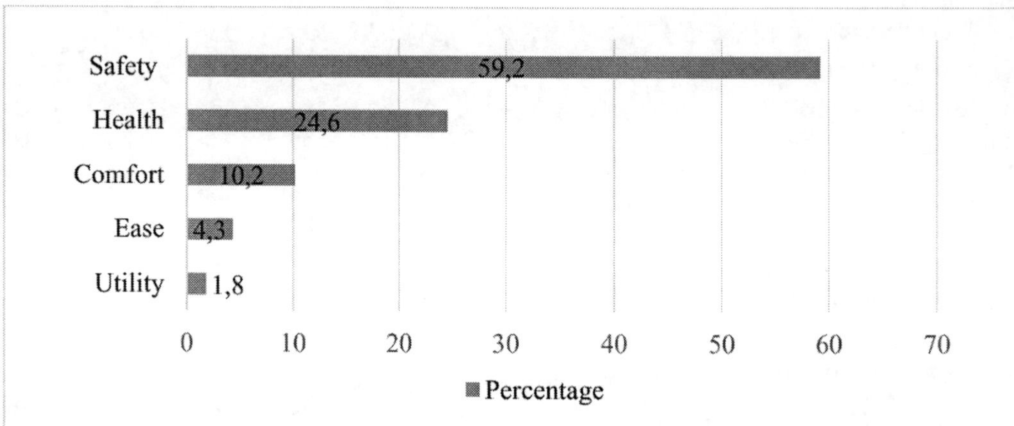

Figure 1. Weighting Variable Results

The safety variable gets the highest weight by 59%. This is because one of the main and important things of the operation and maintenance of a building is safety. The safety variable can be fulfilled well if the health variable can be fulfilled well, therefore the health variable gets the second highest weight of 24.7%. The operation and maintenance of a building is certain to have an impact on the health of the occupants of the buildings around the Directorate General of Human Settlements building and the Directorate General of Highways building within the Ministry of Public Works and Housing. Furthermore, the comfort variable is in the third place at 10.2%. In the operation and maintenance of a healthy building, it can naturally create comfort.

In the fourth place is the ease variable with a weight of 4.3%. Although this criterion is considered important if it is associated with the evaluation of a building, but still less attention is given if viewed from the weight obtained. Meanwhile, the smallest weight is the utility variable, which is needed to ensure the operation and maintenance of a building that meets the applicable regulations and standards, with a weight of 1.8%. With the weight of the five proposed variables, it can be seen that the safety variable is the main variable, since it is the goal of a building infrastructure management, including in building operation and maintenance. Meanwhile, the health and comfort variables still need to be increased again because based on the weight of importance gained in these variables, they do not get much attention from the building manager. The paradigm to improve the health and comfort variables in the provision of building infrastructure must be able to be instilled to all stakeholders, especially the building administrators. Thus, the five proposed variables will get a balanced weight which ranges from ± 20% percentage.

4.1.3. Weighting Sub-Variable Results

After weighting the variables, weighting is done for each sub-variable. The recap of weighting sub-variables can be seen in table 6 as follows:

Table 6. Sub-Variables Weighting

Sub-Variables	Weight (%)
Safety Aspects (X1)	
Strong building structure	28.3%
Fire hazard protection	24.1%
Lightning rod	9.1%
Safety and reliability of electrical installations for buildings that are equipped with electrical installations	17.6%
Building disaster security	20.9%
Health Aspects (X2)	
Ventilation system	15.6%
Lighting system	9.4%
Clean water system	33.1%
Dirty and / or waste water disposal system	19.7%
Sewage and sewage system	12.2%
Rainwater delivery system	10.0%
Comfort Aspects (X3)	
Comfort of space	35.7%
Comfort of air condition in the room	37.6%
View comfort	8.5%
Vibration and noise comfort	18.2%
Ease Aspect (X4)	
Ease of relationship to, from, and inside buildings (means of horizontal relations between spaces / between buildings and facilities of vertical connections between floors in buildings)	27.8%
Evacuation facilities	29.9%
Completeness of infrastructure and facilities for building utilization	42.3%
Utility Aspects (X5)	
Underground and aboveground piping system	7.6%
Underground and aboveground drainage system	18.3%
Low and high volt electrical installations	28.9%
Elevator Installation	20.0%
Gas installation	12.3%
Installation of Waste Water Treatment Plant (WWTP)	12.9%

Based on table 7 above, conclusions can be drawn as follows:

1. In the safety sub-variable, it is known that the highest weight of 28.3% is obtained by the sub-structure of the building structure. This is in accordance with the objective of the operation and maintenance of a building infrastructure that is maintaining the structure of a building, even though the maintenance of a building does not have too large effect when compared to the construction of new buildings or building upgrading or reconstruction.
2. The clean water system sub-variable has the highest weight of 33.1%, which means that the clean water system aspect has the highest importance compared to other sub-variables proposed to measure health variables in buildings that get in the operational and maintenance stages. The clean water system is related to the source or distribution system, the quality and flow of clean / drinking water. Clean water systems are negatively correlated with water shortages, meaning that the better the availability of clean water systems in a building, the better the service level of the building.
3. The comfort of air condition in space variable has the highest weight of 37.6% compared to the other 3 sub-variables proposed. This can be interpreted that according to experts, the comfort of air conditions in the room is considered important to be considered a priority, because this is related to temperature and humidity in the room.
4. The highest weight was obtained by the sub-variable of completeness of facilities and infrastructure of building utilization with a weight of 42.3%. Complete facilities and infrastructure for building utilization are the facilities that are used to facilitate building occupants, such as toilets, parking facilities, worship rooms, lactation rooms, trash bins, communication systems, information systems, and other equipment.
5. Electrical installation sub-variable gets the highest weight by 28.9%. This is because all facilities available in buildings generally use electricity, therefore, electrical installations need to be installed correctly, both low and high voltage because it is one of the sub- variables that can be dangerous for the building and its inhabitants.

5. Conclusion

After observing in the field and analysing the data, the following conclusions are obtained:

1. Through the method of relative importance index, the evaluation of the performance of buildings in the Ministry of Public Works and Public Housing produced 5 (five) filtered variables, including safety variables, health variables, comfort variables, convenience variables and finally utility variables, with 24 (twenty four)) sub-variables.
2. The variable weight of buildings in the Ministry of Public Works and Public Housing consists of a safety variable with a weight of 59.2%, a health variable with a weight of 24.6%, a comfort variable with a weight of 10.2%, a convenience variable with a weight of 4.3%, and utility variable with a weight of 1.8%. Furthermore, the variable results and weights obtained in this study can be used to assess the level of building performance, especially in the Ministry of Public Works and Public Housing.

References

[1] Regulation of the Minister of Public Works and Public Housing Republic Indonesia Number 22 / PRT.M 2018 *Construction of State Buildings*.
[2] Regulation of the Minister of Public Works Republic Indonesia Number 24 PRT.M 2008 *Guidelines for Building Maintenance and Repairs*.
[3] Law of The Republic Indonesia Number 28 2002 *Buildings*.
[4] Suprayitno H & Soemitro R A A 2019 Reflection on basic view of public infrastructure for infrastructure asset management in Indonesia *J. of Infrastructure & Facilities Asset Management* **3** Supplement 1 15-23.
[5] Son L H & George C S Y 2002 *Building Maintenance Technology* (Hong Kong).
[6] Government Regulation of the Republic Indonesia No. 36 2005 *Buildings*.
[7] Presidential Regulation of The Republic Indonesia Number 15 2015 *the Ministry of Public Works and Public Housing*.

IOP Conf. Series: Materials Science and Engineering **930** (2020) 012018 doi:10.1088/1757-899X/930/1/012018

[8] Regulation of the Minister of Public Works and Public Housing No. 27 / PRT 2018 *Certificate of Worthiness of Building Function.*
[9] Regulation of the Minister of Public Works Republic Indonesia Number 16 / PRT.M 2010 *Technical Guidelines for Periodic Building Inspections.*
[10] Antonius & Hermin P 2011 Evaluasi kondisi lingkungan dan struktur pada bangunan publik *Proc. Seminar Nasional Penelitian Masalah Lingkungan di Indonesia, eds Priana Sudjono, I Wayan Arthana, Nieke Karnaningroem* 309-18.
[11] Selowati & Indryani 2012 Analisa kelayakan huni apartemen The Via and The Vue Surabaya dari segi arsitektural *J. Teknik Pomits* **1** 2.
[12] Ery RJ, Raden Roro EM, and Yullianty N 2015 Penentuan indikator kinerja bangunan pendidikan *J. Fondasi* **4** 2.
[13] Regulation of the Minister of Public Works Number 45 / PRT.M 2007 *Technical Guidelines for Building Construction of State Building.*
[14] Regulation of the Minister of Public Works Number 05 / PRT.M 2007 *Technical Guidelines for High-Rise Simple Flats.*
[15] Regulation of the Minister of Public Works No. 30 / PRT.M 2006 *Technical Guidelines for Facilities and Accessibility in Buildings and Environments.*
[16] Daniel CWH, Lo SM, and Edward C Y Y 2008 *The First Guide to Preparing a Building Maintenance and Manual (BMM)* (Hongkong).
[17] Sarachaga J M D, Daniel J E, Alsulami B, and Daniel C F 2016 Evaluation of existing sustainable infrastructure rating systems for their applications in developing countries *J. Ecological Indicator* **71** 491-502.

Study of partnership form to improve idle land based on risk management

B Y Prawiro[1]*

[1]Civil Engineering Department, Engineering Faculty, Universitas Indonesia, Depok, 16424, Indonesia

*Corresponding author's e-mail: bimo.yudha@ui.ac.id

Abstract. Assets for the company are important capital in running the company's business. One form of company assets is land. Land assets are immovable assets provided in the Company's Fixed Assets as long as they are not traded. Generally, land assets can be valuable if it's sold or developed. If developed, land assets can contribute productively to the company, but conversely if these land assets are not developed (idle) will contribute non-productive to the company. Development of land assets can be done by companies itself or partnership with partners. Land assets contributing to non-productive asset will causes some problem such as legal problems and the cost of securing land assets. Legal problems will result in legal costs incurred to solve it. While the costs of securing land assets will continue to arise as long as the land assets are still attached to the Company or have not yet been developed. The purpose of this study is to response those problem. The writer examines the form of cooperation that can be carried out in accordance with regulations, dominant risk variables according to partnership, and then conducts case studies that will encourage risk-based forms of partnership.
Keywords: Asset, Land Asset, Partnership, Risk Management

1. Introduction

Assets are an important capital of a company. Assets are goods or things that have economic value, commercial value or exchange value owned by a business entity, agency or individual. Assets must be managed properly so that productivity can lift the company both in terms of the company's name or in terms of the company's finances because assets are a possible economic benefit obtained or controlled in the future by certain institutions as a result of past transactions or events (Financial Accounting Standard Board (FASB), 1985). In Indonesia has many regulations about asset. One of form asset is land.

Land Assets are generally owned by Companies / Agencies / Agencies both State Owned Enterprises (BUMN) and Private. In the course of assets this form of land can be optimized by developing, selling, or partnership. The Guidelines for Utilizing the Utilization of Fixed Assets of State-Owned Enterprises have been established by the Minister of State-Owned Enterprises (SOE) with the number PER-13 / MBU / 09/2014. In the introduction of this regulation, state-owned enterprises in carrying out their business activities have fixed assets that are less productive and have not been utilized or are not optimally utilized. This can be seen from the Return on Assets (ROA) of SOEs that are still low.

Content from this work may be used under the terms of the Creative Commons Attribution 3.0 licence. Any further distribution of this work must maintain attribution to the author(s) and the title of the work, journal citation and DOI.
Published under licence by IOP Publishing Ltd

Company X is a SOE engaged in the housing sector. Company X has (eight) 8 Region which spread around in Indonesia. In carrying out its business, Company X is equipped with land assets to be utilized. But in the course of Company X there are still undeveloped land assets that are burdening the Company X. in this situation land assets are categorized as non-productive assets. List of Land Productive and Non-Productive is as shown in table 1:

Table 1. Company X's Land Asset

Region	Productive (m^2)	Non-Productive (m^2)	Total (m^2)
Region-1	1.371.564	159.014	1.530.578
Region-2	1.462.910	308.149	1.771.059
Region-3	3.559.851	4.037.963	7.597.814
Region-4	1.700.493	4.299	1.704.792
Region-5	650.149	375.288	1.025.437
Region-6	71.590	83.582	155.172
Region-7	1.601.869	323.151	1.925.020
Reigion-8	30.919		30.919
Total (m^2)	**10.449.345**	**5.291.446**	**15.740.791**
Total (Ha)	**1.045**	**529**	**1.574**
%	**66%**	**34%**	**100%**

Base on table 1 it shown from 8 Regions, Company X has non-productive land asset ± 529 Hectare from total 1.574 Hectare. In this situation, 34% Company X's land asset is not productive yet.

Regarding to Company X's non-productive land asset, its potentially causes problem such as legal problem and cost of securing land asset. Based on data Company X, legal problem and cost of securing land asset are increasing every year. It shown in table 2 as follows:

Table 2. Legal and Securing Land Asset Cost

Description	Year				
	2015 (Million IDR)	2016 (Million IDR)	2017 (Million IDR)	2018 (Million IDR)	2019 (Million IDR)
Legal Problem Cost	16.726	21.801	22.891	23.806	24.520
Securing Land Asset Cost	328	410	432	540	621

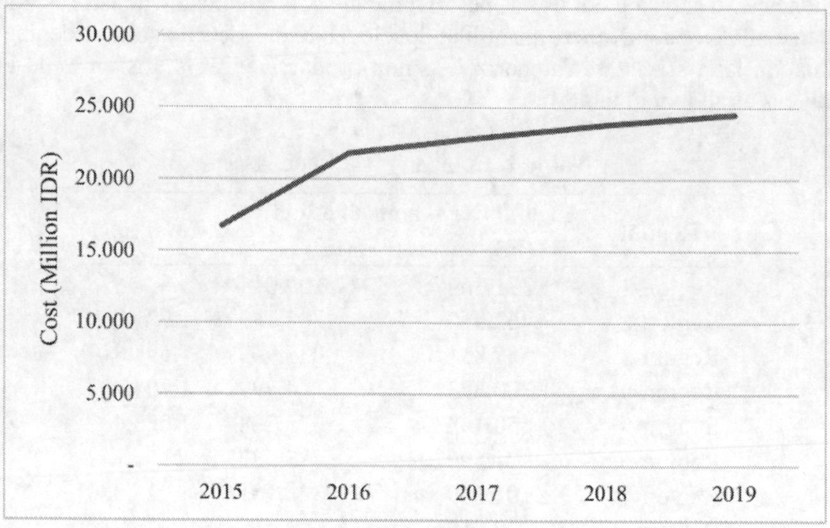

Figure 1. Legal Problem Cost

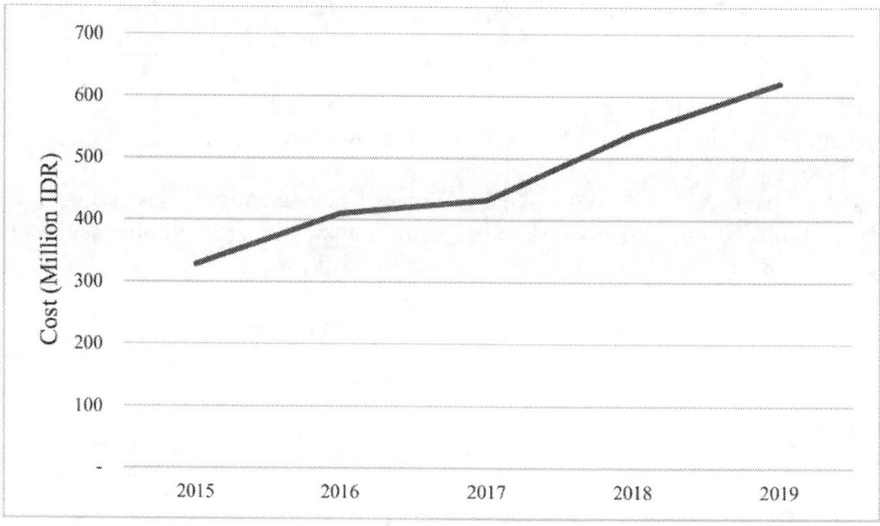

Figure 2. Securing Cost

Base on table 2, figure 1, and figure 2, these issues / problems take cost which increase every year. This condition will impact to Company X if not solve yet. In this research, writer try to identify form of partnership which applicable for Company X based on regulation, and then identify risk variable which probably shown up in partnership based on interview expert, and applied to study case in on of non-productive land Company X.

2. Literature Review

2.1 Public Private Partnership

In a book entitled Public-Private Partnership of Policy and Finance, E.R. Yescombe defines a Public-Private Partnership as a form of cooperation between the government as a public and private party as a private party with the following key elements:

1. Long-term contracts that occur between the government (public) and private (private)
2. For design, construction, financing, and operations carried out by the private sector.

3. Payments during the term of the PPP contract to the private party are carried out by the government and users directly as compensation for the use of infrastructure facilities.
4. Transfer of ownership from the private party to the government at the end of the PPP contract.

With this Public Private Partnership is expected to be able to utilize existing funding effectively and efficiently. Partnership has important meaning for both government and private parties where the horses of these parties will transfer the risks faced. There are two basic theories that underlie the first Public-Private Partnership is the Agency Theory which is based on the transfer of ownership and information rights that are explained in the organizational contract. The most important thing in this theory is the selection of agents according to the criteria of the main tasks and functions. The second theory is the Transaction Cost Theory. This theory focuses on contracts, especially in terms of financial structure, monitor, bonding and residual costs.

2.2 Public Private Partnership Model

According to UNESCAP (2007) the PPP model is clarified into five global categories which generally (not always) increase the involvement and assumptions of the private sector risk. These five categories are:

1. Provision and Management Contract
2. Turnkey
3. Rental
4. Concession
5. Ownership of private assets

The basis of the five global categories of the PPP model is shown in figure 3. Each model has pros and cons and can be suitable for achieving some of the goals of private participation. Specific characteristics of some sectors and their technological development, legal and regulatory regimes, and public and political perceptions about services in the sector can also be a factor in determining the suitability of specific forms of private participation.

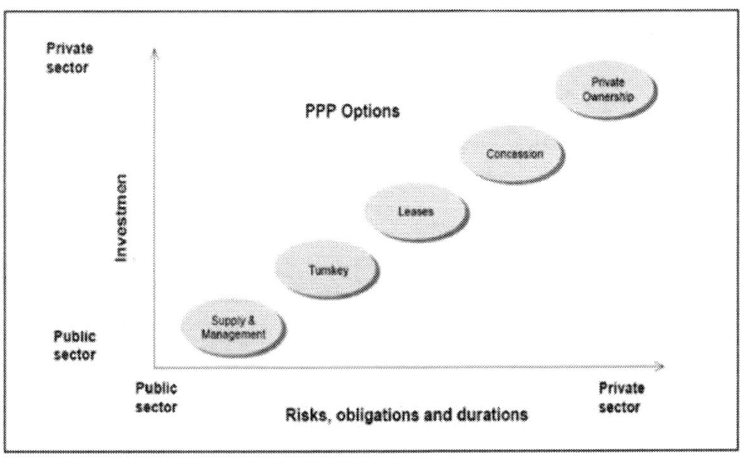

Figure 3. PPP Model [4]

2.3 Regulation

According to the understanding of the Regulation of the Minister of State-Owned Enterprises Number: PER-13 / MBU / 09/2014 Regarding the Guidelines for Utilization of Fixed Assets of State-Owned Enterprises explains that in order to optimize the utilization of fixed assets and at the same time as an effort to improve the performance and value of the company, need to make use of fixed assets. The

utilization of fixed assets must be based on the principles of good corporate governance and the principle of benefits.
In this regulation regulates how the utilization of fixed assets owned by the government (public sector) by:

1. Build Operate Transfer / BGS
2. Build Transfer Operate / BSG
3. Join Operation / KSO
4. Join Venture / KSU
5. Rent
6. Lease

This regulation regulates several objects such as objects, selection of partners, partners, duration, and obligations of partners.

2.4 Risk Management
According to Kezner (1995) risk management is a process of identifying, measuring, developing, selecting and managing options to deal with those risks. Project Risk Management aims to identify and manage risks that are not handled by other parties the project management process. When not managed this risk can cause the project to deviate from the plan and fail to achieve its objectives.The main benefit of risk management is ensuring that the level, degree, type, and visibility of risk management are commensurate with the risk and importance of the project itself for the organization and stakeholders.
R Max Wideman (1992) provides an overview of the integration of risk management with other project management functions on a project as follows:

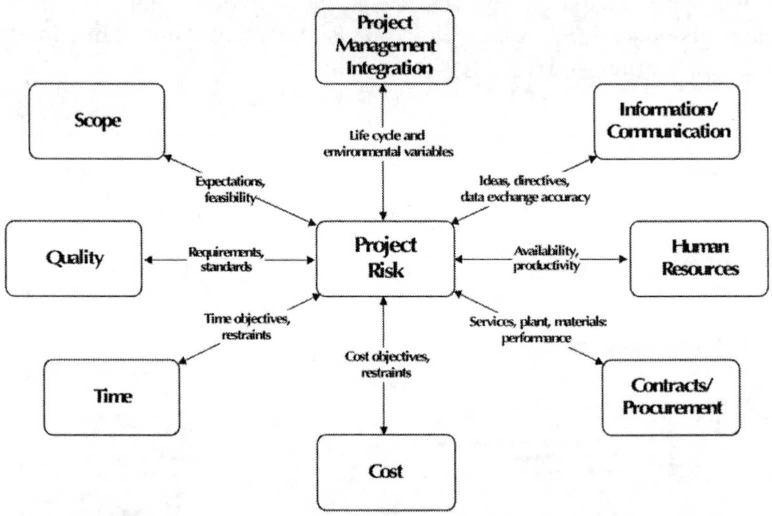

Figure 4. Risk Management Integration on Project

Base on PMBOK 6th Edition Risk Management will describe as follows:

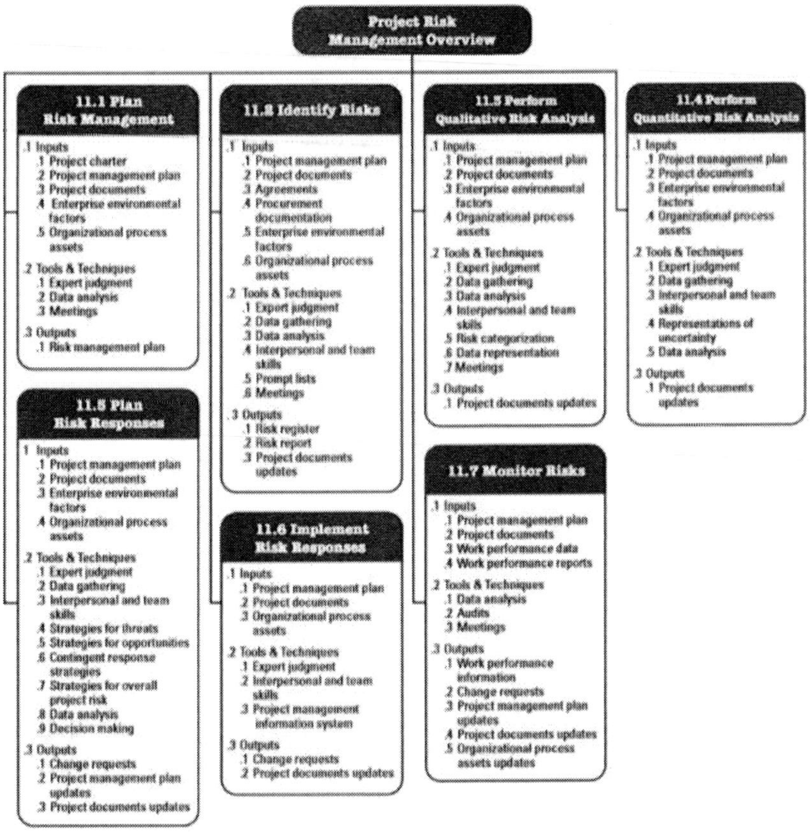

Figure 5. Project Risk Management Overview

3. Research Methodology

This Research is using qualitative, quantitative, and study. First writer will analyse regulation, benchmark and Company data to analysis form of Partnership which applicable for Company X, and then writer will interview important people to get expert judgement about risk variable which probably shown up and impact to Partnership. The risk variable as follows:

Table 3. Risk Variables

No		References	Risk Variable
1			Legal and Contract Risk
	X1	Trismara, 2011	Contract Less Power against Regulation
	X2	Trismara, 2011	Inappropriate Contract selection
	X3	Xu, et al, 2010	Contract violation
	X4	Xu, et al, 2010	delays in land acquisition
	X5	Li, et al, 2012	didn't get permission from the government
	X6	Xu, et al, 2010	change of concession holder
2			Politic Risk
	X7	Xu, et al, 2010	Government Intervention
	X8	Xu, et al, 2010	adverse government intervention
	X9	Wibowo & Mohamed, 2010	abuse power of government

IOP Conf. Series: Materials Science and Engineering **930** (2020) 012019 doi:10.1088/1757-899X/930/1/012019

No		References	Risk Variable
	X10	Xu, et al, 2010	takeover by government
	X11	Li, et al, 2012	slow process of decision making
3			Funding and Financial Risk
	X12	Xu, et al, 2010	interest fluctuation
	X13	Xu, et al, 2010	foreign currencies fluctuation
	X14	Xu, et al, 2010	Cost Changes
	X15	Xu, et al, 2010	Stability funding
	X16	Xu, et al, 2010	tax regulation changes
	X17	Li, et al, 2012	low profit
	X18	Wibowo & Mohamed, 2010	uncertainty demand
	X19	Xu, et al, 2010	market demand changes
	X20	Li, et al, 2012	bad capital structure
	X21	Li, et al, 2012	operator inability
	X22	Li, et al, 2012	slow process of decision making
	X23	Wibowo & Mohamed, 2010	new competitors
4			Design and Development Risk
	X24	Li, et al, 2012	Design not efficient
	X25	Li, et al, 2012	Design Changes
	X26	Li, et al, 2012	slow process of decision making
5			Operational and construction Risk
	X27	Xu, et al, 2010	Closing Project Delays
	X28	Xu, et al, 2010	Material or workers supply
	X29	Li, et al, 2012	cost construction changes
	X30	Li, et al, 2012	Bad Quality
	X31	Xu, et al, 2010	Infrastructure support
	X32	Xu, et al, 2010	Cost Overrun
	X33	Li, et al, 2012	slow process of decision making
6			Force majeure Risk
	X34	Trismara, 2011	bankrupt contractor
	X35	Wibowo & Mohamed, 2010	War
	X36	Wibowo & Mohamed, 2010	unrest condition
	X37	Wibowo & Mohamed, 2010	terrorize

Based on table 3, there are 6 variable which divide into 37 sub variables. After this validated by expert judgement, it will share to respondent which already planned based on experience in PPP project. And then writer will analyse into dominant risk variable which has big probability and impact. And then it will assistance and validate by expert judgement. After that writer will apply to study case in land asset non-productive.

4. Result and Discussion

1. Base on analyse data and regulation, form of Partnership which applicable for Company X are Build Operate Transfer, Build Transfer Operate, Build Owned Transfer, Join Operation, Join Venture, Rent, Lease, License, Management Contract, Private Asset. For the detail will show in table 4 as follows:

Table 4. Applicable Partnership Form for Company X

No.	Regulation/Data	Partnership Form									
		BGS/BOT	BSG/BTO	BOwT	KSO/JO	KSU/JV	Rent	Lease	Licence	Management Contract	Private asset
1	Republic of Indonesia Government Regulation Number 83 Year 2015 Regarding Public Housing (Perum) National Housing Development	√	√	√	√	√	√		√	√	
2	Regulation of the Minister of State-Owned Enterprises Number PER-13 / MBU / 09/2014 Regarding the Guidelines for Utilizing the Fixed Assets of State-Owned Enterprises	√	√		√	√	√	√			
3	Journal, Thesis, etc	√	√				√			√	√

2. Based on analyse questioner which collected and validated by expert and respondents for risk variable in Partnership, 10 dominant risk variables are identified, ranked and divided into each partnership form as follows:

Table 5. Dominant Risk Variable

Dominant Risk Variable		Ranking
X1	Contract Less Power against Regulation	1
X4	delays in land acquisition	2
X6	change of concession holder	3
X3	Contract violation	4
X13	foreign currencies fluctuation	5
X12	interest fluctuation	6
X25	Design Changes	7
X27	Closing Project Delays	8
X15	Stability funding	9
X14	cost construction changes	10

Table 6. Dominant Risk Variable

Partnership Form	Risk Variable Dominant	Total Risk	Expert Validation
BOT	X1; X3; X4; X6; X12; X13; X14; X15; X25; X27	10	√
BTO	X1; X3; X4; X6; X12; X13; X14; X15; X25; X27	10	√

BoWT	X1; X3; X4; X6; X12; X13; X14; X15; X25; X27	10	√
KSO/JO	X4; X12; X13; X14; X15; X25; X27	7	√
KSU/JV	X4; X6; X12; X13; X14; X15; X25; X27	8	√
Sewa	X1; X3; X4; X6; X15; X27	6	√
Pinjam Pakai	X1; X3; X4; X6; X15; X27	6	√

3. Study case for apply this result is in non-productive land asset Company X. in this year (2020) there was initiation Company X Partnership with partner to develop land asset in Region-3. Company X and Partner agreed to use KSO/JO as Partnership Form. Based on last analysis, maybe Company X and Partner choose KSO/JO because it has medium risk dominant, and they can allocate it. Until now there are issues such as land acquisition, change of concession holder and etc. based on result this research will help to response this issue and will register or mitigate risk which not show yet as follows:

Table 7. Risk Response

No	categorize	Issue	Identify Risk	Response Risk
1	Approval Citizen	Approval Citizen not reach minimum percentage based on regulation.	Design Delays Permit Delays Feasibility Study Delays	Government Support, and mediation & socialization to citizen
2	Design	Different interpretation with Partner	Design Delays Permit Delays Partner Willingness to Partnership	Government Support, explanation design to partner, and special offer program for partner
3	Concession holder	Concession holder changes	Amendment Partnership Contract Different Goal and Interpretation Project Partner Willingness to Partnership Project Delays	Explanation Contract existing and Benefit Project
4	Decision	Decision process to long	Project Delays Partner Willingness to Partnership	Monitoring Decision process continually Inform Decision process continually

References

[1] Pemerintah Republik Indonesia 2015 *Peraturan Pemerintah Republik Indonesia Nomor 83 Tahun 2015 tentang Perusahaan Umum* (Pembangunan Perumahan Nasional: Jakarta).
[2] Kementrian BUMN 2014 *Peraturan Menteri Nomor 13 Tahun 2014 tentang Pedoman Pendayagunaan Aset Tetap Badan Usaha Milik Negara* (Jakarta).
[3] Yescombe E R 2007 *Public Private Partnership Priciple of Policy and Finance.*
[4] UNESCAP 2007 *Public Private Partnership in Infrastructure Development an Introduction to Issues from Different Prespective.*
[5] Yunanda Raharjanto 2011 *Model Kerjasama Pemerintah-Swasta Dalam Peningkatan Kapasitas*

Jalur Kereta Api Jabodetabek.
[6] Rin Nurarni Trismara 2011 *Manajemen Risiko Proyek Pembangunan Jalur Kereta Api Batu Bara di Kalimantan Tengah dengan skema KPS (Kerjasama Pemerintah Swasta)* (Indonesia).
[7] Andria Dewi Shinta 2011 *Analisisi Faktor Penentu Keberhasilan Kerjasama Pemerintah Swasta pada Pembangunan Terminal Peti Kemas Palaran* (Indonesia).
[8] Ardhitya 2014 *Analisisi risiko Public Private Partnership dengan Pola Kerjasama Build Operate Trasnfer pada Rest Area Jalan Bebas Hambatan* (Indonesia).
[9] Project Management Institute 2017 *A Guide to Project Management Body of Knowledge, sixth Edition.*

IOP Conf. Series: Materials Science and Engineering **930** (2020) 012020 doi:10.1088/1757-899X/930/1/012020

Adoption and implementation of building information modeling (BIM) by the government in the Indonesian construction industry

M P Sopaheluwakan[1*] and T J W Adi[1]

[1] Departement of Civil Engineering, Faculty of Civil, Planning, and Geo-Engineering, Institut Teknologi Sepuluh Nopember, Surabaya, Indonesia

*Corresponding author's e-mail: mahendra.18031@mhs.its.ac.id

Abstract. Building Information Modeling (BIM) is the development of concepts of collaboration and integration in the construction industry. This concept was introduced in Indonesia formally by the Ministry of Public Works and Public Housing (PUPR) in 2017. Since its introduction, by the Indonesian government the PUPR ministry has made BIM Roadmap, created BIM Team PUPR, and issued ministerial regulations mandating the use of BIM on a state-owned building. BIM provides many benefits that can be utilized by every stakeholder involved in the project. Previous research has focused more on the readiness of the Indonesian construction industry in utilizing BIM. However, studies that focus on the implementation of BIM by the Indonesian government are still very limited. A literature review is used as the main method in this journal to collect information from government publications. This study provides updated information related to government efforts to implement BIM in Indonesia.

1. Introduction

Building Information Modeling (BIM) is a term that defines information modeling technology in the construction industry. BIM is an integration (continuous between the information and the tools used) from various disciplines. The BIM model will produce an intelligent and information-intensive parametric model [1]. The BIM concept supports the process of integration and collaboration. This is the main attraction for its users, including the government as the owner and regulator in the industrial sector in various countries [2]. Other advantages provided by BIM are supervision, interoperability, collision detection in each phase of the project life cycle [3, 4]. Private and public sectors trying to harness the benefits of BIM by using it for various purpose implementations. The public or government sector utilizes BIM to achieve its agenda. For example, the UK is using BIM as part of the 2011 construction strategy which aims to improve the efficiency and quality of the construction industry [5]. Brazil has an agenda to achieve 30% cost savings during the building life cycle. The State of Chile uses BIM to reduce 20% of project costs and save 20% of the costs of the project design process and duration of construction, as well as various other objectives to be achieved by various countries [6].

The Government of Indonesia through the Ministry of Public Development and Public Housing (PUPR) knows the benefits provided by the BIM concept and seeks to utilize them in various government projects. The government as the regulator has the authority and obligation to develop construction equipment and material standards, as well as the development of construction technology following what is mandated in Law No. 2 of 2017 concerning Construction Services. The concept of

Content from this work may be used under the terms of the Creative Commons Attribution 3.0 licence. Any further distribution of this work must maintain attribution to the author(s) and the title of the work, journal citation and DOI.
Published under licence by IOP Publishing Ltd

BIM was formally introduced by the Indonesian government in 2017 by showing a roadmap for the implementation of BIM Indonesia and forming a BIM PUPR Team to accommodate the process of BIM adoption in the Indonesian government, especially the Ministry of PUPR. In addition to creating policies, commitment, and working groups, various events have been organized to deploy BIM concepts in the construction industry in many subjects such as government agencies, university academia, and the practitioners [7].

Research on the topic of BIM in Indonesia has been conducted since 2012 [8]. In that study, it was found that 7 BIM-based studies focus on the dimensions of technology and perspective. The implementation dimension has also been carried out in Indonesia by evaluating the application of the BIM methodology in the construction phase with the subject of contractors already using BIM [9]. BIM practitioners are the main subjects in the study of the relationship between the BIM methodology and the sustainability of construction in Indonesia [10]. BIM practitioners become the main subject of this research with the perspective of investment, risk, and obstacles in adopting the use of BIM [11, 12-15]. Research that focuses on government perspectives about the adoption and implementation of BIM is still very limited.

This paper aims to review the adoption of BIM in Indonesia through a government perspective. This study focused on exploring what the government has done, through various agencies and organizations related to the adoption of BIM in Indonesia.

2. Conceptual background

2.1. Building information modeling
BIM is a parametric data collection of representations of building objects that are designed and built the contain structural, architectural, mechanical, and electrical design information [16]. BIM can be used as a tool in the entire life cycle to facilitate planners, supervisors, K3 (Occupational Health and Safety) management, and contractors. In applying BIM, Azhar [4] found that BIM can be used in the initial phase, the design phase, the pre-construction phase, the construction phase, and the operation phase of the project. Not only serves as a tool, but BIM also provides an opportunity to change rigor in making changes and innovations in the construction sector [3]. Facility management can utilize BIM in careful operations and schedule repairs using real visualizations of buildings in the form of building models [3]. The benefits of interoperability and integration in the BIM concept can be used specifically to oversee various projects owned by one stakeholder. The intended stakeholder is the government who usually has many projects in various locations simultaneously and many assets in the form of buildings or other entities.

2.2. The government's role in the implementation of BIM
The government has a role as a regulator in developing the industrial sector and the construction sector. Smith [17] said that the main success factor in the implementation of BIM nationally was leadership and coordination factors focused on maximizing work efficiency while avoiding arising problems from a segmented implementation approach. The government can become the owner, customer, or regulator in construction projects. In applying the BIM concept, the client/owner in the construction project becomes the main beneficiary [18]. The role of the government is not only limited as a regulator, but the government can also play a role as the initiator, educator, funding agency, demonstrator, and researcher [2].

2.3 BIM implementation
The adoption and implementation of BIM have been carried out by various countries in the world. The United States has become a pioneer country for the adoption of BIM. Various government agencies at every level have compiled a BIM program, set goals for the use of BIM, and made a roadmap for implementation and standards for BIM use [2]. Based on research conducted by Cheng and Lu [2] found that not only America but also many countries in Europe have also implemented BIM significantly. For example, the UK became one of the main BIM implementation countries in Europe by requiring all

government projects to use BIM in 2016. BIM has also entered Asia such as Singapore and Hong Kong to become the country that led to the adoption of BIM in Asia.

3. Methodology

This study used a literature review as the main instrument to find previous research on BIM topics in Indonesia. The information for this study was retrieved mainly through reviewing the literature from journal articles and conference papers; as well as observing relevant reports and guidelines. The literature found will be analyzed to determine the focus of each study, the results of the study, the dimensions of the study where the result will be tabulated. Literature mapping will be conducted to find a gap in BIM research in Indonesia that has not been explored. The next stage in this research is to collect various activities and efforts carried out by the Indonesian government to adopt BIM. The research method used in this paper uses an integrative literature review. An integrative literature review is one type of method in the literature to assess, critique, and locate a topic using available resources [19]. Data sources used in this method come from various sources. The intended data sources are academic journals, construction news, textbooks, industrial journals, government-issued journals, and reports from government institutions. Furthermore, the results of the literature review synthesis to obtain information about government activities related to the implementation process and the adoption of BIM. The main source of information regarding government effort on BIM adoption can be accessed on bim.pu.go.id [7].

4. Finding

4.1 BIM research in Indonesia

BIM research in Indonesia was recorded and published for the first time in 2013 with the topic of the benefits of BIM to streamline construction projects in Indonesia [8]. In a review of BIM research in Indonesia, found that there are seven studies where five of them are focusing on the technological dimension and the two others are focusing on the perspective dimension. Since 2017, BIM research in Indonesia continues to grow. Recent studies in the field of BIM along with their findings and dimensions are attached in table 1. The categorization of each article on the BIM implementation framework dimension is suggested by Jung and Joo [20] following previous research [8]. Literature mapping created to give insight into the article published where the gap of the research can be identified.

Table 1. Summary of BIM research in Indonesia

Theme	Finding	Dimension	Year	Mapping
Design and construction of prefabricated building [8]	BIM has benefit to streamline construction projects in Indonesia	Technology	2013	BIM in Design
Application of BIM for pre-construction [8]	BIM provides information about pre-construction planning and particularly construction logistic planning.	Technology	2014	BIM Technology
Awareness and implementation of BIM among academicians and practitioners [8]	1. High awareness of BIM (70%) low implementation (38%). 2. BIM's main uses are 3-D modeling and visualization	Perspective	2016	BIM Awareness and Perception
Perception of academicians and practitioners of BIM technology implementation in	Practitioners perceive BIM as informative software while academicians perceive BIM as integration software.	Perspective	2016	BIM Awareness and Perception

architecture [8]				
Comparison between Project using BIM and project using a conventional method in term of time efficiency, cost, and human resources [8]	Savings due to using BIM compared to conventional: 1. Project planning time by 50%. 2. Human resources by 26.66%. 3. Personnel cost by 52.25%.	Technology	2016	BIM in Construction
Exploration of technological support for BIM implementation, benefits of BIM, challenges of BIM in construction project [8]	1. The need for technological support for BIM is eminent. 2. BIM benefit is to reduce cost 3. Incompatibility between different BIM software.	Technology	2017	BIM Technology
Challenges of BIM implementation in small-medium enterprise architecture [8]	1. Main benefits of BIM are time efficiency, better communication and coordination, and improved project documentation. 2. Main challenges are lack of skilled BIM user, low demand from a client, high investment cost, and resistance to technological change.	Technology	2017	BIM in Construction
Assessment of BIM in high-rise building construction in Indonesia [9]	Using Building Information Modeling Implementation Index (BIMII) produces a quite achieved score of 67,46% and the Importance Performance Analysis of Building Information Modeling (IPABIM) produced 4 quadrants of importance on BIM achievement.	Technology	2018	BIM in Construction
BIM adoption towards the sustainability of the construction industry in Indonesia [10]	1. Awareness rate of BIM in Indonesia has to reach 67,5 %, although mostly with limited or basic knowledge 2. BIM gives positive impact to sustainability aspects in construction	Perspective	2018	BIM in Construction
Factors that Affects the Maturity Level of BIM Implementation in Indonesia; Case Studies of 5 Construction Key Actors [11]	Factors that affect the maturity level of BIM implementation are: 1. The commitment of the leader 2. System and infrastructure development 3. Quality and capability of human resources	Technology	2018	BIM Technology

Initial Study on Building Information Modeling Adoption Urgency for Architecture Engineering and Construction Industry in Indonesia [12]	6 important aspects of consideration in adopting BIM. Lack of awareness and understanding of BIM in the education sector are key factors that impede BIM adoption. Government and practitioners alike need to develop a strategic roadmap.	Perspective	2018	BIM Awareness and Perception
Investigating Building Information Modelling (BIM) Adoption in Indonesia Construction Industry [13]	1. 60% of respondents already acknowledged and implemented BIM 2. Benefits of BIM have been perceived by the respondent	Perspective	2019	BIM Awareness and Perception
Investment, Returns, and Risk of Building Information Modeling (BIM) Implementation in Indonesia's Construction Project [14]	1. Main priorities in investing BIM, BIM training, updating software, BIM software 2. 3 positive result implementations, and 3 ways to increase BIM returns. 3. 5 risk on BIM implementations	Technology	2019	BIM Technology
The Barrier and Driver Factors of Building Information Modelling (BIM) Adoption in Indonesia: A Preliminary Survey [15]	1. Barrier factors to BIM are adoption lack of BIM experts, lack of understanding, resistance to change 2. Driver factors on BIM adoption are willingness/intention, technological factors, organizational culture	Technology	2019	BIM Awareness and Perception

Based on the finding in table 1, research of BIM from 2012 until 2017 concluded that the progress of BIM in Indonesia is still slow [8]. The low number of the international journal shows that BIM research is not yet mature in Indonesia. This situation had changed from 2018 – 2019. There are 5 publications on the BIM topic in Indonesia. A previous study reported that the low number of research on BIM was published in 2018 [8]. BIM used on high-rise building construction in Indonesia measured by [9] gives an average value of 67,46% (quite achieved) using Building Information Modeling Implementation Index (BIMII) measurement. The instant benefit received from applying BIM is considered to have slightly more impact for the practitioner, rather than the longer-term benefits. The main benefit of BIM has already been understood by practitioners, even though out of 40 respondents, only 2 said that they had competence in understanding BIM, while 25 others described that their knowledge of BIM was limited or just basic [10]. The understanding and level of BIM use in Indonesia are clarified through the discovery that states that BIM has still considered as a technology, not an integrated system, causing the use of BIM is not collaborative among stakeholders [11]. BIM as a tool to encourage sustainability in the construction industry could be achieved if stakeholders had a better understanding of BIM [10] [12]. The education sector is a major obstacle in the BIM adoption process which could be resolved by including BIM in the AEC curriculum. The government and practitioners alike must develop strategic roadmap, regulations, protocols, and standardization in the collaboration and implementation of BIM for the success of its process [12]

Investigation on BIM adoption in Indonesia conducted in 2019 discovered that the current project practice must follow standardization, a collaboration between stakeholders, complex bureaucracy, uses of the conventional method, and system plague effort on BIM adoption [13]. these problem enhanced the previous finding [12]. Recommendation to speed up BIM adoption directed to the government, industry players, and roles of university align with the references given on the previous study. BIM uses in Indonesia, who started firstly in 2012 until 2019, were deemed to be less maximal. The study conducted on industry player who have already adopted BIM resulted in the need for investment in the form of BIM training, new or upgrade hardware, and BIM software, while the return of BIM implementation presumed by industry player is a positive impact on marketing, improving the result of the project process, and increase productivity [14]. The key risks are identified by [14] as follows technical risk, human resource risk, financial risk, management risk, and another risk. The main risk on each type is the rapid renewal of the BIM technology, lack of knowledge and capability of BIM employees and technicians, high short-term investment costs, previous management patterns, and lack of industry standards for BIM application. Barriers and driver factors of BIM adoption study carried out by [15] using an expert as a study object to identify 16 barrier and 18 driver indicators stem from the literature were considered relevant to measure barrier and driver factors of BIM adoption in Indonesia.

A study on the BIM topic in Indonesia used practitioners (13 out of 14 literatures) as the main source of study and information to discover, analyze, and measure of BIM adoption. Academics/Universities place second on object study who (5 out of 14 literatures) who try to extract information mainly perspective of academic, current curriculum of BIM, expertly composed of academic and practitioner, become the main object study of BIM research. Government place last as object study (2 out of 14 literature) despite [11-14] said that the lack of industry standardization, regulation, financing policy, complex bureaucracy, and conventional method used in the industry. This problem can be addressed by the government whose role as regulator and owner of many projects in Indonesia. Literature mapping provided in figure 1 depict BIM research mapping and study object in Indonesia. Collaboration between government, academics/universities, and industry players are needed in the process of adopting BIM in Indonesia. Each construction stakeholder in Indonesia must perform its role following their respective sectors. The study [14] recommended what shall be done by the government, industry players, and academics/universities to ensure BIM adoption in Indonesia.

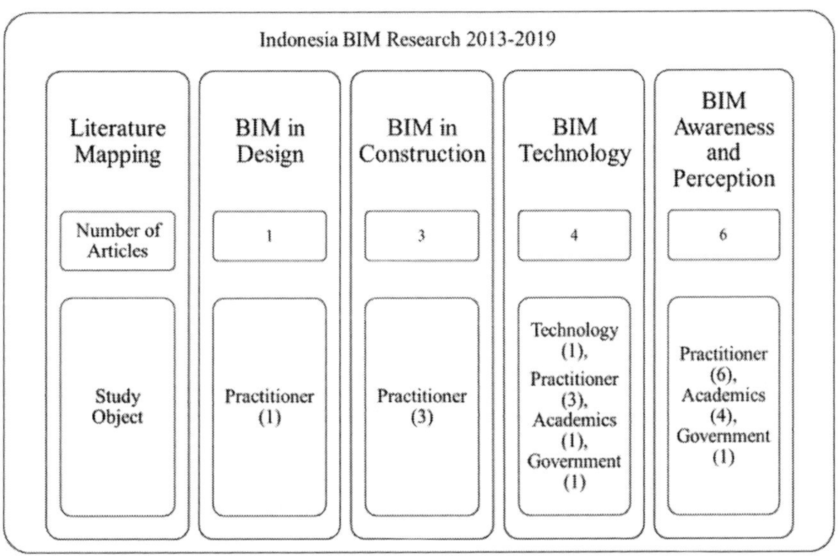

Figure 1. Literature mapping on bim research in indonesia

A literature review on conceptual background shows that the adoption and implementation of BIM have been carried out by various countries. The government's role as a regulator in the process of adoption and implementation BIM in the construction industry is one of the main driving forces. Indonesia became one of the countries that also began to adopt BIM. Research in the field of BIM in Indonesia has been running since 2013 - until now. Based on the literature mapping, industry players, academics, and BIM practitioners have become the main sources in the research. The government as a regulator in Indonesia has been highlighted as research study but limited as a source to find out the adoption and implementation of BIM in Indonesia.

4.2 Roadmap for BIM implementation

The implementation of BIM in Indonesia began formally at the "Digital Construction Day", an international conference held by PT PP in Jakarta on October 4, 2017. The conference invited many speakers from practitioners, government agencies, BIM authoring software, and many researchers/academics. In this conference, it was conveyed that the benefits and use of BIM carried out both by PT PP. Conference the topic discussed at the conference was the achievements of the government, represented by the Ministry of PUPR reports on the implementation of BIM in China and Singapore, experience in using BIM by various consultants and demonstrations of the use of BIM by authoring software and introduction Indonesian BIM Institute as an organization has the aim of developing the implementation of BIM in Indonesia. One of the main results of this conference was the first time public roadmap presentation for the implementation of the Indonesian BIM, which was then called the Indonesian digital construction roadmap by the Ministry of PUPR. In this roadmap, the BIM implementation process is divided into 4 phases which will be implemented in 7 years from 2017 to 2024. The phases submitted by the Ministry of PUPR contain the stages and objectives to be achieved by implementing BIM in Indonesia. Indonesia's digital construction roadmap can be seen in table 2.

Table 2. Indonesian BIM roadmap, source: [22]

2017			2024
Adoption	Digitalization	Collaboration	Integration
• Construction Stakeholder adopted BIM • Compilation of National BIM Standard (SNI) • BIM as a curriculum and competency standard for universities and professional associations	• Licensing • Monitoring and supervision • Commencement of a digital market for the construction sector	• Standard rules related to the construction industry collaboration • Implementation of Virtual Design and Lean Construction • BIM Implementation (3D to 7D)	• Implementation of cloud construction management • Integration of construction process systems (licensing, claim, commissioning, handover, etc.)

4.3 PUPR BIM team

The PUPR BIM Team is a workgroup team formed through Decree of the Director-General of BALITBANG PU (Research and Development Public Works and Housing) No. 32/KPTS/KL/2017 on 12 December 2017 as a follow up to the government's commitment in implementing BIM and the BIM roadmap for Indonesia. This team was formed with 3 main tasks, namely formulating roadmaps as well as BIM implementation strategies, preparing national guidelines/standards about how to apply BIM, and conducting socialization and workshops for BIM. The focus of the work area of the BIM PUPR team is to implement BIM especially within the body of the Ministry of PUPR. Not limited to implementing BIM in the ministry, the PUPR BIM team also has the task to monitor and evaluate pilot projects that apply the BIM concept [7].

One of the first products from the PUPR BIM Team was the "Adoption of BIM in the organization" guidelines, aimed at guiding by helping organizations adopt and implement BIM. These guidelines have been prepared using a variety of sources, namely BIM Essential Guide of Adoption in an Organization by Building and Construction Authority (BCA) issued in August 2013, BIM for Transport and Main Road: A Guide to Enabling the BIM on Road Infrastructure Project, the Queensland government, published in May 2017, the National BIM Guide for Owners, National Institute of Building Sciences, published in January 2017, BIM Guide Series 01 - overview, US General Services

Administration, published in May 2007. This guideline was prepared through collaboration between the BIM PUPR Team and IBIMI. In addition to make guidance products for BIM implementation, the PUPR BIM Team works together with BALITBANG PU and DJBK PU (Directorate General of Construction) to make policies, ministerial policies, standards and protocols, SNI (Indonesian Nationals Standards), SKKNI, adoption guidelines, and what is needed to accommodate the BIM implementation. Figure 2 provides timelines for various activities that have been carried out by the BIM PUPR team in collaboration with various government institutions, industry players (practitioners) and involve academics/universities as a government effort to the adoption of BIM in the construction industry in Indonesia. Activities start in 2017 by the creation of the BIM PUPR team, BIM Roadmap, and IBIMI introduction on public in may to become one of the collaborative teams who help PUPR BIM team on various occasions. PUPR BIM team activities in 2018 marked by BIM implementation commitments and agreed by various government institutions and ministerial regulation of PUPR no 22/2018 was set on this year. BIM team activities starting from 2017 and reaching peak performance in 2019 with many socialization agenda, training, and preparation of standard and guideline as to the main focus in 2019. Other significant activities this year are preparing the BIM institutional agreement at the national level, led by Bappenas in the form of the National BIM Workshop. BIM training for civil servants (ASN), conducted in collaboration between the PUPR BIM team, state institutions, and external parties (IBIMI, academics/universities), was held in August – November to increase the knowledge and understanding of BIM in the internal government.

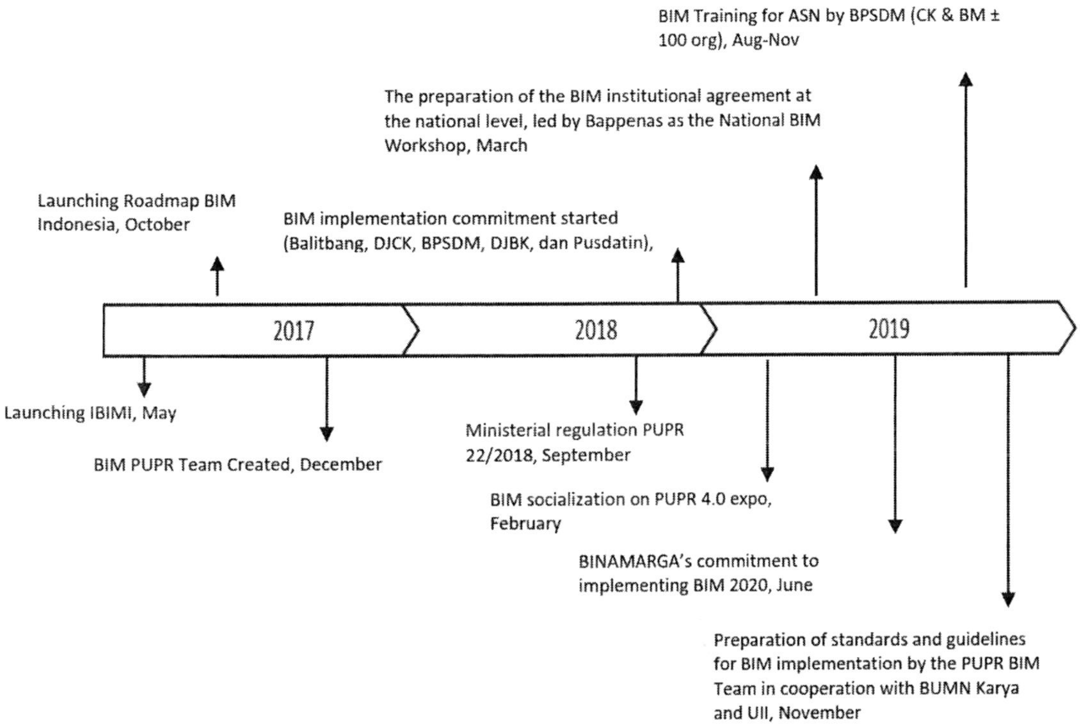

Figure 2. Compilation of PUPR BIM team activities 2017-2019, source : [7]

4.4 Policies – collaboration

One of the various ways in starting the implementation of BIM nationally is by making BIM an obligation or inculcating the obligation to use BIM in regulations, policies, government mandates, statutes, and special policies of the construction sector as a concept to accelerate infrastructure development. This stage in the process of implementing BIM nationally has been applied by several countries, not only regulating regulations and issuing national guidelines or standards but also using the BIM concept as a foothold to achieve national construction goals. Utilizing BIM in achieving the national goals implemented by Brazil to achieve cost reductions of up to 30% in the building life cycle, reducing the cost and time of building planning and construction processes in the country of Chile, streamlining the construction industry, and making interoperability data during the project life cycle in China [6]. The Indonesian government also used the method to implement BIM one of them translated in the direction of the minister PUPR containing 5 breakthroughs for pushing acceleration of infrastructure development, where one of the five directives that contain the implementation of research and construction technologies presented in the Indonesia Infrastructure Finance Forum 2017. These directives had been one of the basic implementation strategies in the infrastructure development of the Ministry of PUPR.

After the formation of the PUPR BIM team and the Indonesian digital construction roadmap in 2017, the Ministry of PUPR also launched PUPR ministerial regulation No. 22 / PRT / M / 2018 on the construction of state buildings. In the PUPR ministerial regulation, one of the attachments, namely No. 13 concerning construction planning service providers, states that the use of BIM is mandatory in non-existent state buildings with area criteria above 2000 m2 (two thousand square meters) and above 2 (two) floors. This regulation clearly states that the use of BIM as one of the obligations for construction service providers of the state buildings construction, therefore starting from 2018 all state-owned buildings that are in the planning and construction phases are obliged to use BIM. Although the use of BIM is required based on PUPR Regulation No. 22 / PRT / M / 2018, there are still no regulations that regulate how far the uses of BIM are and only regulate the final result of BIM operation which is ASME (Architecture, Structure, Mechanical, Electrical), which produced drawing, volume breakdown work, and budget plan. Collaboration between the Indonesian and the UK government by adapting 3 UK infrastructure planning and development methodologies is more effective and sustainable. One of the programs adopted is BIM. The duration of this collaboration is for 2017 – 2021. Further discussion regarding collaboration between Indonesian and the UK government resulted in the Global Infrastructure Program: BIM National strategy workshop on 19-20 March 2019.

5. Conclusions

The implementation of BIM in Indonesia has been started since 2017. This is marked as the publication of the roadmap for the implementation of BIM Indonesia or the Indonesian digital construction roadmap. Not only the BIM roadmap the Indonesian government has also formed a BIM team to implement BIM in government although the main focus is still on the internal needs of the Ministry of PUPR. These tools, formed by the government, are the beginning in producing/developing various tools to implement BIM, one of which is the adoption of BIM in the organization, as well as Minister of PUPR Regulation No. 22 / PRT / M / 2018 regarding the construction of state buildings. Collaborating with various parties in Indonesia, both from academia, practitioners, and various internal institutions, the Indonesian government has carried out the initial stages needed to create space and foster conditions to accelerate the process of adoption and implementation of BIM. These initial stages are needed to further produce guidelines, conduct research, and develop national standards in the use and implementation of BIM to achieve the acceleration of infrastructure development in Indonesia.

Acknowledgements

The authors implied a gratefully acknowledges the financial support given by Institut Teknologi Sepuluh Nopember under freshgraduate scholarship. The author would like to thank of Mr. Tri Joko Wahyu Adi as co-author and academic advisor which has provided support for this research activity.

References

[1] Migilinskas D, Popov V, Juocevicius V and Ustinovichius L 2013 The benefits, obstacles, and problems of practical bim implementation *Proc. Eng.* vol **57** pp 767–74
[2] Cheng J C P and Lu Q 2015 A review of the efforts and roles of the public sector for BIM adoption worldwide *J. Inf. Tech.. Constr.* **20** 442–78
[3] Ghaffarianhoseini A 2015 Building information modelling (BIM) uptake: clear benefits, understanding its implementation, risks, and challenges *Renew. Sustain. Energy Rev.* **75** 1046–53
[4] Azhar S, Khalfan M and Maqsood T 2015 Building information modelling (BIM): now and beyond *Constr. Econ. Build* **12** 4 15–28
[5] Atkinson L, Amoako-Attah J and B-Jahromi A, 2014 Government's influence on the implementation of BIM *Comput. Civ. Build. Eng. Proc. 2014 Int. Conf. Comput. Civ. Build. Eng.* vol **79** pp 423–30
[6] Hore A, McAuley B and West R 2017 BICP global BIM study: lessons for ireland's BIM programme *Construction IT Alliance (CitA) Limited* (Dublin; Irlandia)
[7] Tim BIM Kementrian PUPR 2017 BIM Kementrian Pekerjaan Umum dan Perumahan Rakyat [Online] Available: *http://bim.pu.go.id/materi.html*
[8] Telaga A S 2018 A review of BIM (building information modeling) implementation in indonesia construction industry *IOP Conference Series: Materials Science and Engineering* vol **352** p 1
[9] Khasani R R and Hidayat A 2018 *Assessment of BIM in high-rise building construction in indonesia Int. J. Sci. Technol. Res.* **7** 7 112–7
[10] Zhabrinna, Davies R J, Pratama M M A, and Yusuf M 2018 BIM adoption towards the sustainability of construction industry in indonesia *MATEC Web Conf.* vol **195** no August p 06003
[11] Larasati D, Andini Willis F, Hanifah Y, Alfi A F and Suhendri S 2018 Actors that affects maturity level of BIM implementation in indonesia *Case Studies of 5 Construction Key Actors* 673–81
[12] Agirachman F A, Putra I F, and Angkawijaya A 2018 Initial study on building information modeling adoption urgency for architecture engineering and construction industry in indonesia *MATEC Web Conf.* vol 147 pp 0–5
[13] Utomo J, Hatmoko D, Fundra Y and Wibowo M A 2019 *Investigating Building Information Modelling (BIM) Adoption in Indonesia Construction Industry* vol 2006
[14] Yanda G, Amin M and Soehari T D 2019 Investment, returns, and risk of building information modeling (BIM) implementation in Indonesia's construction project *Int. J. Eng. Adv. Technol.* **9** 1 5159–66
[15] F. R. Utomo and M. A. Rohman 2019 The barrier and driver factors of building information modelling (BIM) adoption in indonesia: a preliminary survey *IPTEK J. Proc. Ser.* vol **0** no 5 p 133
[16] Deshpande A, Azhar S and Amireddy S 2014 A framework for a BIM-based knowledge management system *Procedia Eng.* vol **85**, pp 113–22
[17] Smith P 2014 BIM implementation-global strategies *Procedia Eng.* vol **85** pp 482–492
[18] Bin Y J and Chou H Y 2018 Mixed approach to government BIM implementation policy: an empirical study of taiwan *J. Build. Eng.* **20** 337–43
[19] Torraco R J 2005 Writing integrative literature reviews: guidelines and examples *Hum. Resour. Dev. Rev.* **4** 3 356–67
[20] Jung Y and Joo M 2011 Building information modelling (BIM) framework for practical implementation *Autom. Constr.* **20** 2 126–33
[21] Zhabrinna, Davies R J, Abdillah P M M and Yusuf M 2018 BIM adoption towards the sustainability of construction industry in indonesia *MATEC Web Conf.* vol **195**
[22] BIM PUPR and IBIMI 2018 Panduan adopsi BIM dalam organisasi 1st edition jakarta *Pusat Litbang Kebijakan dan Penerapan Teknologi*

IOP Conf. Series: Materials Science and Engineering **930** (2020) 012021 doi:10.1088/1757-899X/930/1/012021

Analysis of the category of variation order in x project at XYZ ltd

A A Suprapto[1*] and I P A Wiguna[1]

[1] Civil Engineering Department, Faculty of Civil, Planning, and Geo Engineering, ITS

[*]Corresponding author's e-mail: ariefianto.203116@mhs.its.ac.id

Abstract. The construction project is one of the jobs which has a high level of complexity, dynamic and high risk with various uncertainties. Uncertainty which often arises in construction projects of jobs procurement increases or decreases or often mentioned as variation orders. Variation order is considered difficult to be anticipated and have a large impact on construction projects related to the increase in achievement of work progress. Therefore the impact of variation order which is caused by construction costs is greater followed by the higher frequency variation orders will have an impact on the construction project. The purpose of this study was to determine the type of variation order and cost value on X project at XYZ Ltd. This research method was a qualitative study with a literature study to obtain research variables. Secondary data was obtained from the project documents in the form of contract instructions which contained the causes and impacts which were resulted from variation order. Furthermore, Data was classified based on variables from the literature review. The results showed that the percentage of preparing design aspect was 31.50% of the total cost of variation order, site condition was 17.62% of the total cost of variation order, client change was 2.8% of the total cost of variation order and the flexibility contract was 48, 03% of the total cost of variation order. The frequency of variation orders occurance due to preparing designs was 49% of the total frequency, site conditions was 25% of the total frequency, client change was 13% of the total frequency and flexibility contract was 13% of the total frequency.

1. Introduction

Variation order is the impact and risk of uncertainty of a work which experiences the addition or deletion from the original scope of work according to the contract so that it changes the entire contract value and even the time of work completion. [16] explains that construction projects often experience changes commonly called as variation order, even rarely construction projects do not change until the project is completed so that variation order constraints on a construction project process, especially shopping centers, are recommended to be minimized or even avoided because it causes losses to the project. Variation order in a project is generally caused by differences opinion or misunderstanding in communicating between service users and service providers. This misunderstanding can lead to disputes that have an impact on the obstruction of project implementation so that the project fails. Variation order is considered difficult to be anticipated and have a large impact on construction projects related to the increase in achievement of work progress, Therefore the impact of variation order which is caused by construction costs is greater followed by the higher frequency variation orders will have an impact on the construction project [13]. This research uses project X, project X is the construction of mall and apartments worked by XYZ Ltd, XYZ Ltd is a private company engaged in construction, as the object of research. The implementation of X project was assessed as having various obstacles that caused

Content from this work may be used under the terms of the Creative Commons Attribution 3.0 licence. Any further distribution of this work must maintain attribution to the author(s) and the title of the work, journal citation and DOI.
Published under licence by IOP Publishing Ltd

changes in various conditions which are often called variation orders. Constraints in the form of variation order is one of the things that construction companies often experience in working on a project and variation order gives a significant impact related to changes in costs and time and even threaten project failure. Analysis of the main causes of variation order in a X project and similar projects needs to be done in order that the construction company and various other parties who are involved in doing the construction of a construction project know the main causes and are often experienced which become trigger the variation order occurance. This selection of X project at XYZ Ltd. was caused by the number of variation orders is cosidered the most if it's compared to other types of projects which were undertaken by XYZ Ltd., X project often experienced problematic variation order such as price negotiations that were difficult to achieve, administrative processes that were convoluted to obstructed payment processes so that resulting project objectives were constrained. Variation order mapping was important to be found as an analysis of the causative factors of variation order occurance in a construction project. Based on the various explanations have been described above, it is important to analyze the causes of variation order as one of the risks in the process of construction work in order to be able to be assessed and used as a consideration analysis in doing construction work for the actors who are involved in the construction work process both from the service users and service providers in construction field. The title of this research is Analysis of the Category of Variation Order in Project X at XYZ Ltd.

2. Literature

2.1. Variation order

Variation order is proposed of changes in writing between the owner and contractor to change some of the conditions of the initial contract documents, such as adding, reducing work, changes, this can change the specification of contract costs, payment schedules and project schedules. According to [3] variation order can be defined as a modification of the original contract whereas physically, the variation order is a work order letter to confirm revisions of the plan, and the amount of cost compensation to the contractor which occurs during construction implementation, after the signing of the contract between owner and contractor. Variation order is considered as official documents which is signed by both of parties to compensate the contractor for changes, additional work, delays or other consequences of the joint agreement that's written in the contract [12]. Variation orders can also be defined as an agreement signed by the contractor, architect and owner after the initial contract is made and then modified by some scope of work that adjusts to the cost and time [12]. Variation order is considered instructions from the owner or owner of agent which changes the terms and conditions of the contract. This is a written agreement between the contracting parties that represents the addition, deletion, or revision of contract documents, identifying changes in price and time, and describing the nature of the involved work. Variation order is changes in writing between the owner and the contractor to change the conditions of the initial contract documents, by adding or subtracting work. The existence of these changes can change contract costs and project implementation schedules. Variation order can be defined as modifications to the original contract [5]. Changes during the project implementation period often occur due to the wishes of the owners that arise during the implementation of the construction project, this is caused among others because of existence of changes in the scope of work, changes in specifications, changes in material types, changes in architectural planning, changes in work methods, and acceleration of work implementation. The purposes of the variation order are: (1) To change the contract plan in the presence of a special method of payment. (2) For administrative purposes, in determining the method of paying extra work or adding it. (3) To follow the adjustment to the unit price of the contract if there is a change in specifications. (4) For the submission of a reduction in the cost of the incentive proposal there is a change in the value engineering proposal. (5) To adjust the project schedule due to changes. (6) To avoid disputes between the contractor and the owner.

2.2. Causes of variation orders

Causes of Variation Order include the following :

Table 1. Causes of variation order

	Variable	Source
	Rules and Regulation Change	
1	Variation Order is caused by corrupt practices	[3]
2	Variation Order is caused by inconsistent government regulations	[4]
3	Variation Order is caused by ineffective coordination and integration	[4]
4	Procedural unavailability and manual rules for undertaken projects in accordance with government policy	[14]
	Preparing Design	
5	Material that turns into a significant cause of Variation Order	[17]
6	The change of detail and pattern of the image becomes the causes of a significant Variation Order	[18]
7	Design is not construcible	[2]
8	Understanding of the design that is not clear and not well understood	[4]
	Flexibility Contract	
9	The problem of contract adjustment is the causes of the variation order	[8]
10	Poor contract management gives rise to variation order in project implementation	[3]
11	The contract is considered multiple interpretation, this triggers a variation order.	[1]
	Contractor Initiated Change	
12	Consultant costs tend to be minimal	[11]
13	Project designers have minimal experience	[5]
14	Contractors often instruct variation orders	[5]
15	The contractor's supervisor has no experience of the undertaken construction project	[5]
	Client Change	
16	The project owner instructs to modify the design	[18]
17	The project owner failed to make a decision in reviewing documents in a timely manner	[15]
18	Unavailability of licensed experts from the side of the project owner	[18]
	Bad Contractoral Procedure	
19	Budget contract problem becomes the cause of variation order	[8]
20	Unavailability of overall planning project	[15]
21	Failures of construction project contract consultants	[1]
	Improper Coordination Contract	
22	Failure to provide contract information related to project tenders	[19]
23	The decision regarding project contract are considered unsuitable with the initial agreement.	[19]
24	Incompatibility contract with environmental, technical, legal and project budget issues	[7]

3. Research method

This study uses a qualitative approach to determine category that cause variation orders. The discussion of this research is emphasized a case study [20] namely the analysis of the discussion related to the main problems that occur in construction projects related to variation orders, factors and cause sources of variation order. This research begins with the process of elaborating the background then looks for the formulation of the problem and objectives, then proceed by studying literatures to find the factors that cause variation orders, consisting of rules change, preparing design, flexibility contract, client change, contractor change, bad contractoral procedure and improper coordination contract. Factors that cause variation order in secondary data is in the form of contract instructions that contains the incurred causes and costs that are adjusted with the results of literature studies. The re-classification is continued based on the order variation sequence, which causes the greatest cost and the most common frequency. Secondary data in this study are in the form of X construction project data in Surabaya and the used data is contract instruction of X construction projects from the beginning of the project until May 2020.

4. Results and discussion

4.1 Classification of variation order

Various factors about the causes of variation orders are taken from journal, and then crosschecked with field data. Field data which is form of contract instruction was containing the causes of variation orders and costs incurred due to the variation order. So that known the factors which cause variation orders in journal and accordance with the construction project problems that occur in X project namely preparing design, flexibility contract and client change, while site condition was one of the aspects that causes variation order which were found in the field data but not in the journal. Then each cause of variation orders appears cost incurred due to the variation order is taken from contract instruction form. Total costs incurred due to the variation order amounting to Rp 4,636,219,709. While the project X contract value is IDR 261,000,000,000.

Tabel 2. Classification of Variation Orders by Costs

No	Variables	Cost	Frequency	Cost of overall VO / Per Variable	Cost of Contract/ Per Variable
1	Rules and Regulation Change				
2	Preparing Design	Rp 1,460,536,494	49%	31.50%	0.56%
3	Flexibility Contract	Rp 2,226,876,579	13%	48.03%	0.85%
4	Contractor Initiated Change				
5	Client Change	Rp 131,869,796	13%	2.84%	0.051%
6	Bad Contractoral Procedure				
7	Improper Coordination Contract				
8	Site Condition	Rp 816,936,840	25%	17.62%	0.31%

Based on the table above, it is known that the percentage of each aspect of preparing design is 31.50% of the total cost of variation order, site condition is 17.62% of the total cost of variation order, client change is 2.8% of the total cost of variation order and flexibility contract is 48.03% of the total cost of variation order while the percentage value of variation order compared to the contract value is 1.776%. The frequency of occurrence of variation orders due to preparing designs is 49% of the total frequency, site condition is 25% of the total frequency, client change is 13% of the total frequency and flexibility contract is 13% of the total frequency. Aspects or factors that cause dominant order variations based on frequency are preparing designs while flexibility contracts are dominant in terms of the largest costs in causing variation order. This proves that the flexibility contract and preparing designs are as the cause of variation orders which cause the project is constrained thereby increasing construction costs. Previous research often discusses government projects such as school buildings, roads, dams, public facilities. This study discusses private projects in the form of apartment buildings and shopping centers which show the causes of variation orders are dominated by flexibility contracts and preparing designs.

5. Conclusions

Based on the data analysis process, it can be concluded the results of the study are as follows:The percentage of each aspect of preparing design is 31.50% of the total cost of the variation order, site condition is 17.62% of the total cost of the variation order, client change is 2.8% of the total cost of the variation order and the flexibility contract is 48.03 % of the total cost of the variation order. The frequency of occurrence of variation orders due to preparing design is 49% of the total frequency, site condition is 25% of the total frequency, client change is 13% of the total frequency and flexibility contract is 13% of the total frequency. Aspects or factors that cause dominant order variations based on frequency are preparing design while flexibility contracts are dominant in terms of the largest costs in causing variation order.

Acknowledgements

I am deeply indebted to my supervisor, Putu Artama Wiguna, Phd. and Department of Civil Engineering, for warm support, inspiration and thoughtful guidance.

References

[1] Akpan O P and Igwe O 2001 Methodology for determining price variation in project execution *J. Constr. Eng. M..* **127** 5 367-73

[2] Al-Momani A H 2000 Construction delay: A quantitative analysis *Intl. J. Proj. Mgmt..* **181** 51–9

[3] Alnuaimi A S, Taha R A, Mohsin M A and Al-Harthi A S 2010 Causes, effects, benefits and remedies of change orders on public construction projects in Oman *J. of Constr. Eng. Mgmt* **136** 5 615-22

[4] Arain and Low S 2005 The potential effects of variation orders on institutional building projects *Emerald Group Pub. Ltd. 0263-2772 DOI 10.1108/02632770510618462* **23** 11/12 496-510

[5] Assaf S A and Al-Hejji A 2006 Causes of delay in large construction projects *Intl. J. Proj. Mgmt.* **24** 349–57

[6] Bassioni H and Hamza N 2005 Sources of change orders in Kuwait building construction, MSC thesis, department of construction dan building engineering, *Arab Academy for Sci., Tech. dan Maritime Transp.* (Alexdanria: Egypt)

[7] Brown T L, Potoski M and Van S D 2015 Managing complex contracts: A theoretical approach j. publ. administration res. theory **26** 2 294–308 *https://doi.org/10.1093/jopart/muv004*

[8] Charkhakan and Heravi 2012 Identification of changes formation scenarios in construction projects based on changes occurrence paths analysis *Constr. Res. Congr. 2012 © ASCE 2012*

[9] Dosumu O S and Clinton O A 2017 Effects of variation on project of selected building projects in Lagos State *ICCREM 2017*

[10] Enshassi A, Arain F and Al-Raee S 2010 Causes Of variation orders in construction projects in the Gaza strip *J. of Civ. Eng. and Mgmt.* **16** 4 540-51

[11] Goudreau H 2001 The five key elements of a construction contract forget them and you are in trouble *www.hgassociates.com/articlecontracts*

[12] Hale D R, Shrestha P P, Gibson G E Jr and Migliaccio G C 2009 Empirical comparison of design/ build and design/bid/build project delivery methods *J. of Constr. Eng. Mgmt.* **135** 7 579-87
[13] Ibbs C W, Wong C K and Kwak Y H 2001 Project change management system *J. of Mgmt. in Eng.* **17** 3 159-65
[14] Kaming P, Olomlaiye P, Holt G and Harris F 1997 Factors influencing construction time and cost overruns on high-rise projects in Indonesia *Constr. Mgmt.. Econ.* **15** 83–94
[15] Memon Z A 2004 Remedial Measure For Delays At Construction Stage. *J. Mehran Univ. (Eng. Tech.)* **23** 1 9–20
[16] Oladapo A 2007 A quantitative assessment of the cost dan time impact of variation orders on construction projects *J. of Eng., Design dan Tech., Emerald Group Pub. Ltd. 1726-0531 DOI 10.1108/17260530710746597* **5** 1 35-48
[17] Oyewobi, Richard L O J and Bashir O R 2016 Analysis of causes dan impact of variation order on educational building projects *J. of Facilities Mgmt. Emerald Group Pub. Ltd. 1472-5967 DOI 10.1108/JFM-01-2015-0001* **14** 2 139-64
[18] Russell, Simon M H, Min L and Brad W 2014. Causes of time buffer dan duration variation in construction project tasks: Comparison of perception to reality *J. Constr. Eng. Mgmt.* 140
[19] Sambasivan M and Soon Y W 2007 Causes and effects of delays in Malaysian construction industry *Intl. J. Proj. Mgmt.* **25** 517–26
[20] Sugiyono 2014 *Metode Penelitian Pendidikan Pendekatan Kuantitatif, Kualitatif, dan R&D* (Bandung: Alfabeta)

Identifying factors affecting schedule and cost performance on building project

E D Widowati[1*] and F Rachmawati[1]

[1] Department of Civil Engineering, Sepuluh Nopember Institute of Technology, Indonesia

[*]Corresponding author's e-mail: farida_rahma@ce.its.ac.id

Abstract. The project cost and schedule are two major factors which contribute to the project performance. Therefore, it is important to identify factors that affect project cost and schedule performance, which will, in turn, affect the whole project performance. Previous studies have listed many factors have a large impact on time and cost performance, however the discussion related to the relationship between influencing factors and performance parameters is still limited. This research identified and refined 15 influencing factors through literature review. Those factors were grouped into 3 group factors: resource, environmental and management. Then, further literature review further was conducted to determine the relationships between influencing factors with project time and cost performance. The study focuses on factors influencing project cost and time performance based on the physical progress and actual cost. The result shows that resource group is the most variable influencing physical progress, direct cost and indirect cost. While environment and management group influencing physical progress and indirect cost.

1. Introduction

A country's economy gets significant contribution from the construction industry. However, performance measurements in the construction industry are still lagging behind when compared to other industries such as manufacturing [3, 11]. Consequently, it becomes important to measure in order to make the industry sustainable. The construction performance is affected by three main attributes: time, cost and quality. These three attributes are correlated one and each others [10].

Previous studies claimed that lots factors have a big impact on time and cost outcomes. Time and cost performance in construction project are influenced by some variables such as external conditions, project characteristics, design team characteristics , project team performance, procurement system, contractor characteristics and client characteristics [7]. The project management is needed to coordinate and to synchronize the planned cost and schedule to achieve the project goals.

The character of the project, such as unique, uncertain, long-duration nature, and complex make it difficult to predict time and cost projects [17]. Large number of performance indicators that could be measuring and evaluating project performance that could be related to various dimensions [6]. Earned value management (EVM) is one of project controlling tools which integrate cost and schedule aspects [21]. It approaches provides ideas to performance along with variance indexes which should be used to quantify specific construction achievement and situation. Moreover, that can be used to determine the efficiency and effectiveness of past actions of a project about time and cost [11].

EVM consists of three factors including actual cost (AC), earned value (EV), and planned value (PV). The first factor was actual cost of tasks implemented or WBS share performed based on actual

Content from this work may be used under the terms of the Creative Commons Attribution 3.0 licence. Any further distribution of this work must maintain attribution to the author(s) and the title of the work, journal citation and DOI.
Published under licence by IOP Publishing Ltd

physical progress. While EV was the amount of the task implemented or portion of a work breakdown structure and PV was planned budget for performing a task or part of the WBS including details of performance. EVM uses time and cost performance indicator, called as schedule performance index (SPI) and cost performance index (CPI). CPI is derived by dividing EV and AC, while SPI could be determined using the ratio EV to PV.

Several studies on project performance based on EVM have been conducted, including [13] which measure the time, cost and quality of the project but without analyzing the influencing factors. As well as [21] with the gray number method. Another case were done by [3, 4, 7, 12], they identified factors that influence project performance without analyzing the impact of the relationship between influencing factors and performance parameters themselves. Project performance has a complex and dynamic character, so research is needed to identify factors that can affect performance and its relationship with measurements using EVM.

The main objectives of this paper are to identify factors influencing time and cost performance in building projects and to analyze their relationship with SPI and CPI as the indicators of the project performance. The results from this study will offer a preferable understanding of the critical factors and help practitioners implicating building projects to take measures to ensure the attainment of project schedule and cost objectives.

To achieve the objective, this study was done by these subsequent phases. Initially, the factor affecting project time and cost performance was analyzed by an extensive journal reconsider. Then, a further literature review was conducted to determine the relationships between influencing factors with project time and cost performance.

2. Research methods

To fulfill the objectives of this study, that is to identify general factors influencing time and cost performance in the building project. 21 journal articles and research reports were reviewed. Journal articles are more than half of whole paper, and the rest is research reports.

2.1. Study process

The reviewed papers were derived from four main databases: Google Scholar, Scopus, Emerald and Science Direct as listed in figure 1. All journal papers have been revised, than the important substance was summarized through each of them. This substance consist the tittle of the journal, case study, the year of the study, identification of factors contributing to the project performance, data analysis techniques and relationship between factors.

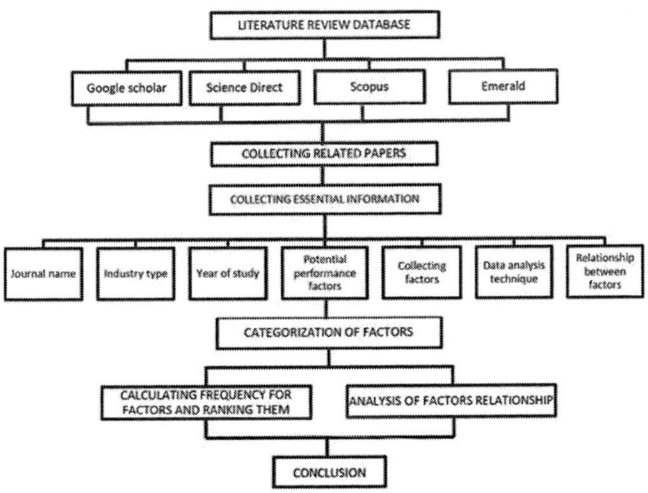

Figure 1. Methodology process

2.2. Journal name

Project performance issues including time and cost indicators have been checked in 13 different journals, and table 1 determine the classification of the publications based on the origin. As shown in that table, 11 different journals name known had appeared the majority of papers on this topic (52 percent of whole articles). The Engineering, Construction and Architectural Management was the majority with 8 articles, and counted for 38 percent of the whole articles. It was continued by Journal of Facilities Management, with 2 papers.

Table 1. Frequency of publication by journals on schedule and cost performance

Journal Name	Frequence	%
Engineering, Construction and Architectural Management	8	38
Journal of Facilities Management	2	9
Other Journals[a]	11	52
Total	21	100

Note: [a]Other journals were those wich had one frequence also *Journal of Civil Engineering and Management, International Journal of Project Management* and *Journal of Economic and Administrative Sciences,* etc.

2.3 Project classify

Figure 2 describe the classification of publication bestow on their case study field: building project, infrastructure project or various construction projects type. Separate the writers who looked for performance factors in the detail industry type, all construction project type was the biggest part of projects, with 62 percent. Building projects was the second reflecting 29 percent, while infrastructure projects th third shown 9 percent of whole projects.

2.4 Publication period

As illustrated in figure 3, publications throughout the last 14 years were classified into five-year periods between 2006 and 2019, and were considered. As illustrated in this illustration, after 2015, there was an upturn in the total of publication papers reported around project performance. With 14 publication papers presented between 2016 and 2019, it shown the biggest number of performance-related researches in all five-year targeted period. Based on this study, there were fewer project-identifying factor studies done before 2010.

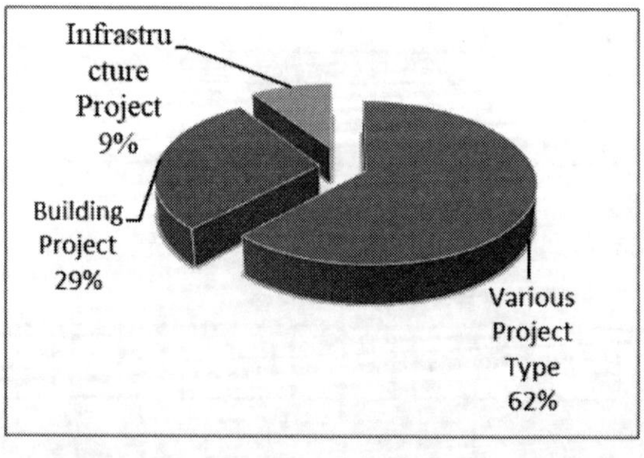

Figure 2. Distribute of journals article in accordance with the type of project

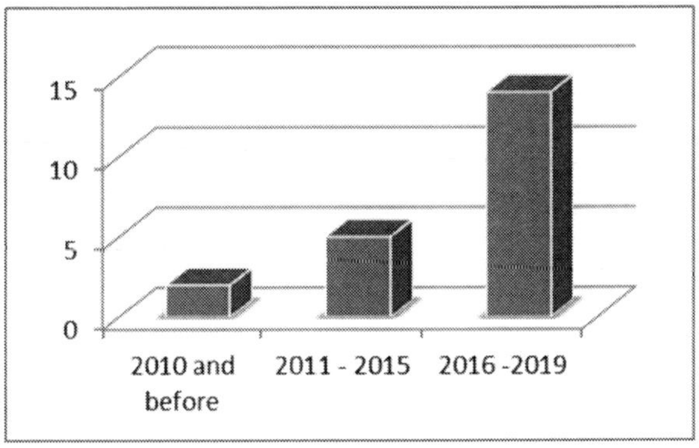

Figure 3. Distribute of journals article in accordance with year of published

2.5 Data processing methods

Some methods were used to assessment the data obtained from the questionnaires, severity index (SI), relative important index (RII), regression, T-test, factor analysis, data classification, etc. Figure 4 shows the classification of data processing methods used in identified journal papers. RII and data classification were the first rank among the processing methods with 29 percent, respectively. The third technique was factor analysis. Other methods equal 16 percent of all utilized techniques.

By a deeper review of previous research that were related with identifying factors of construction time and cost performance, 20 possibility affecting factors were found on table 2.

Show an elaborated empirical study of affecting factors of time and cost performance in the building project done by preceding study. Despite of factors from distinct empirical research are given different names, several of them are identical or overlap. Therefore, factors with few literary journal have combined and then refined them to form the final 15 influencing factors with grouped as follows resource, environmental and management.

Besides the variables found from study literature to get the factors influencing time and cost performance, an analysis is carried out to get the influenced variables. The influenced variables are obtained from the calculation of SPI and CPI equations. Figure 5 explains the steps of identifying variables from the EVM method equation. Where planned value and budget at completion (BAC) obtained from project planning step. So variables that can fickle because influenced by various factors including physical percent complete, direct cost and indirect cost.

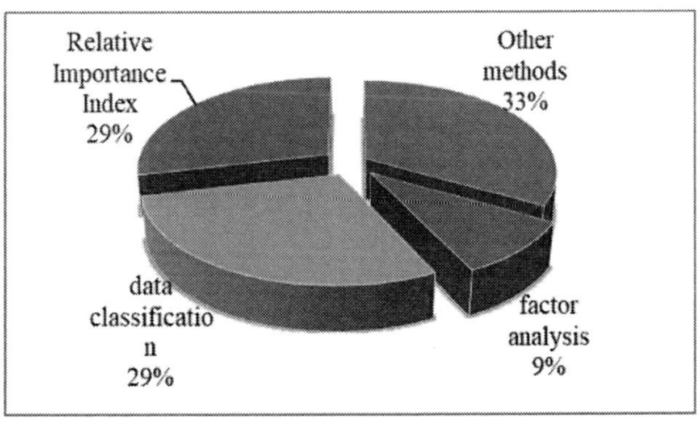

Figure 4. Classification of papers according to data analysis technique

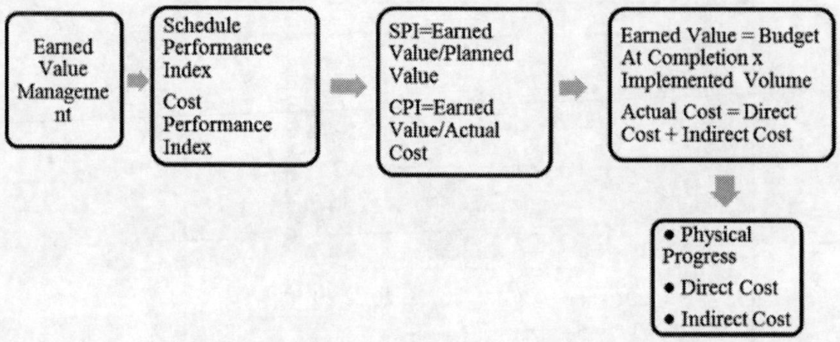

Figure 5. The conceptual model

After obtaining the factors influencing project time and cost performance, the process is followed by identifying the relationship between variables based on the journals that have been reviewed and other journals.

3. Result

As aforementioned, earned value analysis is indicated by SPI and CPI. These two indicators are derived from the ratio earned value and planned value as well as actual cost. Earned value indicator is determined by physical progress and the planned budget. Planned budget and planned schedule are considered as fixed variable, which has been given in the beginning of the project. While physical progress and actual cost are very fluctuate. Therefore, this study only focuses on factors influencing project performance from the perspective of physical progress and actual cost.

3.1. Factors influencing physical percent complete (physical progress)

In general, physical progress is influenced mostly by three major factors; they are environment, resource and management. Environment factor includes weather condition, public facilities and site condition.

Meanwhile, resource factor consists of Materials and labor procurement, availability of materials and equipment on site, skills and productivity of labor, financial condition of owner and contractor, consultant experience level, subcontractor experience level and economic environmental.

Material is one of the factors influencing performance productivity. The availability of sufficient material reduces the risk of waiting by workers [23]. According to [3], shortage of labor and strike by the labor force can influence building project performance. Contractors, owners, and consultants need qualified staff to manage the project they are working on. Deficiency of staff causes a huge material is one of the factors influencing performance productivity. The availability of sufficient material reduces the risk of waiting by workers [29]. According to [3], shortage of labor and strike by the labor force can influence building project performance. Contractors, owners, and consultants need qualified staff to manage the project they are working on. Lack of staff causes a huge burden on the staff at work. In practice, reducing staff can increase responsibility, control, cause poor quality work, improve lower construction management causing construction delays [9, 11]. Moreover, the financial difficulties of the owner can result in late payment by the owner and even closing [10, 11, 26].

Table 2. Schedule and cost performance factors

Variables	Literature
Resources	
Materials and labor procurement	[4], [2], [3], [6], [9], [11], [15],[18], [21]
Availability of materials and equipment on site	[4], [2], [3], [8], [9], [11], [15], [20], [21]
Skills and productivity of labor	[3], [6], [8], [9], [12], [13], [15], [18], [21]
Financial condition of owner and contractor	[8], [5], [7], [6], [14], [17], [18], [20], [21]
Consultant experience level	[2], [6], [7], [11], [15], [16], [20]
Subcontractor experience level	[2], [6], [9], [15]
Economic environmental	[4], [6], [9], [16], [20]
Contractor technology	[11]
Environmental	
Weather conditions	[8], [3], [5], [6], [8], [9], [11], [13], [19], [21]
Public facilities	[3], [13], [20], [21]
Site condition	[4], [2], [3], [5], [6], [9], [11], [13], [19], [21]
Management	
Rework	[6], [9], [13], [17], [21]
Related to design	[8], [6], [7], [8], [11], [16], [18], [19], [21]
The presence of supervising consultant staff	[6], [8], [10], [13], [15], [16], [18]
Slow decision making	[8], [6], [13], [21]
Related to contract	[8], [6], [8], [10], [12], [17]
Project Duration	[12], [14]
Poor coordination	[6]
Safety	[13], [5]
Quality	[15], [19]

The results of the correlation analysis of time overruns are the same as [3] by the known factors as influencing project performance reveal that economic, financial and political clusters have a matter connection with time overrun. The geographical location of a project, soil conditions and weather conditions are examples of the most general environmental effects. Nevertheless, manager construction work will consider physical effect when planning management strategies for avoid the extremes that can make use of what is available resource to get the project goals on time.

Project completions on time, standard of quality and within the assigned budget are the main goals of construction projects. However, there are several unprecedented circumstances that might affect the project performance. To manage building projects efficiently, the coordination process must be applied as an important one function in project management [4]. The more project entities, the more coordination needed. The poor coordination will lead to ineffective project performance, such as project overrun and late schedule [12]. Therefore, it needs such a coordination meeting to support the coordination process [4].

3.2. Factors influencing direct cost

According to the direct cost definition, resource factor is the most dominant factor influencing direct cost. As construction projects take a long time to complete with their complexity and dynamic nature, they are economically risky [6], and have a long history of cost escalation [23]. Total costs can be classified into direct costs related to performing the work, and indirect costs. Direct costs include material, labor, equipment, and subcontractor costs [24]. They can contribute the direct cost influenced by resources, such as not available of material; apparatus and professional workforce impose lots of barrier to a better achievement [9].

According to [10], Material related factors affecting project performance can be associated with material shortages, transportation, material manufacturing, impairment and material changes and delivery. As mentioned by [22], in some developing countries such as Indonesia, the escalation of material price is able to influence the project schedule as the contractor needs to wait until the price. In factors environmental clusters, economic, financial and unexpected price rise of materials had major impact to building project cost performance [3].

Equipment, especially capital equipment, has characteristics and is different requirements for bulk material procurement. [25] State if procurement of capital apparatus has higher waiting times and bigger unit incur costs, and usually for assembly needs special technology. Lack of apparatus, along with apparatus that is not properly maintained, mainly throughout the construction season, could cause project defeat or lead it to diverge.

3.3. Factors influencing indirect cost

Indirect cost, according [14], two major element of it that defined which determine in detail are construction overheads, management costs. Construction overhead can be defined as the total cost incurred by the contractor at the construction site related to delivery work, but is not included in standard labor costs, materials and of construction equipment. Management costs are incurred to maintain functional units and personnel work that carry out general administrative activities carry out supervisory functions, as well as control, register, report and are related to instructions and other functions related to the organization, management and administration of construction companies.

Similar to physical progress, indirect cost is mostly influenced by management, resource and environment factors. Various factors influencing overhead and management cost. One of the most common factors that causing the final project cost to exceed the initial estimate (overhead) is force majeure. Even though the contractor is usually required to obtain an insurance policy against this, if such an event does occur without this provision it may cause delays and further affect costs [5]. Legal compliance and environmental issues are also considered important in influencing project cost performance [8]. Project managers, in addition to their traditional functions, must set up processes for scanning the environment, identifying potential problems, and trying to build power relationships that can assist in managing the factors underlying successful implementation [3].

Construction work managers and supervisors have to apply a suitable leadership style in project management to minimize pointless cost overruns [3]. Inexperienced stakeholders or have new comers can cause delays and cost overruns [11, 16]. According to [10] in [17] factors related to contractor site management, factors related to design and documentation, factors related to financial management can influence cost overruns.

The distribution of relationship between influencing factors with time and cost performance index is illustrated in figure 6. Most portion of the factors were distributed to resource group, indicating the importance of this group, especially in the procurement and availability of material, labor and equipment. Labor productivity and contractor's economic condition also have high importance. The resource factor also has a large impact on time and cost performance, indicated by its effect on implemented volume, direct cost and indirect cost. The second place was management group with having five factors, than the third rank was environment group having three factors.

4. Conclusion

This paper presents a literature study on influencing factors of time and cost performance on building projects based on the EVM method. This paper identifies 15 influencing factors associated with project time and cost performance. Then, these factors grouped into 3 cluster, environmental, resource and management. According to finding in this paper, noted that environmental variables can influence implemented volume and indirect cost. Resources variable can influence implemented volume, direct cost and indirect cost, while management variable also can affect implemented volume and indirect cost.

The finding of this study can find out what factors can affect the project's time and cost performance to provide a better understanding of practitioners to take steps to ensure the achievement of the project schedule and cost objectives.

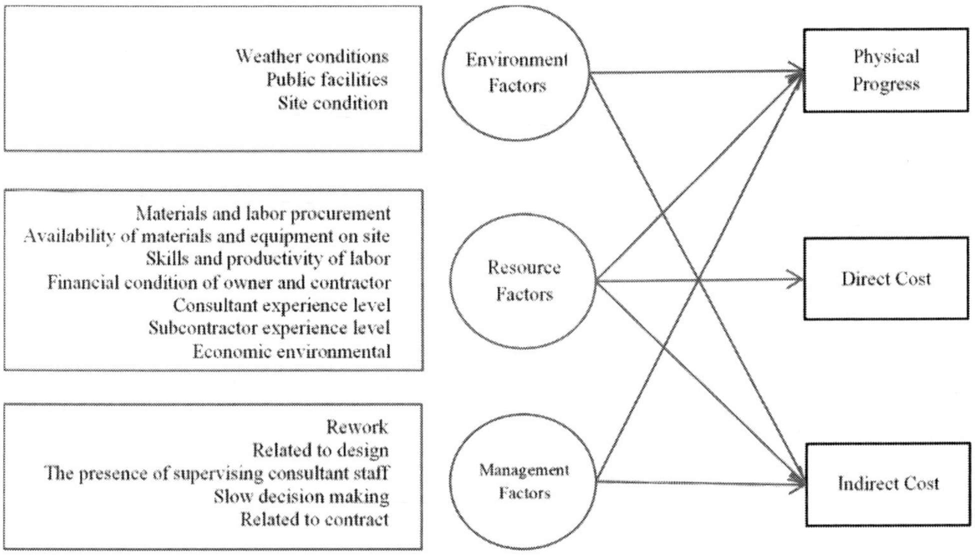

Figure 6. The identified relationship of schedule and cost performance

Acknowledgements
The results of the paper are mainly derived by literature review, so the results still need validation by quantitative research method based on the survey. It is also unknown how big the contribution of each variable influences the project schedule and cost performance index. To validate the findings in next study, then could apply more specific case studies and conduct interviews. So that it can be implemented directly and compared with real conditions on the site.

References

[1] Adam A, Josephson P B and Lindahl G 2017 Aggregation of factors causing cost overruns and time delays in large public construction projects: Trends and implications *Eng., Constr. and Archt. Mgmt.* **24** 3 393 – 406
[2] Adnan E, Sherif M and Saleh A 2009 Factors affecting the performance of construction projects in *J. of Civ. Eng. and Mgmt.* **15** 3 269–80
[3] Akkani P O, Oke A E and Akpomiemie O A 2014 Impact of environmental factors on building *Housing and Bldg. National Res. Center*
[4] Alaloul W S, Liew M S and Zalawi N A W 2016 Identification of coordination effecting building projects performance *Alexandria Eng. J.*
[5] Ali A S, Azmi N F and Baaki T K 2018 Cost performance of building refurbishment works: the case of Malaysia *Intl. J. of Bldg. Pathology and Adaptation*
[6] Creedy G D, Skitmore M and Wong J K W 2010 Evaluation of risk factors leading to cost overrun in delivery of highway construction projects *J. of Constr. Eng. and Mgmt.* **136** 5 528-37
[7] Das D and Ngacho C 2015 Critical success factors influencing the performance of development project: An empirical study of constituency development fund project in Kenya *IMB Mgmt. Rev.* **29** 276-93
[8] Doloi H 2013 Cost overruns and failure in project management: understanding the roles of key stakeholders in construction projects *J. of Const. Eng. and Mgmt.* **139** 3 267-79
[9] Enshassi A, Mohamed S and Abushaban S 2009 Factors affecting the performance of construction projects in the Gaza strip *J. of Const. Eng. and Mgmt.* **15** 3 269-80
[10] Habibi M, Kermanshachi S, 2018 Phase-based analysis of key cost and schedule performance causes and preventive strategies: Research trends and implications *Eng., Constr. and Archt. Mgmt.* 0969 – 9988

[11] Hwang B G, Tan H F and Satish S 2013 Capital project performance measurement and benchmarking in Singapore *Eng., Constr. and Archt. Mgmt.* **20** 2 143 – 59

[12] Hwang B G and Yang S 2014 Rework and schedule performance: A profile of incidence, impact, causes and solutions *Eng., Constr. and Archt. Mgmt.* **21** 2 190 – 205

[13] Khesal T, Saghaei A, Khalilzadeh M, Galankashi M R and Soltani R 2018 Integrated cost, quality, risk and schedule control through earned value management (EVM) *J. of Eng., Design and Tech.*

[14] Leśniak A 2013 The application of artificial neural networks in indirect cost estimation. *11th Intl. Conf. of Numerical Analysis and Applied Mathematics*

[15] Lindhard S and Larsen J K 2016 Identifying the key process factors affecting project performance *Eng., Constr. and Archt. Mgmt.* **23** 5

[16] Lei Z, Tang W, Duffield C and Zhang H F 2017 The impact of technical standards on international project performance: Chinese contractors' experience *Intl. J. of Proj. Mgmt.* **35** 1597–607

[17] Lu W, Hua Y and Zhang S 2017 Logistic regression analysis for factors influencing cost performance of design-bid-build and design-build projects *Eng., Constr. and Archt. Mgmt.* **24** 1 118 – 32

[18] Manley, K., Chen, L., 2016 The impact of client characteristics on the time and cost performance of collaborative infrastructure projects *Eng., Constr. and Archt. Mgmt.* 23 4 511 – 32

[19] Memon A H and Rahman I A 2013 *Analysis of cost overrun factors for small scale construction projects in Malaysia using PLS-SEM method Modern Applied Science* vol. 7 no. 8 pp 78-88

[20] Mohamad R, Ismail S and Said J M 2018 Performance indicators for public private partnership (PPP) projects in Malaysia *J. of Econ. and Admin. Sci.*

[21] *Nadafi S, Moosavirad S H and Ariafat S 2019* Predicting the project time and cost using EVM based on gray number *Eng., Constr. and Archt. Mgmt. 0291*

[22] *Omran A, Abdalrahman s and Pakir A K 2012* Project Performance in Sudan construction industry: A case study *Global J. of Acct. and Econ. Res.*

[23] Soekiman A A, Pribadi B W and Wirahardikusumah R D 2011 Factors Relating to Labor Productivity Affecting the Project Schedule Performance in Indonesia *The Twelfth East Asia-Pacific Conf. on Struc. Eng. and Constr.*

[24] Tarore H, Tjakra and J Walangitan 2 2012 Analisis Optimalisasi Waktu dan Biaya Dengan Program Primavera 6.0 *Jurnal Sipil Statik* **1** 1.53-9

[25] Yang H 2010 A critical review of performance measurement in construction *J. of Facilities Mgmt.* **8** 4 269-84

[26] Zailani S, Iranmanesh M 2016 The moderating effect of project risk mitigation strategies on the relationship between delay factors and construction project performance *J. of Sci. and Tech. Policy Mgmt.* **7** 3

Risk allocation of PPP waste to energy projects in Indonesia: A research framework

W P Utama[1*], A Wibowo[2], D Y Jumas[1], E Rita[1], M Peli[1], and Yulcherlina[1]

[1] Bung Hatta Center for Construction Management and Quantity Surveying Studies, Faculty of Civil Engineering and Planning, Universitas Bung Hatta Jl. Sumatera Ulak Karang, Padang 25133– Indonesia
[2] The Ministry of Public Work and Housing, Republic of Indonesia Jl. Pattimura No. 20 Kebayoran Baru, Jakarta Selatan 12110 – Indonesia

*Corresponding author's e-mail: wahyudi@bunghatta.ac.id

Abstract. Public-Private Partnership (PPP) scheme has been widely adopted in both developed and developing countries to fulfill the need of infrastructure facilities such as toll roads and water supply. In Indonesia, another sector that would be feasible to be developed using the scheme is the energy sector through Waste to Energy (WTE) projects. However, the lack of knowledge and experience of public sectors on the PPP field particularly risk-allocation issues may barrier the project success. This paper presents a research framework to develop a model of risk allocation decision making for PPP WTE projects in Indonesia. A combination of qualitative (literature review and semi-structured interview) and quantitative (Delphi survey) research method is utilized to identify critical risk factors in PPP WTE projects. To develop a risk allocation model, an Adaptive Neuro-Fuzzy Inference System (ANFIS) with two outputs (public and private risk share) is proposed. A set of surveys to collect data for training and checking will be distributed to PPP experts representing government and private sectors in Indonesia, while selected WTE projects are used for data testing. The proposed model is a novel approach in risk allocation PPP research and would be useful for local authorities in Indonesia in proposing the WTE project using the PPP scheme.

1. Introduction

The availability of infrastructure facilities is considered as one of the national economic development pillars. World Economic Forum (WEF) uses the infrastructure as a global competitiveness measure of one country and it empirically correlates to economic growth positively and poverty gap negatively [1]. Electricity as one of the major energy sources plays a vital and strategic role in industrial development. It constitutes a fundamental energy infrastructure which is an essential need for the trade and manufacturing sectors to create high productivity factors [2]. Inadequate supply of electrical energy becomes an obstacle to the growth of investment for driving the economy of a region. Failure to meet such a facility threatens the quality of life and health of the community and impedes economic growth [3].

In reality, the need for electricity in Indonesia has not been fulfilled entirely with around 2,510 villages still unelectrified throughout Indonesia. Based on the Electricity Supply Business Plan (RUPTL) 2018–2027 of PT. PLN, a State Electrical Company, the demand has been estimated at 56,000 MW with the annual projected growth of 6.86%. However, PT. PLN can only supply approximately 30% of the total need. Meanwhile, Government Regulation No. 79/2014 that sets the national energy

Content from this work may be used under the terms of the Creative Commons Attribution 3.0 licence. Any further distribution of this work must maintain attribution to the author(s) and the title of the work, journal citation and DOI.
Published under licence by IOP Publishing Ltd

policy also has the objective to reduce the dependence of fuel oil as a source of energy. In line with this policy, PT. PLN plans to increase the use of alternative sources for electrical power generators, including renewable ones such as wind, solar, and waste.

In terms of waste as a source of energy, current technology has afforded to transform garbage into electricity [4]. According to Zhao et al. [5] and Wang and Zhang [6], waste to energy (WTE) incineration constitutes the most effective technique to manage municipal solid waste. Many countries have successfully adopted and implemented the WTE program, such as Denmark [7], Japan [8], and China [6]. In the southeast Asia region, WTE plants have thrived in Thailand, Singapore, Vietnam, Philippine, and Indonesia [9]. WTE is a solution for solving the energy crisis and at the same time may tackle the urban waste problem [10].

By regulation, PT. PLN is not obligated to develop such plants but is mandatory to buy the electrical power from WTE plants. To promote the WTE implementation, the Indonesian government has set a relatively high price for the WTE-generated electricity i.e., USD 13.3 per kWh based on Presidential decree No 35/2018. Local authorities are therefore strongly urged to actively contribute to developing the plants in their regions. Nowadays, the Indonesian government fosters WTE projects development in several cities such as Jakarta, Tangerang, Bandung, Surabaya, and Makassar (see figure 1). Though the provision of infrastructure facilities is fully under government responsibility, the ability to realize them could not be fulfilled solely. Given the importance of public facilities and the limitedness of public sources, the government strongly encourages the involvement of private sectors.

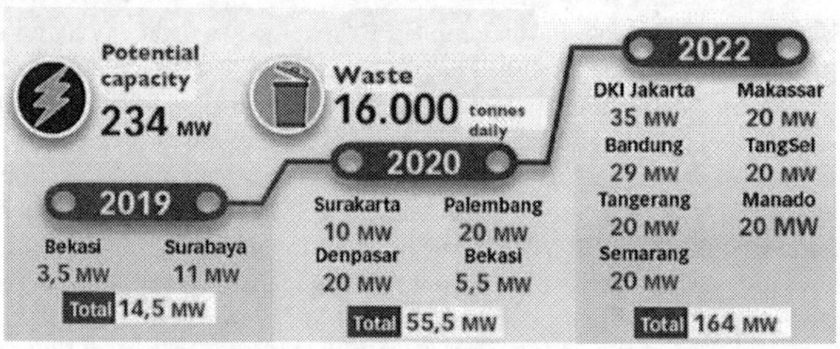

Figure 1. The development plan of Waste to Energy plants in Indonesia [11]

As with other greenfield infrastructure projects, WTE projects are typically capital intensive investments. Examples include the Surakarta 10 MW WTE project, which is worth USD 57.82 million, the Denpasar 15-20 MW of USD 120 million, the Jakarta 35 MW of USD 345.8 million, and Palembang 20 MW of USD 120 million [12]. However, attracting local government participation in WTE plant development features a common problem arising from the fact that the ability to finance the huge investments differs from authority to authority. Local authorities such as Jakarta, Surabaya, and Makassar appear not to have significant obstacles in realizing the WTE plant. On the contrary, cities with a limited budget have to seek private investors to support financing the plants. In connection with the private sector participation in the public sectors, the Indonesian government has regulated the public-private partnership (PPP) infrastructure provision under Presidential Regulation No. 38/2015.

2. Problem statement

The success of PPP projects heavily relies on the advisability of management and risk-sharing. Poor risk management can cause serious problems [6]. There exist many important risk events and project failures in WTE projects implemented under the PPP scheme [13-15]. Therefore, the government and private sectors should understand PPP specific risks regarding the country and type of project, which is WTE in this case.

The basic principle of the PPP scheme is risk-sharing. Risk allocation has a pivotal role in PPP contracts and is the key to the success of the PPP scheme [16]. Distribution of risk means determining which party bears negative consequences if the risk occurs in the future [17]. Furthermore, structuring risk allocation between the government and the private sector in the PPP scheme is often not easy. Chung et al. [18] opine that the intrinsic risk in the PPP scheme is the process of decision making for risk allocation itself. Ameyaw and Chan [19] argue that "The difficulty of a risk-allocation decision becomes the differences in perceptions regarding risk criticality and risk-management capability of each of the parties involved". Thus, an effective decision on risk allocation does not depend on preferred options and mainstream opinions. To tackle this issue, several studies propose decision frameworks to support risk allocation decision-making for PPP projects.

Despite the significance of the risk allocation-related issue in WTE projects, research in this area in particular within the Indonesian context is not well established in the existing body of literature. This research is therefore focused on the model development of risk allocation decision-making for PPP-WTE projects. The results of this research are expected to facilitate decision-makers, especially local governments, in designing policies and specific strategies in managing project risks and at the same time to attract the interest of the private sector.

3. Research and objectives

The research aims at developing a dependable and practical model of risk allocation decision making in PPP waste to energy projects in Indonesia. The specific objectives to achieve the research aim are:

- To identify the characteristic and risk environment in WTE projects;
- To identify and evaluate risk factors of PPP WTE projects;
- To determine the most important risk allocation criteria in PPP WTE projects;
- To develop a risk assessment model for PPP WTE projects;
- To develop a decision-making model of risk allocation for PPP WTE projects.

The present paper is dedicated to describing the research framework that will be adopted to meet this research's aim and objectives.

4. Research framework of the study

This research adopts a sequential combination of research methods encompassing both qualitative and quantitative techniques to enhance data validity and reliability [20]. This research begins with a comprehensive literature review of WTE projects and its risk environment, the PPP applications for WTE projects, and risk management procedure riveted on risk factors and risk allocation. This stage is directed to achieve the research objective 1 and objective 2.

The next stage is the collection of primary data that involves a two-round Delphi survey and personal interviews with industrial practitioners in the WTE projects in Indonesia. This stage is designed to attain research objective 3 and objective 4. The collected data will be statistically analyzed using mean score ranking and factor analysis. The ultimate objective of this research agenda, which is to develop a decision-making model of risk allocation will be achieved by employing the Adaptive Neuro-Fuzzy Inference System (ANFIS). The design of the research framework for this study is illustrated in figure 2.

4.1. Literature review

This research systematically reviews academic publications from scientific journals, proceedings, dissertations, textbooks, institutional reports, and internet sources. This review aims to explore the research transformation in the area of interest. It is firstly directed to investigate the development of incineration WTE projects in Indonesia. The primary objective of the review is to identify relevant risk factors concerning WTE projects and the decision-making approaches applied for risk allocation strategy. This review also assists to define the gaps in the research.

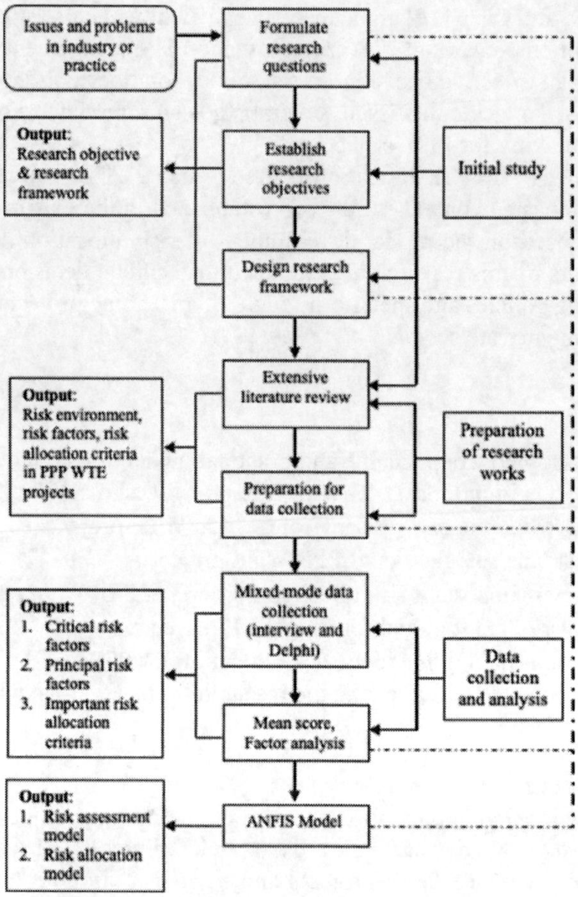

Figure 2. Research framework of the study

4.2. Inteviews

The purpose of the interview is to solicit clarification and elaboration [21, 22], which apprehend the issues from the respondents' perspectives. The qualitative information seized during the interviews enriches the data by extending the insights [23]. After the literature review is done, the research will proceed with interviews with industrial practitioners e.g., WTE operators or investors and concerned governmental bodies e.g., BAPPENAS, PT. PLN, and local authorities. This stage is performed to obtain their viewpoints on WTE projects and their risk milieu. Along with the literature review outcomes, the information gathered from the interviews will be used to prepare for the Delphi questionnaire survey.

4.3. Delphi survey

Delphi is a technique for arranging a process of group communication to deal with a complex problem that needs building up consensus or convergence of thought [24]. The Delphi technique employs a series of questionnaires iterations to collect data from selected experts. This technique has been more appropriate than a simple questionnaire for item prioritization [25] and can minimize the biases resulting from the dominant panelist, unrelated conversation, and group insistence [24].

Among the benefits of the Delphi technique are panelists' agreement achieved through repeated surveys and suitable for studies in which past data are shortcoming for application of other approaches [26]. Given that the information on WTE projects and the number of WTE plants in Indonesia are scanty, the use of the Delphi as a compatible data collection technique for the current study is very reasonable. Thus, this method will be used to evaluate and assess risk factors in PPP WTE projects in Indonesia.

Regarding the number of repetitions, this research will administer the Delphi survey in two rounds. A two-round exercise lessens Delphi panelists' exhaustion and attrition biases than the reiteration rounds [27].

Regardless of the panelist's criteria, determining the suitable size of experts involved is still in disagreement or debatable. Because PPP is new and emerging in Indonesia, identifying panelists having adequate expertise and knowledge also poses another challenge. Two strategies namely purposive sampling and a semi snowballing method will be used to solve this problem, as advised by Ameyaw [26]. The collected data in this stage will then be analyzed using three tests, including Kendall's coefficient of concordance (W), Spearman's rho correlation, and the Kruskal-Wallis test. The expected results from this analysis are the identified critical risk factors and important risk allocation criteria (RAC) in PPP WTE incineration projects. RAC is used to assess each risk factor, which is required for the ANFIS-based model development.

4.4. ANFIS model for risk allocation

Most of the models related to risk allocation are however addressed to general PPP infrastructure projects and water sectors. A hybrid of fuzzy evaluation and statistical mean and neural networks is often used to develop risk allocation decision-making models [26]. The unique characteristics of WTE projects compared with other public facilities e.g., road and water supply may curb the effectiveness of general PPP risk allocation design from previous studies.

This research proposes the ANFIS algorithm to make a major and specific contribution to PPP risk allocation in WTE projects in Indonesia. The process of making decisions in a complex environment involving multi-risk factors, the degree of risk and uncertainty, and unrefined information like WTE projects may result in the troublesome problem [28]. Besides, realistically, the relationship among the risk factors is typically non-linear and some factors unidentified. To deal with the limitations, the use of mathematical concepts of Fuzzy Logic (FL) and Neural Network (NN) is adopted by many researchers. Both FL and ANN have merits and demerits, and the amalgamation of the two algorithms bring advantages to existing ones. ANFIS is a kind of integration of FL and NN.

The fundamental concept of ANFIS is to generate the specified input-output pairs through constructing a set of fuzzy if-then rules. In this research, ANFIS is employed to create an input-output network structure. The input (I) consists of risk factor scores obtained from RAC assessment on identified risk factors, while the output (O) is the result of the fuzzy inference system embedded into the structure of adaptive networks. In this context, government and private risk allocation of each identified risk factors are set as outputs. The design of the ANFIS model for risk allocation of PPP WTE projects can be shown in figure 3.

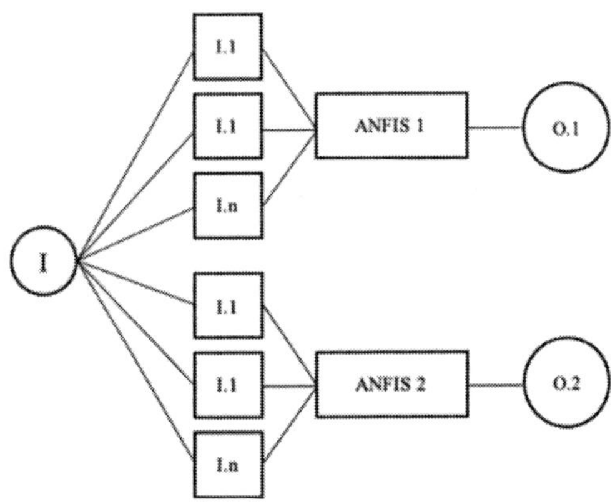

Figure 3. ANFIS network for risk allocation of PPP WTE projects

ANFIS model needs several data for training, checking, and testing. A set of questionnaires will be designed and distributed to PPP experts in Indonesia representing government and private partners.

Selected WTE projects will be used for testing the model. To validate the model network, the common indicators i.e., Root Mean Square Error (RMSE) to measure the performance and the correlation coefficient (R) to view the efficiency will be applied.

5. Conclusions

Development of Waste to Energy (WTE) plants is one of the solutions to solve the municipality solid waste problem and at the same time to increase the electrical power supply. Correspondingly, by Presidential Regulation No. 38/2015, the Indonesian government boosts cooperation between the local authorities and private sector participation to provide public facilities through the PPP mechanism. Due to the WTE projects vary from other public facilities, the PPP implementation such as sharing the risks among partners in such a project would not be synchronized with other public facilities. This paper presents a research framework to develop a model of risk allocation decision making for PPP WTE projects in Indonesia.

The research aim has been established which is to develop a dependable and practical model of risk allocation decision making in PPP WTE projects in Indonesia. Accordingly, specific objectives have been defined to reach the research aim namely A research framework was designed to address the research objectives involving mixed-mode research approaches embracing literature review, semi-structured interviews, and the Delphi survey. The literature review serves the identification of risk environment and risk factors in PPP WTE projects, while semi-structured interviews to gain deep insight into the practice of general PPP and WTE projects in Indonesia. Delphi survey assists to define critical risk factors and the importance of risk allocation criteria. Finally, the research proposes the ANFIS algorithm to develop decision-making model for risk allocation of PPP WTE projects in Indonesia. A better understanding of the mechanism of and the use of reliable risk allocation decision-making techniques may help Indonesian local authorities to attract the participation of private sectors in energy facilities through WTE projects.

Acknowledgments

Authors thank Bung Hatta Center For Construction Management and Quantity Surveying Studies, Faculty of Civil Engineering and Planning, and Universitas Bung Hatta for their positive support.

References

[1] Wibowo A 2016 Perkembangan terkini dalam pembiayaan infrastruktur yang melibatkan partisipasi badan usaha *Prosiding Konferensi Nasional Teknik Sipil X* pp 26-7
[2] Guasch JL and Straub S 2009 Corruption and concession renegotiations: Evidence from the water and transport sectors in Latin America *Utilities Policy* **17** 2 185-90.
[3] Kessides I 2004 Reforming infrastructure: Privatization, regulation, and competition *The World Bank*
[4] Cheng H and Hu Y 2010 Municipal solid waste (MSW) as a renewable source of energy: Current and future practices in China *Bioresource Tech.* **101** 11 3816-24
[5] Zhao XG, Jiang GW, Li A and Wang L 2016 Economic analysis of waste-to-energy industry in China *Waste Manag.* **48** 604-18
[6] Wang L and Zhang X 2017 Critical risk factors in PPP waste-to-energy incineration projects *Intl. J. of Archt., Eng. and Constr.* **6** 2 55-69.
[7] Hulgaard T and Inger S 2018 Integrating waste-to-energy in Copenhagen, Denmark *Proc. of the Inst. of Civ. Engineers-Civil Eng.* (London: ICE Publishing 2018) **171** 5 pp 3-10.
[8] Jung CH, Matsuto T, Tanaka N and Okada T 2004 Metal distribution in incineration residues of municipal solid waste (MSW) in Japan *Waste Manag.* **24** 4 381-91
[9] Hasnan, L 2020 Today's Waste is Tomorrow's Energy [Internet] Kuala Lumpur: The Asean Post; 2019 [updated 2019 Aug 8; cited 2020 Jul 6]. Available from: ttps://theaseanpost.com/article/todays-waste-tomorrows-energy
[10] Faizal M, Wardah YH, Husna MA, Amirah A and Tan YH 2018 Energy, economic and environmental impact of waste-to-energy in Malaysia *J. of Mech. Eng. Res. and Dev.* **41** 3 97-101

[11] Koran Jakarta 2019 Listrik Tenaga Sampah - Pilihan Teknologi Mesti Disesuaikan Karakteristik Sampah, Percepatan PLTSa Butuh Dukungan Semua Sektor [Internet] Jakarta: Koran Jakarta; 2019 [updated 2019 Jul 19; cited 2020 Jul 6]. Available from: http://www.koran-jakarta.com/percepatan-pltsa-butuh-dukungan-semua-sektor/

[12] Wiratmini, N.P.E. Tahun Ini 11 MW PLTSa Terpasang [Internet]. Jakarta: Bisnis.com; 2019 [updated 2019 Aug 8; cited 2020 Jul 6]. Available from: ttps://ekonomi.bisnis.com/read/20190808/44/1133886/tahun-ini-11-mw-pltsa-terpasang

[13] Belevi H and Langmeier M 2000 Factors determining the element behavior in municipal solid waste incinerators 2 *Lab. Experiments Envr. Sci. & Tech.* **34** 12 pp 2507-12

[14] Menard Y, Asthana A, Patisson F, Sessiecq P and Ablitzer D 2006 Thermodynamic study of heavy metals behaviour during municipal waste incineration *Process Safety and Envr. Protection* **84** 4 290-6

[15] Mills E, Kromer S, Weiss G and Mathew PA 2006 From volatility to value: analysing and managing financial and performance risk in energy savings projects *Energy Policy* **34** 2 188-99

[16] Wibowo A and Mohamed S 2008 Perceived Risk Allocation in Public-Private-Partnered (PPP) Water Supply Projects in Indonesia *First Intl. Conf. on Constr. in Developing Countries* **4** 349

[17] Ameyaw EE and Chan AP 2015 Risk allocation in public-private partnership water supply projects in Ghana *Constr. Manag. and Econ.* **33** 3 187-208

[18] Chung D, Hensher DA and Rose JM 2010 Toward the betterment of risk allocation: Investigating risk perceptions of Australian stakeholder groups to public–private-partnership toll road projects *Res. in Transp. Econ.* **30** 1 43-58.

[19] Ameyaw EE and Chan AP 2016 A fuzzy approach for the allocation of risks in public–private partnership water-infrastructure projects in developing countries *J. of Infrastructure Systems* **22** 3 04016016

[20] Abowitz DA and Toole TM 2010 Mixed method research: Fundamental issues of design, validity, and reliability in construction research *J. of Constr. Eng. and Manag.* **136** 1 108-16

[21] Abdul-Aziz AR and Wong SS 2010 Exploring the internationalization of Malaysian contractors: The international entrepreneurship dimension *Constr. Manag. and Econ.* **28** 1 51-61

[22] Utama WP, Chan AP, Zahoor H, Gao R. 2019 Preferred Entry Mode Choices and Transformation of Indonesian Contractors' Strategy in International Markets *J. of Constr. in Developing Countries* **24** 2 173-88

[23] Sandelowski M 2000 Combining qualitative and quantitative sampling, data collection, and analysis techniques in mixed-method studies *Res. in Nursing & Health* **3** 246-55

[24] Walker A, Ameyaw EE and Chan AP 2015 Evaluating key risk factors for PPP water projects in (Ghana: a Delphi study) *J. of Facilities Manag.* **13** 2 133-55

[25] Zahoor H, Chan AP, Gao R and Utama WP 2017 The factors contributing to construction accidents in Pakistan *Eng., Constr. and Archt. Manag.* **24** 3 463-85

[26] Ameyaw EE 2014 Risk allocation model for public-private partnership water supply projects in Ghana [Dissertation] (Hong Kong: The Hong Kong Polytechnic University)

[27] Zahoor H, Chan AP, Utama WP and Gao R 2015 A research framework for investigating the relationship between safety climate and safety performance in the construction of multi- storey buildings in Pakistan *Procedia Eng.* **118** 581-9

[28] Utama WP, Chan AP, Zahoor H, Gao R and Jumas DY 2019 Making decision toward overseas construction projects *Eng., Constr. and Archt. Manag.* **26** 2 285-302

The effect of m-sand and waste marble for strength of concrete

R Hidayawanti[1*], Yuhanah[2], D Mayasari[3] and B Wicaksono[4]

[1,2,3,4] Department of Civil Engineering, Institut Teknologi PLN, Indonesia

*Corresponding author's e-mail: ranti@itpln.ac.id

Abstract. Concrete is a construction material composed of cement, water, and aggregate in the form of sand and gravel. Furthermore, Indonesia experiences a yearly increase in the need for building material, including concrete, due to the rise in development. Smooth and coarse aggregate materials are required to manufacture concrete with the inappropriate disposal of waste marble, adversely impacting environmental pollution. Therefore, tremendous efforts need to be carried out to reduce environmental problems associated with the disposal of waste marble to provide adequate economic value for construction and reduce the collection of rough aggregates in the river. A lot of industrial companies are needed to produce a coarse aggregate with various size variants, thereby, manufacturing waste, also known as M-Sand. The stone ash is one of the numerous coarse aggregate materials that are less desirable in concrete production. Therefore, this study was carried out to determine the compressive strength and water absorption content between fine and coarse concrete using a variety of coarse aggregate replacement materials from the waste marble with a 40% M-Sand substitute. The results showed waste marble variations of 0%, 25%, 50%, 75%, 100% substitution of coarse aggregate with 40% M Sand substitution of fine aggregate. The optimum compressive strength was found in variation 3 (40% M Sand + Waste Marble 50%) of 35.801 MPa at 28 days, with the optimum water absorption obtained at variation 1 (40% M-Sand + 0% Waste Marble) at 1.77%. By analysing the waste marble result of R close to 1, it is concluded that this variation can act as an alternative to the composition of a concrete mixture.

1. Introduction

1.1. Utilization of Concrete Waste

Indonesia experiences a yearly increase in the need for building material, including concrete, due to the rise in development. The use of concrete as a construction material is increasingly widely, as seen from the rise in its production, especially in the ready mix industry. Approximately 70% of building material compilers in the construction world use concrete [1] because it is a relatively inexpensive with a large compressive strength, easy to work with and can be formed according to the desired shape and corrosion resistant [2]. One of the material needs for making concrete is coarse aggregate, which is taken from rivers. The continuous extraction of this material from river causes damages to it, therefore, the use of other resources is expected to minimize and reserves this aggregate sources. One of such alternatives is the use waste of marble. Furthermore, many industrial stone companies are needed in the construction

Content from this work may be used under the terms of the Creative Commons Attribution 3.0 licence. Any further distribution of this work must maintain attribution to the author(s) and the title of the work, journal citation and DOI.
Published under licence by IOP Publishing Ltd

world to produce coarse aggregates of various sizes known as M-Sand or stone ash, which is used an alternative material [3]. M-Sand is a less desirable material used as a substitution of natural sand in concrete production. Therefore, this study examines the manufacture of M-Sand as an alternative to fine aggregate and waste marble as an alternative to coarse aggregate in concrete production [4]. This research was also carried out to determine the compressive strength and absorption rate of the remaining waste marble and M-Sand.

1.2. The Problem to be addressed

The increasing difficulty in obtaining raw materials for coarse aggregate (gravel) and fine aggregates (sand), to minimize the use of marble and M-Sand are some problems to be addressed in this study. The waste needs to be tested against its compressive strength to determine its ability to fulfil the SNI standard requirement for normal concrete. The study utilized the composition waste marble variations of 0%, 25%, 50%, 75% 100% for coarse aggregate and 40% M-Sand of fine aggregate to determine the value that produces optimum compressive strength. M-Sand and waste marble is an environmentally friendly concrete material[5], with more values as a substitute for sand and gravel[6]. Utilization of M-Sand and waste marble has a better compressive strength and absorption value than conventional concrete. In addition, to determine the proportion of the mixture of waste marble with M Sand 40% and to determine the comparison of the compressive strength and absorption of concrete using a mixture of M-Sand and the variation of waste marble with conventional concrete.

2. Methodology

2.1. Proportion of Mixed Concrete used for Mix Design of M-Sand and Waste Marble

Table 1. Material requirements needed for the variation of m-sand and waste marble test

No.	Sample	Material Composition					
		Water (Kg)	Cement (Kg)	Sand (Kg)	Gravel (Kg)	M-Sand (Kg)	Waste Marble (Kg)
1.	Normal Concrete	11.6107	24.7038	48.7398	73.7458	0	0
2.	Variation 1 (M-Sand 40% + Waste Marble 0 %)	11.611	24.704	29.244	73.746	19.49593288	0
3.	Variation 2 (M-Sand 40% + Waste Marble 25%)	11.611	24.704	29.244	55.309	19.496	18.436
4.	Variation 3 (M-Sand 40% + Waste Marble 50%)	11.611	24.704	29.244	36.873	19.496	36.873
5.	Variation 4 (M-Sand 40% + Waste Marble 75%)	11.611	24.704	29.244	22.124	19.496	51.622
6.	Variation 5 (M-Sand 40% + Waste Marble 100%)	11.611	24.704	29.244	0	19.496	73.746
	Total	69.664	148.223	194.959	261.797	97.48	180.677

2.2. Research process

This study determines the correlation value between the use of M-Sand and waste marble as a substitution of fine and coarse aggregate materials in manufacture of conventional concrete. Materials used are water, cement, fine aggregate (sand), coarse aggregate (gravel), M-Sand and waste marble.

Figure 1. Several materials for testing specimen of concrete

3. Result and Discussion

3.1. Laboratory testing result

The aggregate that will be used as concrete mixture material on the test specimens is checked for parameters including gradation, specific gravity, sludge content, organic, absorption.

3.1.1. M-sand gradation analysis test result

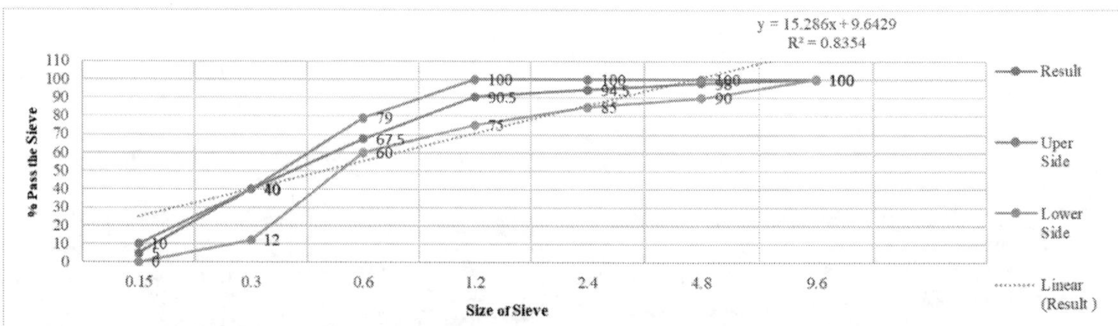

Figure 2. M-Sand gradation analysis test results (source of analysis 2020)

Figure 2 shows the category of M-Sand testing entered in the fine sand (zone III) with a fineness modulus of 3.045, which meets the refinement requirements stipulated in ASTM C-33[7]. The refinement modulus of 2.3 - 3.1, enables the utilization of this broken rock waste as a concrete mixture material.

3.1.2. Specific gravity of m-sand and water absorption

Table 2. Specific gravity data for m-sand and water absorption

No.	Description		Result
1.	Weight of Laboratory (S)		500 gr
2.	Weight of Laboratory after drying (A)		495 gr
3.	Water Weight + pycnometer (B)		872 gr
4.	Pycnometer weight + 500 gr Laboratory + Water (C)		1193 gr
5.	Dry Specific Gravity	$\left(\frac{A}{(B+S)-C} \right)$	2.76
6.	*Saturated Surface Dry* (SSD)	$\left(\frac{S}{(B+S)-C} \right)$	2.79
7.	% Absorption	$\left(\frac{S-A}{A} \times 100\% \right)$	1.01%

The results of testing the specific gravity of M-Sand and water absorption show that the dry specific gravity in the test was 2.76 and the specific gravity under SSD (Saturated Surface Dry) conditions obtained was 2.79. Meanwhile, the test results of water absorption of 1.01%.

3.1.3. Mud content in m-sand

Table 3. Result of mud content data in m-sand

No.	Description		Result
1.	Container Weight (W_1)		306 gr
2.	Weight of the test specimens + container (W_2)		806 gr
3.	Weight of the test specimens (W_3)		500 gr
4.	Dry weight of test specimens after oven + container (W_4)		796.50 gr
5.	Dry weight of test specimens after oven (W_5)		490.50 gr
6.	*Saturated Surface Dry* (SSD)	$\left(\frac{W_3-W_5}{W_3} \times 100\% \right)$	1.90 gr

3.1.4. M-sand water content

Table 4. M-sand water content data result

No.	Description	Result
1.	Container Weight (W_1)	306 gr
2.	Weight of the test specimens + container (W_2)	806 gr
3.	Weight of the test specimens (W_3) = W_2 - W_1	500 gr
4.	Weight of the test specimens after oven + container (W_4)	800 gr
5.	Weight of the test specimens after oven (W_5) = W_4 - W_1	494 gr
6.	% Sand Water Content = $\frac{W_3 - W_5}{W_5} \times 100\%$	1.21%

The examination results of the M-Sand water content showed a test percentage of 1.21%. Furthermore, the water content in the aggregate is strongly influenced by the amount contained in the aggregate. The greater the difference between the original aggregate weight before and after drying the oven, is determined.

3.1.5 Waste marble gradation analysis

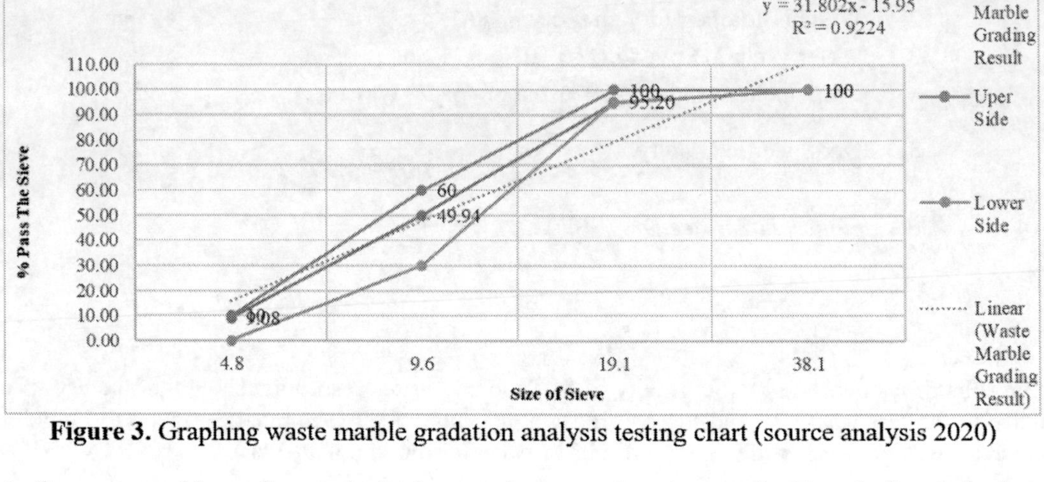

Figure 3. Graphing waste marble gradation analysis testing chart (source analysis 2020)

In the waste marble grading analysis, the experiment needs to be tested with a similar method as the coarse aggregate grading analysis experiment using a set of sieves with 38.1 mm, 25.4 mm, 19.1 mm, 15.9 mm, 9.92 mm, 9.42 mm, 4, 76 mm, 2.4 mm, 1.2 mm and pan. After testing, the results of the gravel left in the sieve are weighed. Therefore, based on the above test, the results showed that the gradation of marble waste used falls into the same category as gravel, at a nominal size of 38.10 mm. The results for the fineness modulus obtained was 6.64, and the modulus value of refinement as a gross aggregate substitution fulfilled the requirements according to SNI 03-2461-2014 [8], which is does not exceed 7%. Figure 3 shows that the Linear Regression method can be applied because R = 0.9224 is close to 1 [9]

3.2. Slump test result

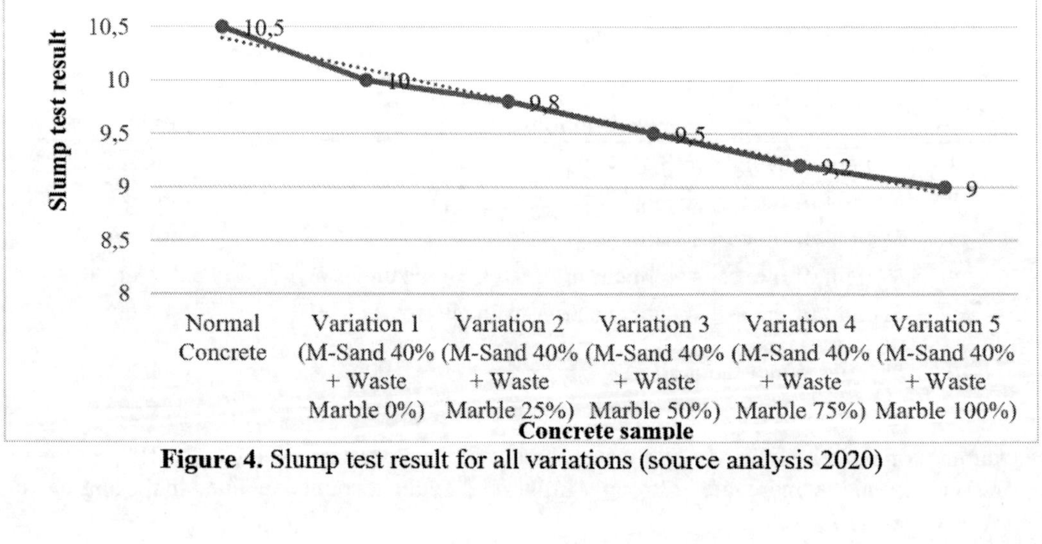

Figure 4. Slump test result for all variations (source analysis 2020)

Figure 4 shows that the slump test obtained is following the specified values ranging from 80 mm to 130 mm and in accordance with National Indonesia standards [10].

3.3. Compressive strength test result

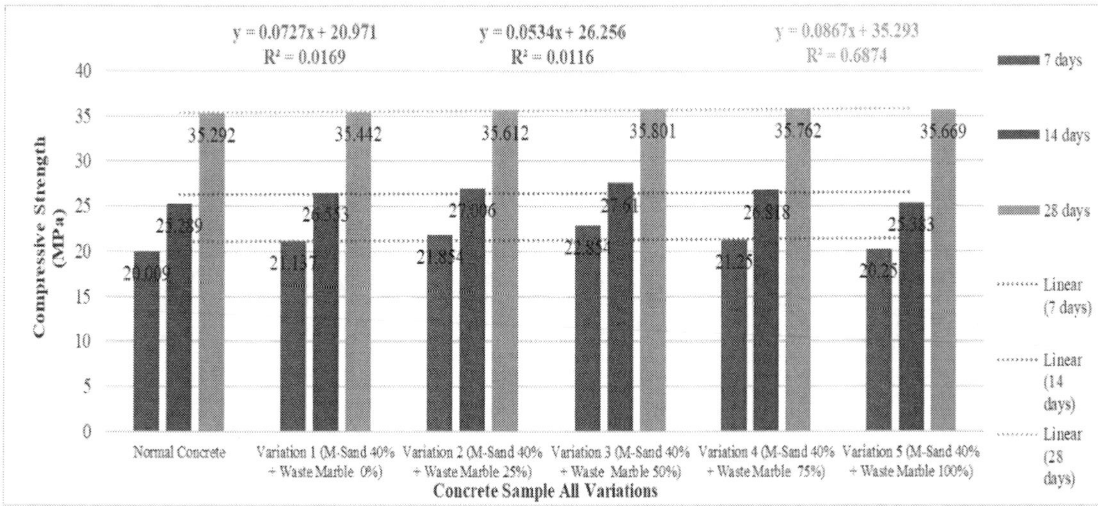

Figure 5. Compressive strength test result for all variations (source analysis 2020)

Figure 5 showed that the highest compressive strength for concrete samples with variation 3 is obtained, with a mixture of 40% M-Sand and 50% waste marble, thereby producing a compressive strength of 27.61 MPa in 28 days. The concrete variation 3 gained an increase with a difference of 2.321 Mpa when compared with the normal variation concrete with compressive strength of 25.289 MPa.

3.4. Concrete Absorption Test Result

Absorption testing is carried out when the concrete age is 28 days, as shown in figure 6.

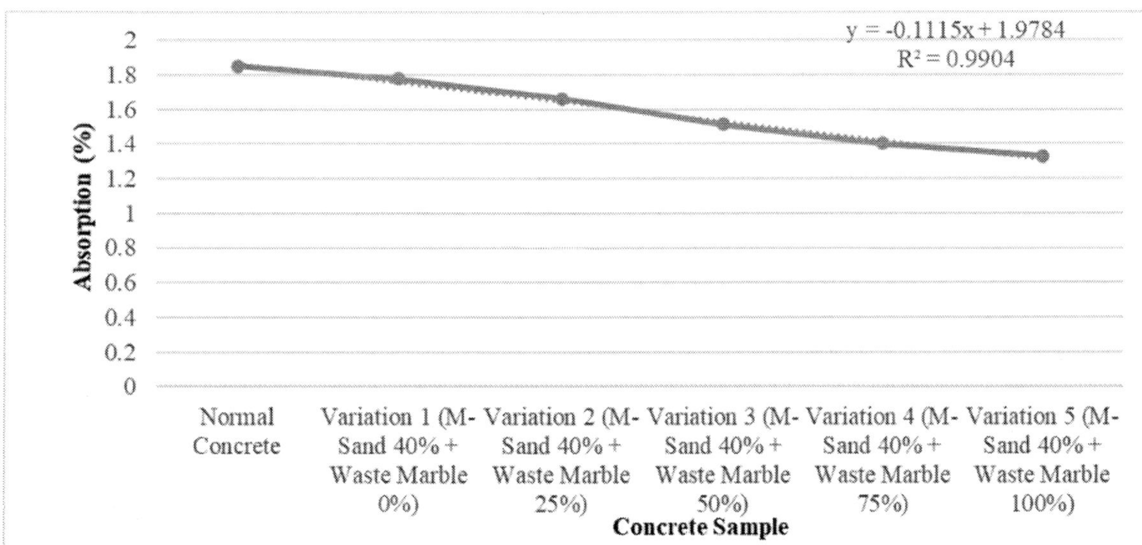

Figure 6. Test results for absorption of all variations (source analysis 2020)

Figure 6 shows that the highest value of the absorption test is normal concrete. The value of variation 1 (M-Sand 40% + Waste Marble 0 %) is closed to the optimum in the alternative use substitutes.

4. Conclusion

The research and analysis results showed that the effect of waste marble at variations of 0%, 25%, 50%, 75%, 100% substitution of coarse aggregate (gravel) and 40% M-Sand produced fine aggregate (sand) with the greatest compressive strength in Variation 3 (M Sand 40% + 50% Waste Marble) of 35.801 MPa in 28 days with a percentage value of 16.76%. This is because coarse aggregate composition is still dominant to help the compressive strength value, with a decrease in compressive strength due to the use of fewer coarse aggregate such as variations 4 and 5.

Furthermore, the optimum value of concrete water absorption was found in concrete variation 1 (M-Sand 40% + 0% Waste Marble) at 1.77% because the addition of material prolongs the water absorption rate. This happens for additional variables that significantly affect water. The more waste marble, the smaller the absorption and impact on workability. However, an R-value close to 1 shows that it can be used for concrete mixes.

Acknowledgments

The authors are grateful to the Chairperson of the Institut Teknologi PLN for the Internal Beginner Lecturer Research Grant No. SK. 0422.SK/A0/2019 dated December 20, 2019.

References

[1] Hidayawanti R, Legino S Sangadji I and Widodo R 2019 *Japan J. GEOMATE.* **16**, pp 95-100
[2] Mahesh J, Nagamani K and Jarin T 2019 Corrosion of reinforcement in concrete with fly ash and manufactured sand *J. Mater. Res. Innov.* **23** 7 pp 413-21
[3] Pauline T and Ashok V 2016 Experimental investigation on properties of concrete containing manufactured sand and recycled aggregates *J. Chem. Pharm. Sci.* **9** 3 pp 1779–83
[4] Suseela K and Baskaran T 2017 Strength analysis on concrete with M-sand as a partial replacement of fine aggregate *Int. J. Civ. Eng. Technol.* **8** 12
[5] Singh M, Choudhary K Srivastava A Singh Sangwan K and Bhunia D 2017 A study on environmental and economic impacts of using waste marble powder in concrete *J. of Building Engineering* pp 87-95
[6] Tugrul Tunc E 2019 Recycling of marble waste: a review based on strength of concrete containing marble waste *J. of Environmental Management.* **231** pp 86-97
[7] ASTM C33 / C33M-18 2018 *Sta. Spec. for Conc. Aggre.*
[8] SNI 03-2461-2014 *The Pro. for Mak. a Norm. Conc. Mix Plan*
[9] Hidayawanti R, Purnama D Iduwin T Legino S and Wachid F 2019 *Proc. Intl. Conf.* 5th *Science Eng. & Env. (Bangkok)* pp 439–45
[10] SNI 1972:2008 *Proc. Conc. Slump Test*

IOP Conf. Series: Materials Science and Engineering **930** (2020) 012025 doi:10.1088/1757-899X/930/1/012025

Compressive strength of geopolymer concrete with fly-ash from Paiton Steam Power Plant and variations of substitution sodium silicate (Na_2SiO_3) with natural zeolite

I N Amini[1,*], Krisnamurti[1] and W T Wahyuningtyas[1]

[1]Department of Civil Engineering, University of Jember, Indonesia

*Corresponding author's e-mail: iqlimana4@gmail.com

Abstract. Geopolymer concrete is one of the technological innovations of concrete without cement. It was prepared using fine aggregates, coarse aggregates, fly-ash, sodium hydroxide (NaOH), and sodium silicate (Na_2SiO_3). In the production of geopolymer concrete, sodium silicate and sodium hydroxide serve as an alkaline activator. One of the natural materials that can be used as an alternative material is natural zeolite. Zeolit contains high silica and alumina, that expected to increase the compressive strength of geopolymer concrete. This research aims to develop geopolymer concrete technology by understanding the setting time behaviour and compressive strength by substitution sodium silicate (Na_2SiO_3) with natural zeolite. The percentage of zeolite that substitute for sodium silicate is 0%, 2.5%, 5%, and 7.5 %. The molarity of NaOH is 12 M. Meanwhile, the ratio of sodium hydroxide and sodium silicate is 1:2.5. Compressive strength testing is carried out on cylindrical specimens 10 cm x 20 cm. The results show that natural zeolite can accelerate the setting time of geopolymer paste. The fastest initial setting time is 48 minutes. Meanwhile, the final setting time is 60 minutes in geopolymer paste with substitution sodium silicate using 7.5 % zeolite. In addition, for compressive strength results, zeolite can increase compressive strength. The highest average of compressive strength results is 46.07 Mpa in substitution of Na_2SiO_3 with 7.5 % zeolite.

1. Introduction

Cement production is one of the main contributors to the greenhouse effect. One ton of cement production can produce one ton of carbon dioxide (CO_2) too into the atmosphere [1]. As the main material of concrete, cement is one important material for concrete construction. This encourages new research about alternative materials for replacing cement which more environmentally friendly.

Geopolymer concrete is an innovative construction material that not uses cement as a binder. The main material used is fine aggregates, coarse aggregates, fly-ash, and alkaline activator. Fly-ash is pozzolanic material from power plant waste. It used with alkaline activator as geopolymer binder for the alternative of portland cement. Its contains 45.59 % silica (Si) and 25.08 % alumina (Al) [2]. Fly ash is reacted with an alkali-activator (NaOH and Na_2SiO_3) to form a polymerization reaction. Sodium hydroxide and sodium silicate greatly affect the mechanical properties of geopolymer concrete, especially in terms of mass ratio and molarity. The mass ratio between sodium silicate and sodium hydroxide is affected for setting time. The higher the mass ratio makes the initial setting longer but

Content from this work may be used under the terms of the Creative Commons Attribution 3.0 licence. Any further distribution of this work must maintain attribution to the author(s) and the title of the work, journal citation and DOI.
Published under licence by IOP Publishing Ltd

speeds up the final setting. Meanwhile, the higher molarity of sodium hydroxide used makes setting time and compressive strength increase [3].

Geopolymer material has a chemical composition similar to natural zeolite but has an amorphous microstructure [4]. Zeolite is crystalline aluminosilicates with structure composed of tetahedral units of silicon-oxygen (SiO4) and alumunium oxygen (AlO4) [10]. Its contains 60.9 % of silica (Si), 8.7 % of alumina (Al), and 11.7 % of Calcium (Ca) [5]. Zeolite is often used as a substitute for cement because it is rich in active SiO_2 and Al_2O_3, such as other pozzolanic materials [6]. With similar chemical composition, zeolite is expected to be one of the materials that can be used in the manufacture of geopolymer concrete. Besides having similar chemical composition, zeolites are also able to increase the compressive strength of concrete up to 15% at 28 days [6].

Zeolite reaction with alkali can form polymerization of inorganic materials/geopolymerization [7]. Alkaline concentrations can affect the dissolution of Si^{4+} and Al^{3+} from zeolites to form sodium-aluminosilicate. For geopolymer concrete production, alumino-silicate is the main material [9]. This compound makes the compressive strength increase. In addition, a large content of Calcium (Ca) in zeolites can also react to form CSH (calcium silicate hydrate) which is the same as the reaction of portland cement forming [7].

This study aims to determine the mechanical properties of geopolymer concrete based on compressive strength and setting time generated on geopolymer concrete with substitution of Na_2SiO_3 with zeolite. With its chemical composition and properties, zeolites are expected to be an alternative use for geopolymer concrete.

2. Material and method

2.1. Material

The material used in this research is fly-ash, alkaline solution (NaOh and Na_2SiO_3), fine aggregates, coarse aggregates, and natural zeolite.

Table 1. Material used for specimen test

Type	Type of Fly-Ash	Zeolite (%)	NaOH (Molarity)	Na_2SiO_3 / NaOH
1	F	0	12	2.5
2	F	2.5	12	2.5
3	F	5	12	2.5
4	F	7.5	12	2.5

2.2. Method

The mix design used in this research is referred to Pavithra et al., 2016 "*A mix design procedure for geopolymer concrete with fly ash*". The specimens used were cylinders Ø10 x 20 cm totalling 36 test specimens. The curring process is wrapped with plastic within 7 days in ambient temperature. The test carried out are setting time and compressive strength.

3. Result and analysis

3.1 Material testing

The purpose of material testing is to find out the specifications of the material used. Material test results must suitable with the material specification requirements that regulated in ASTM. Materials that tested are fly-ash, fine aggregates, and coarse aggregates.

Table 2. Physical properties of fine and coarse aggregate

Aggregate	Specific Gravity	Particle Density (g/cm3)	Water Absorption (%)
Fine Aggregates	2.65	1.45	2.28
Coarse Aggregates	2.76	1.30	1.97
Fly-Ash	2.59	1.11	-

Figure 1. Material testing

3.2 Setting Time

Vicat needle apparatus is used to find out the setting time of geopolymer paste. The test of setting time is carried out for all proportion used. Geopolymer paste prepared according to the standard universal testing method guidelines. The result of Initial setting time is showed when vicat needle dropped at 25 mm. Meanwhile, the final setting time is showed when the vicat needle dropped at 0 mm (figure 3). The initial setting and final setting that occurred in geopolymer paste with substitution of sodium silicate with natural zeolite are shown in figure 2.

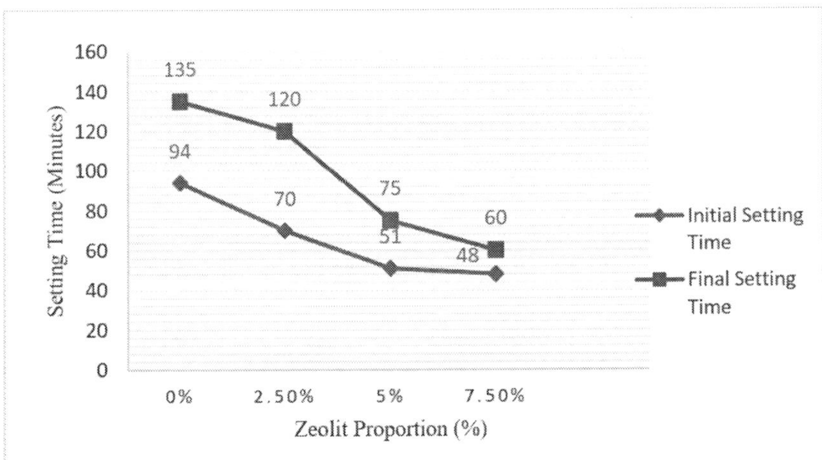

Figure 2. Initial setting and final setting of geopolymer paste with substitution of sodium silicate with natural zeolite.

Figure 3. Setting time test

The fastest initial setting time occurred in the proportion of Na_2SiO_3 substitution with 7.5 % of zeolite is 48 minutes. Meanwhile, the longest initial setting time is normal geopolymer paste with 0% substitution of zeolite which is 94 minutes. The longest final setting time occurs in geopolymer paste with substitution of 0 % zeolite is 135 minutes while the fastest final setting time occurs in geopolymer cement with substitution of 7.5 % zeolite which is 60 minutes.

In this case, the setting time of geopolymer paste is getting faster, linear with increasing substitution sodium silicate with natural zeolite. According to Wang Ru and Wang Gaoyong [11], zeolite can accelerate the setting time because it could transform become C-S-H gel. In addition, according to Ismuyanto [7], C-S-H from zeolite is form by reaction of calcium oxide with silicate compound.

3.3 The proportion of material mixture
The proportion of the material mixture is designed for normal concrete (Table 3).

Table 3. The proportion of material geopolymer concrete

Type	Zeolite	Zeolite	Fly-ash	NaOH	Na_2SiO_3	Coarse Aggregate		Fine Aggregate
	%	kg	kg	kg	kg	1-5 mm (kg)	5-10 mm (kg)	kg
1	0	0.00	6.50	0.97	2.42	8.04	8.04	15.43
2	2.5	0.06	6.50	0.97	2.36	8.04	8.04	15.43
3	5	0.12	6.50	0.97	2.30	8.04	8.04	15.43
4	7.5	0.18	6.50	0.97	2.24	8.04	8.04	15.43

3.4 Slump test
Fresh concrete testing performed for normal concrete is slump testing using an abrams cone (figure 4). The result of the slump test for all proportion is shown in table 4. The slump test result is getting smaller along with the increase of substitution of zeolite. It is due to the fast setting time of proportion that can affect the workability of concrete.

Figure 4. Abrams cone

Table 4. The result of the slump test

Type	Proportion of Zeolit	Slump Test (cm)
1	0%	12
2	2.5%	12
3	5%	11
4	7.5%	10.5

3.5 Compressive strength test

The curring process is done in ambient temperature for 7 days with plastic as a membrane to resist evaporation of the concrete (Figure 5).

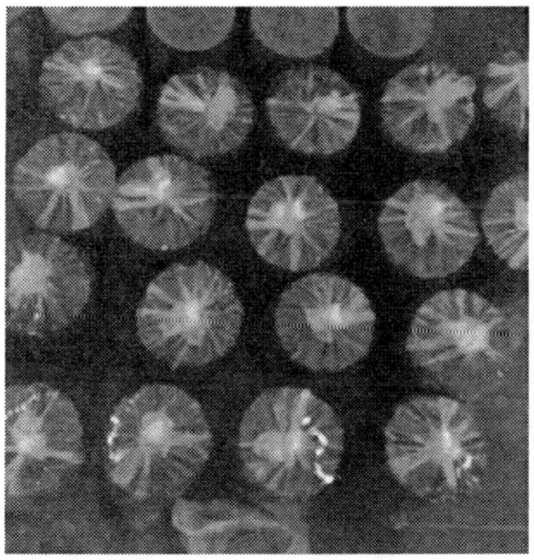

Figure 5. Curring process in ambient temperature

Figure 6. Compressive strength test

The compressive strength test is carried out for all specimens at 7, 14, and 28 days (figure 7).

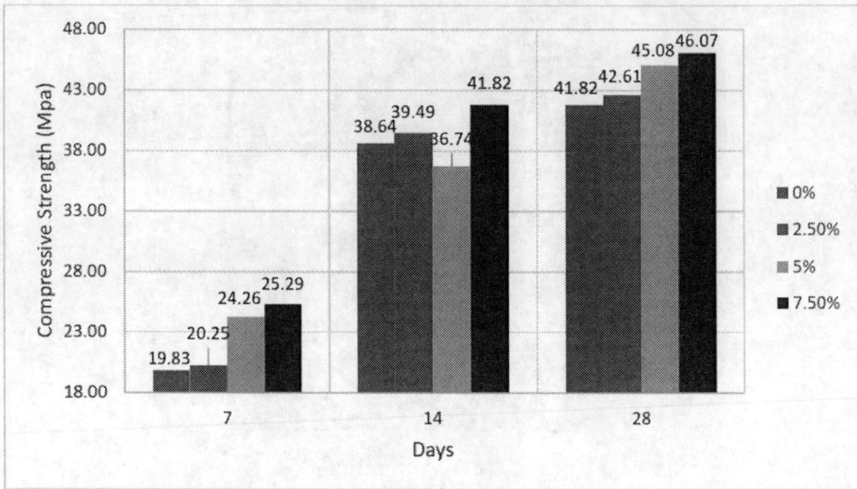

Figure 7. Compressive strength of geopolymer concrete in 7, 14, 28 days.

After curring process for 7 days, the compressive strength test is do for specimens test. Based on the result (Figure 7), for 7, 14, and 28 days, it can be seen that geopolymer concrete without zeolite is the lowest average compressive strength. The largest average compressive strength value is shown by geopolymer concrete with 7.5 % zeolite as subtitution for sodium silicate, which is 46.07 Mpa.

According to Ismuyanto [7], zeolite reaction with sodium hydroxide form sodium-alumino silicate, it can increase the compressive strength of geopolymer concrete. In addition, according to Emam and Yehia [12], zeolite reaction with calcareous material can form calcium silicoaluminate hydrates which contribute for compressive strength result of concrete. It showed that zeolite affects compressive strength for geopolymer concrete and can react with fly-ash and alkaline solution. In this research, zeolite can replace Na_2SiO_3 up to 7.5 %.

3. Conclusion

1. Zeolite substitution for sodium silicate can accelerate the setting time and increase the compressive strength.
2. Zeolite substitution in sodium silicate accelerates the setting time. The fastest setting time occurs in geopolymer paste with 7.5% zeolite as substitution for sodium silicate, i.e 48 minutes for initial setting time and 60 minutes for final setting time.
3. The average of compressive strength results from the proportion of zeolite substitution is getting bigger along with the increase of zeolite substitution.

References

[1] Putra A K, Wallah S E and Dapas S O 2014 Kuat tarik belah beton geopolimer berbasis abu terbang (Fly Ash) *J. Sipil Statik* **2** pp 330-336

[2] Umniati B S, Risdanareni P and Zein F T Z 2017 Workability enhancement Of geopolymer concrete through the use of retarder *Proc. Green Construction and Engineering Education for Suistanable Future* pp 1-9

[3] Ekaputri J J, Triwulan and Damayanti O 2007 Analisa sifat mekanik beton geopolimer berbahan dasar fly ash jawa power paiton sebagai material alternatif *J. Pondasi* **13** pp 124-134

[4] Sumajouw M D and Dapas S O 2013 *Elem. Struk. Beton Bertulang Geopolymer* (Yogyakarta: Andi)

[5] Hidayat B R and Wardhono A 2019 Pengaruh varian distribusi zeolit terhadap abu terbang pada kuat tekan benda uji geopolymer dengan kondisi 12 molar 14 molar dan W/S 0 30 *J. Rek. Tek. Sipil* **1** pp 1–9

[6] Nagrockiene D and Girskas G 2016 Research into the properties of concrete modified with natural zeolite addition *Cons. Build. Mat.* **113** pp 964–69

[7] Ismuyanto B 2017 Mineral zeolit sebagai kandidat pozolan alami untuk pembuatan pasta polimer anorganik *Proc. Inovasi Teknologi dan Rekayasa Ramah Lingkungan untuk Pembangunan Berkerlanjutan* pp 21–4

[8] Pavithra P, Srinivasula Reddy M, Dinakar P, Hanumantha Rao B, Satpathy B and Mohanty A 2016 A mix design procedure for geopolymer concrete with fly ash *J. Clean. Prod.* doi: 10.1016/j.jclepro.2016.05.041.

[9] Nikolov A, Rostovsky I and Nugteren H 2017 Geopolymer materials based on natural zeolite *Cas. Stud. Const. Mat.* **6** pp 198–205

[10] Villa C *et al.* 2010 Geopolymer synthesis using alkaline activation of natural zeolite *Const. Build. Mat.* **24** 11 pp 2084–90

[11] Wang R and Wang G 2016 Influence and mechanism of zeolite on the setting and hardening process of styrene-acrylic ester/cement composite cementitious materials *Const. Build. Mat.* **125** pp 757–65

[12] Emam E and Yehia S 2017 Performance of concrete containing zeolite as a supplementary cementitious material *Int. Res. J. Eng. Tech.*

IOP Conf. Series: Materials Science and Engineering **930** (2020) 012026 doi:10.1088/1757-899X/930/1/012026

The effect of parsial granulated blast furnace slag (GBFS) substitution and ashes of the boiler crust of the palm oil shells on paving block

T Yuhanah[1], R Hidayawanti[2*], D Mayasari[3] and B Wicaksono[4]

[1,2,3,4] Department of Civil Engineering, Institut Teknologi PLN, Indonesia

*Corresponding author's e-mail: ranti@itpln.ac.id

Abstract. Paving block is a building material made of cement, water, and fine aggregate used as an alternative cover or hardening for land surface. It is commonly used for various purposes such as, sidewalks, parking lots, and road pavement. The increasing demand of cement-forming materials is due to the rise in the number of cement users, therefore, an environmentally friendly material is needed to be produced from industrial waste. Generally, Indonesia has extensive oil palm plantations, as a result, excess wastes are produced from the burning of oil palm shells namely, boiler scale ash (contains silica oxides (SiO_2), pozzolan, and other compounds present in cement). Smelting and metal refining industries produce wastes from Granulated Blast Furnace Slag (GBFS), which contains silica oxide as their major chemical composition. Therefore, this study aims to determine the compressive strength and the water absorbing capacity of paving blocks using 10% of Palm Boiler Crust Ash as a substitute for fine aggregate /sand, and the percentage variations of Granulated Blast Furnace Slag (GBFS) waste as a cement replacement material. The results showed that the paving block variation V_2 had the highest compressive strength (22.83 MPa), while the highest water absorption (5.45%) was found in variation V_1, which was higher than the normal paving block (2.39%).

1. Introduction

Currently, paving blocks have many uses which include, their utilization in parking plazas, hotels, recreation areas, historic sites, for terminals, footpaths, and pavement environments in residential complex. They are also called concrete or cone blocks and contains similar constituents present in cement[1]. The increasing demand of cement in the construction industry has led to the various attempts made in searching for raw material with the same chemical compounds present in cement[2]. One of which is the utilization of waste products from industries[3].

Generally, Indonesia has extensive oil palm plantations, as a result, excess wastes are produced from the burning of oil palm shell namely, boiler scale ash (contains silica (SiO2), pozzolan, and other cement constituents)[4]. Smelting and metals refining industries produce wastes from Granulated Blast Furnace Slag (GBFS), which contains silica oxide as their major chemical composition[5].

Therefore, this study aims to determine the compressive strength and the water absorbing capacity of paving blocks using 10% Palm Boiler Crust Ash as a substitute for fine aggregate /sand, and the percentage variations of Granulated Blast Furnace Slag (GBFS) wastes as a cement replacement material.

Content from this work may be used under the terms of the Creative Commons Attribution 3.0 licence. Any further distribution of this work must maintain attribution to the author(s) and the title of the work, journal citation and DOI.
Published under licence by IOP Publishing Ltd

1.1 The Problem to be addressed

The increasing search for cement's raw materials has led to the optimization of industrial wastes, especially those from Granulated Blast Furnace Slag (GBFS) and Palm Boiler Crust Ash (BCA). These wastes are tested for compressive strength in accordance with the SNI standard for normal paving blocks[6]. The percentage composition of the Granulated Blast Furnace Slag (GBFS) wastes used for cement replacement were, 0%, 25%, 50%, 75%, and 100%, while 10% of Palm Boiler Crust Ash (BCA) for fine aggregate /sand substitution. Therefore, the various percentage compositions of GBFS cement with 10% BCA fine aggregate /sand substitution on the paving block produced an optimum compressive strength.

Furthermore, the GBFS and BCA are environmentally friendly materials used for cement and sand substitution, due to their higher compressive strength and water absorbing capacity compared to the conventional paving blocks.

2. Methodology

The Granulated Blast Furnace Slag (GBFS) waste and the Palm Boiler Crust Ash (BCA) were proportionally mixed.

Table 1. Material Requirements needed for GBFS and BCA Paving Block variations.

Percentage / Days	Material	7 Days	14 Days	28 Days	Total
	Cement	2,28 kg	2,28 kg	4,56 kg	9,12 kg
V_0 = 0% (Normal)	Sand	4,80 kg	4,80 kg	9,60 kg	19,20 kg
	Cement	1,71 kg	1,71 kg	3,42 kg	6,84 kg
	Sand	4,20 kg	4,20 kg	8,40 kg	16,80 kg
V_1 = 10%BCA +	GBFS	0,63 kg	0,63 kg	1,26 kg	2,52 kg
25% GBFS	BCA	0,42 kg	0,42 kg	0,84 kg	1,68 kg
	Cement	1,14 kg	1,14 kg	2,28 kg	4,56 kg
	Sand	4,20 kg	4,20 kg	8,40 kg	16,80 kg
V_2 = 10% BCA +	GBFS	1,26 kg	1,26 kg	2,52 kg	5,04 kg
50% GBFS	BCA	0,42 kg	0,42 kg	0,84 kg	1,68 kg
	Cement	0,57 kg	0,57 kg	1,14 kg	2,28 kg
	Sand	4,20 kg	4,20 kg	8,40 kg	16,80 kg
V_3 = 10% BCA +	GBFS	1,89 kg	1,89 kg	3,78 kg	7,56 kg
75% GBFS	BCA	0,42 kg	0,42 kg	0,84 kg	1,68 kg
	Cement	0 kg	0 kg	0 kg	0 kg
	Sand	4,20 kg	4,20 kg	8,40 kg	16,80 kg
V_4 = 10% BCA +	GBFS	2,52 kg	2,52 kg	5,04 kg	10,58 kg
100% GGBFS	BCA	0,42 kg	0,42 kg	0,84 kg	1,68 kg

2.1 Research Process

The issue reviewed in this study was determining the correlation value between the use of Granulated Blast Furnace Slag (GBFS) as a cement substitution with 10% of Palm Boiler Crust Ash (BCA) as a fine gradation /sand substitution in the manufacture of conventional paving blocks.

3. Laboratory Testing Result

3.1 Laboratory Testing Result

The fine aggregate used in the paving block mixture was examined for its parameters[7], which included the gradation and unit weight.

3.1.1 Test Result Analysis of Palm Boiler Crust Ash (BCA)

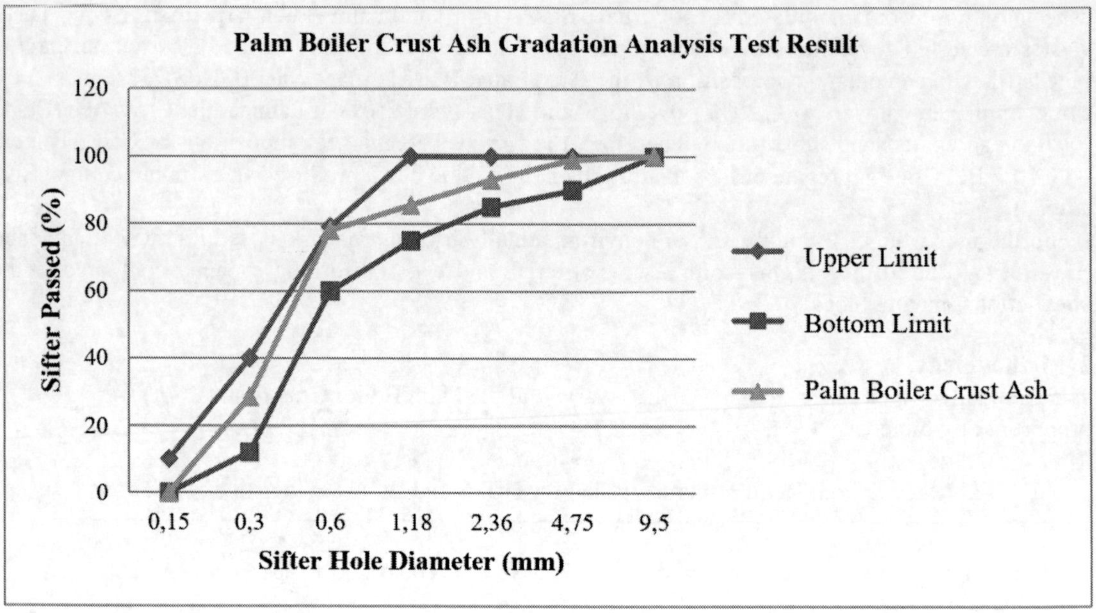

Figure 1. Test Result Analysis of Palm Boiler Crust Ash (BCA) (source of analysis result 2020).

The results showed that Palm Boiler Crust Ash (BCA) belongs to the fine aggregate category (Zone III) with a fineness modulus of 2,158, which meet the requirements stipulated in ASTM C-33[8], namely the modulus refinement of 2.3 - 3.1. Therefore, confirming the use of Palm Boiler Crust Ash waste as a paving block mixture material.

3.1.2 Unit Weight of Palm Boiler Crust Ash (BCA).

Table 2. Unit Weight of Palm Boiler Crust Ash (BCA).

No	Description	Result	Unit
1	Empty Vessel Weight (B_1)	5,88	Kg
2	Verssel Weight + BCA (B_2)	20,24	Kg
3	Sand Weight ($B_3 = B_2 - B_1$)	14,36	Kg
4	Volume of Steel Vessel (V)	0.009916	m^3
5	Unit Weight of Sand (ɣ) $\gamma = \frac{B3}{V}$	1448,16	Kg/m^3

The test result of unit weight of Palm Boiler Crust Ash amounted to 1448,16 kg/m³.

3.1.3 Specific Gravity Analysisi of Granulated Blast Furnace Slag (GBFS).

Table 3. Specific Gravity of Granulated Blast Furnace Slag (GBFS).

No	Examination Test	Result	Unit
1	Weight of GBFS (A)	75	gr
2	Volume of kerosene (B_1)	0,8	ml
3	Volume of kerosene + GBFS (B_2)	22,20	ml
4	Specific Gravity of GBFS = $\frac{A}{B2-B1}$	3,505	gr/ml

3.2 Paving Block Compressive Strength Result

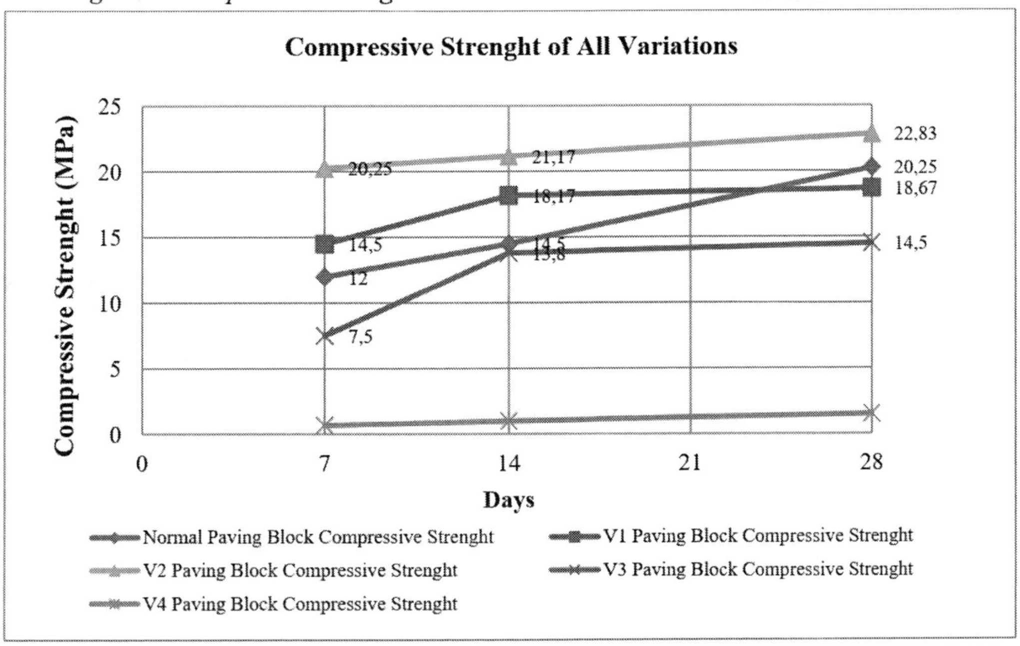

Figure 2. Paving Block Variation Value Testing Results.

Based on figure 2, the compressive strength of variation V2 (10% BCA with + 50% GBFS) had the the highest value compared to the normal paving blocks. While the maximum compressive strength V_2 (10% BCA + 50% GBFS) at 28 days was 22.83 MPa, and higher than the normal variation V_0 (20.25 MPa).

3.3 Paving block Absorption Testing Results

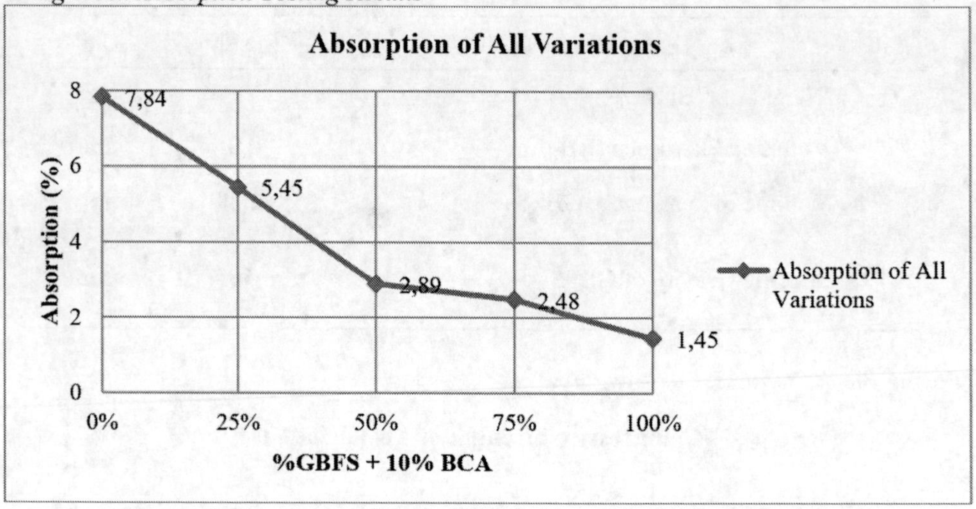

Figure 3. Percentage of Paving Block Absorption in Each Variation

Based on figure 3, the water absorption values of all variations were lower than those of normal paving blocks. The highest absorption value of 5.45% in V1 (10% BCA + 25% GBFS) was lower than the normal paving block V0 of 2.39%. Meanwhile, the value of water absorption was lower with the increasing use of GBFS, as a result, reduced cement use in order for the aggregate binding process of GBFS to occur optimally, which did not cause many pores /cavities in the paving block structure[9].

4. Conclusion

Based on the results of research and analysis, several conclusions were made such as, the highest compressive strength is found in the paving block variation V_2 (10% BCA + 50% GBFS) at 28 days (22.83 MPa), and higher than the normal variation V_0 (20,25 MPa). The water absorption values of all variations are lower than those of normal paving blocks. Therefore, the highest absorption value of 5.45% in V_1 (10% BCA + 25% GBFS) is found to be lower than the normal paving block V_0 of 2.39%.

Acknowledgments

The authors are grateful to the Chairperson of the Institut Teknologi PLN for the Internal Beginner Lecturer Research Grant No. SK. 0422.SK/A0/2019 dated December 20, 2019.

References

[1] Siregar C M A 2016 Komposisi paving block berbasis limbah padat abu boiler pabrik kelapa sawit sebagai bahan pengganti semen. (Bogor: UT- Agro Industrial Engineering) pp 2463

[2] Rahman F and Fathurrahman F 2017 Pemanfaatan hasil pembakaran limbah cangkang kelapa sawit sebagai bahan pengganti pasir pada pembuatan beton normal (Banjarmasin: *Media Ilmu Tek. Sipil* **6** no. 1) pp. 30–40.

[3] Singh G, Das S Ahmed A A Saha S and Karmakar S 2015 Study of granulated blast furnace slag as fine aggregates in concrete for sustainable infrastructure (Amsterdam: *Procedia - Soc. Behav. Sci.*, **195)** pp. 2272–79.

[4] Pa F C, Chik A and Bari M F 2016 Palm ash as an alternative source for silica production *(France: MATEC Web Conf.* **78**) pp 01062.

[5] Walling S A, Bernal S A Gardner L J Kinoshita H and Provis J 2018 Blast furnace slag-Mg(OH)2 cements activated by sodium carbonate (Cambridge:*RSC Adv.* **8** no. 41) pp. 23101–

18.
[6] SNI 03-0691-1996 1996 Bata beton (paving block) (Jakarta: Badan Standar Nasional Indonesia) pp. 1–9.
[7] ASTM C 136-06 IDT 2012 metode uji untuk analisis saringan agregat halus dan agregat kasar (Jakarta: SNI ASTM)
[8] ASTM Standard C33 2003 standard specification for concrete aggregates (Canada: *ASTM Int.*, **vol. i** no. C) p. 11.
[9] Caronge M A, Lando A T Djamaluddin I Tjaronge M W and Runtulalo D 2020 Development of eco-friendly paving block incorporating co-burning palm oil-processed tea waste ash (Washington DC: *IOP Conf. Ser. Earth Environ. Sci.* **419** no. 1) pp 012158.

IOP Conf. Series: Materials Science and Engineering **930** (2020) 012027 doi:10.1088/1757-899X/930/1/012027

The effect of curing in soil and compound on compressive strength of concrete quality fc' 25 MPa

A Karjanto[1*], B Djatmika[1], WH Yoh[1] and PB Susanto[1]

[1]Department of Civil Engineering, Faculty of Engineering, Universitas Negeri Malang, Malang, Jawa Timur, Indonesia.

*Corresponding authors: Adjib.karjanto.ft@um.ac.id

Abstract. Concrete is a material that dominates high-rise building construction materials, because concrete materials are easy to obtain and able to withstand very large loads. In addition, concrete can be made in various forms so that according to needs and can beautify the shape of a building, easy to maintain, durable and relatively cheap prices. But in the implementation of using concrete, there are times when found concrete needs to be perforated on its appearance for the purposes of supporting facilities for building functions such as sanitation, gas, hydrants and others. To increase the compressive strength of concrete so that it meets the desired requirements in addition to the quality of the material, the casting process is also required after the casting. This treatment can be done by several methods, among others: immersion in water, wrapped in wet burlap sacks, watering with water, spraying by compounding, and can also be done by landfilling. Through treatment is expected to be obtained the quality of concrete as planned. Because with the maintenance of concrete, the security of the building structures made of concrete will be assured security. Beside that, concrete is required for minimum usage of 50 years. Therefore the research was carried out with the title "Effect of Soaking (Curing) in Soil and Compound on Compressive Strength of Concrete Quality Fc '25" MPa by using 40 cylindrical concrete specimens and each treatment consisted of 8 test specimens with 300 high dimensions mm and a diameter of 150 mm. Results of a strong test of concrete press performed after the concrete aged 14 days and subsequently converted to a 28-day old concrete produce : Strong high average press that is on concrete with the treatment of using water immersion with a strong press of 23.78 MPa, while the lowest average press is in concrete with the treatment of using a strong burlap sack with a tekasn of 21.88 MPa. Concrete elastic modulus testing using 4 specimens produces an average modulus of elasticity of 24919.59 Mpa.

1. Introduction

Concrete is a material that dominates high-rise building construction materials, because concrete materials are easy to obtain and able to withstand very large loads. In addition concrete can be made various forms so as to suit the needs and can beautify the shape of a building, easy to maintain, durable and relatively inexpensive price. But in the implementation by using concrete, there are times when concrete is found to be in a hole in the container for the purpose of supporting the building functions such as sanitation, gas, hydrants and others.

Based on the research conducted, the average compressive strength and average tensile strength for concrete with normal maintenance are 21,989 MPa and 2,951 MPa. Average compressive strength and

Content from this work may be used under the terms of the Creative Commons Attribution 3.0 licence. Any further distribution of this work must maintain attribution to the author(s) and the title of the work, journal citation and DOI.
Published under licence by IOP Publishing Ltd

average tensile strength of concrete with treatment in the mud is 19,361 MPa and 2,698 MPa or there is a decrease of 0,119% and 0,0856% of normal concrete compressive strength. Average compressive strength and average tensile strength of concrete with treatment in seawater are 18,021 MPa and 2,506 MPa or there is a decrease in 0.180% and 0.151% of the compressive strength and tensile strength of normal concrete [3].

At the age of 7 days, the maximum compressive strength value of PDAM water, sea water and sewage water are 13.522 MPa, 8.759 MPa, 10.296 MPa, respectively. The compressive strength value of PDAM water at the age of 7 days has increased compared to the compressive strength value of sea water and sewage water, while the compressive strength value of sea water has decreased compared to that of sewage water which has increased. So it can be concluded that the highest compressive strength value is PDAM water at the age of 7 days [5].

From the research that has been done to find out how much it affects the curing has on the strength of concrete. Obtained the result that the compressive strength of concrete with the treated (curing) by soaked turns to have a higher strength and end than on left in the open air/without curing. Where the strength of curing concrete turns beyond the compressive strength of the initial plan's compressive strength. As for the uncuring concrete is apparently the force of the press obtained less than the planned compressive strength [11].

According to [3] at the time of execution of regular casting, the lack of perfect compaction, the pouring of its density and imperfect compaction so that the concrete mixture will be not homogeneous. To increase the compressive strength of the concrete firmly so as to meet the desired condition in addition to the quality of material, the implementation of casting is required also after-casting treatment. This treatment can be done by several methods, among others: immersion in water, wrapped in a wet burlap sack, watering with water, spraying by compounding, and also can be done with a filling in the soil.

A curing is a process to keep the humidity and temperature levels ideal to prevent excessive hydration and keep the hydration going on continuously. Curing is generally understood as a concrete treatment, which aims to keep concrete not too fast to lose water, or as an act of maintaining moisture and temperature of concrete, as soon as the concrete finishing process is finished and the total time setting is reached.

Through treatment is expected to be obtained the quality of concrete as planned. Because with the treatment of concrete, the safety of the building structures made of concrete will be assured security. Beside that, concrete is required for minimum usage of 50 years.

2. Research method

The research was conducted using quantitative research methods with an experimental approach based on SNI 2847-2002 and ACI 440-2R-02 in testing. The initial stage in this study was to calculate the compressive strength that occurred on normal concrete cylinders at the age of 14 days. Furthermore, testing for concrete with several treatment methods, namely: treatment immersed in water, treatment in a pile in the soil, and treatment with a compound that is also after the age of 14 days.

Concrete cylinder specimens with diameter size 15 cm height 30 cm where 44 test objects and each treatment was 8 specimens for a compressive strength test and 4 specimens for modulus of elasticity with 25 MPa concrete quality plan. Concrete used is a concrete with the quality of FC ' 25 MPa obtained from the purchase at Varia Usaha Beton Malang ready mix.

Analysis of the data on this research using quantitative descriptive is to take the test result data and describe it in chart form with the test "One Way Anova Analysis". Statistical analysis results in the form of a concrete compressive strength test of the difference in treatment.

3. Result and discussion

3.1 Concrete compressive strength

Test result of compressive strength test of cylindrical concrete using UTM (Universal Testing Machine) conducted at Civil Engineering Laboratory Building D9 State University of Malang, are using cylindrical concrete with the dimensions of the specimens has a diameter of 150 mm, height 300 mm that is tested when the concrete is 14 days old and with a difference in the way of treatment. The result of the 14-day concrete compressive strength that occurred next converted to concrete age 28 days using the Indonesian Concrete Regulation. namely by dividing the caompressive strength that occurs with the value of 0.88 can be seen as in the table 1 up to the table 6 as follows

Table 1. Results compressive strength test of cylindrical concrete with normal treatment (BN).

No	Weight of Specimens	UTM Load Reading	Tools Calibration	Loads	Loads	Cross-sectional Area	Compressive Strength	Compressive Strength Conversion
	(kg)	P (kN)	P (kN)	P (kN)	P (N)	A (mm^2)	fc' MPa	fc' MPa
BN1	11.9	346.2	12.8	333.4	333400	17662.5	18.88	21.47
BN2	12	348.8	12.8	336	336000	17662.5	19.02	21.64
BN3	11.5	359.7	12.8	346.9	346900	17662.5	19.64	22.34
BN4	12.3	355.8	12.8	343	343000	17662.5	19.42	22.09
BN5	11.8	380.6	12.8	367.8	367800	17662.5	20.82	23.69
BN6	12	404.4	12.8	391.6	391600	17662.5	22.17	25.22
BN7	11.7	366.3	12.8	353.5	353500	17662.5	20.01	22.77
BN8	11.7	387.8	12.8	375	375000	17662.5	21.23	24.15
	AVERAGE COMPRESSIVE STRENGTH							22.92

Table 2. Results compressive strength test of cylindrical concrete with immersed in soil (BT).

No	Weight of Specimens	UTM Load Reading	Tools Calibration	Loads	Loads	Cross-sectional Area	Compressive Strength	Compressive Strength Conversion
	(kg)	P (kN)	P (kN)	P (kN)	P (N)	A (mm2)	fc' MPa	fc' MPa
BT1	11.9	364.6	12.8	351.8	351800	17662.5	19.92	22.66
BT2	12.2	376.3	12.8	363.5	363500	17662.5	20.58	23.41
BT3	12.2	346.0	12.8	333.2	333200	17662.5	18.86	21.46
BT4	12.1	399.1	12.8	386.3	386300	17662.5	21.87	24.88
BT5	12.3	387.9	12.8	375.1	375100	17662.5	21.24	24.16
BT6	12.3	401.2	12.8	388.4	388400	17662.5	21.99	25.02
BT7	12.2	326.2	12.8	313.4	313400	17662.5	17.74	20.19
BT8	12.3	405.6	12.8	392.8	392800	17662.5	22.24	25.30
	AVERAGE COMPRESSIVE STRENGTH							23.39

Table 3. Results compressive strength test of cylindrical concrete with compounding (BC).

No	Weight of Specimens (kg)	UTM Load Reading P (kN)	Tools Calibration P (kN)	Loads P (kN)	Loads P (N)	Cross-sectional Area A (mm2)	Compressive Strength fc' MPa	Compressive Strength Conversion fc' MPa
BC1	12	395.6	12.8	382.8	382800	17662.5	21.67	24.66
BC2	11.9	368.2	12.8	355.4	355400	17662.5	20.12	22.89
BC3	11.9	342.8	12.8	330	330000	17662.5	18.68	21.26
BC4	12.1	381	12.8	368.2	368200	17662.5	20.85	23.72
BC5	12	329.9	12.8	317.1	317100	17662.5	17.95	20.42
BC6	11.9	348.5	12.8	335.7	335700	17662.5	19.01	21.62
No	Weight of Specimens (kg)	UTM Load Reading P (kN)	Tools Calibration P (kN)	Loads P (kN)	Loads P (N)	Cross-sectional Area A (mm2)	Compressive Strength fc' MPa	Compressive Strength Conversion fc' MPa
BC7	11.6	362.2	12.8	349.4	349400	17662.5	19.78	22.51
BC8	11.9	396.5	12.8	383.7	383700	17662.5	21.72	24.71
AVERAGE COMPRESSIVE STRENGTH								22.72

Table 4. Results compressive strength test with jute sack covered (BG).

No	Weight of Specimens (kg)	UTM Load Reading P (kN)	Tools Calibration P (kN)	Loads P (kN)	Loads P (N)	Cross-sectional Area A (mm2)	Compressive Strength fc' MPa	Compressive Strength Conversion fc' MPa
BG1	12.1	362.7	12.8	349.9	349900	17662.5	19.81	22.54
BG2	11.9	388.3	12.8	375.5	375500	17662.5	21.26	24.19
BG3	11.9	358.3	12.8	345.5	345500	17662.5	19.56	22.25
BG4	12.1	383	12.8	370.2	370200	17662.5	20.96	23.84
BG5	12.3	316.7	12.8	303.9	303900	17662.5	17.21	19.57
BG6	12.2	327.1	12.8	314.3	314300	17662.5	17.79	20.24
BG7	12.1	330.4	12.8	317.6	317600	17662.5	17.98	20.46
BG8	11.8	353.5	12.8	340.7	340700	17662.5	19.29	21.94
AVERAGE COMPRESSIVE STRENGTH							19.23	21.88

Table 5. Results compressive strength test with soaking in water (BA).

No	Weight of Specimens (kg)	UTM Load Reading P (kN)	Tools Calibration P (kN)	Loads P (kN)	Loads P (N)	Cross-sectional Area A (mm2)	Compressive Strength fc' MPa	Compressive Strength Conversion fc' MPa
BA1	12.3	377.4	12.8	364.6	364600	17662.5	20.64	23.48
BA2	12.3	388.9	12.8	376.1	376100	17662.5	21.29	24.22
BA3	12.1	385.9	12.8	373.1	373100	17662.5	21.12	24.03
BA4	12.2	395.6	12.8	382.8	382800	17662.5	21.67	24.66
BA5	11.9	373.8	12.8	361	361000	17662.5	20.44	23.25
BA6	12.1	376.2	12.8	363.4	363400	17662.5	20.57	23.41
BA7	12.2	392.9	12.8	380.1	380100	17662.5	21.52	24.48
BA8	12.2	364.7	12.8	351.9	351900	17662.5	19.92	22.67
AVERAGE COMPRESSIVE STRENGTH							20.90	23.78

Tabel 6. Average cylinder concrete compression strength test results with each treatment.

No	Cylinder Concrete Test Objects Treatment	Average Compressive Strength (MPa)
1	BN	22.92
2	BT	23.39
3	BC	22.72
4	BG	21.88
5	BA	23.78

Notes:
BN = Compressive Strength Test of Cylindrical Concrete with Normal Treatment
BT = Compressive Strength Test of Cylindrical Concrete with Immersed in Soil
BC = Compressive Strength Test of Cylindrical Concrete with Compounding
BG = Compressive Strength Test with Jute Sack Covered
BA = Compressive Strength Test with Soaking in Water

From the results of the compressive strength of the average cylinder concrete with each treatment obtained graphs as in Figure 3.1 as follows

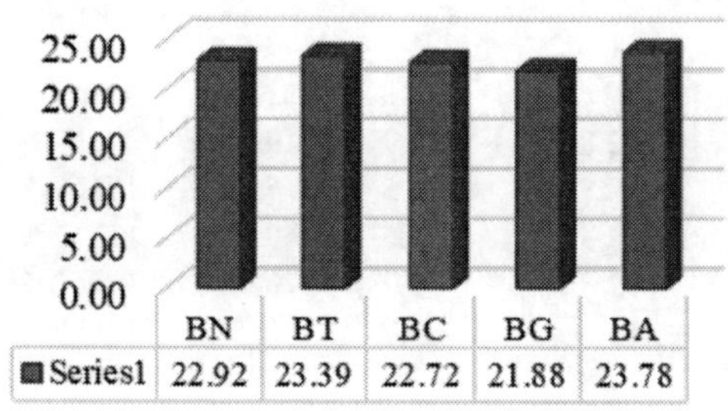

Figure 1. Graph of compressive strength of average concrete with each treatment.

From the result of average compressive strength of cylindrical concrete with each treatment can be concluded the concrete using the treatment soaked in water has the highest average compressive strength of 23.78 MPa, next is a concrete with the treatment using the burlap sacks has a compressive strength the lowest average is 21.88 Mpa.

3.2 Statistical analysis of concrete compressive strength

Hypothesis testing is conducted to determine whether there is a compressive strength difference that occurs in concrete cylinders that have a difference for treatment. The testing of the hypothesis was conducted using the one-way ANOVA analysis using the SPSS V.18 application with the hypothesis formulation in the study is:

H1 : There are differences in the compressive strength of concrete with different treatments
H0 : There is no difference in the compressive strength of concrete with different treatments

Based on the results of the one-way ANOVA testing performed on the compressive strength of concrete cylinder with the difference in the way the treatment obtained result is F = 1.896, df3 = 4, Df4 = 35, then Ftable is 2.64 and the significance level is 0.133. From the calculate F value of 1.896 < 2.64 Ftable, it can be concluded that H0 is accepted and H1 is rejected. While the significance value is 0.133 > 0.05, then Ho is accepted and it can be state there is no significant difference regarding compressive strengt concrete with different treatment.

Post hoc test is done to know the compressive strength of concrete cylinders with different treatments. Because the significance value 0.133 > 0.05 then it can be deduced H0 received or no significant difference in the compressive of concrete with different treatments.

3.3 Concrete elasticity module

The results of the modulus of elasticity testing of cylindrical concrete using UTM (Universal Testing Machine) conducted at the Civil Engineering Laboratory Building State University of Malang, namely by using concrete cylinder with the dimensions of the specimens has a diameter of 150 mm with a height of 300 mm with a total of 4 test objects tested when the cylinder concrete is 14 days old.

Concrete elastic modulus testing was carried out in accordance with ASTM C469. The calculation of modulus of elasticity is obtained by 45% of the maximum load that is subsequently acquired strain and Voltage 1 and subsequently necari strain with a value of 0.00005 according to the regulations and produce voltage and strain 2. Furthermore, the modulus of elasticity is obtained by the stress formula divided by strain. In table 8.1 shows the modulus of elasticity on average after the calculation is 24919.59 MPa. The results of the cylindrical concrete modulus elasticity test can be seen as in table 7 as follows.

In Figure 2 is obtained comparison between stress and strain on concrete after testing of modulus of elasticity. In the concrete cylinder the first test specimen obtainde the smallest stress and strain and the fourth specimen obtained the largest stress and strain. The graph of the stress and strain on the modulus elasticity testing of cylindrical concrete can be seen in Figure 2.

Table 7. The results of the concrete modulus elasticity test.

No	Modulus elasticity (MPa)	Averageof modulus elasticity (MPa)
1	31293.25	24919.59
2	29360.06	
3	10209.91	
4	28815.16	

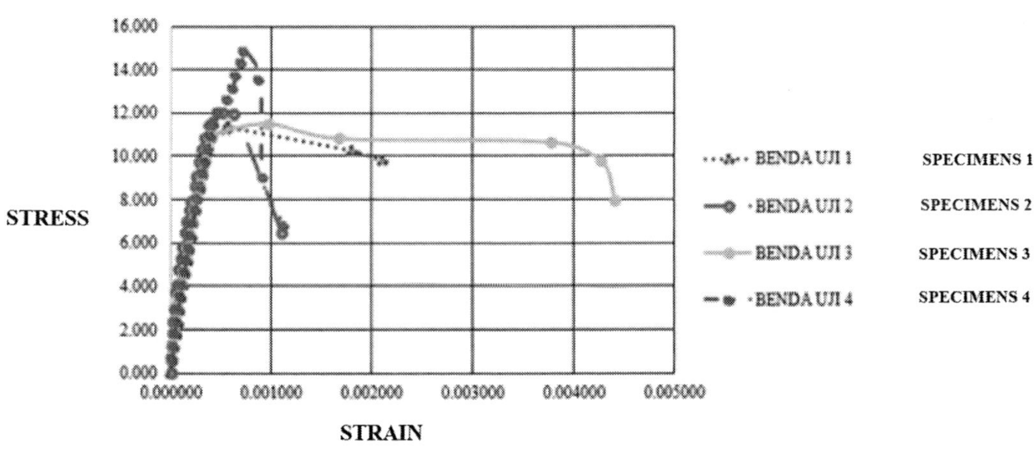

Figure 2. Graph of concrete stress dan strain.

4. Conclusion

Based on the results and discussion in a study entitled "The Effect of Soaking (Curing) in Soil and Compound on Concrete Compressive Strength of Fc '25 MPa" it can be concluded as follows:

1) The results of the concrete compressive strength test carried out after the concrete was 14 days old and subsequently converted to 28-day-old concrete yielded the highest average compressive strength,

namely the concrete treated using water immersion with compressive strength of 23.78 MPa, while the average compressive strength the lowest mean is in concrete with maintenance using burlap sacks with a strong pressure of 21.88 MPa.

2) Based on the data analysis test using "One Way Anova Analysis" using SPSS V.18 application which is performed on compressive strength of cylindrical concrete with different treatment methods, the results are F = 1,896, df3 = 4, df4 = 35, then Ftable is 2, 64 and the significance level of 0.133. From the calculated F value of 1.896 <2.64 Ftable, it can be concluded that H0 is accepted and H1 is rejected. While the significance value of 0.133> 0.05, Ho is accepted and it can be stated that there is no significant difference regarding the compressive strength of concrete with different treatments. From the post hoc test results, the significance value of 0.133> 0.05, it can be concluded that H0 is accepted or there is no significant difference in the compressive strength of concrete with different treatments.
3) Concrete elastic modulus testing using 4 specimens resulted in an average modulus of elasticity of 24919.59 MPa

References

[1] Angjaya and Novi *et al* 2013 *Perbandingan Kuat Tekan Antara Beton dengan Perawatan Pada Elevated Temperature dan Perawatan dengan Cara Perendaman Serta Tanpa Perawatan* (Manado: Teknik Sipil Universitas Samratulangi)

[2] Direktorat Riset dan Pengabdian Masyarakat 2018 Panduan Penelitian dan Pengabdian Kepada Masyarakat XII ed.

[3] Fauzi and Anwar D 2013 *Analisis Kuat Tekan dan Kuat Tarik Beton dengan Perawatan dalam Lumpur dan Air Laut* (Surakarta: Universitas Muhammadiyah)

[4] Oscar N F 2009 Kajian Ekperimental Perilaku Balok Beton Bertulang Tunggal Berdasarkan Tipe Keruntuhan Balok *J. Civil Eng.* **5** (2): 44

[5] Saputra and Dwi R *et al* 2017 Pengaruh Air PDAM, Laut, Comberan Pada Proses Curing terhadap Kuat Tekan Beton fc' 14,53 MPa (Lamongan: Teknik Sipil Universiitas Islam)

[6] Kuhn T 1998 Density matrix theory of coherent ultrafast dynamics *Theory of Transport Properties of Semiconductor Nanostructures (Electronic Materials* vol 4*)* ed E Schöll (London: Chapman and Hall) chapter 6 pp 173–214

[7] SNI (Standart Nasional Indonesia) 03-2847-2002 2002 *Persyaratan Beton Struktural Untuk Bangunan Gedung* (Indonesia: BSN)

[8] SNI (Standart Nasional Indonesia) 03-2847-2013 2013 *Persyaratan Beton Struktural Untuk Bangunan Gedung* (Indonesia: BSN)

[9] Simanjuntak and Johan O *et al* 2015 *Hubungan Peralatan Beton dengan Kuat Tekan* (Medan: Universitas HKBP Nommensen)

[10] Supriyadi, Eko A, Nadia 2013 Studi Analisis Lentur Pada Balok Tumpuan Yang Mengalami Pengeroposan Beton *J. Cons.* **4**

[11] Sutandar and Erwin 2013 *Pengaruh Pemeliharaan (Curing) pada Kuat Tekan Beton Normal* (Pontianak: Universitas Tanjungpura)

IOP Conf. Series: Materials Science and Engineering **930** (2020) 012028 doi:10.1088/1757-899X/930/1/012028

Factors that influence intention in processing organic waste using biopore hole

A Mardiyani[1*], C Utomo[2]

[1]Department of Technology Management, Institut Teknologi Sepuluh Nopember, Indonesia
[2]Department of Civil Engineering, Institut Teknologi Sepuluh Nopember, Indonesia

*Corresponding author's e-mail: anymardiyani.18092@mhs.its.ac.id

Abstract. Biopore hole is needed to reduce organic waste and this depends on a person's intentional behavior that arises due to several factors. Therefore, the purpose of this research is to determine the factors influencing the intention of processing organic waste using biopore hole in adiwiyata schools, Gresik regency. The research employ the extended theory of planned behavior model. Questionnaires were distributed to 393 respondents comprising of principals, teachers and students from 92 adiwiyata schools, The data analysis used structural equation modeling.The results shows that several factors including perceived behavioral control, subjective norms, reward and punishment influence intention in processing organic waste. All three factors possess a significant relationship with intention (P value<0.1) but, subjective norms is the most influencing (β = 3.074, P value<0.1). Based on the result, It was recommended that a review of policies and programs need to be made by government to increase the intention of organic waste processing using biopore hole and will also serve as educational materials to schools.

1. Introduction

Waste exist as a major problem in Indonesia for years and this is comparable to population growth, lifestyle changes, and technological progress. Its volume which incorporate 60% organic and 15% plastic at 65 million tons in a year will continue to increase. Also, food waste contributes significantly to about 40% of the total volume of which 66.39% remains buried in the final processing site (TPA), 19.62% unmanaged and 14% were refined into compost, biogas, and other processed materials [1].

Based on UPT TPA data from Gresik regency environmental office in 2017, the composition of food makes a significant contribution of 42% approximately. Organic waste on the other hand has the capacity of becoming a beneficial to the environment after processing into compost.

Waste management requires legal certainty, clarity of responsibility, authority of the government, role of the community and the business world in order to run proportionally, effectively, and efficiently [2]. The local government made effort to carry out the mandate by involving community participation, this led to the construction of biopore by the regency environmental office. The maintenance of a biopore holes involves filling it with organic waste periodical until they turn into compost [3].
The influence of individual intention leads to the behaviour of processing waste with biopore infiltration hole and the theory of planned behaviour (TPB) is used to measure intention [4]. This theory is widely adopted to predict behavioural intention (BI), which comprises of three variables :

Content from this work may be used under the terms of the Creative Commons Attribution 3.0 licence. Any further distribution of this work must maintain attribution to the author(s) and the title of the work, journal citation and DOI.
Published under licence by IOP Publishing Ltd

attitude (AT), subjective norms (SN), and perceived behavioral control (PBC). According to Ajzen [5], AT is a tendency to respond positively or unpleasantly to an object, person, institution or event. SN is individual perceptions of the expectation of those who are influential in their lives (significant others). PBC is individual perception of how easy or difficult it is to manifest a particular behavior. Also, BI is the tendency of the intention to be involved in a certain behavior. It is assumed that an individual who possesses these factors of BI will have increased intention to perform the behaviour [6]. This study therefore complement previous research by adding two variables, which includes reward and punishment (RP), and perceived usefulness (PU) to the TPB theory, therefore it becomes extended TPB (ETPB). RP is also a factor that influence a person's intention emanated from reinforcement sensitivity theory (RST) [7]. This theory suggests that punishment may reduce undesirable behaviour. On the contrary, rewards may promote favorable behaviours. Perceived usefulness (PU) refers to the extent to which an individual believes that the application of technology enhances job productivity [8]. PU as the second determinant of behavioral intention emerge from technology acceptance model (TAM) by Davis [8]. The proposed hypotheses of this research is revealed in Table 1.

Table 1. Hypotheses definition

Hypotheses	Hypotheses Definition
H1	AT has a significant effect on the intention to process organic waste with biopore hole
H2	SN has a significant effect on the intention to process organic waste with biopore hole
H3	PBC has a significant effect on the intention to process organic waste with biopore hole
H4	RP has a significant effect on the intention to process organic waste with biopore hole
H5	PU has a significant effect on the intention to process organic waste with biopore hole

The findings of this research will be useful to the government as guidelines to review policies and program or as educational materials in schools to change their intention towards processing organic waste.

2. Method

This research builds an ETPB model with 6 latent variables and 19 observed variables as shown in Figure 1. The variables used in this research are latent variables consisting of endogenous latent variables and exogenous latent variables. Endogenous latent variable is the intention in processing organic waste. While the exogenous latent variables are AT, SN, PBC, RP, and PU. A questionnaire is used as the instrument for data collecting for this research.

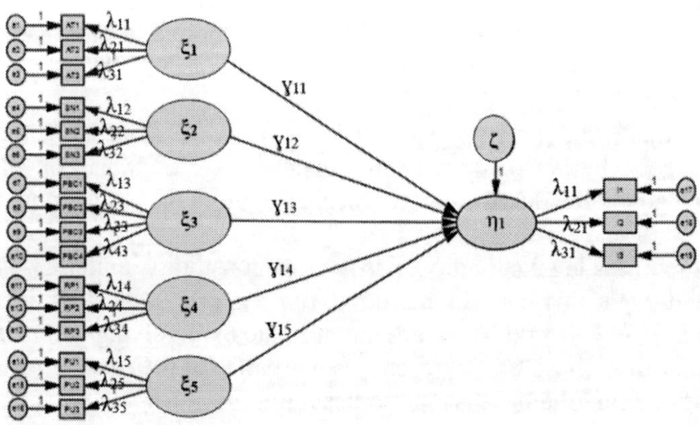

Figure 1. Research model [9-10]

Information :

η = endogenous latent variables

ξ = exogenous latent variables

ζ = Endogenous variable residual error

γ = Direct influence of exogenous latent variables to endogenous latent variables

λ = The relationship between latent and observed variables

e = Measurement error of the latent variable with the observed.

The research were examined and interpreted using descriptive and confirmatory analysis. The descriptive test is used to determine the mean and standard deviation of the results given by respondents and the later uses structural equation modeling (SEM). The research used a quantitative approach based on the test of SEM which was able to verify empirically and especially, the existing relationship between the variables of the proposed model. SEMs are very common analysis technique and are frequently used to test models in different fields of social and behavioural sciences [11]. The objective of the SEMs, as a multivariate statistical analysis technique, is to verify hypotheses concerning the influence of a set of variables on others [11]. The collected data's reliability and validity were tested on the account that the SEM is a first time measurement. Furthermore, model fit assessment, SEM assumption and hypotheses test were conducted.

3. Results and Discussion

All questionnaires were retrieved from the respondents and analysed accordingly. The results of the descriptive statistics, reliability and validity test is presented in Table 2.

Table 2. Descriptive calculation, reliability and validity results

Variable	Indicator	Mean	Std. dev	Loading factor	CR
I	I1; I2; I3	3.87; 3.80; 3.82	0.949; 0.963; 0.981	0.521; 0.554; 0.741	0.7861
AT	AT1; AT2; AT3	3.82; 3.84; 3.78	0.957; 0.975; 1.019	0.557; 0.524; 0.559	0.7290
SN	SN1; SN2; SN3	3.82; 3.69; 3.77	0.970; 1.030; 1.002	0.599; 0.643; 0.575	0.7767
PBC	PBC1; PBC2; PBC3; PBC4	3.74; 3.63; 3.77; 3.75	0.956; 0.991; 0.963; 0.976	0.552; 0.596; 0.608; 0.613	0.8732
RP	RP1; RP2; RP3	3.79; 3.70; 3.77	0.983; 1.021; 0.972	0.550; 0.564; 0.559	0.7357
PU	PU1; PU2; PU3	3.82; 3.70; 3.86	0.996; 1.081; 0.917	0.517; 0.546; 0.523	0.7081

Based on Table 2, The loading factor for each indicator is greater than 0.5, therefore all indicators are valid. Furthermore, the results of the construct reliability (CR) on each variable are greater than 0.70. Therefore it is arguable that the construct reliability test on each variable is fulfilled. The result of the SEM assumption test conducted is shown in Table 3.

Table 3. SEM assumptions test results

Assumptions test	Limit value	Test value
Normality	$-2.58 \leq c.r \leq 2.58$	c.r = 4.526
Linearity (I*AT; I*SN; I*PBC; I*RP; I*PU)	F Linearity	231.541; 234.695; 246.475; 208.805; 190.681
	P-value<0.05	all P-value 0.000
Multicollinearity (AT; SN; PBC; RP; PU)	Tolerance ≥ 0.10	0.427; 0.481; 0.415; 0.413; 0.424
	VIF ≤ 10	2.342; 2.081; 2.410; 2.424; 2.360

Based on Table 3, the critical ratio is outside the boundary value, therefore, the assumption of normality in each variable is not fulfilled. Consequently, it is continued by using the bootstrapping technique [9], [12]. Furthermore, the linearity test has a P-value less than 0.05. As a result the linearity assumption of each latent variable is fulfilled. Table 3 also shows that the tolerance value is greater than 0.10 and the VIF value is less than 10. This indicates that in each exogenous variable, there is no presence of multicollinearity. After several tests, the next measurement is to find a model fit and to test the proposed hypotheses. The model suitability test is measured based on goodness fit parameter as shown in Table 4. Each parameter passed its associated threshold. Furthermore, The hypotheses test is determined from the coefficient and p-value obtained from the regression weight. The results are summarized in Table 5.

Table 4. Model fit test results

Goodness fit parameter	Result	Threshold
Root mean square error of approximation (RMSEA)	0.038	≤ 0.05
Adjusted goodness fit of index (AGFI); Comparative fit index (CFI); Goodness of fit (GFI); Tucker lewis index (TLI)	0.922; 0.965; 0.944; 0.956	≥ 0.90

Table 5. Hypotheses test results

Parameter			Coefficient value (β)	P-value	Results
I	<---	PBC	-2.557	0.056	Significant
I	<---	RP	0.898	0.093	Significant
I	<---	SN	3.074	0.012	Significant
I	<---	AT	-0.327	0.417	Non-Significant
I	<---	PU	-0.260	0.722	Non-Significant

As it is shown in Table 5, three factors influence intention in processing organic waste using biopore hole. They include perceived behavioral control, reward and punishment, and subjective norms. All three factors have a significant relationship with intention (P value < 0.1). However subjective norms possesses a stronger and significant relationship. ($\beta = 3.074$, P value<0.1).

The TPB states that behavioral control is determined by individual beliefs about the collection of resources which consist of equipment, competitiveness, opportunity, and time [5]. An individual with a high perception of control will continue to be motivated and strive for success, this is due to the belief in overcoming difficulties, accuracy and the availability of resources. The results of this research was certified by previous studies which revealed that perceived behavioral control significantly influences behavioral intention [13-14].

Reward and punishment are positive reinforcement that emerge from the behavioristic theory. The theory shows that human behaviour is controlled by reward or reinforcement as they affect individual intention to act. Punishment has a positive effect on the learning process and it is effective in a variety of social settings [15]. This research is supported by previous studies which affirm that reward and punishment have a significant link to individual intention [16]. Gresik government has local regulation on waste management [17] However, there is no imprint of the mechanism of punishment. Therefore, this is a good confirmation of the advantages associated with reward and punishment-based programs to increase the intention of organic waste processing.

Based on this result, every unit increase in subjective norms increases intention towards processing organic waste with biopore hole. Furthermore, subjective norms are actors/environments that influence behavioral intention. These perpetrators include the principal, teachers, parents, friends, and the school environment itself. This is in agreement with the theory of planned behavior [4], which states that subjective norms are decisive factors in implementing some certain behavior. It is also found, to be a significant indicator of intention, confirming to previous studies [13].

4. Conclusion

Several factors influence individual intention in processing organic waste using biopore hole. Based on this research, reward and punishment, subjective norms and perceived behavioral control have influence. Additionally, subjective norms have most influence with the highest coefficient value and positive relationship. Based on the findings of this study, the need for government to review polices and programs is recommended and it will also serve as education materials for schools to increase the intention of processing organic waste.

References

[1] Kementerian Lingkungan Hidup dan Kehutanan Indonesia 2018 *Pedoman Pertanian Perkotaan (Urban Farming)* (Jakarta: Kementrian Lingkungan Hidup dan Kehutanan Indonesia)
[2] Pemerintah Indonesia 2009 *Undang-Undang No. 18 Tahun 2008 : Pengelolaan Sampah* (Jakarta: Pemerintah Indonesia)
[3] Kementerian Lingkungan Hidup Indonesia 2009 *Peraturan Menteri Lingkungan Hidup No. 12 Tahun 2009 : Pemanfaatan Air Hujan* (Jakarta: Kementrian Lingkungan Hidup Indonesia)
[4] Ajzen I 1991 The Theory of Planned Behavior *Orga Beha Hum Decis Process* **50** p 179–211
[5] Ajzen I 2005 *Attitudes, Personality, and Behavior 2nd ed.* (USA: McGrawHill)
[6] Ajzen I 1985 *From Intentions to Actions : A Theory of Planned Behavior* J. Kuhl & J. Backmann (eds.) Action Control : From Cognition to Behavior p 11-39 (Berlin:Springer Heidelberg)
[7] Ferster C B, and B F Skinner 1957 *Schedules of Reinforcement* (New York: Appleton-Century-Crofts)
[8] Davis F D 1989 Perceived usefulness, perceived ease of use, and user acceptance of information technology *MIS Quarterly* **13** p 319–340
[9] Ghozali I 2014 *Model Persamaan Struktural, Konsep dan Aplikasi Dengan Program AMOS 22* (Semarang: Badan Penerbit Universitas Diponegoro Semarang)
[10] Waluyo M 2016 *Mudah Cepat Tepat Penggunaan Tools Amos Dalam Aplikasi (SEM.* UPN "Veteran" Jawa Timur
[11] Tommasetti A, Singer P, Troisi O and Maione G 2018 Extended Theory of Planned Behavior (ETPB) : investigating customers' perception of restaurants' sustainability by testing a structural equation model *Sustainability* **20** p 2580
[12] Widhiarso W 2012 *Pemodelan Persamaan Struktural (SEM) pada Data Yang Tidak Normal* (Yogyakarta: Fakultas Psikologi UGM Yogyakarta)
[13] Gusti A, Isyandi B, Bahri S dan Afandi D 2015 Faktor Determinan Intensi Perilaku Pengelolaan Sampah Berkelanjutan pada Siswa Sekolah Dasar *Andalas J. Public Health* **9** 2 p 65-72
[14] Leeuw A, Valois P, Ajzen I and Schmidt P 2015 Ssing the theory of planned behavior to identify key beliefs underlying pro-environmental behavior in high-school students : implications for educational interventions *J. environ. Psychol.* **42** p 128-138
[15] Corr P J, A D Pickering and J A Gray 1997 personality, punishment, and procedural learning : a test of J. A. Gray's anxiety theory *J. Pers. Soc. Psychol.* **73** 2 p 337–44
[16] Amini F, Ahmad J and Ambali A R 2014 the influence of reward and penalty on households' recycling intention *Proc. APCBEE* (Singapore: Elsevier)
[17] Pemerintah Kabupaten Gresik 2017 *Peraturan Daerah Kabupaten Gresik No. 5 Tahun 2017: Perubahan Atas Peraturan Daerah Kabupaten Gresik No. 9 Tahun 2010 Tentang Pengelolaan Sampah* (Gresik: Pemerintah Kabupaten Gresik)

IOP Conf. Series: Materials Science and Engineering **930** (2020) 012029 doi:10.1088/1757-899X/930/1/012029

Flood inundation numerical modelling due to dam break of Way Sekampung Dam

Z. Nadida[1*], U Lasminto[1]

[1] Civil Engineering Department, Sepuluh Nopember Institute of Technology, Surabaya, Indonesia

[*]Corresponding author's e-mail: zafira@pu.go.id

Abstract. Dam construction holds a major disaster risk in when the dam is failure. The occurrence of dam breach can make water reservoir in the dam flow to the downstream at high speed, so it can cause infrastructure damage and the worst is, victim. This study analyses the extent of flooding in the resulting dam break with piping failure mode. The simulation was conducted using HEC-RAS 5.07, with input data consist of Digital Elevation Model (DEM), precipitation data, and dam's technical specifications. The research object is Way Sekampung Dam. The result of simulation show that after dam breach water flowing to the downstream with depth is 48,6 m and flood innundation area in Pringsewu District is 43.17 km^2.

1. Introduction

Dam construction is a technology to retain water on land. It has purposes for water supply, irrigation, hydroelectric power, etc. The most dam construction type in Indonesia is embankment dam. This is chosen because it can be built on almost all soil conditions. The construction material for dam body can be taken from the soil around it (Sosrodarsono & Takeda, 1977). The embankment dam weakness is being not able to hold water runoff and may cause a landslide on the downstream slope that can make the dam destroyed.

When the dam construction has been operating, it can cause serious disasters. The large volume water in reservoir will flow at high speed, so it can destroy the buildings in the downstream. Various studies for the dam breach have been conducted to minimize the risk occurrence (Xiong, 2011). Many causes that triggered the failure of the dam also became an interesting study to be examined and simulated (Gourbesville, Cunge, Caignaert, 2012). Ministry of Public Works and Housing (PUPR) as an administrator agency issued Ministerial Regulation No. 27/PRT/M/2015 (2015) regarding Dam. The state regulation for dam operation has been made with an emergency plan document that contains study about dam breach.

The object of this study takes place in Way Sekampung Dam located in Pringsewu, Lampung. Way Sekampung Dam location is about 14 km from the capital of Pringsewu regency, which is a populated \ area. The study about flood inundation is important to give a location estimation that can possibly be affected by a dam breach. The purpose of this study is to simulate flood inundation due to a dam breach in Way Sekampung Dam using numerical model.

Content from this work may be used under the terms of the Creative Commons Attribution 3.0 licence. Any further distribution of this work must maintain attribution to the author(s) and the title of the work, journal citation and DOI.
Published under licence by IOP Publishing Ltd

2. Literature review

2.1. Breach parameter

Flood wave prediction caused by a dam breach has high uncertainty (Dewals, et.al. 2014) due to limited information about a dam breach process, time and trigger. Froehlich (2008) collected 74 dam breach cases in embankment dam to get formulation of breach dimension, consisting of average width ($\bar{B}$), side slope ratio (z), height (H_b), water volume above breach bottom (V_w), and formation time of the breach (t_f). K_0 is constant of overtopping failure (1,3) and piping failure (1).

$$\bar{B} = 0{,}27K_0.V_w^{0,32}.Hb^{0,04} \quad (1)$$

$$t_f = 63{,}2\sqrt{\frac{v_\omega}{gh_b^2}} \quad (2)$$

$$B_t = \bar{B} + zH_b \quad (3)$$

$$B_t = \bar{B} - zH_b \quad (4)$$

2.2. HER-RAS for dam break study

HEC-RAS can be used to simulate a dam breach and analyse flood hydrograph by using Saint Venant, Diffusion Wave, or pool routing method. Brunner (2014) stated that the most accurate model technic to capture height and overflow of long narrow reservoir pond is full dynamic wave (unsteady flow). Unsteady flow will be more accurate for with and without breach scenarios. In this method, cross section needs to be created along the reservoir or to create a reservoir pond as a two-dimensional flow. Furthermore, the dam is modeled with the inline structure option.

There are two dam breach failure modes that can be simulated by HEC-RAS: piping and overtopping failure modes. Information that is required for dam breach simulation is area location, failure mode, shape, time, trigger mechanism, weir and piping coefficient, failure location, failure mode (Bruner, 2014).

HEC-RAS has been used for the number of dam breach research such as Ikrom & Wardhna (2020), Yi Xiong (2011). The results of these studies are the determination of high, velocity, and time lag for three scenarios of a dam breach.

3. Research methods

3.1. Study area

Way Sekampung dam is located in Bumi Ratu village, at the right side of the river is Banjero village and at the left side of the river is Pringsewu, Lampung province. The inflow of Way Sekampung dam comes from Way Sekampung River and the discharge of Batutegi Dam. The type of the dam is a rockfill dam. Volume capacity of this dam is 68 million m^3 with effective volume is 33.46 million m^3. This dam has an overflow spillway and a two-gate spillway. The technical data of Way Sekampung dam are shown as follows:

- Catchment area : 346 km^2
- Max Water Level : + 130
- Normal Water Level : + 124
- Height / Length : 55 m / 362 m
- Width : 10 m
- Elevation of overflow spillway : + 124
- Elevation of gate spillway : +116

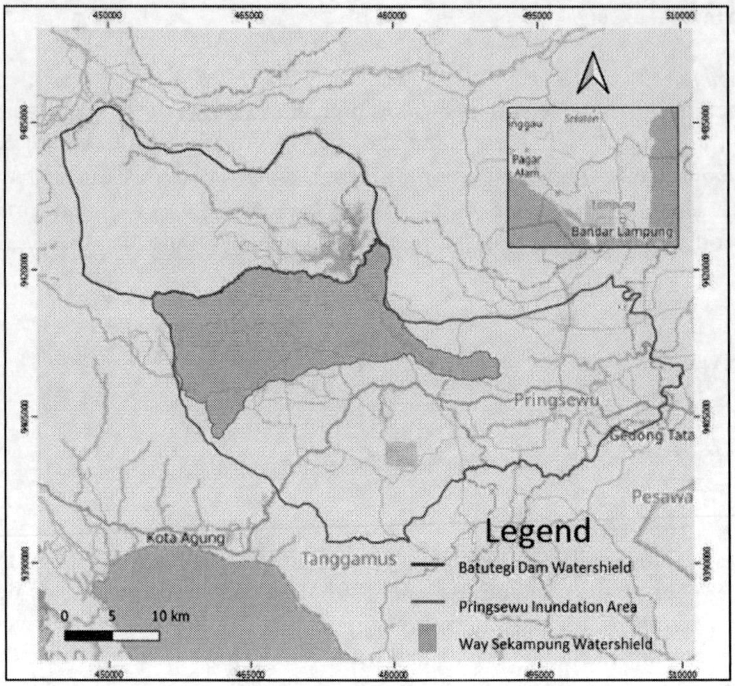

Figure 1. Study area

3.2. Hydrology analysis

Hydrology analysis has purpose to show the water debit that would come to dam in PMF condition. PMF calculation for Way Sekampung Dam is carried out in 3 stages: (1) hydrograph PMF from Way Sekampung Dam watershed; (2) hydrograph from reservoir routing when PMF outflow occurs; (3) calculation of the superposition of both of the hydrographs.

The PMF from the Sekampung Way Watershield is calculated using the Hearsfield formulation. The unit hydrograph is calculated using nakayasu unit hydrograph. Reservoir routing data of Batutegi Dam are obtained from PT. Indra Karya Wilayah 1 (2016). Then, the superposition from the hydrograph is shown in Figure 2.

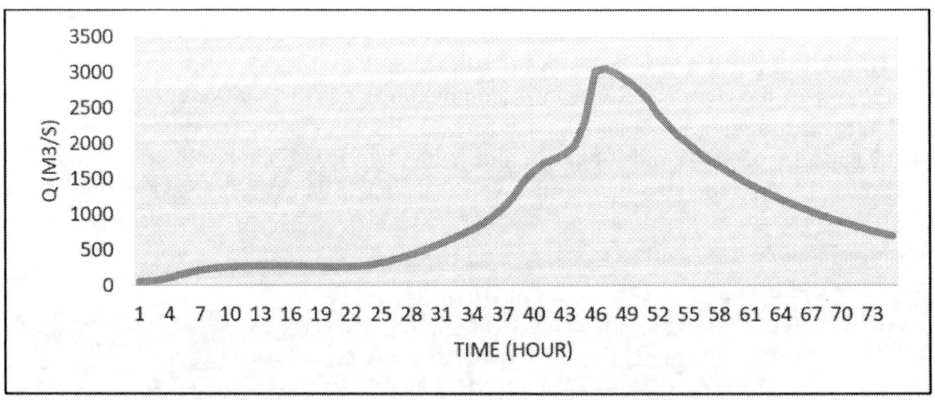

Figure 2. Superposition Hydrograph of Way Sekampung Dam

3.3. Spatial analysis

Spatial analysis is conducted to get contour information for research location, and can be used to simulate basic geometry. Digital Elevation Model (DEM) from DEMNAS with a scale of 1:50.000 is used for this process *(http://tides.big.go.id/DEMNAS/DEMNAS.php)*.

3.4. Model setup
Modeling used HEC-RAS 5.07 of full dynamic wave. Cross section is created along the upstream side, while for the downstream area, it is created with a combination of 1D and 2D. Model combination is made with cross-section connection between 2D area with a lateral structure. The 2D area location was chosen in low elevation area.Hydrograph flow with PMF debit is used for unsteady flow boundary condition in the upstream river, while normal depth is used for the downstream river. Water gate boundary used gate with controlled elevation.

3.5. Boundary condition
Boundary condition of this research involves:
1. Gate spillway does not work;
2. The simulation does not cover lateral inflow in the downstream area;
3. Only captured flooded areas in Pringsewu district;

3.6. BreachpParameter
The dam breach parameter would follow equation made by Froehlich (2008), and failure mode is caused by piping in the crest dam in elevations +120. Inline structure breach is filled with formula (1), (2), (3), (4), with results as shown in Table 1.

Table 1. Breach Parameter

Parameter	Value
$\bar{B}$	100.14 m
t_f	1.25 hour
B_t	126,04
B_b	74,24
k	1
z	0,7

The dimensions of the dam breach calculated in Table 1 cannot be directly entered in HEC-RAS because the dimensions of dam breach cause scour on the valley topography. Meanwhile, this formulation doesn't representative with these conditions. Therefore, we need to adjust the dam breach dimensions as can be seen in Figure 3.

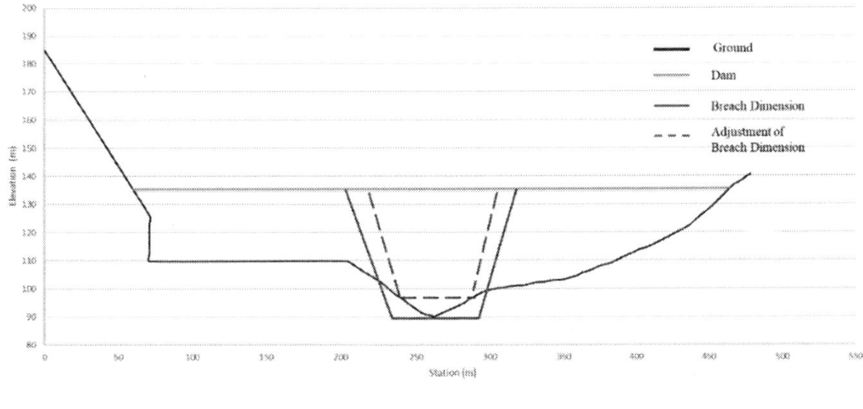

Figure 3. Dam Breach Dimension

4. Result and discussion
Simulation results show that HEC-RAS could simulate the dam breach with the conducted settings. Dam condition before and after dam breach are shown in Figure 4.

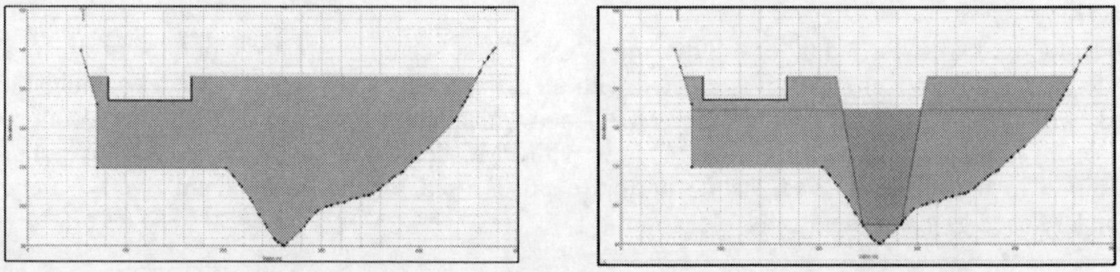

Figure. 4. Comparison of Way Sekampung dam before and after the breach

Figure 5. Present situation of inundation area from RAS-mapper.The closest location with high population is Pringsewu district. There are 5 districts affected by the flooding namely Ambarawa, Pringsewu, Gadingrejo, Sukoharjo and Banyumas. Table 2 shown the total inundated area at Pringsewu District is 43.14 km^2 with Gadingrejo becomes the area with the most extensive flood inundation.

Table 2. Flood Inundation Area In Pringsewu District

Sub District	Area (km^2)
Ambarawa	7,48
Pringsewu	14,11
Gadingrejo	19,83
Sukoharjo	0,63
Banyumas	1,09
Total	**43,14**

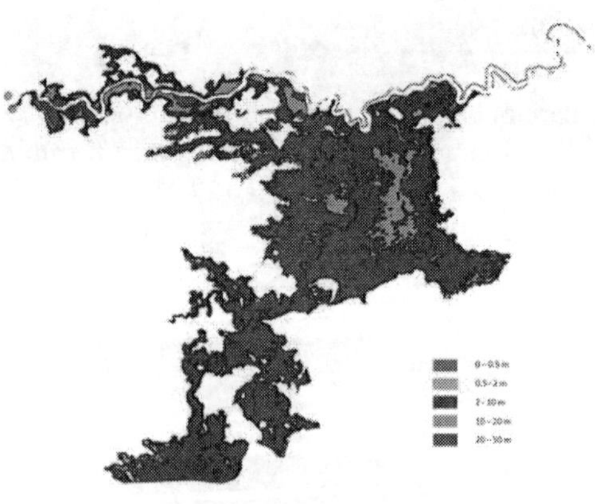

Figure 5. Pringsewu Regency Flood Inundation Map

To see the danger of flood disaster, the hazard area classification is based on inundation height. The categories are based on the Wirustyastuko and Nugroho (2013) which divided flood inundation into 3 categories. However, the average inundation height in Pringsewu district exceeds 2 m and the highest inundation is 49 m, therefore the addition of the classification is made to 5 categories (Table 3).

Table 3. Depth Classification

Category	Depth (m)	Area (km2)
1	0 – 2	7.34
2	2 – 5	12.75
3	5 – 10	17.26
4	10 – 15	4.66
5	20 – 50	1.13
Total		**43.14**

5. Conclusion

- The inundation area that occurred due to the Way Sekampung dam breach caused a flood inundation of 43.17 km^2 in Pringsewu district.
- The largest inundation area is in Gadingrejo district with an area of 19,83 km^2.
- Maximum depth is 49m, with the largest area in category 3 (5 – 10 $km^{2)}$ of 17.26 km^2.

References

[1] Gourbesville P, Cunge J and Caignaert G 2012 Dam Break Flow Modelling with Uncertainty Analysis *Advances in Hydroinformatics Springer Hydrology* (Singapore: Springer) p 107-16
[2] http://tides.big.go.id/DEMNAS/DEMNAS.php
[3] I A Iqbal and W P Nandi 2020 *Hydrodinamic simulation of a dam breach of Cipanas Dam using Hec-Ras 5.05* (IOP Conference Series : Earth and Environment Science)
[4] Ministry of Public Works and Housing 2015 *Peraturan Menteri Pekerjaan Umum dan Perumahan Rakyat No. 27 tentang Bendungan* (Jakarta: Kementerian Pekerjaan Umum dan Perumahan Rakyat)
[5] PT. Indra Karya Wilayah I 2016 *Laporan Hidrologi Pekerjaan Sertifikasi Desain Bendungan Way Sekampung (Sukoharjo)* (Bandar Lampung : PT. Indra Karya Consulting Engineers)
[6] Sosrodarsono S and Takeda K (1977) *Bendungan Type Urugan* (Jakarta: PT. Pradnya Paramita)
[7] W Dandi and N Joko 2013 Analisis wilayah tergenang dan perilaku banjir pada simulasi kegagalan Bendungan Ciawi *J. Civ. Eng.* **20** 2 p 121 – 32.
[8] Xiong Yi 2011 A dam break analysis using hec-*ras J. Water Resource Prot.* **3** p 370-9

IOP Conf. Series: Materials Science and Engineering **930** (2020) 012030 doi:10.1088/1757-899X/930/1/012030

Structural Equation Model (SEM) between risk, safety control systems of risk, and OHS programs on OHS costs in rusunawa projects

A S Harianja[1*] and Y Latief[2]

[1] Post Graduate Student, Department of Civil Engineering, Faculty of Engineering, University of Indonesia, Depok, West Java 16424, Indonesia
[2] Department of Civil Engineering, Faculty of Engineering, University of Indonesia, Depok, West Java 16424, Indonesia

*Corresponding author's e-mail: apriadi.simon@ui.ac.id

Abstract. There is a continuous hampering of Occupational Health and Safety (OHS) in the construction budget due to its consideration as an extra cost. This means decisions on OHS are usually made based on the economics aspect instead of focusing on the ethics and basic rights to a safe workplace. This research was, therefore, conducted to analyze the relationship between risk, its safety control systems, OHS programs, and OHS costs, and also to determine the factors with the dominant effects on OHS costs using the Structural Equation Model - Partial Least Square (SEM-PLS) with the use of Rusunawa projects as a case study

1. Introduction

In recent years, safety in construction has become a global problem and a serious concern for the government. This is mostly reflected in the absence of enough data or correct information to assess the safety of a construction site [1].

One of the risk factors considered to be very influential to the performance of a company in the construction sector is the risk of work accidents. A more complex construction design has been discovered to have a higher risk as well as the OHS costs required to prevent accidents. Therefore, OHS Construction Risk is defined as the measure of the possibility of loss associated with public safety, property, human life, and the environment due to certain sources of hazards during construction [2]. There is, however, a continuous hampering of Occupational Health and Safety (OHS) in the construction budget due to its consideration as an extra cost. The decisions on this concept are, therefore, usually made based on the economics aspect instead of focusing on the ethics and basic rights to a safe workplace [3].

This research was, therefore, conducted to analyze the relationship between risk, its safety control systems, OHS programs, and OHS costs, and also to determine the factors with the dominant effects on OHS costs [4].

2. Literature review

Content from this work may be used under the terms of the Creative Commons Attribution 3.0 licence. Any further distribution of this work must maintain attribution to the author(s) and the title of the work, journal citation and DOI.
Published under licence by IOP Publishing Ltd

2.1. OHS cost for rusunawa project
Indonesia is a developing country with low per capita income and the second-highest number of workers in the world. Moreover, Southeast Asia has also been placed second highest among the regions with the occupational accident fatality rate [6]. This is mainly due to the non-inclusion of safety or accident prevention costs in the project budget in order to reduce the expenses for the contractor. Meanwhile, some of the costs affiliated with accident prevention are as follows [7]:

- First Aid Facilities
- Personal Protective Equipment (PPE)
- Safety Training
- Safety Promotion
- Security Personnel

Statistics showed the highest mortality rate is in the construction industry sector and this majorly due to the poor management of OHS by contractors because of the inadequate value placed on the need to comply with the regulations. Meanwhile, the items required to be included in the cost evaluation [8] include:

- Worker Insurance
- OHS Inspection
- OHS Meeting
- OHS Training
- Payement to safety officers/consultans
- Safety Tool
- Personal Protective Equipment
- Supporting equipment for administration, management, and documentation

Several construction site's employees are being killed or injured every year while others suffer from poor health due to causes related to work. The risk is also not limited to the employees as observed with the frequent deaths and injuries usually recorded in a community when construction activities are not properly monitored. Therefore, some of the items required to be monitored to ensure basic health and safety include [9]:

- Access Location
- Location Limits
- Welfare Facilities
- Sanitation
- Washing Facilities
- Rest Facilities
- Personal Protective Equipment (PPE)
- Storage Areas and Waste Materials
- Lighting
- Emergency Procedure
- First aid
- Reports of Accidents and Dangerous Events

2.2. OHS risk in rusunawa project
According to Flanagan & Norman (1993), the risks in construction projects include [10]:

- Failed settlement according to the design or construction time
- Failure to obtain drawings and planning/permit details on time.

- Unexpected Soil Conditions
- Very Bad Weather
- Labor Strike
- Unexpected increase in price for labor and materials
- Accidents at locations causing injuries
- Damage caused to the structure due to poor work methods
- Unexpected events such as floods, earthquakes, etc.
- Claims from contractors due to loss as well as the costs incurred due to production delays caused by the design team
- Failure to complete the project with a predetermined budget.

2.3. Safety control systems of risk
According to ISO 45001: 2018, the control hierarchy in the safety, health, and occupational health system include [11]:

- Elimination
- Substitution
- Engineering Control
- Administrative Control, and
- Personal Protective Equipment (PPE)

2.4. OHS programs
The details of the Activities of Implementing a Construction Safety Management System based on the Circular Letter of the Minister of PUPR Number 21/PRT/M/2019 are presented as follows [12]:

- Preparation of RKK
- Socialization, promotion, and training
- Work Protective Equipment (APK) and Personal Protective Equipment (PPE)
- Insurance and Licensing
- OHS construction personnel
- Medical facilities, infrastructures, and devices
- Signatures needed
- Consultation with experts regarding construction safety according to the scope of the field works
- Miscellaneous related to construction safety risk control

2.5. The theory of the relationship between risk and OHS costs
Risk Management is defined as the process of identifying, measuring, and financial control of a risk threatening the assets and income of a company or project and with the ability to cause damage or loss. Meanwhile, OHS risk management is a comprehensive, planned, and structured effort to manage OHS risk in order to prevent unwanted accidents in a good system. It is majorly related to hazards and risks with the potential to cause harm in the workplace [13].

According to the PUPR PERMEN Number 21 of 2019 concerning the Construction Safety Management System Guidelines, Article 26, the Construction Safety Risks are classified into [12]:

- Small
- Moderate
- Big

This, therefore, means there is an effect of OHS risk on real-life implementation costs of the concept in construction projects.

2.6. The theory of the relationship between OHS programs and OHS costs

The Occupational Safety & Health Plan is a key document which is usually used as a reference to ensure job safety performance on projects with the focus on workers and other personnel in the field. It contains general regulations on reducing accidents and protecting assets and properties and the estimated costs which are expected to be included in the project budget include [14]:

- Registration and administration fees
- Training costs
- Promotion fee
- OHS operational costs

This means the OHS program to be implemented in the Rusunawa project has the potential to affect the OHS Cost.

2.7. The theories of the relationship between risk safety controls system and OHS costs

The hierarchy of control is one of the things highly considered in OHS risk assessment activities and this is due to the importance of the choice made in providing effective and efficient benefits by reducing the risks to an acceptable level which is also known as the acceptable risk in an organization. Meanwhile, the first control hierarchy is believed to provide higher effectiveness than the second, and this is associated with two basic ideas which are reducing the probability as well as the severity of an accident or exposure. [11]

3. Methodology

This research was conducted with the use of a quantitative survey and this involved taking a sample of at least predetermined 200 respondents from one population while questionnaires were used to obtain primary data.

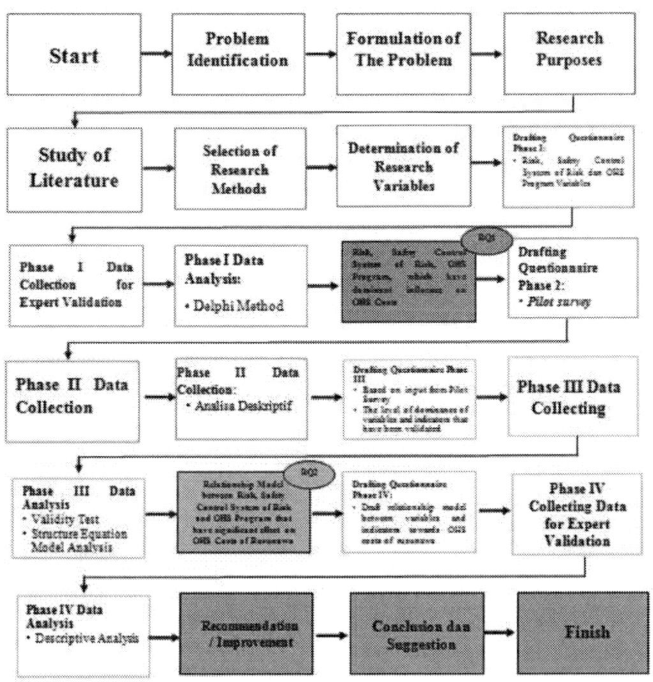

Figure 1. Research Methodology

The first stage involved determining the outputs expected to be used as indicators from each variable. This was followed by the second stage which was validating indicators using expert judgment by applying google docs. The third stage was the distribution of the questionnaires to the respondents while the last step was processing data obtained from the third stage using the SEM PLS program. The process is graphically represented in the Figure 1.

4. Result

The primary data was directly sourced using questionnaires and based on certain criteria one of which is requiring the respondents are construction project professionals with a minimum of 1year experience in Indonesia and results obtained are presented in the following table.

Table 1. The Relationship Between Variable based on SEM-PLS Result

Relationship	Original Sample (O)	T Statistics	Information
X1[b] -> Y1[c]	0,104	1,712	Not Significant
X1 -> Y2[d]	0,096	1,822	Not Significant
X2[e] -> Y1	0,196	2,253	Significant
X2 -> Y2	0,028	0,361	Not Significant
X3[f] -> Y1	0,497	6,491	Significant
X3 -> Y2	0,666	10,690	Significant

[a] Relationship between X and Y Variable which obtained from SEM PLS Program
[b] Risk Variable
[c] OHS General Cost Variable
[d] OHS Spesific Cost Variable
[e] Safety Control Systems of Risk Variable
[f] OHS Programs Variable

The results for the relationship between the variables are attached at the end of this paper after it has been tested by an expert in OHS.

5. Conclusion, limitation, and recommendation

It is recommended that all the indicators for the variables be comprehensively developed to achieve the best result in the next research based on the regulation of the Minister of Public Works and Housing regulations (PUPR) requiring the OHS cost is well presented in constructing the Rusunawa projects.

This research is limited by the inadequate numbers of experts needed to test the validity of the results. It is also possible to develop it by adding more indicators or variables influencing the OHS costs in Rusunawa Projects in order to identify the competency gaps required to be improved.

6. Appendix

Table A. Outer Loading Value Result for Intern Relationship (Between Indicators and Variables)

RELATIONSHIP	ORIGINAL SAMPLE	T STATISTICS	SIGNIFICANCY[A]
X1.1 <- Risk	0,785	13,251	Significant
X1.2 <- Risk	0,938	38,986	Significant
X1.3 <- Risk	0,660	8,100	Significant
X2.1 <- Safety Control System of Risk	0,574	8,493	Significant
X2.2 <- Safety Control System of Risk	0,723	12,887	Significant
X2.3 <- Safety Control System of Risk	0,816	23,093	Significant
X2.4 <- Safety Control System of Risk	0,764	20,923	Significant
X2.5 <- Safety Control System of Risk	0,774	23,751	Significant

X3.1 <- OHS Program	0,609	8,634	Significant
X3.2 <- OHS Program	0,649	12,063	Significant
X3.3 <- OHS Program	0,716	14,878	Significant
X3.4 <- OHS Program	0,583	10,009	Significant
X3.5 <- OHS Program	0,781	30,642	Significant
X3.6 <- OHS Program	0,747	19,830	Significant
X3.7 <- OHS Program	0,778	22,742	Significant
X3.8 <- OHS Program	0,661	12,130	Significant
X3.9 <- OHS Program	0,735	16,884	Significant
Y1.1 <- OHS General Cost	0,822	19,093	Significant
Y1.2 <- OHS General Cost	0,826	20,783	Significant
Y1.3 <- OHS General Cost	0,877	47,643	Significant
Y1.4 <- OHS General Cost	0,822	30,422	Significant
Y2.1 <- OHS Specific Cost	0,737	17,042	Significant
Y2.2 <- OHS Specific Cost	0,780	18,007	Significant

[A]Significant if the T Statistics value > 1,96

Table B. Path Coefficient Significance Relationship Between Variables

No	Relationship Between Variable	Original Sample(O)	T Statistics (|O/STDEV|)	Significance
1	Risk → Safety Control System of Risk	0,369	5,383	Significant
2	Risk → OHS Program	0,157	2,091	Significant
3	Risk → OHS General Cost	0,104	1,712	Not Significant
4	Risk → OHS Specific Cost	0,096	1,822	Not Significant
5	Safety Control System of Risk → OHS Program	0,541	9,216	Significant
6	Safety Control System of Risk → OHS General Cost	0,196	2,253	Significant
7	Safety Control System of Risk → OHS Spesific Cost	0,028	0,361	Not Significant
8	OHS Program → OHS General Cost	0,497	6,491	Significant
9	OHS Program → OHS Specific Cost	0,666	10,690	Significant

Acknowledgements
The author appreciates the financial support provided by the University of Indonesia through PUTI Proceedings of the Fiscal Year 2020 with the contract number: NKB-1184 / UN2.RST / HKP.05.00 /

IOP Conf. Series: Materials Science and Engineering **930** (2020) 012030 doi:10.1088/1757-899X/930/1/012030

2020 managed by the Directorate of Research and Community Engagement (DRPM) Universitas Indonesia.

References

[1] Gurcalini G E and Mungen U2009.An occupational safety risk analysis method at construction sites using fuzzy sets *Int. J. Ind. Ergon.* **39** 2 p 371-387

[2] Brody et al.1990 *An Indirect Cost Theory of Work Accident Prevention* (French: Laval University)

[3] H Lingard and S Rowlinson 2009 *Occupational Health and Safety Management in Construction Project Management* (New York:Spoon Press)

[4] Emre G G, Bilir S and Merve S 2015 *Activity Based Risk Assessment and Safety Cost Estimation for Residential Building Construction Projects* (Turkey: Elsevier)

[5] Ministry of Public Works and Public Housing 2019 *Pedoman Sistem Manajemen Keselamatan dan Kesehatan Kerja (SMK3) Konstruksi Bidang Pekerjaan Umum* (Jakarta:Kementerian Pekerjaan Umum)

[6] World Employment 1995 *An ILO Report*.(United States: Washington Branch)

[7] Ikpe, Hammon and Oloke.2008.*Cost Benefit Analysis for Accident Prevention in Construction Project* (UK: Wolverhampton)

[8] Hamid, Singh and Saleh.2014.*Cost Of Compliance With Health and Safety Management Sistem Among Contractor in Construction Industry*. (Malaysia: Universitas Tun Hussein Onn Malaysia) p 6

[9] Health and Safety Executive.2006.*Health and Safety in Construction* (South Africa: Pretoria) p 2-9

[10] Flangan and Norman.1993:114.*Risk Factors in Cost Estimation : Building Contractor's Experience* (Oxford: Blacwell Scientific)

[11] *ISO45001.2018.Occupational health and safety Vol 54* (Switzerland: Geneva) p 28-31

[12] Ministry of Public Works and Public Housing 2019 *Pedoman Sistem Manajemen Keselamatan dan Kesehatan Kerja (SMK3) Konstruksi Bidang Pekerjaan Umum* (Jakarta: Kementerian Pekerjaan Umum)

[13] Ramli and Soehatman 2010 *Fire Management Practical Guidelines (Fire Management)* (Jakarta: Dian Rakyat)

[14] Department of Public Works.2007:28,30,65.*Construction Training and Competence Training Agency* (South Africa: Pretoria)

[15] AS/NZS 4360.2004. *Risk Management Guidelines*.(Sydney: Standards Australias/Standards New Zealend) p 52-55

[16] Sarwono and Narimawati. 2015:19.*Membuat Skripsi, Tesis dan Disertasi dengan Partial Least Square SEM (PLS-SEM)* (Yogyakarta: ANDI)

[17] Ghozali 2012 *Multivariate Analysis Application with the SPSS Program* (Semarang:Publisher Agency UNDIP)

Updating the slope-movement data on the Batu-Kediri road network using digital information

E Wahyudianto[1*]

[1] Road and Bridge Engineering, Road and Bridge Infrastructure Office, East Java Provincial Government, Jalan Gayung Kebonsari 167 Surabaya, Indonesia.

[*]Corresponding author's e-mail: info@binamarga.jatimprov.go.id

Abstract. Updating the landslide data or slope movement on the road is needed for the safety of traffic movements. Landslide events on the road must be compiled to measure their characteristics, including the causes thereof. In identifying and assessing landslide events, detailed data related to trigger parameters, time of occurrence, impact, or material loss are absolutely necessary. The utilization of digital media ultimately provides very useful disaster data. However, these data sources have different levels of information depth. To obtain sufficient data to analyze landslides, the data must be sorted according to the depth classification of information. The classification is based on landslide coordinates, impacts on traffic movement, time of occurrence, fatalities, triggers, information providers, and types of landslide material. By utilizing the digital media, updating the record of landslides between 2008-2020 resulted in landslide density values as follows: Batu Municipal 9.93 (n/km), Pujon 3.44 (n/ km), Kasembon 1.46 (n/km) and Ngantang (0.91 n/km).

1. Introduction

The rapid growth of online media users or social media is a special phenomenon in the industrial era 4.0. Internet-based media can change behaviours as well as human phenomena in sharing information. Before internet-based media was used, existing information was obtained through media giant companies, so now in the 4.0 era, information can be sourced from anywhere without going through information filters. The positive and negative sides of the phenomenon are dynamics that can be utilized according to the objectives of the users.

Today, disaster occurrence becomes a concern to the community. Both in the forms of occurrence are triggered by natural and non-natural disasters such as technology failure. In all phases of a disaster occurrence, starting from the pre-disaster stage to the post-disaster phase, communication is a determining factor for the success of risk / impact reduction efforts. Rudianto (2012) underlines the accuracy of disaster information. Miscommunication actually worsens the condition of all parties impacted and involved in disaster relief.

The disaster risk reduction program is a top priority as a result of changes in the dynamics of regional growth. Development of the city, the pace of urbanization, and the level of poverty are the foremost challenges faced. Disaster management is more multidisciplinary, a multi-approach, as well as using leading technologies. Disaster institutions are known to be generous in presenting results and data analysis. These data when combined with information that is spread online and social media will be very useful if applied as a source of research information for mitigation activities.

Content from this work may be used under the terms of the Creative Commons Attribution 3.0 licence. Any further distribution of this work must maintain attribution to the author(s) and the title of the work, journal citation and DOI.
Published under licence by IOP Publishing Ltd

IOP Conf. Series: Materials Science and Engineering **930** (2020) 012031 doi:10.1088/1757-899X/930/1/012031

Slope movement on roads has a significant social and economic impact. Damage and loss are generally quite large because the road is an economic link chain. Landslides on the road these days can be easily recorded through uploading activities of the community and institutions through social media. The data obtained are generally above 2008, where online media have begun to stabilize and are well established in presenting the information.

2. Literature review

Prajarto (2007) states that the enthusiasm of the media in broadcasting disaster events has two sides of an impact on society. On the one hand, the disaster information system seeks to be strengthened by the role of the media in the disaster information system. However in practice, instead of getting integrated information in a complete disaster information system, the public will only get a collection of pieces of information about disasters. Wahyudianto (2019) also stated that Crowd-source as well as online media in this case also becomes an important part of the disaster information systems. Those data need to classify and be grouped into several technical criteria. In addition to providing information in a fast tempo, the information conveyed by these media becomes a cover for weakness when official sources do not release the official information. Data that can be traced through an internet search engine is useful as an alternative source of disaster information for analytical purposes. Wahyudianto (2017) used the internet as an alternative source of information to analyze landslides on roads since 2007-2016. Likewise, Floris et al. (2011) conducted a landslide vulnerability assessment using online data in the Veneto region, Italy.

2.1. Slope-movement mitigation on roads

Mitigation of landslides on roads has been carried out by many researchers. Wahyudianto (2019) made a landslide hazard map on the Ponorogo-Pacitan road network, East Java, and validated the results using a landslide inventory field survey on the impact of slope movements due to the Cempaka Cyclone disaster. Guzzetti (2005) adopted Pierson & van Vickle (1993) in assessing the hazard of falling rocks on a road. Likewise, Prina (2004), Parise (2002), Gaurav (2009), and Liu (2006) formulated the vulnerability of vehicle movements based on the probability of the existence of vehicles spatially on a road segment. Approaches related to calculating rockfall risk are standardized by AGS (2010) in Australia, while Dorren (2009) who conducted research on the national road network in Switzerland approached the risk value of vehicle movements based on certain financial values based on the probability of fatalities on road users. Wahyudianto (2017) uses a combined approach, which maps the frequency analysis of landslide events using the method of Huang (2015) which was modified as a result of zero short-duration rain data, and replaces it with daily rainfall. Wahyudianto (2017) calculates the boundary line of the rain trigger for the landslide using the modified Huang (2015) method as a result of zero short-duration rain data and replaces it with daily rainfall.

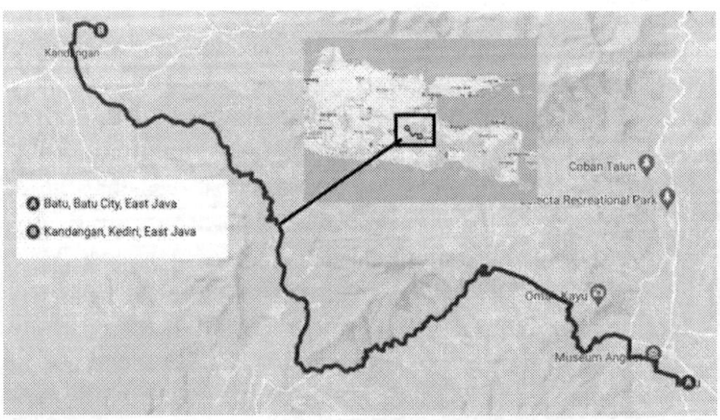

Figure 1. Batu-Kediri provincial road network

2.2. Slope-movement data search

Collecting landslide data on internet-based media has its own advantages, given the ability of search engines to track information is quite well in the past few years. The data generated via the internet have varying degrees of accuracy. There are data that record slope movement through the date of the occurrence, time period, even some of them are very detailed in recording the time of the disaster event, the type of slope movement material, a brief analysis of the cause of the event, as well as the impact caused by the slope movement occurrence. The process of searching data uses the internet search engine as figured in Figure 2.

Wahyudianto (2017) validated internet information sources related to landslide events on his research road as many as 65 sources, with details of 50 days and 92 landslides. The level of accuracy varies. As much as 67% of information was managed to record the time of the occurrence, while the remaining 33% did not know the time of the disaster occurrence.

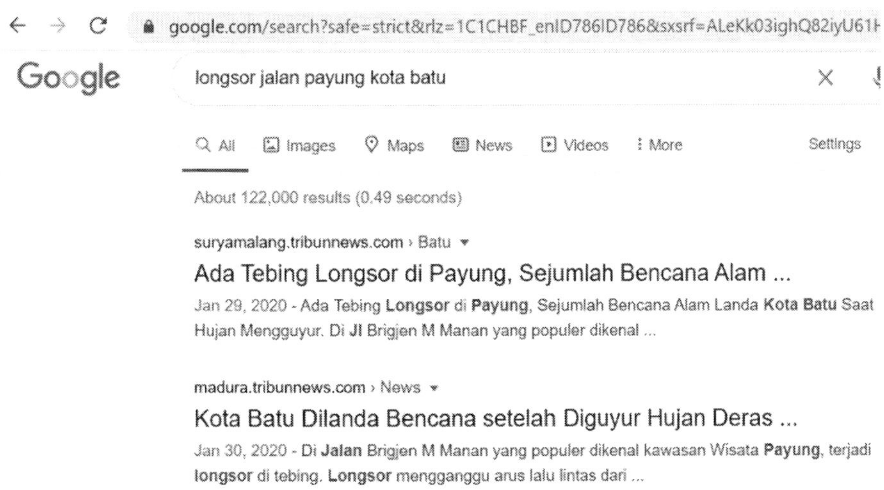

Figure 2. Search for landslide information

3. Research methods

Information that has been collected through the internet is filtered to find important information related to the slope movement. The information includes the day of events, hours, location, quotation of information sources, the impact of traffic movements, type of material, cubic of material, and other information deemed necessary for further analysis. Collecting records of landslide events in this study continues the research of Wahyudianto (2017) and Wahyudianto (2019) as an effort to update the record of occurrence. Wahyudianto (2017) collected data between 2007-2016, whereas Wahyudianto (2019) made an effort to update the data completed until March 2019. From the renewal efforts carried out by Wahyudianto (2019), there were obtained 84 data sources with details of the occurrence of 69 days and 114 landslides. Updating data in this research is completing search results until June 2020.

4. Results and discussions

Mapping slope movement using internet media has an easy source of information. From the data updating process until June 2020, it can be concluded that the pattern of landslide occurrence based on the distribution of the month has slightly changed. If previously the majority of landslides occurred in January and February contributed 80.71% for 12 years of recorded events (Wahyudianto, 2019), then for 13 years of the occurrence, contributions in January and February decreased. The contribution of occurrence in January and February to the end of the search in June 2020 was 76.43% as listed in Table 1. There are four most important rainy months which contributed a lot to slope movements. The months are December, January, February, and March. Wahyudianto (2019) calculated that these months contributed 93.04% of occurrence. Meanwhile, of the results of the renewal process, these months contributed as much as 91.87% of occurrences as listed in Table 1 and Figure 4.

Update of slope movement occurrence record data from 2019 until June 2020 is 9 sources of information, with details of 9 days of occurrence and 9 landslide points. The percentage of hours of the occurrence of all collected data is 30% for the unknown and 70% for the known as listed in Table 3. Of the distribution of landslide events on a yearly basis, most of the disaster events occurred in 2013 and 2014 or amounted to 49.59% as listed in Table 4. The slope movement data that have been recorded from 2007 until June 2020 are 93 sources, with the details of the occurrence are 78 days and 123 location points. The slight change of the slope movement occurrence is assumed as the function of the change of rainy season and rainfall distribution during the months.

5. Conclusions

The results of landslide occurrence that are positioned in the administrative area of the sub-district obtained values of landslide density representing the hazard level of a road segment. The results of the landslide density data updated by Wahyudianto (2019) were in Batu Municipal of 9.27 (n / km), Pujon of 3.08 (n / km), Kasembon (1.46 n / km) and Ngantang (0.85 n / km). From the update record of occurrence between 2007-2020, the largest landslide density value is in Batu City, 9.93 (n / km), followed by Pujon (3.44 n / km), Kasembon (1.46 n / km), and Ngantang (0.91 n / km) as listed in Table 2. From the results of the analysis, it can be concluded that the update data landslide point is the location of the old landslide point in the slope cutting zones. This proves that the influence of the occurrence of a mass movement on the road slope is more contributed by slope cutting than the cumulative rain.

Table 1. Distribution of slope movements on the Batu-Kediri road network during 2007-2020

No	Month	D, days of Occurrence	n, number of landslide points	% D	% n
1	Jan	24	49	30.77	39.84
2	Feb	27	45	34.62	36.59
3	Mar	13	14	16.67	11.38
4	Apr	2	2	2.56	1.63
5	May	0	0	0.00	0.00
6	Jun	1	2	1.28	1.63
7	Jul	2	2	2.56	1.63
8	Aug	1	1	1.28	0.81
9	Sep	0	0	0.00	0.00
10	Oct	1	1	1.28	0.81
11	Nov	2	2	2.56	1.63
12	Dec	5	5	6.41	4.07
	Σ	78	123	100.00	100.00

Table 2. Distribution of slope movements by sub-districts during 2007-2020

Sub-district	Batu	Pujon	Ngantang	Kasembon	Σ Sum
n, number of landslide occurrence	45	48	14	16	123
n (%)	37%	39%	11%	13%	100%
n per year	3.46	4.36	1.27	1.45	11.18
Length, L (km)	4.53	13.97	15.34	10.94	44.78
n/L, density (n/km)	9.93	3.44	0.91	1.46	2.75

Table 3. Level of depth information on slope movement data during 2007-2020

Hour of occurrence	n	n (%)
Unknown	37	30%
Known	86	70%

Table 4. Distribution of slope movements by years of occurrence during 2007-2020

Year	D, day of occurrence	n, number of occurrence
2020, Jan- June	5	5
2019	5	5
2018	9	11
2017	9	10
2016	10	12
2015	2	2
2014	11	35
2013	16	26
2012	4	4
2011	2	5
2010	3	6
2009	0	0
2008	1	1
2007	1	1
Σ	78	123

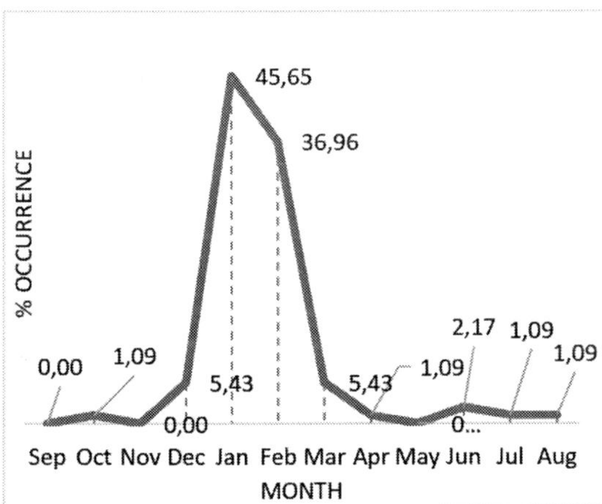

Figure 3. Landslide distribution based on the months of occurrence (2007-2017) by Wahyudianto (2017)

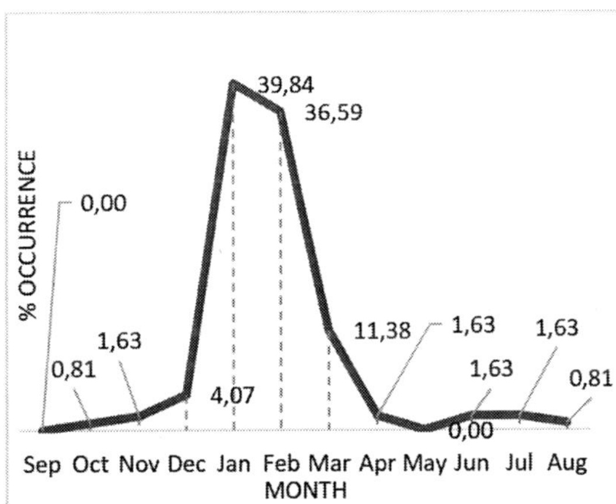

Figure 4. Landslide distribution based on the months of occurrence (2007-2020)

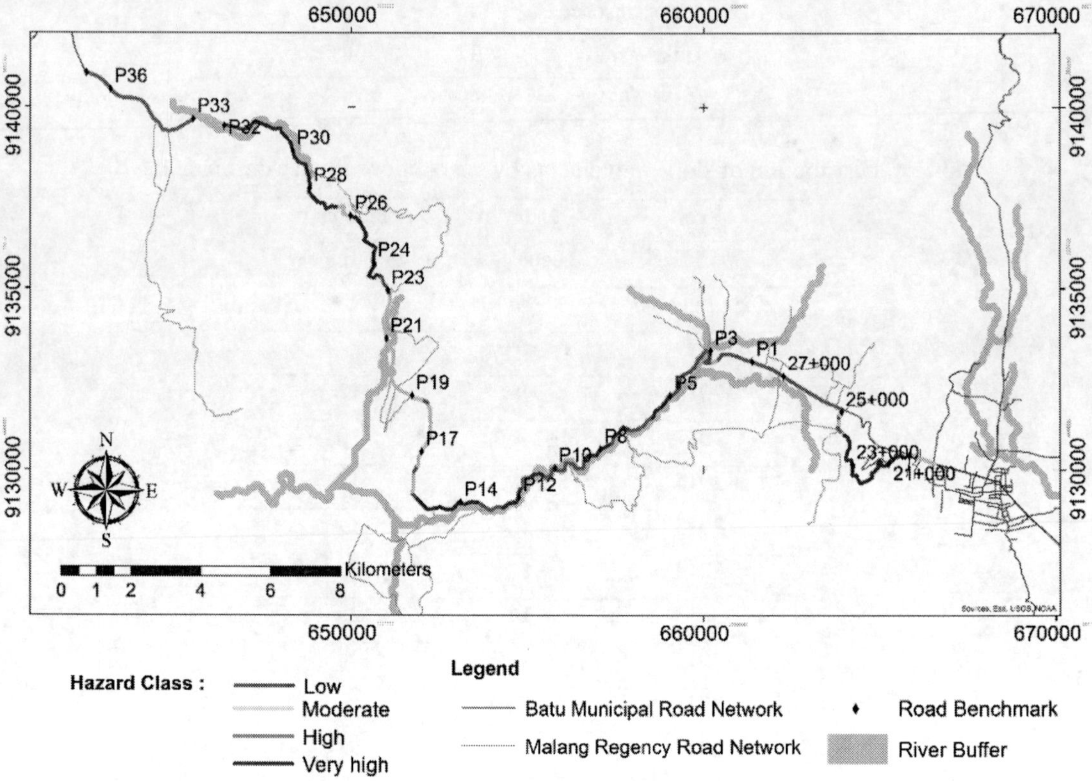

Figure 5. Hazard map on the Batu-Kediri road section from an internet search that was validated using a field-based landslide inventory (Wahyudianto, 2017)

Concrete cracking is known to have a knock-on effect on the durability performance and long-term

Acknowledgements
Special thanks to the Road and Bridge Office, East Java Province, for the permission and opportunity have given in conducting the research.

References

[1] Rudianto 2015 Komunikasi dalam Penanggulangan Bencana *Jurnal Simbolika* **1** 52 L 2
[2] Prajarto N 2007 Bencana, informasi dan keterlibatan media *JSP* **11** 3 (Yogyakarta: UGM)
[3] Wahyudianto E 2019 Pembangunan infrastruktur jalan dalam era teknologi industri 4.0, inventarisasi bahaya longsor jalan pada fase pasca bencana (studi kasus siklon cempaka pada jalan provinsi di Kabupaten Pacitan) *Prosiding Konferensi Nasional Teknik Jalan ke- 10* 513 L 13
[4] Guzzetti F, Dikau R and Glade T 2005 *Landslide hazard and risk assessment* (Bonn: Mathematich-Naturwissenschaftlichen Fakultät Rheinischen Friedrich-Wilhelms-Univestität Bonn)
[5] Pierson L A and Vickle R V 1993 Rockfall hazard rating system participants manual *Report FHWA-SA-93-057*
[6] Prina E, Bonnard C and Vulliet L 2004 Vulnerability and risk assessment of mountain road crossing landslide. *Riv. Ital. di Geotec.* **2** p 76
[7] Parise M 2002 Landslide hazard zonation of slopes susceptible to rock falls and topples *NHESS* **2** p 42

IOP Conf. Series: Materials Science and Engineering **930** (2020) 012031 doi:10.1088/1757-899X/930/1/012031

[8] Gaurav K, Saran S, Stein A and Das I 2009 *Stochastic modelling of land cover (dynamic) elements* Thesis ed. Dehradun (India: Indian Institute of Remote Sensing (IIRS) and International Institute for Geoinformation Enschede, The Netherlands)

[9] AGS 2000 *Landslide Risk Management Concepts and Guidelines* (Sydney: Australian Geomechanics Society: Sub Commitee on Landslide Risk Management)

[10] Dorren L, Sandri A and Raetzo H 2009 *Landslide Risk Mapping for Entire Swiss National Road Network* (Bern, Switerland: Project "Naturgefahren auf National Strassen" NHNR National Hazard on National Roads FEDRO (Federal Road Office) & FOEN Federal Office for the Environment)

[11] Wahyudianto E 2018 Analysis and risk study on landslide hazard frequency at road corridor of Batu City–Kediri Regency border *JCEF* **4** p 3

[12] Wahyudianto E 2019 Pembangunan infrastruktur jalan dalam era teknologi industri 4.0, asesmen bahaya longsor pada jalan raya berbasis data crowd-source dan media online (studi kasus ruas jalan Kota Batu-batas Kab. Kediri*) Prosiding Konferensi Nasional Teknik Jalan ke- 10* 64 L 2

[13] Huang J, Ju N P, Liao Y J and Liu D D 2015 Determination of rainfall thresholds for shallow landslides by a probabilistic and empirical methode *NHESS* **15** p 2715-21

Inspection and evaluation of bridge structures for earthquakes risk

N R Setiati[1*]

[1] Directorate General of Highways, Ministry of Public Works and Housing, 40294 Bandung Jawa Barat, Indonesia

[*]Corresponding author's e-mail: retno.setiati@pusjatan.pu.go.id

Abstract. The 2017 Earthquake Source and Hazard Map Update in 2017 showed a significant change to the previous earthquake map. An increase in the number of active faults and high magnitude earthquake events indicates an increased potential for earthquake risk. The availability of qualified seismic manual is a must to support the direction of government policy in the provision of infrastructure, especially roads and bridges in a sustainable manner. This study aims to determine the value of the vulnerability of existing bridges against earthquakes and alternative reinforcement that can be done. The reinforcement philosophy is based on performance, similar to the philosophy used in the design of new bridges. Performance criteria are examined using earthquakes with a probability of exceeding 7% in 75 years. Performance criteria are determined based on the level of importance and remaining life of the bridge service. An example of a case study was made by examining and evaluating a prestressed girder bridge located in the earthquake zone 4 of the Jakarta area. Analysis and evaluation of bridge capacity is made using the SAP 2000 program. The 3 (three) dimension model is used to calculate the effect of 3 (three) structural dimensions on the dynamic behavior of the bridge. Earthquake input is made in the form of spectra response according to SNI 2833-2016. For earthquake loading in the X and Y direction the g/R scale factor is used (9.81/5), which is 1.96. The basic shear capacity of the column is calculated according to SNI 2833-2016 which consists of the shear capacity of the contribution of the concrete section and reinforcement. Based on the analysis, it was obtained that the column sliding capacity was 1333.3 kN. The value of the capacity is greater than the ultimate shear load of 519.07 kN so it can be concluded that the sliding capacity of the column still meets the earthquake load so that no structural reinforcement is needed.

1. Introduction

1.1. Background

Earthquakes are natural phenomena that are vulnerable to occur in Indonesia. This is due to Indonesia being in the meeting of 3 (three) active tectonic plates (Eurasia, Australia, and Pacific). In addition, Indonesia is also on the path of the world's active volcano (Ring of fire). The earthquake occurred during the last 5 (five) years and caused significant damage. An earthquake of magnitude (Mw = 6.5) occurred in Maluku in 2019, a year after the earthquake accompanied by tsunamis that hit the cities of Palu (Mw = 7.4) and Lombok with Mw ranging from 6.0 to 7.0 [5]. Figure 1 shows the intensity of earthquake events in Indonesia until 2016.

Content from this work may be used under the terms of the Creative Commons Attribution 3.0 licence. Any further distribution of this work must maintain attribution to the author(s) and the title of the work, journal citation and DOI.
Published under licence by IOP Publishing Ltd

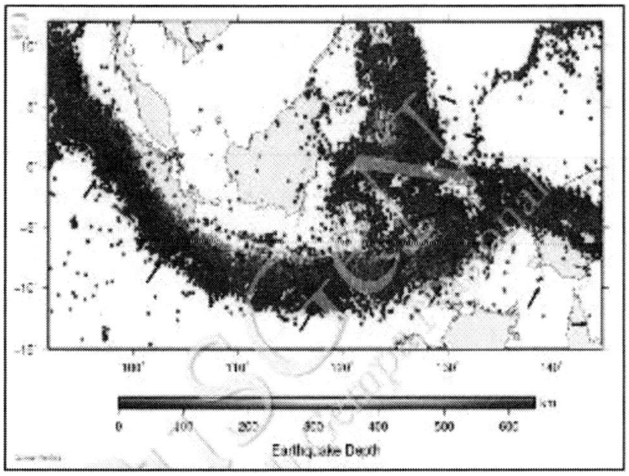

Figure 1. Map of earthquake intensity in Indonesia [4]

The impact of damage caused by the earthquake on road and bridge infrastructure including cracking on expansion joints, bearing shifts, reduction in bridge oprit, sliding cracks on abutments, until the bridge collapses. To avoid sudden collapse of the bridge, periodic checks are needed to determine the value of the existing capacity. The strength of the existing bridge structure needs to be evaluated based on the strength of the earthquake that occurred with a magnitude scale greater than in previous years. After the release of the latest 2017 Earthquake Map, it is necessary to analyze and evaluate the strength of the existing bridges on the earthquake map. This study aims to determine the size of the existing bridge capacity by referring to the 2017 Earthquake Map. If the evaluation results show that the bridge capacity is smaller than the earthquake load, retrofitting, strengthening, or even replacing a new bridge needs to be done.

Bridge structures must be designed to be earthquake resistant, with an emphasis on aspects of good structural ductility and the ability to participate in good energy, so that structures are earthquake resistant. The design of the bridge structure for earthquake loads refers to SNI 2833: 2016 (PerencanaanJembatanterhadap Beban Gempa). Existing bridges built in the decade of the 80s need to be evaluated against the latest earthquake loads. Based on SNI 2833: 2016, the bridge structure is planned with a minimum design life of 75 years and the earthquake force probability is exceeded by 7%, so that the bridge structure is planned to be able to bear the earthquake force with a 1000 year return period. In checking and evaluating existing bridges for earthquake loads, a number of things need to be considered are carrying out an inspection process to identify earthquake hazard risks using structural vulnerability and earthquake hazard data, evaluating bridge capacity for earthquake loads, and selecting appropriate reinforcement techniques to increase bridge resistance to earthquake risk.

2. Methodology

The method of checking and evaluating the existing bridge structure for earthquake loads is shown in figure 2.

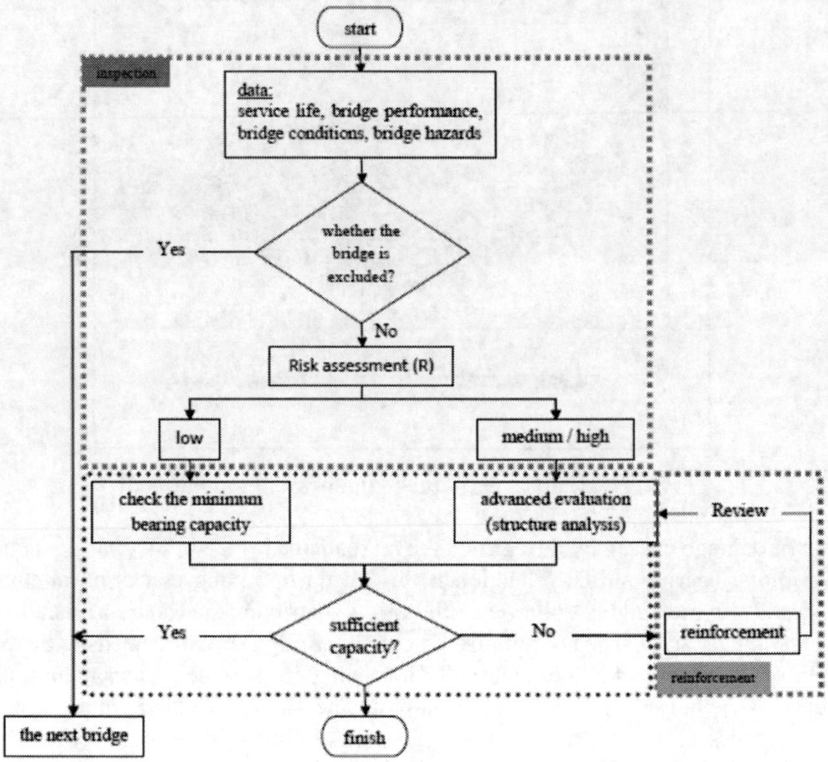

Figure 2. Flowchart for checking and evaluating the structure of existing bridges against earthquake loads

3. Results and discussion

In this research a case study of the inspection and evaluation of bridge structures in the Jakarta area was made. A bridge consists of 3 (three) spans (13.33 m + 26.82 m + 13.33 m) and has a total width of 15 m. The bridge was built in 1989 and has a type of prestressed concrete girders and with soil classification including soft soil. Bridges are included in other categories of bridge importance. The bridge is supported by 3 columns of 1.2 m diameter in each pillar. The length of the existing placement in the pedestal area is 300 mm. Sketch of the elongated and transverse bridge can be seen in figure 3.

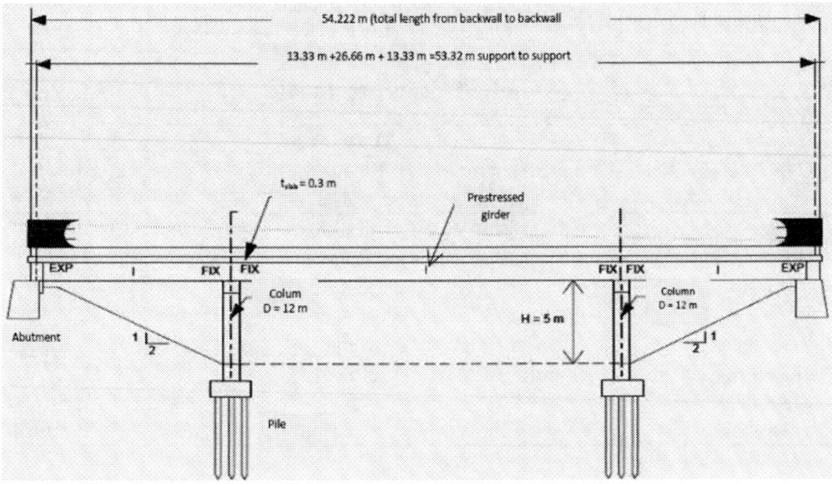

Figure 3. Longitudinal section of bridge

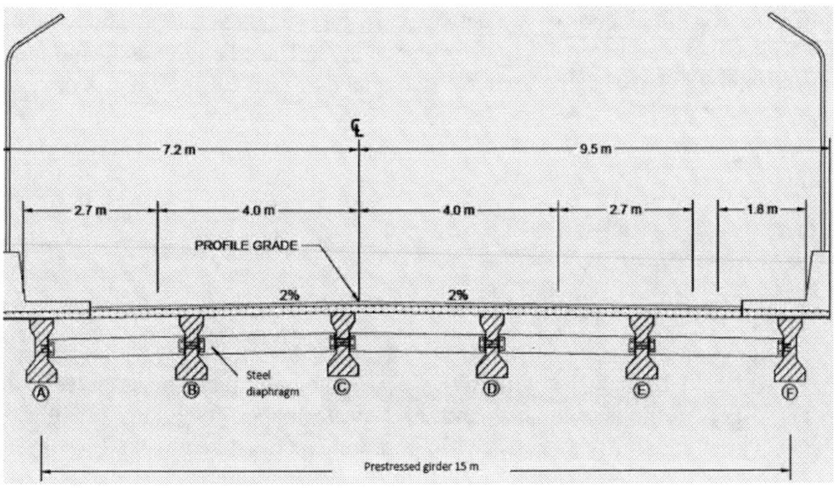

Figure 4. Cross section of bridge

Bridge material data based on figure 3 and figure 4 are shown in table 1.

Table 1. Bridge material

Concrete element:	Strength (MPa)	Modulus of Elasticity (MPa)	Poisson ratio
Girder I	60	36000	0,2
Head beam	30	25000	0,2
Column	30	25000	0,2
Reinforcement element:	Strength of yield (MPa)	Strength of yield (MPa)	Strain min. (%)
Main bar	420	525	14 (d < 19 mm)
Shear bar	420	525	12 (22 < d < 36 mm)

Determination of seismic strength category [1] :
The service life ot bridge is 29 years. The assumed service life is 75 years, so the remaining service life (SUL) = 46. Service life categories are between (16-50) years (SUL 2). The level of performance against earthquakes is included in the TK1 performance level. Based on the 2017 Indonesia Earthquake Map (Pusgen) and SNI 2833: 2016, the level of earthquake hazard for the Jakarta area with soft soil is as follows:

$$PGA_{(g)} = 0.28$$
$$As_{(g)} = PGA_{(g)} \times F_{PGA} = 0.28 \times 1.3 = 0.364$$
$$S_{DS(g)} = S_S \times F_a = 0.55 \times 1.6 = 0.88$$
$$S_{D1(g)} = S_1 \times F_V = 0.23 \times 3.1 = 0.713$$

In accordance with SNI 2833: 2016, with SD1 = 0.713, the earthquake hazard level is included in the earthquake zone 4. For earthquake zone 4 and TK1 performance level, the retrofitting category is included in KPS C (in vulnerability assessment all components must be checked).

Structural joint and bearing [3]
The upper structure is not continuous on the expansion joint, so the bridge is considered not to have good bearing details. The position of continuous bearing in the transverse direction of the bridge is considered not to provide significant damage to the transverse direction of movement of the bridge VT = 5. The minimum length of placement in the longitudinal direction is as follows:

$$L = 13.33 + \left\{\frac{26.82}{2}\right\} = 26.74$$

H = 5 m ; B = 15 m ; S1 = 0.23 ; fV = 3.1 (soft soil)
SD1 = 0.71 (earthquake zone 4 based on SNI 2833: 2016)
α = 0° (no skew).

Minimum length requirements:

$$N = \{0.2 + 0.0017L + 0.0067H\}\{1 + 0.000125\ S^2\}$$
$$N = \{0.2 + 0.0017 \times 26.74\} + \quad 0.0067(5)\{1 + 0.000125\{0\}^2\} = 0.279$$

For earthquake zone 4, the value of N is modified by increasing the percentage of N by 150%, so the value of N becomes N (d) − 1.5 (0.279) = 0.419.

$$\frac{N_{(c)}}{N_{(d)}} = \frac{300}{419} = 0.72 \quad \text{Therefore } 0.5 < \frac{N_{(c)}}{N_{(d)}} < 1$$

Then the value of susceptibility to bearing in the longitudinal direction (VL) is equal to 5. Vulnerability of structure and placement connection (V1).

V1 = max. (VT,VL) = max. (5,5) =5

Determine the value of Q in the shortest and heaviest column (column pillar 2) as follows:

$$Q = 13 - 6\left\{\frac{L_C}{P_S Fb_{\max}}\right\} = 13 - 6\left\{\frac{4}{4(2)1.2}\right\} = 10.5$$

The slope of the skew is less than 20 °, the maximum reduction is 2 so the CVR column vulnerability value = 10.5-2 = 8.5 ≅ 9. While the value of vulnerability in the abutment (AVR) is 0.
Soil conditions from solid to sandy, unsaturated, very dense and gravel. Therefore, soil has a low susceptibility to liquefaction so that the susceptibility to liquefaction (LVR) is 0. The vulnerabilities of other elements of V_2 are as follows :

V_2 = CVR + AVR + LVR = 9+0+0 =9. The overall vulnerability value is the maximum value between $V_1$and V_2. V_e = max. (V_1,V_2) = maks.(5,9) = 9.

Earthquake hazard assessment is determined using earthquake coefficients in a 1 second period (S_{D1}) so that$E = 10S_{D1} = 10(0.71) = 7.1$. Bridge vulnerability rating ($R_V = V_C$ x E = 9 x 7.1 =63.9). Based on the vulnerability rating value where 30 $<R_v<$ 70, then the bridge is included in the medium vulnerability class. The next step is to inspect the bridge capacity quantitatively against earthquake loading. The bridge is then further evaluated for inspection of element capacity and identification of structural deficiencies and their maintenance [2]

Figure 5 shows the 3-dimensional model of the bridge according to the bridge description in figure 3 and figure 4. Beam components, head beams, diaphragms, and columns are modeled with frame elements, then slab are modeled with shell elements. Pile foundations are not modeled specifically because it is assumed that the foundation behaves elastic in an earthquake. In the bridge model, a rigid link is used to connect the upper structure and head beam with different elevations.

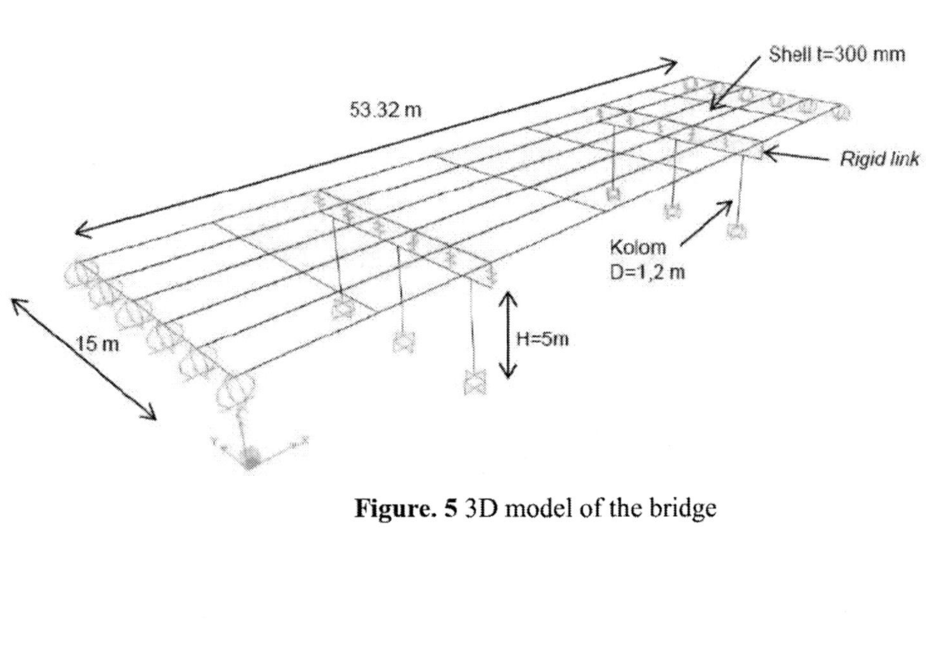

Figure. 5 3D model of the bridge

IOP Conf. Series: Materials Science and Engineering **930** (2020) 012032 doi:10.1088/1757-899X/930/1/012032

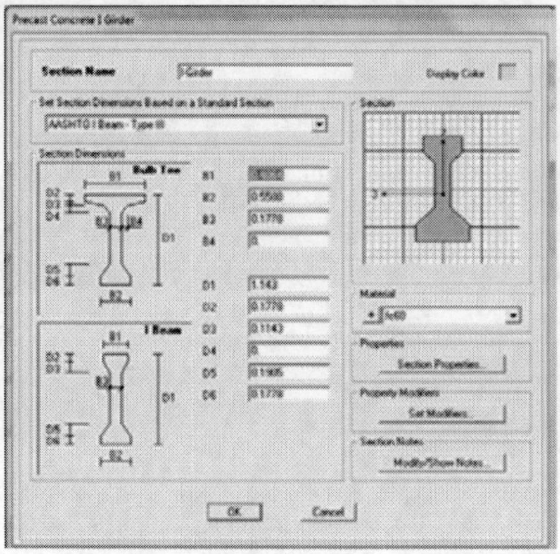

Figure 6. Defining slabs on girder

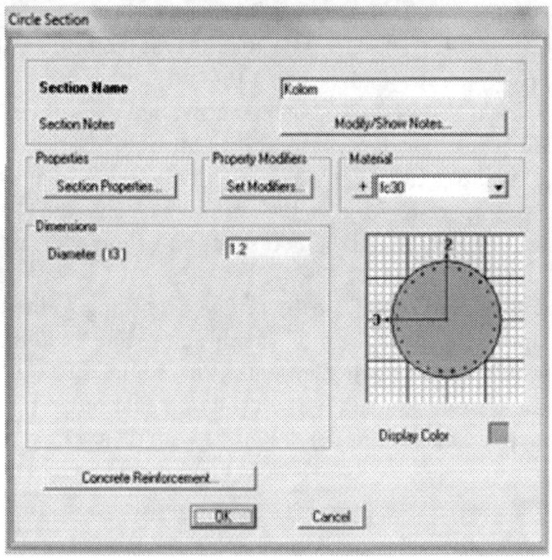

Figure 7. Defining slab in columns

The next step is defining the cross section of the girder and column elements. Figure 6 shows the girder cross-section input while figure 7 shows the column cross-section input. Then proceed with the definition of the slab elements as in figure 8.

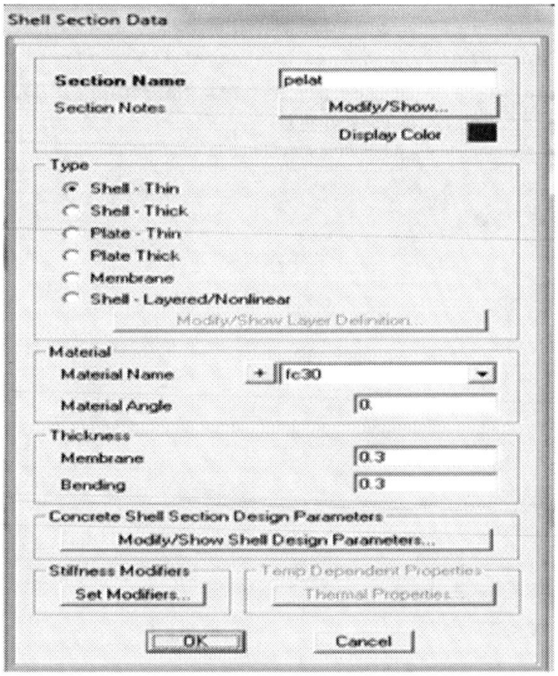

Figure 8. Defining slab in columns

The shell element was chosen because it was assumed that against earthquake loading, the floor could behave like a plate or membrane simultaneously. Plates have a ratio of thickness to width equal to 0.1 so it is considered a thin shell (thin shell).

Earthquake loading

Earthquake input is made in the form of spectra response in accordance with SNI 2833: 2016. The spectral response coordinate points can be inputted in the definition of the spectral response function as shown in figure 9.

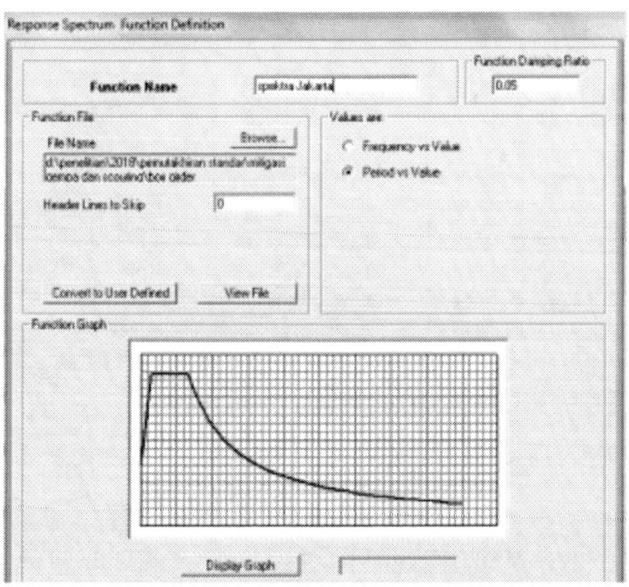

Figure 9. Spectra response of the Jakarta region - soft soil

Based on figure 9, the Jakarta area is a region with a high level of vulnerability to earthquakes, therefore the bridge structure in this region must be designed to withstand earthquakes, with emphasis on aspects of good structural ductility and ability to participate in good energy, so that the structure is able to withstand Jakarta's characteristic earthquake [6].

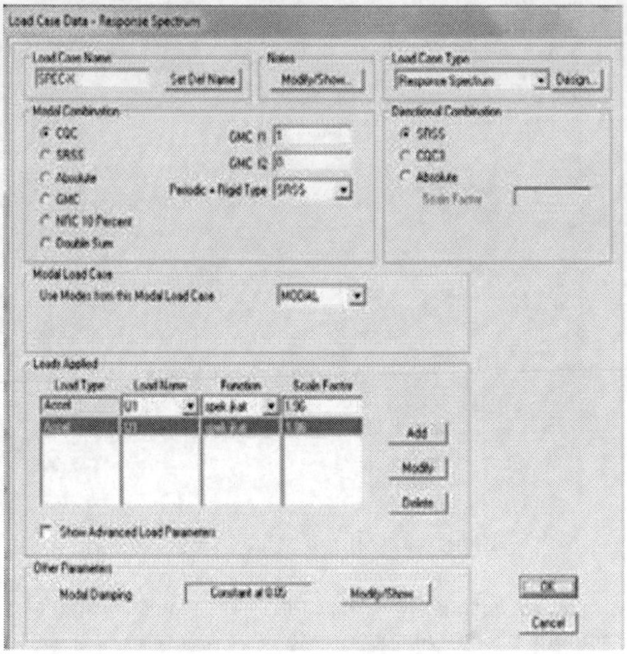

Figure 10. Definition of X-direction response spectrum

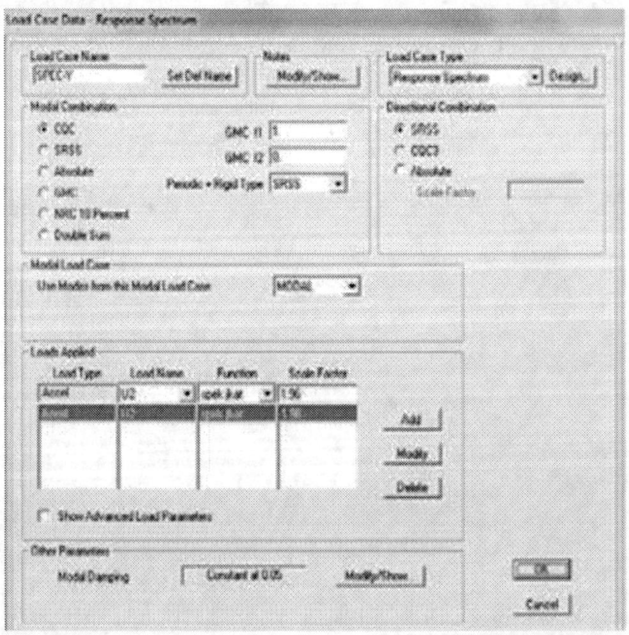

Figure 11. Definition of Y-direction response spectrum

For earthquake loading in the X (Spec-X) and Y (Spec-Y) directions, a scale factor of g/R is used where g is gravity and R is the response modification factor. For compound columns on other bridges the value of R is taken equal to 5 according to SNI 2833: 2016[5], so the scale factor is used 9.81/5=1.96. The combination of earthquake loading includes :

1. Combination 1 = DL+EQx
2. Combination 2 = DL+EQy
3. Combination 3 = DL+EQx +0.3EQy

The analysis tool discussed in this paper can provide practising engineers and researchers withthe ability to assess detailed behavioural response of structural elements subjected to unforeseen loading conditions such as earthquake. It could also be a useful tool in the assessment of structural elements with unusual detailing, particularly if constructed in seismic prone area.

Check column capacity

Column capacity is checked with the help of column interaction diagrams. In this example, a response is taken in the second column of the first pillar (K2P1). Based on structural analysis, the internal forces are obtained as shown in table 2Earthquake loading.

Table 2. The forces in the K2P1 column element

	Internal forces in the column				
Load	M_{3-3} (kNm)	M_{2-2} (kNm)	V_{3-3} (kN)	V_{2-2} (kN)	P (kN)
DL	463.19	0.00	0.00	233.80	1709.00
Eqx	879.75	0.00	0.00	285.27	0.73
Eqy	0.00	788.80	302.31	0.00	0.00
Comb-1	1342.94	0.00	0.00	519.07	1709.73
Comb-2	463.19	788.80	302.31	233.80	1709.00
Comb-3	1342.94	236.64	90.70	519.07	1709.73
Comb-4	727.12	788.80	302.31	319.38	1709.21

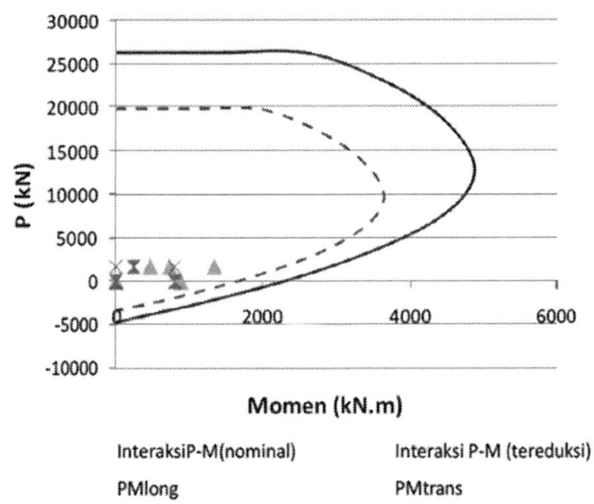

Figure 12. Definition of load combination

Based on figure 12, it can bc seen that the internal forces in the column due to earthquake are still within the column capacity area, so that the column meets the capacity against earthquake loading. Column shear capacity can be calculated in accordance with SNI 2833: 2016 which consists of 2 (two) components, namely the shear capacity of contributions from the cross section of concrete Vc and shear contributions from reinforcement Vs.

Shearcapacity of the concretecolumn section $V_C = 0.083\beta\ f_c^{'} b_v d_v$ (1)

with b_v is the dimension of the column, dv is the effective sheardepth.
Vc = 0,083(2) 30(1200)(0,75*1200)
Vc = 981957 N
ThereforeP_u< $0.1A_g f'_c$ then

$$V_C^{'} = \left\{\frac{P_u}{0.1A_g f_c^{'}}\right\} V_C = \left\{\frac{1709000}{0.1(1130400)(30)}\right\} = 494858\ N \quad (2)$$

Shear capacity of shear reinforcement

$$V_S = \frac{\pi}{2}\left\{\frac{nA_{SP} f_{yh} D}{s}\right\} \quad (3)$$

with Asp is the area of shear reinforcement, fyh is the yield shear reinforcement, D is the diameter of the concrete core, and s is the shear reinforcement.

$$V_S = \frac{\pi}{2}\left\{\frac{2(132.66)(420)(1200-2(50))}{150}\right\} = 1282981\ N$$

$\phi V_n = \phi(V_{c'} + V_s)$ = 0.75(1777839) = 1333379 N = 1333.3 kN>V_u = 519.07 kN. So it can be concluded that the column sliding capacity meets the earthquake loading.

4. Conclusion

Based on the results of the structural evaluation it can be concluded that the column capacity still meets the requirements for earthquake loading so that no structural retrofitting is needed. To prevent sudden collapse of existing bridges against earthquake loads, periodic bridge inventory and inspection efforts should be undertaken. From the results of the inspection, if the capacity of the existing bridge is smaller or equal to the earthquake load, it is necessary to take retrofitting or strengthening the bridge.

Acknowledgements

The author would like to thank Mr. Ir. Deded P. Sjamsudin, M.Eng.Sc. as the Head of Institute Road of Engineering and Mr. Joko Purnomo which has provided support for this research activity. Furthermore, the authors also wish to thank Mr. Wawan Witarnawan, Mr. Abinhot Sihotang, and Mr. Samun Haris as a resource person who has given the science is very helpful to the author.

References

[1] AASHTO 2011 *Guide specification for LRFD seismic bridge design 2nd edition*
[2] Bridge Safety Assurance (BSA) 2004 Seismic vulnerability manual *New York State Dept. of Transp.*
[3] FHWA 2006 Seismic retrofitting manual for highway structures: Part 1-bridges *Dept. of Transp.* no FHWA-HRT pp 6-32 (Virginia: McLean)
[4] PuSGeN 2017 Pedoman perencanaan jembatan terhadap beban gempa berbasis simpangan *Peta Sumber dan Bahaya Gempa Indonesia*
[5] Pasau G, Ferdy and Tamuntuan 2017 Pengamatan seismisitas gempa bumi di wilayah pulau Sulawesi menggunakan perubahan nilai a-b *Jurnal MIPA Unsrat Menado online* **6** 1 31-5
[6] SNI 2833: 2016 Perencanaan jembatan terhadap beban gempa
[7] Silitonga D Z and Imran I 2019 Penilaian kerentanan seismik pada jembatan box girder beton prategang menerus bentang majemuk eksisting melalui pengembangan kurva fragilitas analitik *Jurnal Teknik Sipil*

Preliminary 3D numerical pushover analysis of laterally loaded pile groups

A Yuwono[1*], W A Prakoso[2] and Y Lase[2]

[1] PhD Student, Civil Engineering Department, Faculty of Engineering, Universitas Indonesia, Depok, West Java 16424

[2] Lecturer, Civil Engineering Department, Faculty of Engineering, Universitas Indonesia, Depok, West Java 16424

*Corresponding author's e-mail: amelia.yuwono@ui.ac.id

Abstract. This paper reports the results of 3D numerical pushover analyses of laterally loaded pile groups. The effects of pile structural models, soil strengths, and foundations systems are evaluated. Single piles, single pile groups, and multi pile groups are considered in the analyses. The main objective of this paper is to examine important parameters to be considered further. The findings include the overestimation of pile-soil interactions (lower efficiency factors), as well as the highly complex pile internal forces warranting a further examination in the context of high seismic demands in Indonesia.

1. Introduction

Pile foundations – typically used in pile groups – are widely used for construction in Indonesia and worldwide. One of the design criteria is lateral performance of the pile groups. This topic is still examined further in many recent international publications [e.g., 1-7]. The approach to address this topic includes the pushover analysis and the use of non-linear pile-soil interactions [e.g., 8-12].

The Indonesia geotechnical code SNI 8460:2017 [13] stipulates that the maximum lateral displacement of laterally load single piles is 12 mm for nominal seismic loads and 25 mm for strong seismic loads. There have been discussions whether these limits would be compatible with the seismic demands for infrastructures in Indonesia as stipulated in the Indonesia seismic code SNI 1726:2019 [14].

This paper reports the results of a series of preliminary 3-dimensional numerical pushover analyses of laterally loaded pile groups. Different effects are to be evaluated, namely the effect of pile structural models, the effect of soil strengths, and the effect of foundations systems. Single piles, single pile groups, and multi pile groups are considered in the analyses. The objectives of this paper are to assess the suitability of the numerical models used and to examine important parameters to be considered further.

2. Numerical models

The assumed geotechnical conditions consist of two soil layers. The thickness of the upper soil layer is 7.5 m, overlying a 9 m thick, lower soil layer; the basic sketch is shown as figure 1. The undrained shear strength s_u of the upper soil layer is varied to evaluate the effect of the strength on the behavior

Content from this work may be used under the terms of the Creative Commons Attribution 3.0 licence. Any further distribution of this work must maintain attribution to the author(s) and the title of the work, journal citation and DOI.
Published under licence by IOP Publishing Ltd

of laterally loaded pile groups, while the undrained soil modulus E_u is 150 × s_u. The s_u values considered are 20 kPa, 40 kPa, 60 kPa, 80 kPa, and 100 kPa. The s_u and E_u of the lower soil layer are 200 kPa and 50 MPa, respectively. The undrained friction angle of both soil layers is zero, and the groundwater is not modeled explicitly. The soil constitutive model used is the Mohr-Coulomb model. The assumed piles are 500 mm square piles, driven to the top of lower soil layer; the length is then 7.5 m, and the pile tip may numerically rotate freely. The modulus of the piles is 30 GPa. In some analyses, the elastoplastic piles are assumed to have an ultimate bending moment capacity of 400 kN-m. The piles are modeled numerically as embedded piles.

In each pile group, there are nine (9) piles spaced 1.5 m center-to-center, and the pile cap is 4.5 m × 4.5 m, as shown in figure 2. The stiffness of the concrete pile caps is determined based on a thickness of 1.0 m. For the multi pile group model as shown in figure 3, there are 0.25 m × 0.75 m concrete tie-beams connecting the nine pile caps, and there are 0.25 m thick concrete floor slabs connected to the pile caps and tie-beams.

The series of analyses was conducted using Plaxis 3D [15], and the so-called "Undrained C" type of analysis (total stress analysis) was implemented. The soil elements were 10-node tetrahedral elements. The embedded pile and the tie-beam elements were 6-node beam elements, while the pile cap and floor slab elements were 15-node thin shell elements. All the elements for 40 m × 40 m × 16.5 m meshes were automatically generated by Plaxis 3D as shown in figure 1; the number of nodes was between 215,000 and 244,000 nodes, while the number of soil elements was between 157,000 and 178,000 elements. The lateral loads were applied at the pile cap level.

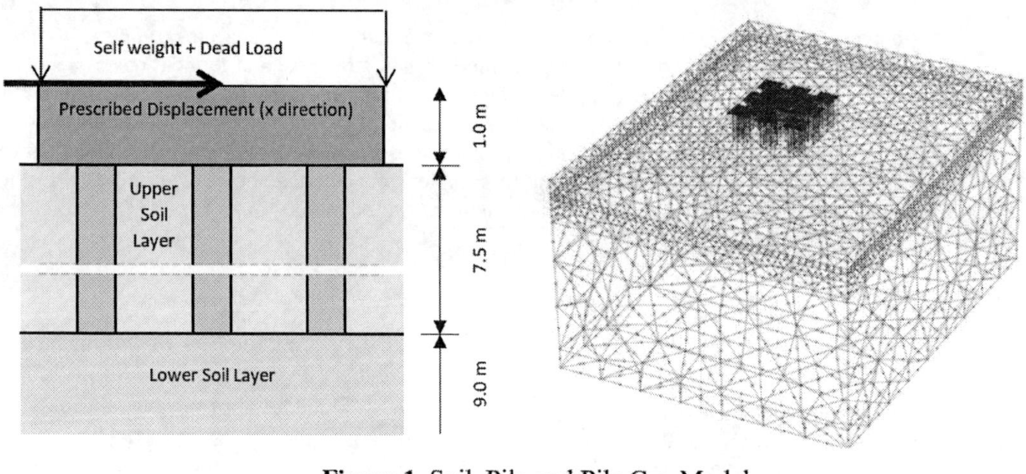

Figure 1. Soil, Pile and Pile Cap Model.

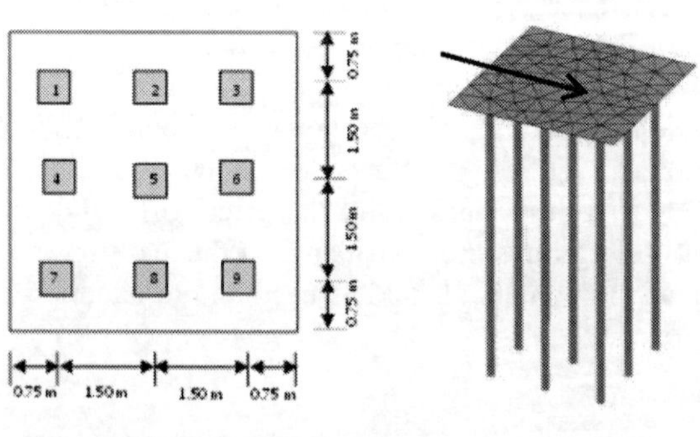

Figure 2. Single Pile Group Model (1PC9).

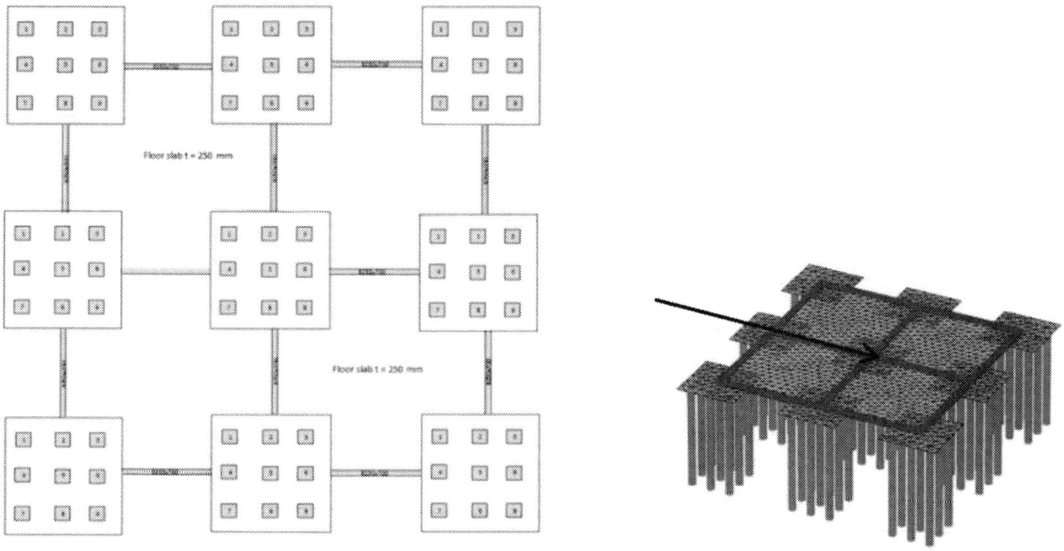

Figure 3. Multiple Pile Group Model (9PC9).

3. Results and discussion

Different effects were evaluated in three series of analyses, namely the effect of pile structural models, the effect of upper soil layer properties, and the effect of foundations systems. For the analysis of pile structural models (elastic piles and elastoplastic piles), the single pile group 1PC9 model was modified by removing the upper soil layer. In addition, the pile tips were restrained for translational movement, but were free for rotational movement. The pile cap was then laterally displaced by an amount of 0.1 m.

The results are shown in figure 4 for both structural models. For the elastic 1PC9, the lateral force increases linearly with an increase in lateral displacement. However, for the elastoplastic 1PC9, the lateral force increases linearly up to a lateral force of 336 kN, and it subsequently starts to behave non-linearly and becomes plastic at a lateral force of about 440 kN. By comparing both results, one could conclude that the elastoplastic pile structural model could provide elastoplastic pile structural behavior.

The effect of upper soil layer properties on the behavior of laterally loaded single piles are analyzed and shown as figure 5. The elastoplastic single piles have no pile caps, and the pile heads were laterally displaced by an amount of 0.1 m. As expected, the initial stiffness of the single pile responses increases linearly with an increase in undrained shear strength s_u.

Referring to [13], the lateral loads at 25 mm displacement for s_u values of 20 kPa and 80 kPa are 125 kN and 520 kN, respectively. However, the overall response curves would become more non-linear with an increase in s_u. The lateral loads at 50 mm displacement for the above s_u values are about 240 kN and 735 kN, respectively. All these indicate that the soil-pile interaction becomes more complex for greater soil shear strength and modulus.

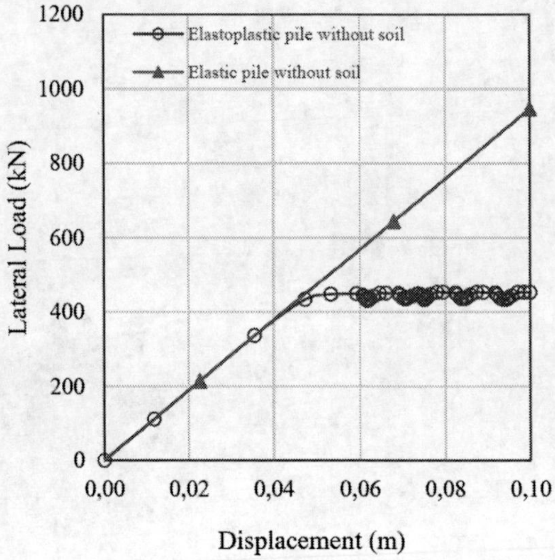

Figure 4. Lateral load vs displacement for single pile group (1PC9) without upper soil layer

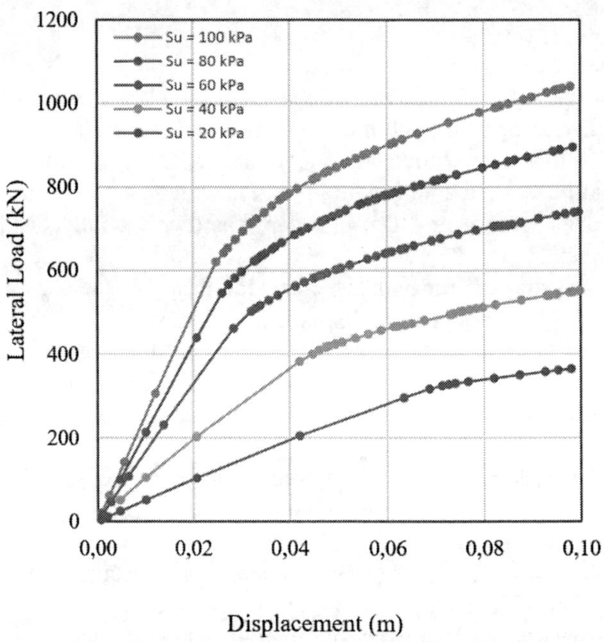

Figure 5. Lateral load vs displacement for single 500 mm square piles (P1) for various s_u (elastoplastic pile, no pile cap)

The behavior of laterally loaded single pile group 1PC9 is analyzed and shown as figure 6. Note that there is a gap between the pile cap and the upper soil layer. The pile heads were laterally displaced by an amount of 0.1 m. Referring to [13], the lateral loads at 25 mm displacement for s_u values of 20 kPa and 80 kPa are about 550 kN and 2340 kN, respectively. The lateral loads at 50 mm displacement for the above s_u values are 1000 kN and 4050 kN, respectively. The efficiency factor for 1PC9 with upper soil layer s_u of 20 kPa at 25 mm lateral displacement is about 0.49 [= 550 kN / (9 ·

125 kN)], while that with upper soil layer s_u of 80 kPa is about 0.50 [= 2340 kN / (9 · 520 kN)]. The efficiency factor for 1PC9 with upper soil layer s_u of 20 kPa at 50 mm lateral displacement is about 0.46 [= 1000 kN / (9 · 240 kN)], while that with upper soil layer s_u of 80 kPa is about 0.61 [= 4050 kN / (9 · 735 kN)]. In addition, the 25 mm and 50 mm efficiency factors for 1PC9 with upper soil layer s_u of 40 kPa are 0.48 and 0.52, respectively, while those for 1PC9 with upper soil layer s_u of 40 kPa are 0.50 and 0.63, respectively. For a lower lateral displacement, the efficiency factor appears to be relatively independent of the soil shear strength and modulus, while for a higher displacement, the efficiency factor appears to increase with an increase in s_u. It is highlighted that these efficiency factors tend to be lower than the factors suggested by Reese et al. [16], indicating the models tend to overestimate the interaction among the piles in 1PC9.

The behavior of laterally loaded multi pile group 9PC9 is analyzed and shown as figure 7. Note that there is a gap between the pile cap and the upper soil layer. The pile heads were laterally displaced by an amount of 0.1 m. Referring to SNI 8640:2017, the lateral loads at 25 mm displacement for s_u values of 20 kPa and 80 kPa are about 4 MN and 13 MN, respectively. The lateral loads at 50 mm displacement for the above s_u values are 8 MN and 23.5 MN, respectively. The efficiency factor for 9PC9 with upper soil layer s_u of 20 kPa at 25 mm lateral displacement is about 0.39 [= 4 MN / (81 · 125 kN)], while that with upper soil layer s_u of 80 kPa is about 0.30 [= 13 MN / (81 · 520 kN)]. The efficiency factor for 9PC9 with upper soil layer s_u of 20 kPa at 50 mm lateral displacement is about 0.41 [= 13 MN / (81 · 240 kN)], while that with upper soil layer s_u of 80 kPa is about 0.39 [= 23.5 MN / (81 · 735 kN)]. These efficiency factors were lower than the factors for single pile groups, indicating complex interactions among pile groups. It is highlighted that these efficiency factors again tend to be lower than the factors suggested by Reese et al. [16], indicating the models tend to overestimate the interaction among the pile groups in 9PC9.

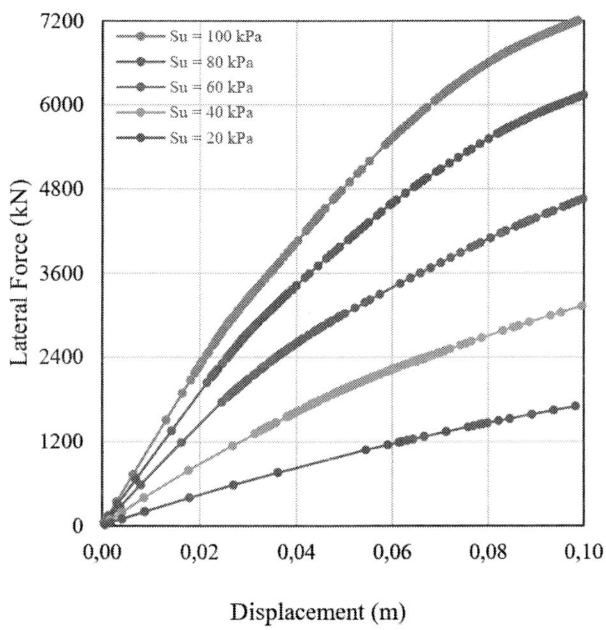

Figure 6. Lateral load vs displacement for single pile group (1PC9) for various s_u (elastoplastic pile, gap beneath pile cap)

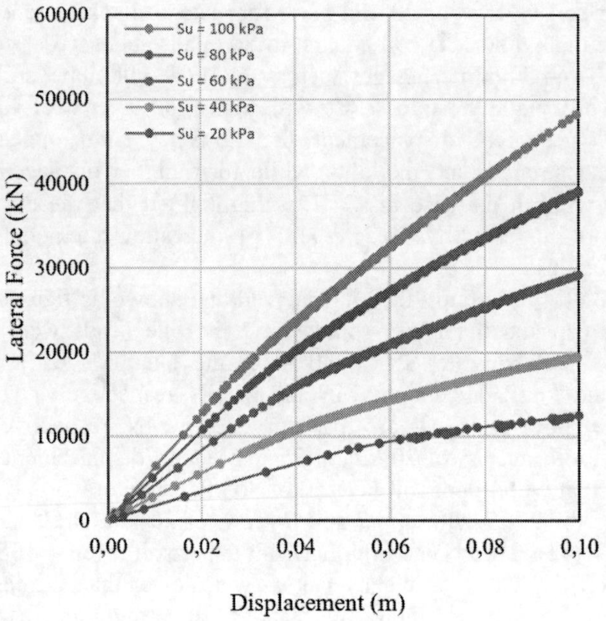

Figure 7. Lateral load vs displacement for multi pile group model (9PC9) for various s_u (elastoplastic pile, gap beneath pile cap)

The pile lateral displacement, bending moment, and shear force for 1PC9 with upper soil layer s_u of 20 kPa at 100 mm lateral displacement are shown in figures 8, 9, and 10, respectively. It can be seen that the pile lateral displacement in figure 8 is not exactly uniform, caused by the complex interaction among the piles. The pile bending moment diagrams shown as figure 9 in general are different among the piles, and the maximum pile bending moments occurred in the connection between the piles and the pile cap. The trailing piles (piles in the back) have the highest pile bending moments (375 – 385 kN-m). The leading piles (piles in the front) have the second highest bending moments (311 – 341 kN-m), while the middle piles have the lowest bending moments (295 – 363 kN-m). For lateral displacement smaller than or equal to 10 mm, the leading piles (piles in the front) have the highest bending moments and the trailing piles (piles in the front) have the second highest bending moments, while the middle piles have the lowest bending moments. The piles would not experience pile structural plastification at the same time, an important feature warranting a further examination in the context of high seismic demands in Indonesia [14].

The pile shear force diagrams shown as figure 10 in general are different among the piles, and the maximum pile shear forces occurred expectedly in the connection between the piles and the pile cap. The average maximum pile shear forces for the leading and trailing piles is about 256 kN and 210 kN, respectively, while that for the middle piles is about 134 kN. This difference in pile shear force is also an important feature warranting a further examination in the context of high seismic demands in Indonesia [14].

4. Conclusions

The results of numerical pushover analyses of laterally loaded pile groups using Plaxis 3D were reported in this paper. The effects of pile structural models (elastic and elastoplastic models), soil strengths, and foundations systems (single piles (P1), single pile groups (1PC9), and multi pile groups (9PC9)) were evaluated. The highlighted observations are as follows:

- The pile structural models could provide the expected elastic and elastoplastic pile foundation system behavior
- The single pile behavior is a function of soil shear strength and modulus, and the behavior would change for more complex soil-pile interactions.
- The pile group efficiency factor for single pile groups appears to be relatively independent of the soil shear strength and modulus for a relativel low lateral displacement, but it would change change for more complex soil-pile interactions. The efficiency factors observed in the 3D models appear to be lower than the factors typically suggested in literature.
- The pile group efficiency factor for multi pile groups was lower than the factor for single pile groups, indicating complex interactions among pile groups. The efficiency factors observed in the 3D models again appear to be lower than the factors typically suggested in literature.

The pile internal forces (bending moments and shear forces) are highly complex. The pile structural plastification and the pile shear ultimate behavior would not occur at the same time, important features warranting a further examination in the context of high seismic demands in Indonesia.

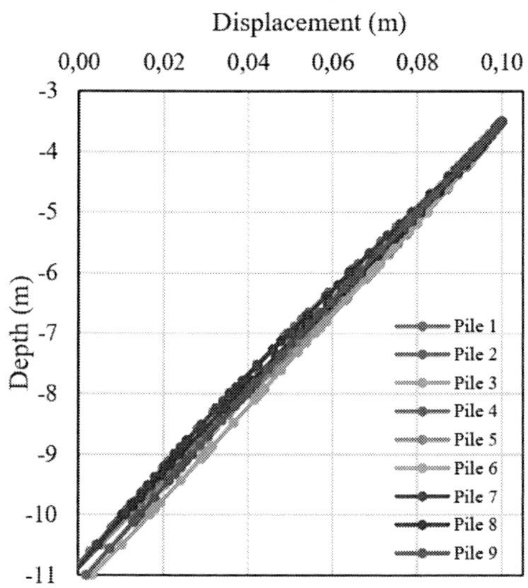

Figure 8. Lateral displacement diagram for single pile group (1PC9) at lateral displacement of 100 mm (s_u = 20 kPa, elastoplastic pile)

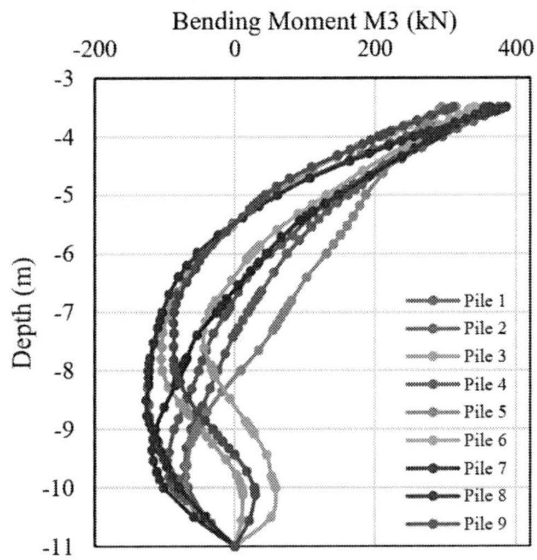

Figure 9. Bending moment diagram for single pile group (1PC9) at lateral displacement of 100 mm (s_u = 20 kPa, elastoplastic pile)

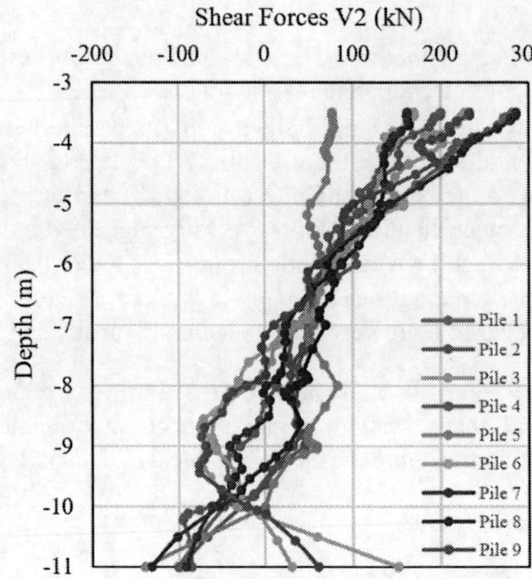

Figure 10. Shear force diagram for single pile group (1PC9) at lateral displacement of 100 mm (s_u = 20 kPa, elastoplastic pile)

5. References

[1] Gao S, Gong J, and Feng Y 2016 Equivalent damping for displacement-based seismic design of pile-supported wharves with soil–pile interaction *Ocean. Eng.* **125** p 12-25

[2] Mayoral J M, Pestana J M, and Seed R B 2016 Multi-directional cyclic p–y curves for soft clays *Ocean. Eng.* **115** p 1-18

[3] Blanco G, Ye A, Wang X, and Goicolea J M 2018 Parametric pushover analysis on elevated RC pile-cap foundations for bridges in cohesionless soils *J. Bridge. Eng.* **24** 1 04018104 (*Preprint* gr-qc/04018104)

[4] Rahmani A, Taiebat M, Finn W D L and Ventura C E 2018 Evaluation of p-y springs for nonlinear static and seismic soil-pile interaction analysis under lateral loading *Soil Dyn. Earthq. Eng.* **115** p 438-47

[5] Dehghanpoor A, Thambiratnam D, Taciroglu E and Chan T 2019 Soil-pile-superstructure interaction effects in seismically isolated bridges under combined vertical and horizontal strong ground motions *Soil Dyn. Earthq. Eng.* **126** 105753 (*Preprint* gr-qc/105753)

[6] Sakellariadis L, Marin A and Anastasopoulos I 2019. Widening of Existing Motorway Bridges: Pile Group Retrofit versus Nonlinear Pile-Soil Response *J. Geotech. Geoenviron.* **145** 12 04019107 (*Preprint* gr-qc/04019107)

[7] Su L, Wan H P, Dong Y, Frangopol D M and Ling X Z 2019 Seismic fragility assessment of large-scale pile-supported wharf structures considering soil-pile interaction *Eng. Struct.* **186** p 270-81

[8] Liu J, Wang W and Dasgupta G 2014 Pushover analysis of underground structures: Method and application *Sci. China Technol. Sci.* **57** 2 p 423-37

[9] Sun L and Zhang C 2004 Improvement of pushover analysis taking account of pier-pile-soil interaction *Proc. 13th World Conf. on Earthq. Eng.* (Vancouver:Canada)

[10] Mardfekri M, Gardoni P and Roesset J M 2013 Modeling laterally loaded single piles accounting for nonlinear soil-pile interactions *J. Eng.* **2013**

[11] Taïeb B and Sofiane B 2014 Accounting for ductility and overstrength in seismic design of reinforced concrete structures *Proc. 9th Inter. Conf. on Struct. Dyn.* EURODYN 2014 (Porto: Earthquake Engineering Research Institute)

[12] Wang X, Ye A, He Z and Shang Y 2016 Quasi-static cyclic testing of elevated RC pile-cap foundation for bridge structures *J. Bridge Eng.* **21** 2 04015042 (*Preprint* gr-qc/04015042)

[13] Badan Standarisasi Nasional 2017 *Persyaratan Perancangan Geoteknik SNI 8460:2017* (Jakarta: BSN)

[14] Badan Standarisasi Nasional 2019 *Tata Cara Perencanaan Ketahanan Gempa untuk Struktur Bangunan Gedung dan Nongedung SNI 1726:2019* (Jakarta: BSN)

[15] Brinkgreve R B J, Zampich L M and Ragi Manoj N 2019 *Plaxis Connect Edition 2019* Delft p 60

[16] Reese LC, Isenhower W M and Wang S T 2006 *Analysis and Design of Shallow and Deep Foundations* (Hoboken: Wiley)

IOP Conf. Series: Materials Science and Engineering **930** (2020) 012034 doi:10.1088/1757-899X/930/1/012034

Vs and CPT based evaluation of location with high liquefaction damage during 2018 Palu earthquake

H A Rahmawati[1*], W A Prakoso[1], and A Rahayu[2]

[1] Department of Civil Engineering, University of Indonesia,Depok 16424
[2] Department of Civil Engineering, University of Tadulako, Palu 94118

*Corresponding author's e-mail: hanif.audina81@ui.ac.id

Abstract. Palu earthquake occurred on 28 September 2018 caused severe liquefaction damages for the city. Based on field surface damage survey by Indonesia Geology Agency, a location was classified within the high liquefaction damage area. The pre-event soil characteristics of this location was provided by resutls of MASW (Multichannel Analysis of Surface Wave) and CPT (Cone Penetration Test) surveys. The soil deposit could characteristized by relatively low shear wave velocity (Vs) and low of cone resistance (qc) values. The potential liquefaction presented by safety factor (0.6 to 0.9) obtained based on both methods was consistent to depths from 2.0 m to 5.0 m. This safety factor appears to be consistent with the high liquefaction damage reported by the Agency.

1. Introduction

The 28 September 2018 Palu earthquake with magnitude of 7.4 caused severe damages in the City of Palu. Liquefaction is a phenomenon in which sandy soils lose their shear strength due to the increase in porewater pressures induced by seismic shaking (e.g., [1]). Ishihara [2] presented the liquefaction phenomenon with a simulation of vibration of sand and water in an aquarium; the results included the compaction of sand and the increase in the water level.

It has been shown (e.g., [3]) that the liquefaction phenomenon depends on the type of soil, relative density, initial confining pressure, as well as the intensity and duration of seismic vibration. The soil type most prone to liquefaction is low density sandy soils in low confining pressure conditions subjected to a large magnitude earthquake. In addition, poorly graded sands tend to be more vulnerable compared to well graded sands. Futhermore, it has been shown (e.g., [4]) that a higher fines content would provide a higher resistance to liquefaction.

A number of field tests have been suggested to evaluate the liquefaction potential, including standard penetration tests (e.g., [3]), cone penetration tests (e.g., [5]), and shear wave velocity (e.g., [4]). Figure 1 shows the relationship between the normalized shear wave velocity and the liquefaction resistance, represented by the cyclic resistance ratio [4]. It is noted that shear wave velocity could be measured using the MASW (Multichannel Analysis of Surface Waves) geophysical method.

The Indonesia Geology Agency has developed the liquefaction potential map based on the post 2018 event field survey [6]. The red, orange, and yellow colored areas are areas with very high, high, and medium liquefaction potential, respectively. The white areas were not mapped during the survey.

Content from this work may be used under the terms of the Creative Commons Attribution 3.0 licence. Any further distribution of this work must maintain attribution to the author(s) and the title of the work, journal citation and DOI.
Published under licence by IOP Publishing Ltd

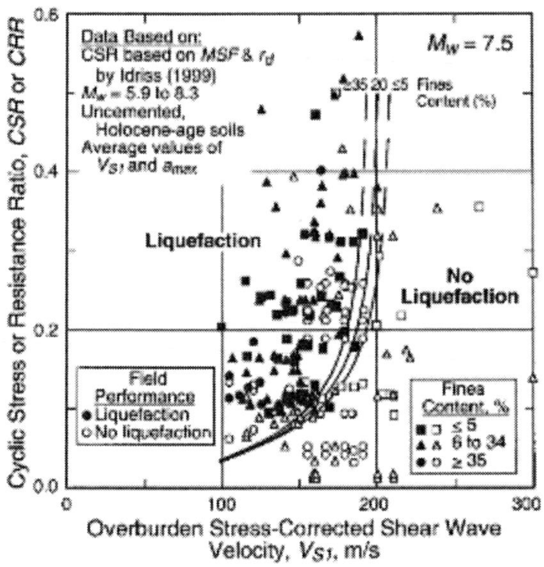

Figure 1. Liquefaction potential based on shear wave velocity [4]

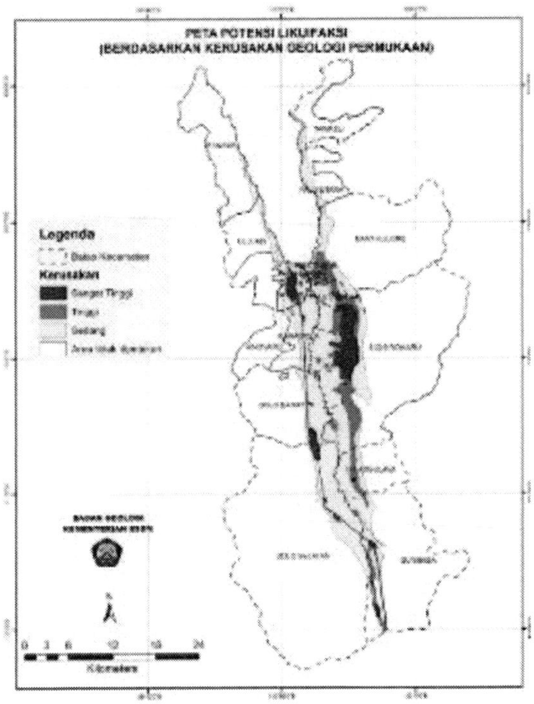

Figure 2. Liquefaction Potential Map based on field survey [6]

Table 1. Classification of Damages [6]

Colour Code	Classification	Remark
Yellow	Medium	Liquefaction not observed Crack less than 1 cm Settlement not observed Lateral spreading not observed Soil surface bulging not observed
Orange	High	Local liquefaction observed Crack less than 5 cm Settlement less than 10 cm Lateral spreading less than 5cm Soil surface bulging observed
Red	Very High	High density liquefaction observed (more than 25% locations surveyed) Crack more than 5 cm Settlement more than 10 cm Lateral spreading more than 5 cm Soil surface bulging observed extensively

Concrete cracking is known to have a knock-on effect on the durability performance and long-term service life of reinforced concrete structures [1, 2]. This phenomenon is primarily attributed to complex loading conditions that may arise from external loading under day-to-day service in addition to environmental exposures, creating even more complex physical and chemical changes in concrete properties [3]. This will cause not only localised distress and stiffness degradation of structural members but also poor resistance of penetration from aggressive agents (e.g. alkali-silica reaction and chloride transport) which will result in further significant deterioration [4, 5].

Figure 3. Research Location

2. Research Method

The research method employed is summarized as a flowchart shown as Figure 4. The shear wave velocity data (V_s) and the cone resistance data (q_c) are used to calculate their respective Cyclic Resistance Ratio (CRR) using the methods summarized in [7]. Both the V_s data and the q_c data are subsequently normalized to 1 atm confining pressure, as the methods employ the V_{s1} data and the q_{c1} data. It is noted that the groundwater level was about 2.0 m below the ground surface.

The seismic demand represented by the Cyclic Stress Ratio (CSR) is also calculated based on the method suggested in [7]. The CSR is a function of soil vertical effective and total pressures, as well as the peak ground acceleration (PGA) of the considered earthquake. In this paper, the PGA value is estimated based on the average PGA calculated based on the methods proposed by the number of PGA was 0.462, calculated everage empirical equation by Boore et al. [8] and Campbell and Bozorgnia [9].

The liquefaction potential is then determined based on the safety factor (SF), defined as the ratio of CRR to CSR. SF less than or equal to 1.0 would indicate that liquefaction would occur, while SF greater than 1.0 would indicate otherwise.

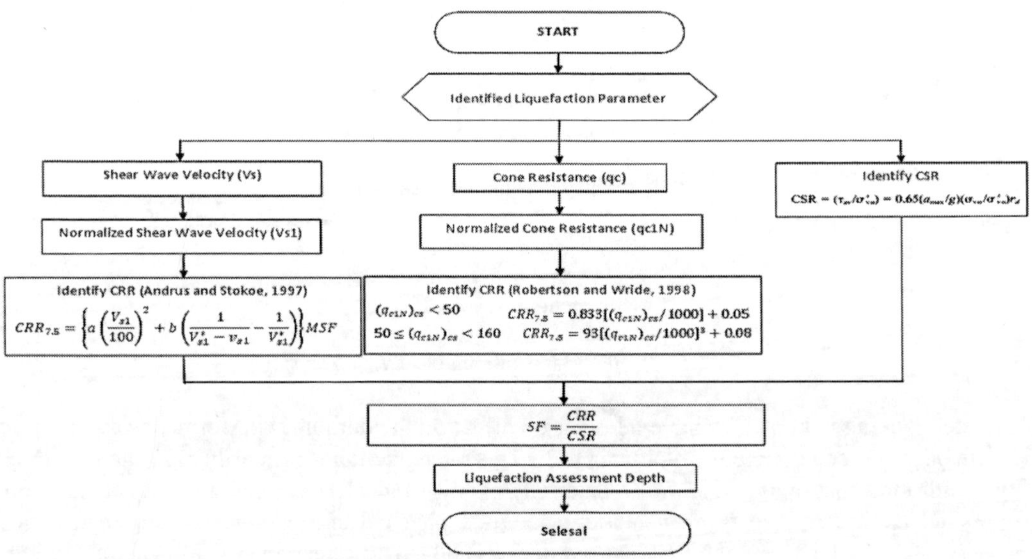

Figure 4. Research method flow chart.

3. Liquefaction Assessment Result

Figure 5a presents the distribution of V_s with depth. The V_s values in general increase with depth, indicating an increase in soil compaction level with depth (e.g., [10]). Figure 5b presents the distribution of q_c with depth. In general, the q_c values were less than 10 MPa to a depth of 10.0 m, and they tend to increase with depth as well. It is noted that, based on data from a nearby bore hole, the particle size distribution of soil in this location is predominantly coarse grained sands.

Figure 6a presents the cyclic resistance ratio (CRR) from the PKA10 MASW data and the CPT7 data. Figure 6a alos shows the cyclic stress ratio (CSR) for the research location. The values of q_c based CRR are between 0.08 and 0.60, while those of V_s based CRR are between 0.09 and 0.60. The values of CSR above ground water level are lower than those below the ground water level, because of the difference in effective soil svertical stresses. The values of CSR decreases with an increase of soil depth.

Overall, the liquefaction safety factor for the research location is between 0.30 and 1.20, and the depths of which SF < 1.0 are between 1.0 m and 12.0 m as shown in Figure 6b. It can be seen that, to a depth of 5.0 m, the SF based on q_c is similar to that based on V_s. However, for depths greater than 6.0 m, the SF based on q_c is much lower than that based on V_s.

Based on liquefaction potential analysis proposed by Andrus and Stokoe [4], liquefaction would occur at depths from 2.0 m to 12.0 m. A comparison between Figures 5a and 6a, the liquefaction SF is much less than 1.0 for depths up to 6.0 m, and the SF is about or greater than 1.0 for V_s values of 180 m/s or greater.

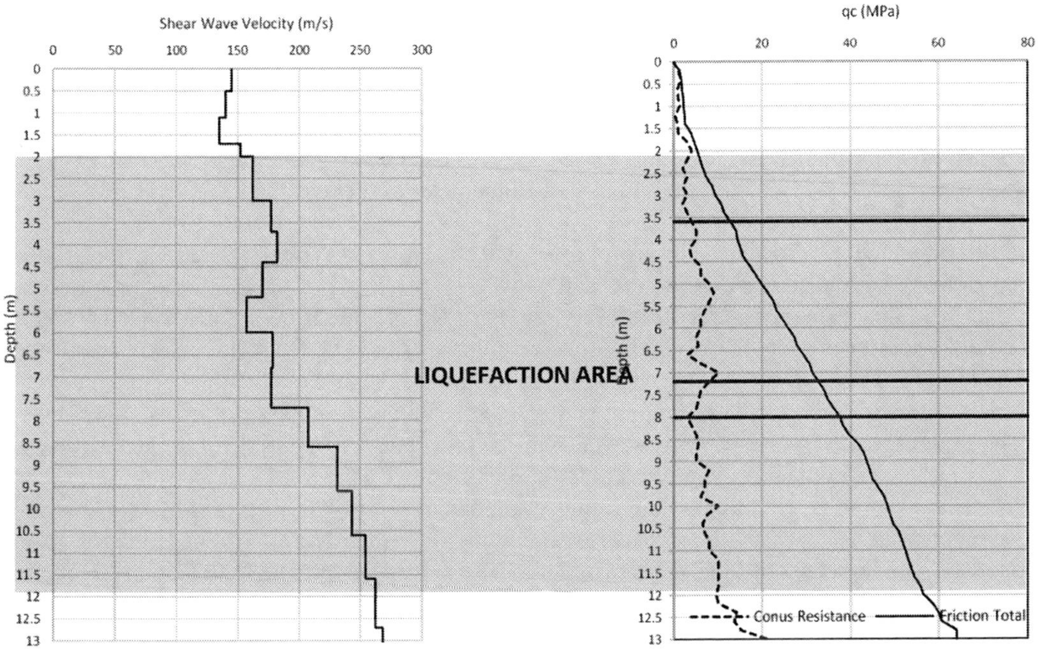

(a) (b)

Figure 5. Liquefaction potential by depth (a) Graph of Shear wave velocity of PKA 10 (b) CPT Survey Result of CPT07.

IOP Conf. Series: Materials Science and Engineering **930** (2020) 012034 doi:10.1088/1757-899X/930/1/012034

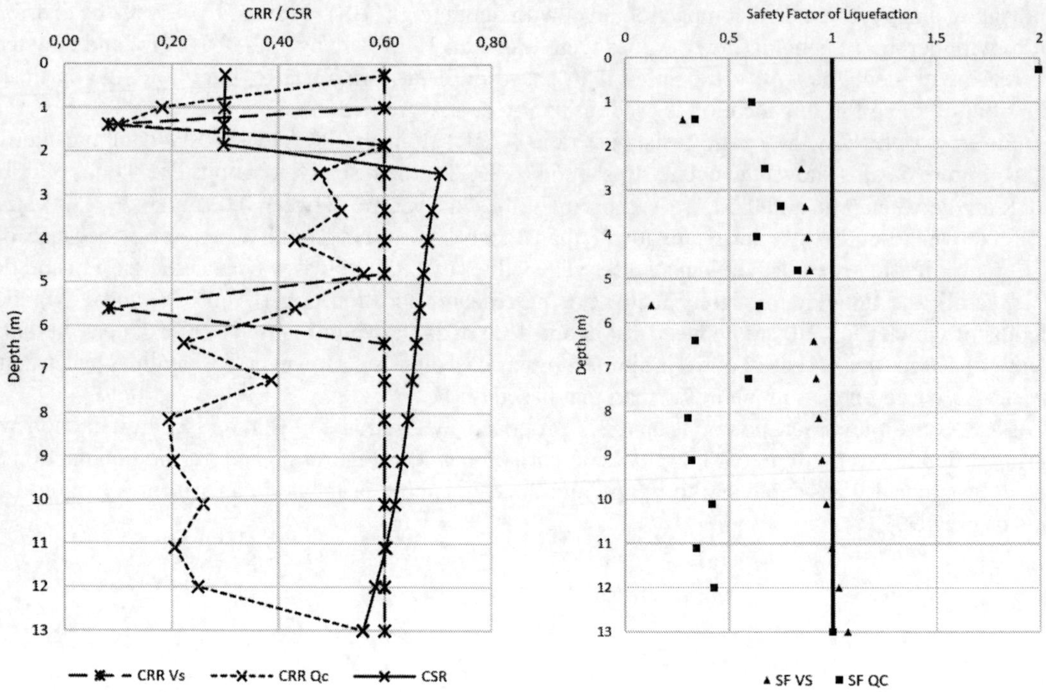

(a) (b)

Figure 6. (a) Cyclic Resistance Ratio (based on V_s and q_c) and Cyclic Stress Ratio Charts and (b) Safety Factor of Liquefaction Potential

Based on liquefaction potential analysis proposed by Robertson and Wride [5], liquefaction would occur at depths from 2.0 m to 12.0 m. A comparison between Figures 5b and 6b, the liquefaction SF is much less than 1.0 for q_c values of about 10 MPa or less. Furthermore, the friction ratio of CPT7 was used to estimate the soil types using the method proposed by Robertson [11]; the soil types are suggested to be silts and sands. This is in accordance with Ishihara [2] proposal, that liquefaction occurs in silt and sand materials which would be affected by pore water pressure below ground water level.

The liquefaction SF is consistently lower than 1.0 (between 0.6 and 0.9) for depths between 2.0 m and 5.0 m based on both tests. This appears to be consistent with the results of the field survey reported in [6]. This location is within with high liquefaction damage area, characterized by crack more than 5 cm, settlement more than 10 cm, and lateral spreading more than 5 cm. Although this is an indirect comparison, the liquefaction field survey [6] could be used to validate the liquefaction SF obtained by the methods based on the shear wave velocity and the cone resistance.

4. Conclusions

This paper reported liquefaction analysis results employing the pre-event shear wave velocity V_s obtained from MASW survey and the pre-event cone penetration q_c data for a location in Balaroa area, Palu. This location is within the high liquefaction damage area as reported by the a field survey after the 28 September 2018 Palu earthquake. The liquefaction safety factor indicated by both data sets was consistent for depths up to about 5.0 m, and the safety factor was between 0.6 and 0.9. The safety factor indicated by the V_s data was higher for depths greater than 6.0 m. This low liquefaction safety factor appears to be consistent with the reported high liquefaction damage class, suggesting a good correlation between the analysis and the field survey results.

Acknowledgements
The authors gratefully acknowledge the research grant PUTI provided by University of Indonesia for the budget year of 2020 (Contract No. NKB-1188/UN2.RST/HKP.05.00/2020). The authors also acknowledge the contribution of the Provincial Public Works Agency and Mr. Dandung Sri Harninto for CPT data of Palu.

References

[1] Kramer S and Elgamal A 2001 *Modeling Soil Liquefaction Hazards for Performance Based Earthquake Engineering* (Berkeley: Pacific Earthquake Engineering Research Center. University of California)
[2] Ishihara K 1985 Stability of natural deposits during earthquakes *Proc. of The Eleventh ISSMFE* (San Francisco: CA, USA) p 321-376
[3] Seed H B and Idriss I M A 1970 *Simplified procedure for evaluating soil liquefaction potential* (California: University of California)
[4] Andrus R D and Stokoe I I K H 2000 Liquefaction resistance of soils from shear-wave velocity *J Geotech. Geoenviron.* **126** 11 p 1015–25
[5] Robertson P K and Wride C E 1998 Evaluating cyclic liquefaction potential using the cone penetration test *Can. Geotech. J.* **35** 3 p 442–59
[6] Badan Geologi 2018 *Di Balik Pesona Palu* (Bandung: Badan Geologi, Kementrian ESDM)
[7] Youd T L and Idriss I 2001 Liquefaction resistance of soils *J. Geotech. Geoenviron. Eng.* **127** p 297–313
[8] Boore D M, Joyner W B and Fumal T E 1993 Estimation of response spectra and peak accelerations from western North American earthquakes: an interim report. *USGS Open-File Report 503* p 72
[9] Campbell K W and Bozorgnia Y 1993 Empirical analysis of strong ground motion from the 1992 Landers, California, earthquake *Bull. Seismol. Soc.* **84** 3 p 573–88
[10] Building Seismic Safety Council 1995 *A New Technical Explanation of the 1994 NEHRP Recommended Provisions* (Wasington DC: Building Seismic Safety)
[11] Robertson P K 1990 Soil classification using the cone penetration test *Can. Geotech. J.* **27** 1 p 151–8

IOP Conf. Series: Materials Science and Engineering **930** (2020) 012035 doi:10.1088/1757-899X/930/1/012035

The comparison of empirical and finite element methods to evaluate soil foundation capacities for coastal dike

Suyuti[1*], Zulkarnain K.Misbah[1], and Zhafirah M. Ardhana[2]

[1] Lecturers at Study Program of Civil Engineering, Faculty of Engineering, Universitas Khairun, North Maluku, Indonesia
[2] Undergraduate Student at Study Program of Civil Engineering, Faculty of Engineering, Universitas Khairun, North Maluku, Indonesia

[*] Corresponding author's e-mail: suyuti@unkhair.ac.id

Abstract. There are over 6,300km of coastlines in the North Maluku province of Indonesia due to tidal waves which originate from the pacific ocean. The waves deposit finely loosed materials and light soil formations on coastlines, therefore causing abrasion. A common example occurred in Toseho, Tidore, and Archipelago districts. Moreover, the local residence has been faced with this challenge for up to thirty years. For this reason, deliberations to construct a dike became significant, although, the soil properties are vulnerable. This reasearch, therefore, compares the empirical calculation to the finite element method (FEM) to ascertain soil foundation capacities. The results showed the safety factor, Fs, for dike height, H_d, of 3.5 m and width bottom, B_1, of 12.5m appeared below 2.0 for soil friction φ_s of 10~ 15°, while estimates above 2.0 depicts soil friction φ_s of 20°. Furthermore, the soil stresses beneath the dike were determined empirically at q_{max} of 5.97 t/m^2 and σ_s of 4.57 t/m^2, while FEM evaluated the deformation elastic at 12.5 cm.

1. Introduction

North Maluku is a province in Indonesia characterized by extensive coastlines where over 40% suffered abrasion due to natural disasters [1]. This is triggered by the existence of soft clay, and/or loose sand deposits. However, several structures, including residences, roads, public offices etc, owned by government and private entities are situated in this district, although, these properties are constantly threatened by unfavorable wave action.

As one of the counter measures to prevent the natural disaster, tall dikes have been erected. These were built without any physical significance to geotechnical standards. This study reports the comparison of the calculation method in the design of coastal dike on silty sand soil. Typical cases include Toseho village, Tidore, and Archipelago district in Halmahera Islands, North Maluku.

These local communities are known to settle in traditional houses along coastlines built with conventional construction methods for 40 years. Figure 1 revealed these locations, in recent development, no longer provide safe habitation due to high risk of wave disturbance.

Content from this work may be used under the terms of the Creative Commons Attribution 3.0 licence. Any further distribution of this work must maintain attribution to the author(s) and the title of the work, journal citation and DOI.
Published under licence by IOP Publishing Ltd

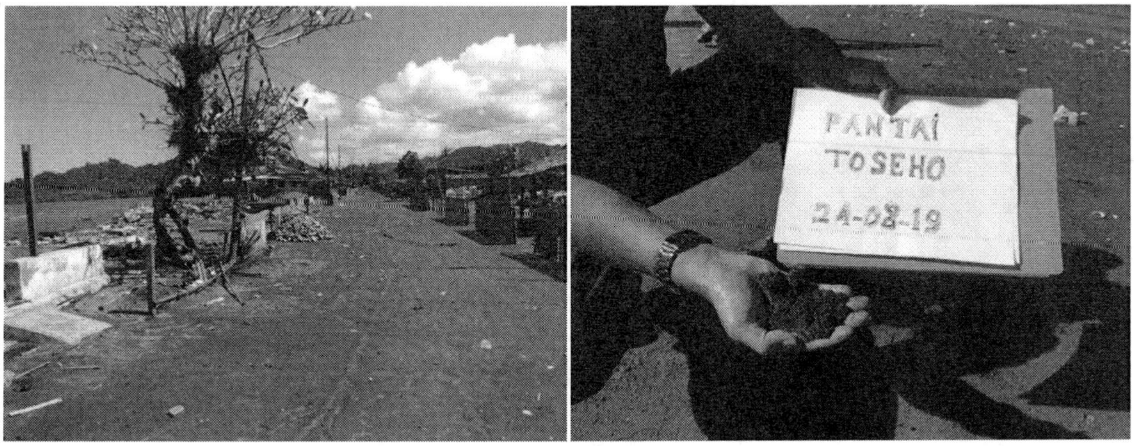

a) Recently land situation of the local houses   b) Sand material views at the ground surface

Figure 1. Photo recently situation of the local residential at Toseho coastline (Photographer Suyuti, 24.08.2019).

However, the unfortunate breakdown of these residences was attributed to coastal abrasion as only trees were used as protection mechanism. In addition, over 15 families are known to currently reside on Toseho coastline and figure 2 captures the present condition.

a) Houses before collapsing   b) Houses after damaged by sea waves

Figure 2. Photo recently situation of the local residential at Toseho coastline (Photographer Suyuti, 24.08.2019).

Figure 2 also shows the traditional semi-permanent houses built with materials such as bamboo used as reinforcement for the basic wall, sand for wall, wood for buttress, midrib of sugar palm for wall, etc.

For a long time, the people settled for conventional methods of coastal protection such as planting trees, coral stones construction etc [2], hence, the need to undertake modern dike construction. This study, therefore, aims to provide design calculations for this purpose.

2. Methodology of Research

The empirical and FEM approach compares the nature of wave forces on the dike. In addition, the required stability criterion of soil foundation is calculated by following flow chart (figure 3).

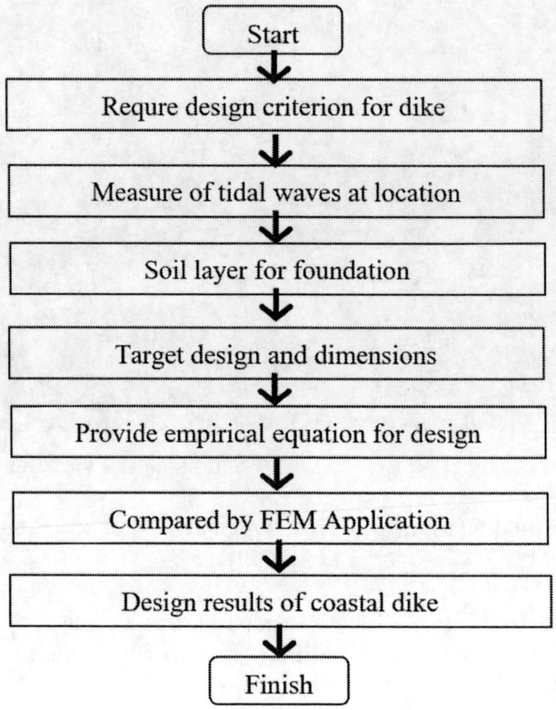

Figure 3. Flow chart of design calculation for dike.

2.1. Measure tidal waves

Three points (A, B and C) were marked on the coastline at a length of 483 m at Toseho to measure the tidal waves. Furthermore, by using global positioning system (GPS), the coordinate point surveyed on April 21st, 2020 generated 0°20'57,91"U and 127°39'17,30"T. The coordinates using Google Earth application are shown in figure 4.

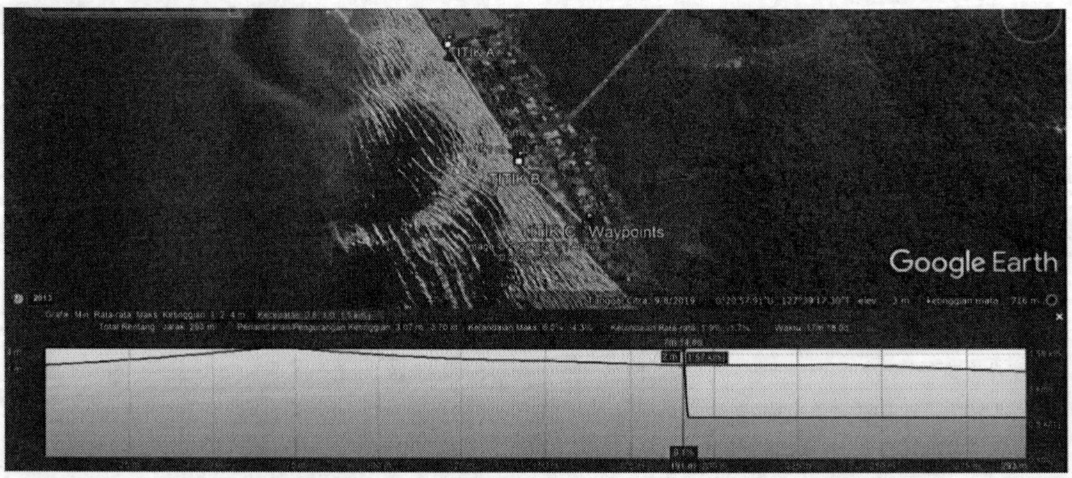

Figure 4. Photo of three points of the coastline situation at Toseho Village (Photographer Zhafirah, 21.04.2020).

Figure 4 also indicates the various wave elevations obtained 1.05 ~ 2.0 meters at a length of 20 m on the cross section [2].

2.2. Soil layer for foundation

The soil layer was investigated by a simple method to express the parameters for the foundation, e.g. cohesion (c, t/m^2), internal friction (ϕ, °), and unit weight of the saturated soil (γ_{sat}, t/m^3). Meanwhile, the surface layer was tested by plugging a hole on the ground surface.

2.3. Design and dimensions of the dike

The design was prepared without any reinforcement using dimensions such as width at top (B_0, m), height (H_d, m), width at bottom (B_b, m), gradient slope at the front side (n_1), and gradient slope behind the structure (n_2). However, the dike was constructed with a foundation depth (D_f, m), and design height in respect to design water level of waves (DWL is High water level added by global warming) (figure 5) [3].

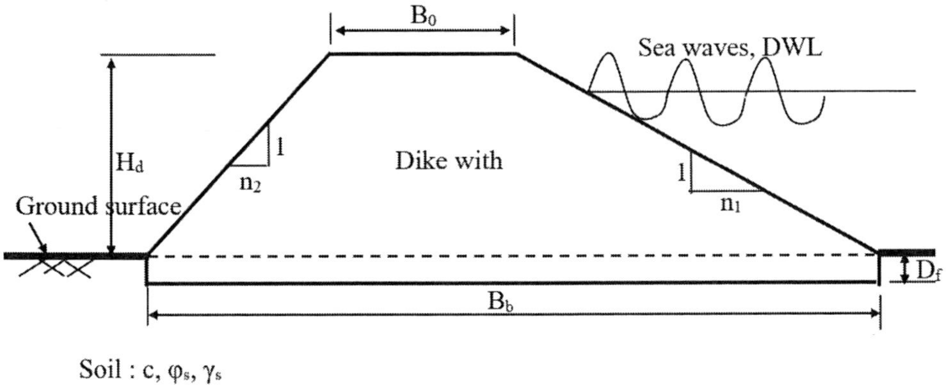

Figure 5. The prototype of dike on the subsoil use for coastal.

Figure 5 shows that width of the bottom B_b used to make simplified for calculation, it is based on assumption footing of dike as rigid footing foundation with depth D_f (m).

3. Fundamental Theory

The required stability criterion was prepared with allowable bearing capacity of soil foundation and the vertical loading pressure provided as follows:

3.1. Empirical Equations and formulas

The factor of the vertical loading pressure of the dike to allowable bearing capacity of soil Fs, is written as: [4]

$$Fs = \frac{q_{us}}{p_d} > 2.0 \quad (1)$$

The foundation pressure beneath the coastal dike construction (without waves force) p_d (t/m^2) is calculated by:

$$p_d = \gamma_d H_d \quad (2)$$

By simplifying, the classical theory of Terzaghi and Peck (1948) may be used to predict ultimate bearing capacity of soil q_{us} (t/m^2) [5].

$$q_{us} = cN_c + qN_q + \tfrac{1}{2} B_b \gamma'_s N_\gamma \quad (3)$$

The allowable bearing capacity of soil q_{as} (t/m^2) is calculated by.

$$q_{as} = \frac{q_{us}}{Fs} \tag{4}$$

where c = the cohesion of soil (t/m^2), q = the pressure at ground surface (t/m^2) ($q = \gamma_{sat} \times D_f$), D_f = foundation depth (m), γ'_s = the effective unit weight of soil (t/m^3) ($\gamma'_s = \gamma_{sat} - \gamma_w$), γ_{sat} = the unit weight of saturated soil (t/m^3), γ_w = the unit weight of water (t/m^3), N_c, N_q, N_γ are factor of bearing capacity respected to the internal friction of soil (φ_s).

This section investigates the internal friction of the soil φ_s (°) by a simple method to observe the surface density condition.

Load stress of the dike construction on the soil ground surface q_{max} (t/m^2) due to wave forces is calculated by: [6]

$$q_{max} = \frac{\sum_{i=1}^{n} V}{B_b} \times \left\{ 1 \pm \left(\frac{6e}{B_b} \right) \right\} \tag{5}$$

where $\sum V$ = total of vertical load (ton), B_b = width bottom of the dike (m)

The eccentricity of dike construction e (m) due to the moment of coastal wave forces is predicted by: [6]

$$e = \frac{B_b}{2} - \left(\frac{\sum_{i=1}^{n} (MV - MH)}{\sum_{i=1}^{n} V} \right) < \frac{B_b}{6} \tag{6}$$

where $\sum MV$ = total moments of dike vertical load (t.m), $\sum MH$ = total moments of wave horizontal load (t.m), $\sum V$ = total vertical load (ton).

3.2. Finite Element Method-2D

Finite Element Method (FEM) is applied by following the plane strain model in order to compare empirical calculation. The global relation between force (F, kN), stiffness of materials (K, kN/m^2), and displacement (U, m), is shown by equation [7]

$$\{F\} = [K]\{U\} \tag{7}$$

In the soil parameter using Mohr-Coulomb material model for FEM-2D simulation, the relation between Young's Modulus (E, kN/m^2), shear modulus (G_{ref}, kN/m^2), Eodometer modulus (E_{oed}, kN/m^2), and Poisson's ratio of soil (υ) are written as [8],

$$G_{ref} = \frac{E_s}{2(1+\upsilon)} \tag{8}$$

$$E_{oed} = \frac{(1-\upsilon)E_s}{(1-2\upsilon)(1+\upsilon)} \tag{9}$$

3.2.1. Making geometry model

The geometry of dike on the ground estimated to produce the FEM-2D simulations of the coastal dike on a silty sand layer. Geometry model is made coordinates of points (x, y) as shown in figure 6.

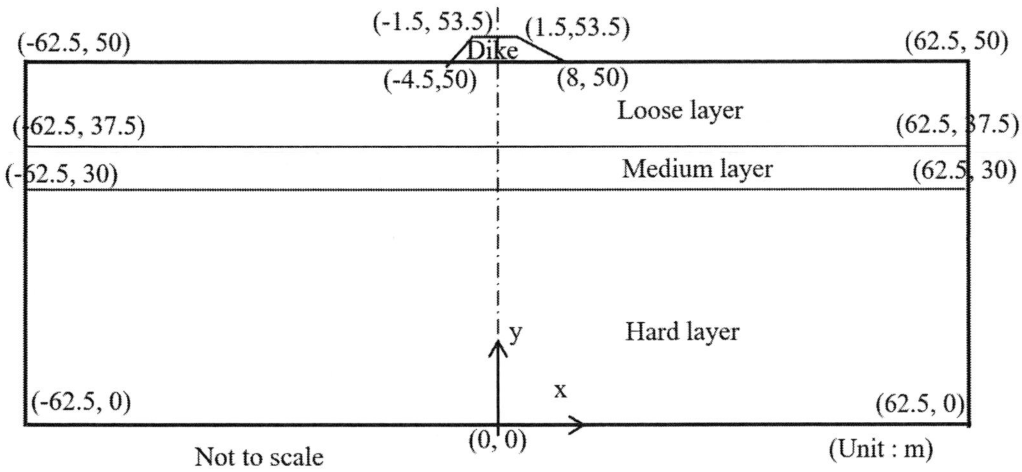

Figure 6. Modeling geometry of dike on the ground for FEM-2D.

3.2.2. Soil parameters of material model

Soil parameters and dike simulations were provided to calculate the stability criterion. The soil properties at the ground are roughly predicted that based on field investigation. It is found loose to medium layers of silty sand [2, 4]. The results of input parameters of soil and dike are listed in table 1 and table 2, respectively.

Table 1. Input soil parameter of soil [2, 4, 9].

Soil layer		Loose layer	Medium layer	Hard layer
Depth (m)		0 ~ 12.5	12.5 ~ 20	20 ~ 50
Material type		Drained	Drained	Drained
Parameters of soil:	Unit			
Unsaturated (γ_{unsat})	(kN/m^3)	12.5	15.0	18.0
Saturated (γ_{sat})	(kN/m^3)	17.5	19.0	21.0
Permeability k_x	(m/day)	1.0	2×10^{-2}	2×10^{-4}
Permeability k_y	(m/day)	1.0	2×10^{-2}	2×10^{-4}
Cohesion (c)	(kN/m^2)	1.0	1	1
Internal friction $Ø_s$	(°)	20	25	30
Young Modulus E_s	(kN/m^2)	1.23×10^4	4.5×10^4	9.8×10^4
Poisson's ratio v_s	-	0.35	0.3	0.3
Dilantancy ψ	(°)	0	0	0
G_{ref}	(kN/m^2)	4.56×10^4	1.73×10^4	3.76×10^4
E_{oed}	(kN/m^2)	$1.974 \text{ X } 10^4$	6.5×10^4	1.32×10^5
Velocity v_s	m/s	59.76	106.3	143.3
Velocity v_p	m/s	124.4	198.9	268

Table 2. Input soil parameter of dike [2, 9].

Dike construction		
Material type		Drained
Parameters of dike:	Unit	
Unsaturated (γ_{unsat})	(kN/m^3)	19.0
Saturated (γ_{sat})	(kN/m^3)	22.0
Permeability k_x	(m/day)	100
Permeability k_y	(m/day)	100
Cohesion (c_d)	(kN/m^2)	1
Internal friction $Ø_d$	(°)	40
Young Modulus E_d [9]	(kN/m^2)	1.2 x 10^5
Poisson's ratio v_d	-	0.3
Dilantancy ψ	(°)	10
G_{ref}	(kN/m^2)	4.6 x 10^4
E_{oed}	(kN/m^2)	1.6 X 10^5
Velocity v_s	m/s	154.3
Velocity v_p	m/s	288.7

The soil and dike parameters are modeled in Mohr-Coulomb and drained material type. It is setting up as materials elasto plastic in calculation for stability and settlement criteria.

3.2.3. The application program of FEM-2D
The FEM application was employed commercial software to calculate the stress and displacement at every point. Meanwhile, the gradient slopes n_1 and n_2 were placed at the ocean side and behind the dike, respectively. Furthermore, dimension geometry was applied based on design results to implement empirical calculaton.

4. Results and Discussions

Based on simple tests in field, soil parameters for the calculation of bearing capacity were assumed such as internal soil friction φ_s of 10 ~ 20°, unit weight γ_{sat} = 17.5 kN/m^3[10]. It is shown in Figure 7.

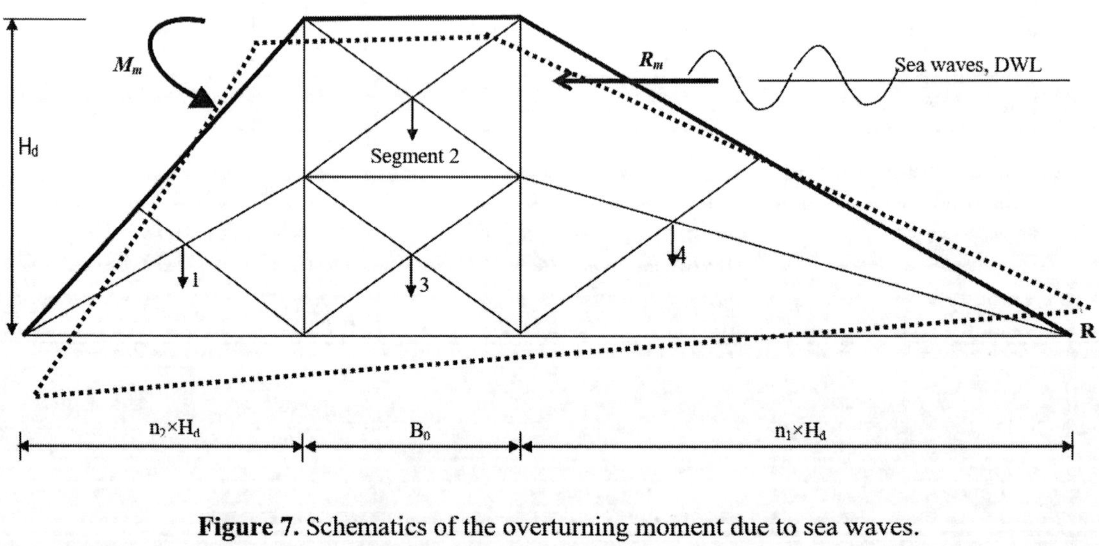

Figure 7. Schematics of the overturning moment due to sea waves.

Figure 7 shows the dimensions of the structure were used by height H_d of 3.5 m, width the top B_0 of 2.0 m, n_1 = 2, n_2 = 1, bottom B_1 of 12.5 m.

The calculation results of factor of safety Fs are listed in table 3.

Table 3. Calculation results of factor of safety F_s (dike $\gamma_d = 2.5$ t/m^3).

Soil internal friction φ_s (°)	Bearing capacity q_{us} (t/m^2)	Factor of safety Fs	Remark
10°	8.3	0.95 < 2.0	No. Safe
15°	16.2	1.84 < 2.0	No. safe
20°	31.2	3.56 > 2.0	Safe

The calculation parameters for soil stress beneath the coastal dike included the dynamic force of waves R_m of 2.07 ton, eccentricity e of -0.37 m < 2.08 m (B_b/6). Then, calculation of statis force due to waves were presented as total moments of dike vertical load MV of 438.12 (tm), total moments of wave horizontal load MH of 17.95 (tm), and total vertical load V of 63.42 (ton).

The factor of safety calculation results (Fs) are found less than Fs of 2.0 or no safe for internal friction φ_s of 10° and 15°, and more than 2.0 for for internal friction φ_s of 20°. Therefore, this calculation results of Fs for the internal friction φ_s of 20° is compared by FEM-2D as explained below. The steps for FEM calculation are as follows:

- Carry out soil and dike modeling,
- Input parameter of the material model of soil and dike,
- Prepare fixed soil ground,
- Generate mesh for ground and dike,
- Set up boundary conditions of the construction,
- Calculate initial pore pressure of the soil,
- Set up calculation of elasto-plastic with stage construction (t = 0 day),
- Run application program, and
- Check the calculation results of soil stress and displacement.

The calculation results of soil the stresses beneath the dike using an empirical equation and FEM-2D are listed in table 4.

Table 4. Calculation results of soil stress beneath dike.

Soil stress beneath the dike	Calculation result (t/m^2)		Allowable bearing. capacity, q_{as} (t/m^2)	Remark
	Empirical	FEM-2D		
Maximum stress, q_{max}	5.97	$\sigma_s = 4.57$	15.6	Safe
Minimum stress, q_{min}	4.17			

Table 4 shows that calculation result of soil stress beneath the dike with FEM-2D is found fitted to the empirical calculation method in equations of 5 and 6. Moreover, the allowable bearing capacity of soil foundation is obtained more than maximum stress (q_{max}) of 5.97 t/m^2. Those input soil and dike parameters above that applied in the empirical and Finite Element Methods are reasonable to used in the simulation.

5. Conclusions

The research summarized the following:

- The safety factors (Fs) of soil the bearing capacity is predicted for internal friction φ_s of 20°.
- Calculation results for soil stress beneath the dike with the empirical mehod are fitted to FEM-2D by using the Mohr-Coulomb model with drained type material
- Calculation result for vertical displacement is evaluated at 12.5 cm, and
- Soil parameters are reasonable to use in simulations.

Acknowledgments
The authors are grateful to the LPPM of Universitas Khairun in year 2020 and also the Soil Mechanics at the Department of Civil Engineering, Faculty of Engineering, Universitas Hasanuddin for supporting FEM.

References

[1] Ministry of Public Works 2014 Coastal protection in Ternate and Sofifi
[2] Zhafirah M Ardhana 2020 Design construction of coastal dike by using FEM-2D *StudyProgram of Civil Engineering Faculty of Engineering Universitas Khairun*
[3] Triatmodjo B 1999 *Offshore Engineering* (Yogyakarta: Universitas Gadjah Mada) pp 1-201
[4] Coduto D P 1994 *Foundation Design: Principles and Practices* Prentice-Hall International Edition pp 44–91
[5] Das B M 1979 *Introduction to Soil Mechanics* (The IOWA State University Press) 1st edition pp 229–53
[6] Das B M 1999 *Principles of Foundation Engineering* (California State University: PWS Publishing) 4th edition pp 152–97
[7] Suhendro B 2000 *Finite Element Method and Application* (Yogyakarta: Gadjah Mada University) pp 85-122
[8] Plaxis 2019 *Material Model Manual* pp 1-135
[9] Bowles J E 1989 *Soil Properties and Geotechnics* 2nd edition (Jakarta: Erlangga) pp 25-273
[10] Craig R F 1987 *Soil Mechanics* (Jakarta: PT. Erlangga) pp 1–31

IOP Conf. Series: Materials Science and Engineering **930** (2020) 012036 doi:10.1088/1757-899X/930/1/012036

Comparison of ultimate bearing capacity based on empirical method, interpretation of loading pile test and finite element

Y Lastiasih[1*] **and P T K Sari**[1]

[1]Civil Engineering Departement, Institute Technology of Sepuluh Nopember Surabaya, Indonesia

[*]Corresponding author's e-mail: yudhi.lastiasih@gmail.com

Abstract. Civil Engineers in designing the ultimate brearing capacity of the pile usually uses several empirical methods. Then usually the loading pile test will be conducted to confirm the ultimate carrying capcity in the field. Based on conditions that often occurs in the field, an empirical method of research is carried out to find out which method is closest to results of the field test. So that in designing the ultimate bearing capacity of the deep foundation will be closer to the field conditions. The empirical methods used in this case are the method of Bazara and Luciano Decourt, while the interpretation pf filed results uses the methods of Chin, Mazurkiewiz and Lastiasih. Based on the results of empirical methods and interpretation of the ultimate bearing capacity of the field test, the results of the pile loading test will be analysed using finite element. The result of this research shows Finite Elemet method, Luciono Decourt and Lastiasih are the closest produce produce ultimate bearing capacity with loading pile test until failure.

1. Introduction

Recently, the construction of high rise buildings has strict requirements, especially related to the planning of the foundation. All building plans in Jakarta must carry out loading pile tests as one of the requirements in carrying out foundation work. Planning in several areas of the big city has also begun to use loading pile test requirements. The problem that is often encountered in the field is the result of the interpretation of the ultimate bearing capacity of the loading pile test in the field that is not comfortable with empirical theories or finite element-based support programs. Therefore, the author will discuss the ultimate bearing capacity of each method, so it can be concluded which method is the closest to the results of the field test pile. Several studies have conducted numerical analysis to compare the results of interpretation of field test piles and see the relationship between soil and piles as has been done by KRASIńSKI[1], Zein [2] and Lastiasih [3]

KRASIńSKI, et al [1] conducted numerical analysis for piles with a diameter of 1500 m with length L = 19.7 m, where the results of the load distribution on the pile can be seen in figure 1. The numerical analysis results showed that around 81% of the total load was received by friction bearing capacity and 19% held by the base bearing capacity of pile, while the field test results show that the total load received by pile is 68% received by friction bearing capacity and 32% is held by the base bearing capacity of pile.

Content from this work may be used under the terms of the Creative Commons Attribution 3.0 licence. Any further distribution of this work must maintain attribution to the author(s) and the title of the work, journal citation and DOI.
Published under licence by IOP Publishing Ltd

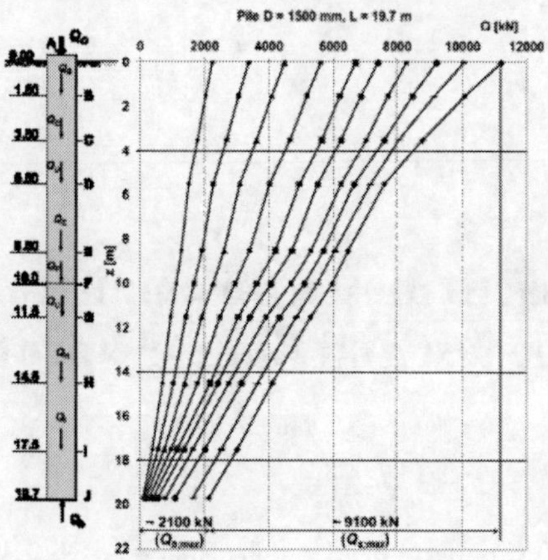

Figure 1. Load distribution in numerical modeling of pile foundation [1]

Zein et al. [2] present an evaluation of the bearing capacity of bored pile which is measured directly and obtained from the interpretation method of selected load test data based on different failure criteria. Zein [2] conducted an experiment on a field scale, which was to install a bored pile in alluvial soil deposits and load by static load until it failure. The measured failure load varies from 195 to 520 kN and pile movements of 1.3 to 2.7% D are needed to mobilize the load at the time of testing. The friction bearing capacity contributes more than 83% of the total bearing capacity and is mobilized when the movement is 0.2 to 0.8% D. Total load receive by the friction bearing capacity 83%. These result inidicate that the quantity is almost same ast the experiment from KRASIńSKI [1]. Zein [2] also interpreted the results of the loading test using several methods. As for the results obtained for the Fuller and Hoy methods show the best overall performance in estimating the ultimate load, the De Beer and Van der Veen methods produce a good evaluation while the Davisson interpretation method shows relatively poor results. The method that produces the highest ultimate carrying capacity estimate among the methods used by Zein [2] is Chin. The results of Zein et al [2] are in line with the results from Lastiasih et al [3]. Lastiasih [3] compared the Chin [4] method with Mazurkiewizc [5], Brinch Hansen [6] and Van der Veen [7] and obtained the highest estimate ultimate bearing capacity with Chin method. But Lastiasih also interpreted using the Decourt Extrapolation [8] method, and when compared with the Chin [4] method it turned out that the highest Decourt Extrapolation [8] method produced the ultimate carrying capacity.

Existing interpretation methods also still produce varying values, so Kaniraj [9] and Lastiasih [3] try to develop new interpretation methods. The ultimate bearing capacity interpretation method based on the results of the pile test in the form of a load-settlement curve that has been mathematically extrapolated was also developed by Kaniraj [9] as a hyperbolic equation. This concept has been effectively used to estimate the ultimate carrying capacity of large diameter piles in several projects (Sanjay Gupta & Ravi Sundaram, [10]). The equation used for the hyperbolic model is as follows:

$$\frac{S'}{Q'} = a + bS' \tag{1}$$

Where :
$S' = s/D$ s = settlement D = Pile diameter
$Q' = Q/Qr$ Q = Applied Load Qr = Reference Load
a and b are constants determined from a linear regression analysis of S'/Q' and S'.

In addition to the formula developed by Kaniraj [9]., Lastiasih [3] also established a formula to interpret the ultimate carrying capacity with quadratic hyperbolic, using the following equation:

$$Q = \frac{a\left(x^2 + bx\right)}{x^2 + cx + d} \tag{2}$$

Where :
Q = Ultimate load
x = displacement
a,b, c dan d = Constanta

Kaniraj [9] and Lastiasih [3] used hyperbolic equations as an effort to estimate the ultimate bearing capacity based on the results shown in the load vs. decrease curve. The results obtained show a hyperbolic shape for both single pile and group pile according to Eurocode 7 [11], as seen in figures 2 and 3.

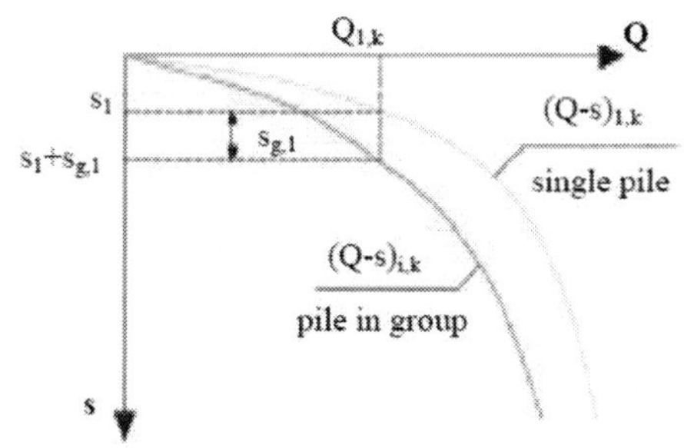

Figure 2. Load-settlement curves for single piles and group piles [11]

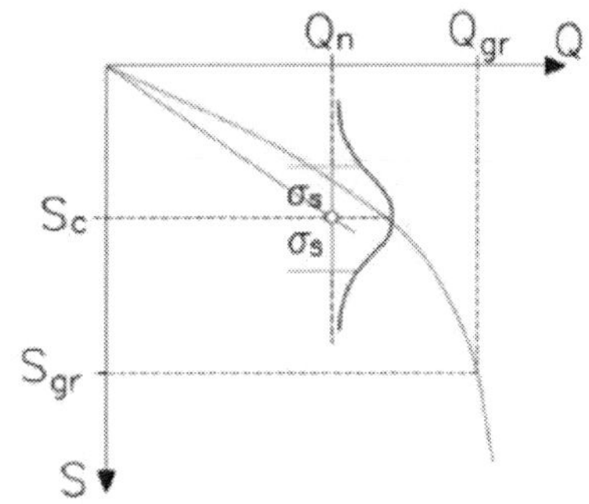

Figure 3. Schematic diagram to determine the Q-s characteristics of the "i" pile in a group [11]

Based on existing conditions in the field that the results of several previous researchers who produced a different reality between the results of field testing and the results of the ultimate bearing capacity calculation nterpretation, the researchers tried to compare the ultimate bearing capacity of interpretant results with empirical theories and finite elements, to determine which method was more appropriate to use. The empirical method used is Luciano Decourt and Meyerhoff Bazara, because the average projects are using that method. Likewise, the selection of interpretation methods with Chin was also carried out in many projects, and 2 other methods were chosen namely Mazurkiwiz and quadratic hyperbolic from Lastiasih.

2. Research data

This study uses data from loading pile test results, and soil test results data in one of the projects in the Dharmawangsa Jakarta area, where the piles tested with a diameter of 100 cm in 5 locations. Pile data, depth, and ultimate load tested on the pile can be seen in table 1

Table 1 Pile data tested

Pile location	ϕ	L	Qult load test
	m	m	ton
BP98		30	1200
BP145		26,03	700
BP190(F)	1	24,97	700
BP334		25,77	800
BP355 (F)		24,32	265,625

IOP Conf. Series: Materials Science and Engineering **930** (2020) 012036 doi:10.1088/1757-899X/930/1/012036

Soil data of 3 points in the research location are dominated by clayey silt layers, as shown in table 2 to table 4.

Table 2 Soil parameters BH 1

Depth m	Soil consistency	N_{SPT}	Cu t/m2	γsat t/m3	E Mpa	ν	ϕ
0,00 - 6,50	Silty Clay	7,00	4,20	1,66	7,20	0,20	12,60
6,50 - 13,50	Silty Clay	15,00	9,00	1,68	12,00	0,30	27,00
13,50 - 15,00	Sandy silt	24,00	14,40	1,96	17,40	0,35	30,00
15,00 - 21,00	Clayey silt	30,50	18,30	2,00	21,30	0,30	30,00
21,00 - 27,50	Sand	37,50	0,00	1,75	25,50	0,35	35,26
27,50 - 29,00	Clayey silt	37,50	22,50	2,00	25,50	0,33	30,00
29,00 - 32,50	Sandy silt	37,50	22,50	2,00	25,50	0,30	30,00
32,50 - 36,50	Clayey silt	37,50	22,50	2,00	25,50	0,33	30,00
36,50 - 40,50	Silty sand	30,00	0,00	1,75	21,00	0,30	35,26
40,50 - 45,00	Clayey silt	23,75	14,25	1,93	17,25	0,30	30,00
45,00 - 51,50	Sandy silt	29,33	17,60	2,00	20,60	0,35	30,00
51,50 - 57,50	Sand	37,50	0,00	1,75	25,50	0,38	35,26
57,50 - 58,50	Silty sand	37,50	0,00	1,75	25,50	0,30	35,26
58,50 - 60,50	Clayey silt	37,50	22,50	2,00	25,50	0,35	30,00

Table 3 Soil parameters BH-2

Depth m	Soil consistency	N_{SPT}	Cu t/m^2	γsat t/m^3	E Mpa	ν	ϕ
0,00 - 5,50	Silty Clay	2,33	1,40	1,53	4,40	0,20	4,20
5,50 - 9,50	Clayey silt	10,50	6,30	1,70	9,30	0,30	18,90
9,50 - 15,50	Clayey silt	37,50	16,03	1,75	25,50	0,35	30,00
15,50 - 21,50	Sand	37,50	0,00	1,75	25,50	0,38	35,26
21,50 - 37,00	Sandy silt	37,50	22,50	2,00	25,50	0,35	30,00
37,00 - 47,00	Clayey silt	18,25	10,95	1,71	13,95	0,33	30,00
47,00 - 52,50	Sand	31,33	11,30	1,92	21,80	0,30	30,00
52,50 - 60,50	Clayey silt	27,50	19,50	2,00	19,50	0,35	30,00

Table 4 Soil parameters BH-3

Depth m			Soil consistency	N_{SPT}	Cu t/m^2	γsat t/m^3	E Mpa	ν	ϕ
0,00	-	11,50	Silty Clay	3,50	2,10	1,64	5,10	0,30	6,30
11,50	-	14,50	Clayey silt	10,50	6,30	1,70	9,30	0,30	18,90
14,50	-	35,50	Sand	37,40	0,00	1,74	25,44	0,38	35,22
35,50	-	37,50	Sandy silt	19,00	12,20	1,73	14,40	0,35	30,00
37,50	-	43,50	Clayey silt	27,33	16,40	1,87	19,40	0,35	30,00
43,50	-	55,50	Sand	34,17	0,00	1,71	23,50	0,33	34,63
55,50	-	60,50	Clayey silt	21,33	12,80	1,85	15,80	0,30	30,00

The soil data are used for modeling on plaxis and input phisik parameter to calculated ultimate bearing capacity by Luciano Decourt and Meyerhof Bazara. The soil data is used to find the ultimate carrying capacity with diameter and depth according to the pile test in the field as seen in table 5

Table 5 Location soil test and loading pile test

Location Loading Pile Test	Location Soil Test
BP98	BH-1
BP145	BH-2
BP190(F)	BH-3
BP334	BH-1
BP355 (F)	BH-1

3. Calculation Method

The interpretation method used in this study is the method of Chin [4], Mazurkiewizc [5], Lastiasih [3], while the empirical method used is the Luciano Decourt [12] method and Meyerhof-Bazara method [13][14]

3.1. Chin Method (1971)

The procedure to determine the ultimate load using this method is as follows:

- Figure s / Q for s, where s is a settlement, and Q is the load applied.
- Ultimate load (Q) ult = 1 / C1.

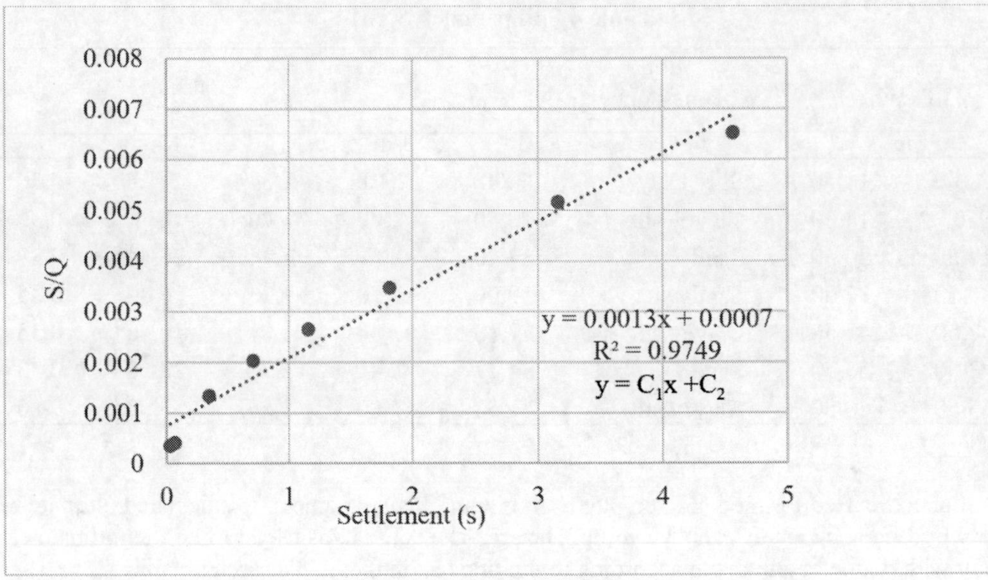

Figure 4. Chin Method

3.2. *Mazurkiewizk Method*

The steps to find the ultimate carrying capacity of the results of the load settlement curve using the Mazurkiewizc method are as follows:

Plot the load- settlement curve, select the amount of settlement value in the curve and draw a vertical line that intersects the curve. Then draw a horizontal line from this point of intersection on the curve until it intersects the load axis. From the intersection of each curve, draw a 45° line to cut the next load line. The intersection is approximately located on a straight line. The point obtained by the intersection of this line extension on the vertical axis (load) is the ultimate load.

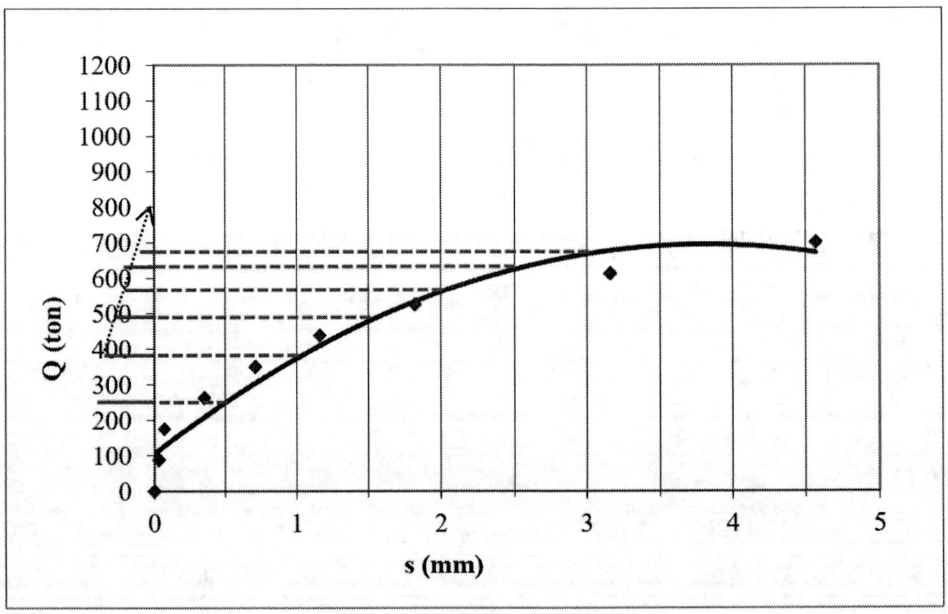

Figure 5. Mazurkiewicz Method

3.3. Quadratic Hyperbolic Method [3]
As seen in Eq.2 where to get constants a, b and c can be done using the help of the MatLab program.

3.4. Luciano Decourt Method [12]
Luciano Decourt Method is valid for all soil types (cohesive & non-cohesive), the pure data input from N_{SPT} value.

$$Q_P = \alpha q_P \text{ x } A_P = \beta(\check{N}_P \text{ x } K) \text{ x } A_P \quad (3)$$

where :
q_P = stress on based foundation in t/m^2, A_P = Sectional area of the base foundation, α= base coefficient = 1 (driven pile), $\check{N}_P$ = The average of N_{SPT} around 4B above until 4B below the base of pile foundation (B = Pile diameter) = $\sum_{i=1}^{n} \frac{Ni}{n}$

K = coefficient of characteristic soil: 12 t/m^2, for clay; 20 t/m^2, for clayey silt; 25 t/m^2, for sandy silt; 40 t/m^2, for sand

$$Q_S = \beta \text{ x } q_S \text{ x } A_S = \beta(\check{N}_S/3 + 1) \text{ x } A_S \quad (4)$$

where : q_S = friction stress in t/m^2 , $\check{N}_S$ = The average of N_{SPT} along foundations are embedded, with restrictions: $3 \leq N \leq 50$, A_P= Circumference x length foundation which sets , β = shaft coefficient =1

3.5. Meyerhoff- Bazara Method [14][13]
Bearing capacity by Meyerhoff Bazara follows the equation below :

$$P_{ult} = C_n.A_{base} + \sum Cl_i.As_i \text{ or } P_{ult} = 40\overline{N}.A_{base} + \sum_i^n \frac{N_i}{2or5} \times As_i \quad (5)$$

Where :
Cli = resistance of friction pile at segmen i = fsi; Asi = friction area of segment i = O_i x h_i; O_i = perimeter of pile; $\overline{N}$ = The average of N2, 4D along the end of the pile down s/d 8D along the end of the pile to the top

$$N_1 = 15 + \tfrac{1}{2}(N - 15) \text{ or } N_1 = 0{,}6N \quad (6)$$

$$N_2 = \frac{4N_1}{1 + 0{,}4\sigma'_o} \quad (7)$$

$$N_2 = \frac{4N_1}{3{,}25 + 0{,}1\sigma'_o} \quad (8)$$

3.6. Finite Element Method
The Finite Element Method (FEM) is a method of approximation the behavior of continua. In this numerical technique the system is discretecized into many meshes or element, then the equability and compatibility of each element, and whole system will be examined. In geotechnical engineering, PLAXIS is one of the most widely used finite element softwares. To model the pile used in this study into PLAXIS 2D, a work area with a width of 50 m and a depth of 60.5 m was used and the geometry was simulated using the axisymmetric model in which the pile was positioned along the axis of symmetry. With the aid of standard fixation boundary conditions, concrete piles with a total length vary

according to table 1 modeled. The pile has a diameter of 1.0 m; then it is defined as a column with a width of 0.5 m. Both the soil and the piles are modeled with 15-node elements.

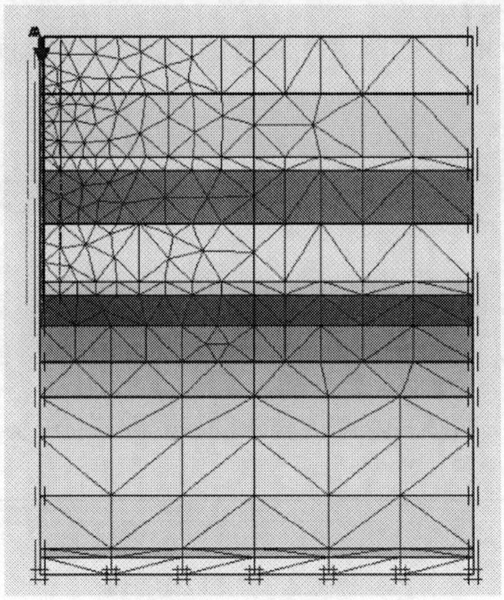

Figure 6. Modelling FEM

4. Result and Analysis

Based on the results of interpretation using the Chin [4], Mazurkewizc [5], Lastiasih [3], Finite Element methods and on the results of loading pile test results are obtained as follows:

Table 6 Comparison Ultimate Bearing Capacity of Pile with Vary Method

Pile Location	The pile load test	Loading Pile Test interpretation methods result			The pile bearing capacity using empirical methods		Finite Element
	field	Chin	Mazurkiewizc	Lastiasih	Meyerhoff Bazara	Luciano Decourt	Plaxis
BP 98	1200	2000	1550	1650,79	2177,09	985,3	1620
BP 145	700	1111	1080	846,17	2104,05	903,6	1029,5
BP 190 (F)	700	769	800	709,98	1233,49	889,6	1381,2
BP 334	800	1429	1050	1650,79	1998,52	1444,3	1185,8
BP 355 (F)	265,625	417	360	263,90	1234,85	843,8	540

Based on the table 5, it can be seen that the results of the pile test that failure are closer to the results of the interpretation method using the method developed by Lastiasih [3]. Whereas both empirical methods used in this study, the results are closer to the reality in the field is using the method of calculation of the bearing capacity of Luciano Decourt [12]. But when compared with finite elements, the results of finite elements are closer than empirical methods. Chin's interpretation method yields an estimate of the ultimate carrying capacity that is relatively greater compared to other methods. These results are almost the same as the results of research conducted by Zein[2]. The Chin [4] and

Mazurkiewiz [5] method is more graphic so that the results obtained may differ between planners from one another. Therefore in the case of choosing the right approach, it is better to use a specific formula so that whoever does it will get the same results. While for the empirical method Luciano Decourt [12] method was chosen because the carrying capacity produced was more pessimistic than the Bazara - Meyerhoff [13][14]method.

5. Conclusion

Based on research conducted, it can be concluded as follows

1. Analysis by interpreting the ultimate bearing capacity based on the results of the field test pile has been carried out in this study. The temporally results that obtained in this study namely data interpretation in the form of a relationship curve between load and settlement is most appropriate using the quadratic hyperbolic method by Lastiasih. As for the empirical formula, it is recommended to use the Luciano Decourt method.
2. Based on the results of interpretations that are still varied it is necessary to do a lot of data so that correction factors can be obtained for the methods of Luciano Decourt and Meyerhoff Bazara, so the results of empirical calculations are close to the results of field testing
3. To get the ultimate bearing capacity that is right in the field by using finite elements, axysymmetry modeling is carried out and the piles are assumed linearly elastic soils rather than plates. And smoothing the mesh around the pile and Mohr Coulomb modeling undrained conditions for the soil layer.

Reference

[1] Adam KRASIńSKI and Mateusz Wiszniewski 2017 Static Load Test on Instrumented Pile - Field Data and Numerical Simulations *Studia Geotechnica et Mechanica* **39** 3 17–25.

[2] Abdul Karim M Zein and Enass M Ayoub 2016 Evaluation of Measured and Interpreted Failure Loads of Bored Piles in Alluvial Soil Deposits *International Journal of GEOMATE* **10** 1 1636–43.

[3] Yudhi Lastiasih and Indra Djati Sidi 2014 Reliability of Estimation Pile Load Capacity Methods *Journal of Engineering and Technological Sciences* **46** B 1 1–16.

[4] Chin F K 1970 Estimation of the Ultimate Load of Piles from Tests Not Carried to Failure *Proceedings of Second Southeast Asian Conference on Soil Engineering* (Singapore City) 81-92.

[5] Mazurkiewicz B K 1972 Test Loading of Piles According to Polish Regulations *Preliminary Report No. 35 Commission on Pile Research* (Royal Swedish Academy of Engineering Services, Stockholm).

[6] Hansen J B 1963 Discusson on Hyperbolic Stress-Strain Response: Cohesive Soils *Soil Mech. Found. Div. ASCE, 89(SM4).*

[7] Van der Veen C 1953 The Bearing Capavity of Pile *Proceeding 3rd Internation Conference on Soil Mechanics and Foundation Engineering* (Zürich).

[8] Federation of Piling Specialists 2006 *Handbook on Pile Load Testing no 47.*

[9] Kaniraj S R and Samantha S 1996 Interpretation of Safe Load from Pile Loading Tes *Seminar on Piles-IGC-Delhi Chapter* (New Delhi) 97-102.

[10] Sanjay Gupta and Ravi Sundaram 1999 Pile Foundations for Flyover (ROB) in Jaipur - Geotechnical Aspects *Proceedings Indian Geotechnical Conference - IGC – 99* (Calcutta).

[11] EN 1997-1:2004 *Eurocode 7: Geotechnical Design - Part 1: General Rules* (Brussels, Belgium) 171.

[12] Decourt L 1999 Behavior of foundations under working load conditions *Proceedings of the 11th Pan-American Conference on Soil Mechanics and Geotechnical Engineering* (Foz-DoIguassu, Brazil) **4** 453-88.

[13] Bazaraa A R and Kurkur M M 1986 N-values used to predict settlements of piles in Egypt *In Proceedings of In Situ '86* (New York, American Society of Civil Engineers: New York, NY, USA) 462–74.

[14] Meyerhof G G 1976 Bearing Capacity and Settlement of Pile Foundation *J. Geotech. Eng. Div.* 102 197–228.

Circular lining behaviour due to earthquake load in MRT Jakarta underground tunnel area CP-106

F Muhammad[1]*, W A Prakoso[1]

[1] Department of Civil Engineering, University of Indonesia, Depok 16424

*Corresponding author's e-mail :Fathurrizal.muhamamd@ui.ac.id

Abstract. One solution to traffic congestion in Jakarta is to build a mass transportation system in the form of Mass Rapid Transit (MRT), especially the application of underground tunnel structures. This study examines the modeling of tunnel behavior on the effect of earthquake loads. This study calculates circular tunnel behavior in static and dynamic conditions due to earthquake effects. Static condition analysis using Muir Wood's theory and dynamic conditions using the theories of Wang (1993) and Panzien (2000) as well as supported by empiric and numeric calculations. The depth of the MRT tunnel in the CP-106 area is at a depth of 11 meters by diverting by clay with NSPT 3-20. The results show that in static condition has tunnel deformation that works at 8.06 mm in empiric and 15.12 mm in numeric. In the dynamic analysis with earthquake acceleration in 300 cm/s^2 produces oval deformation at 15.95 mm in empiric and 15.20 mm in numeric. The maximum deformation limit given is 20 mm with a maximum earthquake acceleration value of 400 cm/s^2. In conclusion, MRT Jakarta area CP-106 in static condition has a lower deformation than dynamic conditions, but both conditions fulfill the deformation tunnel requirements.
Keywords: Transportation, Static, Dynamic, Earthquake Load, Oval Deformation

1. Introduction

Large cities in developing countries are often faced with transportation problems. Jakarta, the capital of the State of Indonesia, is an example of a large city in a developing country that has problems in the transportation sector. One of the solutions implemented by the Jakarta Government to overcome transportation problems is to build a mass transportation system in the form of Mass Rapid Transit (MRT).

Most of the Jakarta MRT projects especially the tunnel structure are in northern Jakarta. Specifically in the central Jakarta CP-106 area that has already been completed (e.g., [1]). Broadly speaking, the soil base in Jakarta is a volcanic clay deposition that is approximately 2500 years old. The land surface in the northern part of Jakarta consists mostly of soft clay (e.g., [4]). Besides, large- scale earthquakes will also propagate earthquake vibrations so that they experience amplification or magnification of the shock of the structure above or below. According to the Indonesian earthquake threat map published by USGS (2012), the city of Jakarta is in the Peak Ground Acceleration zone between 0.245 g - 0.326 g (Marlihat and Mangape, 2009). Seeing the importance of planning in the construction of earthquake-resistant tunnels in Jakarta, especially in the initial lane of the CP-106 tunnel area, research on modeling the behavior of tunnels to the effects of earthquake loads and the impact on surrounding buildings is needed.

Content from this work may be used under the terms of the Creative Commons Attribution 3.0 licence. Any further distribution of this work must maintain attribution to the author(s) and the title of the work, journal citation and DOI.
Published under licence by IOP Publishing Ltd

IOP Conf. Series: Materials Science and Engineering **930** (2020) 012037 doi:10.1088/1757-899X/930/1/012037

Figure 1. Research Location at first phase (Jakarta MRT, 2015)

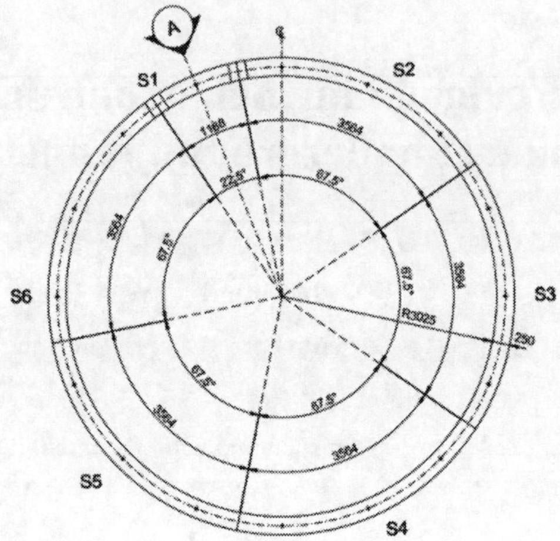

Figure 2. Typical precast and joint lining segments between segments (Jakarta MRT, 2015)

In the first phase, Jakarta MRT construction starts from Lebak Bulus towards the 15.7-kilometer Bunderan HI, including the construction of 13 stations. Figure 1 shows the Jakarta MRT line CP-106 area starts from Dukuh Atas as a research location with a tunnel diameter of 7.3 m which was dug below the surface of the ground using the Tunnel Boring Machine (TBM) using the shield tunneling method. The tunnel consisting of precast segments can be used in single or multiple tunnels to support soil or rock and water pressure. A tunnel construction requires 7 precast lining segments in a full circle. Figure 2 shows the dimensions of the precast segment and the connections between the precast lining segments.

2. Research Method

First of all, this research examines the static and dynamic conditions in the tunnel. This analysis uses several standards to determine the circular tunnel behavior in the Jakarta MRT tunnel and this standard reference has never been used by the Jakarta MRT project. In planning tunnel lining, there are things to consider, namely the behavior of circular tunnel lining. Muir Wood's (2002) theory provides a method for analyzing static conditions of tunnels by considering Axial Force, bending moment, and lining deformation. Static analysis of circular tunnel behavior can be described in the equation:

$$N = \frac{r_o(S_n + 2S_t)\cos 2\theta}{3} + p_w \cdot r_e + N_o \tag{1}$$

$$M = \frac{r_o \cdot r_e(2S_n + S_t)\cos 2\theta}{6} \tag{2}$$

$$U = \frac{r_e \cdot r_o^{\,3}(2S_n + S_t)\cos 2\theta}{18E_{lin} \cdot I} \tag{3}$$

The effects of earthquakes on underground structures can be grouped into two categories, namely ground vibration and soil failure. Ground vibration refers to the deformation of the soil produced by seismic waves that propagate through the earth's crust. The main factors affecting ground vibration are soil shape, dimensions and depth of the structure, the nature of the surrounding soil or rocks, the nature of the structure, and the weight of the soil under review (Zahrah, 1987). The tunnel behavior approach refers to the elastic beam model for deformation caused by the surrounding soil. Three types of seismic response deformation in underground structures are axial compression - extension, longitudinal moment, and ovaling. Axial deformation of the tunnel resulting from seismic waves produces motion parallel to the axis of the tunnel and causes compression and stress. The tunnel deformation mode due to seismic waves can be seen in figure 3 (Wang, 1993).

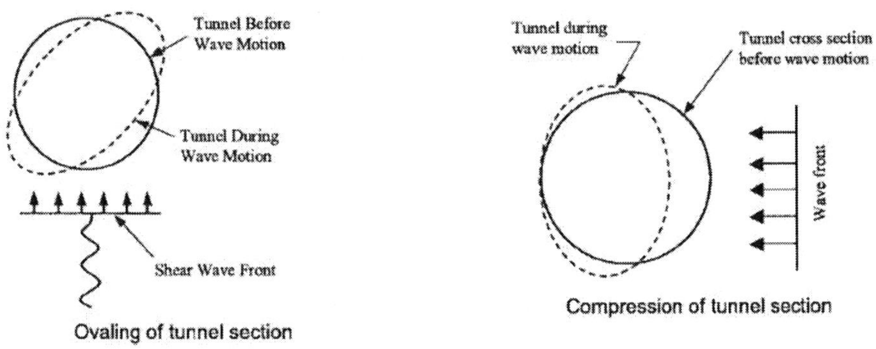

Figure 3. Mode of deformation of a tunnel due to seismic waves (Owen and Scholl, 1981)

The equation to determine the ground shear strain caused by dynamic loads is as follows:

$$\gamma_{max} = \frac{\tau_{max}}{G_m} = \frac{\left(\frac{PGA_M}{g}\right)\sigma_v.R_d}{\frac{E_m}{2(1+\upsilon_m)}} \quad (4)$$

with,
PGA_M = the acceleration of the peak land adjustments influences the site class
σ_v = earth vertical stress (kN/m2)
τ_{max} = earth shear stress (kN/m2)
Rd = stress reduction factor
Em = modulus of soil elasticity (kN/m2)
Gm = earth shear modulus (kN/m2)
υ_m = soil Poisson ratio

The forces and moments on a tunnel caused by seismic waves act on the axis of the tunnel. Maximum friction between the tunnel wall and the surrounding soil results in axial strain on the tunnel wall. The lining response coefficient can be obtained using a graph between the compressibility ratio and the flexibility ratio. Lining resistance response due to ground pressure with earthquake load is calculated by Equation (4) - (6).

$$C = \frac{E_m(1-\upsilon_l^2)r}{E_l.t(1+\upsilon_m)(1-2\upsilon_m)} \quad (5)$$

$$F = \frac{E_m(1-\upsilon_l^2)r^3}{6E_l I(1+\upsilon_m)} \quad (6)$$

$$K_l = \frac{12(1-\upsilon_m)}{2F+5-6\upsilon_m} \quad (7)$$

with,
C = compressibility ratio (response due to soil pressure)
F = ratio of replacement (response layer to withstand soil pressure) I = moment of inertia (m^4)
r = lining radius (m)
t = lining thickness (m)

K = lining response coefficient

For the bending moment equation, axial force and oval deformation due to earthquake loads on the cross-section of the tunnel as a function of maximum flexural strain are described as follows :

$$M_{max} = \frac{1}{6}K_1 \frac{E_m.r^2.\gamma_{max}}{1+\upsilon_m} \tag{8}$$

$$N_{max} = K_2.\tau_{max}.r = K_2 \frac{E_m.r.\gamma_{max}}{2(1+\upsilon_m)} \tag{9}$$

$$\Delta d = \frac{D.K_1.F.\gamma_{max}}{3} \tag{10}$$

The underground construction research site is located on the CP 106 section which is viewed from the Upper Dukuh pedestrian bridge (STA 13k + 000) to the Upper Dukuh Station (STA 13k + 916). In this study, primary and secondary data are used as input data in the modeling of circular lining tunnel behavior analysis in the Jakarta MRT tunnel construction project. Primary data in the form of results of land investigations that have been carried out previously as well as the results of surveys and documentation. Secondary data in the form of soil layers interpretation and tunnel design.

3. MRT Jakarta Tunneling Result

In this study, the soil layer in the Jakarta MRT tunnel from Upper Dukuh Bridge to Upper Dukuh Station were taken in one section is BR 21. Soil conditions at the construction site are dominated by soft clay at a depth of less than 10 m from the ground surface and silt layer to sand with medium density at a depth of 15 to 20 m.

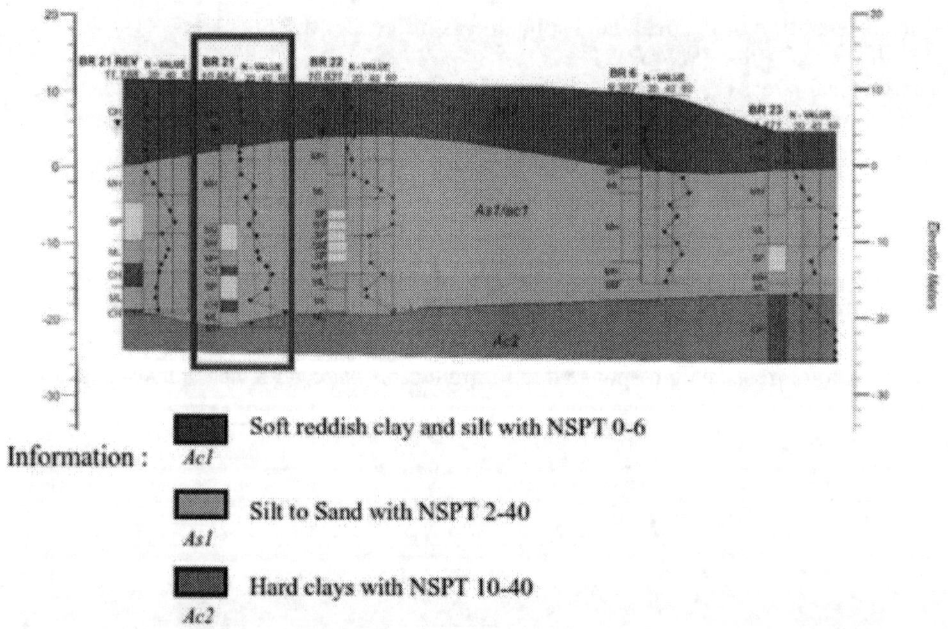

Figure 4. Interpretation of the Jakarta MRT tunnel layer CP 106 section

For numerical calculations using Plaxis-2D, earthquake loads are given in the time acceleration calculation format. In this study, variations in earthquake load applied at 300 cm/s^2 at Southern Sumatra, Indonesia obtained from USGS. Soil movements that occur in the subsoil or solid layer, its characteristics are regulated using the maximum soil acceleration parameters. The maximum value of ground motion for dynamic analysis is more effective using the initial earthquake record (Nuttli, 1979). In table 1 the parameters of the earthquake data are described in the analysis and the acceleration of Southern Sumatra can be seen in figure 5.

Table 1. Details of earthquake load parameters

Acceleration (cm/s^2)	Earthquake Source	Magnitude	Peak time (s)
300	Southern Sumatra, Indonesia, 2007	8.4	36.80

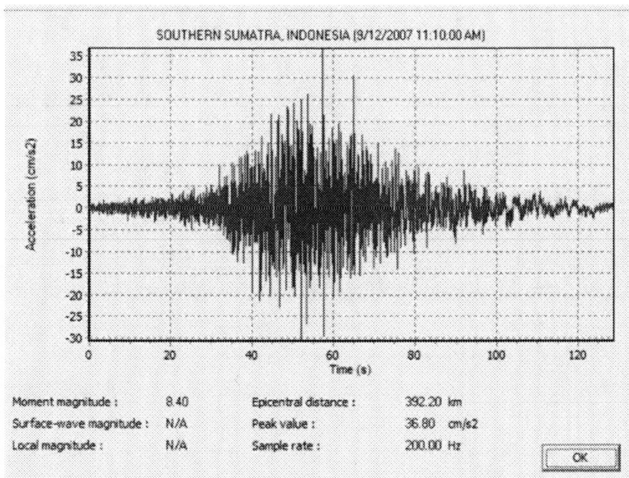

Figure 5. Southern Sumatra earthquake acceleration recording (USGS, 2012)

3.1. Calculation Phase

The analysis phase carried out in this study was 3 stages. In general, the analysis phase can be seen in figure 6. For more details, the analysis phase will be described as follows.

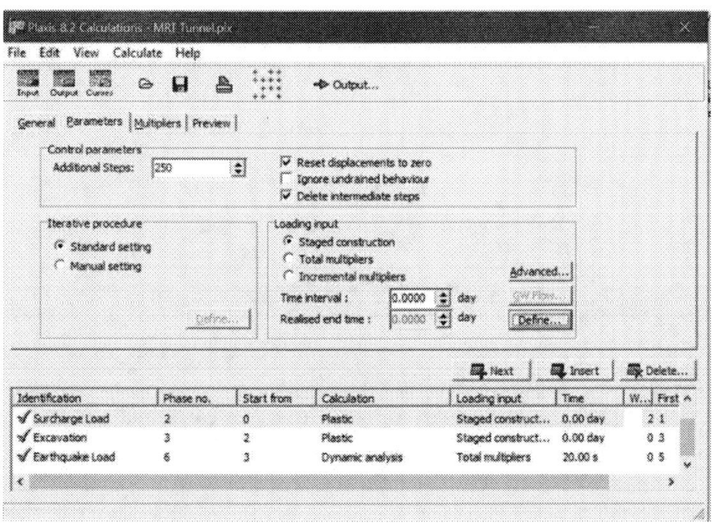

Figure 6. Calculation Stage Settings

3.1.1. First Phase (Surcharge Load)

The first phase applies static loads to the ground before tunnel excavation is carried out. This type of calculation in this phase uses plastic calculations by loading the input construction stage. The static load application is set from the sheet tab-define parameter. The structural model in the first phase calculation can be seen in figure 7.

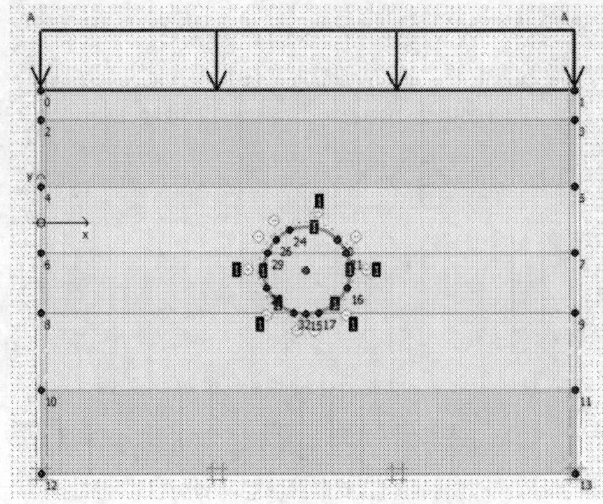

Figure 7. Structure model in the first phase calculation

3.1.2. *Second Phase (Tunnel Excavation)*

Behavior applied to the model in the second phase is by activating cluster tunnels and soil excavation. This type of plastic analysis calculation by loading input stage construction is still used in the second phase. The structural model in the second phase calculation can be seen in figure 8.

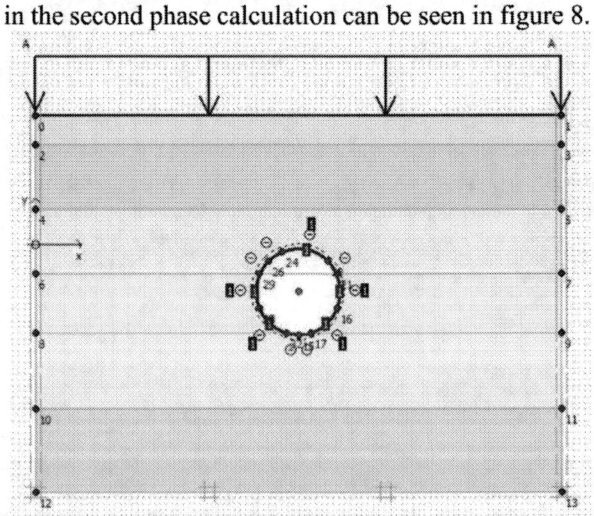

Figure 8. Structure model in the second phase calculation

3.1.3. *Third Phase (Application of Earthquake Load)*

In the last phase or third phase, dynamic load in the form of an earthquake load is given to the tunnel model. Setting the type of dynamic analysis calculation, by loading the input total multipliers. On the parameters tab, the settings are performed on an iteration procedure using a manual system with dynamic sub-steps of 3. Earthquake load is entered by opening the multiplier -ΣMdispl tab sheet. The earthquake load format used uses the SMC (Strong Motion CD) data format from USGS. The data chosen in this analysis is the acceleration type of data. The last place is to set the dynamic time of earthquake load. For more details, can be seen in figure 9.

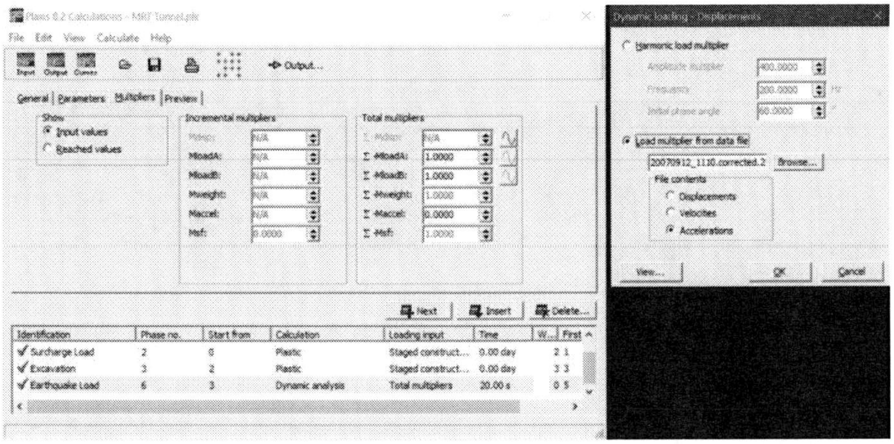

Figure 9. Application of earthquake loads in the spent phase calculation

3.2. Static Load Analysis

The final result of static load analysis based on empirical analysis refers to Muir Wood's theory in the Land Transport Authority (2002) and numerical analysis using Plaxis-2D. Overall lining strength based on the interaction diagram is still within safe limits, with a deformation lining tunnel below 20 mm (Mott Mc Donald, 2013). For an overall comparison of the deformation analysis, axial forces and bending moments can be seen in figure 10 in the diagram of internal forces.

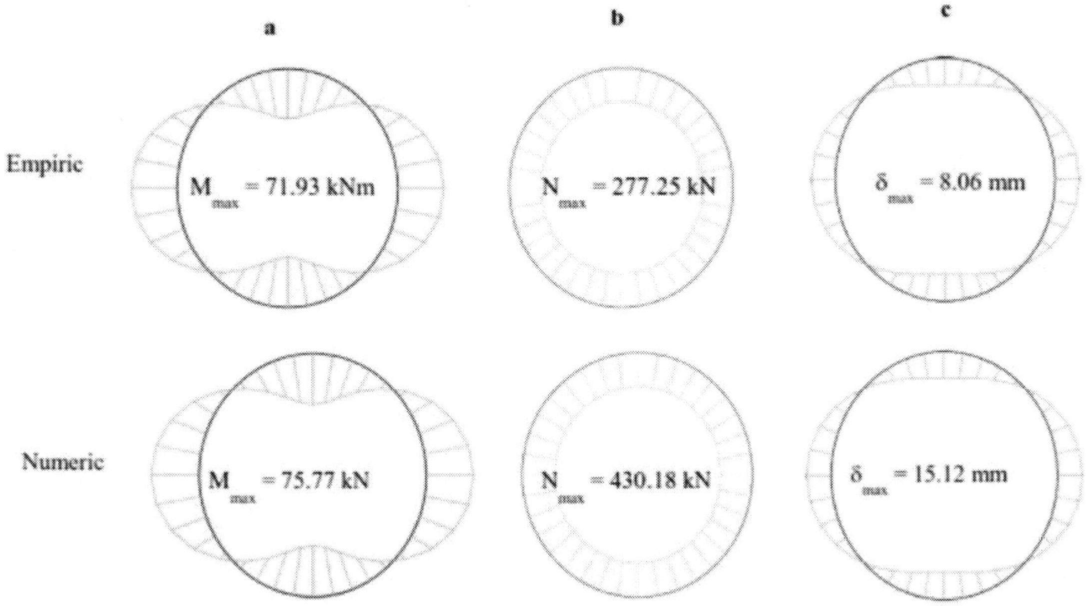

Figure 10. Comparison of the behavior of circular tunneling static load analysis tunnel. (a) Bending Moment, (b) Axial Force, (c) Deformation.

3.3. Dynamic Load Analysis

For dynamic load analysis, the calculation of the empirical method uses the methods of Wang (1993) and Panzien (2000) as well as numeric using Plaxis-2D. This method is valid and has a high level of accuracy if the results are the same. Circular tunnel behavior in overall dynamic load analysis can be seen in figure 11.

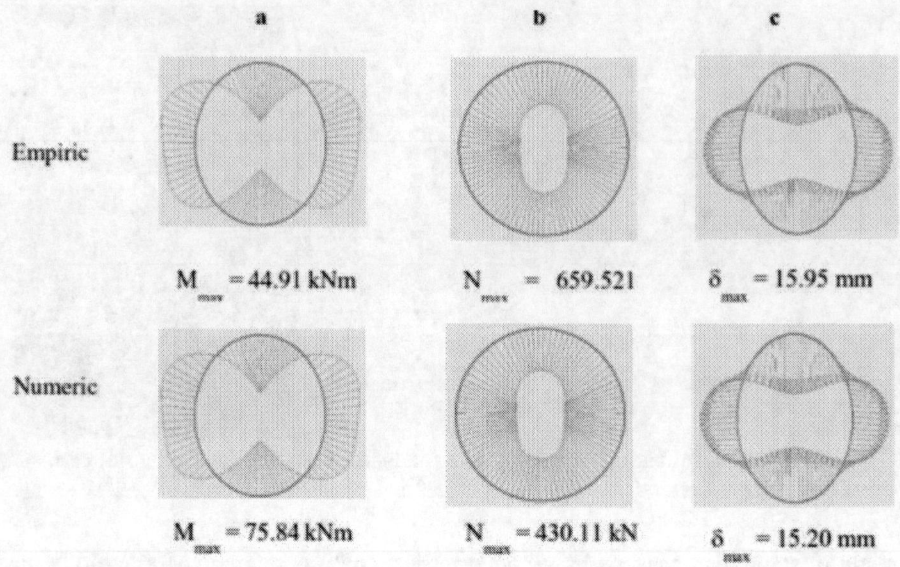

Figure 11. Comparison of the behavior of circular tunneling dynamic load analysis. (a) Bending Moment, (b) Axial Force, (c) Deformation.

The largest oval deformation is based on the empirical method of 15.95 mm and based on the numerical method of 15.20 mm. From the deformation graph, dynamic analysis produces the largest deformation value due to the earthquake acceleration of 300 cm/s^2, but the deformation of the lining tunnel is still within the given limit.

4. Conclusion

The lining ability which is assessed based on axial force, bending moment, and oval deformation due to static and dynamic loads is carried out in section BR 21 in the CP-106 area. Deformation values in the analysis of static conditions have smaller results than dynamic conditions due to dynamic conditions considering the peak ground acceleration (PGA) value of 300 cm/s^2 and do not consider friction between the lining and the ground. So this can make the tunnel deformation greater than the static condition which only considers overburden loads and surcharge loads. But from these two conditions, tunnel deformation is still within the safe limits because it has a deformation value below 20 mm.

References

[1] U.S. Geological Survey 2012 Earthquake Hazard Program *USGS Website.*
[2] Marlihat B dan Mangape I 2009 Analisis Hazard Gempa dan Usulan Ground Motion pada Batuan Dasar untuk Kota Jakarta *Journal Teknik Sipil, ISSN 0853-2982* **16** No 3 (Institut Teknologi Bandung: Bandung).
[3] Associated General Construction of America 2007 *Petition497, AGCA* (California, USA).
[4] Land Transport Authority 2002 *Civil Design Criteria for Road and Rail Transit System.*
[5] Federal Highway Administration 2009 *Technical Manual for Design and Construction of Road Tunnels* (United States Department of Transportation).
[6] Zahrah T F 1987 A Seismic Design of Underground Structures *Tunneling and Underground Space Technology* **2** No 2 165-97.
[7] Wang J N 1993 Seismic Design of Tunnels *Quade and Douglas Inc* (New York).
[8] Owen G N dan Scholl R E *1981 Earthquake Engineering of Large Underground Structures: Prepared for the Federal Highway Administration and the National Science Foundation* FHWA/RD-80/195 January.
[9] Hashash 2001 Seismic Design and Analysis of Underground Structures *Tunneling and Underground Space Technology* **16** 247-93.

IOP Conf. Series: Materials Science and Engineering **930** (2020) 012037 doi:10.1088/1757-899X/930/1/012037

[10] Kuesel 1969 Earthquake Design Criteria for Subways *Journal of Structural Division* ASCE ST6 1213-31.
[11] MRT Jakarta 2015 *Basic Engineering Design* (Jakarta: Jakarta Metro Engineering Consultant).
[12] Nuttall O W 1979 *The Relation of Sustained Maximum Ground Acceleration and Velocity to Earthquake Intensity and Magnitude* Miscellaneous Paper no S-73-1 Report 16 (US Army Corps of Engineers Waterways Experiment Station, Vicksburg, Mississippi).
[13] Anonim 2012 Tata Cara Perencanaan Ketahanan Gempa Untuk Struktur Bangunan dan Non Gedung *SNI 1726:2012* (Badan Standarisasi Nasional).
[14] MacDonald M 2013 *Geotechnical Interpretative Report* (Jakarta: SOWJ-JV).

The stabilization of soft soil using admixture of palm oil boiler ash and *MATOS*

Ma'ruf M. A.[1*], Rusliansyah[1], Ritonga F. I. D.[1], Azizah B.[1]

[1] Department of Civil Engineering, Lambung Mangkurat University, Banjarmasin, Indonesia.

[*]Corresponding author's e-mail: afief.maruf@ulm.ac.id

Abstract. Banjarmasin has an alluvial area with soft soil thickness of 28-42.4m. Therefore, the land is not structurally feasible for construction due to the low California Bearing Ratio (CBR). One of the solutions used to overcome this problem is soil stabilization. This study aims to examine the effect of adding palm oil boiler ash and *MATOS* in accordance with CBR value to stabilize soil conditions. The laboratory CBR method was used to determine the original and stabilized conditions of the soil. The initial and stabilized soils used an admixture of 2% *MATOS* as an additive and palm oil boiler ash with percentage variations of 6%, 9%, and 12% of the dry weight soil. This was further kept for 7 days curing before being tested. The result showed that the highest CBR value is obtained at the variation of 6% palm oil boiler ash + 2% MATOS with an increase in CBR value of 7.542% from the original soil condition.

1. Introduction

Soft soils are low bearing capacity and high compressibility soils such as those commonly found in Banjarmasin, an alluvial area with a thickness of 28-42.4m [1]. The area is not structurally feasible for construction due to the low California Bearing Ratio (CBR) value, low permeability value, and high water saturation.

Handling soft soil in Banjarmasin is currently carried out by physical treatment, such as adding the gleam pile reinforcement, thereby, increasing the total shear strength. However, it is also important to determine the soils micro-characteristics, chemical components, and shear strength during construction. Current technological developments show that soil stabilization is not only done physically but also chemically. The success of chemical stabilization is seen from the existence of accurate initial information about the micro and chemical soil that will be stabilized. Therefore, chemical and physical characteristics on a micro scale are also important to know, especially for soft soil in Banjarmasin.

One of stabilization form is by using palm oil boiler ash and *MATOS*. Palm oil boiler ash, a waste product produced from processing palm oil, is used as a filler due to the pozzolan nature described as Siliceous, which functions as a binder of soil particles. *MATOS*, which is one of the most recent stabilization materials, is further added as an additive from industrial products. Its chemical characteristics are uniform, thereby, making it easier to use on a large scale because it is an industrial product. These chemicals function to solidify and stabilize soil in the form of a fine powder consisting of metals and inorganic mineral composition such as flour and salt. Previous studies showed that the addition of *MATOS* as an additive improves the physical and mechanical properties of clay soil.

Content from this work may be used under the terms of the Creative Commons Attribution 3.0 licence. Any further distribution of this work must maintain attribution to the author(s) and the title of the work, journal citation and DOI.
Published under licence by IOP Publishing Ltd

Therefore this study examines the effect of adding *MATOS* and palm oil boiler ash as an additive and filler for changes in the characteristics of soft alluvial soil in the Banjarmasin area.

2. Literature Review

Boiler ash is one of the waste materials widely available from the palm oil processing industry. Previous studies have been carried out to examine boiler ash's characterizations from the waste of the palm oil industry [2]. Furthermore, the chemical and physical characterizations were analyzed using X-Ray Fluorescence (XRF), X-Ray Diffraction (XRD), particle size distribution, Fourier Transform Infrared Spectroscopy (FTIR) and Scanning Electron Microscope (SEM). From the XRF analysis, the major component of boiler ash such as silica oxide (SiO_2), calcium oxide (CaO), and ferum oxide (Fe_2O_3) are determined. However, in order to produce geopolymer, the raw material need to contain 3.7% Al_2O_3, 40.60% SiO_2 and CaO. The median particle size of boiler ash was compared with fly ashes, and its morphology showed irregular particle size and shape which looked similar to calcined kaolin, although no plate-like structure was observed.

MATOS® is an additive that functions to solidify the physical-chemical stabilizer of soils [3]. It is a fine powder material consisting of inorganic mineral composition, which tends to freeze and stabilize the soil's physical-chemical processes. The characteristics include no odor, a pH of 8.37, a volume unit weight of 2.35gr/cm^3, and water solubility of 1:3 [4]. *MATOS* needs to be dissolved in water at a 10% molarity rate before usage. The various components negatively influence the function of topsoil, thereby reducing its level. Furthermore, the calcium cation (Ca++) in the added particles sticks directly to the soil surface. Moreover, it eliminates the inhibitory effect of ionic bonds make soil particles more easily charged with negative ions (anions). The Ca++ cation binds easily with the soil particles and helps to supply more substitute ions forming aluminum silica acid compounds. Therefore, this tends to form 3 dimensional honeycomb structures between the soil particles. When exposed to groundwater or rainfall, the added particles containing sulfur (SO_3) with soil in the absence of *MATOS* produce sulfuric acids, which causes cracks.

Conversely, in the presence of *MATOS*, the added particles bond with soil particles and dries due to dehydration reaction. In addition, crystals form, which appear between the cement mixtures that bind the soil particles. These crystals are needle-like structures multiplied and enlarged to micron cavities that absorb water (porosity), thereby preventing cracking. *MATOS* works to improve the quality of road construction and also reduces the cost requirements. It reacts with soil and hydrated cement to produce complex binding particles with a strong framework and stable layer. Also, its use reduces the impact hazardous impact of dust on the environment and makes the surface waterproof in all weather.

MATOS's application procedure is very simple and starts by dredging and stirring with the soil until it becomes a homogeneous mixture. This process crushes large grains into smaller pieces and reduces the moisture content of the soil. *MATOS* is further added and stirred again to ensure a thorough mixture. Furthermore, the needed quantity of water is added to the mixture to achieve Optimum Moisture Content (OMC), making the chemical process work. Compaction is an important aspect that has to be thoroughly carried out with the appropriate equipment to achieve maximum compaction. The soils' density increases with MATOS's addition, which shows the results before the application's completion. The reaction of *MATOS* with soil takes place within the first 24 hours, and in the first 7 days, most chemical reactions have already been completed. Furthermore, the chemical reactions tend to continue for 365 days due to the nature of *MATOS* and based on hydration.

In terms of soil stabilization, [5] conducted a study on increasing the free compressive strength (In terms of soil stabilization, a study was carried out on increasing the Unconfined Compressive Strength (UCS) of clay soil after adding *MATOS* to stabilize sand and cement [5]. The results showed that the UCS of clay before sand and cement stabilization was 3.51kg/cm^2 at 0% *MATOS,* while the stabilization of a mixture 28% sand and 6% cement was 9.10kg/cm2. The addition of 2%, 4%, 6%, 8% and 10% *MATOS,* corresponded to the UCS values, which increased to 10.92kg/cm^2, 11.18kg/cm^2, 11.44kg/cm^2, 13.78kg/cm^2, and 14.56kg/cm^2 respectively. The percentage of *MATOS* used as an additive for this study was 2%.

3. Methodology

3.1. Sampling and Preparing Test Materials
The soft soil was sampled at Jl. Gubernur Syarkawi, Gambut Village, Banjar District, South Kalimantan. They were taken in disturbed and undisturbed conditions to obtain uniformity. However, this study employed the use of disturbed samples, as shown in figure 1. The test sample were prepared in the field and molded until it was ready to be tested in the Soil Mechanics Laboratory, Lambung Mangkurat University, at Banjarbaru.

The palm oil boiler ash used in this study was obtained from PT. Perkebunan Nusantara XIII Pelaihari Palm Oil Mill, Jl. Soekarno Hatta No. 1A Pelaihari, Tanah Laut District, South Kalimantan. Meanwhile, *MATOS* used was purchased from PT. Watukali Capita Ciptama as the manufacturer and distributor of the *MATOS* product brand, as shown in figure 2.

3.2. Laboratory Testing and Analysis
The laboratory test study was carried out in the Soil Mechanics Laboratory of Civil Engineering Department, Lambung Mangkurat University, Banjarbaru, South Kalimantan. The CBR test was used to determine soils in its original or initial conditions and those that have been stabilized or with additives. This test aimed to determine the CBR value of the initial and stabilized soil conditions. Then, the laboratory sequence was as follows:

- The soil samples were divided into 4 container boxes, each weighing 20 kg, with one initial and three stabilized conditions. The soil was further checked for its dry weight unit.
- The initial soil conditions were further tested for CBR to obtain the number of blows that produced the maximum CBR graph.
- The next step was preparing the stabilized soil condition samples by mixing the three container box left with palm oil boiler ash and *MATOS*.
- Palm oil boiler ash was initially mixed with soil samples that were crushing into smaller pieces by a percentage of 6%, 9%, and 12% of the dry weight soil in its initial condition. Each percentage was placed in a different box.
- *MATOS* was dissolved with water to obtain the optimum moisture content (OMC) of the initial soil sample condition. It was further mixed in the composition from the previous step using 2% of the dry weight soil.
- The soil sample mixed with *MATOS* further cured for 7 days, and the CBR testing was carried out with the number of blows according to the maximum graph of its initial condition.

After the laboratory test was completed, it was analyzed by comparing the CBR test result of soft soils before and after mixing the admixture of palm oil boiler ash and MATOS. This final step was carried out to determine the change of behavior in CBR value before and after the stabilization process and cured for 7 days, in order to examine the effects of adding admixtures.

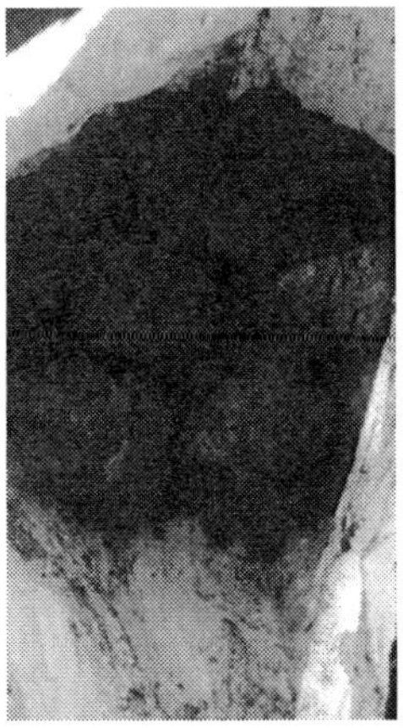

Figure 1. Soft soil samples.

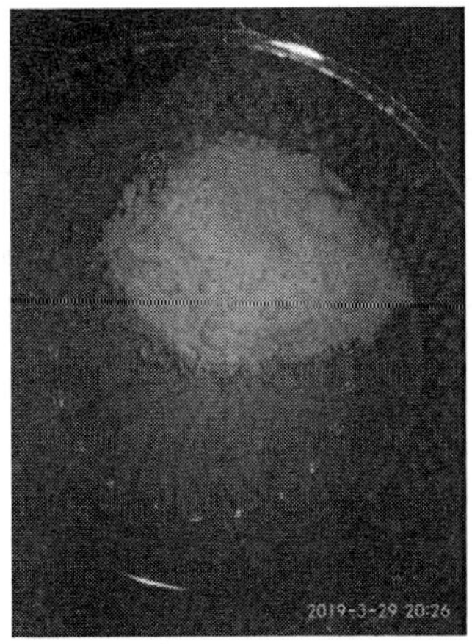

Figure 2. *MATOS* soil stabilizer.

4. Results And Discussion

4.1. Initial Sample Condition

The first step was to determine the optimum moisture content (OMC) of the initial soil condition with a laboratory test used to obtain the water value and open moisture contents at 70.35% and 28.76%, respectively. This means that the initial soil condition was too moist for *MATOS* to react because the requirement was dissolved in the OMC condition. However, irrespective of this condition, this study also examined the effect of high water content.

The initial and stabilized samples were examined using CBR Test, as shown in figure 3. Furthermore tables 1 and 2 show that the values for each variation of blows are 10, 25, and 56, with a graph used to determine the maximum stabilized conditions. The graph shows that the maximum conditions obtain from 56 blows, so the stabilized conditions will use 56 blows.

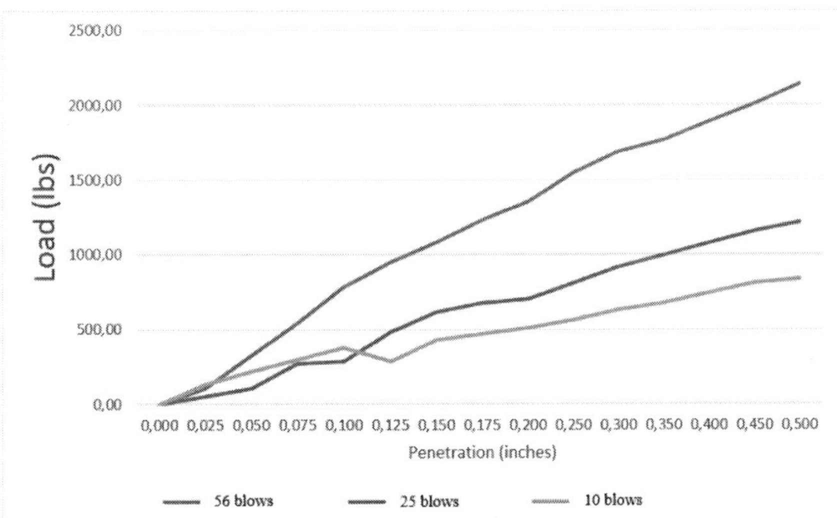

Figure 3. Result of CBR test in initial condition.

Table 1. Result of CBR test in initial condition.

Elapsed Time (Minute)	Penetration		56 Blows		25 Blows		10 Blows	
	(0.01 Mm)	(Inch)	Proving Ring 10^{-4} Inch	Load Lbs	Proving Ring 10^{-4} Inch	Load Lbs	Proving Ring 10^{-4} Inch	Load Lbs
0.00	0.00	0.000	0.00	0.00	0.00	0.00	0.00	0.00
0.50	63.50	0.025	4.00	107.70	2.00	53.85	5.00	134.63
1.00	127.00	0.050	12.00	323.11	4.00	107.70	8.00	215.40
1.50	190.50	0.075	20.20	543.90	10.00	269.26	11.00	296.18
2.00	254.00	0.100	29.10	783.53	10.50	282.72	14.00	376.96
2.50	317.50	0.125	35.20	947.78	18.00	484.66	10.50	282.72
3.00	381.00	0.150	40.20	1082.41	23.00	619.29	16.00	430.81
3.50	444.50	0.175	45.50	1225.11	25.00	673.14	17.50	471.20
4.00	508.00	0.200	50.30	1354.35	26.00	700.06	19.00	511.58
5.00	635.00	0.250	57.50	1548.22	30.00	807.77	21.00	565.44
6.00	762.00	0.300	62.50	1682.84	34.00	915.47	23.50	632.75
7.00	889.00	0.350	65.60	1766.31	37.00	996.24	25.00	673.14
8.00	1016.0	0.400	70.00	1884.79	40.00	1077.02	27.50	740.45
9.00	1143.0	0.450	74.40	2003.26	43.00	1157.80	30.00	807.77
10.00	1270.0	0.500	79.50	2140.58	45.00	1211.65	31.00	834.69

Table 2. Result of CBR Value in initial condition.

Blows	Load at Penetration (Lbs)		CBR Value (%)	
	0.1"	0.2"	0.1"	0.2"
10	377	512	12.57	11.37
25	283	700	9.42	15.56
56	784	1354	26.12	30.10

4.2. Stabilized Sample Condition

The soil sample that mixed with palm oil boiler ash and *MATOS* was cured for 7 days. This was followed by examining the CBR Test with 56 blows each on 3 variations of sample as shown in figure 4 and Table 3. The result shows that the combination of 6% palm oil boiler ash and 2% MATOS provided the best result after being tested with a 7 days curing period.

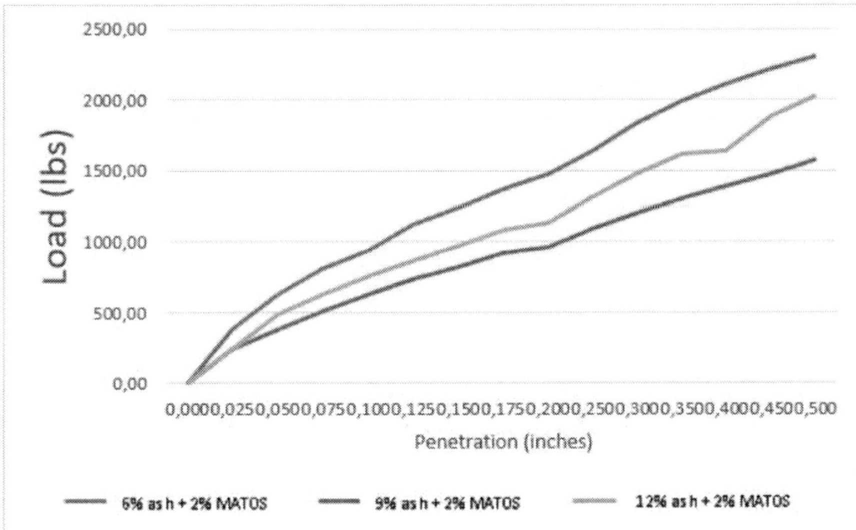

Figure 4. Result of CBR test in stabilized condition.

Table 3. Result of CBR test in stabilized condition.

Elapsed Time (Minute)	Penetration (0.01 mm)	Penetration (Inch)	6% + 2% Proving Ring 10^{-4} Inch	6% + 2% Load Lbs	9% + 2% Proving Ring 0,0002 Cm	9% + 2% Load Lbs	12% + 2% Proving Ring 0,0002 Cm	12% + 2% Load Lbs
0.00	0.00	0.000	0.00	0.00	0.00	0.00	0.00	0.00
0.50	63.50	0.025	14.00	376.96	9.00	242.33	9.00	242.33
1.00	127.00	0.050	23.50	632.75	14.00	376.96	18.00	484.66
1.50	190.50	0.075	30.00	807.77	19.00	511.58	23.50	632.75
2.00	254.00	0.100	35.00	942.39	23.50	632.75	28.00	753.91
2.50	317.50	0.125	41.50	1117.41	27.50	740.45	32.00	861.62
3.00	381.00	0.150	46.00	1238.57	30.50	821.23	36.00	969.32
3.50	444.50	0.175	51.00	1373.20	34.00	915.47	40.00	1077.02
4.00	508.00	0.200	55.00	1480.90	35.50	955.86	42.00	1130.87
5.00	635.00	0.250	61.00	1642.46	40.50	1090.48	49.00	1319.35
6.00	762.00	0.300	68.00	1830.93	44.50	1198.18	55.00	1480.90
7.00	889.00	0.350	74.00	1992.49	48.50	1305.89	60.00	1615.53
8.00	1016.0	0.400	78.50	2113.65	51.50	1386.66	61.00	1642.46
9.00	1143.0	0.450	82.50	2221.35	55.00	1480.90	70.00	1884.79
10.00	1270.0	0.500	85.50	2302.13	58.50	1575.14	75.00	2019.41

Table 4. Result of CBR Value in stabilized condition.

Admixture	Load at Penetration (Lbs)		CBR Value (%)	
	0.1"	0.2"	0.1"	0.2"
6% + 2%	843	1456	28.09	32.37
9% + 2%	577	996	19.22	22.13
12% + 2%	739	1277	24.64	28.40

4.3. Comparison and Analysis

The result obtained from the CBR Test was compared with the graph results of the initial and the stabilized condition, as shown in figure 5 and table 5. The comparison showed that the combination of 6% palm oil boiler ash and 2% *MATOS* graph were above the initial CBR graph, while the other admixtures were below.

This unique trend was mostly caused by the initial sample, which is too moist, while the requirement for *MATOS* to work efficiently was by OMC conditions. The admixture combination of 9% and 12% palm oil boiler ash showed that the moisture used for the reaction was too high, thereby reducing the density below the OMC condition. Conversely, the 6% palm oil boiler ash combination makes the moisture content closer to the OMC condition, thereby obtaining the best CBR value, which further increased to 7.542% from the original soil condition.

From this study, it is concluded that of the combination of palm oil boiler ash and *MATOS* as an admixture for the soft soil, changes the CBR value, which is also dependent on the initial moisture content. A moisture too far from the OMC condition tends to vary or fall below the initial condition.

This study does not discuss the effect of the chemical solution on groundwater as no test was carried out due to the limitations of time and funds. Therefore, further study is recommended to consider using *MATOS*, with a test carried out to investigate its effect on groundwater.

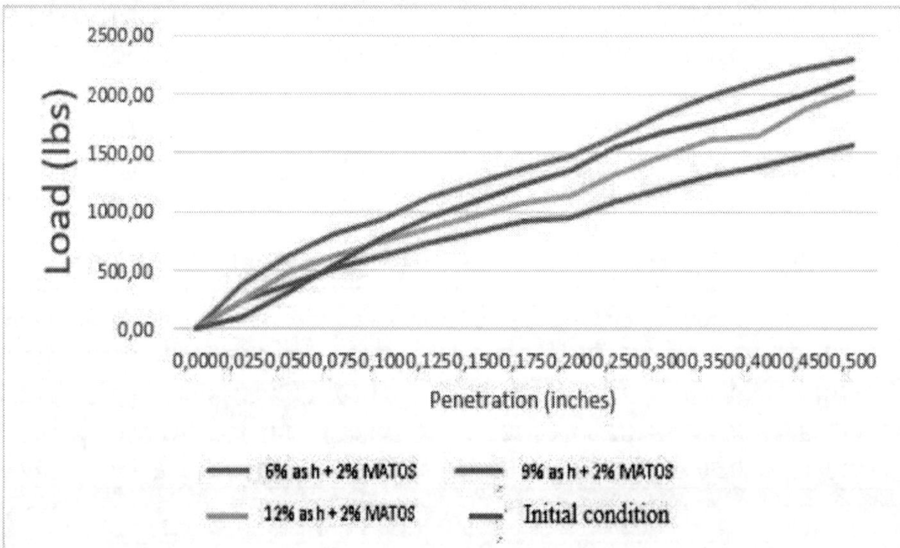

Figure 5. Comparison between initial and stabilized condition.

Table 5. Comparison of CBR Value in initial and stabilized condition.

Admixture	Load at Penetration (Lbs)		CBR Value (%)	
	0.1"	0.2"	0.1"	0.2"
Initial	784	1354	26.12	30.10
6% + 2%	843	1456	28.09	32.37
9% + 2%	577	996	19.22	22.13
12% + 2%	739	1277	24.64	28.40

5. Conclusion

In conclusion, the highest CBR value was obtained at the variation of 6% palm oil boiler ash and 2% MATOS, with an additional increase to 7.542% from the initial soil condition. Therefore, adding the combination of palm oil boiler ash and *MATOS* as an admixture for the soft soil changes CBR value behaviour. In addition, the study shows that when the soil's initial moisture is too far from the OMC condition, it falls below the initial condition.

Acknowledgement

The authors are grateful to the University of Lambung Mangkurat for supporting this study by providing a grant, through the Institute for Research and Community Service in the scheme of the Program Dosen Wajib Meneliti Research Scholarship.

References

[1] Ma'ruf M A, Rusliansyah, Fitriati U and Rachman A A 2019 Digital Mapping of Hard Soil Depth in Banjarmasin City *International Journal of Engineering and Technology* **11** 5 316.

[2] Yahya Z, Al M, Abdullah M M, Al Bakri, Kamarudin H, Nizar K, Sandu A V, Vizureanu P and Razak R 2013 Chemical and Physical Characterization of Boiler Ash from Palm Oil Industry Waste for Geopolymer Composite *Revista de Chimie -Bucharest- Original Edition-* **64** 1408.

[3] PT. Watukali Capita Ciptama 2016 *MATOS® Book* (Yogyakarta: PT. Watukali Capita Ciptama)

[4] Universitas Gajah Mada 2010 *Laporan Hasil Uji Laboratorium untuk MATOS* (Yogyakarta: Universitas Gajah Mada)

[5] Muda A 2016 Analisis Kuat Tekan Bebas Pada Penambahan Matos Terhadap Stabilisasi Tanah Lempung Dengan Semen *Media Ilmiah Teknik Sipil* **5**(1) 30-9.

IOP Conf. Series: Materials Science and Engineering **930** (2020) 012039 doi:10.1088/1757-899X/930/1/012039

The effect of skirt footings for road settlement on peat soil

D A W Gautama[1*], B S Supanji[2] and W Rahayu[2]

[1]Postgraduate student, Universitas Indonesia, Civil Enginering Department, 16424 Depok, Indonesia

[2]Lecturer, Universitas Indonesia, Civil Enginering Department, 16424 Depok, Indonesia

*Corresponding author's e-mail: danarariangga@gmail.com

Abstract. The construction of peat soils is a big challenge in the construction world coupled with the very special nature of peat soil. Road construction on peat soil is a major obstacle because the deformation that occurs on the road is so large. Skirt footings are one of the shallow foundations commonly used for building offshore oil refineries. With the principle of making soil that has a low bearing capacity into a single unit to be able to withstand the burden on it. Skirt footings are placed under the embankment with a parallel position stretching to the width of the road plan. Among the skirt, footings are connected with steel straps so that during the loading process the skirt footings decrease can be controlled. The purpose of this research is to make skirt footings a shallow foundation under a road built on peat soil. Tests were also carried out on two models of shallow foundation to determine the effect produced by skirt footings on the soil. The other two types of shallow foundation models are un-skirted footings and mini piles. Comparison of the results of tests between skirt footings with un-skirted footings, it can be concluded that skirts under circular footings act as a buffer against the effective stress of the soil. While from the comparison of test results between skirt footings with mini piles, it can be concluded that circular footings play a role in reducing the effect of the differential settlement on the soil. The results of this study are that skirt footing can withstand differential settlement patterns and able to increase road stability.

1. Introduction

The Republic of Indonesia is a country located on the equator with a tropical climate, consisting of 17,504 islands with a total area of 5,455,675 km^2. Peat in Indonesia is spread over 8 provinces with a total area of 14.9 million hectares of peat soil. Indonesia contributes 8% of total global peat soil. Borneo island has a land area of 4.8 million hectares of peat [8]. The construction on the territory of the peat becomes quite a difficult challenge due to the nature of peat which has a high-value permeability along with major consolidation that took place very quickly. Several solutions can be applied in the handling of existing peat soil such as excavation, layer change, transferring, or pre-compression on peat soils. The use of peat soil improvement methods has grown rapidly over the past decade, starting from the use of geotextile, geogrid, to changing the composition of peat soil composition by mixing certain materials aimed at reducing the water content in peat soil pores.

Content from this work may be used under the terms of the Creative Commons Attribution 3.0 licence. Any further distribution of this work must maintain attribution to the author(s) and the title of the work, journal citation and DOI.
Published under licence by IOP Publishing Ltd

Skirt Foundation or skirt footings are known to increase bearing capacity on shallow foundations on soils that have the low bearing capacity. Skirt footings are commonly used in offshore work areas. Circular skirt footings or bucket foundations or suction caissons are commonly used in the offshore oil and gas industry [2-3]. Skirts can increase the capacity of additional moments, which increases due to the suction produced by the skirt during loading. Skirt footings work system is to make the soil as a unit to channel the burden of the structure above it to the stronger ground and increase the ability to withstand overturning loads [4-6].

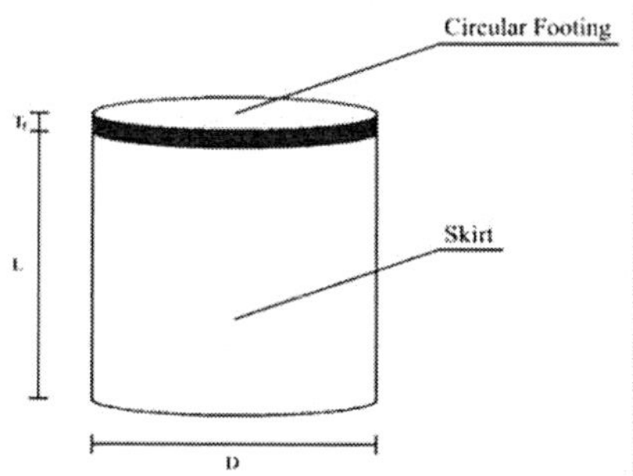

Figure 1. Circular skirt footings

Figure 2. The application circular skirt footings in the field [7]

Pre-construction methods on peat soil include work methods with replacing peat, replacing peat methods on-site (*in situ*). The pre-loading method is a method that aims to accelerate the process of primary consolidation in peat soils and also to create conditions for Over-Consolidation Ratio (OCR) on peat soils. The peat soil matrix will change under load which will reduce the permeability and compressibility with the resultant value of the shear strength increasing [1]. According to the construction and building guidelines by public work service regarding the Planning for Road Embankment Construction on Peat, the Pre-Loading Method has two main functions against decrease including:

1. Adding additional weight to minimize primary degradation during life service.
2. Adding additional weight to minimize secondary settlement during life service.

2. Methods

2.1. Material properties

Material properties come from the drill log soil data of the Bukit Rawi area, Central Borneo. The drill log digging to a depth of 18 meters, but the data used in PLAXIS only the majority of soil properties. These are sand until the depth of 2 meters, peat soil until the depth of 10 meters, and the deep sand until 14 meters. The peat soil in the drill log sampling location has fibrous peat criteria with the value of cohesion is 73.89 kN/m^2 and the angle of friction's value is 11.42°.

2.2. PLAXIS modeling

The research uses assistance from PLAXIS 2D software with Soft Soil Creep modeling. To find out the effect of skirt footings on the deformation that occurs, axisymmetric modeling is used in this study. Axi-symmetry modeling aims to take into account the structural forces that occur at the foundation. The calculation of axisymmetric modeling is based on the existing load and not based on the displacement of forces. There are 16 variations of skirt footings with variations in size (D) and length (L) skirt footings. The variation between length (L) and diameter (D) used are (L / D) 0.25; 0.33; 0.5; 0.5; 0.67; 0.75; 1; 1; 1; 1; 1.33; 1,5; 2; 2; 3; 4. The embankment load is changed to a static load by multiplying the height of the embankment by the density of the soil. To help stabilize it better, a meter soil deep down under the embankment is raised in OCR value so that the foundation can work optimally withstand the load. An embankment test is carried out to determine the behavior of the soil in bearing the load, the test using a plane-strain model with all the same soil parameters. The results obtained are the amount of soil deformation that occurs and the stability factor of the embankment.

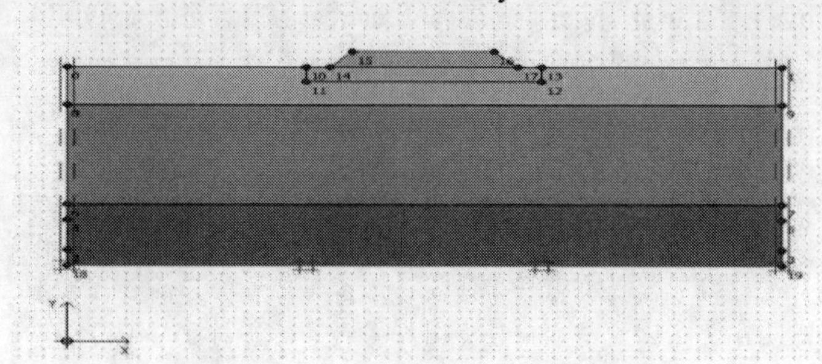

Figure 3. Fieldwork of embankment.

Table 1. The soil condition calculation.

U 5 days (mm)	U 365 days (mm)	SF
137	159	1,372

As a comparison to the function of the length of skirt footings (L) and skirt footings diameters (D), an experiment was conducted with the other type of shallow foundation. The other shallow foundations are un-skirted footings and mini pile models. To make sure that the result of all type shallow foundations can be compared properly, several adjustments are used such as, the model un-skirted footings using the diameter (D) equal to the skirt footings and mini pile models use length (L) same as the skirt footings.

1. Consolidation of 5 days, short-term consolidation.
2. Consolidation of 365 days, to see the behavior of embankment in a year.
3. Consolidation of minimum pore pressure.
4. Calculate the stability value of the embankment.

3. Results and discussion

3.1 Skirt footings

Skirt footings were tested with 4 variations in length (L) and 4 variations in diameter (D). It aims to see the effect of the length and diameter of the skirt footings on the deformation that occurs.

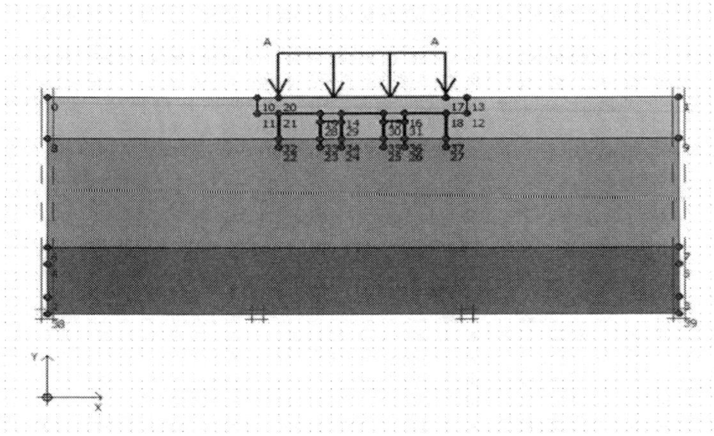

Figure 4. Skirt footings fieldwork.

Table 2. The result of skirt footings.

D (m)	L (m)	L/D	U 5 days (mm)	U 365 days (mm)	Average U 5 days (mm)	Average U 365 days (mm)	SF
1	1	1	190	212	111,5	177,5	2,883
	2	2	70	107			2,921
	3	3	83	162			2,881
	4	4	103	229			2,883
2	1	0,5	189	204	95,75	122,5	2,835
	2	1	60	77			2,829
	3	1,5	64	92			2,834
	4	2	70	117			2,835
3	1	0,33	208	233	102,75	143,25	2,863
	2	0,66	64	91			2,864
	3	1	67	110			2,859
	4	1,33	72	139			2,861
4	1	0,25	186	201	91	108,75	2,874
	2	0,5	59	72			2,869
	3	0,75	59	77			2,873
	4	1	60	85			2,877

Table 2. shows a variety of results from skirt footings tests. The variations in the value of soil deformation resulted from each diameter of skirt footings, in general, the results obtained are the bigger the diameter of the skirt footings (D) used, the smaller the soil deformation that occurs, it is directly proportional to the fact that the smaller comparison value of the long skirt footings to the diameter of the skirt footings (L / D) then deformation results obtained are also getting smaller. If the parameters used are grouping by diameter (D) then the deformation of the skirt footings chart obtained with a different length. From the graph with the same diameter (D), it can be seen that the pattern is almost the same but looks different if the wider diameter (D) skirt footings are used, the deformation that occurs tends to be smaller. From the graph, it is found that from 4 different diameter types (D), it shows that the deformation of the soil in skirt footings with the length (L) of 2 meters is the smallest, namely 60 mm at D 2 meters and 59 mm at D 4 meters.

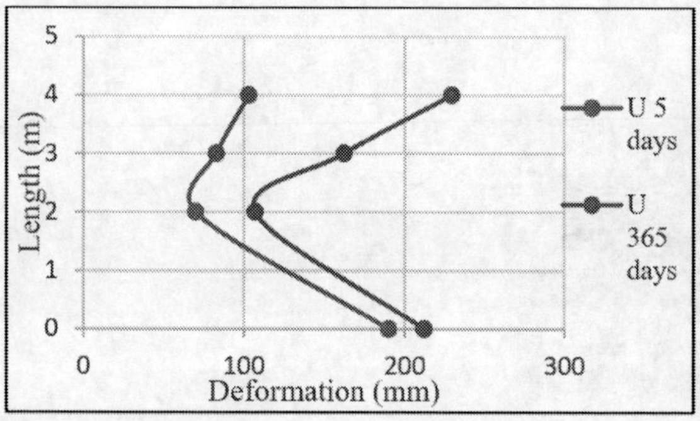

Figure 5. Skirt footings D 1 meter.

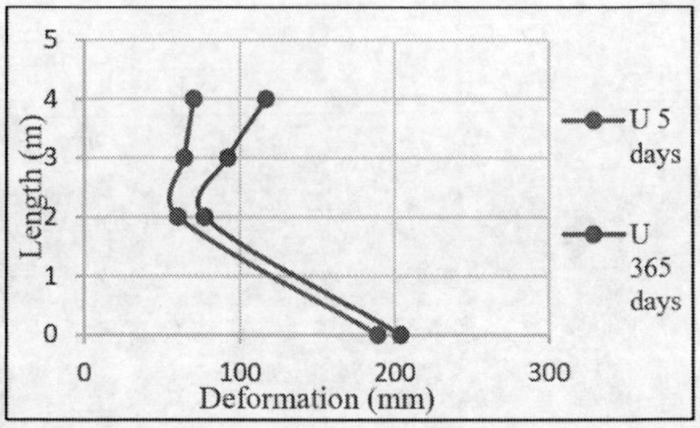

Figure 6. Skirt footings D 2 meter.

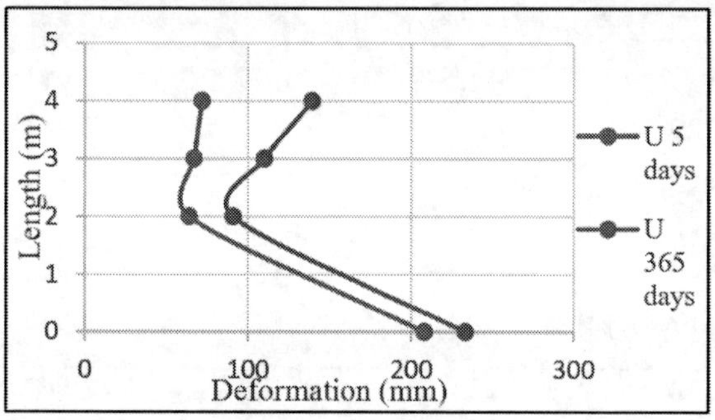

Figure 7. Skirt footings D 3 meter.

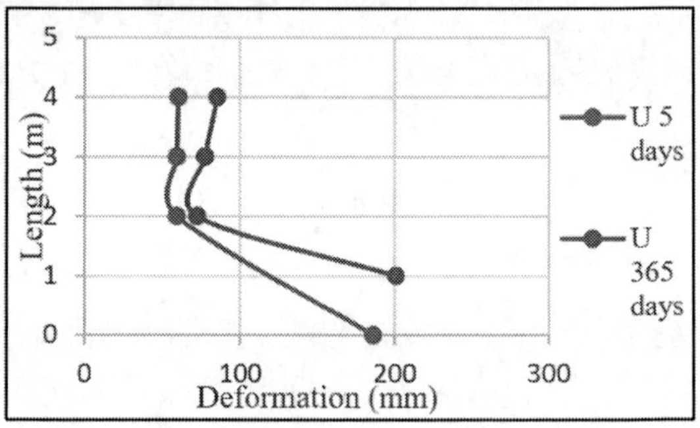

Figure 8. Skirt footings D 4 meter.

When seen from another viewpoint the length of the skirt footings (L), there are 4 variations in the length of the skirt footings, the deformation that occurs also changes. The length of the skirt footings influences reducing the deformation of the soil that occurs. From the 4 types of skirt footings with different diameters (D), it can be seen that skirt footings with a diameter of 4 meters have a small deformation when viewed from each different length of skirt footings. At 365 days the deformation is much more noticeable pattern changes than the deformation of 5 days at each skirt footings (L), the most obvious change is the deformation that occurs at D 1 meter with 4 meters L. In the consolidation of 5 days, the deformation was 103 mm while the consolidation 365 was deformed as far as 229 mm.

IOP Conf. Series: Materials Science and Engineering **930** (2020) 012039 doi:10.1088/1757-899X/930/1/012039

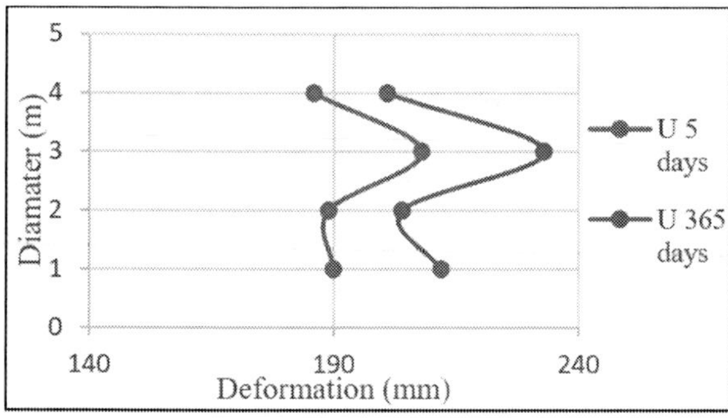

Figure 9. Skirt length 1 meter.

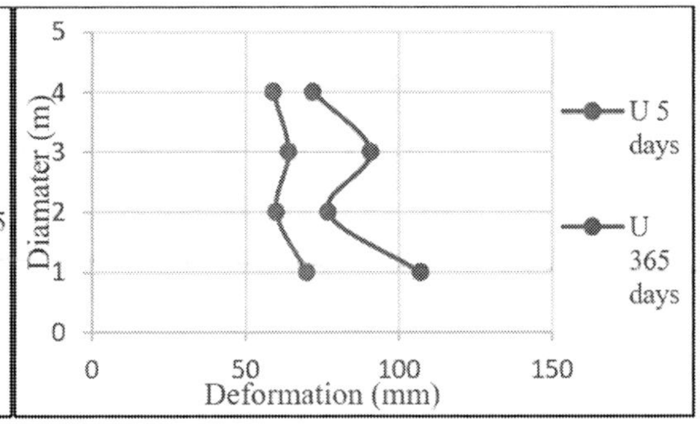

Figure 10. Skirt length 2 meters.

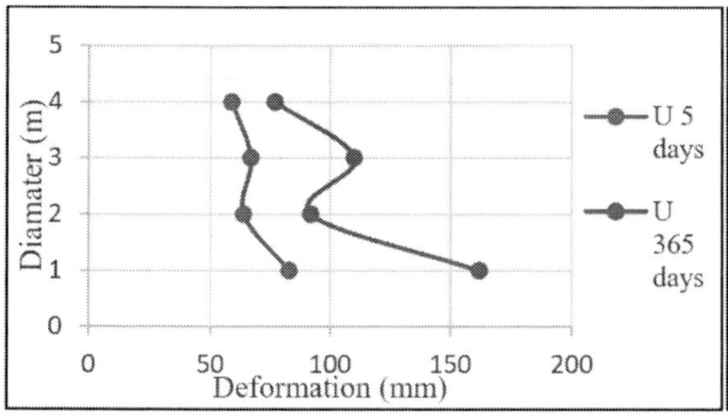

Figure 11. Skirt length 3 meters.

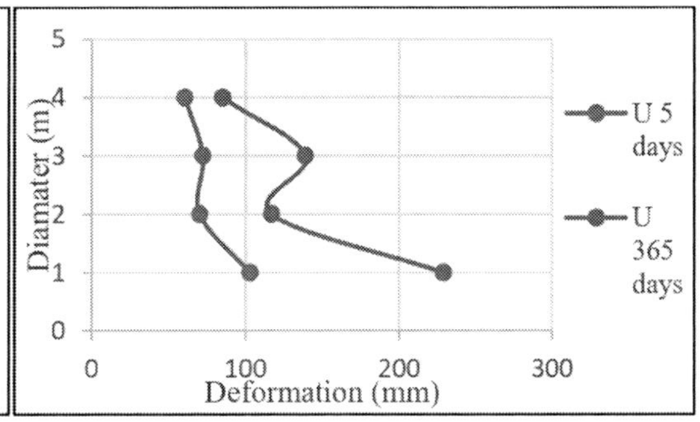

Figure 12. Skirt length 4 meters.

3.2 Un-skirted footings

To determine the effect of the length of skirt footings (L) on soil deformation, an un-skirted footings test was performed. Equation tests performed on un-skirted footings with skirt footings are using the same diameter (D) and the same test stages. The only difference lies in the work area of un-skirted footings that do not use the skirt wall under the circular footing.

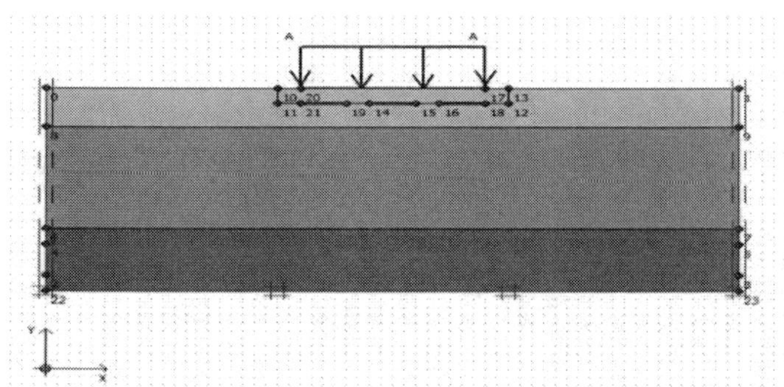

Figure 13. Fieldwork un-skirted footings.

Table 3. Un-skirted footings test result.

D (m)	U 5 days (mm)	U 365 days (mm)	SF
1	190	200	2,886
2	203	213	2,835

IOP Conf. Series: Materials Science and Engineering **930** (2020) 012039 doi:10.1088/1757-899X/930/1/012039

3	216	231	2,856
4	196	207	2,867

Figure 14 and figure 15 show the difference deformation at consolidation 365 days and consolidation5 days, it means the skirt wall under circular footings has a major effect in reducing the deformations. There are big gaps between these two shallow foundations. When the diameter is 1 meter, the difference quite small. The biggest difference at deformation measure from both type of footings lies in diameter (D) 2 meters where the deformation decrease almost 100 mm for both of consolidation.

Figure 16 and figure 17 show the effective stress at both types of footings. The effective stress indicates the deformations of soil, at un-skirted footings the effective stress value far less than at skirts footings. Stress that occurs in the soil directly leads to the embankment without being blocked by the foundation. It means the capability of un-skirted footings to reduce the deformations value not as good as skirt footings (soil deformations tends bigger in un-skirted footings).

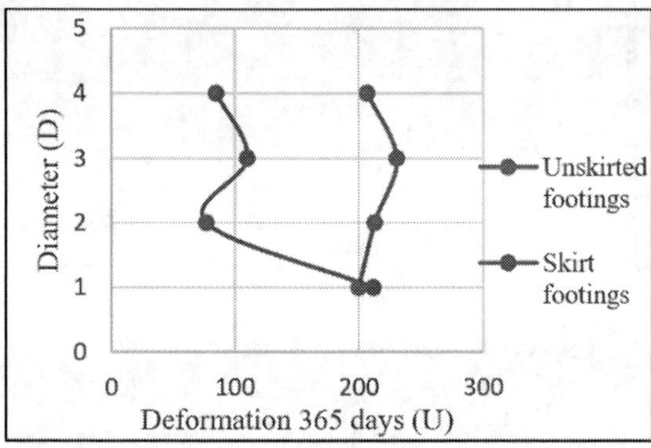

Figure 14. Skirted vs un-skirted footings 365 days.

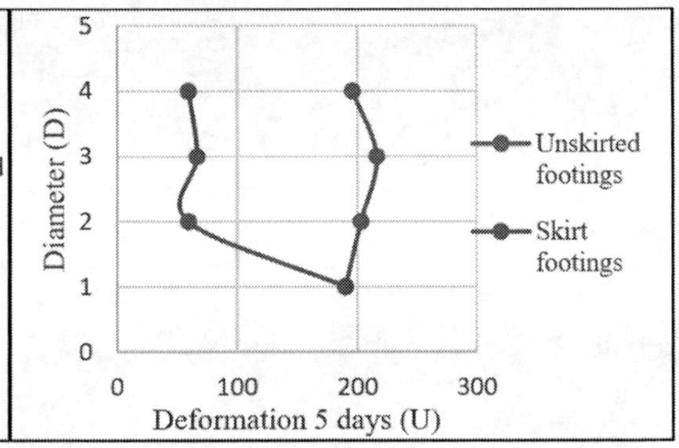

Figure 15. Skirted vs un-skirted footings 5 days.

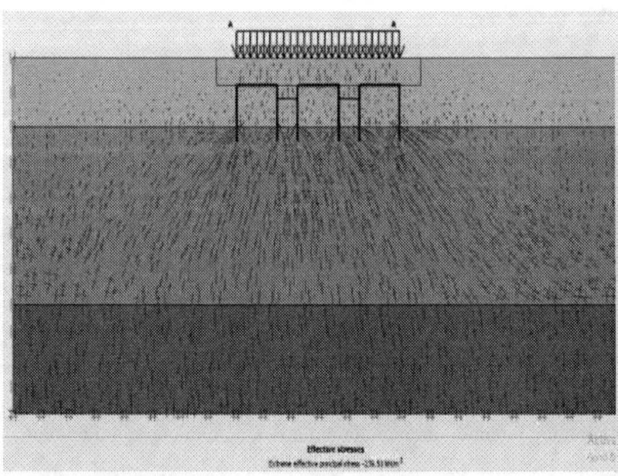

Figure 16. Effective stress at skirt footings.

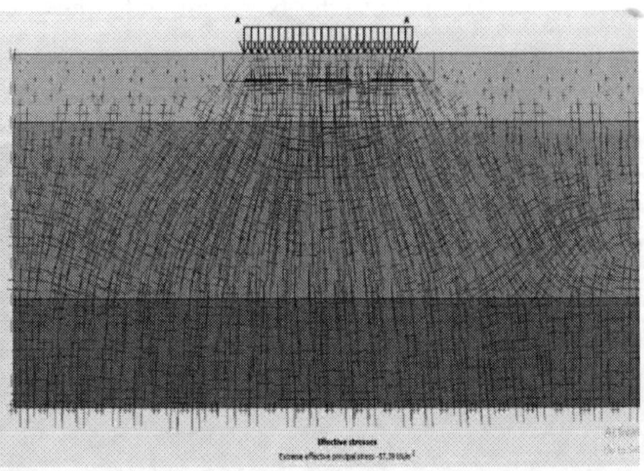

Figure 17. Effective stress at un-skirted footings.

3.3 Mini pile

The method of soil improvement of peat soils for road embankment construction was issued by the Public Works Department with the number Pd T-11-2005-B, stabilization using cement and wooden pile. To compare the results of the skirt footings method with wooden piling and cement methods from the Public Works Department, wooden piling was replaced with the mini pile to adjust to current field conditions, and the cement was not used as a stabilizing material but was replaced by the consolidation ratio (OCR) of the original soil increased. The length of the mini pile used in the study matches the length of the skirt footings (L) to see the effect of the diameter of the skirt footings (D) on the possible soil deformation. The distance between the mini piles also follows the distance rules on the skirt footings, which are 1 meter. Mini pile calculation steps are the same as skirt footings calculation steps, the difference lies in the mini pile that does not have circular footings.

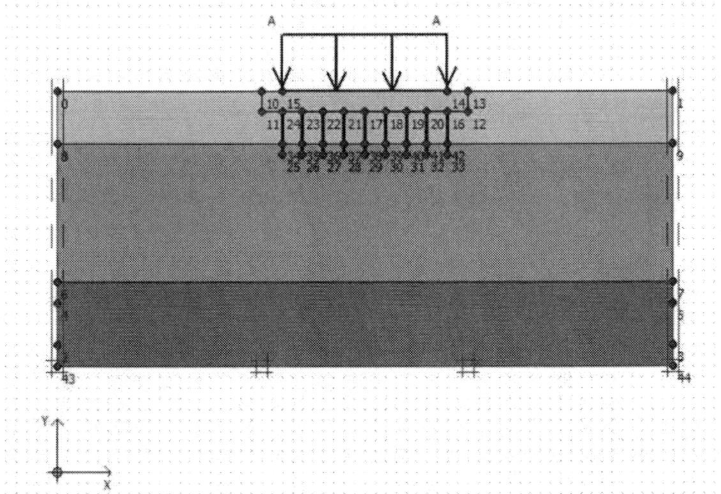

Figure 18. Fieldwork mini pile.

Table 4. The test result of the mini pile.

L (m)	U 5 days (mm)	U 365 days (mm)
1	20	21
2	13	16
3	19	30
4	27	58

There is a striking difference in deformation between the two types of shallow foundations. Figure 19 and figure 20 show the deformation that occurs in the mini pile is relatively small when compared with skirt footings with an average deformation stands at 27 mm for mini pile and 107 mm for the skirt footings, but the concern is the pattern of soil subsidence occurred. From figure 21 and figure 22. shows the vertical deformation results of research using a mini pile as a shallow foundation and skirt footings, soil deformation that occurs in the mini pile are very small but it can be seen that the deformation pattern decreases unevenly on the soil (differential settlement). The vertical deformation pattern of the mini pile appears to be spread deep into all layers, on the other hand, the vertical deformation pattern of skirt footings centered on stabilizing soil. This shows the function of circular footings (D) in holding deformation fixed in one area (stabilization soil).

IOP Conf. Series: Materials Science and Engineering **930** (2020) 012039 doi:10.1088/1757-899X/930/1/012039

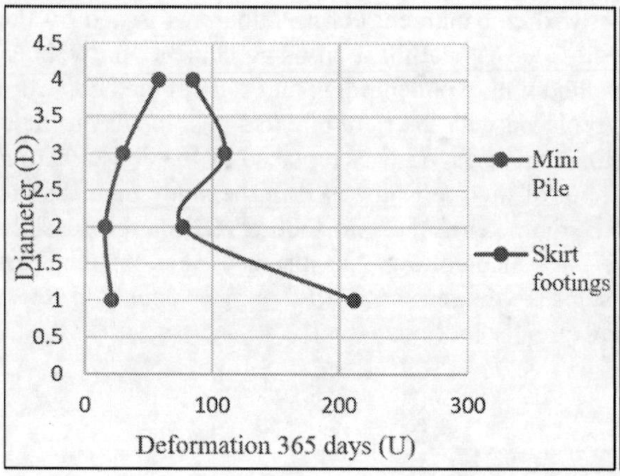

Figure 19. Skirt footings vs mini pile 365 days.

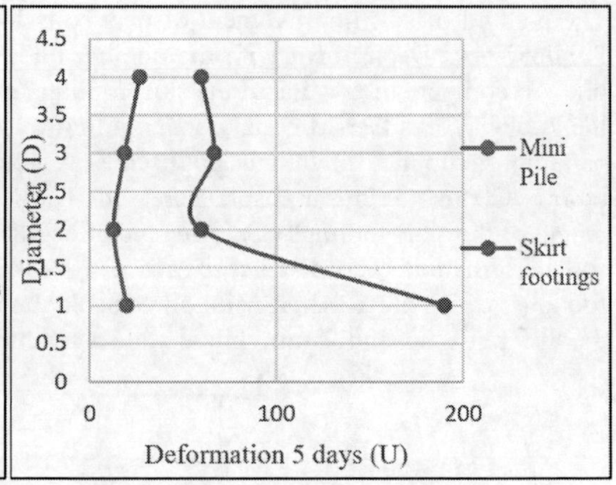

Figure 20. Skirt footings vs mini pile 5 days.

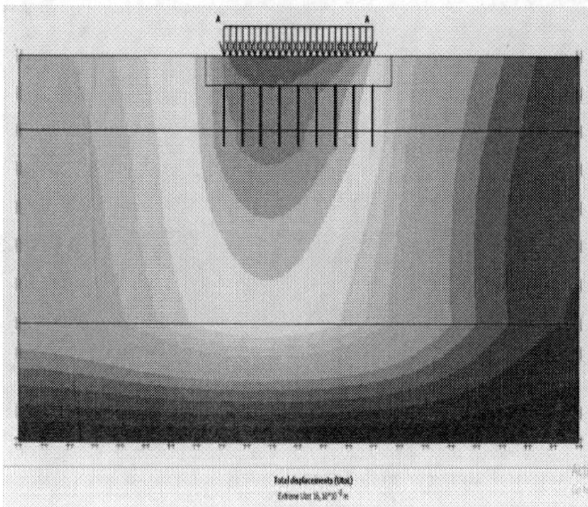

Figure 21. Vertical deformations at the mini pile.

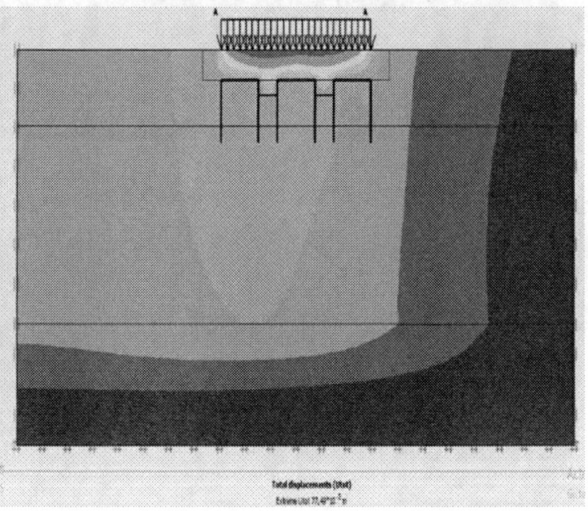

Figure 22. Vertical deformations at skirt footings.

4. Conclusion

The main purpose of this research is to investigate the effect of skirt footings on differential settlement and stability which the results show the embankment stability value goes up with the help of skirt footings. The function of the skirt wall under the footings is to hold the load from the embankment and distribute evenly to the soil around the skirt. The function of circular footings is to help the settlement of the soil evenly and to place the load area around the stabilization soil.

From the tests conducted, skirt footings are better at holding loads than un-skirted footings, this can be seen from the effective stress received by the pile is bigger when the skirt footings are used than the pile using un-skirted footings. Skirt footings are also better at resisting the effects of uneven decrease (differential settlement), this is evident from the vertical deformation pattern on the skirt footings concentrating the load on an area of land that has been stabilized while on the mini pile the distribution pattern is not centered on an area of land.

References

[1] Munro R 2004 *Dealing with bearing capacity problems on low volume roads constructed on peat* (The Highland Council, Transport, Environmental & Community Service, HQ, Glenurquhart Road, Inverness IV3 5NX: Scotland.)

[2] Wang Y, Lu X, Wang S and Shi Z 2006 The response of bucket foundation under horizontal dynamic loading *Ocean. Eng.* **33**(7) 964–73.

[3] Randolph M F, Gaudin C, Gourvenec S M, White D J, Boylan N and Cassidy M J 2011 Recent advances in offshore geotechnics for deepwater oil and gas developments *Ocean. Eng.* **38**(7) 818–34.

[4] Kellezi L, Hofstede H W L and Hansen P B 2005 Jack-up footing penetration and fixity analyses *Proceedings of the 1st international symposium frontiers in offshore geomechanics (ISFOG)* (Perth: Australia) 559–65.

[5] Kellezi L, Kudsk G and Hofstede HW L 2008 Skirted footings capacity for combined loads and layered soil conditions in Foundations *Proceedings of the 2nd British Geotechnical Association international conference on foundations eds M. J. Brown, M. F. Bransby, A. J. Brennan and J. Knappett* (Dundee UK: HIS Press) **1** 923–35.

[6] Jostad H P and Andersen K H 2006 Potential benefits of using skirted foundations for jack-up platforms *Proceedings of the offshore technology conference* (Houston: TX) OTC 18016.

[7] Rogers M 2018 *Framo technology contributes to safe and secure offshore wind foundations* Retrieved from https://energynorthern.com/2018/12/17/framo-technology-contributes-to-safe-and-secure-offshore-wind-foundations/

[8] Ritung S, Wahyunto K, Nugroho, Sukarman, Hikmatullah, Suparto dan Tafakresnanto C 2011 *Peta Lahan Gambut Indonesia Skala 1:250.000 (Indonesian peatland map at the scale (1:250,000)* (Balai Besar Penelitian dan Pengembangan Sumberdaya Lahan Pertanian: Bogor, Indonesia).

IOP Conf. Series: Materials Science and Engineering **930** (2020) 012040 doi:10.1088/1757-899X/930/1/012040

Structural behavior of open truss FRP bridge without side support

I Sidik[1*], R Irawan[1]

[1]Directorate General of Highways, Ministry of Public Works and Housing, Jl. A.H. Nasution 264, Bandung 40294, Indonesia

*Corresponding author's e-mail: indra.sidik@pusjatan.pu.go.id

Abstract. Lightweight makes the advantage of Fiber Reinforced Polymer (FRP) as one of the choices of truss bridge without side support materials. The simple shape of the structure and bridge connection system is believed to be the reason the bridge with this type of structure can use in emergencies that require rapid mobilization and installation. However, with the limited choice of FRP structural element profiles available, combined profile preparation was carried out in the structure to be used as a truss bridge structure profile. This paper describes the structural response identification of the use of a combined profile as seen from static and dynamic load testing of a full-scale bridge structure with a 16-meter span that planned to use as a pedestrian and light vehicle bridge. The load is applied in stages to be able to observe changes in the response of the structure and connection system. The results show that the bridge without side support has a load-bearing capacity that is strongly influenced by the stability of the compressive members and connection systems, which is very visible in the dynamic characteristics of the bridge in the lateral direction which is quite far when compared with the tests in other references.

1. Introduction

Emergency bridges are an essential part of maintaining mobility and logistics connectivity in the event of a disaster. The latest preparation of Bailey steel bridge emergency bridge-type by the Directorate General of Highways was carried out in 2018 and 2019 to meet the priority needs of disaster management in 22 areas of the national road implementing offices in Indonesia. That preparation shows that there is a significant need to prepare an emergency bridge used to serve light operational disaster relief vehicles, which can be transported quickly without the need for heavy equipment during installation in remote areas, and which can be done by workers without high skills [1].

Based on these needs to be able to increase the speed of installation and long-term performance of emergency bridges, the Institute of Road Engineering was trying to develop a bridge structure made from Fiber-Reinforced Polymer (FRP). The application of composite materials such as FRP has considerable potential in developing bridge structures. FRP has suitable physical and mechanical properties, such as the ratio of strength to weight, high strength and stiffness, high impact strength, non-magnetic and non-thermal conductive properties, as well as high resistance to corrosion and chemicals [2]. It can also be an effective solution with the life cycle of a structure to reduced installation and maintenance costs for corrosion and chemicals [4]. Increasing mass production as applied to the aircraft industry can anticipate high initial production costs and uncertain long-term costs such as maintenance to protect against fire, and dismantling and disposal of FRP materials [6]. The flexibility of the direction of installation of the fiber to accommodate the main force planned to bear is one of the economic benefits of using FRP elements [6].

Content from this work may be used under the terms of the Creative Commons Attribution 3.0 licence. Any further distribution of this work must maintain attribution to the author(s) and the title of the work, journal citation and DOI.
Published under licence by IOP Publishing Ltd

The development of bridges made from FRP in Indonesia can follow what the USA develop for pedestrian and light vehicles bridges since the late 1980s [7]. Better and more comprehensive knowledge about the fundamental properties of the composite, and their long service life [2] related to the complex structural behavior that is strongly influenced by composite materials and jointing methods [3] can obtain through that development. Things to note are that the main ingredient of FRP elements, especially glass fibers, experience deterioration due to increased temperatures, especially if burdened with compression and shear [8] and also due to UV radiation for more than 3000 hours [9] which very decisive in a tropical country like Indonesia.

Research and development of bridges made from FRP in Indonesia began in 2013 has produced two prototype bridge frames for pedestrians and light traffic respectively in 2016 and 2019. FRP profiles formed by the pultruded method, which is estimated to be a cost-effective composite manufacturing technique [3], can use for the development of this bridge in the future. The limited FRP pultruded profile which produces for the structure of building roofing in Indonesia, applies as bridge truss structure elements with some modification.

The study began by identifying the physical and mechanical parameters of the pultruded FRP material made from glass, which showed a significant effect of aging testing and an increase in temperature on changes in mechanical properties [10]. The research continued by making a test model truss system in 2014, and a prototype of an FRP pony-truss Warren-type bridge with side support in 2015-2016 using two closed C channel pultruded-FRP-profiles combined into a box-like profile (Figure 1). The results obtained from these tests show that there was a failure in the connection area between the joint profile and the steel plate used to connect the FRP pultruded profile frame.

Figure 1. Prototype pony-truss Warren-type 8 meters span bridge with side support in 2016 using a combined of closed two closed C channel pultruded-FRP-profiles

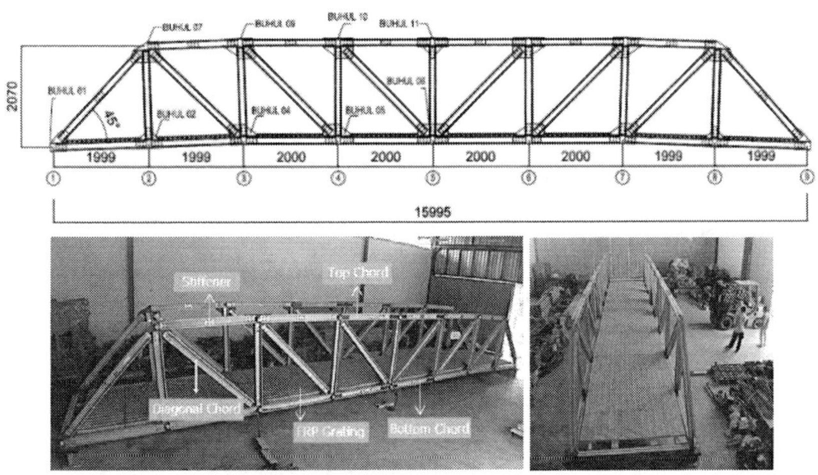

Figure 2. Views and parts of the 16 meters span FRP bridge

Based on this, finally, in 2019, the purpose of this study is to determine the structural performance of the FRP bridge prototype development result. A pony-truss FRP bridge prototype was developed to identify the structural response using the same FRP pultruded combined profiles. Static and dynamic load testing carried out on a full-scale bridge structure that planned to utilize as a pedestrian bridge and an emergency light vehicle.

The bridge prototype is a truss bridge with 16 meters long and 2.5 meters wide which has a length of 2 times that of the 2016 test bridge (Figure 1). The frame uses a combined two open C channel pultruded-FRP-profile with a depth of 150 mm and a thickness of 7 mm (Figure 3) combined with bolts and gusset plates and coupling plates in the center of each compression chord to minimize the occurrence of buckling [10].

2. Experimental and numerical model

The FRP bridge prototype that studied is a bridge system of Pratt type form pony-truss without using side support on vertical trunks as is commonly applied to be the choice in the design of pedestrian bridges and light vehicles [1] because of its simple shape [12]. The truss bridge also has high resistance in the longitudinal direction but low resistance to shear [1].

The bridge consists of two main components; namely, the truss system rests on two trusses and the floor system. The truss system consists of a compressed member, tension member, and a diagonal member, the flexible floor beam being a buffer to provide stability [13]. Stringers and floor beam supported the deck system where the type of pony-truss type bridge structure like this is one of three types of structural systems developed into pedestrian bridges in the USA [7].

One of the planning criteria for pultruded FRP bridges in the form of pony-truss in 2019 refers to the "Guide Specifications for Design of FRP Pedestrian Bridges, AASHTO 2008". The planned load is pedestrian load of 3 kPa. The bridge also allows light vehicles to be able to cross in emergencies such as ambulances or other logistical transport vehicles so that it is assumed to be 5 tons, and the wind load for the open frame bridge system is 1.68 kPa. The conditions that must be fulfilled are deflection not exceeding 1/200 span length, fundamental vertical frequency not less than 5 Hz and lateral frequency fundamental not less than 3 Hz.

2.1. FRP pedestrian bridge prototype

Testing performed on a prototype FRP bridge with a width of 16 meters, and a width of 2.5 meters, where the span range is an effective span for FRP truss bridges that range from 9 meters to 30 meters with a width of between 1.5 meters and 3 meters [7] as shown in **Error! Reference source not found.**.

2.2. Member profiles, deck and connections

The profile used for the truss component was C channel profile 150.65.20, with a profile thickness of 7 mm, and the deck system uses a 50 mm pultruded grating as shown in Figure 3 dan Figure 4. The estimated total weight of the truss bridge reaches approximately 5 tons including bridge truss elements, connection systems, and pultruded grating for bridge deck systems. Details of truss elements weight are shown below.

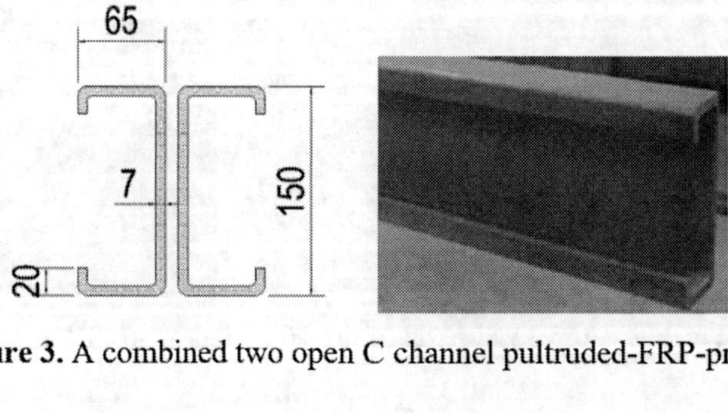

Figure 3. A combined two open C channel pultruded-FRP-profile

Figure 4. **Pultruded FRP grating for bridge deck**

Table 1. Bridge truss element

No.	Element	Number of Element (pcs)	Length (m)	Volume (m3)	Weight (kg)	∑ Weight (kg)
1	Top Chord 1	12	1.890	0.007659	13.786	165.434
2	Top Chord 2	8	1.890	0.007659	13.786	110.290
3	Top Chord 3	16	1.890	0.007659	13.786	220.579
4	Diagonal Chord 1	8	2.395	0.009713	17.483	139.867
5	Diagonal Chord 2	12	2.443	0.009713	17.483	209.801
6	Floor Beam	30	1.800	0.007293	13.127	393.822
7	Stringer	11	2.320	0.009365	16.857	185.427
8	Vertical Chord 01	4	1.800	0.007293	13.127	52.510
9	Vertical Chord 02	14	1.800	0.007293	13.127	183.784
	Specific Gravity = 1800 kg/m^3		Total Chord Weight (kg)			1,662

At the connection, the bridge test model uses ASTM A325 16 mm diameter bolts and 10 mm thick gusset plates and addition, the coupling plate stiffener is connected to shorten the buckling length and bind bridge chord compression members as shown in Figure 5.

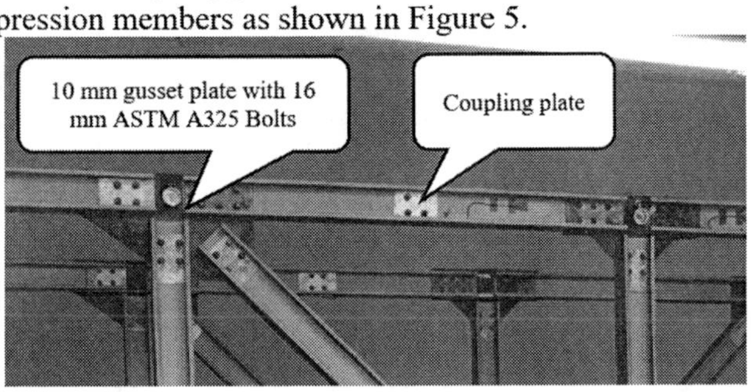

Figure 5. Connections system on FRP Bridge Prototype

2.3. FRP material characteristics

The composition of the content of heterogeneous FRP constituent materials makes FRP elements anisotropic and brittle [6] will affect the design process of FRP elements or connection systems [3]. The mechanical properties of PFRP materials must be tested beforehand because of the fiber composition and stiffness that varies from manufacturer to manufacturer [14], which shows a significant difference compared to the American Society of Civil Engineers reference, 2010. Pre-Standard for Load & Resistance Factor Design (LRFD) of Pultruded Fiber Reinforced Polymer (FRP) Structures especially in the minimum thickness and shear strength requirements that seen in Table 2 [10].

Table 2. The minimum FRP characteristics based on ASCE 2010 and 2013 test results [14]

	ASCE 2010	Fiber-beam (FRP)
Fiber type	Min. *E*-glass	*E*-glass
Minimum thickness/element	Min 9,6 mm	7 mm
Tensile strength	207 Mpa	370,4 ± 18,5 MPa
Compressive strength	207 Mpa	115 ± 12,9 MPa
Flexural strength	207 MPa	498,0 ± 42, 1MPa
Shear strength	31 MPa	40,3 ± 7.7 MPa
Modulus of elasticity	20684 MPa	20600 MPa

2.4. Numerical model
A numerical model created to verify the results of the response test. FRP bridge behavior is simulated with structural analysis software with fundamental frequency as a reference. At this stage, trial and error are carried out on the modulus of elasticity to obtain the same fundamental frequency.

3. Test setup
Identification of the structural response characteristics carried out with dynamic and static testing of the FRP bridge prototype. The things observed in the FRP bridge prototype performance test include the visual condition of the elements and structure of the bridge, the strain on the chord, structural deformation, and the natural frequency of the bridge structure. Dynamic tests carried out in the vertical and lateral directions of the test model. The vertical structural response obtained by passing people walking and running on the bridge. The transversal response obtained by giving a boost to the frame with the help of human force. The sensor used is an accelerometer placed in the middle of the span (Figure 6). The direction of the accelerometer is positioned according to the testing requirements, namely vertically and transversely of the bridge.

Static tests are carried out by overloading the bridge with a maximum load of planned 12 tons. At each load increase, a visual inspection conducted to observe changes in the bridge behavior. In addition to visual observations, strain measurements and deformations are also carried out. Strain measurement using strain gauge is mounted on top and bottom chord, while deformation measurement uses Total Station with a point of view on each gusset plate. The instrumentation performance test layout of the FRP bridge prototype shown in Figure 6 and Figure 7. Complete loading order and weight on each segment presented in Table 3 which indicated in I, II, III, and IV number for each load increment. Loading segment numbering indicated by two joints individual connected one span.

4. Test result

4.1. Dynamic test
Dynamic test results on a natural vertical frequency were 8.06 Hz for people walking and 7.39 Hz for people running. Meanwhile, the natural frequency of the lateral direction is 1.46 Hz, which means it is still lower than the AASHTO regulation, which is not less than 3 Hz.

4.2. Static test
In static testing, the FRP bridge test model has collapsed close to loading 9 tons. Test visualization from the beginning to bridge collapse seen in Figure 8.

Error! Reference source not found. shows the results of deflection measurements on the FRP bridge prototype. The maximum deflection measured at 9 tons load is 15,83 mm measured by the Total Station and the deflectometer calibrated before testing. Meanwhile, **Error! Reference source not found.** is a strain sensor recording on the lower and upper FRP members installed in the middle of the range where negative value indicated compression stress, and the positive value indicated tension stress.

Measurement in vertical deflection and lateral deformation of the top chord also carried out. Figure 9 shows the change in the position of the top chord towards the inside of the bridge until it loads 6 tons.

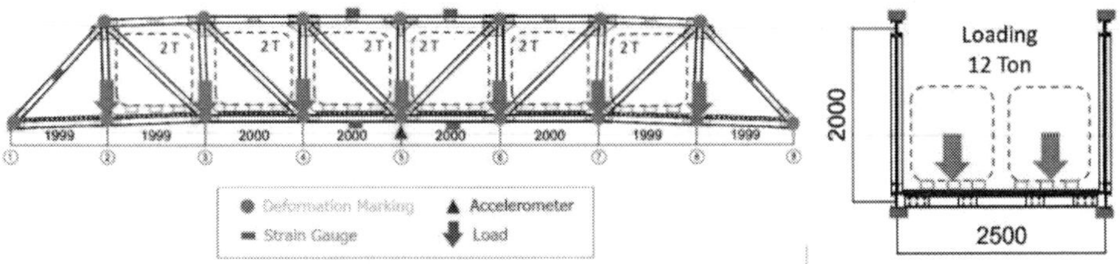

Figure 6. Instrumentation layout for bridge loading

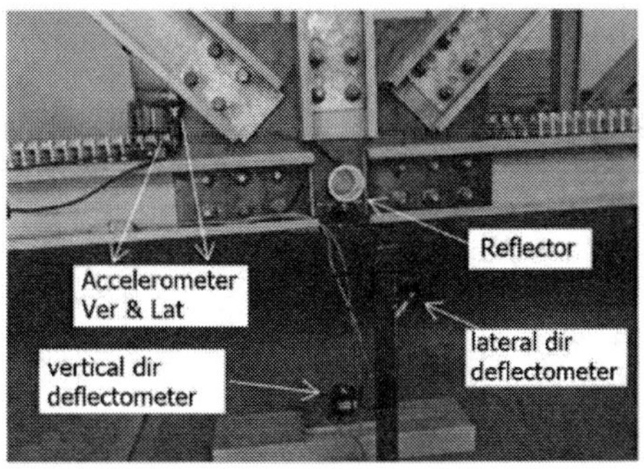

Figure 7. Typical instrumentation attached in gusset plate

Table 3. Planned sequence load in each span

Span indication loading (ton)	1-2	2-3	3-4	4-5	5-6	6-7	7-8	8-9	Total load (ton)
Load I	-	0.5	0.5	0.5	0.5	0.5	0.5	-	3.0
Load II	-	1.0	1.0	1.0	1.0	1.0	1.0	-	6.0
Load III	-	1.5	1.5	1.5	1.5	1.5	1.5	-	9.0
Load IV	-	2.0	2.0	2.0	2.0	2.0	2.0	-	12.0

Table 4. Displacement of FRP bridge prototypes at each point static test

Displacement (mm)	1	2	3	4	5	6	7	8	9
Load I[a]	-0,90	-3.40	-4.60	-5.80	-6.30	-5.60	-4.70	-3.90	-0.70
Load II[a]	-0.90	-5.10	-8.10	-10.10	-11.20	-10.40	-8.30	-5.50	-0.20
Load III[b]	-	-	-	-	-15,83	-	-	-	-

[a] Displacement Loads I and II are obtained from Total Station measurements
[b] Displacement Load III is obtained from the deflectometer sensor just before prototype collapse

Table 5. Strain of the FRP bridge prototype during static testing

Load Combination	Bot 01[a] με	Bot 02[a] με	Top 01[b] με	Top 02[b] με
I (3 ton)	216.3	243.8	-158.8	-173.4
II (6 ton)	348.3	411.4	-339.3	-374.1
III (9 ton)	495.6	544.7	-420.9	-532.8

[a] Bot: Bottom Chord

Load Combination	Bot 01[a] με	Bot 02[a] με	Top 01[b] με	Top 02[b] με

[b] Top: Top Chord

(a) (b)

(c)

Figure 8. FRP bridge prototype static load testing process (a) initial condition, (b) loading process, (c) collapse at Load III

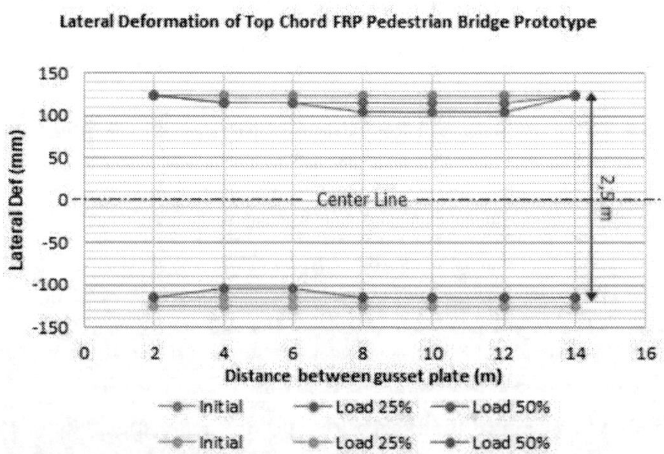

Figure 9. Lateral deformation of top chord of the FRP bridges

5. Analysis and discussion

5.1. Numerical model

Fundamental frequencies referred to according to measurement results are 7.39 Hz and 8.06 Hz. Tuning model results obtained modulus of elasticity of 20830 MPa (for f=7.39 Hz) and 24930 MPa (for f=8.06 Hz) from the initial assumption of 20600 MPa.

Next, the comparison between deflection values generated by the simulation on the numerical model with the values measured in the test shown in Table 6.

Table 6. **Experimental and numerical analysis deflection values in every joint**

Displacement (mm)		**1**	**2**	**3**	**4**	**5**	**6**	**7**	**8**	**9**
Load I	Exp [a]	-0,90	-3.40	-4.60	-5.80	-6.30	-5.60	-4.70	-3.90	-0.70
	Num [b] $f_{7.39}$	0.00	-2.45	-4.58	-6.15	-6.82	-6.15	-4.58	-2.45	0.00
	Num [b] $f_{8.06}$	0.00	-2.06	-3.85	-5.17	-5.73	-5.17	-3.85	-2.06	0.00
Load II	Exp [a]	-0.90	-5.10	-8.10	-10.10	-11.20	-10.40	-8.30	-5.50	-0.20
	Num [b] $f_{7.39}$	0.00	-4.91	-9.16	-12.30	-13.65	-12.30	-9.16	-4.91	0.00
	Num [b] $f_{8.06}$	0.00	-4.12	-7.69	-10.34	-11.47	-10.34	-7.69	-4.12	0.00
Load III	Exp [a]	-	-	-	-	-15,83	-	-	-	-
	Num [b] $f_{7.39}$	0.00	-7.36	-13.73	-18.46	-20.47	-18.46	-13.73	-7.36	0.00
	Num [b] $f_{8.06}$	0.00	-6.18	-11.54	-15.51	-17.20	-15.51	-11.54	-6.18	0.00

[a] Exp: Experimental
[b] Num: Numerical analysis

Figure 10 illustrates the deformation of the FRP bridge prototype obtained from numerical and experimental simulations at each loading stage. Relative deflections of the bridge model obtained as L/1010 (for experimental models), L/781 (for numerical models f=7,39 Hz), dan L/930 (for numerical models f=8.06 Hz), where L is the total length of the bridge.

Those ratios were obtained in 9 tons loading, which still under L/200 in the AASHTO regulation. Even though the prototype collapsed at 9 tons, the deflection value still fulfilled the requirements. The results obtained indicate that the FRP bridge structure prototype has adequate structural stiffness.

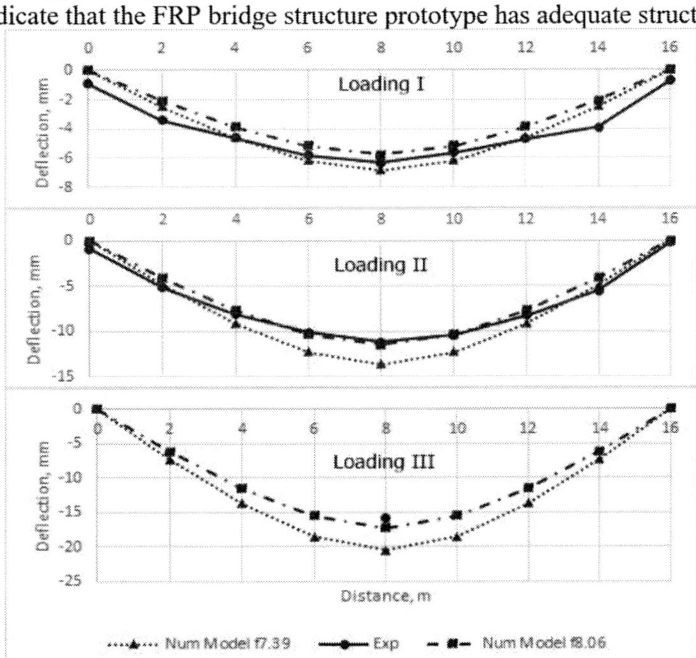

Figure 10. Deflection of prototype FRP pedestrian bridge subjected to different static loading stages

5.2. Analysis of the causes of collapse

Although the bridge stiffness in the vertical direction is quite adequate, there are still other parts of the structure that have weaknesses. Figure 9 illustrates the deformation that occurs in the top chord due to the 6 tons loading with a maximum deformation value of 20 mm. Deformation, as in figure 8, allows the top chord to transfer compressive force and moments. When the load reaches 9 tons, the test model collapses at the top of the chord, as shown in Figure 11. This condition indicates that the top chord is a weak point in this prototype.

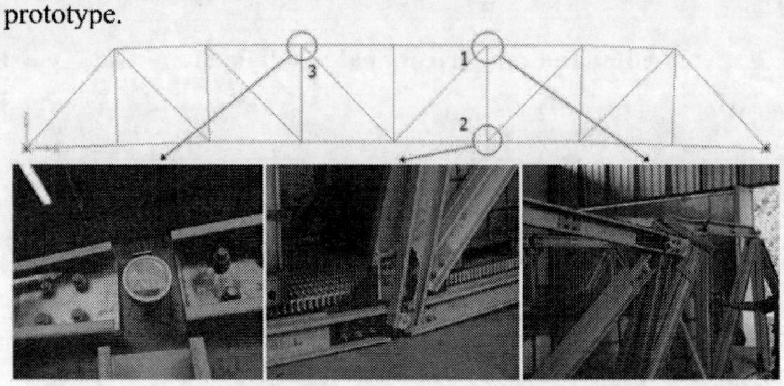

Figure 11. The condition of the bridge prototype after collapse

No. 1 in Figure 11 shows the first collapsed connection, then affects the nearest subsequent connection and members as shown in No. 2 connection. On the other side, No. 3 also has already shown damage, which is the occurrence of cracks which generally occur (Figure 12).

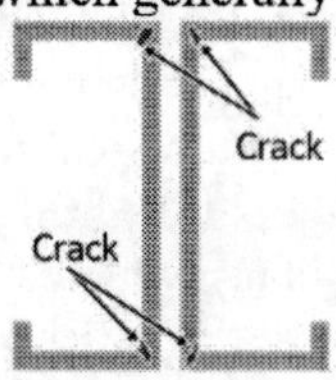

Figure 12. Crack location in no. 3 joint in figure 11

If strain results on one of the top chords analyzed, the stresses occur while the prototype collapses were:

$$\sigma_{fail} = E \times \varepsilon_{fail}$$
$$\sigma_{fail} = 20600 \times 532{,}8 \times 10^{-6}$$
$$\sigma_{fail} = 10{,}9 \text{ MPa}$$

The compressive stress that occurs is still lower than the compressive strength, but the bridge has already failed. It is likely that the top chord not only holds the axial force but holds another force that occurs on its weak axis. An open truss system without this side support requires considerable stability in the compressive area.

The first prototype is only able to span less than 10 meters. When the first prototype was made to exceed that, with the bridge's own weight, the prototype had already suffered damage to the FRP profile and was less stable. After development, the second prototype is able to reach a span of 16 meters and has sufficient vertical stiffness and is stable enough to withstand its dead load and live load even though it still has shortcomings that must be evaluated.

6. Conclusion and recommendations

This article explains the identification of structural responses using a combined two open C channel pultruded-FRP-profile with a depth of 150 mm and a thickness of 7 mm due to dynamic and static load testing on a full-scale bridge planned used as an emergency pedestrian and light vehicle bridge. The prototype is a pony-truss bridge with 16 meters span and 2.5 meters wide. Dynamic test results showed this prototype is rigid in the vertical direction but not rigid enough in the transverse direction of the bridge. The bridge prototype has been static tested until it collapses during the loading phase III (9 tons load), which planned to consist of four loading stages (maximum load of 12 tons). This FRP bridge prototype has adequate vertical stiffness because deflections obtained from static tests still meet permissible deflection requirements.

For further development of this research, the stiffness of the transverse direction of the bridge must increase. The recommendations is to increase the capacity of members which accept a compressive force by replacing FRP profiles that have a thicker web and flange profiles.

Acknowledgements
The author would like to thank Dr. Winarputro Adi Riyono and Bagus Aditya Wardhana for all their dedication as pioneers of FRP bridge research in the Ministry of Public Works and Housing. This work was supported by the Institute of Road Engineering (IRE) Research Grant F.Y. 2019, Ministry of Public Works and Housing, Republic of Indonesia.

References

[1] Sedlacek G 2004 Innovative developments for bridges ssing FRP composites *Adv. Poly. Comp. Struc. Appli. Const.*

[2] Bacinskas D 2017 Structural analysis of GFRP truss bridge model *Mod. Build. Mat. Struct. Tech. MBMST* **172** pp 68-74

[3] Kim S Y, Yoo J H Kim H Shin K Y and Yoon S J 2018 Failure modes of single and multi-bolted joint in the ultruded fiber polymer composite members

[4] Zhang X 2014 Life cycle sssesment (LCA) of fibre reinforced polymer (FRP) composites in civil application *Eco Effic. Const. Build. Mat.* pp 565-91

[5] Haak A J 2018 Life-cycle-cost Evaluation of Bridges with Fiber Reinforced Polymers (FRP) University of Rhode Island

[6] Potyrala P B 2011 Use of fibre reinforced polymer composites in bridge construction *Sta. Art Hyb. All-Comp. Struc.*(Barcelona: Universitat Politécnica de Catalunya)

[7] Bank L C 2006 Application of FRP composites tp bridges in USA *Int. Coll. Appli. FRP Brid.* (Japan Society of Civil Engineers) pp 9-16

[8] Correia J R, Gomes M M Pires J M and Branco F A 2013 Mechanical behaviour of pultruded glass fibre reinforced polymer composites at elevated temperature: experiments and model assessment *Comp. Struc.* **98** p 303–13

[9] Bazli M, Jafari A Ashrafi H Zhao X L Bai Y and Singh R K 2020 Effects of UV radiation, moisture and elevated temperature on mechanical properties of GFRP pultruded profiles *Cons. Build. Mat.* **231** p 117-37

[10] Riyono W A and Nasiri J 2019 Characterisation of C-shape pultruded fibre reinforced polymer for bridge structures *Spri. Nat. Jour.*

[11] Bakht B and Csagoly P F 1976 Lateral buckling of pony truss bridges *Trans. Res.* pp 14-8

[12] Butler M A, Swanson J A Rassati G A and Dues E F 2018 High resolution modeling and modeling of connections in pony truss bridges *Trans. Res. Rec. Jour. Trans. Res. Boar.* pp 1-10

[13] Singh S B and Chawla H 2018 An investigation of material characterization of pultruded FRP H- and I-beams *Mech. Adv. Mat. Struc.* **25** 2

[14] Adi W 2013 Karakterisasi pultruded fiber reinforced polimer (PFRP) (Bandung: Pusat Penelitian dan Pengembangan Jalan dan Jembatan)

IOP Conf. Series: Materials Science and Engineering **930** (2020) 012041 doi:10.1088/1757-899X/930/1/012041

Analytical modeling of 3D-printed reinforced concrete beams

J Chandra[1*], P Pudjisuryadi[1], A Antoni[1] and H Wibowo[2]

[1] Civil Engineering Department, Petra Christian University, Siwalankerto 121-131, Surabaya, Indonesia
[2] Department of Civil, Construction, and Environmental Engineering, Iowa State University, 813 Bissell Road, Ames, Iowa, United States of America

*Corresponding author's e-mail: chandra.jimmy@petra.ac.id

Abstract. Three-dimensional (3D) printing for cementitious materials such as concrete has become increasingly popular. Numerous research efforts have been undertaken to fabricate structural elements, such as beams, using 3D-printed concrete. Because the behavior of 3D-printed reinforced concrete (RC) beams is not well understood, research in this area is still ongoing to investigate the behavior and compare it with conventional RC beams. In this paper, the results of an analytical study using finite element software on 3D-printed RC beams are presented. The main challenge was to determine a constitutive material model of the 3D-printed concrete for nonlinear analysis that was quite different from normal concrete. The developed model was validated using the results from several past experimental tests on 3D-printed RC beams. The results showed that the analytical model can accurately predict the maximum flexural strengths of the 3D-printed RC beams. However, the analytical model overestimated the initial stiffness of the beams. Furthermore, several local failures, such as shear failure of nodal points and bond failure between rebars and concrete, could not be well simulated by the analytical model. Thus, future research is needed to correctly define the constitutive material model for 3D-printed RC beams.

1. Introduction

In recent years, new technology to produce concrete has been developed using three-dimensional concrete printing (3DCP) [1]. 3DCP is done by extruding mortar mixture gradually to form layers of concrete mortar and finally producing the expected shape according to the design. Therefore, in 3DCP, there is no need to build formwork as in conventional concrete construction. 3DCP uses an automatically controlled machine to produce any shape or geometric model of structural elements as desired [2].

3DCP for buildings is an environmentally friendly construction method that provides limitless possibilities for the realization of buildings with complex geometry [3]. With 3DCP, the paradigm of building designers is expanded as each building component can be unique without the need to spend extra cost on complex formwork as well as casting process. Other advantages of 3DCP are automation and timing optimization, lower construction cost, and reliability of the results [4, 5].

However, until now, the manufacturing process of 3DCP still encounters many obstacles, e.g., the concrete mixture condition is still inconsistent and requires a special base material [6, 7]. Moreover, there is limited research on the behavior of structural elements made using 3DCP in which the material

Content from this work may be used under the terms of the Creative Commons Attribution 3.0 licence. Any further distribution of this work must maintain attribution to the author(s) and the title of the work, journal citation and DOI.
Published under licence by IOP Publishing Ltd

properties are quite different than that of conventional concrete [8-10]. Thus, research in this area is still needed.

This paper focuses on studying the behavior of 3D-printed reinforced concrete (RC) beams. The study was done analytically using a finite element computer software [11] by comparing the available experimental results [12] with the numerical results of simulating said experiments in the software. Of particular interest is the load-deflection relationship of the specimens as well as their maximum strengths.

2. Past research on the structural behavior of 3D-printed RC beams

Al-Chaar et al. [12] have studied the flexural behavior of 3D-printed RC beams by performing bending tests on nine specimens with various combinations and types of meshes and bars. The objective of their study was to obtain flexural capacity of these beams whereby the behavior was different from that of conventional RC beams because the interfacial bonding between concrete mortar layers is limited. The beams were 4.88-m long, with a clear span of 4.57 m and shear span ratio of 6 or greater. The beams were truss beams with total depth of 203 mm and width of 140 mm. The longitudinal view of a typical beam is shown in figure 1.

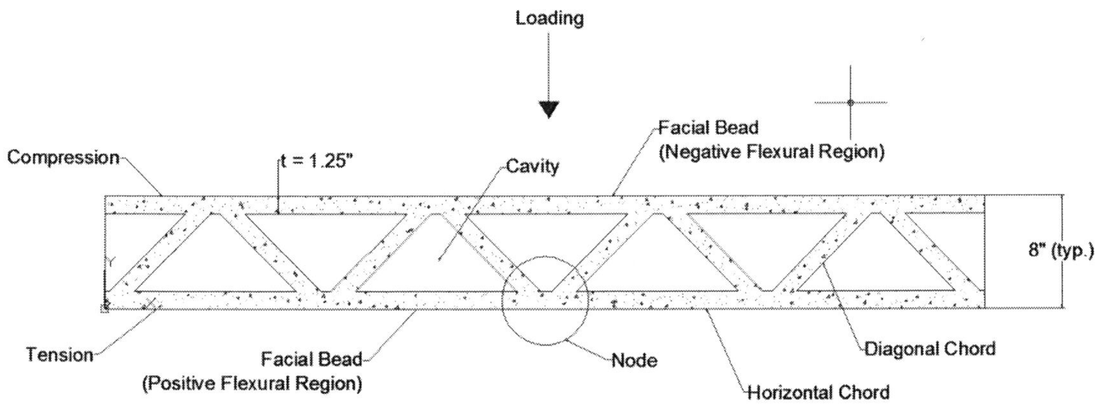

Figure 1. Longitudinal view of 3DCP RC beams tested by Al-Chaar et al. [12].

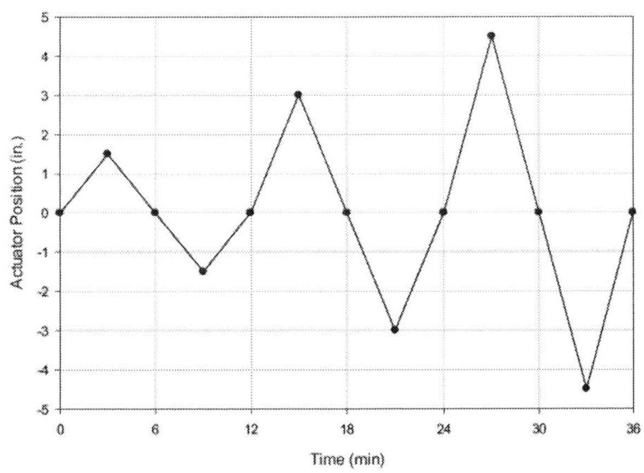

Figure 2. Loading protocol used by Al-Chaar et al. [12] (1 inch = 25.4 mm).

In this paper, only three out of nine specimens tested by Al-Chaar et al. [12] were investigated due to the limitation of the analytical model to simulate the basalt mesh used between the concrete mortar layers. These specimens are 1SR-S-0, 4DR-S-0, and 5DR-B-0. Specimen 1SR-S-0 was a singly-reinforced beam with three 10-mm diameter steel bars in the tension side while specimen 4DR-S-0 was a doubly-reinforced beam with three 10-mm diameter steel bars each in the tension and compression sides. Specimen 5DR-B-0 was a doubly-reinforced beam with three 10-mm diameter basalt bars in the tension and compression sides. Because specimen 1SR-S-0 was a singly-reinforced beam, it was cyclically loaded in one direction only, whereas specimens 4DR-S-0 and 5DR-B-0 were subjected to cyclic loading in both positive and negative directions. The loading protocol for the specimens is displayed in figure 2.

3. Analytical modeling of the 3D-printed RC beams

In this paper, the analytical modeling of the 3D-printed RC beams as tested by Al-Chaar et al. [12] was done using SAP2000 computer software [11]. The truss beams were modeled using frame elements for its tension, compression, and diagonal chords. Furthermore, fiber P-M-M hinge that is available in SAP2000 [11] was used to account for the material nonlinearity. The elevation view of the analytical model of the 3D-printed RC beams can be seen in figure 3.

Figure 3. Elevation view of the 3D-printed RC beams as modelled in SAP2000 [11].

The most important parameter to be defined in this analysis is the constitutive material model for the 3D-printed concrete. According to a study by Le et al. [13], the hardened properties of 3D-printed concrete is different than that of conventional concrete due to anisotropy in the layered concrete mortar. The compressive strength of 3DCP was reduced by 5 to 15 percent, depending on the loading direction, for a straight-line printing structure and the reduction can go up to 30 percent for a curvy-shaped printing structure as compared with conventional concrete. In addition, because of the gap between its layers due to the printing process, the tensile bond strength between the layers of 3D-printed concrete is weaker. However, because the truss beams were modeled using frame elements, the reduction of tensile bond strength was not able to be taken into account in the analytical model.

In this analysis, the Mander's model for unconfined concrete [14] that is available in SAP2000 [11] was used as the constitutive material model for concrete as displayed in figure 4a. For rebars, Park's model for reinforcing bar [15] was used as displayed in figure 4b.

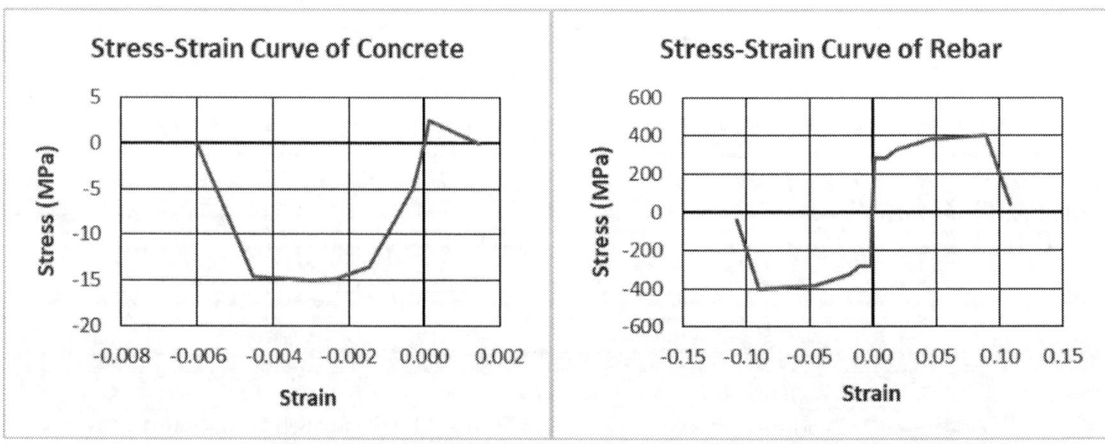

(a) (b)

Figure 4. Stress-strain curve of materials used in the analytical model: (a) concrete and (b) rebar.

The analyses were carried out using nonlinear static analysis with several load cases to simulate cyclic loading conditions as defined in the experiment by Al-Chaar et al. [12]. Displacement control was used to monitor the nodal displacement in order to simulate the loading protocol used in the experiment. From the analysis results, reactions in the supports were combined to obtain the total force applied to the 3D-printed RC beams. Subsequently, force-displacement curves of the beams obtained from the analyses were compared with those obtained from the experiment by Al-Chaar et al. [12] in order to see whether the analytical model could well simulate the behavior of 3D-printed RC beams.

4. Analysis results

The analysis results were compared with experimental results from a study by Al-Chaar et al. [12] in terms of the load-displacement curves of the specimens and their maximum strengths. The comparison of maximum strengths between experimental and analytical results can be seen in table 1 while the comparison of load-displacement curves between experimental and analytical results are displayed on Figures 5, 6, and 7 for specimens 1SR-S-0, 4DR-S-0, and 5DR-B-0, respectively.

Table 1. Comparison of maximum strengths between experimental and analytical results.

Specimen ID	Maximum Experimental Strength (kN)	Maximum Analytical Strength (kN)
1SR-S-0	12.881	11.898
4DR-S-0	14.154	12.924
5DR-B-0	7.958	8.416

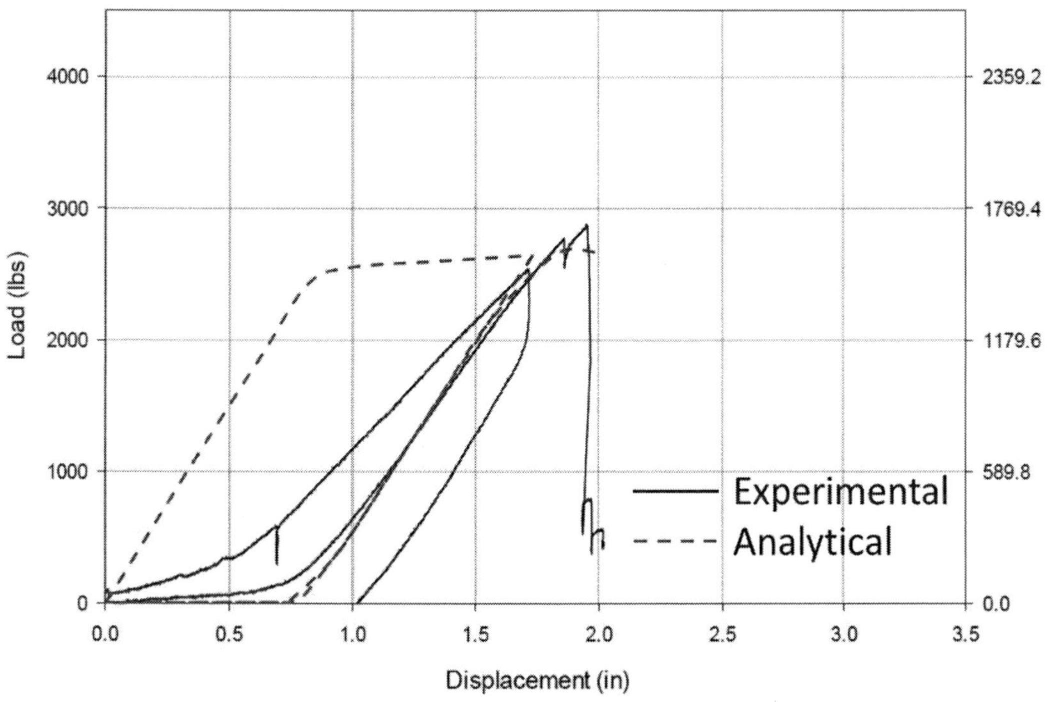

Figure 5. Load-displacement curves of specimen 1SR-S-0.

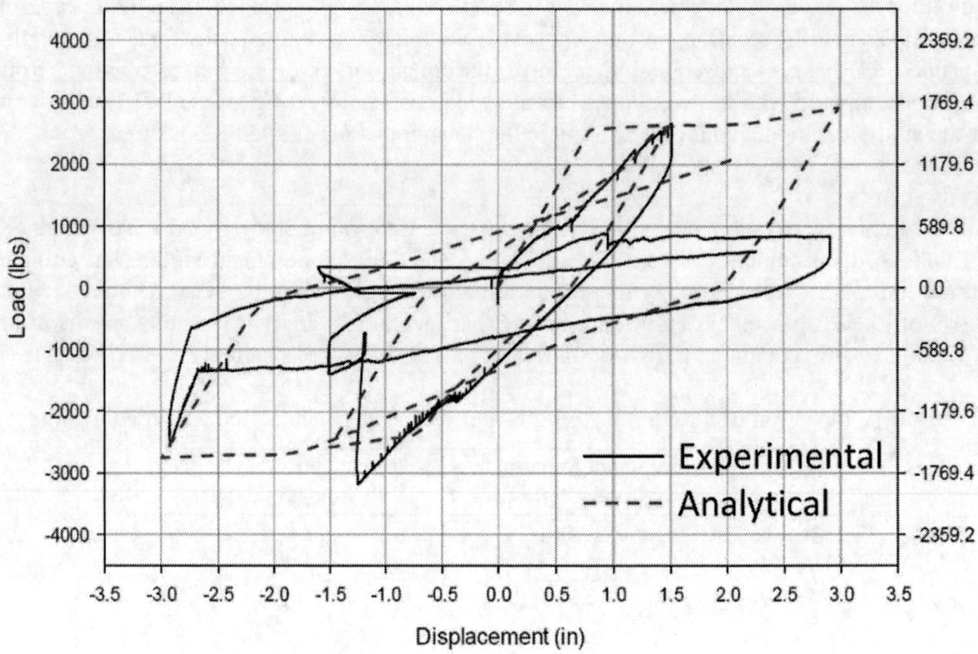

Figure 6. Load-displacement curves of specimen 4DR-S-0.

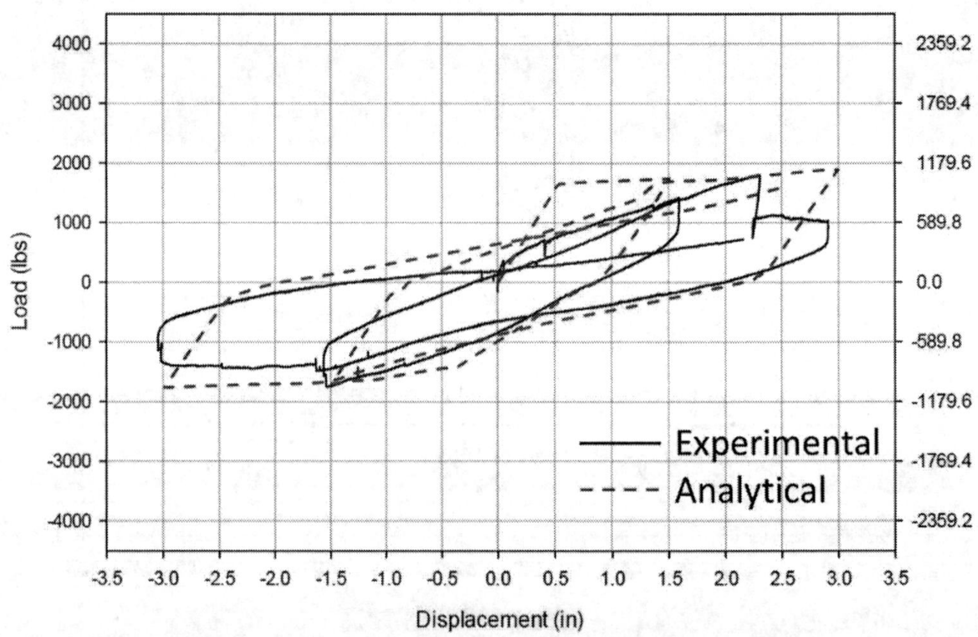

Figure 7. Load-displacement curves of specimen 5DR-B-0.

The observations and comparison for each specimen modeled analytically are presented as follows:

1. Specimen 1SR-S-0: For the load-displacement curve, it can be seen in figure 5 that the analytical model predicts higher initial stiffness compared with the experimental result. The experimental results, however, did not follow common behavior as there was a significant increase in the beam stiffness starting from displacement of 0.7 inches onwards. Usually, for RC beams, the

stiffness will decrease as the displacement increases because of crack propagation. For the second cycle, it can be seen in figure 5 that the analytical model can predict the reloading curve quite accurately, including the maximum strength. Nevertheless, for the degrading branch, the analysis could not proceed further due to convergence problems after the beam strength started to decrease. From table 1, it can be seen that the maximum experimental and analytical strengths for specimen 1SR-S-0 do not differ much, only by 7.63%.

2. Specimen 4DR-S-0: For the load-displacement curve, it can be seen in Figure 6 that the analytical model predicts a higher initial stiffness compared with the experimental result. Moreover, from the hysteretic behavior, it can be seen that the analytical model shows a ductile flexural response, whereas for the experimental result there was a significant reduction in strength and stiffness after the beam reached its maximum strength in negative direction. This was due to shear failure of the nodal points near the supports (complete detachment of the nodes from the horizontal chords). Nevertheless, this failure mode could not be simulated well by the analytical model that uses frame elements. From table 1, it can be seen that the maximum experimental and analytical strengths for specimen 4DR-S-0 only have 8.69% difference.
3. Specimen 5DR-B-0: For the load-displacement curve, similar to the other cases, it can be seen in figure 7 that the analytical model predicts a higher initial stiffness compared with the experimental result. Furthermore, the analytical model shows a ductile flexural response, whereas for the experimental result there was a reduction in strength after the beam reached its maximum strength in positive direction. This was due to bond failure between basalt rebars and concrete. However, this behavior could not be simulated well by the analytical model that assumed a perfect bond between the rebars and concrete. From table 1, unlike other cases that underestimate the experimental strengths, for this specimen, the analytical model predicts slightly higher strength with a difference of 5.76%.

5. Concluding remarks

From this initial study, the following remarks can be made:

- The analytical model consistently predicts a higher initial stiffness compared with the experimental result. This can be due to the stiffness of the 3D-printed RC beams being less than that of the conventional RC beams. The constitutive material model used for concrete in the analysis is the Mander's model for unconfined concrete [14], which is actually for conventional concrete and, thus, it results in higher initial stiffness of the beams. Furthermore, another possibility is that the analytical model assumed perfect geometry and no damage in the specimen before loading, whereas in the experiment there might be some imperfections in geometry as well as slight damage in the specimen before testing, as seen in specimen 1SR-S-0 that did not follow common material behavior (increase in beam stiffness as the displacement increasing).
- The analytical model can accurately predict the maximum flexural strengths of 3D-printed RC beams. The comparisons show that the difference between maximum experimental and analytical strengths are below 10% for all specimens.
- From the hysteretic behavior, it can be concluded that the analytical model that used frame elements with fiber P-M-M hinges could not predict some modes of failure of the 3D-printed RC beams, i.e., shear failure of the nodal points and bond failure between rebars and concrete. It is highly recommended to use a more advanced analytical model (i.e., solid element) in order to simulate the cyclic behavior of 3D-printed RC beams correctly.
- Future research is needed to define correctly the constitutive material model for the 3D-printed concrete which is different than that for conventional concrete, especially in terms of the stress-strain curve and its hysteretic behavior, the interfacial bonding between concrete mortar layers, and the bonding between rebars and concrete.

Acknowledgments
The authors are grateful for the funding provided by the Deputy for Research Empowerment and Development, Ministry of Research and Technology/National Research and Innovation Center, Republic of Indonesia under PDUPT scheme no. 002/SP2H/LT-MULTI/LL7/2020.

References

[1] Schutter G D, Lesage K Mechtcherine V Nerella V N Habert G and Agusti-Juan I 2018 Vision of 3D printing with concrete technical, economic and environmental potentials *Cem. Conc. Res.* **112** 25-36
[2] Gosselin C, Duballet R Roux P Gaudillière N Dirrenberger J and Morel P 2016 Largescale 3D printing of ultra-high-performance concrete - a new processing route for architects and builders *Mat. Design* **100** 102-9
[3] Duballet R, Baverel O and Dirrenberger J 2016 Classification of building systems for concrete 3D printing *Auto. Const.* **83** 247-58
[4] Hager I, Golonka A and Putanowicz R 2016 3D printing of buildings and building components as the future of sustainable construction *Proc. Eng.* **151** 292-9
[5] Buswell R A, d Silva W R L Jones S Z and Dirrenberger J 2018 3D printing using concrete extrusion: A roadmap for research *Cem. Conc. Res.* **112** 37-49
[6] Kazemian A, Yuan X Cochran E and Khoshnevis B 2017 Cementitious materials for construction-scale 3D printing: Laboratory testing of fresh printing mixture *Cons. Build. Mat.* **145** 639-47
[7] Ma G, Li Z and Wang L 2018 Printable properties of cementitious material containing copper tailings for extrusion-based 3D printing *Cons. Build. Mat.* **162** 613-27
[8] Lim S, Buswell R A Le T T Austin S A Gibb A G F and Thorpe T 2012 Developments in construction-scale additive manufacturing processes *Auto. Cons.* **21** 1 262-8
[9] Bos F, Wolfs R Ahmed Z and Salet T 2016 Additive manufacturing of concrete in construction: Potentials and challenges of 3D concrete printing *Virt. Phys. Prot.* **11** 3 209-25
[10] Perrot A, Rangeard D and Pierre A 2016 Structural built-up of cement-based materials used for 3D-printing extrusion techniques *Mat. Struc.* **49** 4 1213-20
[11] Computer and Structures Inc. 2019 *SAP2000 Structural Analysis Program* (Computer and Structures, Inc.)
[12] Al-Chaar G K, Stynoski P B and Banko M L 2018 Structural behavior of layer-printed reinforced concrete beams *Open. Cons. Build. Tech. Jour.* **12** 375-88
[13] Le T T, Austin S A Lim S Buswell R A Law T Gibb A G F and Thorpe T 2012 Hardened properties of high-performance printing concrete *Cem. Conc. Res.* **42** 558-66
[14] Mander J B, Priestley M J N and Park R Theoretical stress-strain model for confined concrete *ASCE Jour. Struc. Eng.* **114** 3 1804-26
[15] Park R and Paulay T 1975 *Reinforced Concrete Structures* (Canada: John Wiley & Sons, Inc.)

Effect of reinforced concrete beam confinement under cyclic loading on ultimate drift ratio

L S B Wibowo[1*] and M S D Cahyono[1]

[1]Department of Civil Engineering, Widya Kartika University, Surabaya, Indonesia

*Corresponding author's e-mail: leonarduswibowo@widyakartika.ac.id

Abstract. The purpose of this paper is to evaluate the effect confinement on ultimate drift ratio of reinforced concrete beam specimens using normal concrete. All beam specimens were tested under cyclic loading. The measured compressive strength of the concrete is in the range of 25 – 41.8 MPa, longitudinal reinforcement yield strength is in the range 350 – 570 MPa and transverse reinforcement yield strength is in the range 275 –570 MPa. Test parameters include s/d_b ratio, ρ_t, and $V_s,a/M_n$ ratio. The results showed that $V_s,a/M_n$ ratio have a significant effect on the ultimate drift ratio. Increase the nominal shear strength by confinement leads to an increase of ultimate drift ratio. In other hands, spacing of transverse reinforcement not exceeding $8d_b$ can achieve minimum 3.00% ultimate drift ratio.

1. Introduction

In the previous research, the amount of transverse reinforcement has an influence on the performance of reinforced concrete beams under cyclic loading and has an effect on shear strength and ultimate drift ratio. One of ACI318-14 [1] requirement for spacing of hoops is eight times the diameter of the smallest longitudinal bar enclosed. Based on ASCE 41-13 [2], drift limit for reinforced concrete beams to achieve life safety condition are between 2.0% to 2.5%. Test results by Panagiotou et al. [3] shows beam with reduced transverse reinforcement spacing was capable of more displacement cycles and larger displacement amplitude than beam with wider transverse reinforcement spacing. Ultimate drift ratio is defined as the ratio of maximum lateral displacement to the total of shear span. The analysis presented seeks to establish a direct relationship between, s/d_b ratio, transverse reinforcement ratio, and $V_s \times a/M_n$ ratio on ultimate drift ratio. This study focused on reinforced concrete beams under cyclic loading.

2. Experimental Database

The beams were subjected to cyclic loading and single curvature. 15 specimens were tested in a vertical position and 11 specimens were tested in a horizontal position. The test setup is presented in Fig. 1. The shear span, measured from the center of load application to the top of the concrete base block. Data of 26 beam specimens were collected from various sources (Kinugasa and Nomura [4], Fang et al. [5], Ou et al. [6], Vu [7], Cheng and Giduquio [8], Panagiotou et al. [3], Marefaat et al. [9], Walker and Dhakal [10], Ou and Chen [11], Tanarslan [12], and Jin et al. [13]). Details of the beam data including dimensions, material properties, and ultimate drift ratios. The ultimate drift capacity ratio of the beams is defined as the displacement corresponding to the maximum shear strength (V_{max}) or 20% drop of the maximum shear strength (V_{max}).

Content from this work may be used under the terms of the Creative Commons Attribution 3.0 licence. Any further distribution of this work must maintain attribution to the author(s) and the title of the work, journal citation and DOI.
Published under licence by IOP Publishing Ltd

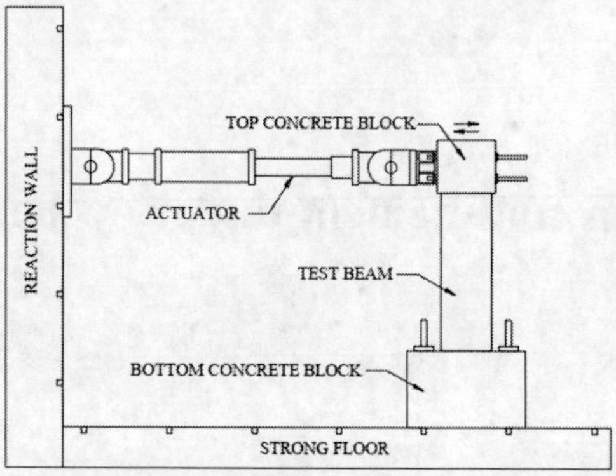

(a) Vertical position

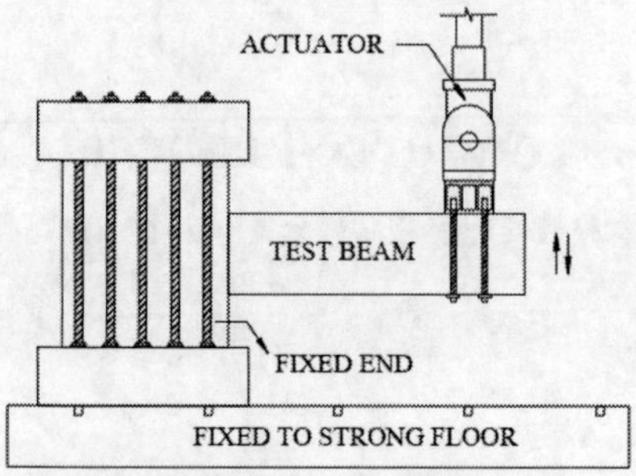

(b) Horizontal position

Figure 1. Test setup

The parameters considered in the study included concrete compressive strength (f_c'), yield strength of the longitudinal reinforcement (f_y) and transverse reinforcement (f_{yt}), longitudinal and transverse reinforcement ratio, and aspect ratio. The following criteria were considered in establishing database are 1) Only cantilever beams subjected to standard cyclic testing procedure as in figure 1 and figure 2) Only beams which has the ratio of top reinforcement is the same as the ratio of bottom reinforcement ($\rho = \rho'$), 3) Only beams for which the complete data are known were used. table 1 presents the range of properties of the beams included in the database. A summary of beam specimen and material properties is provided in table 2.

Table 1. Range of Properties in the selected beam database.

Variable	Units	Minimum	Maximum
f_c'	MPa	25	41.8
f_y	MPa	350	570
$\rho = \rho'$	%	0.42	2.16
f_{yt}	MPa	275	570.2
ρ_t	%	0.19	1.43
a/d	-	2.6	5.0
s/d_b	-	2.62	25

Table 2. Summary of beam specimen and material properties.

Reference	Beam name	f_c' (MPa)	b (mm)	d (mm)	f_y (MPa)	ρ	f_{yt} (MPa)	ρ_t	s/d_b	a/d	Type*
Kinugasa and Nomura (2006)	A1	28.2	200	170.0	352; 361	0.95%	366	0.75%	3.93	2.94	V[a]
	A2	26.9	200	170.0	352; 361	0.95%	366	0.75%	3.93	2.94	V
	B	33.3	200	170.0	402	1.75%	366	0.19%	9.43	2.94	V
Fang et al. (1993)	B1-6	33.1	200	332.0	489	2.16%	310	1.43%	2.62	2.56	H[b]
Ou et al. (2011)	B-0	25	300	426.2	448	0.78%	448	0.48%	6.29	3.52	H
Vu (2013)	At-0	30.36	300	433.0	524	1.49%	570.16	0.84%	3.48	2.77	H
	Bt-0	39	300	433.0	524	1.49%	431.88	0.84%	3.48	2.77	H
Cheng and Giduquio (2011)	SP1	37	250	540.0	485; 421	0.72%	420	0.46%	7.86	3.89	V
Panagiotou et al. (2013)	BEAM 1	36.8	762	1115.8	503; 455	0.65%	455	0.28%	22.00	3.42	H
	BEAM 2	36.8	762	1115.8	503; 455	0.65%	455	0.51%	12.00	3.42	H
Marefat et al. (2009)	PN-CS4	27.8	150	265.0	356	0.51%	310.27	0.19%	25.00	3.02	V
Walker and Dhakal (2009)	A1 (rev)	41.8	250	349.9	350	1.71%	445	0.36%	6.94	4.00	H
	A2 (rev)	41.8	250	349.9	350	1.71%	445	0.63%	3.97	4.00	H
	D1 (rev)	25.6	410	354.5	570	0.42%	560	0.22%	10.94	3.95	H
Ou and Chen (2014)	Bt-0	38	300	432.5	444	1.53%	432	0.88%	3.45	2.77	H
Tanarslan (2011)	Beam 1	25.2	200	320.0	414	1.47%	275	0.38%	3.75	5.00	H
Jin et al. (2018)	CB-1-i	35.5	80	170.0	405.97	1.66%	297.89	0.94%	8.33	4.00	V
	CB-1-ii	35.5	80	170.0	405.97	1.66%	297.89	0.94%	8.33	4.00	V
	CB-2-i	35.5	160	345.0	392.1	1.14%	405.97	0.66%	8.00	4.00	V
	CB-2-ii	35.5	160	345.0	392.1	1.14%	405.97	0.66%	8.00	4.00	V
	CB-3-i	35.5	240	540.0	399.13	1.24%	392.1	0.63%	6.25	4.00	V
	CB-3-ii	35.5	240	540.0	399.13	1.24%	392.1	0.63%	6.25	4.00	V
	CB-4-i	35.5	320	735.0	402	1.16%	392.1	0.63%	8.33	4.00	V
	CB-4-ii	35.5	320	735.0	402	1.16%	392.1	0.63%	8.33	4.00	V
	CB-5-i	35.5	400	930.0	374.84	0.65%	392.1	0.50%	8.33	4.00	V
	CB-5-ii	35.5	400	930.0	374.84	0.65%	392.1	0.50%	8.33	4.00	V

[a]vertical position
[b]horizontal position

3. Result and Analysis

The experimental-to-nominal flexural strength ratio ($M_{exp.}/M_n$) is between 0.71 and 1.61 for all test specimens as shown in Table 3., where $M_{exp.}$ is the average peak flexural strength from the two loading directions and M_n is determined per ACI318-14[1] with test material properties. Ultimate drift ratio (d_u) is defined at the point when one of the following two criteria is first met: 1) the load dropped 20% from the peak on the envelope curve; or 2) the load dropped more than 20% in the repeated cycles at the same target drift level. Ultimate drift ratio (d_u) is the average from the two loadings direction. As shown in Figure 2, most of the specimen can achieve an ultimate drift ratio more than 2.0% and only one specimen has ultimate drift ratio below 2.0%.

Table 3. Summary of Analysis Results

Reference	Beam name	d_u (%)	$M_{exp.}/M_n$	$V_s \times a/M_n$
Kinugasa and Nomura (2006)	A1	5.42	1.15	2.573
	A2	5.03	1.11	2.586
	B	2.39	0.90	0.325
Fang et al. (1993)	B1-6	5.96	0.96	1.242
Ou et al. (2011)	B-0	4.66	1.19	2.309
Vu (2013)	At-0	4.68	0.99	1.828
	Bt-0	4.99	1.06	1.378
Cheng and Giduquio (2011)	SP1	4.53	1.18	2.321
Panagiotou et al. (2013)	BEAM 1	3.70	1.00	1.492
	BEAM 2	5.46	1.08	2.735
Marefat et al. (2009)	PN-CS4	1.74	0.71	1.006
Walker and Dhakal (2009)	A1 (rev)	3.04	0.98	1.165
	A2 (rev)	4.29	1.01	2.038
	D1 (rev)	3.57	0.99	2.164
Ou and Chen (2014)	Bt-0	4.99	1.27	1.734
Tanarslan (2011)	Beam 1	3.84	1.36	0.925
Jin et al. (2018)	CB-1-i	5.42	1.61	1.813
	CB-1-ii	4.72	1.59	1.813
	CB-2-i	5.66	1.41	2.561
	CB-2-ii	6.96	1.45	2.561
	CB-3-i	7.71	1.17	2.131
	CB-3-ii	8.53	1.13	2.131
	CB-4-i	5.57	1.18	2.234
	CB-4-ii	6.33	1.18	2.234
	CB-5-i	4.01	1.06	1.933
	CB-5-ii	4.35	1.07	1.933

ρ_t is ratio of area of distributed transverse reinforcement to gross concrete area perpendicular to that reinforcement. To calculate transverse reinforcement ratio is presented in eq. (1)

$$\rho_t = \frac{A_{st}}{b \cdot s} \qquad \text{Eq. (1)}$$

where, A_{st} is total area of transverse reinforcement, b is width of beam, s is spacing of transverse reinforcement.

V_s is nominal shear strength provided by transverse reinforcement. To calculate nominal shear strength is presented in eq. (2)

$$V_s = \frac{A_{st} \times f_{yt} \times d}{s} \quad \text{Eq. (2)}$$

where, A_{st} is total area of transverse reinforcement, f_{yt} is yield strength of transverse reinforcement, d is effective depth of beam, and s is spacing of transverse reinforcement.

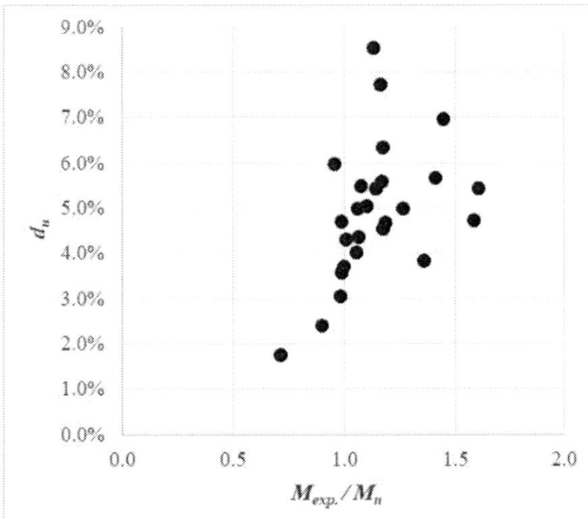

Figure 2. Relationship between $M_{exp.}/M_n$ and d_u

Table 4. Correlation coefficients, R, for confinement and ultimate drift ratio

Confinement parameter	s/d_b	ρ_t	$V_s \times a/M_n$
Correlation coefficient, R	0.437	0.535	0.584

A series of linear regression analyses were performed to identify the most influential parameters, on beam ultimate drift ratio. Correlation coefficients R, for the complete database of 26 specimen test for various parameters are presented in table 4. Parameters $V_s \times a/M_n$ produce the highest correlation coefficients with beam ultimate drift ratio, with R being 0.584.

3.1 Effect of s/d_b ratio

In the database, there are two specimens that have s/d_b larger than $20d_b$. Ultimate drift ratio (d_u) of specimen BEAM 1 (Panagiotou et al., 2013) and PN-CS4 (Marefat et al., 2009) is 3.70% and 1.74%, respectively. Results s/d_b presented in figure 3 show that the ultimate drift ratio between that two specimens are different, even though their s/d_b almost the same. The other specimens that have s/d_b less than $15d_b$ can achieve ultimate drift ratio more than 2.0%. Based on figure 3, the relationship between s/d_b and ultimate drift ratio is not clear. Most of the specimen that have s/d_b less or equal than $8d_b$ can achieve 3.00% ultimate drift ratio or larger.

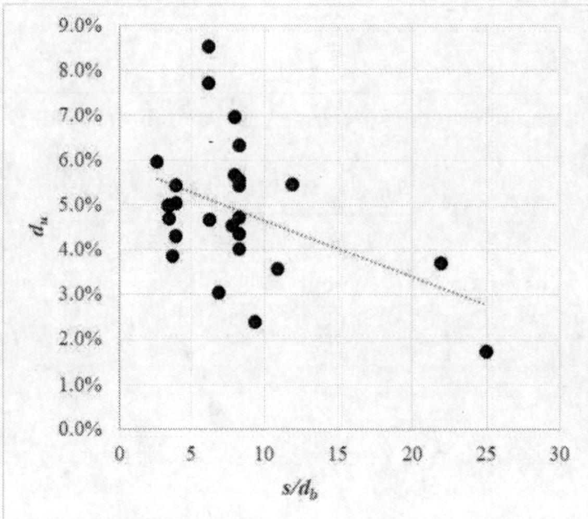

Figure 3. Relationship between s/d_b and d_u

3.2 Effect of transverse reinforcement ratio (ρ_t)

Figure 4 and table 3 indicate that ultimate drift ratio is correlated with ρ_t, with ultimate drift ratio varying between 1.74% and 8.53% as ρ_t increase from 0.19% to 1.43%. Specimen B (Kinugasa and Nomura, 2006), specimen PN-CS4 (Marefat et al., 2009), and specimen D1 (Walker and Dhakal, 2009) which have a transverse reinforcement ratio below than 0.25%. As shown in figure 4, only specimen D1 can achieve ultimate drift ratio larger than 3.00% compare to other two specimens.

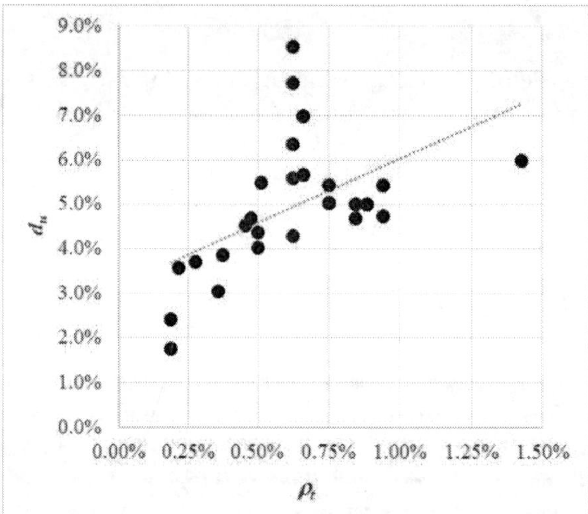

Figure 4. Relationship between ρ_t and d_u

3.3 Effect of $V_s \times a/M_n$

As noted earlier, $V_s \times a/M_n$, has a significant impact on ultimate drift ratio, where V_s is nominal shear strength, a is shear span and M_n is nominal moment capacity. As shown in figure 5, specimen B (Kinugasa and Nomura, 2006) and specimen PN-CS4 (Marefat et al., 2009) cannot achieve 2.50% drift, even though specimen PN-CS4 (Marefat et al., 2009) has $V_s \times a/M_n$ equal to 1.00. Most of the specimens have $V_s \times a/M_n$ larger than 1.00 can achieve ultimate drift ratio larger than 3.00%.

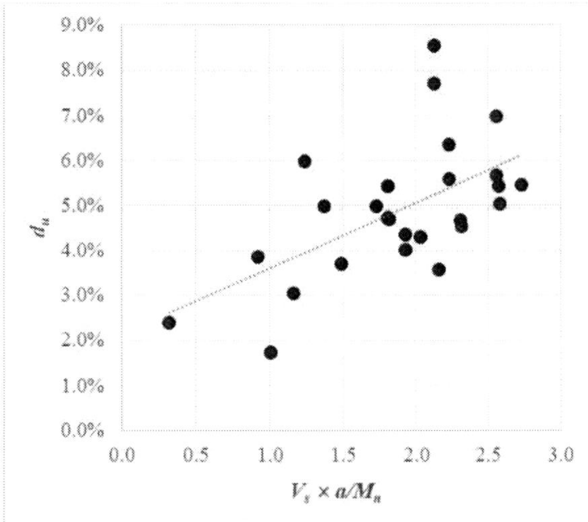

Figure 5. Relationship between $V_s \times a/M_n$ and d_u

4. Conclusions

Based on the findings of this study, the following conclusions with regards to ultimate drift ratio of cyclic beams can be drawn:

1. The experimental-to-nominal flexural strength ratio (M_{exp}/M_n) larger or equal to 1.00 can achieve 3.00% ultimate drift ratio.
2. A maximum transverse reinforcement spacing not exceeding $8d_b$ can achieve minimum 3.00% ultimate drift ratio.
3. The study showed that $V_s \times a/M_n$ increased followed by the ultimate drift ratio.

Acknowledgments

The authors thank to the Ministry of Research, Technology, and Higher Education of the Republic of Indonesia for funding this research through Hibah Penelitian Dosen Pemula (PDP) and Universitas Widya Kartika.

References

[1] ACI Committee 318-14 2014 *Building Code Requirement for Structural Concrete (ACI 318-14) and Commentary (ACI 318R-14)* (Farmington Hills, MI: American Concrete Institute)

[2] ASCE/SEI 41-13 2013 *Seismic Evaluation and Retrofit of Existing Buildings (ASCE 41-13)* (Reston, Virginia: American Society of Civil Engineers)

[3] Panagiotou M et al 2013 *Effect of hoop reinforcement spacing on the cyclic response of large reinforced concrete special moment frame beams* (Berkeley: Department of Civil and Enviromental Engineering, University of California)

[4] Kinugasa H and Nomura S 2006 Deterioration mechanism of shear-resisting system in RC beam subjected to reversed cyclic loading after flexural yielding *ACI Special Publication* **237** 1 1-14

[5] Fang I K, Yen S T Wang C S and Hong K L 1993 Cyclic behavior of moderately deep HSC beams *J. Struct. Eng.* **119** 9 2573-92

[6] Ou Y C, Tsai L L and Chen H H 2012 Cyclic performance of large-scale corroded reinforced concrete beams *Earthq. Eng. Struct. Dyn.* **41** 593-604

[7] Vu N N 2013 *Residual shear strength and ductility evaluation of corroded RC beams* (Taiwan: National Taiwan University of Science and Technology)

[8] Cheng M Y and Giduquio M B 2014 Cyclic behavior of reinforced concrete flexural members

using high-strength flexural reinforcement *ACI Struct. J.* **111** 1 1-10
[9] Marefat M S and Shirazi S M H 2009 Cyclic Response of Concrete Beams Reinforced by Plain Bars *J. Earthq. Eng.* **13** 4 463-81
[10] Walker A F and Dhakal R P 2009 Assesment of material strain limits for defining plastic regions in concrete structures *Bull. NZ. Soc. Earthq. Eng.* **42** 2 86-95
[11] Ou Y C and Chen H H 2014 Cyclic behavior of reinforced concrete beams with corroded transverse steel reinforcement *J. Struct. Eng.* **140** 9 1-10
[12] Tanarslan H M 2011 The effects of NSM CFRP reinforcements for improving the shear capacity of RC beams *Constr. Build. Mat.* **25** 2663-73
[13] Jin L, Yu W Xiao S Zhang S Du X Han J and Li D 2018 Effect of cross-section size on the flexural failure behavior of RC cantilever beams under low cyclic and monotonic lateral loadings *Eng. Struct.* **156** 567-86

IOP Conf. Series: Materials Science and Engineering **930** (2020) 012043 doi:10.1088/1757-899X/930/1/012043

Determination of LRFD environmental load factor in the java sea for unbraced monopod structures using reliability analysis

Paramashanti[1*] and D A Larasati[2]

[1] Ocean Engineering Department, Bandung Institute of Technology, 10 Ganesha Street, Bandung
[2] Ocean Engineering Student, Bandung Institute of Technology, 10 Ganesha Street, Bandung

*Corresponding author's e-mail: parama@ocean.itb.ac.id

Abstract. Case studies were carried out for unbraced monopod structures located in the Java Sea by reviewing the value of the reliability index. This sea has a depth of 33 meters and a 100-yr wave height of 6.29 meters. The structures are designed optimally with a unity check ratio of 1.00 using API RP 2A WSD and API RP 2A LRFD standards under storm condition, assumed to be unmanned. For LRFD criteria, the value of the environmental load factor is varied, namely 1.35, 1.20, 1.10, and 1.00. Collapse analysis using static pushover analysis of the model produces a minimum RSR value meeting the minimum criteria for un-manned structures of 1.6. The reliability analysis process uses the First Order Reliability Method II (FORM II). The structural reliability index value has met the minimum criteria issued by DNV of 3.71 for the structure that is designed using the environmental load factor of 1.35. It can be concluded for unbraced monopod structures designed in the Java Sea, the environmental load factor value of 1.35 is following the minimum reliability index value proposed by DNV.

1. Introduction

The process of oil and gas field exploration requires an infrastructure that designed optimally using applicable design standards to be able to withstand the equipment and environmental loads later. Structures designed using Load Resistance Factor Design (LRFD) standards are much more reliable compared using Working Stress Design (WSD) criteria due to the use of load factor, based on the geographic location [1]. WSD criteria consider the uncertainty components related to resistance and load variable using the safety factor by reducing the allowable strength. While LRFD criteria include the uncertainty of resistance and load variable using the statistics of the resistance and load variable. The WSD criteria became uneconomical because it doesn't consider the uncertainty factor [8].

Commonly used standards are published by the American Petroleum Institute (API) which adapts water conditions in the Gulf of Mexico. The environment conditions in the Gulf of Mexico are relatively more extreme than in Indonesia, so the application of environmental load factors in Indonesia needs to be reviewed. Indonesia has relatively calmer environmental conditions and more shallow water fields. The use of monopod structures will be more economical for exploiting these conditions. The partial load factors in API were also calibrated for jacket platforms [10]. The load transfer between jacket platform and monopod platforms are also different, where monopod platforms distribute both vertical and horizontal loads to one member. It is necessary to study the use of monopod structures.

The calculation of offshore structure analysis is the subject of uncertainty. Uncertainty can be faced by considering the uncertainty parameters of the resistivity and load variables. The values are analyzed based on the amount of information about the statistical parameters that contain uncertainty [1].

Content from this work may be used under the terms of the Creative Commons Attribution 3.0 licence. Any further distribution of this work must maintain attribution to the author(s) and the title of the work, journal citation and DOI.
Published under licence by IOP Publishing Ltd

Accordingly, uncertainty modeling is essential to perform reliability analysis towards offshore structures.

Studies about the determination of environmental load factors in nearby regions produce different values. These values significantly depend on the geographic conditions of each region [9]. Kurian et. al. (2014) proposed the environmental load factor of 1.25 for jacket platform structure in Malaysian waters. This value is obtained from the system reliability of all jacket components. Meanwhile, Ronalds (2007) suggested the environmental load factor of 3.00 for concrete monopod structures in the Australia North West Sea. Reliability analysis was conducted by reviewing the Reserve Strength Ratio (RSR) of the structure's main leg with a reliability index target of 3.3. This target is converted from the probability of failure (*PoF*) of 5×10^{-4} for unmanned structures (ISO 19902).

2. Methodology

2.1 Structure Modelling

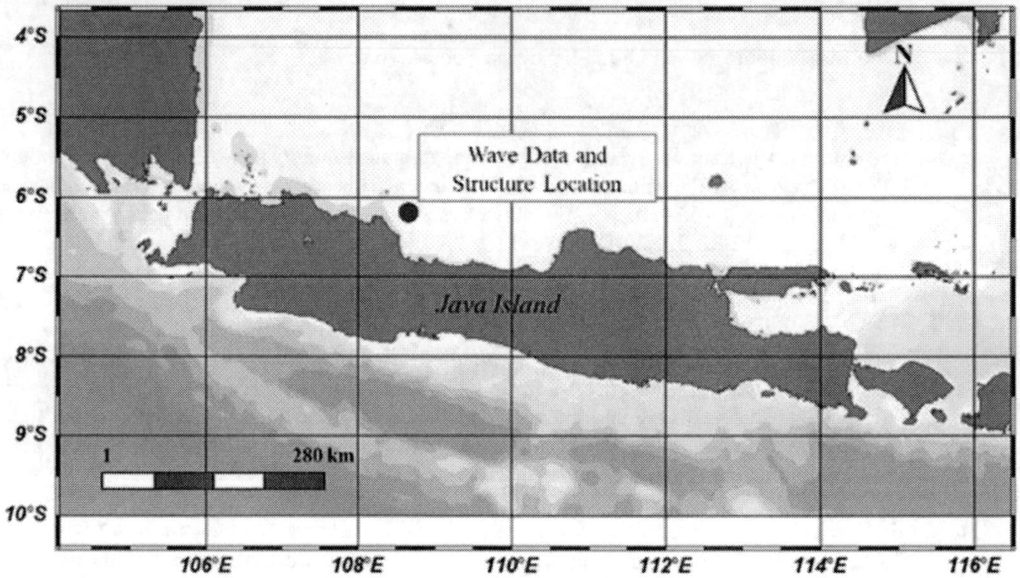

Figure 1. The wave data and the structure is located in the Java Sea.

This study took place in the Java Sea (Figure 1) with a depth of 33 m. With the same environmental and topside load conditions, the optimum design (UC = 1.00) is carried out by using API RP 2A WSD and API RP 2A LRFD, which may produce more than one different structure. For LRFD criteria, variations are applied for the environmental load factor values (γ_E), namely 1.35, 1.20, 1.10, and 1.00. A total of five structures are managed by changing the thickness of the leg, while keeping the diameter for all structures at the same value, using SACS.

The deck is located 12 m above Stillwater and modeled as a simple beam with a weight of 1000 kN and length of 10 m. The structure is assumed to be unmanned with zero marine growth. The whole structure is made of steel, assumed to have a yield stress of 248 MPa. The 100-yr return period storm load is used for design. The environmental loading only includes waves, other environmental loadings such as current and wind loading are neglected. The whole monopod configurations considered in this study are summarized in Table 1.

Table 1. Monopod Configuration.

Parameter	Symbol	Units	Value
Water Depth	h	m	33
Leg Diameter	D	cm	86
Thickness	t	cm	1-1.5
Topside Weight	W	kN	1000

The structure is assumed to fail by bending in the main leg. Besides bending, axial compression is also subjected to the main leg due to the weight of the beam.

2.2 *Wave Data Analysis*

Hourly significant wave data set for 60 years from the Java Sea is used as the raw data for this study. The raw data will be processed to obtain the relationship between the wave height and wave period, and also the 100-yr return period wave height. The statistic parameters and distribution of the data set are firstly determined before conducting the main process. Normal, Lognormal, Log Pearson III, and Gumbel distributions are selected for distribution fitting tests. These distributions are tested using the Kolmogorov-Smirnov test.

The wave period value follows the wave height. To obtain the relationship between the wave period and wave height, all hourly significant wave data are plotted against its significant wave period. A regression analysis took place to determine the equation that closely matches with the relationship between the two variables. Later, the equation will be used to determine the period of the 100-yr return period wave height.

Extreme value analysis is performed to obtain the 100-year return period wave height. The maximum significant wave height from each year is collected and sorted from the lowest to the highest value. Then the 100-year return period wave height is calculated by using the extreme analysis equation from the selected distribution. Since the data set is in the form of significant wave data, it needs to be converted into maximum wave height data. Chakrabarti (1996) proposed the relationship between significant wave heights and maximum wave height in equation (1) below.

$$H_{max} = 1.86\ H_{1/3} \tag{1}$$

The 100-yr return period wave height and the calculated wave period obtained from extreme value analysis will be used in designing the model. While the maximum significant wave height of each year will be used for static in-place analysis to determine the wave base shear equation.

2.3 *Static Pushover Analysis*

Static pushover analysis is performed on each monopod to obtain the structure's maximum base shear value right before collapsing ($BS_{collapse}$). This value is used for the reliability index value calculation. The Reserve Strength Ratio (RSR) value of the structure must exceed the minimum value of 1.60 for unmanned structures with high consequence of failure.

2.4 *Static In-place Analysis*

This analysis is performed to obtain the wave base shear function. The wave base shear function (BS_{wave}) is defined as the relationship between the structure's maximum base shear and the maximum wave height that have been converted from maximum significant wave height data for 60 years. The maximum base shear value of the structures is obtained from the in-place analysis, then it is plotted against the maximum significant wave height.

Equation (2) shows the wave base shear expression:

$$L = BS_{wave} = aH^b \tag{2}$$

IOP Conf. Series: Materials Science and Engineering **930** (2020) 012043 doi:10.1088/1757-899X/930/1/012043

Where *H* is the wave height, while coefficients *a* and *b* are determined from curve fitting.

2.5 Reliability Analysis

The reliability analysis uses maximum wave height data statistical parameters. First Order Reliability Method (FORM) II is used for reliability analysis. The collapse failure mode is chosen for performance failure. Safety margin between resistance and load is indicated by limit state equation as follows:

$$g = R - L \leq 0 \tag{3}$$

Where g is the limit state function. Here, the collapse base shear acted as the resistance variable (*R*), while the wave base shear acted as the load variable (*L*). By inserting equation (2), equation (3) becomes equation (4).

$$g = BS_{collapse} - aH^{b} \leq 0 \tag{4}$$

Reliability analysis is performed using the performance function in equation (3) where the wave height carries the uncertainty value. So, the overall system uncertainty is strongly dependent on the wave height. The obtained reliability indexes need to be compared with the target values. DNV proposed a recommended target reliability index (β) of 3.71, or equivalent with a probability of failure of 10^{-4} for a non-redundant structure.

3. Results and Discussion

3.1 Design Wave Height

The Kolmogorov-Smirnov test results are shown in Table 2.

Table 2. The Kolmogorov-Smirnov test result.

Distribution	Maximum D_n
Normal	0.0695
Lognormal	0.0299
Rayleigh	0.0683
Gumbel	0.0236

The test result above shows that Gumbel distribution is the closest distribution form to the 60 years of hourly wave data distribution. The critical D_n^{α} for 2358 data with significance level of 15% is 0.0242. It is concluded that the Gumbel distribution can be accepted to represent the wave data. The probability density of data and each distribution can be seen in Figure 2.

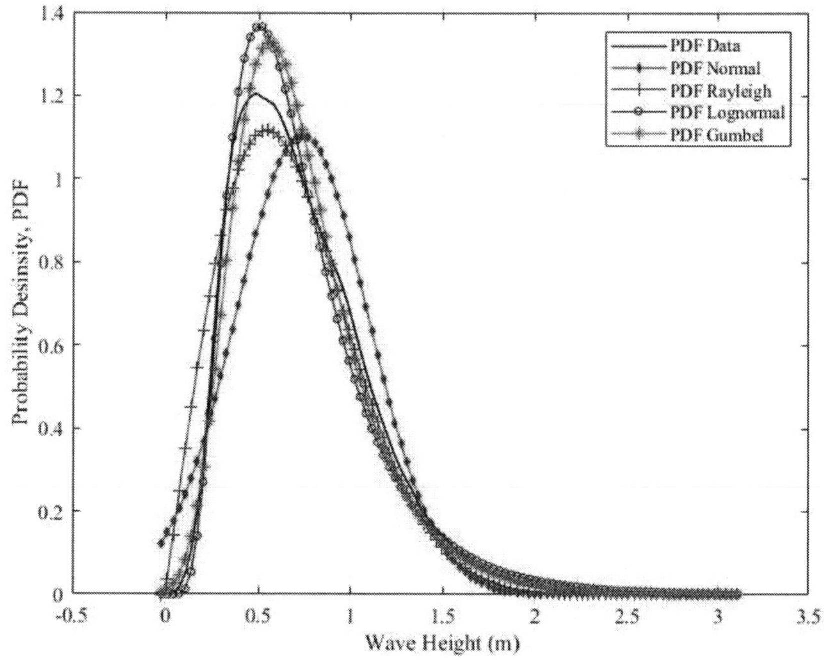

Figure 2. The probability density function of wave data and test distributions.

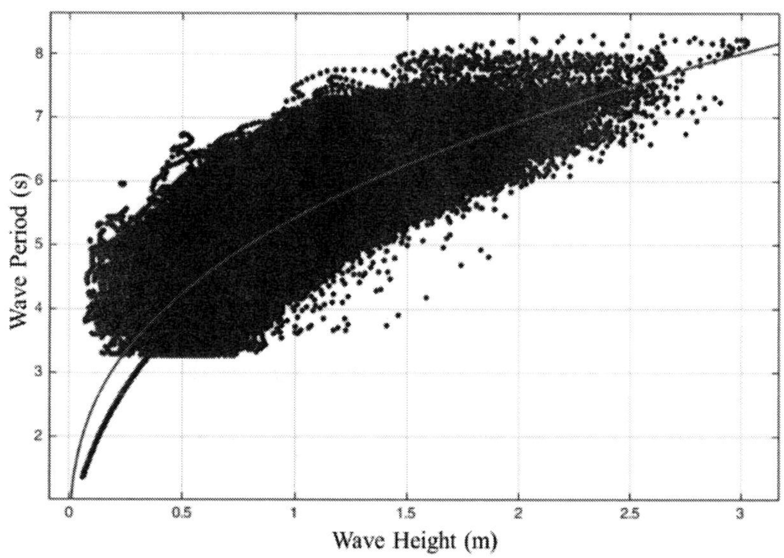

Figure 3. Hourly significant wave data set plotted against its period.

Figure 3 shows the relationship between the wave period and wave height that can be expressed in the equation (5) below.

$$T(H) = 6.241\, H^{0.4126} \tag{5}$$

From the extreme value analysis using Gumbel distribution, we obtain the design wave height of 6.29 m, with a wave period of 11.47 s. These values will be used as the environmental load for structure design.

3.2 Optimized Structure Design

Five structures are obtained after optimally designed using different environmental load factor values that are shown in Table 3. One structure is designed based on WSD criteria, and the rest are designed using LRFD criteria with a different environmental load factor.

Table 3. Optimum structure dimensions for each criterion.

Criteria	γ_E	*D (cm)*	*t (cm)*
WSD	1.00	86	1.375
LRFD	1.35	86	1.350
	1.20	86	1.280
	1.10	86	1.235
	1.00	86	1.190

The UC value of each structure is checked using allowable combined stress in equation (6) and equation (7) for cylindrical members from API RP 2A WSD and API RP 2A LRFD respectively. Axial compression and bending stress are subjected in the leg member. A cylindrical member under combined compression and bending loads should be proportioned to satisfy the following requirements at all cross-sections along their length.

$$\frac{f_a}{F_a}+\frac{C_m\left(f_{bx}{}^2+f_{by}{}^2\right)^{1/2}}{\left(1-\frac{f_a}{F_e'}\right)F_b}\leq 1.0 \tag{6}$$

where:

C_m	=	Reduction factor, for restrained end members, 0.85
F_a	=	Allowable axial compressive stress, ksi (MPa)
f_a	=	Acting axial stress, ksi (MPa)
F_b	=	Allowable bending stress, ksi (MPa)
f_{bx}, f_{by}	=	Bending stress corresponding to the member x and y-axis respectively, ksi (MPa)
F_e	=	Critical elastic buckling stress, ksi (MPa)

$$\frac{f_c}{\phi_c F_{cn}}+\frac{1}{\phi_b F_{bn}}\left(\left[\frac{C_{my}f_{by}}{\left(1-\frac{f_e}{\phi_c F_{ey}}\right)}\right]^2+\left[\frac{C_{mz}f_{bz}}{\left(1-\frac{f_e}{\phi_c F_{ez}}\right)}\right]^2\right)^{1/2}\leq 1.0 \tag{7}$$

where:

C_{my}, C_{my}	=	Reduction factor corresponding to the member y and z-axis respectively, for restrained end members, 0.85
F_{bn}	=	Nominal bending stress, ksi (MPa)
f_{by}, f_{bz}	=	Bending stress corresponding to the member y and z-axis respectively, ksi (MPa)
F_{cn}	=	Nominal axial compressive strength, ksi (MPa)
f_c	=	Axial compressive strength due to factored loads, ksi (MPa)
F_{ey}, F_{ez}	=	Euler buckling stress corresponding to the member y and z-axis respectively, ksi (MPa)
ϕ_b	=	Resistance factor for bending stress, 0.95
ϕ_c	=	Resistance factor for axial compressive strength, 0.85

The applied stress subjected to the main leg member due to external loads and its allowable stress for each structure are shown in Table 4.

Table 4. The applied stress value and allowable stress value for each structure.

Criteria	γ_E	D (cm)	t (cm)	Applied Stress			Allowable Stress			
				Axial Stress (MPa)	Bending Stress Y (MPa)	Bending Stress Z (MPa)	Axial Stress (MPa)	Euler Stress (MPa)	Bending Stress Y (MPa)	Bending Stress Z (MPa)
WSD	1.00	86	1.375	25.45	0	92.05	60.7	60.7	233.08	233.08
LRFD	1.35	86	1.350	28.44	0	126.46	74.23	87.33	282.22	282.22
	1.20	86	1.280	29.77	0	118.26	74.35	87.48	278.6	278.6
	1.10	86	1.235	30.7	0	112.18	74.43	87.57	276.07	276.07
	1.00	86	1.190	31.7	0	105.67	74.51	87.66	273.36	273.36

The ratio of the applied stress against allowable stress and the UC value for main leg member of each structure are summarized in table 5.

Table 5. Ratio of the applied stress against allowable stress and the UC value for main leg of each structure.

Criteria	γ_E	D (cm)	t (cm)	Ratio			UC
				Axial Stress	Bending Stress Y	Bending Stress Z	
WSD	1.00	86	1.375	0.419	0	0.395	1.00
LRFD	1.35	86	1.350	0.383	0	0.448	1.00
	1.20	86	1.280	0.400	0	0.424	1.00
	1.10	86	1.235	0.412	0	0.406	1.00
	1.00	86	1.190	0.425	0	0.387	1.00

The table above shows that the bigger the environmental load factor, the bigger the bending stress ratio.

3.3 Pushover Analysis

The D/t ratio values of all structures are greater than 60 so local buckling due to axial compression should be considered in pushover analysis. Both elastic and inelastic local buckling should also be considered for all structures due to the value of thickness more than 6 mm. As shown in table 6, the greater the thickness value, the greater the base shear value needed to make the structure collapse. The value of the corresponding base shear can be called a collapse base shear ($BS_{collapse}$).

Table 6. Pushover analysis results of each structure consisting of RSR and $BS_{collapse}$ value.

Criteria	γ_E	D (cm)	t (cm)	RSR	$BS_{collapse}$ (kN)
WSD	1.00	86	1.375	2.96	47.44
LRFD	1.35	86	1.350	2.89	46.38
	1.20	86	1.280	2.72	43.77
	1.10	86	1.235	2.62	42.11
	1.00	86	1.190	2.52	40.44

IOP Conf. Series: Materials Science and Engineering **930** (2020) 012043 doi:10.1088/1757-899X/930/1/012043

The RSR values of all structures are above the recommended value of 1.60 for unmanned structure with high consequence of failure. All structure's failure is caused by local buckling in the main leg as figure out in structure deformation in figure 4.

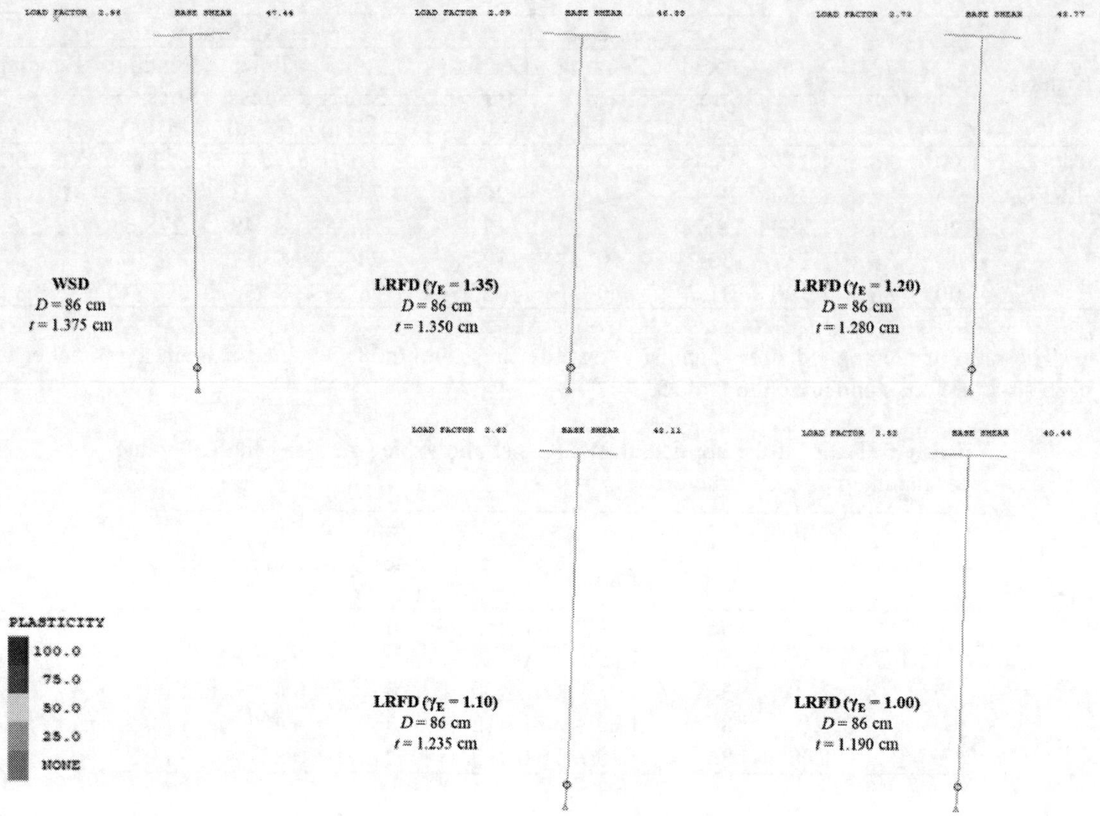

Figure 4. Structure deformation in pushover analysis for each design criterion.

3.4 Wave Base Shear

Figure 5 shows the relationship between wave base shear and wave height values. The result shows that same maximum base shear values are produced from all environmental load factor values. This indicates that the maximum base shear value caused by the wave is only affected by the diameter of the structure.

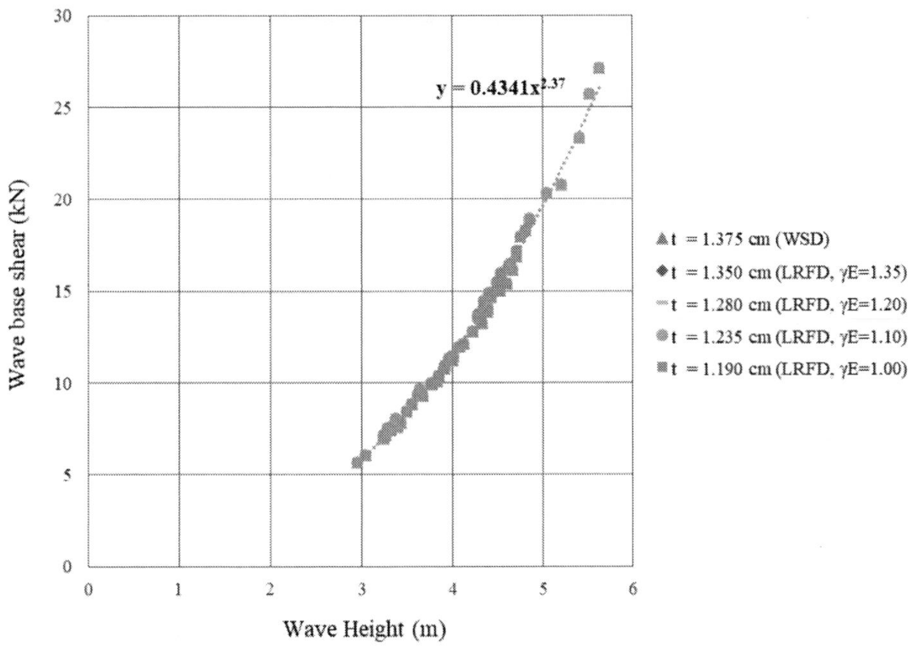

Figure 5. Maximum base shear plotted against wave height. Equation y shows the relationship between the two variables.

From the figure above, we obtain the wave base shear equation in equation (8).

$$BS_{wave} = 0.4341\ H^{2.37} \tag{8}$$

This equation is used as the performance function component for reliability analysis.

3.5 Reliability Index

The performance function for reliability analysis can be seen in equation (9).

$$g = BS_{collapse} - 0.4341\ H^{2.37} \leq 0 \tag{9}$$

When conducting reliability analysis, collapse base shear ($BS_{collapse}$) performed as the deterministic variable, while wave height (H) is the uncertainty variable. Uncertainty parameters used in this analysis is the maximum wave height statistical parameter. The maximum wave height data distribution follows the Gumbel distribution. Statistical parameters such as mean and standard deviation are used in the FORM II calculation. The calculation results of the reliability index using FORM II are shown in Table 7.

IOP Conf. Series: Materials Science and Engineering **930** (2020) 012043 doi:10.1088/1757-899X/930/1/012043

Table 7. Reliability Index (β) and the probability of failure for each design criteria using FORM II.

Criteria	γ_E	D (cm)	t (cm)	β	PoF
WSD	1.00	86	1.375	3.77	8.087×10^{-5}
LRFD	1.35	86	1.350	3.71	1.033×10^{-4}
	1.20	86	1.280	3.55	1.905×10^{-4}
	1.10	86	1.235	3.45	2.830×10^{-4}
	1.00	86	1.190	3.34	4.231×10^{-4}

The results above show that the reliability index of the unbraced monopod structure designed with an environmental load factor of 1.35 meets the recommended reliability index value proposed by DNV for a non-redundant structure of 3.71. The size of the structure may give contributions to the structure's safety.

4. Conclusion

The reliability index of the structure designed with API RP 2A LRFD environmental load factor value of 1.35 meets the requirement proposed of 3.71 for non-redundant structure. This indicates that the environmental load factor is sufficient for the unbraced monopod structure. However, this research uses only one topside load value and dimensions that are not too diverse. Further studies using varied topside weight values, dimensions, and more resistivity parameters need to be conducted.

References

[1] Adduri P R 2006 *Robust Estimation of Reliability in the Presence of Multiple Failure Modes* (India: Wright State University)

[2] American Petroleum Institute 1993 *Recommended Practice for Planning, Designing and Constructing Fixed Offshore Platforms – Load and Resistance Factor Design, API RP 2A-LRFD* (Washington: American Petroleum Institute Publications and Distribution Section)

[3] American Petroleum Institute 2000 *Recommended Practice for Planning, Designing and Constructing Fixed Offshore Platforms – Working Stress Design, API RP 2A-WSD* (Washington: American Petroleum Institute Publications and Distribution Section)

[4] Chakrabarti S K 2005 *Handbook of Offshore Engineering* (Amsterdam: Elsevier)

[5] Cossa N J, Potty N S Idrus A B Hamid M F A and Nizamani Z 2012 Reliability analysis of jacket platforms in Malaysia environmental load factors. *Research J. App. Sci. Engineering, and Technol.* **4** pp 3544-54

[6] Efthymiou M, Graaf J W Tromans P S and Hines I M 1997 Reliability-based criteria for fixed steel offshore platforms *Jour. Offshore Mechanics and Arctic Eng.* **119** pp 120-124

[7] Haldar A and Mahadevan S 2000 *Probability, Reliability, and Statistical Methods in Engineering* (Singapore: John Wiley & Sons, Inc)

[8] International Standards Organization 2007 *Petroleum and Natural Gas Industries – Fixed Steel Offshore Structures, ISO 19902* (Switzerland: International Standards Organization)

[9] Nizamani Z, Kurian V J and Liew M S 2014 Determination of environmental load factors for ISO 19902 code in offshore *Ocean Engineering* **92** pp 31-43.

[10] Ronalds B F, Cosson D G Pinna R and Cole G K 2007 Towards monopod design criteria for reliability. *Jour. Petro.Sci. and Eng.* **57** pp 8-15

IOP Conf. Series: Materials Science and Engineering **930** (2020) 012044 doi:10.1088/1757-899X/930/1/012044

The effect of Street canyon width towards wind flow in between high-rise buildings

Y Sanjaya[1*], D Priambodo[1], P W Sarli[1], & H D Setio[1]

[1]Programmed Study of Civil Engineering, Faculty of Civil and Environmental Engineering, Institut Teknologi Bandung, Bandung, Indonesia

*Correponding author's e-mail: Pw.sarli@itb.ac.id

Abstract. Wind load is often defined as a static uniform distributed load in the design code, especially on small structures such as billboard, pedestrian cross bridge, etc. However, as a fluid, wind can deform from a laminar flow to a turbulence flow or it can also be amplified if enough disturbance around the object exists. If these small structures are not designed for a dynamic load or the corrected amplified load, it will collapse. This study aims to investigate the effect of street canyon width towards wind flow by observing a simplified model of 4 symmetrical high-rise buildings with changing street canyon width. The numerical model uses CFD with RANS in OpenFOAM v1912 which is validated with experimental results using a low wind speed wind tunnel and Particle Image Velocimetry (PIV). An Experiment has been performed in BBTA3 using a model with 1:420 scale, there are 2 models with different street canyon width. The result showed there is an increment of the mean wind speed about 5.92-20.20%.

1 Introduction

The growth of high-rise buildings in Indonesia is increasing rapidly, especially in major cities such as Jakarta and it can cause local wind flow behavior. As a fluid, wind can easily transform, with enough disturbance it can be amplified or transformed from laminar flow to turbulent flow. In recent years there were some incident of a billboard and pedestrian cross bridge failure due to high wind speed. Therefore, the study of wind flow around buildings is important. In this study we will examine the fluid behavior due to the change of street canyon width.

Blocken et.al [1] observed a significant amplification factor of up to 1.4 of wind speed at the passage centerline. The amplification factor is defined as wind speed at measurement point around buildings compared to wind speed at measurement point without any buildings. It can be an indicator of the effect of the buildings to wind speed. The amplification caused by the Venturi effect at the pedestrian level it is only about 8%, so there is a need of further investigation to what causes the increase of wind speed.

Kuo et.al [2] researched about pedestrian level wind environment in a street canyon between a high-rise building with a podium and a low level attached house. Kuo et.al [2] modelled the street canyon width (S): 0.375 D; 0.75 D; 1.25 D; and 2.25 D, where D is high-rise building width. The effect of the width canyon resulted in 3 different wind flow, these are: resistance flow for S = 0.375D, interaction flow for S = 0.75 D and 1.25 D, and isolation flow for S = 1.75 D and 2.25 D. These types of flows are explained further in a research by Blocken et.al [1] research. The resistance flow has the maximum amplification factor is located at the passage center line just beyond the passage entrance, the interaction

Content from this work may be used under the terms of the Creative Commons Attribution 3.0 licence. Any further distribution of this work must maintain attribution to the author(s) and the title of the work, journal citation and DOI.
Published under licence by IOP Publishing Ltd

flow has the maximum amplification is located somewhat further downstream compared with the resistance flow type, and the passage jet extends to considerable distance downstream passage exit. While for the isolated flow the maximum amplification factor is no longer situated in passage center line but at the corner streams near the passage entrance.

CFD is a numerical solution to model a wind flow, which still needs more development. CFD can be cheapest solution to study wind flow behavior, but as a numerical approach, CFD have some limitations and error, therefore the development of a numerical model by CFD must be validated by experimental result. CFD is a numerical solution to model wind flow which still needs more development.

Mittal et.al [3] has reviewed that the RANS standard k-ε turbulence model overestimates about 10% for the strong wind region (U/U_{reff} >1) and underestimate about 5 times or more for the weak wind region (U/U_{reff} <1). RANS standard k$-\varepsilon$ cannot produce wind back flow on the top of the roof and has over predict turbulence kinetic energy in separated flows around windward corners of buildings. Steady RANS method is the best choice economically, however for improving accuracy of RANS standard k-ε, the effect of different model closure coefficient needs to be investigated in detail. Yoshie et.al [4] observed that the wind speed difference between experiment and CFD using RANS standard k-ε turbulence model was several tens of % at most if it was limited to the highest wind speed region. R. Yosie et. al [5], defined the maximum difference between CFD and experiment results is 15%.

In order to get better understanding about wind flow around buildings, non-intrusive measurement is needed, therefore in this study measurement using PIV was used. [3] PIV has high spatial resolution and directional sensitivity, but it is expensive and could be dangerous due to the usage of the laser and this method is not suitable for the cluster of buildings. In this study we found it hard to get enough particle in the pedestrian level wind area so other level measurement area was chosen.

This study aims to have a better understanding about the effect of street canyon width to wind flow behavior around high-rise buildings. In this study 2 models of 4 symmetrical high-rise buildings with difference of street canyon width are used, RANS standard k-ε turbulence model is used in OpenFOAM v1912 software to validate the experiment results.

2 Methodology

2.1 Experimental Setup

The experiment was performed by using an open circuit low wind speed wind tunnel in BBTA3 laboratory. The wind tunnel can generate wind speed up to 45m/s. The wind tunnel has 1.25m of length and $0.5\ m \times 0.5\ m$ on the inlet and $0.51\ m \times 0.51m$ on the outlet. The wind tunnel used in this study is shown on **figure 1.**

Figure 1 Low wind speed wind tunnel

The experiment started with determine the wind profile that satisfy the log law approach with $\alpha = 0.28$. To create wind profile for the turbulence generator as shown at **figure 2**, the turbulence generator is made in triangle shape. There are 4 triangles with height (h) = 21 cm and width (l) = 3 cm located in the middle, and 2 triangles with h = 29.5 cm and l = 4 cm. located at the sides. The turbulence generator has 4 cm space each, and 5 cm to side of test section.

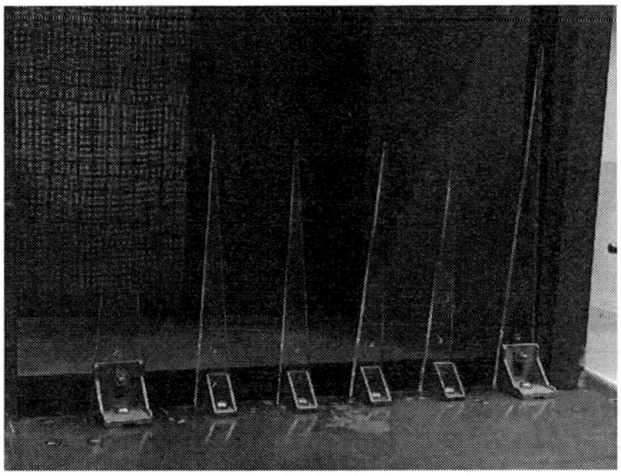

Figure 2 Turbulence generator

The measurement of wind profile generated, by the turbulence generator was measured by using PIV at 1 m from the test section inlet when the test section is empty. 10-time frame data is taken and plotted as shown at **figure 3**. The picture contains 10 frame data, average of the data, power law wind profile, and CFD simulation using OpenFOAM. The velocity at the free stream is 15 m/s that started in reference height z_{reff} = 300 mm.

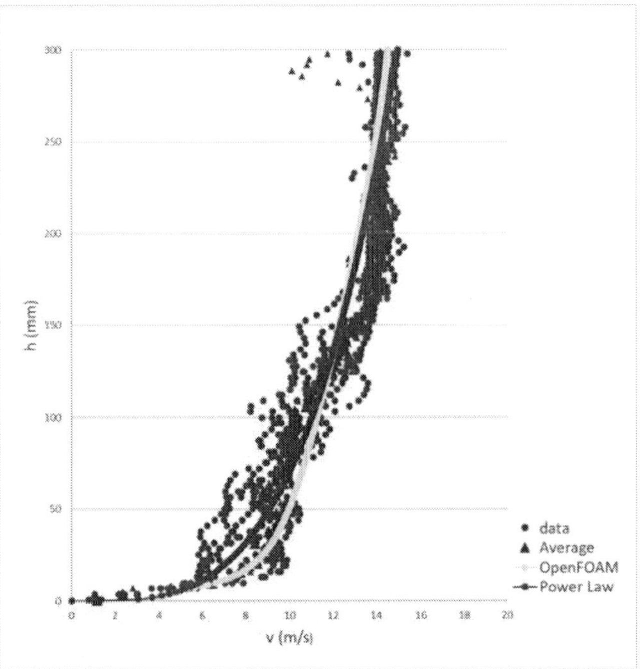

Figure 3 Wind profile

IOP Conf. Series: Materials Science and Engineering **930** (2020) 012044 doi:10.1088/1757-899X/930/1/012044

2.2 PIV (Particle Image Velocimetry)

PIV is a nonintrusive measurement method by using a laser and super CCD camera to capture the particle movement. A laser sheet creates a plane of measurement and on the normal of the plane there's CCD camera to capture the particle displacement with a determine time frame (Δt). With the known displacement and time, the PIV software will simulate the particle movement as a velocity vector. The working principle of the PIV is shown at **figure 4.**

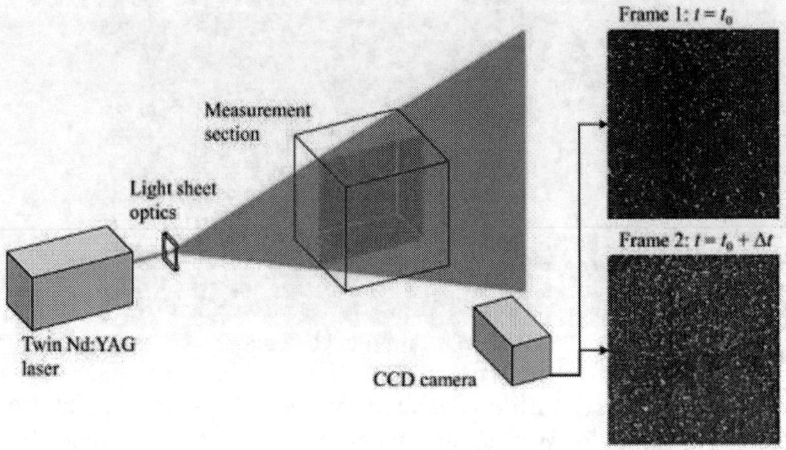

Figure 4 Working principle of PIV

2.3 Experiment Model

In this study 2 different models were used; the models contain 4 symmetrical high-rise buildings with difference street canyon width and 1:420 scale. The 3D schematic of the model is shown at **figure 5.** The wind generated in x axis direction with 0° of angle attack is illustrated in **figure 6**. The measurement point is taken at z = 10 cm is shown at **figure** 7. The measurement height z = 10 cm chosen due to lack of particle that can be measured by the PIV system in the laboratory. Model's parameters are detailed in **table 1.**

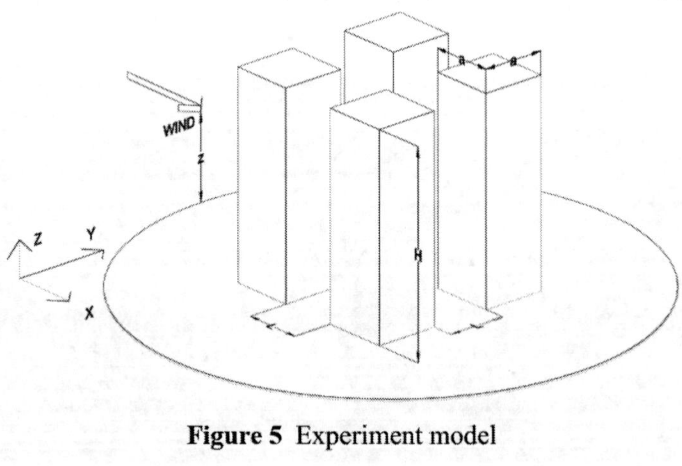

Figure 5 Experiment model

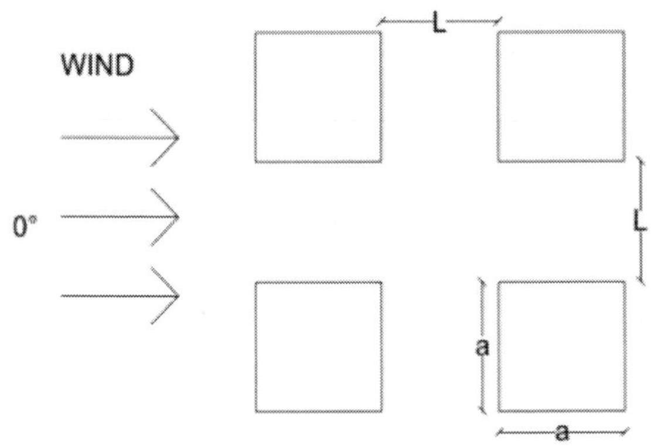

Figure 6 Top view of the model

Figure 7 Measurement points

Table 1 Experiment model parameters

Parameters	
Street canyon width (L)	72 mm, 167 mm
Height of buildings (H)	238 mm
Width of buildings (a)	77 mm
Angle of attack	0°
Wind Speed at freestream	15 m/s
Measurement height (z)	10cm

2.4 Numerical simulation

In this study the simulation is performed by using OpenFOAM v1912 with RANS standard k-ε turbulence model. OpenFOAM is an open source software and use C++ as library. In operating OpenFOAM cases 3 directories are needed:

- A constant directory that contains full description of the case mesh and specifying physical properties
- A system directory for setting parameters associated with the solution procedure itself, it contains a least 3 files: **controlDict** for controlling run parameters, **fvSchemes** for determine the discretization schemes that being used, **fvSoultion** contains equation solvers, tolerances, and other algorithm set for the run.
- The 'time' directories containing individual files of data for particular fields. The data contains initial value and boundary conditions, and the solution is written in 'time' directories. [5]

In R Yoshie et.al [4] stated that the small horizontal computational domain (width) create significant increase for wind speed result, therefore the horizontal computation domain should be extended outside the rim of surrounding object. The standard vertical computational domain is 7.2 H (object height), but there is no substantial problem when the vertical domain 3H is used, therefore the computational vertical domain suggested is 3H or higher. The difference between the calculation result of coarse grid and finer grid was considered small, the standard grid would be satisfactory between buildings suggested should be divided into 10 portions or more. Tominaga et.al [6] defined that computational domain must be have blockage ratio less than 3%, with the horizontal computational domain about 5H from each sides of the buildings, 5H for vertical domain, 10H for computational domain behind the buildings. In Jorg Franke et.al [7] researched the maximum computational domain for behind the buildings is 15H. The mesh that used in this study can be seen at **figure 8.**

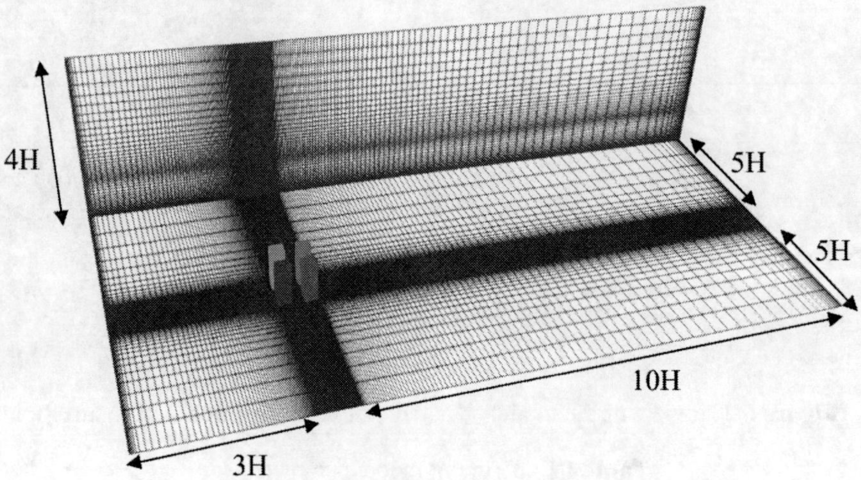

Figure 8 Mesh for numerical simulation

After the mesh is generated the initial value of fields and boundary condition is determined in 0 time dictionary. The boundary condition being used in this study is summarized in **table 2.**

Table 2 Boundary condition

Top, Bottom, Buildings, sides wall	noSlip condition
Inlet	Power Law with wind speed at free stream is 15 m/s
outlet	Zero gradient

To simulate the wind flow using steady RANS standard k-ε, SIMPLE (Semi-Implicit Method for Pressure Linked Equation) algorithm is used. SIMPLE scheme solves the momentum equation by giving an initial guess value of the field, the velocity result is then used for correcting the pressure equation and the corrected pressure and velocity value is used to solve the transport equation until the solution converge [8]. In D.A. Jones et.al [9] research the SIMPLE algorithm can become unstable, so to improve the stability a relaxation factor is needed. The recommended relaxation factor for the pressure equation is 0.3 and the remaining equation is 0.7.

For the divergence schemes the linear upwind schemes is used, this schemes is second order accuracy but, as it is more complex than linear interpolations and can produce unbounded solutions, the latter is a better choice [10]. To summarize the case setup for solving the CFD simulation shown at table

Table 3 CFD case setup

Application solver	simpleFoam (steady state)
Divergence schemes	Linear Upwind schemes for U, k, ε
Turbulence model	RANS standard k-ε
Number of iterations	3000

IOP Conf. Series: Materials Science and Engineering **930** (2020) 012044 doi:10.1088/1757-899X/930/1/012044

3 Results and Discussion

3.1 PIV result

The measurement was taken at the level of (z) =10 cm above the ground surface of test section, as shown in **figure 3** the wind speed at 10 cm is 10.295 m/s (U_{reff}) and the calculated turbulence intensity by dividing standard deviation of wind profile data at z =10 cm, by the average value of wind speed at z = 10 cm is 9.11%. The vector result from the PIV measurement is shown at **figure 8(a)** for model 1 and **figure 8(b)** for model 2. The figure shows that model 1 produce higher wind speed in the center of passage line.

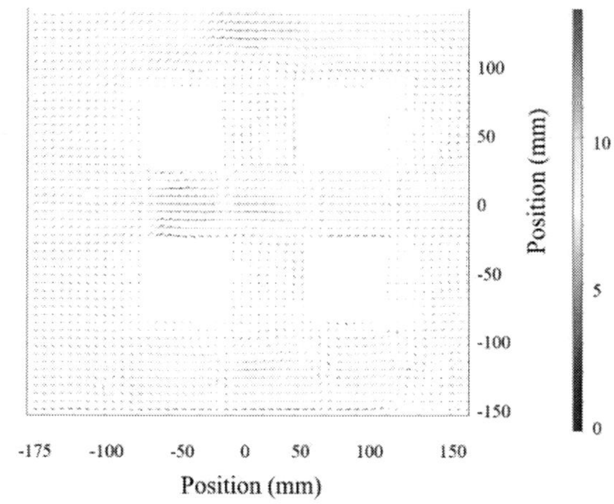

Figure 9 PIV output for vector field model 1

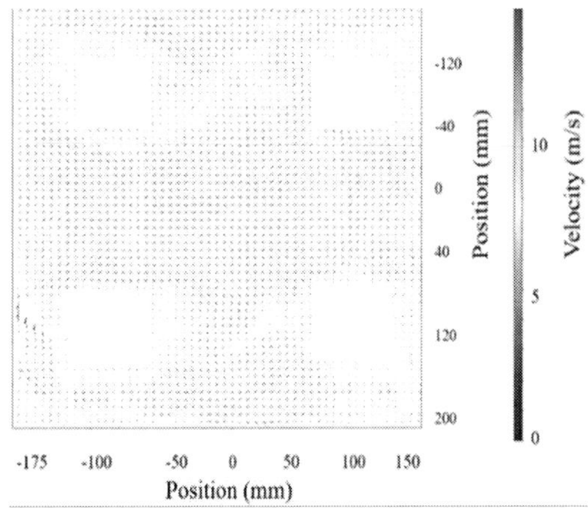

Figure 10 PIV output for vector field model 2

3.2 OpenFOAM result

The flow field results at z = 10cm from OpenFOAM is shown in **figure 11** for model 1 that converge ater 2200 iteration and **figure 12** for model 2 that converge after 2400 iterations. In the figure shows that occurs high wind speed occur at the corner of the front buildings, and the amplification of the wind speed occur is higher in the model 1 simulation.

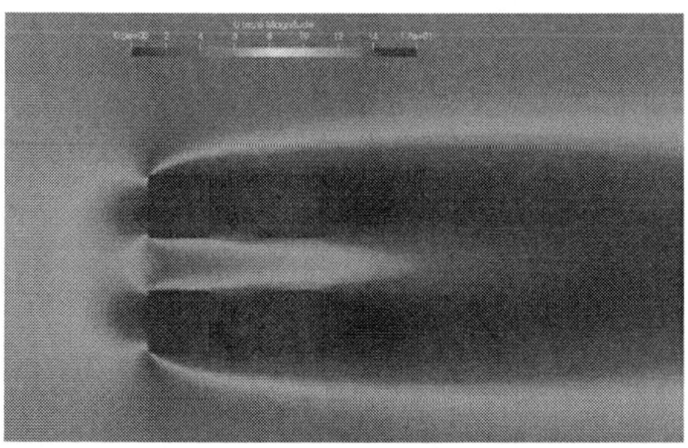

Figure 11 OpenFOAM flow field output of model 1

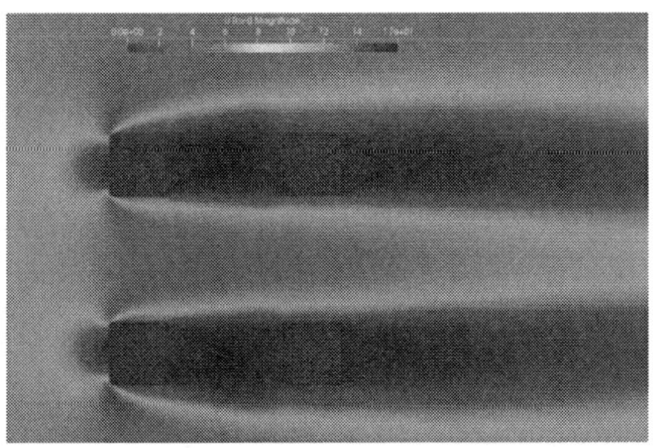

Figure 12 OpenFOAM flow field output of model 2

3.3 Comparison of CFD simulation and PIV results

For better understanding of the simulation and experiment result, the vector of wind speed has been plotted in graph shown at **figure 13** for model 1 results and **figure 14** for model 2 results. The data is only extracted on the center line of the passage due to the limited resource in this research. From the graph we can see the effect of the street canyon width to the wind speed and the validation of CFD results with experiment results.

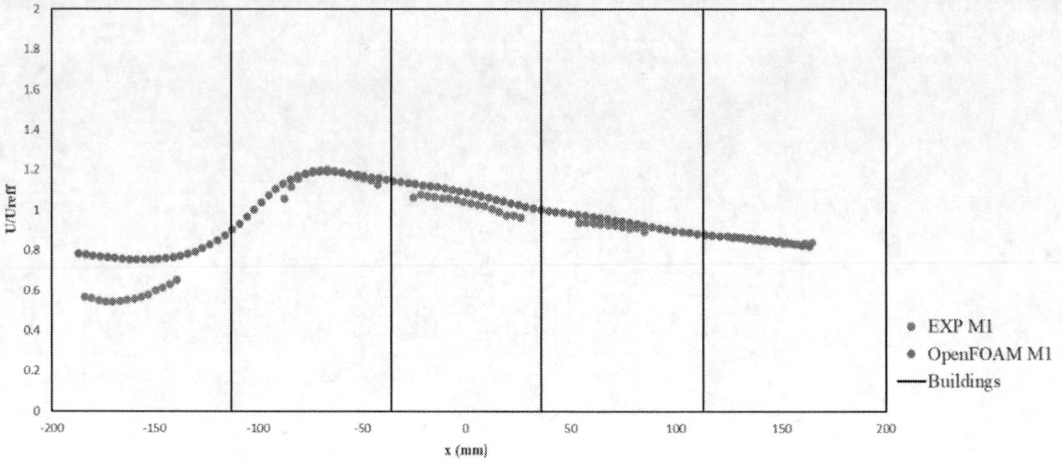

Figure 13 Experiment and OpenFOAM results for model 1

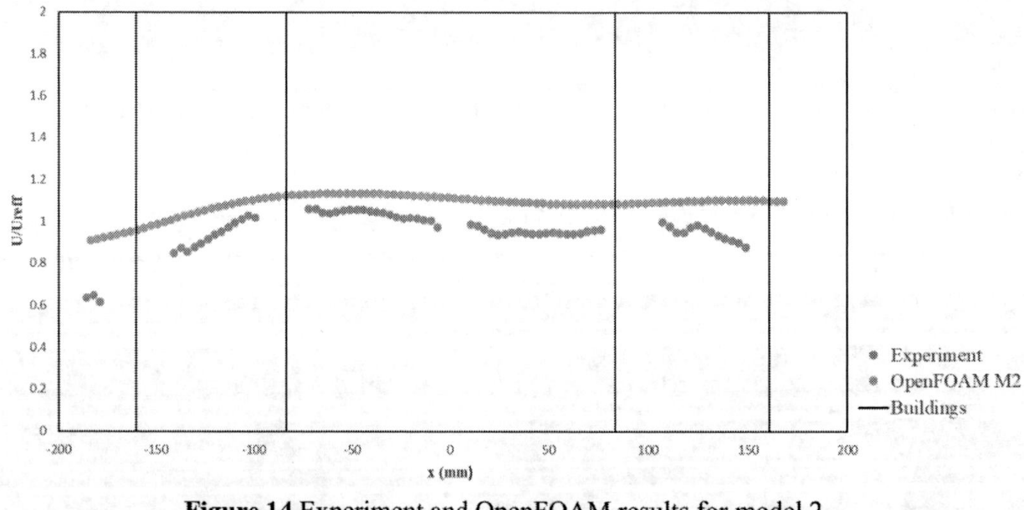

Figure 14 Experiment and OpenFOAM results for model 2

The error occurs in strong wind (U/U_{reff} >1) for the model 1 is small has shown at **figure 15** and for the model 2 is shown at the **figure 16.** The error occurs from model 1 and model 2 is considered good, both model has error below 15%, if referring to R Yoshie et.al [5] research, where the maximum error occurs when simulating CFD using RANS standard k-ε turbulence model is 15%.

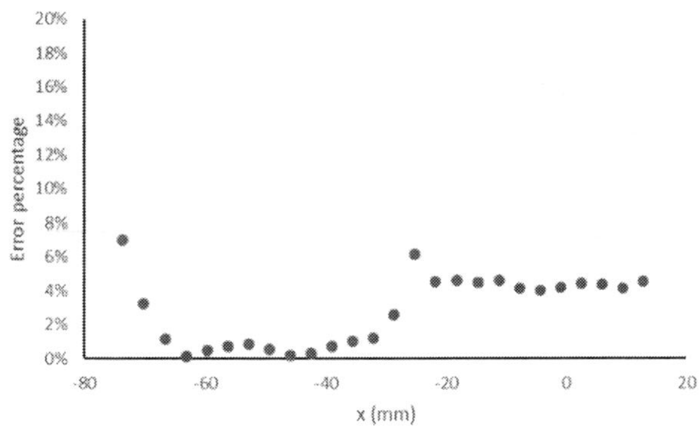

Figure 15 Error on the strong wind region of model 1

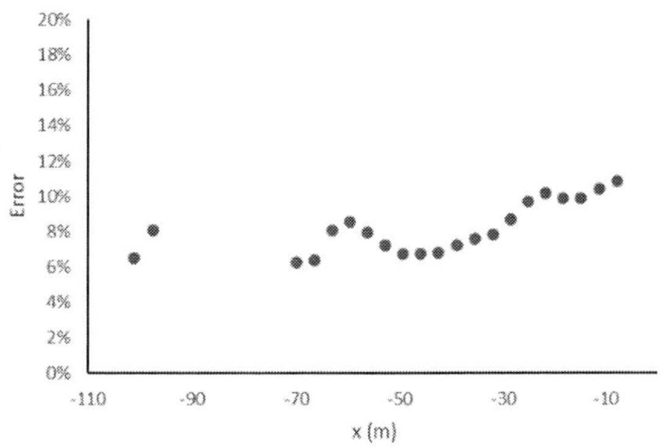

Figure 16 Error on the strong wind region of model 2

3.4 Effect of street canyon width

The result of wind speed ratio (U/U_{reff}) on z =10 cm, shows different amplification factor in the center of passage line. From the graph above, 20.20% amplification of the wind speed occurs in the model 1, and 5.92% occurs in the model 2.

Referred to Blocken et.al [1] and Kuo et.al [2], the effect of street canyon width to the wind flow of the model 1 with the ratio of street canyon width and building width, L/a =0.94 is categorized interaction flow due to the amplification occurs at the center passage line located further from the upstream because of the merge of the corners stream of the front buildings. The wind flow of the model 2 with the ratio of street canyon width and building width, L/a =2.17 is categorized as isolation flows due to the maximum amplification is not occurs in the center line of the passage because there is no interaction between corners flow in the passage.

4 Conclusion

The behavior of wind between 4 high-rise building in this study has been validated. To summarize the results of the following study, here are the conclusions made:

1. Amplification occurs for model 1 is 20.20% and model 2 is 5.92%, the amplification on model 1 is higher due to the smaller street canyon width.
2. The change of the street canyon width affects the flow behavior, for model 1 the flow is categorized as interaction flow, while model 2 is isolated flow.
3. RANS standard k-ε model is more suitable to validate the strong wind flow region with U/U_{reff} > 1

IOP Conf. Series: Materials Science and Engineering **930** (2020) 012044 doi:10.1088/1757-899X/930/1/012044

Future research

In the next step of the study is to study about effect of unsteady wind flow around high-rise building to structures arounds the buildings.

Acknowledgement

This research is done by the help of the Program Penelitian, Pengabdian kepada Masyarakat dan Inovasi ITB (P3MI-ITB) Grant scheme 2017-2019 from ITB. The experiment was performed in the low wind speed wind tunnel laboratory of BPPT-BBTA3

References

[1] Blocken B, Carmeliet J and Stathopolous T 2007 CFD evaluation of wind speed conditions in passages between parallel buildings - effect of wall function roughness modifications for the atmospheric boundary layer flow *J. Wind. Eng. Ind. Aerod.* **95** 9-11 941-62

[2] Kuo C Y, Tzeng C T Ho M C and Lai C M 2015 Wind tunnel studies of a pedestrian-level wind environment in a street canyon between a high-rise building with a podium and low-level attached houses (Energies)

[3] Mittal H, Sharma A and Gairola A 2018 A review om the study of urban wind at the pedestrian level around buildings *J. Build. Eng.* **18** 154-63

[4] Yosie R, Mochida A Tominaga Y Kataoka H and Yoshikawa M 2005 Cross comparisons of CFD prediction for wind environment at pedestrian level around buildings part 1 (Seoul: The Sixth Asia-Pasific Conference on Wind Engineering)

[5] OpenFOAM Ltd (ESI Group) 2019 OpenFOAM The open source CFD toolbox user guide (Bracknell:OpenFOAM Ltd (ESI Group)),

[6] Tominaga Y, Mochida A Yoshie R Kataoka H Nozu T Yoshikawa M and Shirasawa T 2008 AIJ guidelines for practical application of CFD to pedestrian wind environment aroud building *J. Wind. Eng.* 1749-61

[7] Franke J, Hellsten A Schlunzen H and Carissimo B 2007 Cost best practice guideline for CFS simulation of flows in the urban environment (Hamburg: University of Hamburg)

[8] Tu J, Yeoh G H and Liu C 2008 Computational fluid dynamics a practical approach *Elsevier*

[9] Jones D A, Chapuis M Liefvendahl M Norrison D and Widjaja A R 2016 RANS simulations using OpenFOAM software (Australia: Maritime Platform Division DST Group Science and Technology Group)

[10] Ferziger J H and Peric M 2002 Computational methods for fluid dynamics (Berlin: Springer)

Review on 3D printed concrete as structural beam members

P Pudjisuryadi[1], **A Antoni**[1] **and J Chandra**[1]

[1] Civil Engineering Department, Petra Christian University, Siwalankerto 121-131, Surabaya, Indonesia

Abstract. Recently, Three-Dimensional Concrete Printing (3DCP) has gained its popularity as construction material. It offers several advantages over conventionally casted concrete such as absence of formwork, reduction of construction equipment transportation, greater safety, reduction of labour cost, and many others. However, the technology also comes with many challenges. Researches in this area can be classified into two issues which are fresh and hardened states of the printed concrete. Investigation of fresh concrete focuses on the optimal design of rheological properties which determines the pumpability, extrudability, and buildability. While mechanical properties of hardened 3DCP are usually investigated by adopting standard tests for conventionally casted concrete. However, due to the complexity of printing process, it is often that 3DCP behaviours cannot be predicted by its mechanical properties obtained from the tests. This paper reviews some studies done by others to highlight potential manufacturing process related weak points of 3DCP as structural beam members.

1. Introduction

In this rapidly growing technology era, one concrete construction method has been relatively favoured to be researched and developed, namely the Three-Dimensional Concrete Printing (3DCP) [1]. 3DCP is a technology to build concrete by extruding plastic mortar against a casting bed through a nozzle opening which movement is controlled by a robotic arm system [2]. The extruded mortar filaments will shape certain desired 3D body of concrete by adding subsequent layers on top of previous ones, without the use of any formwork. Aside from freedom of formwork, 3DCP has many other advantages, some of them are reduction of construction equipment transportation, reduction of construction time and expenses, greater safety, and saving labour cost [3]. However, to fully utilize those benefits, complexity of 3DCP process rises many challenges. Characteristic of both fresh and hardened states of the concrete should be well investigated.

Fresh concrete for 3DCP should possesses good rheological properties such as pumpability, extrudability, and buildability [4]. Those can be achieved if the mix design of the fresh concrete is well collaborated with the chosen printing process. In other words, different printing process may require different mix concrete design for a successful 3DCP. Consideration of the fresh concrete itself may include the initial yield strength and its development, and the setting time. The yield stress should be low enough that the concrete can be pumped and extruded, but its strength development should be quick enough that the extruded layer could sustain the load from the subsequent layers above. Setting time should be designed that the fresh concrete remains workable during printing process. Logically, in the design process of those properties, the printing process should already be included. Planar path of each layer, relative height of subsequent concrete layer, and speed of printing head (deposition rate) should be balanced that the extruded material could remain stacked, bonded in layers and sustain the weight of other layers that are deposited on top. This challenging relationship between fresh concrete and the manufacturing process of 3DCP has attracted many researchers [5-6].

Content from this work may be used under the terms of the Creative Commons Attribution 3.0 licence. Any further distribution of this work must maintain attribution to the author(s) and the title of the work, journal citation and DOI.
Published under licence by IOP Publishing Ltd

The next thing that becomes the concern of engineers after successfully constructs 3DCP without collapse, is the quality of the resulting hardened concrete. Commonly, researchers adopt standard tests of conventionally casted concrete to measure the quality of hardened 3DCP. Le et al. [7] conducted experiments that can give some idea about 3DCP hardened properties. Due to gently vibrated hopper and small pressure during extrusion, good 3DCP tends to have higher density than that of conventional concrete. However, it should be noted that the higher density observed were measured from specimens which were cut from 3DCP slab which was relatively easy to print. Compressive strength showed variations (72 MPa to 102 MPa) if the test samples were cut from different printing specimen, even with the same fresh concrete. Specimens which were cut from printed slab showed higher strength (and lower anisothropic behaviour) than that which were cut from curvy printed shapes (more difficult printing process). From flexural strength test, it was observed that lower layers tended to have higher strength since they had higher density and lower w/c ratio. Inter-layer tensile bond strength highly depended on the gap time between layers. Gap time is the time interval of printing process between two subsequent layers at the same position. It can be concluded that due to the nature of printing process, it is logical that the hardened 3DCP properties is highly anisotrophic. Bonding strengths between adjacent layers and filaments play very important role in determining the quality hardened concrete because they are potential weak points of 3DCP. Horizontally, the mechanical properties in the direction of filament path is different to its perpendicular direction. Vertically, the mechanical properties of different layers are also different since lower layers have higher density due to the pressing weights of layers above them.

Furthermore, applicability of 3DCP as structural members raises challenges for engineers even more. There are many things should be considered for assessing the hardened 3DCP quality, which may not be necessary for conventionally casted concrete. Adopting standard assessment for conventionally casted concrete is not enough to give accurate information of how good a 3DCP is. Structural member can show highly non-linear behaviour from the very beginning, which cannot be predicted by standard tests of concrete material. Some reviews of this concern are presented in this paper.

2. Research significance

This paper raises awareness that if 3DCP is to be used as structural members, investigation of 3DCP hardened mechanical properties alone may not be enough to accurately predict the structural performance. Once 3DCP is used to construct certain structural member, the behaviour of the member is not as simple as conventionally casted concrete. Premature non-linear behaviour may arise which is not the case in conventionally casted concrete structural member.

3. Research on 3DCP as structural beam member

The main advantage of 3DCP that complex body forms can be made, has opened opportunities for concrete industry. With 3DCP technology, final weight of structural beam member can be reduced by manufacturing complex shape (introducing hollows) with proper analysis to guarantee the adequacy of the intended application (topology optimization). For example, a beam member can be printed to form a truss like shape (Figure 1), instead of classical prismatic shape. However, to fully utilize this advantage, some manufacturing challenges have to be overcome. Efforts done by others are presented in the following subsections.

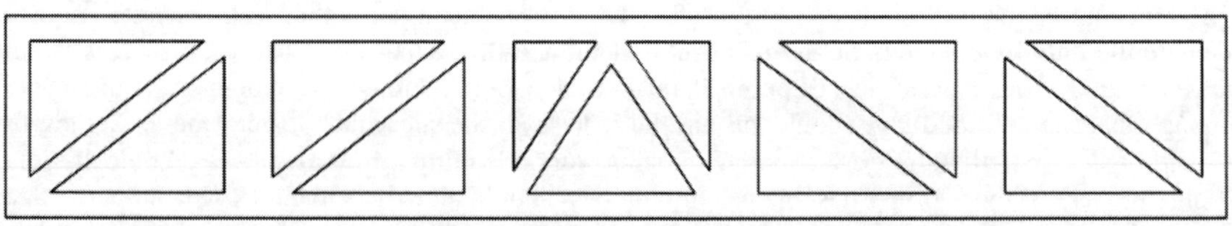

Figure 1. Truss like shape structural beam.

3.1 Non-Segmented 3DCP structural beam

Al-Chaar et al. [8] conducted experiment on truss like 3DCP beams. The beam was 3D printed as one body at one time. Tension reinforcement bars were embedded inside the bottom horizontal chord. A cyclic flexural test was conducted on the beams. It can be seen from the force-displacement relationship (Figure 2), that the beam stiffness started very low and then slightly increased. This behaviour is not normal. It cannot be explained by the material properties alone. There are potential weakening points in 3DCP members in its filament to filament and layer to layer bonding. A numerical modelling by using SAP2000 was done by authors to predict the behaviour of the beams tested by Al-Chaar et al. [8] if treated as conventional homogeneous concrete without considering the weakening of bonding strength. The resulting load-displacement curve is presented in Figure 3. It can be observed that the analytical model possesses stiffness much higher than that of the tested 3DCP beam (Figure 2). This shows that, one should be aware of the potential degradation of 3DCP structural beams due to the potential weakening of the bonding. More complex printing paths will introduce more variation of age differences and qualities between connections.

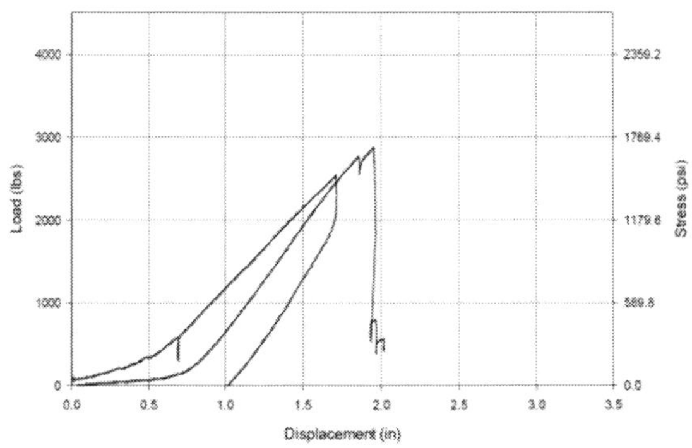

Figure 2. Load-displacement curve of specimen 1SR-S-0: experiment

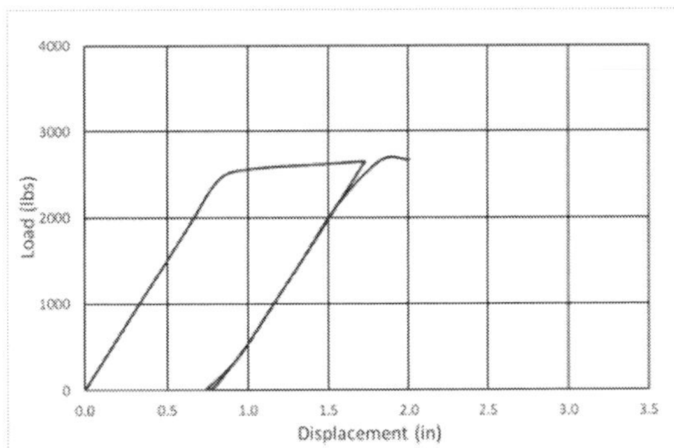

Figure 3. Load-displacement curve of specimen 1SR-S-0: analytical

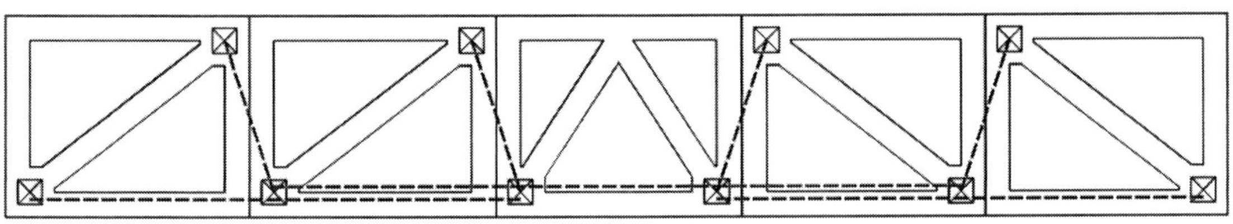

Figure 4. Segmented 3DCP structural beam (adopted from [9]).

3.2 Segmented 3DCP structural beam (non-prestressed connection)

One approach to avoid the weakening problem of the bonding strength, is by printing 3DCP structural members in segments. Making smaller segments reduces the complexity of printing process that consistency of the quality is easier to maintain. Asprone et al. [9] constructed concrete beam from 3DCP segments and conducted bending test. However, the connections needed to assemble the segments raises other challenges. In this case, some hollows in the segments should be made for this connection purpose (see figure 4). Dashed lines in figure 4 represent tension steel bars used to connect the segments. Tension bars mechanically connected and secured with wet joint were used for the connection system. Result of the bending test is presented in figure 5. It was reported that the beam behaved linearly at initial stage (when everything was still intact, with no cracks). But this state was only observed in the very start of the test that it rapidly entered non-linear stage (start of cracking stage, progression of cracking stage, final failure stage). The linear stage (line OA) is comparable with conventionally casted RC beams. But the non-linear stage (line ABC), stiffness reduced quite significantly, not because of degradation of material or wrong optimization but mainly because of the non-effective interface/connection system (the connected segmental system is prone to local damages). This shows that the printed concrete quality may be improved, but the connection system introduces different kind of weakening.

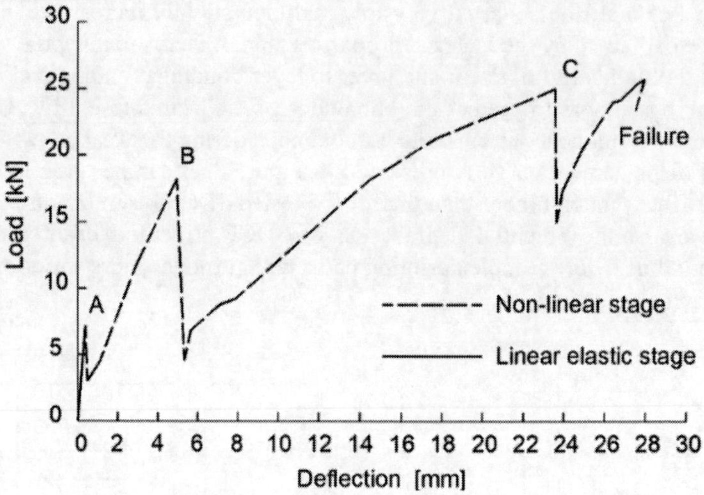

Figure 5. Linear elastic and non-linear stages of load-deflection curve of the beam (adopted from [9]).

3.3 Segmented 3DCP structural beam (prestressed connection)

For segmented 3DCP structural beam to work, the connection system may be improved by using post-tensioning approach. Vantyghem et al. [10] presented prestressed concrete beam design based on topology optimization and manufacturing process with concrete extrusion technology. The illustration of the beam segments and its reinforcement as well as post tensioning tendons is presented in figure 6. Due to the symmetric geometry, only half span of the beam is drawn. Bending test was performed on the manufactured beam to record the load-displacement relationship of upper and bottom chords of the beam. The goal was to verify the service load performance of the optimized shape. A finite element analysis of the beam was also conducted and compared to experimental results.

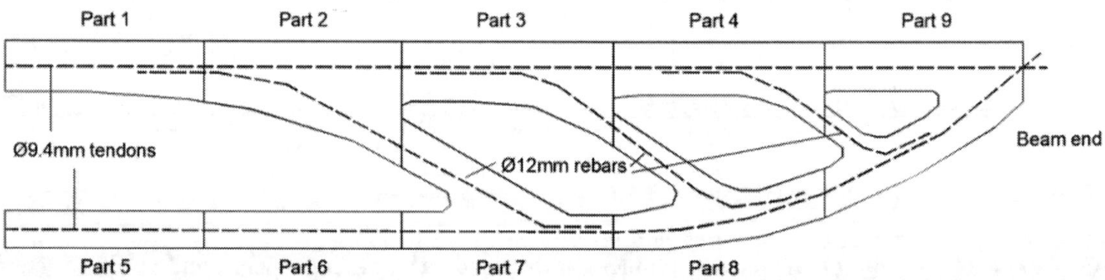

Figure 6. Illustration of the segmented 3DCP structural post tensioned beam (adopted from [10])

Results of the bending test is presented in figure 7. The upper chord showed a good fit (Figure 7a) between the experimental and numerical results. On the other hand, the lower chord shows a much higher deviation (Figure 7b) from the numerical result. Vantyghem et al. stated that this can be attributed to tolerances in the position of the post-tensioning strand within the lower chord, as well as to the sectional assembly. In larger structures, the positioning of the post-tension strand should be easier to control and eliminate this problem. Another factor of uncertainty is the determination of the equivalent material characteristics of the hybrid structure (3DCP segments and grout infill).

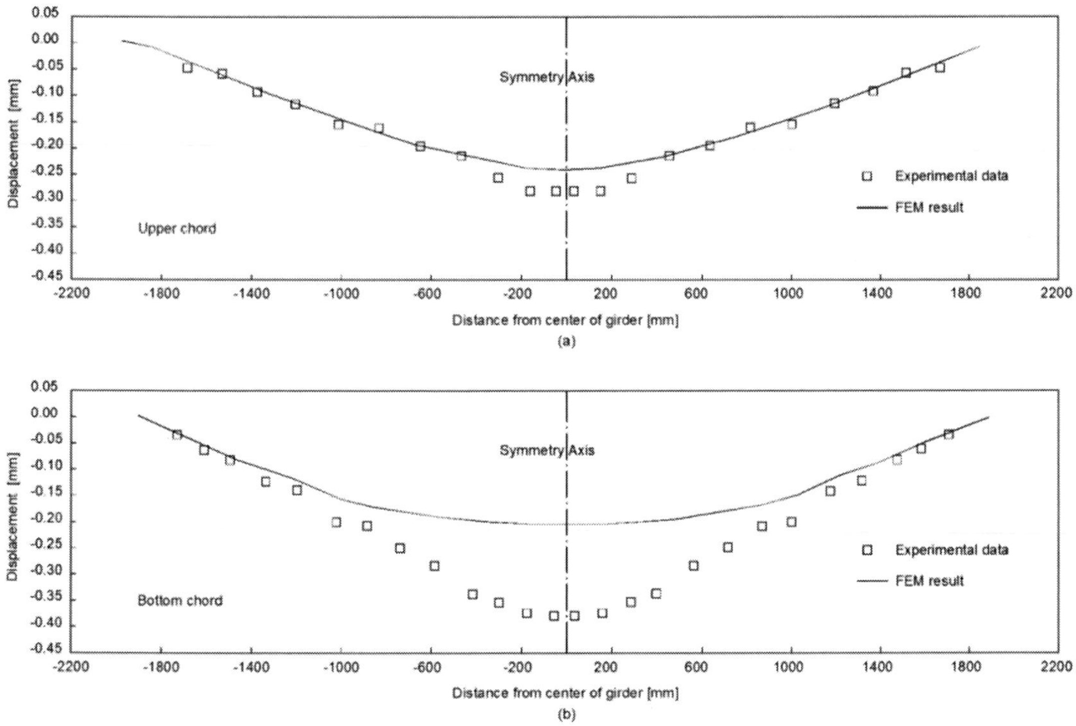

Figure 7. Force-Displacement of upper and bottom chord (adopted from [10]).

4. Discussions and concluding remarks

Reviewing some works by others as well as conducting numerical analysis regarding the use of 3DCP as structural beam members, some discussions and conclusions can be drawn:

- Due to its manufacturing complexity, hardened properties of 3DCP which is commonly assessed by using small specimens, may not be enough for predicting 3DCP structural member behaviour. As a structural member, 3DCP possesses many things to be considered beside of the material hardened properties, that can contribute to its overall behaviour, mainly due to non-homogenous condition introduced by the printing process.
- To overcome weak points introduced by printing large complex shapes, assemblage approach of smaller 3DCP segments are usually taken. Post-tensioning approach to connect the segments is proven to be quite successful. However, in other point of view, assemblage approach is a little bit contradictory to the initial concept of 3D printing which gain its popularity from its one step manufacturing process.

Acknowledgments

The authors are grateful for the funding provided by the Deputy for Research Empowerment and Development, Ministry of Research and Technology/National Research and Innovation Centre, Republic of Indonesia under PDUPT scheme no. 002/SP2H/LT-MULTI/LL7/2020.

References

[1] Schutter G D, Lesage K Mechtcherine V Nerella V N Habert G and Agusti-Juan I 2018 Vision of 3D printing with concrete — technical, economic and environmental potentials *Cem. Concr. Res.* **112** 25–36

[2] Gosselin C, Duballet R Roux P Gaudillière N Dirrenberger J and Morel P 2016 Large-scale 3D printing of ultra-high performance concrete - a new processing route for architects and builders *Mater. Des.* **100** 102–9
[3] Wrangler T et al 2016 Digital concrete: opportunities and challenges *RILEM Tech. Lett.* **1** 67-75
[4] Valkenaers H, Jansen D Voet A Gysel A V and Ferraris E 2014 Additive manufacturing for concrete: a 3d printing principle *Proceedings of the 14th euspen international conference* **1** 139-42
[5] Ma G, Li Z and Wang L 2018 Printable properties of cementitious material containing copper tailings for extrusion based 3D printing *Constr. Build. Mater.* **162** 613–27
[6] Kazemian A, Yuan X Cochran E and Khoshnevis B 2017 Cementitious materials for construction-scale 3D printing: laboratory testing of fresh printing mixture *Constr. Build. Mater.* **145** 639–47
[7] Le T T, Austin S A Lim S Buswell R A Law R Gibb A G F and Thorpe T 2012 Hardened properties of high-performance printing concrete *Cem. and Conc. Res.* **42** 558-66
[8] Al-Chaar G K, Stynoski P B and Banko M L 2018 Structural behavior of layer-printed reinforced concrete beams *The Op. Const. and Build. Tech. J.* **12** 375-88
[9] Asprone D, Auricchio F Menna C and Mercuri V 2018 3D printing of reinforced concrete elements: technology and design approach *Const. and Build. Mat.* **165** 218-31
[10] Vantyghem G, Corte W D Shakour E and Amir O 2020 3D printing of a post-tensioned concrete girder designed by topology optimization *Auto. in Const.* **112** 1-11

Rapid visual assessment of building vulnerability due to earthquake potential hazard in Surabaya

D Irawan [1], B Suswanto [1], A R Amalia [1], D Iranata [1*]

[1] Department of Civil Engineering, Faculty of Civil, Planning and Geo Engineering, Sepuluh Nopember Institute of Technology, Surabaya, East Java, Indonesia.

*Corresponding authors e-mail: data@ce.its.ac.id

Abstract. This article studies about Rapid Visual Assessment using android application RViSITS. The Institut Teknologi Sepuluh Nopember team has developed this application, and the purpose is to conduct a more accessible assessment survey in the digital version instead of using the standard form, applying the same standard called FEMA 154. This building vulnerability research was carried out using the application, due to data requirement for mitigation action needed by Surabaya government, due to two newly discovered faults passing the city, namely Surabaya and Waru faults. These two new faults concealed by the National Earthquake Research Centre (PuSGeN) lead to higher earthquake potential hazards in Surabaya, as stated in the SNI 1726. In this research, a two-phase survey was conducted to a total of 366 buildings. From the development of the result mapping of the application, it is discovered that the buildings built after SNI 1726 regulation issued have a higher level of resistance toward earthquake loads, especially the buildings in Surabaya, where gradual seismic risks keep increasing. Therefore, it is necessary to develop mitigation strategies due to the earthquake potential hazard in Surabaya.

1. Introduction

Indonesian territory is surrounded by complicated tectonic plates. The archipelago of Indonesia represents the meeting of Pacific Plate and Indian-Australian Plate (in Eastern Indonesia), Eurasia Plate, and Indian-Australia Plate (in Western Indonesia), and a smaller plate (Caroline Plate and Philippine Sea Plate). Those surrounding plate boundaries can be estimated by the zone of the active earthquake, the zone of mass movement in the mountain region, the zone of volcanism, zone of magmatism, and zone of hydrocarbon [1]. The zone of active earthquakes has already been verified by the Indonesian Meteorological, Climatological, and Geophysics Agency (BMKG) recorded repository regarding the recent earthquake in its website [2]. Another research about this theme in Sulawesi [3] and Lampung [4] is another support to the evidence of Indonesia's great potential of earthquake hazard.

Earthquake potential hazards have to be standardized by the Indonesian government. Many cities in Indonesia have regulations to deal with earthquake possibilities, which can occur at any time in the future. Surabaya, as the second-largest city in Indonesia, and as the home for approximately 2.9 million people in 2015 [5], faces its own problem regarding earthquake possibilities.

Content from this work may be used under the terms of the Creative Commons Attribution 3.0 licence. Any further distribution of this work must maintain attribution to the author(s) and the title of the work, journal citation and DOI.
Published under licence by IOP Publishing Ltd

Figure 1. Surabaya fault (top line) and Waru fault (bottom line) through Surabaya.

Table 1. Recent earthquake recorded around Surabaya.

No.	Date	Depth (km)	Location	Magnitude
1	July 24th, 1984	33	Bojonegoro	4.7
2	May 14th, 1992	33	Java Sea (South of Madura)	4.7
3	July 28th, 2006	10	Grobokan	4.5
4	January 24th, 2007	35	Bojonegoro	4.3
5	February 28th, 2015	45.06	Java Sea (South of Madura)	4.1
6	June 25th, 2015	9.77	Bojonegoro	4.3
7	November 4th, 2016	17.87	Ngawi	4.7
8	February 20th, 2017	10	Madura	3.7

Table 2. Earthquake variable value in Indonesia national standard of earthquake resistant building.

Name of standard	Value of PGA (g)	Value of SS (g)	Value of S1 (g)
SNI 1726-2002	0.1	0.38	0.24
SNI 1726:2012	0.3 – 0.4	0.6 – 0.7	0.2 – 0.3
SNI 1726:2019	0.4 – 0.5	0.7 – 0.8	0.4 – 0.5

Based on the National Earthquake Research Centre (PuSGeN) released Maps of Earthquake Source and Hazard of Indonesia 2017 [6], it is acknowledged that Surabaya is passed by Baribis-Kendeng Fold-thrust zone, which has 0.05 mm/years slip rate [7]. The Baribis-Kendeng Fold-thrust zone that passes Surabaya consists of two newly discovered faults, as listed as Surabaya fault and Waru Fault [6]. Both of these faults can be seen in Figure 1 [8]. And, a recent earthquake that happened around Surabaya city, which is presented in table 1, confirms that Baribis-Kendeng is potentially posed as an active Fold-thrust zone.

Based on three available Indonesian National Standard (SNI) of Earthquake Resisting Building Design starting from SNI 1726-2002 [9]; SNI 1726:2012 [10] and SNI 1726:2019 [11], there has been gradual seismic increasing risks in Surabaya. And this started from The First Indonesian National Earthquake Standard (SNI 1726-2002). In this regulation, it was stated that Surabaya was categorized to a low-risk hazard zone, with PGA value was 0.1g. It can be seen in table 2. In SNI 1726:2012, it was stated that PGA value was 0.3g - 0.4g and categorized as a middle-risk hazard zone. The newest standard SNI 1726:2019, changed the value to higher number of Surabaya's PGA to around 0.4g – 0.5g. However, from table 2, the S_S value in the RSNI2 1726:2018 was still categorized as middle zone, despite the fact that it was close to high risk hazard zone. Therefore, every construction needs special design requirement. And further, it is also necessary to develop mitigation strategies due to the earthquake potential hazard in Surabaya.

All the newly discovered faults, which are newly published on earthquake hazard maps, or new standards, become the reasons why mitigation strategies have to be accelerated in Surabaya. To propose the mitigation techniques, the database of building assessment will be needed as one of the compulsory requirements. The conventional method of building assessment needs to involve several building and structure experts, and other related fields involved. Therefore, it requires more time and more expenses. Due to the assessment problem, Federal Emergency Management Agency (FEMA) suggests an application of a technique called Rapid Visual Screening (RVS) presented in FEMA 154 [12].

The RVS is a quick and low-cost method that has been selected to evaluate building vulnerability subjected to seismic loading. This method can also be applied to buildings with minimum access as it can be carried out through an observation conducted from the outside surroundings of the buildings. RVS technique has been widely used by several researchers to evaluate building vulnerability due to earthquake potential hazards [13], [14], [15].

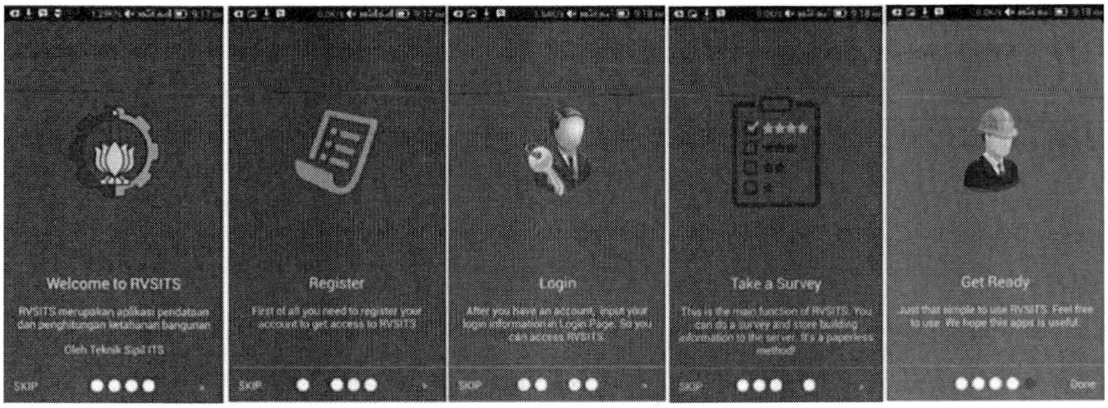

Figure 2. RViSITS application digital android form.

In this research, the RVS android based application, which has been developed by Institut Teknologi Sepuluh Nopember, is called RViSITS (Rapid Visual Survey by Institut Teknologi Sepuluh Nopember). The application of RViSITS has been verified by numerical analysis, which shows the same results as the results of previous studies [13]. Moreover, since RVS method by FEMA 154 still applies several manual paper forms, the proposed RViSITS application can be selected and used as a faster and simple method. The manual paper forms of the RVS method by FEMA 154 can be omitted using RViSITS application, which is based on android systems, as shown in figure 2.

2. The Material and Method

2.1. Research Methodolgy

Since the earthquake potential hazards in Surabaya have increased, as presented in [9], [10], [11], it is necessary to perform building vulnerability assessment using RVS method. The main purpose of this research is to collect and build database of buildings' vulnerability in Surabaya. Therefore, several steps must be carried out to achieve the purpose of the research. The first step is literature study related to the earthquake potential hazard, building assessment and vulnerability, and hazard mitigation and response in Surabaya. Then, the second step is field survey, which is very crucial in this research as to gather building data from several types of buildings in Surabaya. Along with the data collecting process, the software development toward the mapping of the assessment results must be conducted simultaneously Result from mapping and data processing are needed as well, before this paper comes to a discussion and draws a conclusion. The Illustration of the research methodology is shown in figure 3.

2.2. Field Survey

The field survey of this research started with the sequence of permission processing. This field survey obtained two official permission letters from Surabaya Government, the first letter was from Surabaya Planning Board (Bappeko), and the second letter was from the National Unity, Politics and Society Protection Board of Surabaya Government (Bakesbangpol dan Linmas). The field survey was conducted by several numbers of registered surveyors. These registered surveyors had been trained to use the application in advance before carrying out the field survey as the method in FEMA 154 [12] and RViSITS's user guide [16]. Meanwhile, the copyright of the application had already been registered in Indonesia's Ministry of Law and Human Rights, Republic of Indonesia, with registration number 000154073. The phase of the field survey is illustrated in figure 4.

The first field survey phase was conducted on several buildings that belonged to the Government of Surabaya. The duration of this field survey was one month. These government buildings were chosen due to the fact that apart from the function of these buildings, like public buildings, the Surabaya government has the authority to rebuild or renovate the buildings. Therefore, if the results of the survey showed unqualified scores and did not meet the requirements. Surabaya government could take necessary action toward these buildings. In this phase, there were 110 buildings surveyed using RViSITS. And the buildigs were separated to several government building categories such as: schools, health facilities, mosques, churches and district offices.

Figure 3. Research methodology

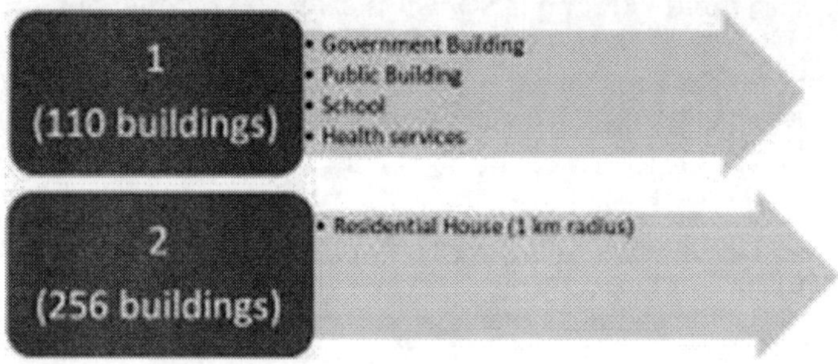

Figure 4. Field survey phase classification.

The initial survey was carried out to all the surveyed buildings located in the areas of where the faults passed. Some of the districts that Surabaya fault passed are: Asemrowo, Sambikerep, Sawahan, Sukomanunggal, Tandes, Benowo and Pakal. The districts near Waru fault are fewer than the districts near Surabaya fault. Waru fault only covers two districts, the two districts are Karangpilang and Wiyung. The results of the first phase survey project were reported to the Government of Surabaya [17]. To compare any actions that the Government of Surabaya performs regarding this case, the report of the results is illustrated and discussed in this article.

The second extension survey was carried out to another 256 different buildings. The field survey duration was just 35 days. It focused on residential buildings, which were built only in a radius of 1 km from the fault estimated locations revealed by the National Earthquake Research Centre (PuSGeN). The residential building survey was carried out to collect information about the housing around the city near to the faults. The results from the survey are expected to be able to be applied and used can be used as consideration for Surabaya Government to plan its urban regional planning, especially the one that is related to building construction regulation.

IOP Conf. Series: Materials Science and Engineering **930** (2020) 012046 doi:10.1088/1757-899X/930/1/012046

However, there were several boundary conditions used as input in the application operated in this study. Each of them is listed below:

1. Surveyors were only for trained people who have the capability or knowledge regarding building, structures, and earthquakes.
2. The names of the buildings were written by the official name or the owner's name.
3. The Addresses and GPS locations were entered as clearly as possible.
4. The construction date entered was the time of the initial construction.
5. The seismic level that is used as the default is set as "High" to perceive the building score well, especially when faced with the current level of potential earthquake hazard. As an example of data comparison, the building constructed before 2012 was set as "Low". While another building constructed in the middle of 2012 and 2019 was set to "Middle".
6. The soil condition in this fault area is categorized as "Soft Soil". Nonetheless, any data needed for East part of Surabaya City, the input then was categorized as "Poor Soil".
7. Other categories, that were not limited to the list, are adjusted due to the real condition in the field, when the survey activities were being conducted by the surveyors.

Figure 5. Java fault (.kmz) line in Goggle Earth.

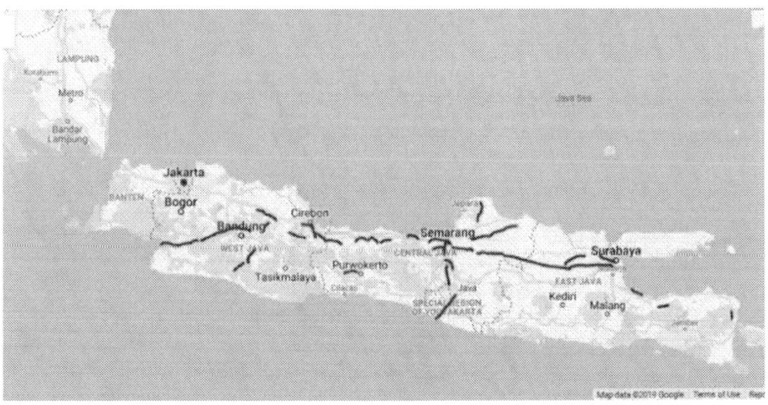

Figure 6. Java fault (.kmz) line in RViSITS website.

2.3. Development of Website Mapping

There are many features available besides data-based collection in RViSITS website. The added features of the website in this research phase are the results of the mapping and uploading google earth line data (.kmz). The mapping results will be discussed later, however kmz extension data that is used in this focused article is Java fault. The comparison of Java fault line opened in Google Earth and in this website can be seen in the figure 5 and figure 6. It is expected that the progress of the development of this website will be able to accomodate google earth line data, thus the results can satisfy and meet the result mapping needed by public use in the future. In addition to this development, the result mapping has a change as well in this research phase. In the previous project, the website can only show the building location captured by GPS. However, in this current research project, both qualified and unqualified results of the survey are properly illustrated on the result map according to the GPS location.

3. Results and Discussion

3.1. Survey Results

Survey result in this section will be separated as a graph that defined numbers of building categories in each surveyed phase. Another result in this article shows the result mapping of the two phases of survey. As seen in the previous figure 4, there are 110 buildings surveyed in the first phase. These 110 buildings are divided in two different fault areas, Surabaya fault and Waru fault. Numbers of different building categories and the survey results are illustrated in figure 7. The bar chart in figure 7 illustrates numbers of building on the first phase survey result that are divided in 4 building categories: 19 buildings of public services (mosques and churches), 30 buildings of government buildings (district offices), 57 school buildings, and 4 buildings of health services.

From the illustrated graph, it is shown that all the health services show a great result as these buildings are categorized to qualified classification. Unfortunately, this result is gained from very limited survey data taken just from 4 surveyed buildings. This is due to the regulation applied in Indonesia, that every health service building that requires a survey must have an official permit from the Ministry of Health. Due to the difficulties of obtaining the official permit from the Minister of Health for health service buildings, this study only managed to survey four heath service buidings, eventhough it had attained an official permision from Bappeko Surabaya. The most number of surveyed buildings are schools. There are 24 buildings, categorised as qualified. Unfortunately, there are more number of buildings categorised as not qualified, these are 33 buildings. The government buildings show similar case as half of the surveyed buildings gain qualified result. Whereas, the other half of the buildings are not qualified. The results obtained form the public services building, including mosques and churches show that 6 buildings are categorised as qualified buildings, meanwhile 13 of them are categorised as not qualified buildings.

In the second phase survey, different results are gained. Figure 8 illustrates the residential house survey results in that phase. The graph shows that 76% of the residential building near to new discovered fault are classified in not qualified category. And, only 24% of them are qualified.

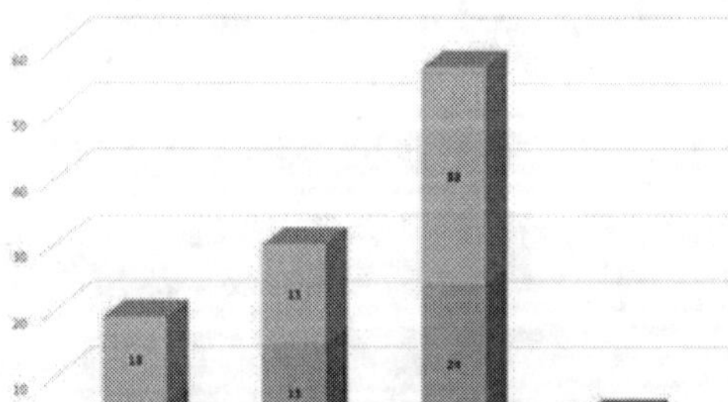

Figure 7. First phase survey result.

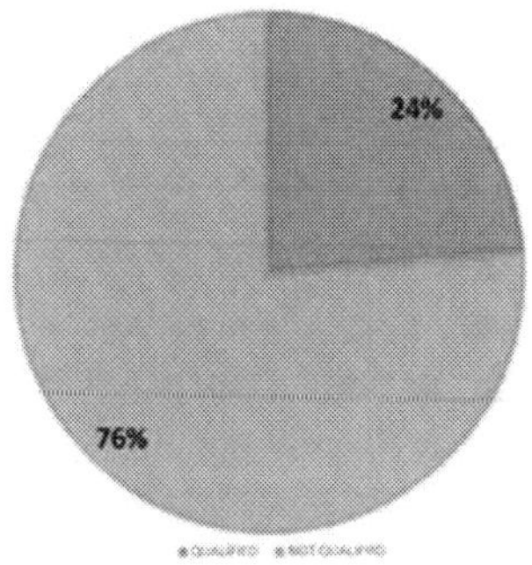

Figure 8. Second phase survey result of residential house.

3.2. Comparison of Seismic Category

Our developed website can map the result in two categories: qualified (represented by green dots) and not qualified (represented by red dots), as depicted in figure 9 to figure 12. The development of the increased capability of this application shows better result in the survey. In this research project, two different seismic categories are compared. The first one is seismic category based on Indonesian National Standard used when the building are constructed. For an example, if a building was built in 2003, and considered using the Indonesian National Standard (SNI 1726-2002) [9], in which the seismic category of Surabaya was classified tas "low", it is then entered into the "Low" zone. Another example if a building was built around 2013, and considered using the Indonesian National Standard (SNI 1726-2012) [10], in which the seismic category of Surabaya was categorized as "Middle", it is then entered in the application into "Middle" zone. Since the potential seismic risks in Surabaya keeps increasing, due to the finding of the existence of the two newly found faults, in this research the seismic category of the area, where the two faults pass are categorized as "High". This modification in seismicity criteria definitely changes the final result scores of the buildings as written in FEMA 154 that used in the RViSITS application. If the result is "qualified", the maps will show green dots. In the opposite, if the result is "not qualified" it will show red dots.

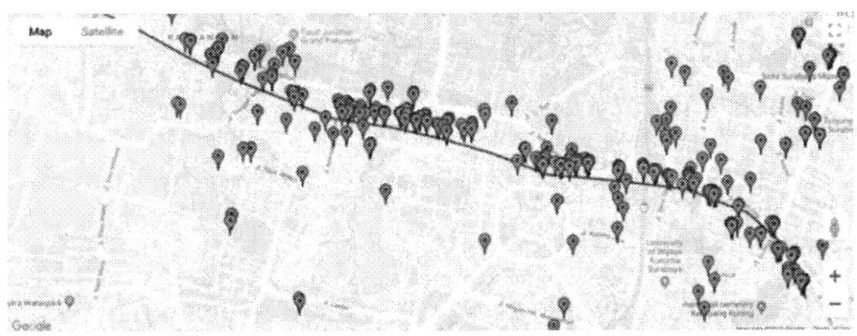

Figure 9. Surabaya fault – RViSITS's maps for seismic criteria follow the standard when the building was built.

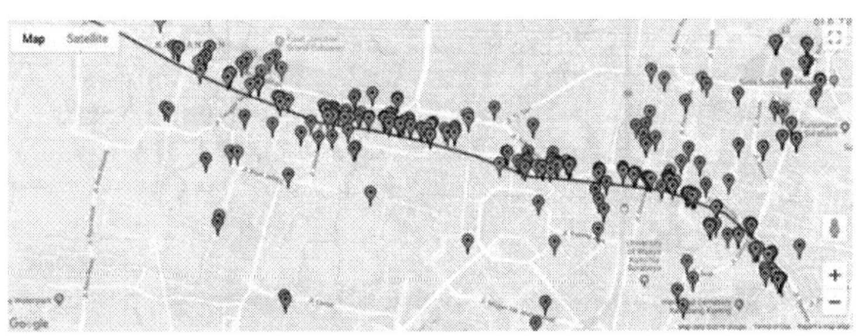

Figure 10. Surabaya fault - RViSITS's maps for high seismic criteria.

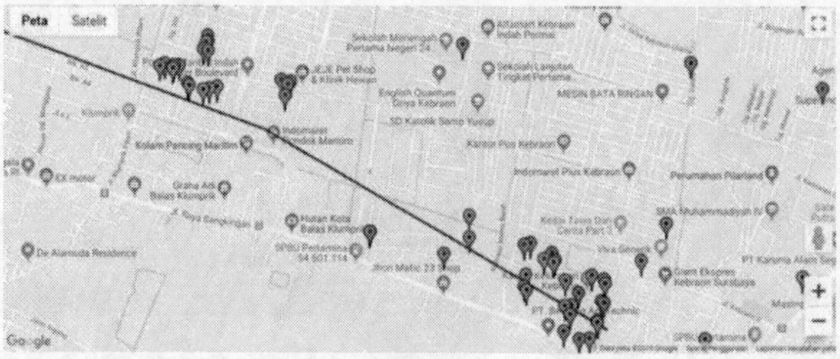

Figure 11. Waru fault – RViSITS's maps for seismic criteria follow the standard when the building was built.

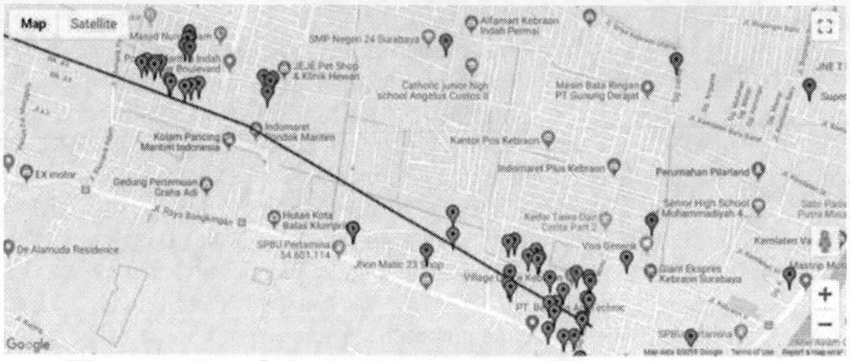

Figure 12. Waru fault - RViSITS's maps for high seismic criteria.

The mapping result are shown in figure 9 to Figure 12. Each of the mapping result illustrates the condition of the building assessment result based on different seismic criteria written above. Figure 9 and Figure 10 show the assessment results of the buildings near Surabaya fault, whereas figure 11 and figure 12 show the assessment results of the buildings near Waru fault. The different results in the maps are shown in figure 9 and figure 10 of Surabaya fault, and figure 11 and figure 12 of Waru fault. All the pictures show that there are changes of the vulnerability score of the buildings, and these lead to different colours in the maps. The vulnerability score using RViSITS of some buildings, which were built from 2003 until 2012 increase, and the building colours in the maps will change from green dots (means: qualified) into red dots (means: not qualified). Hence, Bappeko Surabaya should perform the hazard mitigation policies due to the assessment results using RViSITS based on the increasing seismic hazard in Surabaya.

3.3. Surabaya's Mitigation Strategies

In term of responding to earthquake potential hazard, Surabaya still has low resilience performance index [18]. Hence, it is necessary to develop mitigation strategies due to the earthquake potential hazards in Surabaya. Since the building vulnerability assessment results have been obtained and collected as the database system using RViSITS, the following steps which consist of mitigation strategies in Surabaya should be developed and supported by Bappeko Surabaya. A general procedure of developing mitigation strategies is presented in figure 13. Meanwhile, the buildings which have 'not qualified' result from field survey should be further analysed compared to the other buildings with 'qualified' results. The conceptual design and method of retrofitting solution should be proposed to the 'not qualified' buildings. The retrofitting solution should be evaluated for both short-term and long-term benefits using cost-benefit analysis. The proposed solution should involve some experts in structural building engineering, and Surabaya Government representatives. After the proposed solutions has been approved, the Government representatives, which is Bappeko Surabaya, must issue the practical acting code and regulations regarding the building assessment results as well as the retrofit design and method. This code and regulations must cover the construction of all buildings in Surabaya. Further, a successful strategies of earthquake hazard mitigations requires high public risk awareness and public trust in risk reduction policies. And, this can be best performed by developing risk awareness campaigns, including training programs.

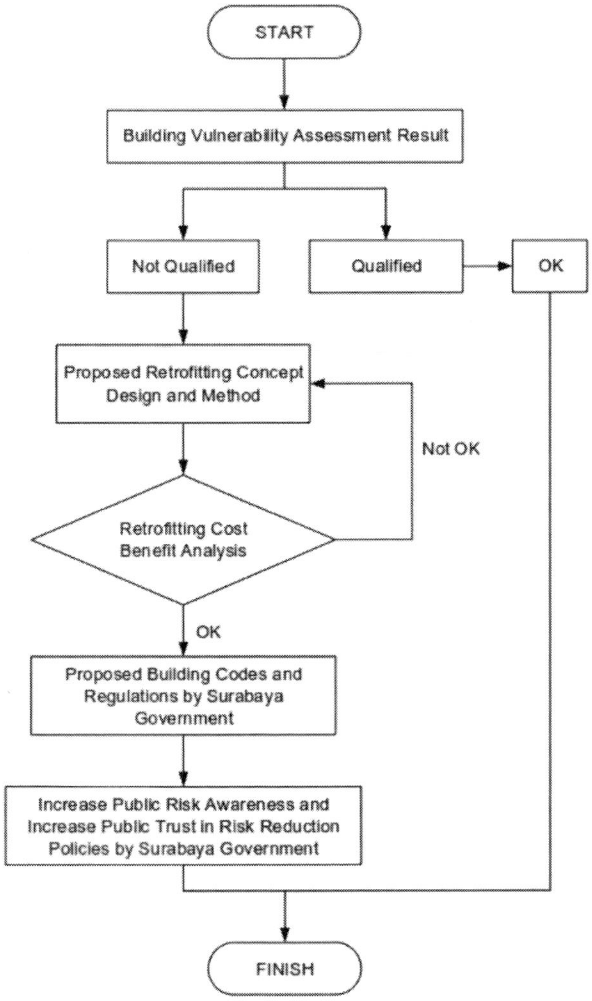

Figure 13. General procedure of developing mitigation strategies.

4. Conclusions

The obtained results of the field surveys using RViSITS show that it requires 65 days to conduct a survey of 366 buildings. That means the use of RViSITS to help rapid visual survey for building assessment are relatively quick in time duration. A building built in accordance with the Indonesian National Standard Criteria (SNI) applied at the time of the construction, will have higher damaged risks when it is faced with increased earthquake potential hazard, especially in Surabaya, which has gradual increasing seismicity risks. In the future, it is expected that the development of other capabilities to generate reliable rapid visual survey results on this RViSITS application can be carried out, including the use of the latest standard and codes in the application. And, it is also expected that other types of categories and number of buildings can be collected and observed for the development of this kind of research in the future.

IOP Conf. Series: Materials Science and Engineering **930** (2020) 012046 doi:10.1088/1757-899X/930/1/012046

5. References

[1] Zakaria Z 1987 Aplikasi Tektonik Lempeng 123–131
[2] B. M. K. dan Geofisika 2019 No Title [Online] Available: http://repogempa.bmkg.go.id/repo_new/repository.php.
[3] Baeda A Y, Husain F, Seismic Z, and Tenggara 2020 A Kajian Potensi Tsunami Akibat Gempa Bumi Bawah Laut di Perairan Pulau Sulawesi *Jurnal Teknik Sipil - Jurnal Teori dan Terapan Bidang Rekayasa Sipil* **19** 75–82
[4] Naryanto H S Analisis Potensi Kegempaan Dan Tsunami 71–78
[5] Surabaya Government 2017 *Kota Surabaya dalam Angka 2017* (Surabaya: Balai Pusat Statistik)
[6] Pusat Penelitian Gempa Nasional 2017 *Peta Sumber Dan Bahaya Gempa Indonesia Tahun* 2017 (Bandung: Pusat Penelitian dan Pengembangan Pemukiman, Kementerian Pekerjaan Umum)
[7] Koulali A, Susilo S, and McClusky S 2016 Crustal strain partitioning and the associated earthquake hazard in the eastern Sunda-Banda Arc *Geophysical Research Letter* **43** 1943–49
[8] Google Earth 2019 *Surabaya and Waru Fault* [Online] Available: https://earth.app.goo.gl/RJTz99
[9] Departemen Pemukiman dan Prasarana Wilayah 2002 *Standar Perencanaan Ketahanan Gempa Untuk Struktur Bangunan Gedung*
[10] Nasional B S 2012 *Tata cara perencanaan ketahanan gempa untuk struktur bangunan gedung dan non gedung* (Indoensia: Badan Standarisasi Nasional)
[11] Nasional B S 2019 *Tata cara perencanaan ketahanan gempa untuk struktur bangunan gedung dan non gedung* (Indonesia: Badan Standarisasi Nasional)
[12] ATC, FEMA 154 2002 *Rapid Visual Screening of Buildings for Potential Seismics Hazards: A Handbook* 2nd ed (Washington DC: Applied Technology Council for Federal Emergency Management Agency)
[13] Wahyuni E, Iranata C, Suswanto B, Nurcahyo C B, and Sutrisno W 2018 Assessment of Vulnerable Buildings Due to Earthquake Loading Using Rapid Visual Screening Smartphone Application," IJASEIT **8** 567–72
[14] Perrone D, Antonietta M, Pecce M, and Rossi F 2015 Rapid visual screening for seismic evaluation of RC hospital buildings *ISTRUC* **3** 57–70
[15] Adam C 2012 A Rapid-Visual-Screening Methodology for the Seismic Vulnerability Assessment of Historic Brick-Masonry Buildings in Vienna in *World Conf. on Earthquake Engineering*
[16] Laboratory of Structure, Civil Engineering Dept., Institut Teknologi Sepuluh Nopember, 2019 User Guide Mobile RViSITS p. 9
[17] Wahyuni E, Amalia A R, and Irawan D 2018 *Laporan Kajian tingkat kerawanan bangunan dan infrastruktur terhadap beban gempa di Kota Surabaya*
[18] Pamungkas A, Ciptaningrum M U, Jaelani L M, and Iranata D 2019 Surabaya Resilience Index for Potential Earthquakes : An Institutional Perspective *Australasian Journal of Disaster and Trauma Studies* **23** 1–11

3D Finite element modeling of circular reinforced concrete column confined with CFRP under different eccentric loads

A B Christianto[1], B Piscesa[1*], M M Attard[2], Faimun[1], and P Aji[1]

[1] Civil Engineering Department, Institut Teknologi Sepuluh Nopember, 60111 Surabaya, Indonesia
[2]School of Civil and Environmental Engineering, The University of New South Wales (UNSW), Sydney NSW 2052, Australia

* Corresponding author: piscesa@ce.its.ac.id

Abstract. This paper presents nonlinear finite element analysis of eccentrically loaded circular Reinforced Concrete (RC) column confined with Carbon Fiber Reinforced Polymer (CFRP). The purpose of this study is to validate the proposed plasticity-fracture model of Piscesa et. al. (2019) to model RC column confined with CFRP under eccentric loading. The software used in the simulation is and inhouse finite element package called 3D-NFLEA. Two available specimens in the literature are investigated and an existing numerical result is also included for comparison purposes. From the analysis and investigation, it was found out that the negative effect of confinement at the outer concrete region was caused by significant difference between the concrete dilation rate at the inner to the outer concrete core region. In addition, From the comparison, close agreement between the numerical models and the available test result are discovered in this study. As for the column ductility, it was found out that the column ductility was higher for column with larger initial load eccentricity.

1. Introduction

Carbon Fiber Reinforced Polymers (CFRP) are a solution for external retrofit applications that have become popular and are widely used in structural retrofitting. Carbon Fiber Reinforced Polymers (CFRP) have advantages such as being lightweight, high strength, and easy to implement in the field [1]. By using CFRP in the potential plastic hinge area region [2], the lateral deformation capacity of RC column can be increased. When the plastic hinge formed in RC column, the column may exhibit reduced load carrying capacity. Increase in the lateral deformation capacity of CFRP confined concrete is caused by the confinement effect of the CFRP jacket which enhance both the concrete core strength and ductility [3-7]. The enhanced lateral deformation capacity of CFRP confined RC column allows the concrete material within the plastic hinge region to remain intact at large deformation cycles [8-11]. Since CFRP is added later to retrofit the column, the CFRP element is considered here as the external confining devices. (see Piscesa et al. [12]).

From the available study, many investigations of CFRP confined RC column have been focused on its behavior under concentric load [6, 13-15]. These studies showed that the retrofitting method was successful in increasing the strength and ductility of RC column. However, more research work is

Content from this work may be used under the terms of the Creative Commons Attribution 3.0 licence. Any further distribution of this work must maintain attribution to the author(s) and the title of the work, journal citation and DOI.
Published under licence by IOP Publishing Ltd

necessary to investigate the behavior of CFRP confined RC column in detail under eccentric loading condition. This eccentric load condition will enable strain gradient in the cross section and thus resulted in more complex behavior of the concrete material under unequal confinement condition.

In this paper, CFRP confined RC columns under eccentric loading are investigated. An inhouse 3D-NLFEA finite element package [16, 17] is used in the numerical simulation. The concrete constitutive model is based on the Piscesa et al. [12, 18] plasticity-fracture model which is restraint sensitive and path-dependent. The model can predict plastic-volumetric compaction which is important to capture the concrete behavior where both high confining pressure and high stiffness of confining devices are present [19]. The constitutive model for the steel solid material and the reinforcing bars obeys the J_2 metal plasticity model. While the FRP material is modeled as an elastic transversely orthotropic material.

Since the loading considered in the analysis is concentric, it is wise to include the 2nd order effect in the analysis. The 2nd order effect analysis will account for the geometrical change in the column due to additional eccentricity. From the previous study [20-22], 3D-NLFEA was successful in capturing the behavior of Concrete Filled Steel Tube (CFST) column under concentric and eccentric loads which involves the 2nd order effect in the analysis. For comparisons purposes, two specimens from Bisby and Ranger [23] are investigated. In addition, numerical result with the same specimens carried out by Lin and Teng [24] are included in the analysis. This paper discusses the lateral modulus contour distribution both in the elastic stage and after the ties is yield. Comparisons of axial load as function of the lateral mid-height displacement curves between the model and the test results will be presented. Finally, ductility evaluation of the CFRP confined RC column is briefly discussed.

2. Constitutive model of material

Figure 1 shows the stress-strain diagram for the CFRP, rebar, and steel loading plate materials. In the finite element model, the CFRP element was modeled as an elastic transversely orthotropic material. An elastic-fracture constitutive model is used for the CFRP element. The softening modulus is set to one-fourth of the hardening modulus to prevent sudden drop of stresses in the CFRP elements. The fiber orientation is set to be oriented in the hoop's direction. The elastic modulus for the CFRP element which consisted only for the resin modulus is set to at least ten percent of the CFRP elastic modulus but not more than 10 GPa. The rebar is modelled using embedded truss element which obey the elastic perfectly plastic model. There is no strain hardening considered and buckling of longitudinal bars are not considered in the analysis. The steel loading plate is modelled as solids element which follows the J_2 metal plasticity model with no strain hardening behavior considered.

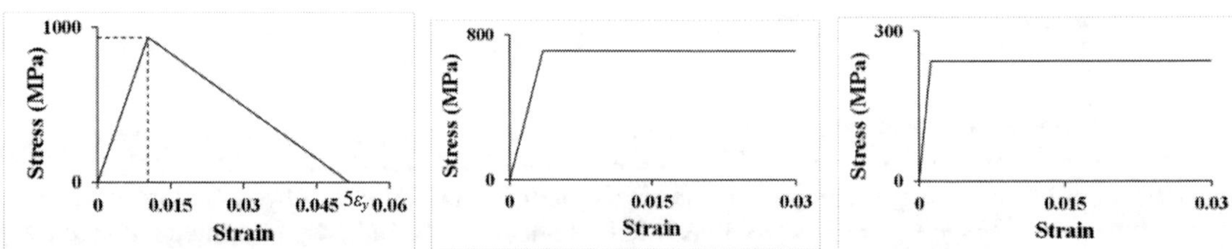

(a) CFRP material (b) rebar material (c) steel plate material

Figure 1. Stress–strains diagram of specimen materials.

The concrete material is based on the plasticity-fracture model developed by Piscesa et al. [12, 18]. In a way, the used plasticity-fracture model is similar in theory with [25] but differs in term of the compression plasticity model. For passively confined concrete, a model which is not only pressure dependent but also restraint sensitive should be used to correctly capture the behavior in the tested specimen. Here, the restraint sensitiveness of the plasticity-fracture model is determined from the lateral modulus value obtained from the analysis during the simulation. From previous study [16], it was shown that the use of plasticity-fracture model [12, 18] can accurately predict the response of CFRP confined RC column under concentric loading. However, further validation for non uniform confining pressure should be carried out. This paper also serves as a validation of the Piscesa et al. [12, 18] plasticity-fracture model to predict the behavior of CFRP confined concrete under eccentric loading. Inside the 3D-NLFEA finite element package, it is important to note that the return mapping algorithm for the plasticity-fracture model was based on two nested algorithm of multi-surface plasticity model [17].

3. Reinforced concrete column model

3.1. Specimen geometry detail

The modeled specimens are obtained from Bisby and Ranger [23] with eccentricity value 30 and 40 mm. This specimen is noted as C30 and C40 in this paper. 3D-NFLEA uses SALOME platform [26] to prepare the geometry of the model and to mesh the model. The meshed model is then extracted as an input in 3D-NLFEA. Figure 2 shows the geometrical interpretation of the specimens. As shown in Figure 2, the specimen diameter is 152 mm while the specimen height is 608 mm. The load eccentricity considered are 30 and 40 mm. The longitudinal bar consisted of four bars with 6.4 mm diameter. The confining bar had the same diameter as the longitudinal bar. The pitch spacing of the longitudinal bar is 100 mm (see Figure 2b for more details). The FRP thickness for one ply is 0.38 mm.

The concrete strength of the specimens is 33.2 MPa for compression and 3.557 MPa for tensile. The modulus of elasticity for the concrete is 20431.1 MPa. The CFRP material have a yield tensile strength of 933 MPa, ultimate tensile strength of 1014 MPa, elastic modulus of 90000 MPa, rupture strain of 1.12 % and the Poisson's ratio of 0.3. The yield strength of the bar is 710 MPa while the Young's modulus is 200 GPa. The steel loading plate is considered to have the yield strength close to that of A36 steel material.

At both ends, a 25 mm thick steel loading plate is used to distribute the knife-edge load from the steel to concrete. This steel plate is required to avoid stress concentration due to sharp displacement control at the top end. A thin line of displacement control is set at the top steel loading plate while at the bottom plate, a thin line of restraint is provided in all direction. Schematic details on the boundary conditions are shown in Figure 2 (c).

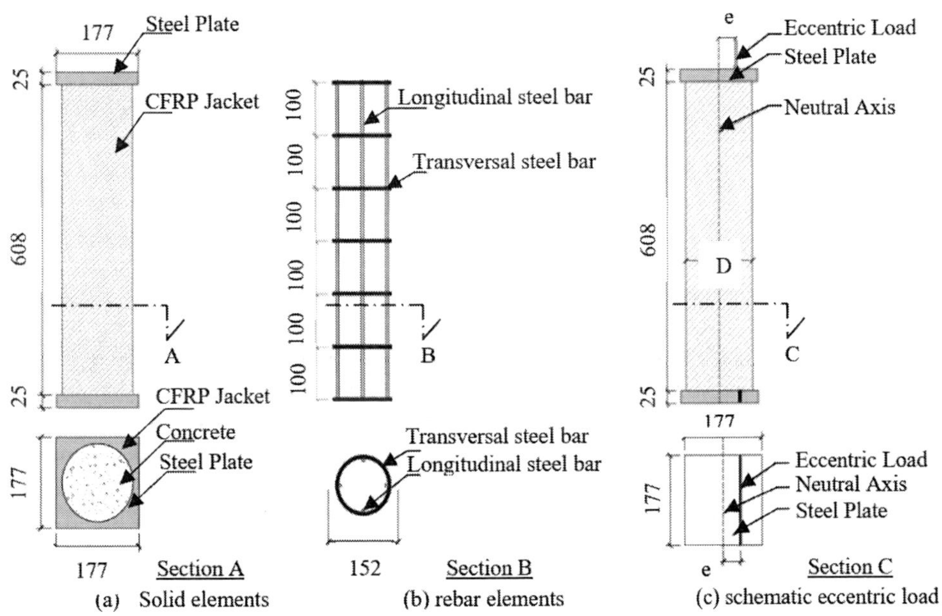

Figure 2. Side view and cross section of specimen.

3.2. 3D Modeling of Specimen

IOP Conf. Series: Materials Science and Engineering **930** (2020) 012047 doi:10.1088/1757-899X/930/1/012047

The 3D model of RC columns is prepared using SALOME 9.3.0 [26] as shown in Figure 3. The total number of solid elements is 27,702 and 27,084 are used to model C30 and C40 columns, respectively. The concrete, CFRP and reinforcing bars elements are shown separately as shown in Figure 3 (a), (b) and (c), respectively. The full meshed element of the specimens as shown in Figure 3 (d). The knife-edge load at the top end, which act as a displacement control during the numerical analysis, can be moved with different eccentricities while at the bottom, a fixed restraint boundary condition is sought.

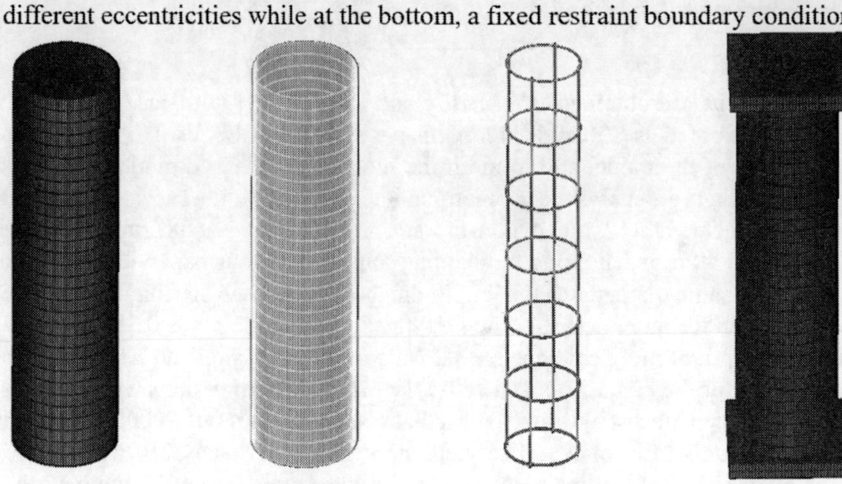

(a) concrete (b) CFRP jacket (c) reinforcing bars (d) full model

Figure 3. Meshed model of a typical column under eccentric loading.

4. Analysis results and discussions

4.1. Lateral modulus contour distribution

In this section, the lateral modulus for RC columns confined with CFRP members is being examined in detail to understand more about the confinement effectiveness. The focus of investigation is on the development of the lateral modulus in the elastic phase and when the transverse steel is yielded. The lateral modulus was found to be the highest at the concrete core due to the circular ties and the CFRP jacket work together to confine the inner concrete core (see Figure 4 (a)). On the other hand, the outer concrete core only confined by CFRP and thus have lower lateral modulus value due to less confined behavior. This indicates that the effectiveness of the CFRP may not be fully functioning before the circular ties is yields.

The 3D surface of the lateral modulus contour at the elastic phase for specimen C30 and when the steel ties are yielded are shown in Figure 4 (a) and (b), respectively. The lateral modulus value in the inner concrete core is higher than that of the outer concrete core. At the ties level, for concrete area under compression, there is a high value of lateral modulus concentration as shown in Figure 4 (a). This shows that the circular ties stiffness is more concentrated than the CFRP. Arching actions between the ties are clearly shown and was associated with the steel ties. Outside the arches, the lateral modulus is lower than the one exerted by the CFRP wrap. This condition would give different concrete dilation rate between the two core regions. Since at the outer concrete core the concrete dilation rate is much higher than the inner core, there will be a negative effect in the outer concrete core. Hence, the value of the lateral modulus at the outer region will be less than the one exerted by the CFRP wrap if it was used as a single confining device. Figure 4 (b) shows when the lateral modulus between the ties where it is drops and only occurs in the compressed region of the concrete. This shows that, the CFRP tubes starts to effectively confine the concrete.

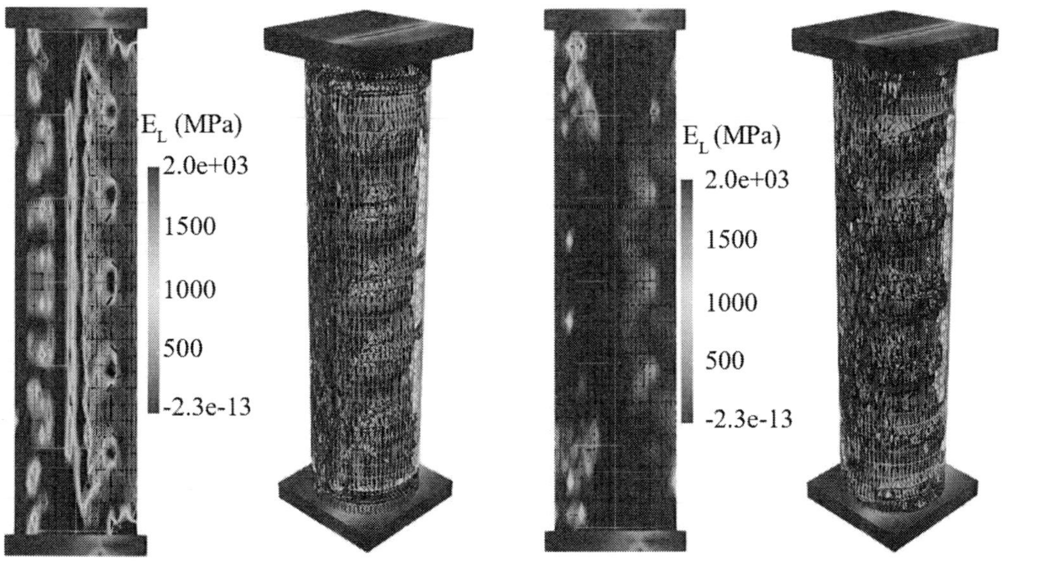

(a) elastic stage (b) after the circular ties is yield

Figure 4. The C30 specimen volume with the lateral modulus contour.

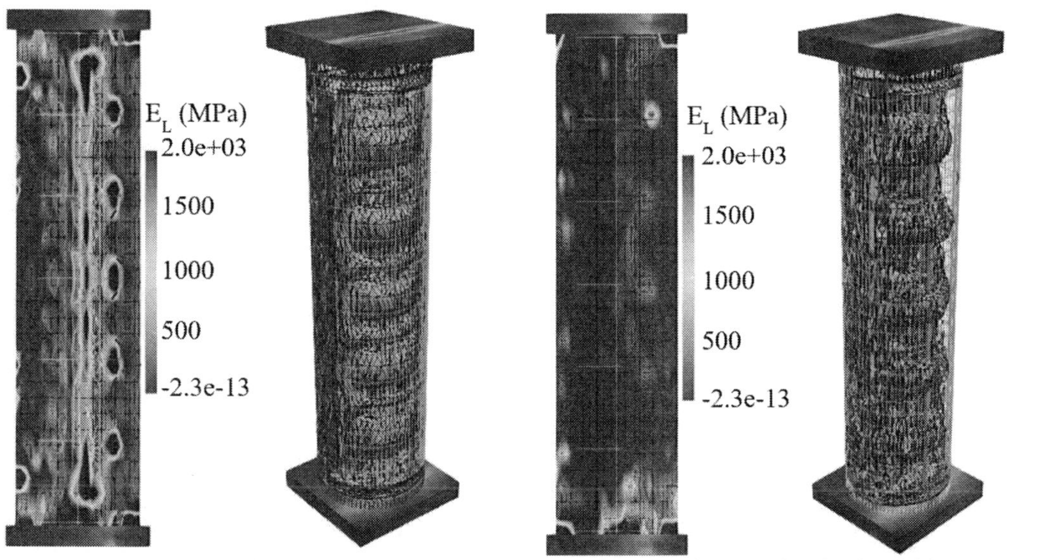

(a) elastic stage (b) after the circular ties is yield

Figure 5. The C40 specimen volume with the lateral modulus contour.

Figure 5 (a) shows the lateral modulus contour and the 3D surface at elastic stage and Figure 5 (b) shows the same condition when the steel ties are yielded for specimen C40. Similar conclusions were obtained from specimen C30 for specimen C40, as shown in Figure 5.

4.2. Comparisons with experimental result

Figure 6 shows the comparisons between the numerical models and the test results for specimen C30 and C40 in terms of axial force versus additional eccentricity at the mid-height for C30. In Figure 6, the FEA results carried out by Lin et al. [24] was executed using ABAQUS finite element software package. The concrete constitutive model in [24] was from [27, 28]. As shown in Figure 6(a), the initial stiffness response from 3D-NLFEA, test result, and from [24] are similar. At lateral displacement about 2 mm, the axial load from both 3D-NLFEA and [24] are higher than the test result. However, at lateral displacement around 11 mm, the axial load from 3D-NLFEA is in good agreement with the test result, while the axial load from [24] was found to be higher than the test result. The same conclusion was found for specimen C40 as shown in Figure 6(b).

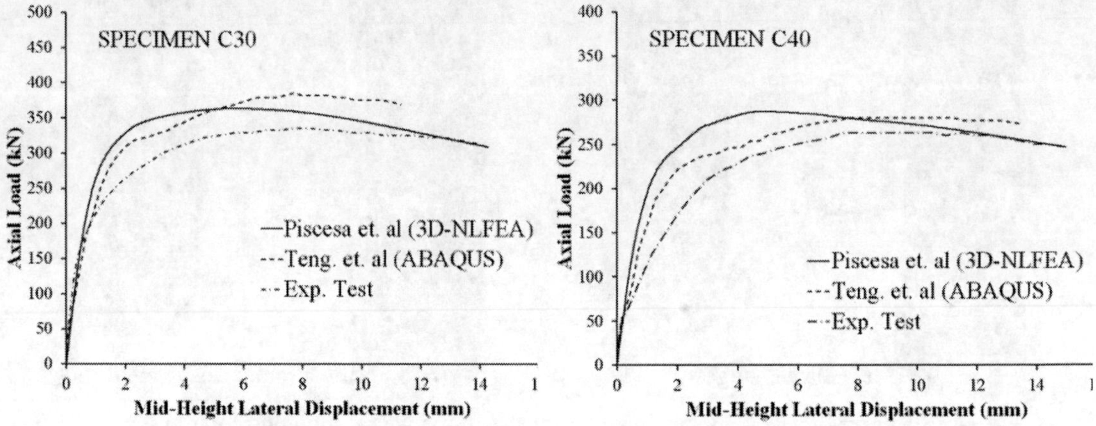

(a) C30 with 30 mm load eccentricities (b) C40 with 40 mm load eccentricities

Figure 6. Comparison of axial load–lateral displacement curves of columns.

4.3. Ductility of column

The evaluation of column ductility is based on the I_{10} ductility index which is based on the energy ratio. The I_{10} ductility index was proposed by Foster and Attard [29] in which often used to estimate the flexural toughness of beam in ASTM C1018 [30]. In the research of Samani et al. [31], the I_{10} ductility index was used to evaluate the ductility of RC columns. This energy is computed as the area beneath the moment-curvature curve (M-φ). The bending moment is obtained by multiplying the axial load (P) with the total lateral mid-height displacement (Δ). By taking the value of the strain that occurs in the mid-height cross section (zero axial strain) of the column then dividing it by the distance between the measured axial strain value to the neutral line of the cross section, the curvature value is obtained from Paraview 5.8 (post-processor, see [32, 33]). To get the value of I_{10}, one can compute the area ratio of the M-φ from the analysis (point OCD) to the area of triangle OAB (see Figure 7).

Table 1 shows the value for the I_{10} ductility index for both columns. As shown in Table 1, the I_{10} ductility index for specimen C30 is lower than specimen C40. As the load eccentricity increased, the ductility of column becomes higher. This can be well understood as the compression in concrete is less and therefore the column behaves more like a beam and therefore more ductile. Several researchers [34-37] have performed ductility tests of concrete columns applied to conventional concrete and CFRP reinforced concrete columns at eccentric loads.

Table 1. Ductility index of RC column.

No.	Specimens ID	I_{10}
1	C30	8.284
2	C40	8.434

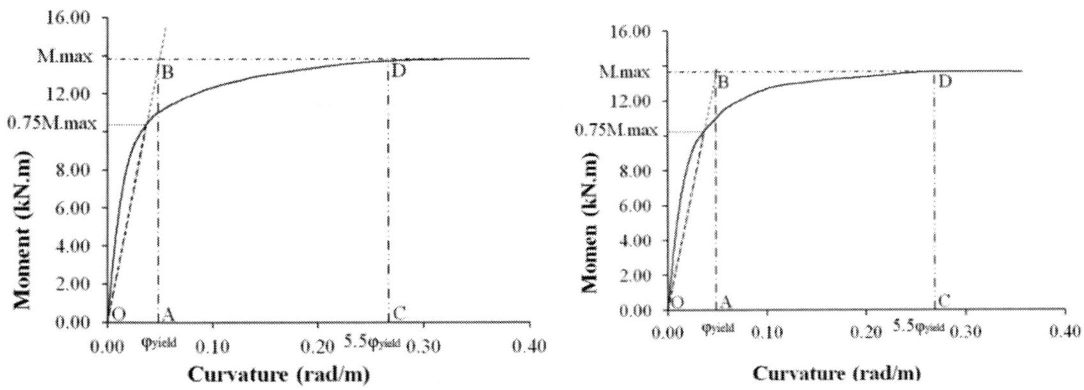

(a) specimen C30 (b) specimen C40

Figure 7. Ductility index calculation of column.

5. Conclusions

This paper presents full three-dimensional nonlinear FEA of circular reinforced concrete column confined with CFRP under different eccentric loads. The FEA are carried out using the 3D-NLFEA program developed by the Piscesa et al. A plasticity-based constitutive model developed by Piscesa et al. is used for the concrete material under compression and fracture-plasticity based constitutive model developed by Piscesa et al. (2019) for concrete material under tension. Two test results available in the literature from Bisby and Ranger (2010) are investigated.

From the analysis and investigation, it was found out that the significant difference between the concrete dilation rate at the inner to the outer concrete core region caused by a negative effect of confinement at the outer concrete region. The negative confinement effect can be attributed to the following: in the ties level for concrete area under compression there is a concentration of lateral modulus (at the elastic phase). However, the concentration of the lateral modulus at ties level disappears when the circular ties is yield. From the comparisons between the models and the test results showed that the 3D-NLFEA prediction agrees well with the test result. As for the column ductility, it was found out that the column ductility was higher for column with larger initial load eccentricity.

Further research should be focused on more comparisons with the experimental data which cover different load eccentricity, dimension, and strength of material properties. It should also be noted that the material data inputted in the analysis is a mean data extracted from the experiments. In practice, there are possibilities of having variations in the material strength which may result in lower material strength. Thus, further research is needed to investigate RC columns confined with CFRP under different eccentric loads and it is necessary to investigate in detail on the confining pressure distribution.

6. References

[1] Binici B 2008 Design of FRPs in circular bridge column retrofits for ductility enhancement," *Eng. Str.* **30** 766-76

[2] Binici B and Mosalam K M 2007 Analysis of reinforced concrete columns retrofitted with fiber reinforced polymer lamina *Compos. Part B: Eng.* **38** 265-76

[3] Xiao Y and Wu H 2000 Compressive behavior of concrete confined by carbon fiber composite jackets *J. Mater. Civil Eng.* **12** 139-46

[4] Samaan M, Mirmiran A, and M Shahawy 1998 Model of confined concrete by fiber composites," *J. Struct. Eng-ASCE* **124** 1025-31

[5] Rochette P and Labossiere P 2000 Axial testing of rectangular column models confined with composites *J. Compos. Constr.* **4** 129-36

[6] Mirmiran A, Shahawy M, Samaan M, Echary H E , Mastrapa J C, and Pico O 1998 Effect of column parameters on FRP-confined concrete *J. Compos. Constr.* **2** 175-85

[7] Harries K A and Kharel G 2003 Experimental investigation of the behavior of variably confined concrete *Cem. Concr. Res.* **33** 873-80

[8] Sheikh S A and Yau G 2002 Seismic behavior of concrete columns confined with steel and fiber-reinforced polymers *Structural Journal* **99** 72-80

[9] Seible F, Priestley M N, Hegemier G A, and Innamorato D 1997 Seismic retrofit of RC columns with continuous carbon fiber jackets *J. Compos. Constr.* **1** 52-62

[10] Saadatmanesh H, Ehsani M R, and L Jin 1996 Seismic strengthening of circular bridge pier models with fiber composites *ACI Struct. J.* **93** 639-738

[11] Iacobucci R D, Sheikh S A, and Bayrak O 2003 Retrofit of square concrete columns with carbon fiber-reinforced polymer for seismic resistance *Structural Journal* **100** 785-94

[12] Piscesa B, Attard M M, and Samani A K 2016 A lateral strain plasticity model for FRP confined concrete *Compos. Struct.* **158** 160-74

[13] Ahmad S H and Shah S P 1982 Complete triaxial stress-strain curves for concrete *ASCE J. Struct. Div.* 108 728-42

[14] Becque J Analytical modeling of concrete columns confined by FRP 2000

[15] Mirmiran A and Shahawy M 1997 Behavior of concrete columns confined by fiber composites *J. Struct. Eng.* **123** 583-90

[16] Piscesa B, Attard M M, and Samani A K 2018 3D Finite element modeling of circular reinforced concrete columns confined with FRP using a plasticity based formulation *Compos. Struct.* **194** 478-93

[17] Piscesa B, Attard M M, Prasetya D, and Samani A K 2019 Modeling cover spalling behavior in high strength reinforced concrete columns using a plasticity-fracture model *Eng. Struct.* **196** 109336

[18] Piscesa B, Attard M, Samani A, and Tangaramvong S 2017 Plasticity Constitutive Model for Stress-Strain Relationship of Confined Concrete *ACI Struct. J.* **114**

[19] Piscesa B 2018 *Modeling confined concrete using plasticity formulation PhD. Thesis, School of Civil and Environmental Engineering* (Australia: The University of New South Wales, Sydney, Australia)

[20] Piscesa B, Attard M, Suprobo P, and Samani A 2017 Investigation on the fiber based approach to estimate the axial load carrying capacity of the circular concrete filled steel tube (CFST) in *IOP Conference Series: Mater. Sci. Eng.* vol 267 no 1: IOP Publishing p 012017

[21] Piscesa B, Attard M M, and Samani A K 2017 Plastic Dilation Rate Characteristic of Concrete Confined with Steel Tube in *XIV International Conference on Computational Plasticity. Fundamentals and Applications (COMPLAS)* (Barcelone)

[22] Piscesa B 2017 Numerical Investigation on The Behaviour of Concrete-Filled-Steel-Tube Column under Eccentric Loading *presented at the The Third International Conference on Civil Engineering Research (ICCER)* (Surabaya)

[23] Bisby L and Ranger M 2010 Axial–flexural interaction in circular FRP-confined reinforced concrete columns *Constr. Build. Mater.* **24** 1672-81

[24] Lin G and Teng J 2017 Three-dimensional finite-element analysis of FRP-confined circular concrete columns under eccentric loading *J. Compos. Constr.* **21** 04017003

[25] Červenka J and Papanikolaou V K 2008 Three dimensional combined fracture–plastic material model for concrete *Int. J. Plast.* **24** 2192-20

[26] *The Open Source Integration Platform for Numerical Simulation* 2017 http://www.salome-platform.org/ [Online] Available: http://www.salome-platform.org/

[27] Yu T, Teng J, Wong Y, and Dong S 2010 Finite element modeling of confined concrete-II: Plastic-damage model *Eng. Struct.* **32** 680-91

[28] Yu T, Teng J, Wong Y, and Dong S 2010 Finite element modeling of confined concrete-I: Drucker–Prager type plasticity model *Eng. Struct.* **32** 665-79

[29] Foster S J and Attard M M 1997 Experimental tests on eccentrically loaded high strength concrete columns *Structural Journal* **94** 295-303
[30] Standard A ASTM C1018-97 *Standard Test for Flexural Toughness and First-Crack Strenght of Fiber Reinforced Concrete (Using Beam with Thrid Point Loading)*
[31] Samani A K, Attard M M, and Foster S J 2015 Ductility in concentrically loaded reinforced concrete columns *Aust. J. Struct. Eng.* **16** 237-50
[32] Ahrens J, Geveci B, Law C, Hansen C, and Johnson C 2005 36-ParaView: An End-User Tool for Large-Data Visualization *The Visualization Handbook* 717
[33] Ayachit U 2015 *The paraview guide: a parallel visualization application*
[34] Csuka B and Kollár L P 2012 Analysis of FRP confined columns under eccentric loading *Compos. Struct.* **94** 1106-16
[35] Yu T, Wong Y, and Teng J 2010 Behavior of hybrid FRP-concrete-steel double-skin tubular columns subjected to eccentric compression *Adv. Struct. Eng.* **13** 961-74
[36] Fam A, Flisak B, and Rizkalla S 2003 Experimental and analytical modeling of concrete-filled FRP tubes subjected to combined bending and axial loads *ACI Struct. J.* **100** 499-509
[37] Bunni N, Scott B, Park R, and Priestley M 1982 Stress-strain behavior of concrete confined by overlapping hoops at low and high-strain rates-discussion *J. Am. Concr. Inst.* **79** 496-98

Connection system of precast beam and column with wet joint for short span bridge

A Karjanto[1*], Nindyawati[1], and P B Susanto[1]

[1]Department of Civil Engineering, Faculty of Engineering, Universitas Negeri Malang, Malang, Jawa Timur, Indonesia.

Corresponding author: adjib.karjanto.ft@um.ac.id

Abstract. The objective of this study is to see whether or not there is a difference in the flexure behavior between three-supported continued segmental precast concrete beam joined with a cast in place. The formation of the bond is 90° vertical combined with 45° against the horizontal reinforcement, making use of the beam as the formwork. The monotonic pressure was applied constantly at one stage until the beams reached the ultimate. The result of the experiment showed that the flexure behavior of the three-supported continued pre-cast beams joined with the cast in place at the condition before cracks (elastic) had the same trend as that of monolith beam at the base area, as well as at the supporting points. However, before passing the yielding condition (the pulling reinforcer yielded), the precast beam had broken into pieces at the supported area, while the monolith beam had a trend in elastic condition, yielding condition, and ultimate condition. The ductility value of the pre-cast beam was 1.33 (at the base area) and 1,5 (at the supported area). The ductility value of the pre-cast beam was 1.0 at the base as well as at the supported area. This means pre-cast beams reach ultimate before the iron reinforcer yield (the ultimate structure is collapse). The cracking patterns showed that at the joined area the bonding failed to happen between the pre-cast beam and the bonding agent.

1. Introduction

Beam flexural behavior can be known from the amount of the load (gravity, lateral) that works on the beam giving rise to the torque, curvature, deflection, rotation, ductility, and cracked patterns. According to Park and Pauly (1975), curvature (φ crack, φ y, φ u) is the response of each concrete strain condition and steel against the effective height of the beam in each load condition (elastic, yield, and Ultimate).

Deflection is a large decrease in the beam due to the load read on the dial gauge. The torque is an analysis of reinforced concrete sections that use data from direct measurements of concrete strain readings and steel strain on the stain meter at each stage of loading. Rotation (angle magnitudes) is round the corners of both edges. The ductility (ductility deflection index) is the ratio of beam deflection when destroyed to beam deflection when the tensile reinforcement reaches melting. A crack pattern through direct observation of the microscope crack detector, with an accuracy of 1 divs = 0.02 mm.

Content from this work may be used under the terms of the Creative Commons Attribution 3.0 licence. Any further distribution of this work must maintain attribution to the author(s) and the title of the work, journal citation and DOI.
Published under licence by IOP Publishing Ltd

IOP Conf. Series: Materials Science and Engineering **930** (2020) 012048 doi:10.1088/1757-899X/930/1/012048

2. Experimental study

2.1 Fabrication of the test specimens

The dimensions of the beams and precast bridge columns are made to observe deflections, torques, curvature, rotation, ductility, and cracked pattern, as can be seen in Figure 1. In addition to the monolith, the method is made as a comparison and the details in Figure 1

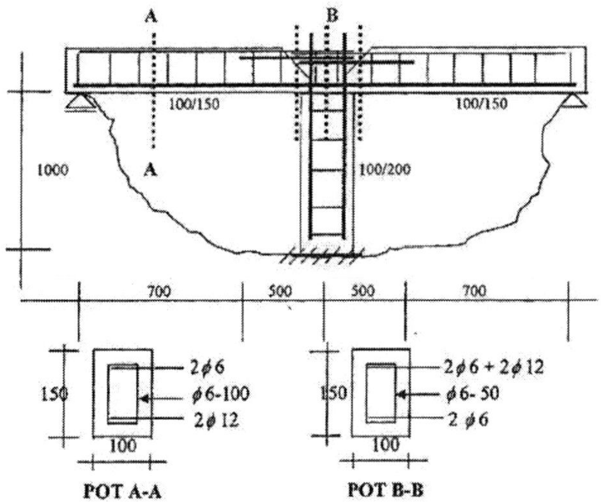

Figure 1. The dimension of the Prefabricated Concrete Bridge Portal

Information:
The spindle in the joint area is made ½ the distance from the precast spindle (see detail. B-B)

2.2 Beam testing

Beam testing can be detected from measuring devices mounted on a beam testing scheme and the placement of the strain gauges and dial gauges shown in Figure 2

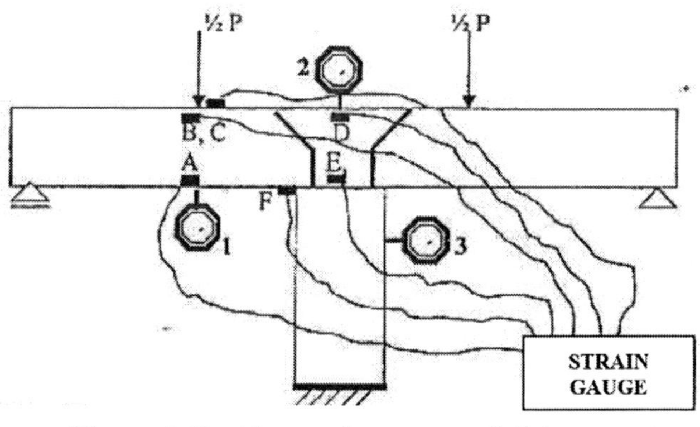

Figure 2. Position strain gauge and dial gauge

Information:
A = B = D = E = steel strain gauge patch in reinforcement
C = F = concrete strain gauge patch in surface concrete
 = Dial gauge

3. Analysis of experimental result

3.1 Load – deflection

The early collapse of beams marked with lending and concrete strain was increased but a constant steel strain and followed by cracks were increasingly widened to enter concrete press areas. The correlation between load and deflection base area for both models of Beams are shown in Figure 3 with an average difference of 6/7.3 = 82%). The collapse of the beams on the focus before the melt pull

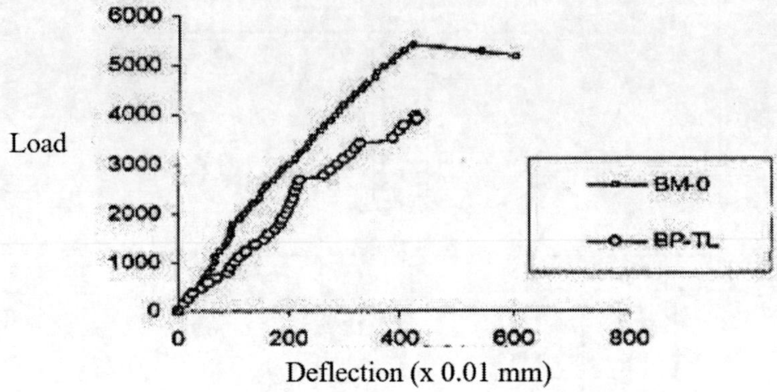

Figure 3. Correlation of Deflection Load in the Base Area

Table 1. Deflection of Beams in the Base and Support Areas

Type of Beam	Location	Deflection (mm)
BM-0	Support	6.00
	Mid Span	8.00
BP-TL	Support	7.30
	Mid Span	8.52

3.2 Moment - curvature

The relation between bending moment and curvature are reviewed in the Center and base areas (under centralized load P) due to centralized loads (P) and analyzed resulting in concrete strain (EC), compressed steel Strain (ES), tensile steel strain (εs), moment (M-kgm) and curvature (Φ-rad/M) shown in Table 2

Tabel 2.a Correlation of Moment – Curvature of Cracked Condition, Elastic

Type of Beam	Location	P (kg)	Ec	εs	εs	Torque (kgm)	ϕ (rad/m)	Percent M to BM-0 (%)
BM-0	Support	611.9979	0.000420	0.000192	0.000140	141.3120	0.004708	100
	Mid Span	611.9979	0.000343	0.000081	0.000173	217.470	0.003969	
BP-TL	Support	557.0540	0.000094	0.000429	0.000120	134.6253	0.004023	89.02220776
	Mid Span	557.0540	0.000095	0.000074	0.000061	207.9250	0.001200	

Table 2.b Correlation of M – Curvature of Cracked Condition, Reinforcement Melting (Yield)

Type of Beam	Location	P	Ec	εs	εs	Torque	ϕ	Percent M to BM-0
		(kg)				(kgm)	(rad/m)	(%)
BM-0	Support	2699.8660	0.002631	0.002149	0.001296	623.4066	0.036769	100
	Mid Span	2699.8660	0.002112	0.001367	0.001796	959.2805	0.030062	
BP-TL	Support	1985.5950	0.001884	0.000687	0.000687	598.4795	0.012131	90.54421760
	Mid Span	1985.5950	0.001884	0.001186	0.001134	885.4954	0.033062	

Table 2.c Correlation of M – Curvature After Cracked, Collapse Condition (Ultimate)

Type of Beam	Location	P	εc	εs	εs	Torque	ϕ	Percent M to BM-0
		(kg)				(kgm)	(rad/m)	(%)
BM-0	Support	2727.3380	0.002634	0.002199	0.001494	629.7499	0.037177	100
	Mid Span	2727.3380	0.002164	0.001477	0.001845	969.0415	0.030838	
BP-TL	Support	2013.0670	0.002106	0.001069	0.000975	594.8228	0.024423	89.81070100
	Mid Span	2013.0670	0.003164	0.001413	0.001134	815.2563	0.033062	

From Table 2.b, it can be seen that the beam with the best ability is a monolith beam with a reinforcement (BP-TL) in the crack condition, elastic (crack), having an ability of 90.54% under the monolith beam (BM-0). This situation is also in the melting and collapse condition.

3.3 Load – rotation

The correlation of load and rotation for all models is shown in Figure 4 and the analysis is shown in Table 3. The tendency of the load and rotation relationships of the two beams as in the load and deflection relationship.

From table 3 It can be seen that the prefabricated beam rotation is greater than the monolith beam. That is, judging from the value of the beam rotation, the monolith beam has a better ability than the precast beam because before the rotation reaches the maximum, the precast beam has collapsed due to the unfavorable connection strength.

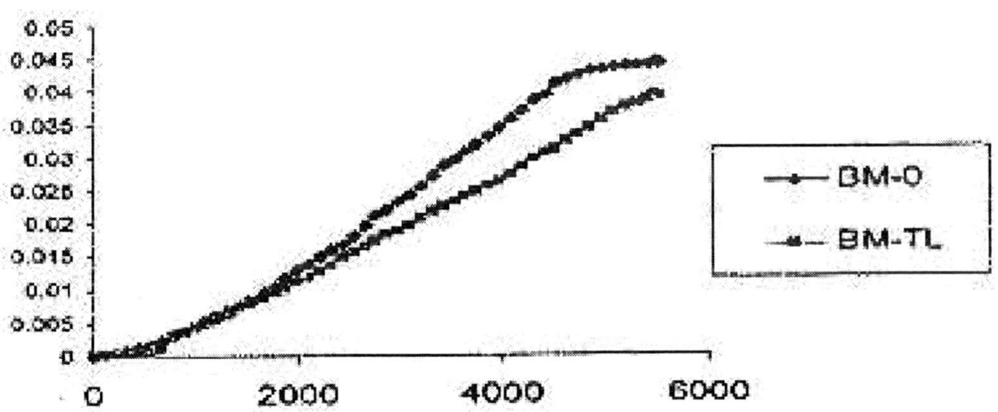

Figure 4. Correlation of Load dan Rotation

Table 3. Correlation of Load dan Rotation Collapse Condition (Ultimate)

Type of Beam	Location	P	ϕ	Percent M to BM-0
		(kg)	(rad/m)	(%)
BM-0	Support	2727,3380	0,044612	100
	Mid Span	2727,3380	0,037006	100
BP-TL	Support	2727,3380	0,039572	88,70266915
	Mid Span	2727,3380	0,039120	85,45347967

3.4.Ductility

The deflection ductility index (μ) or displacement ductility is a comparison of the beam deflection at the time of being crushed against the deflection of the beam when the tensile reinforcement reaches melting point. The magnitude of the experimental test index ductility for each model is given in table 4

Table 4. Deflection Ductility Index (μ)

Type of Beam	Location	μ (x0,01 mm)	μ (x0,01 mm)	μ
		Collapse Beam	Yield Strain	(ductility)
BM-0	Support	300	200	1.5
	Mid Span	400	300	1.33333333
BP-TL	Support	366	360	1.01666666
	Mid Span	208	200	1.04

From Table 4 It is known that failure occurs on precast beams with pass reinforcement because it has the same ductility index equal to 1.0

That is, for precast beams with reinforcement (BP-TL) in the following condition:

1. on the verge of collapse when the steel of reinforcement reaches melting.
2. Structural collapse occurs brittle.
3. The collapse occurs suddenly after the melting steel reinforcement.

3.5.Crack pattern

In Figure 5 There is a bending crack that continues to go near the axis of the beam and towards sliding bending. Due to the addition of the load, the bending shear crack becomes larger and leads to the concrete compressive area and causes the beams to break. It can be concluded that the monoliths beam crack pattern (BM-0) is an ideal model for the comparison of precast beams.

In the precast beam with reinforcement (BP-TL), occurs the same crack with BM-0, which is a bending crack that continues towards the near-axis of the beam and towards the sliding, forming a flexural shear crack. Due to the addition of the load, the bending shear crack becomes larger and leads to the concrete compressive area and caused the beam to become crushed. On the unfractured area, there is no crack, indicating that the reinforcement system that is used does not affect the flexural strength of the beam. Cracks occur in the area of the connection line, starting from the top side, radiating to approach the axis of the beam, toward the horizontal (sliding crack) after the reinforcement is approaching yielding.

Table 5. Beam Crack Width

Type of Beam	Location	Crack Width (mm)
BM-0	Support	0,46
	Mid Span	0,51
BP-TL	Support	0,92
	Mid Span	0,86

From Table 5 It appears that the smallest crack width on the BP-TL type of experimental beam is 0.92 mm (supported) and 0.86 mm (mid span) above the width of the crack beam monoliths (BM-0).

The experimental results above which consist of load and deflection relationships, moment and curvature relations, load and rotation relationships, ductility, crack widths show that:

1. Monolith beams (BM-0) have the level of quality as planned to withstand gravity loads.
2. Beam precast with reinforcement (BP-TL) have an average quality of 87% when compared to the results of plans, theoretical and monolithic beams, and have sufficient ability to withstand gravity loads, when viewed from deflection ductility and crack width. This means that the load that works on precast is quite safe when reduced by 80% of the load acting on a monolith beam.

4. Conclusion

Based on the results of experimental studies on the design of the connection system between the beam and precast column on the structure of the short span bridge with a cross-section connection method when compared with conventional systems (monoliths) it can be concluded as follows.

1. In the system of beams and non-precast columns (monoliths) have a good ability to withstand the gravity load when viewed from the deflection, curvature, rotation and deflection ductility of 1.5 (on the bar area) and 1.33 (on the base area)
2. In precast specimens with the connection system between the beam and precast column on the bridge structure shows behavior that resembles the behavior of monolithic specimens. A striking difference in the amount of load that occurs, where the ability of the precast beams and columns has a capability of 87% when compared to conventional systems and has a deflection ductility of 1.0 (should be > 1)
3. A series of precast beams and columns have conditions: (1) on the threshold of collapse before the reinforcement reaches melting and (2) The collapse of the structure occurs in a way.
4. The failure of the Precast column and beam on the connection area is caused not due to the reinforcement system but due to the connection material because there is no good unity between precast and the connection material (shear cracking occurs). The proposed advice is the need for advanced research on the connection material so that it has a good ability (monoliths) in uniting precast concreteMonolith beams (BM-0) have the level of quality as planned to withstand gravity loads.

References

[1] M R Dicky and Rifani 2001 Kajian eksperimental perilaku sambungan prepack polimer antara balok beton daur ulang dengan normal *J. Tek. Sip. ITB Bandung* **8** 1 p 11 – 20
[2] Nindyawati and Susanto P B 2004 Kajian Eksperimental Kekuatan Lentur Dari Sambungan Balok Beton Pracetak Segmental Bertulang Tunggal Dengan Metode Cor Di Tempat (Wet Joint*) tidak dipublikasikan* TPSDP Universitas Negeri Malang UM
[3] Park R and Pauly T 1975 *Reinforced Concrete Structure* (New York: John Wiley & Sons)
[4] S Siddid 1995 Struktur Bangunan Sistem Rangka Daktail Tahan Gempa dengan Komponen Precast *DPU Seminar Teknologi Pengurangan Dampak Gempa & Tsunami* (Jakarta)

[5] W S Biemo and R M Dicky 1998 Penilaian dan Rehabilitasi Gedung Pasca Kebakaran dengan Menggunakan Beton Polimer *Makalah Seminar Teknologi Beton* BPPS (Semarang: Universitas Diponegoro)

A simplified method to estimate the fundamental frequency of simple span bridges supported on lead rubber bearing

H Sugihardjo[1*], and A B Habieb[1]

[1] Department of Civil Engineering, Sepuluh Nopember Institute of Technology, Surabaya, East Java, Indonesia
[2] Member of Bridge and Tunnel Safety Commision, Ministry of Public Works and Housings, Indonesia.

*Corresponding author's e-mail: hidayat@ce.its.ac.id

Abstract. The impulsive load method applied on the mid-span of a girder is often used to measure the fundamental frequency of the bridge. This method commonly involves a truck that is dropped from a certain height to generate an impulsive load. In high seismicity zones, bridge structures often involve base isolators such as lead rubber bearing (LRB) to reduce the seismic force demand. However, the application of LRBs on the bridge girders may reduce its vertical fundamental frequency. This fundamental frequency may decrease further if the girder is supported on the frames or cantilever beams on high columns. In this study, the fundamental frequency of several bridge spans is computed through a simplified analytical study based on a single degree of freedom assumption. The results show that the application of LRBs does not significantly affect the fundamental frequency of the girder when compared to the identical girder supported on hinge-roller supports, due to the high vertical stiffness of the LRBs. In average, the values of fundamental frequency obtained through the simplified method and the FE modeling are lower than which obtained from the dynamic loading test. It indicates that the stiffness of the existing bridge is higher than the analyses.

1. Introduction

In the evaluation of constructed bridges, static and dynamic test are required. The static test can be performed by placing a row of trucks having an equivalent load minimum 70% of the design load. The deformation of the elements of the bridges, particularly the vertical deflection of the girders shall be less than the design deflection or which required by the code.

In the dynamic test, the simplest method is to perform an impact load on several points of the girders such that a maximum deflection is obtained. In simple span bridges, the impact load P(t) can be in the form of truck axis dropped from the height of 200 mm at the middle span, as seen in Figure 1. The jumper device is presented in Figure 1c: small white wheels are used for the transportation. This loading method presents the implementation of an impulsive ramp load that occurs in a very short duration. Immediately after the impact load, the girders exhibit a damped-free vibration response. The fundamental frequency of the structure can be computed based on the amplitude from crest to crest by means of Fast Fourier Transform (FFT), while the damping ratio can be computed based on the obtained logarithmic decrement.

Content from this work may be used under the terms of the Creative Commons Attribution 3.0 licence. Any further distribution of this work must maintain attribution to the author(s) and the title of the work, journal citation and DOI.
Published under licence by IOP Publishing Ltd

In simple span bridges, in which the supports have an infinite vertical stiffness, (i.e: real hinge-roller support), the fundamental frequency can be estimated using the empiric formula 100/L to 125/L as reported in many classic literatures, where L is the bridge span [1-3].

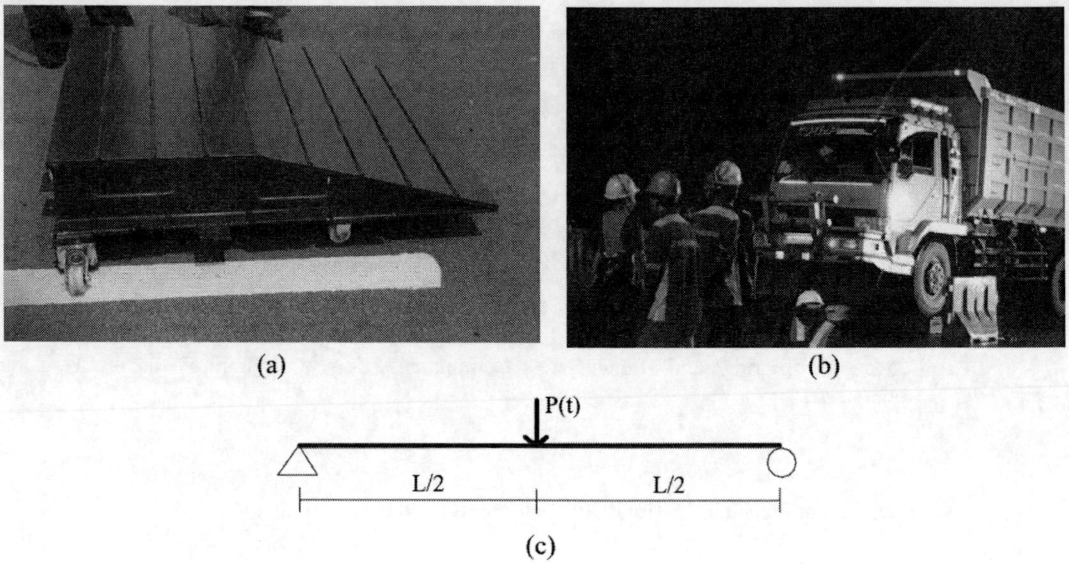

Figure 1. Dynamic loading test using jumper: (a) Steel jumper; (b) Front wheels of the truck on jumper; (c).Model of impulsive truck load P_t on the mid span of the girder

The above empiric formulas may not be suitable for isolated girders because the isolator device also affects the vertical frequency of the girder system. As a matter of fact, in Indonesia nowadays, the application of seismic isolators on bridges becomes much more popular due to the finding of many new active faults that increases the seismic coefficient. The application of seismic isolation system can reduce significantly the seismic base shear so that more economical design can be achieved. Investigations on the application of base isolation on the bridges have been reported in many literatures. In the experimental study on a three-continuous spans bridge having the length of 90 m, an increase of fundamental frequency was observed due to the increase of stiffness of the LRB, caused by the low temperature of -10° C during the test [4]. A parametric study was conducted to evaluate the effect of LRB and ground motion on the response of isolated bridges. The study concludes that for a certain ground motion, the smaller values of maximum isolator displacement and isolator force indicate a better seismic performance [5].

Another study investigates the possibility of buckling and tensile force on high damping natural rubber (HDNR) bearings at particular condition in a three-spans bridge using simple-span I-girders having 30 m of length [6]. That unexpected condition can be avoided by selecting the suitable shape factor of the bearings.

In this paper, the behavior of a simple-span bridge isolated using LRB is investigated. The study focuses on the variation of the fundamental frequency due to the contribution of the LRB vertical stiffness and the stiffness of the substructure. The method used in this study is the simplified analytic method. The analytic approach assumes a single degree of freedom (SDOF) system with a lumped mass in the mid-span of the girder. The girders can have prismatic or non-prismatic shape. The investigation is performed also on the bridges supported on the abutments that have large vertical stiffness and the bridges supported on the non-prismatic cantilever beams that have low stiffness. The later bridge type is usually applied in fly-over structures. The developed analytical method is validated through the experimental dynamic test performed on several spans (60 m each) of Jakarta-Cikampek Elevated Bridges [7].

2. Materials and methods

In Figure 2, the vertical stiffness of the girder is defined as the required force to generate a deflection at the mid-span of one unit deflection. Considering that $E_G I_G$ is the flexural stiffness of the girder, L is the span, and m is the beam mass assumed concentrated at the mid-span, the vertical stiffness of the girder can be computed through the equation (1).

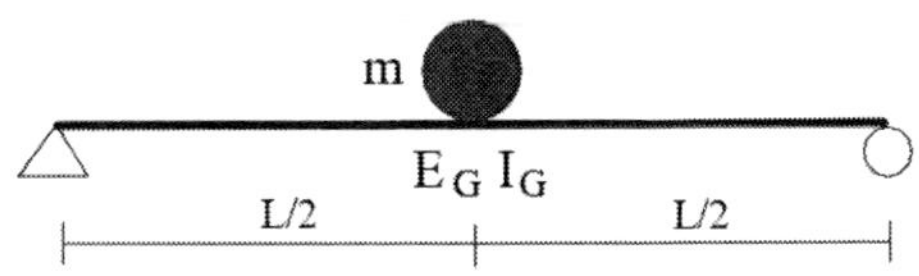

Figure 2. Model for determining the stiffness and frequency of the simple span girder

$$k = \frac{48 E_G I_G}{L^3} \tag{1}$$

While, the Fundamental frequency is expressed as the following:

$$f = \frac{1}{2\pi}\sqrt{\frac{k}{m}} = \frac{1}{2\pi}\sqrt{\frac{48 E_G I_G}{\bar{m} L^4}} \tag{2}$$

where $\bar{m} = m/L$ is the mass per length. In a girder supported on real hinge-roller having infinite vertical stiffness, it can be assumed that the half of the girder mass $\frac{1}{2}\bar{m}l$ is concentrated at the mid-span. The rest of mass is equally distributed to the two supports. Based on this assumption, the frequency can be expressed as $1.56\sqrt{(EI_g/\bar{m}L^4)}$. This value is close to the analytic formula for a simple span girder $f = 1.57\sqrt{(EI_g/\bar{m}L^4)}$ [8]. The value of m in the Equation (2) can be added by the mass of the truck if that additional mass is significant compared to the mass of the girder.

In the steel girder system, the girder is usually nonprismatic in order to obtain economical design, in which the stiffness in the mid-span is larger than that of in the ends of the girder. As seen in Figure 3, using the moment area method, the stiffness of the nonprismatic girder kNP can be derived as expressed in the Equation (3). If the properties of the prismatic bridge α=β=1 are substituted, the Equation (1) is obtained.

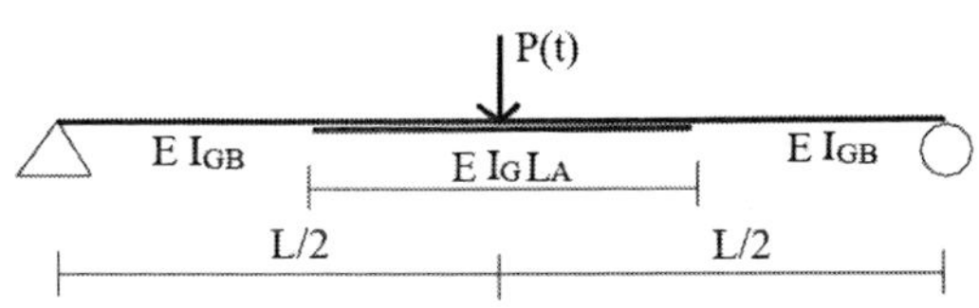

Figure 3. Model for calculating stiffness of non prismatic girder

$$k_{NP} = \frac{48 E_G I_G}{L^3}\left[\frac{2}{\frac{2-6\alpha+6\alpha^2-2\alpha^3}{\beta}+6\alpha-6\alpha^2+2\alpha^3}\right] \tag{3}$$

where: $\alpha = \frac{L_A}{L}$; $\beta = \frac{I_{GB}}{I_G}$

When the LRB isolation system is used, the vertical stiffness of the LRB k_LRB reduces the stiffness of the girder, due to the vertical deformation of the LRB, as illustrated in the dynamic model in Figure 4. In this study, two cases are observed.

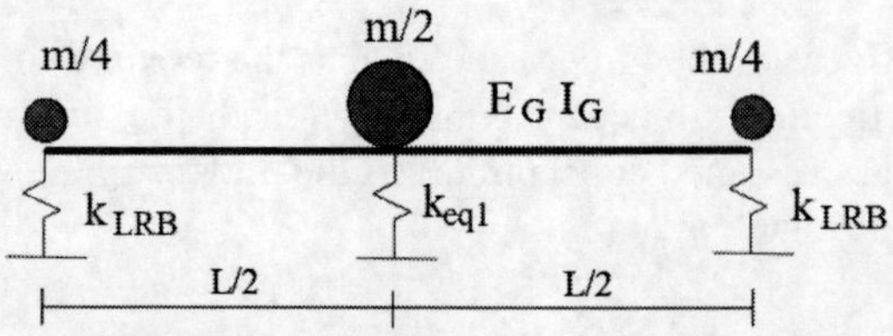

Figure 4. Dynamic models for simple span girder supported by LRBs

2.1. Case 1: girders supported on rigid abutment

Based on the assumption that the girder is rigid and the abutment does not experience axial deformation, the stiffness of the girder decreases due to the contribution of the stiffness of LRB, k_LRB. With the assumption that the stiffness of the girder is parallel to the stiffness o the LRB, the value of girder stiffness k_eq1 can be derived from the unit deflection in the mid-span taking into account the stiffness of the girder and the LRBs, as expressed in the Equation (4). It can be observed that the second term of the denominator in the Equation 4 causes the reduction of the girder stiffness due to the presence of the LRB. If the real hinge-roller support with infinite stiffness is used, that second term becomes zero and Equation (1) is obtained. In order to compute the fundamental frequency through the Equation (4), the mass considered is the total mass of the girder because the stiffness of LRB contributes to the equation of motion.

$$k_{eq1} = \frac{1}{\frac{L^3}{96E_G I_G} + \frac{1}{4k_{LRB}}} \quad (4)$$

2.2. Case 2: girders supported on flexible cantilever

In flyovers, the girders are usually supported on cantilever pierheads having non prismatic shape. The girder stiffness will decrease further due to the vertical deformation of the cantilever. The stiffness of the structure is the equivalent stiffness taking into account the stiffness of the cantilever, LRB, and girder from the Equation (4) in series form. Two conditions are investigated in this study:

1. If the mass of the truck is neglected, no contribution of the column pillar deformation is considered. Thus, the decrease of the stiffness is caused only by the stiffness of the cantilever. Using the moment area method, the stiffness of the non-prismatic cantilever k_{C1} can be computed through the Equation (5) [9]. The stiffness of the cantilever and the LRB forms a series system, so that the equivalent stiffness k_{eq2} can be computed through the Equation (7). The mass of the cantilever in the dynamic equilibrium is assumed as the half-span of the cantilever from the end.

$$k_{C1} = \frac{2E_C I_{CB}}{L_C^3}\left[\frac{\gamma^3}{-2\gamma+\ln(1+\gamma)+\gamma^2}\right]; \gamma = \frac{I_{CA}-I_{CB}}{I_{CB}} \quad (5)$$

$$k_{eq2} = \frac{k_{C1}k_{LRB}}{k_{C1}+k_{LRB}} \quad (6)$$

$$k_{eq3} = \frac{1}{\frac{L^3}{96E_G I_G} + \frac{1}{4k_{eq2}}} \quad (7)$$

Where: L_C is length of cantilever; γ is coefficient of non prismatic cantilever; E_C is modulus elasticity of cantilever; I_{CA} is inertia moment of cantilever at section A; I_{CB} is inertia moment of cantilever at section B where the unit deflection is applied as shown in Fig. 5.

In addition, if more than one concentrated load P(t) are applied in the cantilever system, the value of k_{eq3} is the average values of each girder.

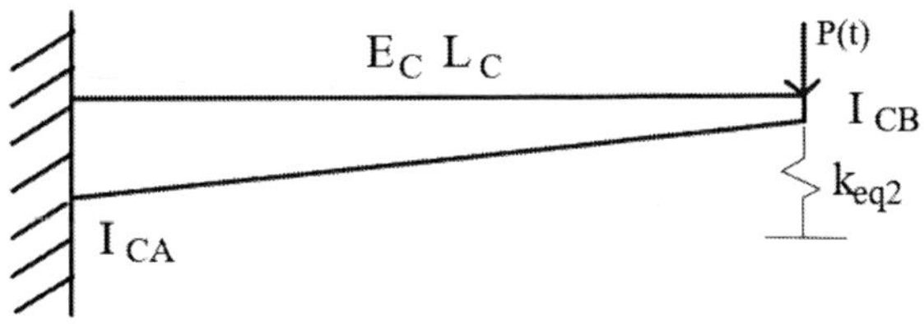

Figure 5. Model for determining the rigidity of non prismatic cantilever

2. If the mass of the impulsive truck is relatively large compared to the mass of the girder, a rotation on the pillar column φ can be generated due to the nonsymmetrical load. This rotation at the top of the column causes a similar rotation at the end of the cantilever, so that a vertical deflection takes place, as seen in Figure 6. The stiffness of the cantilever k_{C2} decreases further due to the presence of the rotation at the top of the pier column. Using the moment area method, the cantilever stiffness can be computed trough Equation (8). The equivalent stiffness of the series system of the LRB and cantilever k_{eq4} is expressed in the Equation (9). The stiffness of the structure k_{eq5} in Equation (10) can be obtained by substituting k_{eq2} in Equation (7) by k_{eq4}. The assumed mass for dynamic equilibrium is one-half of the cantilever mass added with the mass of the truck. The mass of the column is neglected because it does not contribute to the vertical stiffness.

$$k_{C2} = \frac{(2E_C I_{CB})(E_P I_P)n^3}{L_C^3 E_P I_P[-2n+\ln(1+n)+n^2]+L_C^2 h E_C I_{CB}[\ln(1+n)n^2]} \tag{8}$$

$$k_{eq4} = \frac{k_{C2}k_{LRB}}{k_{C2}+k_{LRB}} \tag{9}$$

$$k_{eq5} = \frac{1}{\frac{L^3}{96E_G I_G}+\frac{1}{4k_{eq4}}} \tag{10}$$

Where: $E_P I_P$ is flexural rigidity of pier's column; h is height of pier's column; n is the modulus of elasticity ratio E_s/E_c; E_s is the modulus of elasticy of steel; E_c is the modulus of elasticity of concrete

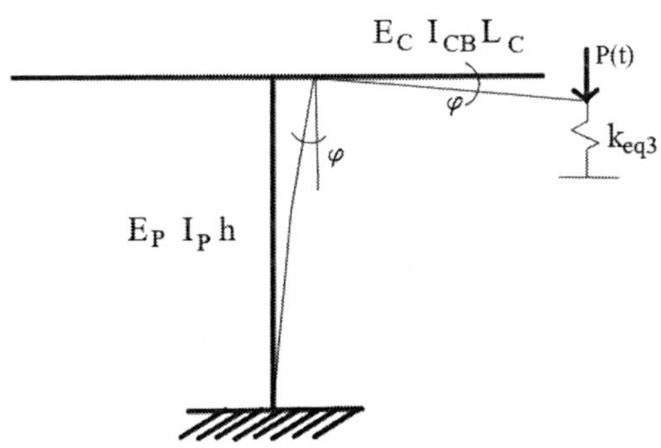

Figure 6. Effect of rotation at top of pier's column to rigidity of non prismatic cantilever

3. Results and Discussions

For validation, the proposed simplified method is implemented in a simple-span bridge in Jakarta-Cikampek elevated bridge. In every three spans, expansion joint connection is used. Meanwhile in the other ends of the girders, link slab connection is used. The section of the bridge is Composite Steel Box Girder having an effective span of 58 meters and supported on the single column piers with cantilever pierheads, as seen in Figure 7 [7]. The girders under study are the one located at the end of the cantilever and the one close to the pier's column. The accelerometers used for loading test are indicated by red circles. The impulsive load applied is the truck load having weight of 30 tonf that is supported by two girders. Thus, a single girder is assumed to carry 15 tonf of impulsive load.

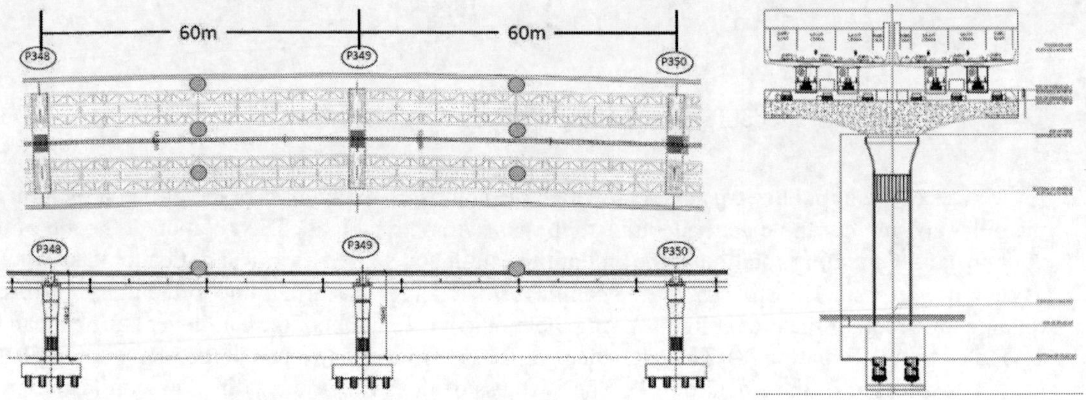

(a) (b)

Figure 7. Elevated steel box girder bridge with span of 60 meter
(a). Plan and elevation; (b). Cross section

3.1. Case 1

The location of the observed girders is between piers P349-P350 and the box girders are located at the end of the cantilever. The simple-span composite nonprismatic girder is divided into five segments consisting of two end-segments having web thickness of 16 mm, length of 11,000 mm made of JIS SM520 steel, and three intermediate and middle segments having web thickness of 20 mm, length of 3x12,000 mm made of JIS SM570 steel. The dimension of the steel box girder can be seen in Figure 8a. The cross section of the bridge as presented in Figure 8b indicates the average concrete slab thickness of 318 mm and width of 11,500 mm (the effective width of a single box girder is 4895,5 mm). With the concrete compressive strength fc' of 35 MPa and elastic-section assumption, the moment inertia of the composite girder is obtained I_G = 3.3177E+11 mm^4. The girder stiffness k_{NP} can be computed through Equation (3) resulting in16272 N/mm.

The mass of the bridge of 204.2 N sec^2/mm is assumed one half of the total span and concentrated at the mid-span. Using Equation (2) the bridge frequency is obtained 1.422 Hz (with assumption of real hinge-roller support) as described in Tabel 1. If the girder is assumed prismatic, a frequency of 1.424 is obtained. Thus, only 0.14% of different is observed. From the technical specification, the vertical stiffness of the LRB is 1489540 N/mm. Thus, using Equation (3) the values of k_{eq1} = 32455 N/mm and fundamental frequency = 1.420 Hz (taking into account the LRB stiffness) are obtained, as reported in Table 1, column 2. That obtained frequency is only 0.14% lower than that of using real hinge-roller assumption. It is cause by the large vertical stiffness of the LRB, which is 91.5 times larger than that of the girder. If the truck mass 15 tonf is taken into account, the frequency of 1.392 Hz is obtained, as seen in Tabel 1, column 3. This value is only 1.97% lower than the frequency without considering the truck mass.ncrete cracking is known to have a knock-on effect on the durability performance and long-term service life of reinforced concrete structures [1, 2]. This phenomenon is primarily attributed to complex loading conditions that may arise from external loading under day-to-day service in addition to environmental exposures, creating even more complex physical and chemical changes in concrete properties [3]. This will cause not only localised distress and stiffness degradation of structural members but also poor resistance of penetration from aggressive agents (e.g. alkali-silica reaction and chloride transport) which will result in further significant deterioration [4, 5].

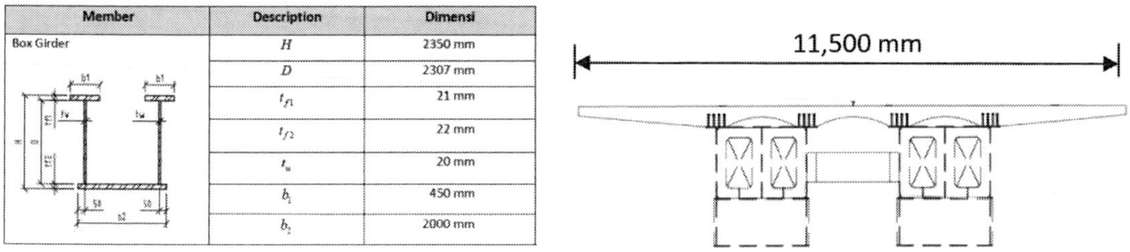

(a) (b)

Figure 8. Properties of the bridge: (a) Steel cross section at intermediate and mid segment (t_w=20 mm); (b) Transverse section of girder

In order to analyze the cantilever stiffness, two conditions are considered:

3.2. Case 2a

The values of the frequency in Tabel 1, column 4, are computed as the following: with I_{CA} = 5.19922E+12 mm^4 and I_{CB} = 9.60097E+11 mm^4 ; from Equation (4) it is obtained that γ= 4.415 and k_{C1} = 1209641 N/mm. The stiffness of the system without the truck mass k_{eq2} through Equation (5) and k_{eq3} through Equation (6) are obtained 667539 N/mm and 32151 N/mm, respectively. The frequency computed through Equation (2) is obtained 1.303 Hz. Furthermore, the frequency of the box girder near the pier column is obtained 1.307 Hz and the averaged values is 1.305 Hz as seen in table 1, column 4. This value is 8.10% lower than the frequency without considering the cantilever stiffness as seen in Tabel 1, column 2. It indicates that the contribution of the cantilever stiffness is significant to reduce the fundamental frequency of the structure.

Tabel 1. Fundamental frequencies of the bridges estimated by the simplified method

Bridge Location	Span L (m)	Frequency (Hz)						
		Real Hinge	With LRB	LRB+Truck Load	Including Pier's Stiffness: No Truck Load	Including Pier's Stiffness: With Truck Load	FEM	Loading Test
		Eq. (3)	Eq. (4)	Eq.(4)+truck mass	Eq. (7)	Eq. (10)		
		1	2	3	4	5	6	7
P349-P350-a	60	1.422	1.420	1.392	1.305	1.283	1.400	1.454
P196-P197-b	60	1.422	1.420	1.392	1.305	1.283	1.360	1.410
P469-P470-c	60	1.422	1.420	1.392	1.305	1.283	1.346	1.340
P586-P587-d	60	1.422	1.420	1.392	1.305	1.283	1.357	1.495
P194-P195-e	60	1.422	1.420	1.392	1.305	1.283	1.362	1.405
P32-P34-f	60	1.422	1.420	1.392	1.305	1.283	1.340	1.430

Note: a-One girder's edge connected with link slab; b-One bearing supported on steel frames; c-Two girders supported on steel frames; d-One girder supported on one cantilever; e-Four girders supported on steel frames; a-One girder's edge connected with link slab

3.3. Case 2b

If the additional mass of truck of 15 tonf is taken into account, having n=7.2, E_pI_p=1.56406E+17 N-mm^2 and height of the column h=11806 mm, it is obtained through Equation (8), (9), and (10) that k_{C2}, k_{eq4} *and* k_{eq5} are 673590, 463837, 31982 N/mm, respectively. The frequency of structure, taking into account the rotation of the pier column due to unsymmetrical load, is obtained 1.280 Hz as reported in Table 1, column 5.

Using the similar method, for the second box girder (near the pier column), the frequency of the structure is obtained 1.286 Hz. Thus, the average frequency is 1.283 Hz, as seen in table 1, column 5. This value indicates that 1.69% reduction of the frequency is resulted if the additional mass of truck 15 tons is considered.

In Table 1, column 6, the results from Finite Element Model (FEM) are reported: a fundamental frequency of 1.400 Hz is obtained. This value comes from the analyses performed by the designer, without considering the mass of the truck, based on the 23rd mode in vertical direction and mass participation of 46%. For the other five spans, P196-P197, P469-P470, P586-P587, P194-P195, and P32-P34, in which the other ends of the girders are supported on steel portal frames having various stiffness, their frequencies are reported in Tabel 1, column 6 (obtained from FEM) and column 7 (obtained from loading test). The average difference of the results between the simplified method (column 4) and FEM is 4.10%. It may be caused by several aspects that cannot be satisfied by the simplified method: for instance the link slab connection at the other end of the girder (even though its stiffness is only 3.95% of the girder stiffness), and the stiffness of the other girder ends because they are supported on the steel portal frames instead of cantilever structure.

As seen in Tabel 1, column 7, the average frequency obtained from the loading test is 1.422 Hz. This value is based on the observation of the accelerometer which is then post-processed through FFT as presented in Figure 9. The frequency obtained from the loading test is 1.305 Hz and 1.361 Hz larger than which obtained from the simplified analytical method and FEM, respectively. This result indicates that the constructed bridge structure is stiffer than the designed structure.

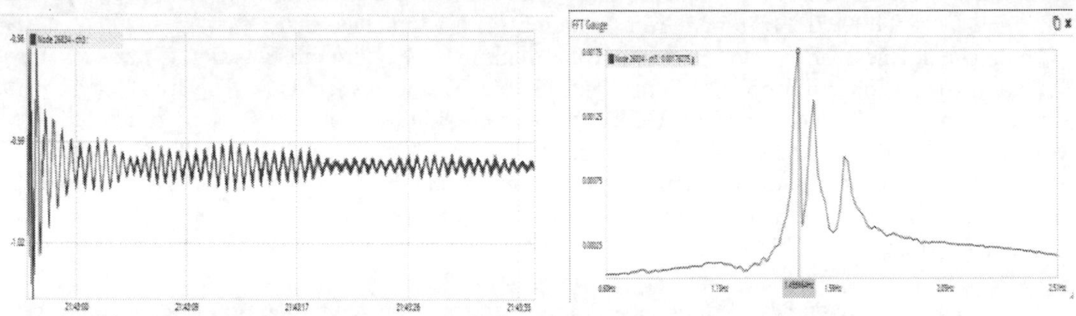

(a) (b)

Figure 9. (a) Time domain acceleration history obtained in the loading test; (b) Frequency domain acceleration obtained through Fast Fourier Transform

4. Conclusions

From the above discussions, several conclusions can be drawn:

1. The difference of frequency between prismatic and nonprismatic simple-span girders is not significant. In this study, with box composite section and 60 m of length, the average difference is only 0.14%
2. The application of LRB as the bridge support results in the frequency that is close to the girder supported on real hinge-roller supports. The difference is only 0.14% that can be neglected. Thus, for simple span bridge using LRB isolation, the fundamental frequency can be computed through the empirical formula for simple span beam supported by real hinge-roller supports.
3. The simplified method to compute the fundamental frequency results in the values that 4.10% lower than which obtained from FEM analyses. As a matter of fact, the simplified method does not take into account the link slab stiffness and assumes that all girders are supported on identical cantilever structures.
4. In the loading test, the fundamental frequency is obtained larger than which obtained from the simplified method and FEM analyses, indicating that the bridge is stiffer than the estimated design.

References

[1] Cantieni R 1983 *Dynamic load tests on highway bridges in Switzerland: 60 year experience of empa* (Swiss: Swiss Federal Laboratories for Material Testing and Research)

[2] Burdett O and Corthay S 1995 Dynamic load testing of Swiss bridges *Proc. of IABSE symposium San Francisco extending the lifespan of structure* **73** 2 p 1123-8

[3] Vaza H 2016 Research on the Improvement of Bridge Management System 1992: Case Of Bridge Condition Assessment in the Decentralized Indonesia *Dissertation Graduate School of Social Systems and Management* (Japan: University of Tsukuba)

[4] Lee S Y, Koh H M, Ha D H, Ok S Y, Ryu H S, Kang H T and Kim Y S 2004 Field Testing of Seismically Isolated Bridge Design Based on Minimum LCC Concept *13th World Conf. on Earthq. Eng.*

[5] Hameed A, Koo M S, Thang T D, and Jeong J H 2008 Effect of lead rubber bearing characteristics on the response of seismic-isolated bridges *KSCE J. Civ. Eng.* **12** 3 p 187-96.

[6] Tubaldi E, Mitoulis S A, Ahmadi H and Muhr A 2016 A parametric study on the axial behaviour of elastomeric isolators in multi-span bridges subjected to horizontal seismic excitations *Bull Earthq. Eng.* **14** p 1285–310

[7] KKJTJ 2019 *Evaluation of static and dynamic loading test on Jakarta Cikampek elevated bridges (from various sources)* (Indonesia: Bridge and Tunnel Safety Commision Ministry of Public Works and Housings)

[8] Paz M and Kim Y H 2019 *Structural dynamics: theory and computation 6th edition* (New York: Springer)

Analysis study of extended end plate connection due to cyclic load using finite element method

Y S Yoganata[1*], B Suswanto[1], D Iranata[1], and D Irawan[1]

[1] Civil Engineering Department, Faculty of Civil, Planning, and Geo Engineering, Institut Teknologi Sepuluh Nopember, Surabaya, Indonesia

[*]Corresponding author's e-mail: budi_suswanto@ce.its.ac.id

Abstract. In steel frame structures, the connection is very important element to design, especially for earthquake resistant structures. How to choose the material and and dimension are very important things in steel connection to receive the load. Nowadays, the application of joints using extended end-plate has become popular because of it easy fabrication, the erection, and precise seismic performance. This connection is used to connect beam to column or connect the two beams. As a study material, we'll make the detail of steel connection due to cyclic load using Abaqus 2017. In this research, we will remodel the connection from previous research with a end-plate modification using finite element method. The dimension of beam is I - 360 x 170 x 12 x 8 and the column is I - 213 x 222 x 10 x 7 with the thickness of end plate is 15 mm and using Grade 8.8 bolts. There are 3 types variation model, one for verification of previous research, and the other two were modified connections. The result of the study is Extended End-Plate can affect the value of the displacement. Connection using Extended End-Plate has smaller displacements, so its better to receive the loads that can cause the collapse. Extended End-Plate also affects the value of lateral load that can be borne. Connection using Extended End-Plate can receive the greater lateral loads, so its more stable and stronger to holding the loads. From the three specimens analyzed, the first model has the best performance where there are extended end plates on both sides.

1. Introduction

Infrastructure in Indonesia must be designed as an earthquake-resistant building because of so many earthquake happened. If an earthquake occurs, the structure will have vertical and lateral movement. Generally, steel has tensile strength and it will be attracted until failure with the specific load limit. Damage can occur on the column/beam or the connecting part, depending on the quality of the material and dimensions used.

When a partial strength connection is used in the moment resisting frame (MRF), the friction occurs in the location of the plastic hinge from the beam or column to the connection. When this friction occurs, it is very important to consider the characteristics of the beam to column connection. The application of partial-strength connection is a common and relatively inexpensive solution to be applied in MRF if we compared to full-strength joint types. Previous studies have shown that detail, this connection can also be an alternative to structures that located in seismic regions, enabling the precise of control and response of dissipative elements.

Content from this work may be used under the terms of the Creative Commons Attribution 3.0 licence. Any further distribution of this work must maintain attribution to the author(s) and the title of the work, journal citation and DOI.
Published under licence by IOP Publishing Ltd

Today, the application of Bolt Extended End-Plate (BEEP) has become popular because of its easy in fabrication, erection and the precise seismic performance. This joint is for connecting a beam to column or connecting two beams together. The BEEP connection is included in ANSI 2010 / AISC 358, and the results of the seismic test is the BEEP can provide the considerable seismic toughness and durability.

Because of he number of this tests is limited, numerical simulations need to be done. A more accurate finite element model for parametric analysis is very important. Therefore, more detailed study of the variables that affecting the end-plate capacity is needed to design steel beam-column joints using extended end plate. This study refers to previous research by Doh, *et al.* For a more detail description of BEEP behaviour, an analysis using finite element method will be done. Displacement of the load, and the rotation caused by the moment will be detailed in this study. So, we get a thorough behaviour about it.

2. Method

The method is to make beam-column joint modelling in the Abaqus program. The modelling is based on the previous research by Doh, *et al* [4] as verification and be determiner to continue the other models.

2.1 Numerical Program

In the Abaqus program, the connection is modelled with the following stages:

a. Parts consist of columns, beams, end plates and bolts.
b. The contact between the beam and the end plate uses Tie-Constraint type. As for the contact between the end-plate with column and bolts with beam and end-plate uses the friction type.
c. The nut is modelled together with a bolt.

Figure 1. Layout of the connection using Abaqus 2017. At this step, the all parts are combined to be one.

2.2 Material Properties

a. The material used for all parts is shown in table 1.

Table 1. Modelling material detail.

Material	Detail	F_y N/mm²	F_u N/mm²	E N/mm²
Column	213 x 222 x 10 x 7	248	483	207000
Beam	360 x 170 x 12 x 8	248	483	207000
End-Plate	Thick of 15 mm	248	483	207000
Bolt	Radius of 20 mm	634	827	210000

The table above is a detailed quality of each material that inputted into the Abaqus in the material properties module.

b. Steel Density value of 7850 kg/m^3 and Poisson ratio of 0.3.

2.3 Load

The load is of two types, the first is a concentrated load of 100 kN. The second is the cyclic load according to AISC regulations. The load is located at the end of the beam.

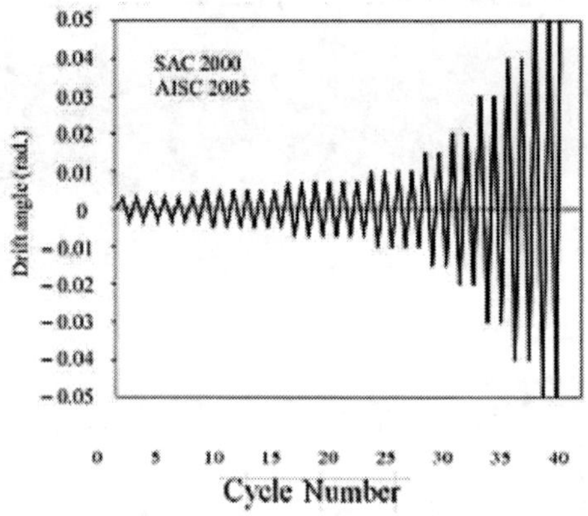

Figure 2. Cyclic Loading System. The load is inputted in the load module with a concentrated load type, whereas the cyclic load is inputted in the boundary condition module.

2.4 Specimen Details

The modelling is in three specimens, there are an extension of the end-plate used for the specimen PS-1 with an end-plate extension on both sides, for the specimen PS-2, there was an extension of the end-plate on the left side, whereas for specimen PS-3, there is no extension on either side of the end-plate. The column height is 1900 mm and the beam length is 1500 mm. The details of each connection model in Abaqus is depicted at figure 3 to figure 5 and listed in table 2.

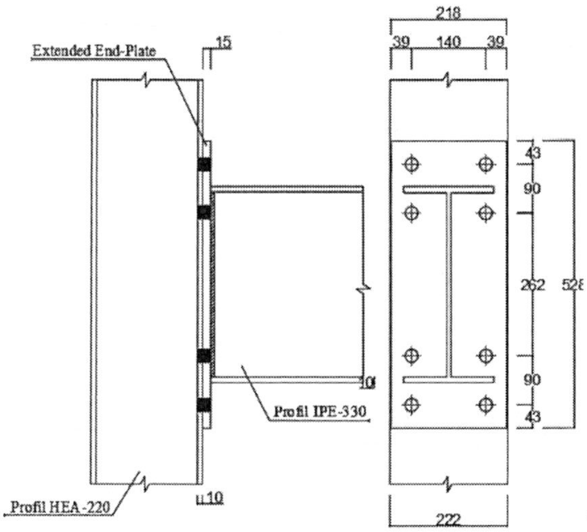

Figure 3. Detail of Specimen PS-1.

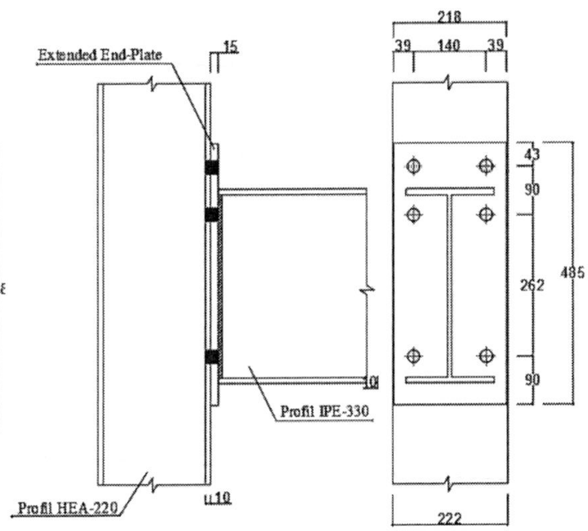

Figure 4. Detail of Specimen PS-2.

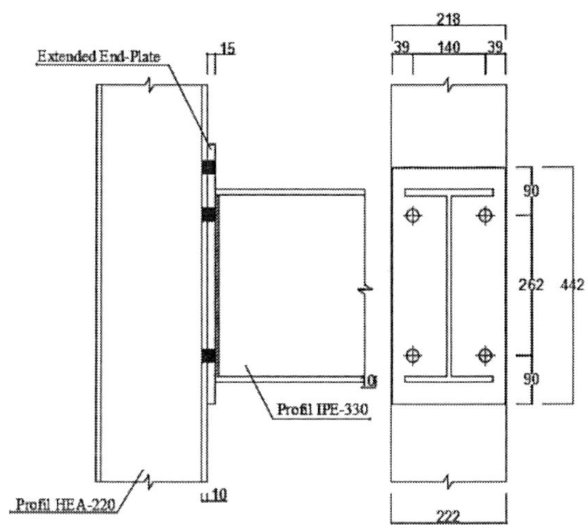

Figure 5. Detail of Specimen PS-3.

In figure 3 to 5 we can see the detail of each material that used, there are column, beam, end plate, and bolts. For the length of column is 1900 mm and 1500 mm for beam. There are differences in end plate size and number of bolts in each specimen. The diameter of bolt is 20 mm for all specimen.

Table 2. Size detail of model.

Part	PS-1	PS-2	PS-3
L Column (mm)	1900	1900	1900
L Beam (mm)	1500	1500	1500
End-Plate (mm)	218x528x15	218x485x15	218x442x15
Number of Bolt	8	6	4

2.5 Finite Element Analysis Method

Finite element method analysis is performed in all three specimen variations. The results of the modelling will be compared to one another, so the most effective connection results are obtained. Finite element method analysis is obtained from the modelling results using Abaqus program. Here are the few indicators to analyze the modelling results:

1. End-Plate Capacity Analysis on joints
   To obtain the ability of end-plate in the beam-column connection to accept the cyclic load, starting from the first yield to the ultimate yield.
2. Ductility
   The ductility factor of structure (μ) is taken from the ratio between the ultimate deviation and the first yield deviation. Ductility value of each specimen needs to be considered in choosing the right method to applied in high-rise buildings.
3. Envelope Curve
   The graph between displacement and lateral load will show which specimen is able to holding the largest lateral load. The envelope curve is obtained from the peak load in one cycle from each cycle of the lateral loading.
4. Hysteretic curve
   The Hysteretic curve will show an increase and decrease displacement while the cyclic load is working. Through the Hysteretic curve, the relative dissipation ratio for each specimen is obtained.
5. Modified Connection Failure Patterns in End-Plate Parts
   After all processes have been completed, a failure pattern is obtained in the connection structure of the entire specimen. Observations on the occurrence of deformation due to loading will be displayed in this parameter.

3. Result and Discussion

a. Verification Of Previous Research

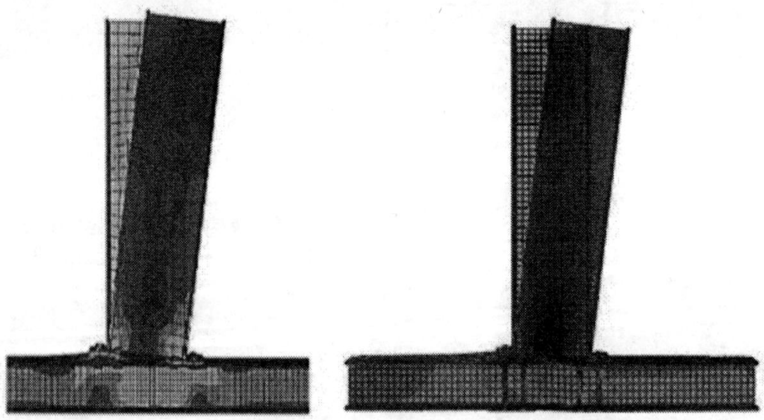

(a) Doh, *et al.* (b) Specimen PS-1

Figure 6. Comparison of S-Mises between experiments with PS-1.

In figure 6, the stress distribution that occurs has a similar pattern. The beam is bending to the right side because of the load at the end of beam by 100 kN.

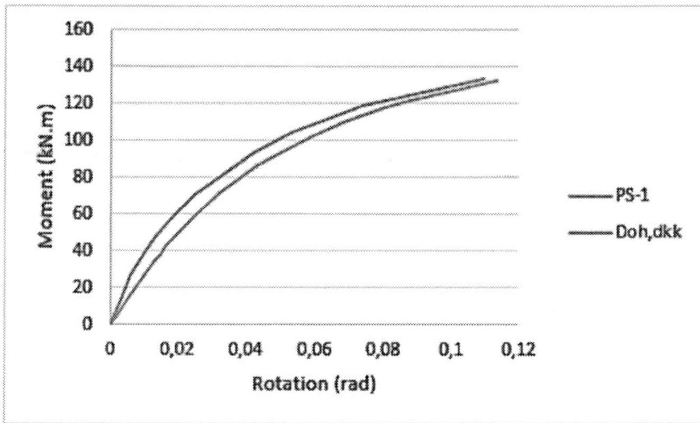

Figure 7. Verification Graph of Moment-Rotation.
Both graphs have the same pattern, although the moment value of the PS-1 modelling results is greater than the results of the Doh, *et al* modelling along the curve. But we can see that the difference is not too large.

Table 3. Verify PS-1 Modelling.

No	Variation	Moment-Rotation	
		Mmax (kN.m)	θ (rad)
1	Doh, *et al.*	132.45	0.1137
2	PS-1	133.65	0.1098
Difference		1.2	0.0039

From the table 3 we can see that the difference is very small. Then the next modelling can be analysis using other variations and adding the cyclic loads.

b. Result and Analysis
Other variations modelling is carried out to obtain the connection behaviour due to cyclic load. Following are the results of numerical analysis using Abaqus.

c. Stress Distribution

(a) PS-1 (b) PS-2 (c) PS-3

Figure 8. Result of S-Mises of the three specimens.

Figure 8 shows the stress distribution in all specimens connection model. Generally, that figure shows the same yielding pattern. Only at the PS-3, yielding is also occurs at the end of the column.

d. Ductility

Table 4. Ductility Factor.

Specimen	Displacement while		Ductility (μ)
	First yield (mm)	Ultimate (mm)	
PS-1	12.196	48.580	3.983
PS-2	17.996	58.749	3.265
PS-3	23.141	70.286	3.037

Specimen PS-1 has the highest ductility value, it means that the extended end-plate can increase the value of ductility if applied to steel structures as listed at table 4.

e. Envelope Curve

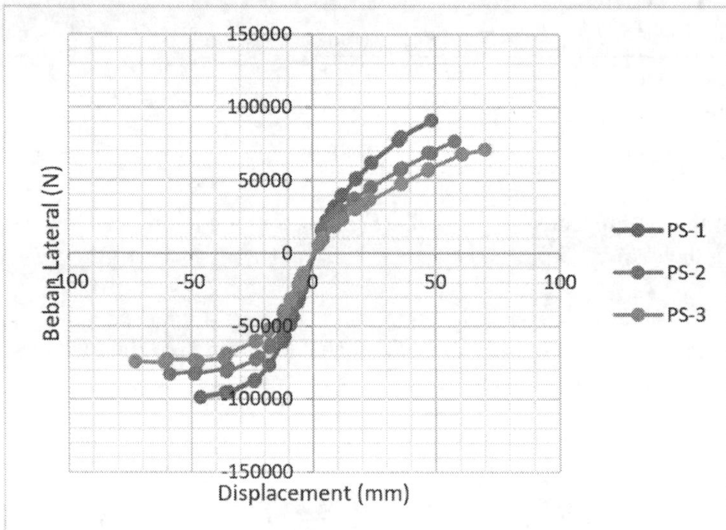

Figure 9. Envelope Curve. Maximum lateral load capacity for PS-1 is greater than PS-2 and PS-3. Therefore, the PS-1 connection is more able to receive the load than the other variations.

f. Rotational Stiffness

Table 5. Value of rotational stiffness.

Specimen	Moment (Nmm)	Rotation (rad)	Rotational Stiffness (N.mm/rad)
PS-1	136398.300	0.0345	3956.869
PS-2	114995.250	0.0417	2756.914
PS-3	106473.900	0.0499	2133.596

Specimen PS-1 has the highest rotational stiffness value. This means the PS-1 is the best connection to receive the force. Even this is also affected by the extended end-plate on both sides as listed in table 5.

g. Energy Dissipation

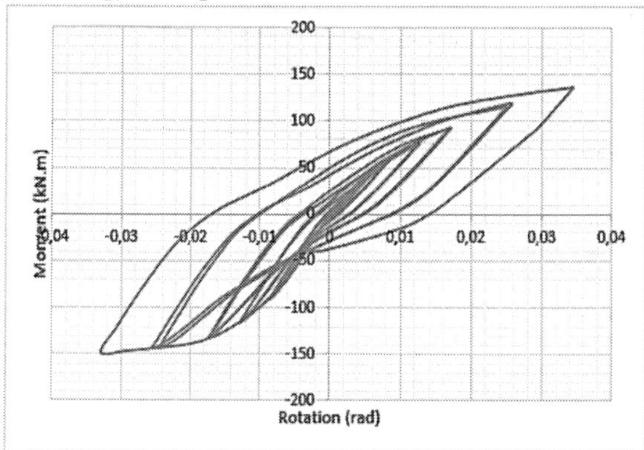

(a) Momen-Rotation Graph of PS-1.

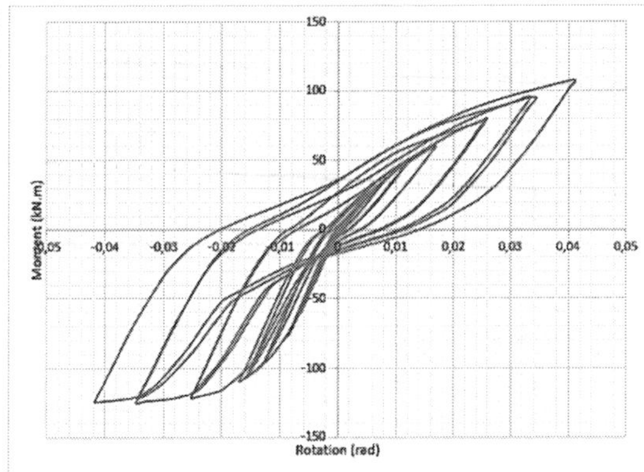

(b) Momen-Rotation Graph of PS-2.

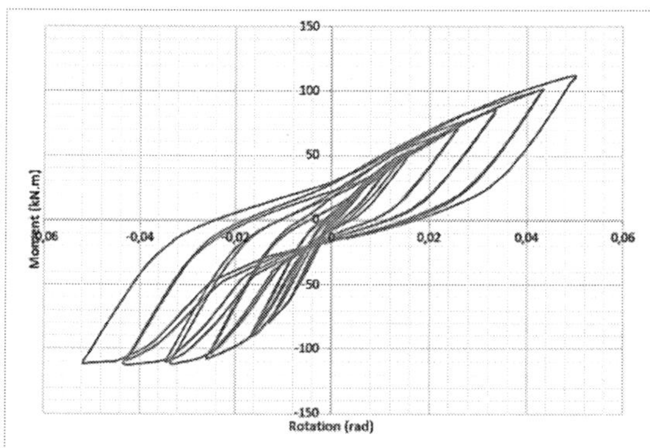

(c) Momen-Rotation Graph of PS-3.

Figure 10. Momen-Rotation Graph.
From the figure (a), (b), and (c) we can see the differences of the pattern. PS-1 has the highest moment value, but smallest rotation. PS-1 is the best connection to be applied.

Table 6. Value of ratio of energy dissipation.

Specimen	Cyclic Area (kN.mm)	Total Area (kN.mm)	Ratio of Energy Dissipation (β)
PS-1	2208.752	8986.635	0.246
PS-2	2075.095	9026.891	0.230
PS-3	2435.136	10701.247	0.228

The relative dissipation energy ratio of the three specimens is greater than 1/8 (0.125), so it is still meets the required criteria or in other words, the structure still can maintain the stability before it collaps. The specimen PS-1 is the most stable in accepting the load at the maximum displacement level as listed in table 6.

h. Failure Pattern

Observation of the failure patterns starts from the beginning to the end of the modelling. For specimen PS-1 the failure occurs in bolts, yield in column body and wings, while for beam it is relatively safe. Whereas for PS-2 and PS-3 the failures occur in bolts, yield in column bodies and ends of beam bodies. This is because there is no extended end-plate either on the right side of the PS-2 or on both sides of the PS-3. In the PS-2 and PS-3, the failure mode of column body is caused by a transverse pressure at the end positive load step then pull out on the right side of bolts, and generally yielding also occurs in the end plate.

4. Conclusion

From the three modelling, we get the conclusion as follows:

a. From the analysed of ductility values, it can be concluded that the PS-1 is the most able to maintain the stability due to cyclic load that cause the first yielding, while maintaining the strength and rigidity. From these explanations it can be concluded that the extended end-plate is able to increase the value of ductility if applied to the steel structures.
b. The PS-1 connection has the highest rotational stiffness value, it is the most rigid when compared to the other two specimens. This means the PS-1 connection is the best connection type to receive the force. Even this is also affected by the extended end-plate on both sides.
c. Based to the results of the analysis, the ratio of relative dissipation energy in the three specimens is greater than 1/8 (0.125). However, when compared between three specimens, specimen PS-1 has the largest relative dissipation ratio (β) value of 0.246, it means the PS-1 connection is the most able to absorb the energy by external loads. The PS-1 connection is the most stable in accepting the load at the maximum displacement level.
d. Extended end plate makes the displacement value to be smaller, but it can receive greater lateral forces.

References

[1] American Institut of Steel Construction (AISC) 360 2016 *Spesification of Structural Steel Buildings* (American National Standards Institute)
[2] AISC 2005 *Seismic Provisions for Structural Steel Buildings* (Chicago, Illinois)
[3] Augusto H, Silva L S, Rebelo C and Castro J M 2017 Cyclic behaviour characterization of web panel components in bolted end-plate steel joints *J. Const. Steel Res.* **133** 310–33.
[4] Bahaz A, Amara S, Jaspart J P and Demonceau J F 2017 Analysis of the Behaviour of Semi Rigid Steel End Plate Connections *EDP Sciences 02058*
[5] Diaz C, Martí P, Victoria M and Querin O M 2011 Review on the modelling of joint behaviour in steel frames *J. Constr. Steel Res.* **67** 741–58
[6] Doh S I, Chin S C and Fakih K A 2018 Behavior of extended end-plate steel beam to column connections *The Open Civ. Eng. J.* 250-62
[7] Eladly 2019 Behaviour of stainless steel beam-to-column bolted connection part 1: simplified FE model *J. Constr. Steel Res.* 164 105784
[8] Elsabbagh A, Sharaf T, Nagy S and ElGhandour M 2019 Behavior of extended end-plate bolted connections subjected to monotonic and cyclic loads *Eng. Struc.* **190** 142–59
[9] Gödrich L, Wald F, Kabeláč J and Kuříková M 2019 Design finite element model of a bolted T stub connection component *J. Constr. Steel Res.* **157** 198–206
[10] Haghollahi A and Jannesar R 2018 Cyclic behavior of bolted extended end-plate moment connections with different sizes of end plate and bolt stiffened by a rib plate *Civ. Eng. J.* **4** 1
[11] Ismail R E S Fahmy, Khalifa, and Mohamed 2016 Numerical study on ultimate behaviour of bolted end-plate steel connections *Latin Americ. J. Sol. Struc.* **13** 1-22
[12] Liu X, Ataei A and Bradford M A 2017 Computational modelling of the moment-rotation relationship fordeconstructable flush end plate beam-to-column composite joints *J. Constr. Steel Res.* **129** 75–92
[13] Silviana 2017 Studi kekuatan sambungan batang tarik pelat baja dengan alat sambung baut *J. Inotera* ISSN 2581-1274 **2** 26-31
[14] Standar SNI 1726-2012 2012 *Standar Perencanaan Ketahanan Gempa Untuk Struktur Bangunan Gedung* (Badan Standarisasi Nasional)
[15] Standar SNI 1729-2015 2015 *Tata Cara Perencanaan Struktur Baja Untuk Bangunan Gedung* (Badan Standarisasi Nasional)

IOP Conf. Series: Materials Science and Engineering **930** (2020) 012050 doi:10.1088/1757-899X/930/1/012050

[16] Standar SNI 03-1729-2002 2015 *Tata Cara Perencanaan Sambungan baut Pada Struktur Baja* (Badan Standarisasi Nasional)
[17] Zhao Y, Bu Y and Wang Y 2019 Study of stainless steel bolted extended end-plate joints under seismic loading *Thin-Walled Structures* **144** 106255

The structural behavior of castellated beam with shape variation using finite element methods

V W Yustisia[1*], B Suswanto[1], D Irawan[1], D Iranata[1]

[1] Department of Civil Engineering, Institut Teknologi Sepuluh Nopember, Surabaya, Indonesia.

*Corresponding author's e-mail: yustisia.vicky16@mhs.ce.its.ac.id

Abstract. Castella steel is an alternative material used in the structure of beams and columns in the construction of a building. Castella web holes in previous studies mentioned the following: hexagonal, honeycomb, octagonal, diamond, and hexagonal. The advantage of using the castella steel beam is that it has a higher web geometry without increasing its weight so that it could add up the inertia value of its profile. The purpose of this study is to analyze castella steel beams with various geometric shapes such as hexagonal, octagonal, and diamond with couple plates on the web weld joints to determine the behavior of castella steel beams when receiving axial loads using the finite element method of the abaqus. Analysis of the behavior of castella steel beams, with or without using of a plate, does not significantly influence the deflection that occurs. However, the addition of the coupled plate has a significant influence on the stress-strain value in the welded joints on the web. Hence, the addition of a plate coupled to the web joints of the castella steel beam profile is highly recommended to support strength resistance as a long-term structure. Analysis of the performance of castella steel beams by the addition of coupled plates for each castella hole with a load equal to 75 kN, resulting in the best capability on the value of the displacement is the hexagonal profile that is 12.839mm compared to other profile displacements of 14.433mm diamond hole profiles and 14.972 octagonal hole profiles.

1. Introduction

Castella steel is an alternative material that can be used in the construction of beam and column structural elements. Castella steel beam is a modification of the I-beam that experiences longitudinal and horizontal cuts in the web of Budi *et al.*. [4]. Castella rods are I-profile rods with hexagonal holes in the web made from cutting a hot rolled steel profile I rod which forms a polygon pattern in which the web is halved and joinned by welding at web position so that the profile increases 40-60% higher than original profile Sonck's *et al.*.[8]. The hole in the web of castella has several geometric shapes that have been done by research, among others: hexagonal, honeycomb, octagonal, diamond, and hexagonal with a cut corner. There are six possible patterns of failure in the castella beam, including the formation of the flexural mechanism, lateral torsion bending, formation of the Vierendeel mechanism, rupture of the welded joint, shear buckling and bending stress on the web of Jamadar and Kumbhar [3].

Castella steel beams have a higher web geometry with their weight that does not increase so it could increase the value of the inertia of the profile in accepting the load, this is an advantage of using castella. Research by Jamadar and Kumbhar [3] about holes in castella beam bodies generally hexagonal, circular, diamond, or square. The purpose of this study is to examine holes with hexagonal, circular, diamond, or square shapes. There are six possible patterns of failure in the beam, among others: the

Content from this work may be used under the terms of the Creative Commons Attribution 3.0 licence. Any further distribution of this work must maintain attribution to the author(s) and the title of the work, journal citation and DOI.
Published under licence by IOP Publishing Ltd

formation of the bending mechanism, lateral torsion bending, the formation of the Vierendeel mechanism, rupture at the welded joint, shear buckling on the web and bending stress on the web. Sonck's research, *et al.*. [8] about the lateral torsion bending resistance of castella steel beams with a modification of the residual stress pattern during the castellated beam manufacturing process will result in a resistance that lies about one bend curve lower than the resistance obtained using the original residual stress pattern. It corresponds to a maximum reduction in buckling resistance of around 13%. Research Budi, *et al.* [4] an optimizing the size and hole spacing of hexagonal geometric castella steel beams, led to the fact that the distance between holes affected the location of the maximum stress concentration. The maximum stress concentration is at the edge of the area of the hexagonal hole for a longer hole distance while the maximum stress concentration moves to the web connection area for shorter hole spacing. The basis of the increased strength analysis for the original profile and the location of the stress concentrations of all the models is 60 degrees, namely the angular size with the hole spacing which results in the optimum model of castella steel beams worth between 0.186ho to 0.266ho. Laboratory test results are approximately the same as FEM analysis with an average ratio of 1.011 and a COV value of 0.069. The method used for the analysis of castella steel beam models with FEM is valid. Yossef and Taher's research [10] regarding cost optimization in a composite floor system with castella steel beams, produces a way to reduce post-bending bodies, the ratio of openings to the web can vary between 19 to 34 and the ratio of openings to widths must be less than 2.28 with an angle (alpha) between 54.8 to 76.5 degrees with an average value of 62.12 degrees. Extending the distance from end to end is not recommended because it does not affect optimizing material values. Partial rigid joints with a restraint factor of R = 0.75 for the main girders make up 10% the price of composite plates compared to simple end connections. Weld becomes very important in the process of making castella steel beams. Understanding welding is the process of connecting the metal by making the parts that are joined merge (coalescence) into a single unit Dewobroto [1].

In this study, the aim is to examine the stress-strain analysis at the connection of the web with a castella couple beam plate with various geometric shapes, hexagonal, octagonal and diamond when receiving axial loads using the Finite Element Method of the Abaqus auxiliary program.

2. Research Method

2.1. General

Analysis of castella steel beams with variations in the geometry hole mounted by a couple plate on the welding joint is applying the finite element method (FEM). The outline of this research is to find out the stress-strain behavior of castella steel beams with a variety of geometric holes mounted by a couple plate on the welded joints of the web when receiving axial loads. The analysis used the 2019 Abaqus program, an analysis program using the finite element method. Logan, D.L [5] describes the finite element method is a numerical way to solve physical-mathematical engineering and problems. There are 3 models with different castella steel hole geometric that will be examined and analyzed using steel material Grade, namely BJ41 with yield strength (fy) 250 MPa and tensile strength (fu) 410 MPa and BJ50 Grade that has a yield strength (fy) 290 MPa and with tensile strength (fu) of 500 MPa. It is explained in the following table 1.

Table 1. Profile of Castella Steel 225x75x5x7mm with a length of 3,000 mm.

No.	Steel Profile (mm)				Mechanism of Steel				Hole of Castella Steel							Hole Type of Castella
	H	B	tw	tf	Type	(fy)	(fu)	(%)	ho	b (mm)	e (mm)	ntotal	e/ho (ratio)	α	s	
1	225	75	5	7	BJ 41	250	410	18	149,74	43,50	39,83	48,00	0,27	60	166,7	Hexagonal
2	225	75	5	7	BJ 41	250	410	18	172,10	43,40	39,83	48,00	0,23	60	166,5	Octagonal
3	225	75	5	7	BJ 41	250	410	18	150,00	75,00	39,83	42,00	0,27	45	189,8	Diamond
4	225	75	5	7	BJ 50	290	500	16	149,74	43,50	39,83	48,00	0,27	60	166,7	Hexagonal
5	225	75	5	7	BJ 50	290	500	16	172,10	43,40	39,83	48,00	0,23	60	166,5	Octagonal
6	225	75	5	7	BJ 50	290	500	16	150,00	75,00	39,83	42,00	0,27	45	189,8	Diamond

Young modulus (E) of 200,000 MPa and a poison ratio of 0.3 and a plastic strain value used was 0.0001 according to Irawan, C. [2]. The initial steel beam profile was WF 150x75x5x7 which was modified into 225x75x5x7 castella steel beam profile as detailed in figure 1, which was adopted from the research of Budi *et al.* [4].

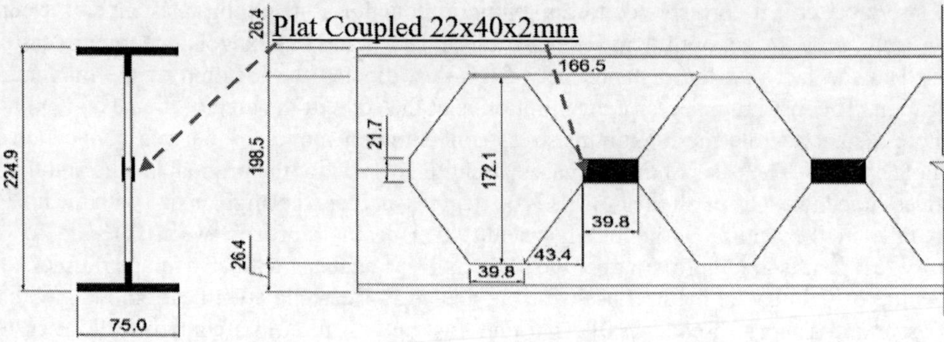

Figure 1. Preliminary details of castella steel beam profiles.

2.2. Calculation Analysis

Perforations made on the castella web greatly affect the structural performance of the beam. Therefore, it is necessary to do a practical analysis to provide optimal perforation on the beam. Optimization of beam dimension profiles and beam hole detail in this study refers to the previous 3 researchers namely the Jamadar AM research, and Kumbhar P. D [3] for diamond holes, in the study of Budi, L., *et al.* [4] for hexagonal holes and in the Soltani MR study, [6] for octagonal holes. It is explained in figure 2, figure 3, and table 2.

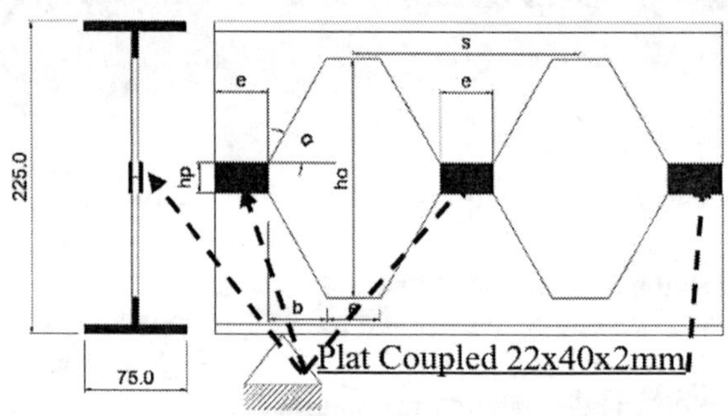

Figure 2. Detail profile of hexagonal and octagonal.

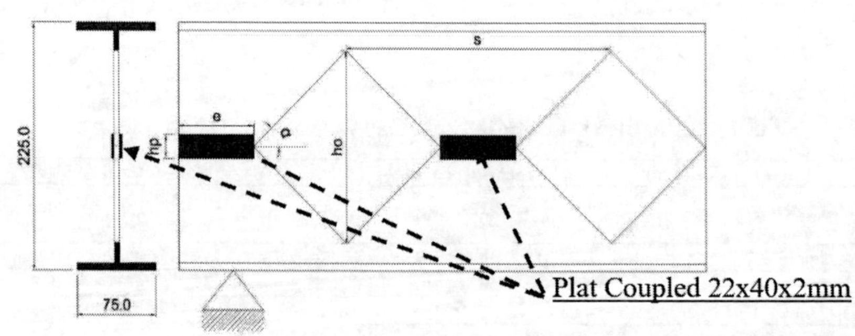

Figure 3. Detail Profile of Diamond.

Table 2. Detail profil of castela steel beam 225x75x5x7.

No.	Hole of Castella Steel								Hole Type of Castella
	ho[a]	b[b] (mm)	e[c] (mm)	n-total[d]	e/ho[e] (ratio)	α[f]	s[g]	hp[h]	
1	150	43,50	39,8	48,00	0,27	60	166,7	-	Hexagonal
2	172	43,40	39,8	48,00	0,23	60	166,5	21,7	Octagonal
3	150	75,00	39,8	42,00	0,27	45	189,8	-	Diamond
4	150	43,50	39,8	48,00	0,27	60	166,7	-	Hexagonal
5	172	43,40	39,8	48,00	0,23	60	166,5	21,7	Octagonal
6	150	75,00	39,8	42,00	0,27	45	189,8	-	Diamond

Note
[a] ho : height of the castle hole
[b] b : projection of horizontal distance in the castella hole
[c] e : width of the horizontal distance on the web. $\left(\frac{L-(n_{total}.2.b}{2.n_{total}}\right)$
[d] n : the total number of castella holes
[e] e/ho : ratio
[f] α : angle of the hole
[g] s : the distance between the hole points, Hexagonal and Octagonal = 2e + 2b; Diamond = ho + b
[h] hp : vertical height between holes = couple plate width
couple profile : 39.8mm long, 21.7mm wide and 2.5mm thick

In this study using the finite element method with modeling that refers to the research profile of Budi *et al.* (2017). Modeling indicators are stated verified by experimental test results if the results of finite element analysis of the compressive strength (displacement) produced no more than 5% of the results of previous studies. Load on castella steel beams as described in figure 4.

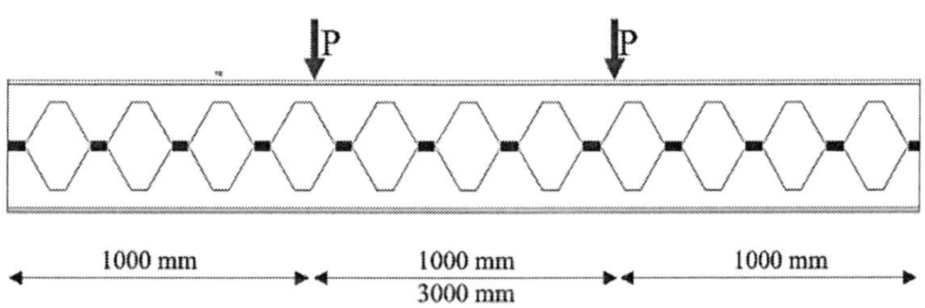

Figure 4. Loading of Castella Steel Beams.

Wang, *et al.* [9] conducted a study on the capacity of vertical shear buckling in the body of castella steel beams with a hexagonal hexagonal body hole. Comparison between bending mode simulation is obtained from the results of experimental tests with the results of bending failure occurring on the body. Prediction of deformation changes using finite element simulation gives very good results.

3. Result of Analysis

In this study, the results of the analysis will be compared with the previous studies. Indicators of this verification stage are load values and experimental test displacement and finite element analysis results as described in table 3.

IOP Conf. Series: Materials Science and Engineering **930** (2020) 012051 doi:10.1088/1757-899X/930/1/012051

Table 3. Verification of analysis results and prior research results.

	Vicky (2020)	Budi et al. (2017) FEM	Budi et al. (2017) Eksperimental	Ratio		
	(a)	(b)	(C)	(b/a)	(c/b)	(c/a)
Force (kN)	74,377	73,958	75,021	0,994	1,014	1,009
Displacement (mm)	12,762	12,772	13,790	1,001	1,080	1,081
Ratio (F/Disp)	5,828	5,791	5,440	0,994	0,939	0,933
			Average	0,996	1,011	1,008
			Deviation	1,478%		0,360%

The results of the modeling analysis show the deviation difference from the results of previous studies is quite good, with a deviation result below 5%. Thereby, the modeling stages are declared feasible to be used in further profile analysis. Figure 5 shows a verification graph showing the results coinciding between previous research data and this study.

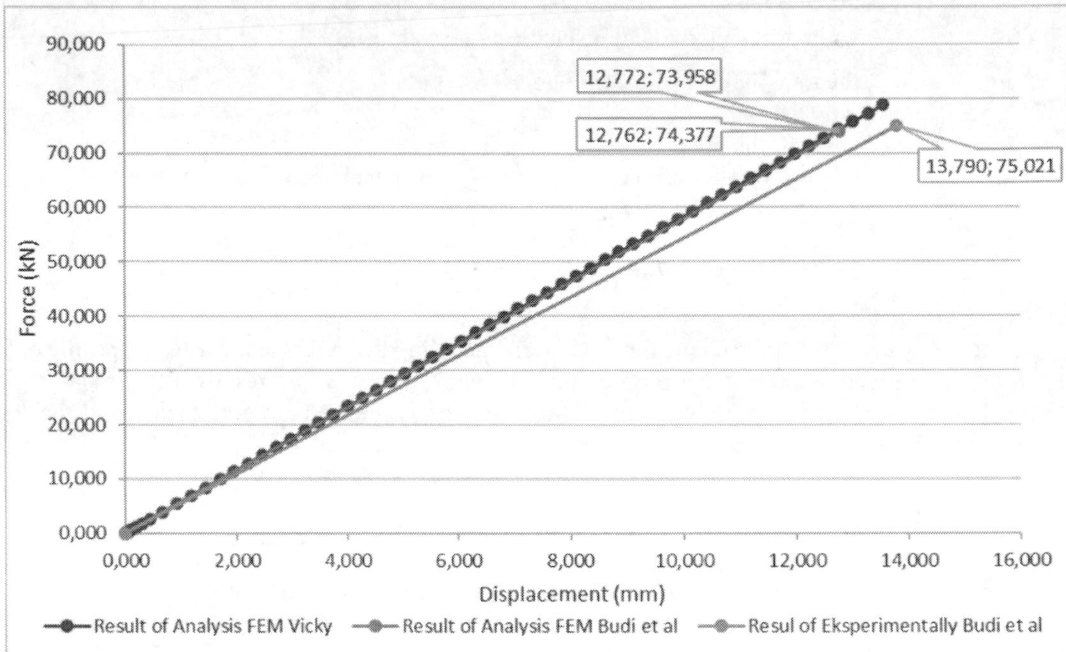

Figure 5. Verification results of modeling analysis.

Figure 6 shows the performance graph profile of each castella steel beam hole with a comparison of the displacement value that occurs to the load (P) which is the same as the hexagonal hole results have the best performance in terms of the smallest displacement value among other hole profiles.

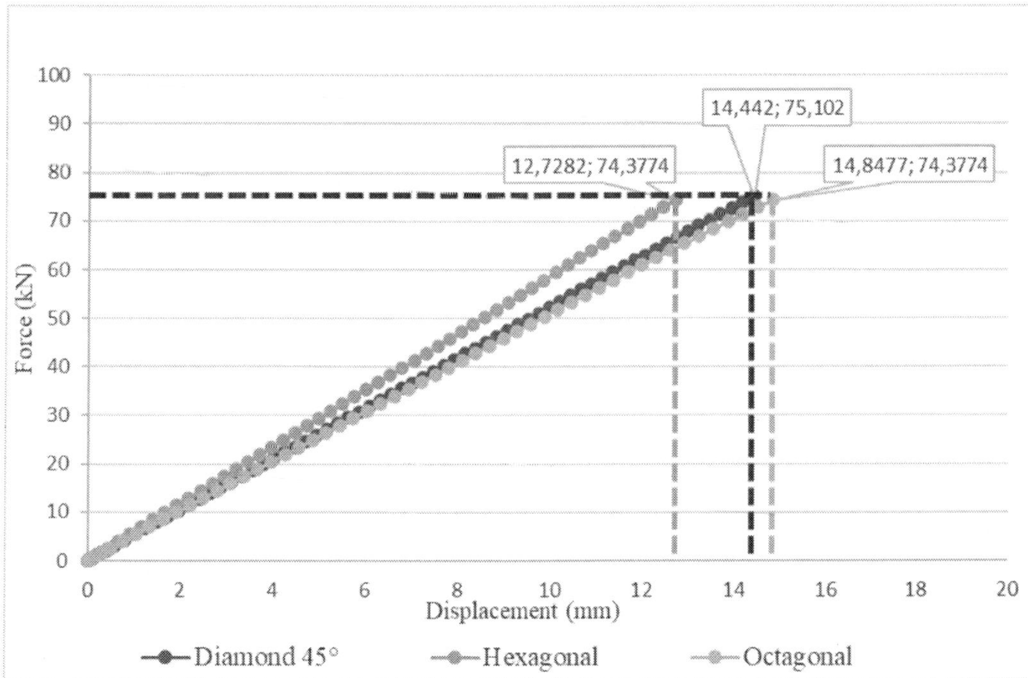

Figure 6. Comparison of displacement values to the load given to each castella hole is diamond, hexagonal, and octagonal hole.

From table 4, it can be concluded that the castella steel beam with the best performance in receiving the same static load is indicated by the smallest displacement (displacement) is on the castella steel beam with the hexagonal hole profile.

Table 4. Results of performance analysis of castella steel beams.

	Diamond 45°	Hexagonal 60°	Octagonal 60°	Type
P (kN)	75	75	75	
Displacement (mm)	14,42	12,84	14,97	Hexagonal 60°
Phase Load Es - Ps	140,38	135,88	104,38	Diamond 45°
Phase Disp Es - Ps	26,06	31,28	25,44	Octagonal 60°

Further review of the ability of castella steel beam material in accepting loads up to the plastic phase has different performance categories between profiles. Even further review is on the stresses that occur in each section of the beam. Figure 7 show the result of the stress values of each castella steel beam hole profile in the middle span position and the pedestal position. The comparison result of the stress values of each castella steel beam in table 5.

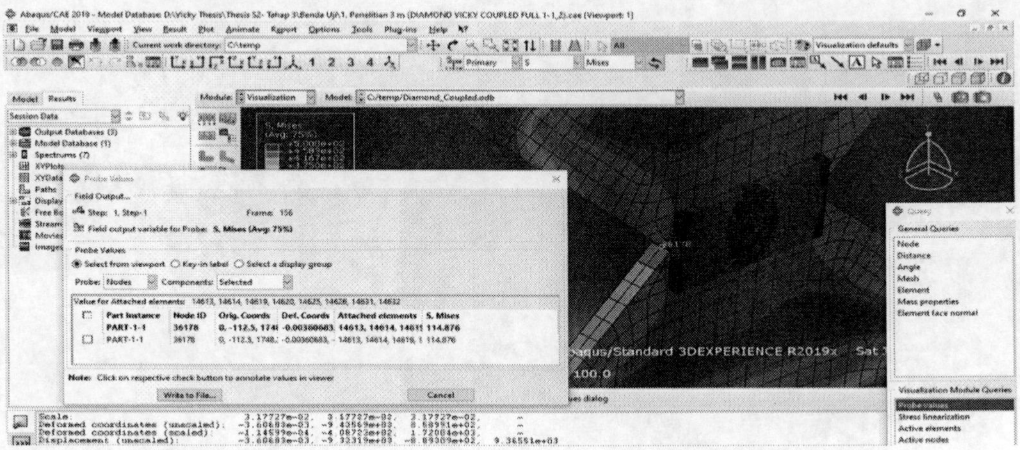

Figure 7. Example of von mises stress castella diamond steel beams in the middle span with BJ-50 steel grade.

Table 5. Results of Comparison Analysis of Castella Steel Beam Stresses.

Grade	Profil Stresses (MPa)					
	Midle Span			Support		
	Hexagonal	**Octagonal**	**Diamond**	**Hexagonal**	**Octagonal**	**Diamond**
BJ41-Coupled	242	407	410	287	256	400
BJ50-Coupled	234	340	115	213	507	403
BJ50-Uncoupled	366	474	256	446	500	424
Average	**281**	**407**	**260**	**315**	**421**	**409**

The table 5 shows that the optimal average stress value is in the profile of castella steel beams with hexagonal orifice profiles. It shows that the hexagonal hole profile has the best structural performance in receiving loads by reviewing the average stress value generated from the stresses in the center of the span and support, average values is 298MPa for the hexagonal hole; 414MPa octagonal hole; and 334MPa of diamond holes.

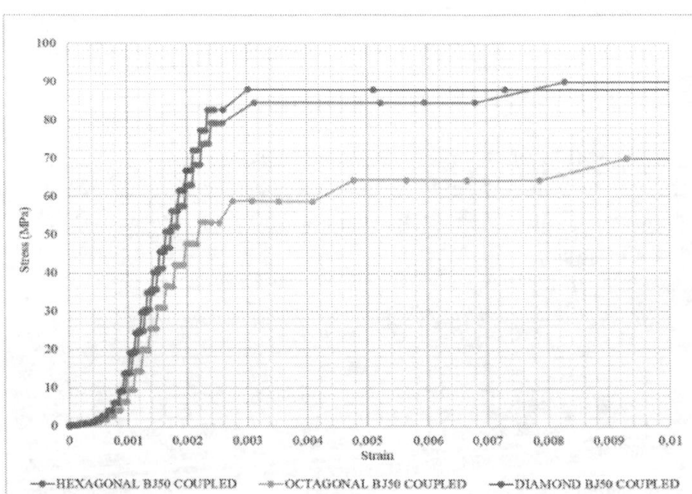

Figure 8. Stress – Strain curve of steel grade BJ-50.

Figure 7 show the steel grade BJ-50 have the best stress - strain values is diamond hole profiles castella steel beam. This shows different behavior according to the grade of steel and the type of castella hole profile used.

4. Conclusion

Based on the results of the analysis and discussion the following points can be summarized:

- Analysis of the behavior of castella steel beams with or without the addition of coupled plates has no significant effect on a deflection that occurs.
- Strain stress analysis on castella steel beams influences the use of coupled plates, and it does not use coupled plates that occur in welded web joints. The value of stressed steel beams with a coupled plate is smaller when compared to steel beams that do not use a coupled plate. Thereby, the addition of a plate coupled to the web joints of the castella steel beam is highly recommended to support the durability of the strength profile as a long-term structure.
- Analysis of the performance of castella steel beams with the addition of coupled plates for various hole profiles with a load of 75 kN, resulting in the best capability on the hexagonal profile in terms of the value of the displacement which occurs is 12.839mm smaller when compared to other profile displacement is 14.433mm for diamond hole profiles and 14,972 for the profile of the octagonal hole.

References

[1] Dewobroto W 2016 *Struktur Baja Perilaku, Analisis & Desain – AISC 2010* 2nd Edition (Tanggerang: Jurusan Teknik Sipil UPH) p 595
[2] Irawan C 2019 Kontribusi Beton Pengisi Terhadap Peningkatan Kinerja Tiang Pancang *Spun Pile* Akibat Beban Lateral Siklik dan Aksial Tekan *Disertasi Dr.* (Surabaya: Institut Teknologi Sepuluh Nopember)
[3] Jamadar A M and Kumbhar P D 2015 Parametric study of castellated beam with circular and diamond shaped openings *Int. Research J. of Eng. and Tech. (IRJET)* **2** 2 715-22
[4] Listiyono B, Sukamta and Partono W 2017 Optimization analysis of size and distance of hexagonal hole in castellated steel beams *Sustainable Civil Engineering Structures Construction 2016 (SCESCM 2016)* vol 171 pp 1092– 9
[5] Logan D L 2012 *A First Course in the Finite Element Method* 5th Edition (Global Engineering: United States of America) p 1
[6] Soltani M R, Bouchair A and Mimoune M 2012 Nonlinear FE analysis of the ultimate behavior of steel castellated beams *J. Constr. Steel Res.* **70** 101-14
[7] Standar Nasional Indonesia 1729-2015 *Spesifikasi untuk Bangunan Gedung Baja Struktural* (Jakarta: Badan Standardisasi Nasional)
[8] Sonck D and Belis J 2017 Lateral-torsional buckling resistance of castellated beams *J. Struct. Eng.* **143** 3 1-9
[9] Wang P, Wang X and Ma N 2014 Vertical shear buckling capacity of web – post in castellated beam with fillet corner hexagonal web openings *Eng. Struct.* **75** 315-26
[10] Yossef N M and Taher S 2019 Cost optimization of composite floor systems with castellated steel beams *Pract. Period. Struct. Des. Constr* **24** 1 1-3

IOP Conf. Series: Materials Science and Engineering **930** (2020) 012052 doi:10.1088/1757-899X/930/1/012052

Influence of link spacing on concrete shear capacity: experimental investigations and finite element studies

W Don[1], K Chong[1], M Aitken[1], A Tambusay[1,2], B Suryanto[1]* and P Suprobo[2]

[1] Institute for Infrastructure and Environment, School of Energy, Geoscience, Infrastructure and Society, Heriot-Watt University, Edinburgh, United Kingdom.
[2] Department of Civil Engineering, Faculty of Civil, Planning and Geo Engineering, Sepuluh Nopember Institute of Technology, Surabaya, East Java, Indonesia.

*Corresponding author's e-mail: B.Suryanto@hw.ac.uk

Abstract. This paper investigates the influence of varying the spacing of shear reinforcement on the response of six geometrically identical reinforced concrete beams. Each beam was designed with differing longitudinal and transverse reinforcement configurations to provide improved understanding of the importance of providing sufficient shear reinforcement following guidance provided in Eurocode 2. Three different cases of varying transverse reinforcements were tested: one without the presence of shear reinforcement, one with shear reinforcement non-compliant to the design provisions and one where the design of the beams fully complies with the design guidance. In order to verify the experimental results, an open source digital image correlation software was utilised to allow for a computational crack mapping, along with the use of nonlinear finite element analysis software to enable a more thorough investigation of the individual beam response. The results obtained highlighted that shear links non-compliant to Eurocode 2 consequently led to unpredictable modes of failure, whereas beams containing shear links at the required amount and spacings displayed significant improvements in the mode of failure and ductility. Beams with the omittance of shear links resulted in the lowest load capacity, due to the sudden, brittle shear failure.

1. Introduction

Concrete is a highly versatile material that possesses high compressive strength but low tensile resistance. When combined with steel reinforcement, however, concrete can provide the ability to withstand significant tensile loads and exhibit excellent ductility, allowing structural elements to fail in a ductile manner with extensive plastic deformation when overloaded. This type of failure is commonly referred to as the flexural failure and is the desired failure mechanism. This mode of failure is generally aimed for in the design of a structural element due to the ability to provide ample warning before failure ultimately takes place. This behavioural response can be predicted to a high degree of accuracy with the utilisation of the current design specifications. Furthermore, the margin of safety of reinforced concrete members can be increased by designing members that are under reinforced in flexure which enables a gradual structural failure to occur.

To avoid a sudden catastrophic shear failure, shear reinforcement are generally provided with the main purpose of increasing the shear capacity of the structural concrete [1]. By increasing the spacing between the shear links, the potential can arise for inclined shear cracks to propagate between the links. In practice, this can be found in structural beam design subjected to relatively low shear stresses

Content from this work may be used under the terms of the Creative Commons Attribution 3.0 licence. Any further distribution of this work must maintain attribution to the author(s) and the title of the work, journal citation and DOI.
Published under licence by IOP Publishing Ltd

commonly utilised in regions with none-to-low seismic activity and in a lightly reinforced, deep transfer beam [2]. In many cases of the design of reinforced concrete elements, where the concrete alone provides enough shear resistance against the low shear stresses imposed, it is good practice to provide minimum shear reinforcement. It is essential that the degree of safety of the elements is analysed and preventive measures are imposed to mitigate any potential catastrophic damage to occupants and society. Hence provisions of maximum spacing are given in current design codes to ensure the presence of at least one shear link in any potential formation of a diagonal shear crack. Furthermore, to ensure that the stirrups are adequately anchored in the concrete (hence effective to prevent a premature shear failure), Eurocode 2 [4], for example, limits the spacing of shear links in the longitudinal axis to be no more than the lesser of 0.75*d* or 600 mm. Other design codes specify more conservative limits; for example, ACI 318-14 [5] limits the longitudinal spacing to the lesser of 0.5*d* or 600 mm (or the lesser of 0.25*d* or 300 mm in members subjected to significant shear stress); JSCE [6] to the lesser of 0.5*d* or 300 mm in areas requiring shear link and the lesser of 0.75*d* or 400 mm in other areas; and CSA A23.3-04 [7] to the lesser of 0.63*d* or 600 mm. This design aspect is obviously very simple to implement yet essential, as if this is overlooked or ignored can potentially lead to a brittle, undesirable shear failure.

The paper provides an analysis of the behaviour of six steel reinforced concrete beams designed with varying tensile reinforcement ratios and varying shear reinforcement spacing. Two beams are designed in accordance with the current design provisions in Eurocode 2, with the remainder of the four beams designed with none to non-complying shear reinforcements. A principal focus in the paper is the effect of the variation of the spacing of the shear links on the overall structural response covering different design aspects from standard to non-complicit design. To this end, digital image correlation (DIC) [8-11] is utilised to allow for a detailed crack development during the experimental testing of the beams to be monitored and verify the experimental behaviour during testing. Alongside this, nonlinear finite element analysis (FEA) is also undertaken to study the underlying mechanisms exhibited by each beam, using the ATENA Science software package [12]. In this paper, it is shown that through the use of these advanced modern-day computational methods (FEA and DIC), more refined understanding of the shear behaviour can be obtained which will increase the understanding of the significance of the shear spacing in the shear design of reinforced concrete members.

2. Experimental Programme

The experimental programme involved the testing of six geometrically identical reinforced concrete beams under four-point loading. All beams had a 150×250 mm rectangular cross-section dimension and a 30mm clear concrete cover to the main bars. The only varying parameter between specimens were the longitudinal and transverse (shear) reinforcement arrangements. The geometry and breakdown of all the reinforcement layouts of the specimens are illustrated in figure 1. The first three beams were reinforced with three H12 bottom bars (denoted L series), whereas the remaining three beams had two H16 bottom bars (denoted H series)– corresponding to longitudinal reinforcement ratios of 1.06% and 1.26%, respectively. Two of the beams in each series had H8 transverse reinforcement provided at a spacing of 300 mm (≈1.4*d*) and 150 mm (≈0.7*d*) over the shear span, corresponding to the shear reinforcement ratios of 0.22% and 0.45%, respectively. No transverse and longitudinal top reinforcements were provided in the remaining two beams (Beams L1 and H1).

2.1. Materials, fabrication and curing

Casting of the test specimens occurred on two separate days, using the same grade of concrete (C25/30) and slump (S2). The concrete was ordered from a local ready-mix company, with each batch used to fabricate three beams. Table 1 provides a breakdown of the mix design for the concrete, together with the corresponding concrete compressive strength for each beam. The concrete strengths were computed from the conversion of the mean strengths of three concrete cube samples with dimensions of 100×100×100 mm at the day of testing. All beams were cured under a wet burlap, wrapped in polyethylene sheeting at ambient laboratory conditions until approximately three days prior to testing. All beams were tested between 23 and 30 days after casting.

IOP Conf. Series: Materials Science and Engineering **930** (2020) 012052 doi:10.1088/1757-899X/930/1/012052

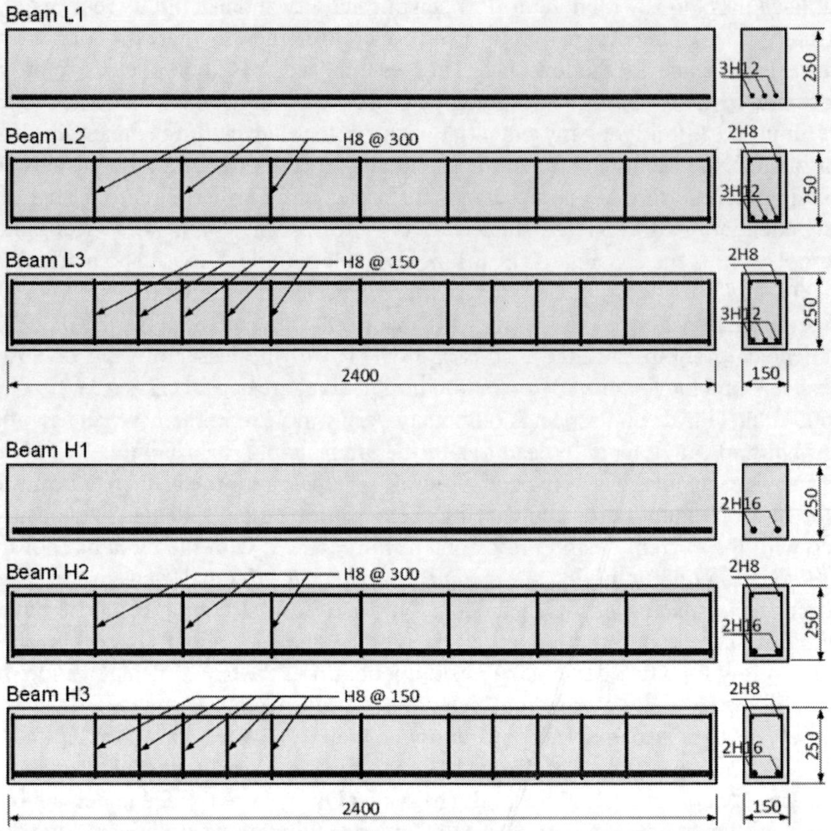

Figure 1. Beam geometry and bar configuration (dimension in mm). In each beam, 30 mm clear cover was provided to the main bars.

Table 1. Summary of concrete mix and strengths. (WRA: BASF Glenium 126).

OPC (kg/m^3)	GGBS (kg/m^3)	4/20 mm (kg/m^3)	0/4 mm (kg/m^3)	WRA (kg/m^3)	w/b	Compressive strength (MPa)					
						L1	L2	L3	H1	H2	H3
150	150	1020	940	1.7	0.6	23.8	25.8	26.5	29.8	31.1	31.6

2.2. Test configurations and measurement

The schematic of the test setup is presented in figure 2. Prior to testing, each beam was simply supported on two solid steel rods placed 1800mm apart, with the support configuration of a pin at one end (by means of a steel rod welded to a pair of steel plates) and a roller at the other end (by means of a steel rod free to rotate between two steel plates). The equipment used to perform the monotonic loading was a 2000kN Losenhausen servo-hydraulic testing machine. The load was applied through a rigid spreader (steel) beam to direct the load from the testing rig to two 50 mm diameter steel rods placed 600mm apart. At all loading and support points, steel plates with dimensions of 100×60×200 mm were provided. The loading imposed on the test beams under a crosshead speed of 0.5mm/min, paused at regular intervals of 10kN to allow for the manual crack mapping of the beam to progress. The imposed loads were recorded using a load cell installed in the test rig, while the beam deflections were recorded using one LVDT (linear variable displacement transducer) placed under the beam at the midspan. All data was recorded using a 16-bit USB data acquisition system at a sampling rate of 5Hz.

IOP Conf. Series: Materials Science and Engineering **930** (2020) 012052 doi:10.1088/1757-899X/930/1/012052

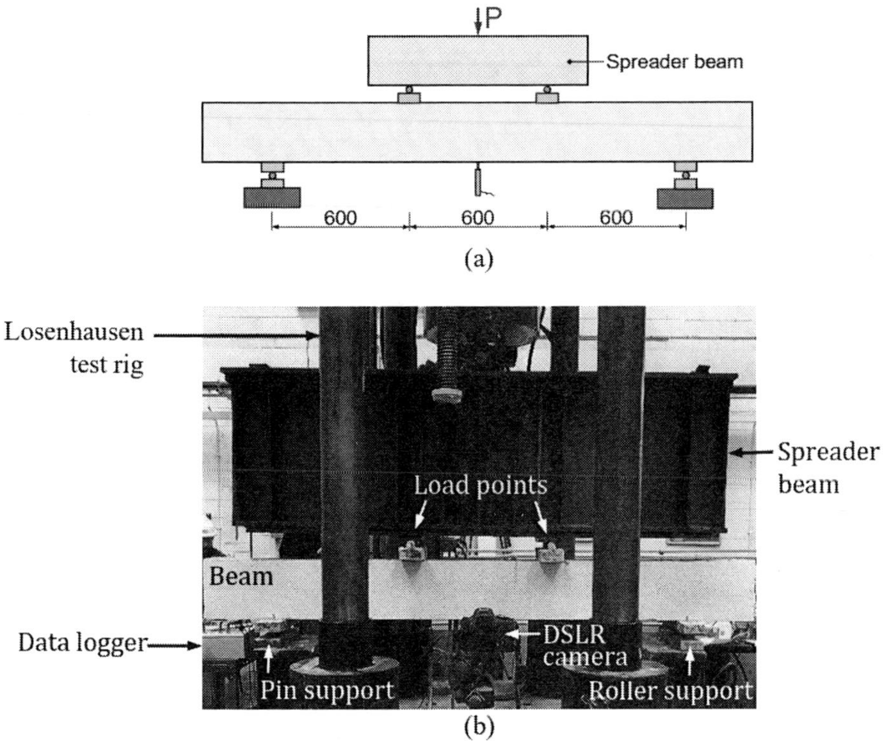

Figure 2. (a) Schematic of test setup; (b) typical view prior to testing.

Prior to testing, the both sides of each beam were whitewashed. Grid lines were then drawn on one side to allow for the manual crack mapping during the testing process. The other side was sprayed with a random black speckle pattern to facilitate the digital crack mapping. This was done by recording the crack development on the beam surface at 2kN increments using a 18MP Canon EOS Rebel T3i digital camera. The acquired digital images were then analysed using an open-source DIC software Ncorr to obtain crack development and failure mode [8-11]. The camera was mounted on a sturdy tripod and controlled remotely through a Desktop PC to avoid inadvertent camera movement throughout the testing process.

3. Nonlinear Finite Element Analysis

Three-dimensional nonlinear finite element analyses were undertaken for each beam using ATENA Science software package, developed by Červenka Consulting [12]. Figure 3(a) displays the typical finite element meshes used to represent the beams. The concrete was modelled using 8-node hexahedral elements with a typical size of 30 mm. The embedded reinforcement was modelled using a one-dimensional element. The plates at the loading and support points were modelled using an elastic tetrahedral element. In this study, the "Cementitious2 User" model was used for the concrete.

Figures 3(b) and (c) display a summary of the compression, tension and shear transfer models implemented in the software package. In the compression model, the compression hardening function is defined using an elliptical relation through the following expression [12,13]

$$\frac{\sigma_c}{f_c'} = f_{co} + (f_c' - f_{co})\sqrt{1 - \left(\frac{\varepsilon_c - \varepsilon_c^p}{\varepsilon_c}\right)^2} \quad (1)$$

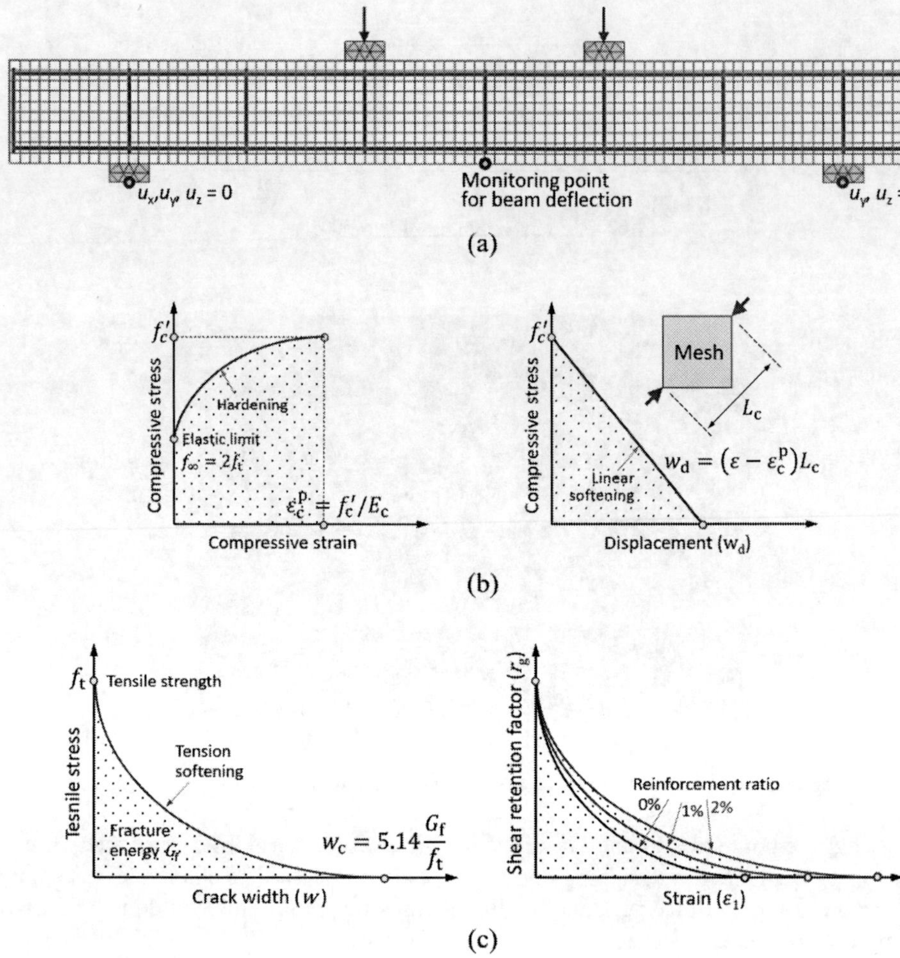

Figure 3. (a) Typical finite element mesh; (b) compression model; and (c) tension and shear transfer models.

where σ_c is the normal compressive stress (MPa); f_c' is the mean cylinder compressive strength (MPa); f_{co} is the compressive stress at post-elastic point (MPa); ε_c is the concrete strain (mm/mm); and ε_c^p is the plastic strain at the peak point (mm/mm). Beyond the peak, the softening relation is determined through an analogical model commonly known as the crush band model (to ensure mesh objectivity) [14]. The assumption was made on the localisation of post-peak compressive stress and energy dissipation in a plane normal to the direction of principal stress. A w_d value of 0.05 mm was used in this study following the recommended value for normal concrete [15].

In the tension model, the exponential softening law is adopted following the formulation of fictitious crack model based on the crack-opening law and fracture energy which are expressed in the following manner [12],

$$\frac{\sigma_t}{f_t} = \left(1 + \left(c_1 \frac{w}{w_c}\right)^3\right) \exp\left(-c_2 \frac{w}{w_c}\right) - \frac{w}{w_c}(1 + c_1{}^3)\exp(-c_2) \tag{2}$$

where σ_t is the normal tensile stress (MPa); f_t is the tensile cracking strength (MPa); w is the crack opening (mm); and w_c is the crack opening at the complete release of stress deriving from fracture energy (mm). Empirical constants $c_1 = 3$ and $c_2 = 6.93$ are considered.

The shear transfer model of cracked concrete is represented by a shear retention factor following the formulation proposed by Kolmar [16]. The shear modulus is related to the strain normal to the crack which is indicative of crack opening. For more detailed information of cementitious user material used in this study, readers are referred to reference [12].

4. Results and Discussion

4.1. Test observations

Figures 4(a) and (b) present the load applied at one side of each beam plotted against the mid-span deflection for L and H series beams respectively. For illustrative purposes, figure 5 displays typical failure crack patterns obtained from the L series beams (comparable crack patterns were obtained from the H series beams). During testing, flexural cracking was first observed at a shear (half of the machine) load of approximately 8kN, which corresponds to a moment value of 4.8kNm and tensile stress in the concrete at the bottom most part of the beam of approximately 3.1MPa. At increasing loads, new flexural cracks formed along the length of the beam, while pre-existing cracks extended vertically and increased in width. As displayed in figures 4(a) and (b), this resulted in a significant reduction in beam stiffness. It is interesting to note that the post-cracking response of the beams in each series was similar to each other and once the cracks had uniformly developed, the response followed an approximately straight line at a slope of approximately 30% of the initial (pre-cracking) stiffness. However, this trend is only true up to the formation of significant diagonal shear cracks, which occurred (initially) at one (weaker)

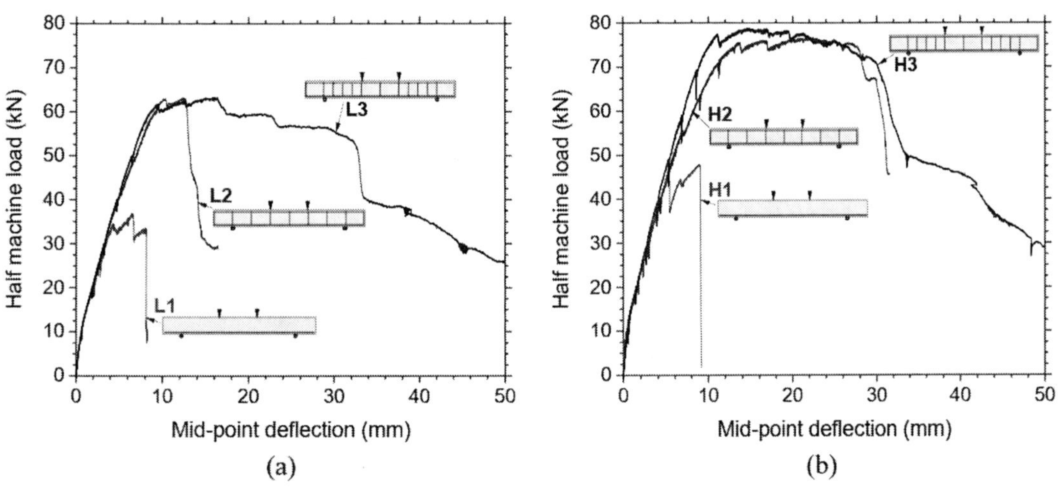

Figure 4. Shear-deflection response: (a) Beams L1–3 and (b) Beams H1–3.

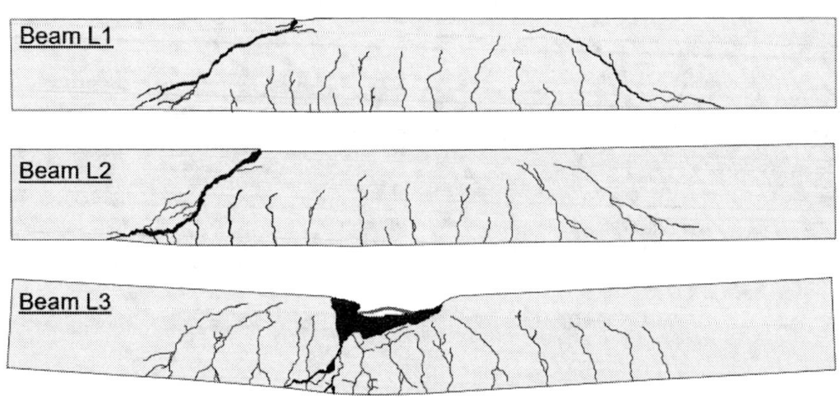

Figure 5. Typical crack patterns after failure.

side of a beam, at an applied shear load of ~35kN for Beam L1 and ~45kN for Beam H1. Failure finally occurred when the diagonal shear crack had extended to the point of load application and propagated down to the support along the top of the bottom reinforcement, leading to the formation of an *S-shaped* failure crack pattern. The failure was brittle, as indicated by the sudden drop in load-carrying capacity.

As the spacing of the transverse reinforcement is decreased from 600 mm (Beams L1 and H1) to 150 mm (Beams L3 and H3), the mode of failure is altered from shear to flexure. This is attributed to the fact that in these beams, the failure load that corresponds to the shear capacity is higher than that of the flexural capacity. It is interesting to note that although this relation also applies to Beams L2 and H2, these beams failed in shear as the spacing of transverse links in these two beams did not satisfy the maximum link spacing outlined in Eurocode 2 (0.75*d* or, in this case ~160 mm). Shear crack can, therefore, develop between two shear links where the shear is resisted only by the concrete. The evidence for this can be seen from figure 5 in which the formation of the diagonal (shear) crack between two shear links is clearly evident, with the crack at the mid-depth forming at an angle of ~70° from the horizontal (or an increase of ~25° from the angle of the critical crack in Beam L1).

With reference to figures 4(a) and (b), it is interesting to note that although Beams L2 and H2 could not develop to their full load capacity (based on the amount of the shear links provided), the shear capacities of these beams exceeded the shear capacity of the concrete itself by a considerable amount (compare, for example, the load capacity of Beams L2 and H2 to Beams B1 and L1). This could be related to the general increase in shear strength with an increase in shear span-to-depth (a/d) ratio [17] (with the increase in this ratio is dictated by the spacing of the shear reinforcement), although the presence of the transverse links might have also prevented the propagation of the critical shear crack and hence makes a contribution to the overall increase in shear capacity. At this moment, however, it is not possible to obtain a quantitative measure of the increase in shear capacity as the tension bars in both beams reached their yield capacity before the beam reached the shear capacity. It is also still not clear as to why Beam H2 was able to continue to resist considerable load after the formation of a diagonal crack at a shear load of approximately 56kN (see the slight drop in load in figure 4(b)). This is possibly due to the presence of the shear links as stated above; however, further investigation is needed to clarify this issue.

4.2. Crack developments

To provide detailed insights into the response of the beams, figure 6 illustratively displays the formation and propagation of concrete cracks in Beams L1 and L2 obtained from the DIC crack mapping.

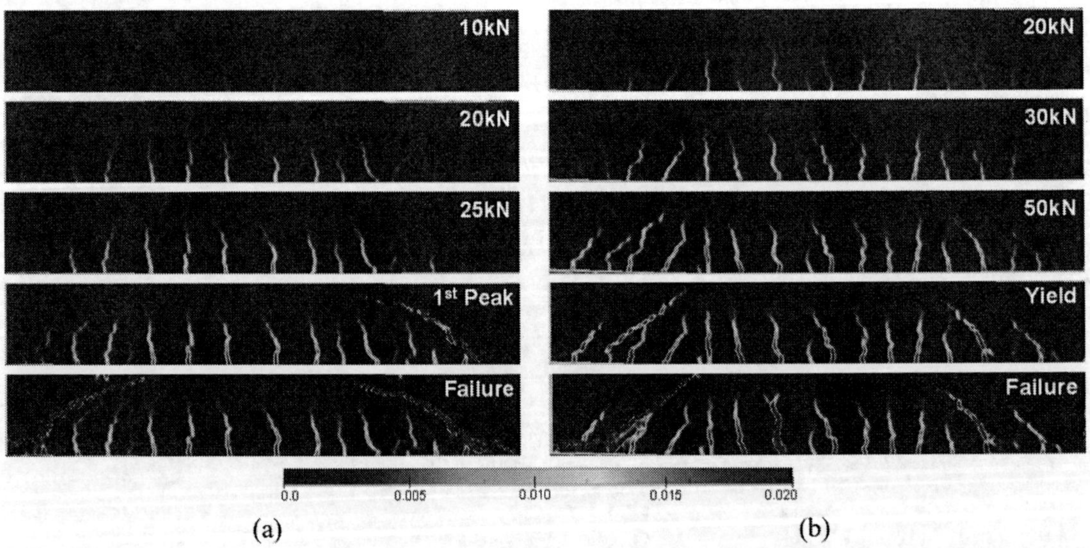

Figure 6. Crack developments at selected load levels during testing: (a) Beam L1 and (b) Beam L2.

It is evident from figure 6 that at early stages of loading, both beams exhibited uniform crack distribution, consisting primarily of flexural cracks which extend vertically with increasing loads. Some of the cracks over the side (shear) span then developed a rotation at their tips and merged with other cracks to develop a critical crack at both ends of the beam. This critical crack then propagated towards the point of load application at the top and down to the top layer of tension reinforcement and continued as a nearly horizontal (splitting) crack along the tension steel. This resulted in a sudden and brittle failure, with no ductility beyond the peak load (see figures 4(a) and (b)).

4.3. Nonlinear finite element analysis

The load (applied at one side of each beam) vs midspan deflection responses for all beams obtained from the finite element analyses are presented in figures 7(a) and (b), and can be compared to experimental results presented earlier in figures 4(a) and (b). As can be seen, there is an excellent agreement between the predicted and observed responses in terms of the overall load-deflection response and mode of failure. It is evident that as the spacing of transverse reinforcement decreases, the load capacity and ductility of the beams increase; this behaviourial response could be well captured by the simulations. The predicted and observed failure crack patterns are also in good agreement (compare, for example, figures 5 and 7). Of interest to note is the successful representation of the increasing angle of the failure crack in beams with intermediate link spacing (i.e. Beams L2 and H2). The crack patterns over the centre span could also be well simulated and are similar to the experimental crack patterns displayed previously in figure 6, highlighting the value and and accuracy of the finite element models employed in this study.

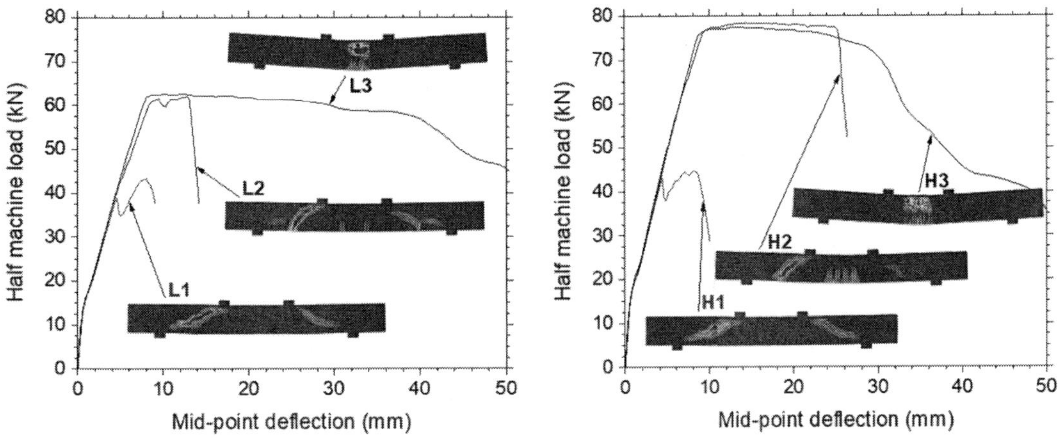

(a) (b)

Figure 7. Predicted shear-deflection response for (a) Beams L1–3 and (b) Beams H1–3.

5. Concluding remarks and conclusions

Six geometrically identical beams produced from two batches to the same specification of concrete mix (C25/30) were tested. Beams L1 and H1 had no shear reinforcement; Beams L2 and H2 were non-compliant with the necessary shear reinforcement spacing detailed in Eurocode 2 and Beams L3 and H3 fully complied with this provision. The purpose of this study was to determine the influence of varying the link spacing on the shear capacity of each concrete section. This was achieved by performing experimental testing, digital crack mapping and nonlinear finite element analyses. The following conclusions can be drawn:

1. Beams L1 and H1 yielded the lowest load bearing capacities. With the omittance of shear reinforcement, the beams, as expectedly, failed in shear at a maximum load of 35kN and 45kN while exhibiting very low ductility. These beams are crucial in highlighting the dangerous, sudden brittle failure which can occur in structural members with no shear reinforcement.

2. Beams L2 and H2 exhibited significantly higher load bearing capacities and deflections than those discussed previously. However, the subsequent differing modes of failure highlights how inadequate spacing of shear links can lead to unpredictable responses. Shear failure occurred regardless of the inclusion of shear linkages due to the lack of sufficient spacing.
3. Beams L3 and H3 yielded load bearing capacities similar to that of L2 and H2 respectively. However, of major significance is the change from shear to flexural failure. As a result of the sufficient spacing of shear links, the brittle premature shear failure could be prevented. Instead, the mode of failure was very ductile, with the gradual propagation of cracks and the beam deflecting significantly more than the previous designs. This behaviour is sought after by structural designers as it allows adequate warning before failure ultimately occurs.
4. Nonlinear finite element analyses show strong agreements on the overall load-deflections and failure modes of the beams tested, validating the experimental behaviours exhibited. Moreover, the use of digital image correlation further exacerbated the accuracy between analyses and testing through detailed crack mapping throughout the testing.
5. It is shown that errors in the design of a structural member can potentially be catastrophic and the use of advanced analysis tools could be useful in this regard; for example, for identifying potential problems or assessing behavioural response influencing the performance and safety of a structure.

References

[1] Suryanto B, Morgan R and Han A L 2016 Predicting the response of shear-critical reinforced concrete beams using Response-2000 and SNI 2847:2013 *Civ. Eng. Dimens.* **18** 1 16–24
[2] Collins M P and Kuchma D 1999 How safe are our large, lightly RC beams, slabs, and footings? *ACI Struc. J.* **96** 4 482–90
[3] Anderson B G 1957 Rigid frame failures *ACI J. Proc.* **53** 1 625–36
[4] BSI, 2015 *BS EN 1992-1-1:2004+A1:2014: Eurocode 2: Design of concrete structures. General rules and rules for buildings* London, British Standards Institution
[5] ACI, 2014 *ACI 318M-14 Building code requirements for structural concrete and commentary* Farmington Hills, American Concrete Institute
[6] JSCE, 2007 *Standard specifications for concrete structures – 2007 design* Tokyo, Japan Society of Civil Engineers
[7] CSA, 2019 *CSA A23.3:19 Design of concrete structures* Toronto, Canadian Standards Association
[8] Blaber J, Adair B and Antoniou A 2015 Ncorr: open-source 2D digital image correlation Matlab software *Exp. Mech.* **55** 6 1105–22
[9] Suryanto B, Tambusay A and Suprobo P 2017 Crack mapping on shear-critical reinforced concrete beams using an open source digital image correlation software *Civ. Eng. Dimens.* **19** 2 93–8
[10] Tambusay A, Suryanto B and Suprobo P 2018 Visualization of shear cracks in a reinforced concrete beam using the digital image correlation *Intr. J. Adv. Sci. Eng. Inf. Tech.* **8** 1 573–8
[11] Tambusay A, Suryanto B and Suprobo P 2020 Digital image correlation for cement-based materials and structural concrete testing *Civ. Eng. Dimens.* **22** 1 6–12
[12] Červenka V, Jendele L and Červenka J 2018 Program documentation ATENA theory Czech Republic, Červenka Consulting
[13] Menétrey P and Willam K J 1995 Triaxial failure criterion for concrete and its generalization *ACI Struc. J.* **95** 3 311–8
[14] Mier J G M V 1986 Multiaxial strain-softening of concrete part 1: fracture *Mat. Strut.* **19** 111 190–200
[15] Červenka J, Červenka V and Laserna S 2018 On crack band model in finite element analysis of concrete fracture in engineering practice *Eng. Fract. Mech.* **197** 27–47
[16] Kolmar W 1986 Specification of shear transfer across cracks in nonlinear finite element analysis of reinforced concrete structures *PhD Thesis* Darmstadt University of Technology Germany
[17] Collins M P, Bentz E C and Sherwood E G 2008 Where is shear reinforcement required? review of research results and design procedures *ACI Struc. J.* **105** 5 590–600

IOP Conf. Series: Materials Science and Engineering **930** (2020) 012053 doi:10.1088/1757-899X/930/1/012053

Temperature dependence and activation energy of electrical conduction in an engineered cementitious composite

B Suryanto[1]*, D Saraireh[1] and A Tambusay[1,2]

[1] Institute for Infrastructure and Environment, School of Energy, Geoscience, Infrastructure and Society, Heriot-Watt University, Edinburgh, United Kingdom.
[2] Department of Civil Engineering, Faculty of Civil, Planning and Geo Engineering, Sepuluh Nopember Institute of Technology, Surabaya, East Java, Indonesia.

*Corresponding author's e-mail: B.Suryanto@hw.ac.uk

Abstract. The electrical impedance of an engineered cementitious composite was measured over the temperature range 7–60°C and data are presented over the frequency range 20Hz–1MHz. The work has highlighted that the resistive and reactive components of the complex impedance both displayed an inverse relationship with temperature. When presented in an Arrhenius format, the bulk resistance displayed a linear response over the entire temperature range studied and the activation energy, E_a, for electrical conduction was evaluated as 21.7 kJ/mol (0.22 eV/ion).

1. Introduction

Engineered Cementitious Composite (ECC) is a family of cement composite that is capable of exhibiting strain hardening response in tension, with a tensile strain capacity in the order of a few percent commonly achieved [1]. This unique property is attributed to the ability of this group of cement composite to form multiple micro-cracks of small widths (typically <0.1 mm) [2, 3], thereby making it particularly attractive for applications where cracking and long-term durability are critical. Two areas of application that have received increasing attention are durable cover and repair material [4, 5]; therefore, it is not surprising that the durability and long-term performance of ECC in the natural environment have been actively researched by various laboratories around the world [6-9].

Recently, there has been a growing interest in exploiting the self-monitoring capabilities of ECC, through monitoring of its electrical properties [10-12]. Due to the vagaries of weather in the natural environment, however, ECC will certainly be subjected to constantly changing load and environmental (i.e. temperature and humidity) conditions which would, in turn, affect its electrical response. The latter is particularly relevant for ECC applied in the near-surface region, as if used as a cover or repair material. As far as electrical based monitoring is concerned, there is a need to develop a systematic approach that can be used to treat the effects of every contributing factor on electrical response (i.e. temperature, moisture, cracking/damage) in isolation. It is set against this background that this paper presents the findings of a testing programme that investigates the influence of temperature on bulk electrical properties. Data are presented in the Arrhenius format, thereby allowing evaluation of the activation energy for electrical conduction. The knowledge gained from this work will be used for developing an activation energy-based temperature correction protocol which can be used to remove temperature effects from field measurements.

Content from this work may be used under the terms of the Creative Commons Attribution 3.0 licence. Any further distribution of this work must maintain attribution to the author(s) and the title of the work, journal citation and DOI.
Published under licence by IOP Publishing Ltd

2. Experimental Programme

2.1. Materials and mix proportion

The mix proportion for the ECC used in this experimental programme is presented in table 1. The binder comprised Portland cement CEM I 52.5N to BS EN197-1 [13] and fine fly-ash (Superpozz SV80, supplied by ScotAsh) at a cement/fly-ash ratio of 1:1.8 (or a replacement level of approximately 64%). The typical oxide analysis of these materials is presented in table 2. Fine silica sand with a mean particle size of 120 μm was added into the mix at a sand/cement ratio of 0.6:1 by mass. The fibres used in the mix were Kuralon K-II REC15 polyvinyl alcohol fibres, which were added at a dosage of 2% by volume. The fibres had an average diameter of 39 μm, length of 12mm, tensile strength of 1.6GPa, and tensile strain capacity of 6.5%. They were supplied with a proprietary coating agent (1.2% by mass) by the manufacturer to reduce excessive chemical bonding with the surrounding matrix. A high-range water-reducing (HRWR) admixture (BASF MasterGlenium ACE499) was added at a dosage rate of 1% by mass of binder to improve mix workability and fibre dispersion. The water/binder ratio was set at a relatively low value (=0.28) to produce an ECC matrix of desired strength.

Table 1. Summary of ECC mix and compressive strengths.

OPC (kg/m^3)	FA (kg/m^3)	S (kg/m^3)	HRWR (kg/m^3)	PVA (kg/m^3)	w/b	Compressive strength (MPa)		
						28 days	90 days	180 days
454	818	273	4.54	2.6	0.28	51.9	63.7	66.6

Notes: OPC: CEM I 52.5N; FA: fly ash; S: sand; HRWR: high-range water-reducing admixture; and PVA: polyvinyl alcohol.

Table 2. Typical oxide analysis of materials (+ = not determined).

Properties	OPC	FA
Chemical analysis		
SiO_2	19.9	52.7
Al_2O_3	4.8	26.6
Fe_2O_3	3.1	5.6
K_2O	+	+
CaO	62.4	2.4
MgO	2.2	1.2
Na_2O equivalent	0.54	1.7
SO_3	3.0	0.3
Free CaO	+	0.03
Total phosphate	+	0.5
Loss on Ignition (LOI)	+	<2.0
Physical properties		
Specific gravity	3.15	2.20
Surface area (m^2/kg)	375	1300
Fineness (% retained on 25 μm)	+	<25

2.2. Test specimens, fabrication and curing

In this experimental programme, a total of 14 ECC specimens were fabricated. As illustrated in figure 1, three, 40×40×160 (long) mm prismatic specimens were used for electrical measurements (see figure 1(a)); five, dog-bone (DB) shaped specimens of geometry and dimensions in accordance with the JSCE recommendation [14] for tensile testing (see figure 1(b)); and six, 50 mm cuboidal specimens for compressive strength tests.

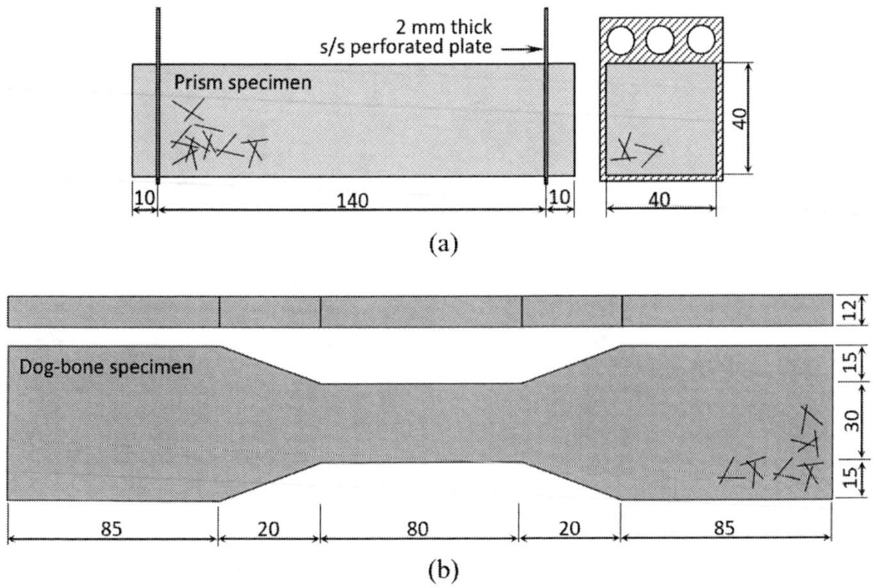

Figure 1. Schematic of test specimens: (a) prism; and (b) dog-bone shaped (DB) specimen. All dimensions are in mm.

A 10-litre Hobart planetary motion mixer was used to prepare the mix in a single batch. Specimens for electrical measurements were cast in polystyrene moulds, with a pair of 45×65×2 (thick) mm stainless steel (316L) perforated electrodes provided at the time of casting at 140 mm ceantres to facilitate electrical measurements. The perforations on the electraodes were 10 mm in diameter and arranged in a staggered pattern with a 15 mm pitch. The relatively large hole size was chosen to allow for the ECC slurry to flow easily through the perforations during casting. The use of perforated electrodes also ensured intimate bonding with the ECC and reduced any possible interfacial problem at the electrode/sample contact region [15]. The dog bone and cube specimens were cast in custom-made Plexiglas and standard three-gang steel moulds, respectively. All these moulds were treated with a proprietary release agent prior to casting.

After casting, all specimens were covered with thick cling film and allowed to cure for 24 hours. The specimens were then demoulded and stored in a curing tank at 20±2°C until they were required for testing (28 days for the dog-bone specimens; 28, 90 and 180 days for the cubes; and 360 days for the prisms). With regard to the latter, hydration could be considered as having ceased and, as such, at this extended period of time, the electrical response would be purely due to the changing in temperature.

2.3. Specimen conditioning and electrical measurements

Prior to testing, each prism specimen was removed from the curing tank and then placed in a thermostatically controlled circulating water bath (initially at ≈7°C). The water was then raised to the temperature at which electrical measurements were required, which, in the case of the work presented, at 10 to 60°C in 10°C increments (with one additional measurement at 15°C). At the required temperature, temperature was held for ~5 minutes in order to minimise thermal gradient within the specimen. This was confirmed by monitoring the internal temperature within each specimen (through a thermistor which was embedded at the time of casting). Once thermal equilibrium was reached, electrical impedance measurement was then undertaken on each specimen using an E4980AL Keysight high-precision LCR meter. The LCR meter was operated at a constant signal amplitude of 350mV and controlled by a desktop PC using LabVIEW virtual instrument. A logarithmic sweep was made over the frequency range 20Hz–1MHz, with the impedance recorded at 20 spot frequencies within this range. The connection to the LCR meter was by means of four individually-screened (short) coaxial cables, with connection to the electrode by means of alligator clips.

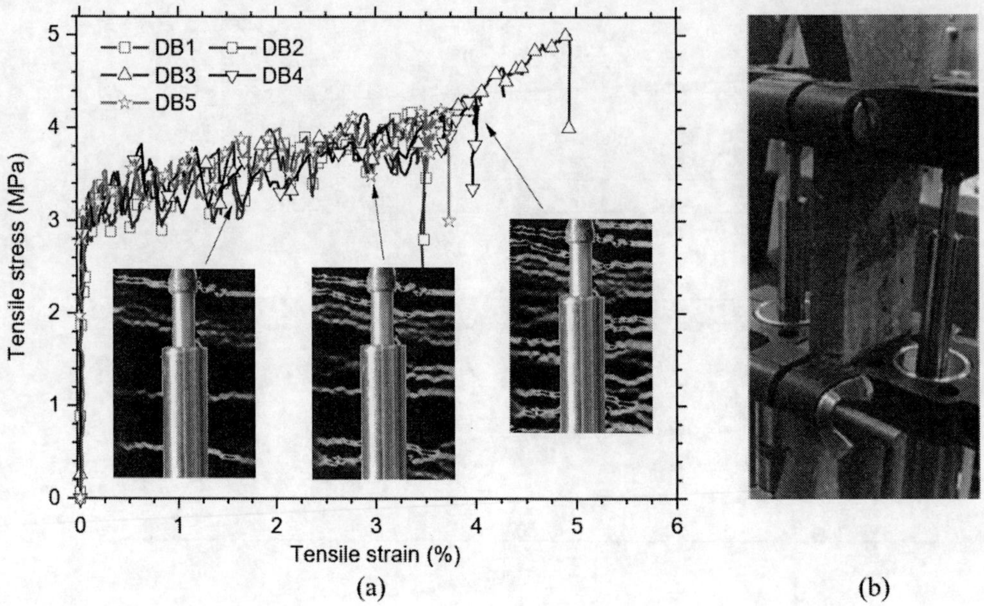

Figure 2. (a) Typical tensile stress-strain response obtained from dog-bone specimens; and (b) specimen during tensile testing showing two displacement transducers for strain measurements.

Although the focus of this paper is on the electrical properties of an ECC, it was considered appropriate to confirm the tensile strain hardening properties of the material and, for completeness, check the strength development with time. The dog-bone specimens were therefore tested using a 100kN 4206 Instron at 28 days after casting. Each specimen was held at both ends using pneumatic grips and loading was then applied under a crosshead speed of 0.5 mm/min. Figure 2(a) presents the tensile stress-strain responses obtained from the dog-bone specimens, together with an example of the crack patterns obtained using an automated crack mapping technique [16, 17]. It is evident that all specimens displayed a tensile strain hardening response with notable fluctuations in stress with increasing strain, highlighting clearly its distinctive tensile strain-hardening property. The fluctuations in stress can be attributed mainly to progressive development of micro-cracks and the evidence for this can be seen from the strain plots displayed in the figure, which were obtained from the central (narrower) section of the dog-bone. The strength development of the ECC over a 180-day period was determined by testing the ECC cubes using a 3000kN Avery-Denison testing machine, with the load applied at a rate of 50kN/min. The compressive strengths obtained on the 28th, 90th and 180th days of curing are presented in table 1.

2.4. Preliminaries and data presentation

The impedance, $Z(\omega)$, of a cementitious system (in Ohm, Ω) subjected to a small-signal sinusoidal electric field at an angular frequency, ω, can be written in rectangular form as,

$$Q(\omega) = Z'(\omega) - iZ''(\omega) \tag{1}$$

where the real component, $Z'(\omega)$, is the resistive component and the imaginary component, $Z''(\omega)$, is the reactive component (both in Ω). These two parameters are commonly presented in the Nyquist format, with $Z''(\omega)$ plotted against $Z'(\omega)$ over a wide frequency range (see, for example, [18]). For clarity, a typical impedance response obtained using the two-point electrode configuration is presented in figure 3. Two distinct regions are evident: a spur at the right-hand side and a semi-circular arc whose centre is depressed below the real axis at the left-hand side. The intercept of the low-frequency end of the arc with the real axis (i.e. near or at the junction between the electrode spur and the arc, or commonly referred to as the *cusp* point) is easily identifiable and could be considered as the bulk resistance of the test specimen, R (Ω), and will be related to ionic conduction processes [19].

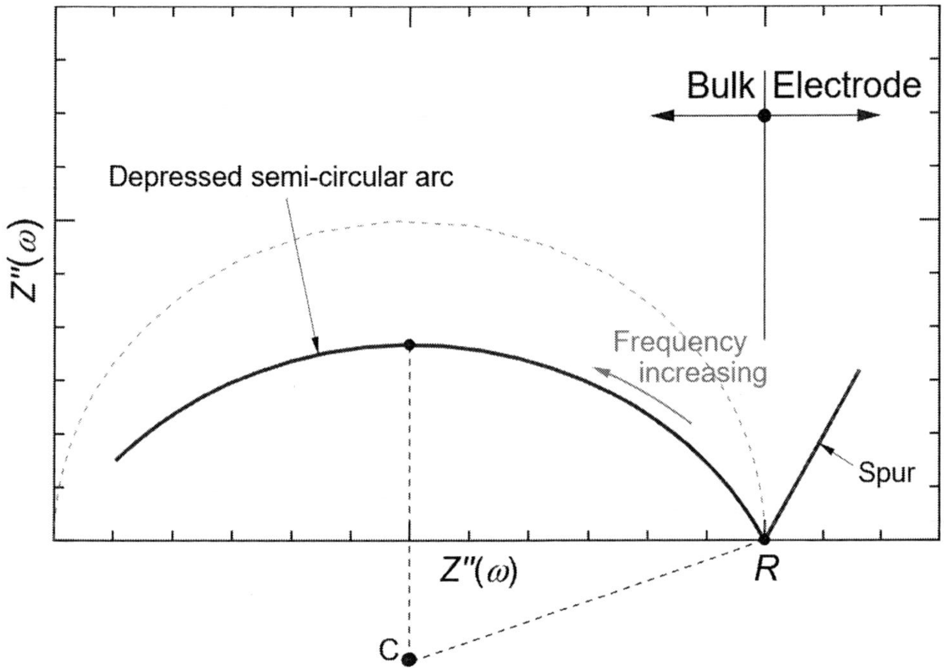

Figure 3. Schematic diagram of the complex impedance response of a cement-based material.

3. Results and Discussion

3.1. Complex impedance and bulk resistance

The typical complex impedance spectra under varying temperatures are displayed in Nyquist format in figure 4(a), with frequency increasing from right-to-left across the curve over the frequency range 20Hz–1MHz. Although measurements were obtained at 20 spot frequencies per decade, for clarity, only selected frequencies are highlighted with data markers. In general terms, a similar overall response to the schematic shown earlier in figure 3 is evident, comprising a depressed semi-circular arc and a weakly developed spur at the low-frequency end; the full extent of the spur would only become apparent at frequencies considerably smaller than the lower frequency limit considered in this study (i.e. $\ll 20$ Hz) [20]. The data clearly show that the real and imaginary components of the impedance are both affected by temperature at all test frequencies. As the specimen temperature increases, there is a progressive displacement of the entire impedance response to the left-hand side toward the origin, together with a reduction in the radius of the bulk arc which indicates an overall reduction in specimen impedance. Increasing the specimen temperature has also the effect of shortening the length of the bulk arc, with the end of the arc shifting toward the cusp-point.

Taking the cusp-point as the bulk resistance of the ECC specimen, figure 4(b) displays the variation in resistance as the sample temperature is increased from approximately 7°C to 60°C. Best-fit line to the test data is plotted in the figure in solid line through the measurement points, with the resulting fitting equation displayed on the figure. The results obtained from the two prismatic specimens indicate good repeatability and it is evident that over the temperature range considered, the bulk resistance exhibits an inverse relationship with temperature, with resistance decreasing with increasing temperature, although the changes in resistance at higher temperatures are not as sensitive to changes in temperature as those at lower temperatures. This natural temperature dependence is well-documented and could be attributed primarily to the changes in ionic mobility within the pore network (hence changes in pore-fluid resistance) [21]. In this work, it is anticipated that the influence of other factors such as dissolution of ions and microstructural changes [22] is much limited in extent.

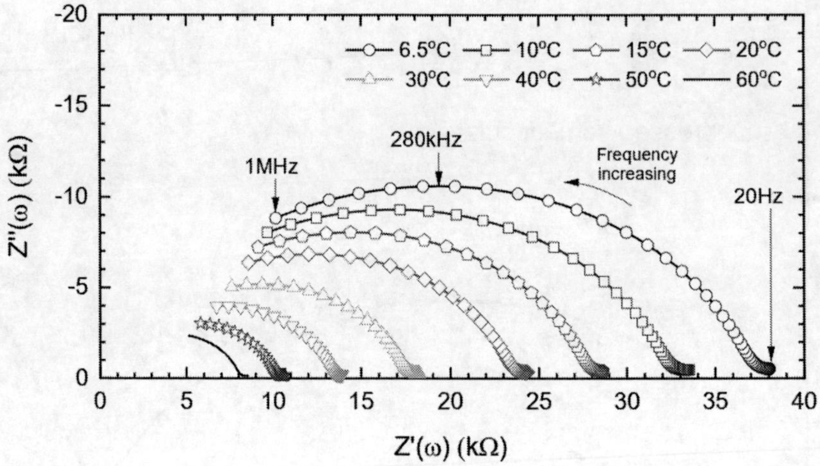

(a)

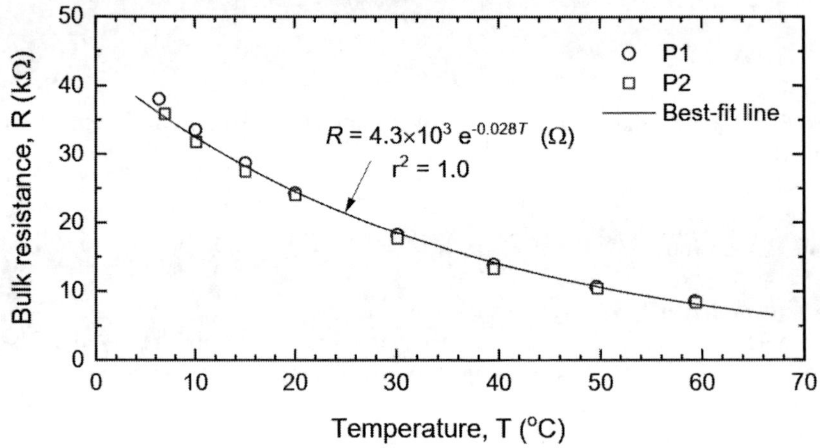

(b)

Figure 4. Variations in (a) impedance and (b) resistance with temperature.

3.2. Arrhenius presentation

Given that ionic conduction in ECC is a thermally activated process, the Arrhenius relationship is applied [22, 23],

$$R = R_o e^{\left[\frac{E_g}{R_g T_k}\right]} \tag{2}$$

where R is the bulk resistance (Ω) at temperature T_k (Kelvin); R_o is the pre-exponential constant and represents the nominal resistance at infinite temperature (Ω); E_a is the activation energy for the conduction process (J mol^{-1}); R_g is Universal Gas constant (8.3141 J mol^{-1} K^{-1}). Figure 5 presents the natural logarithm of bulk resistance, $ln(R)$, plotted against $1000/T_k$. In this format of presentation, the sample temperature increases from right-to-left along the horizontal axis; the data markers represent the measured resistance, while the solid line represents the linear fit from which the slope could be used to obtain the apparent activation energy (i.e. multiplying the slope by R_g will obtain E_a in kJ mol^{-1}). From the data presented, the activation energy for conduction processes over the temperature range 7–60°C was obtained as 21.7 kJ mol^{-1} (0.22 eV per ion) which is in agreement with the value reported in [24].

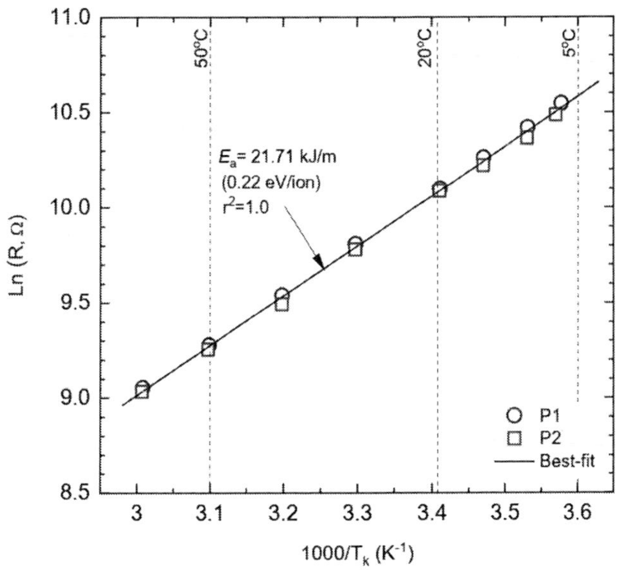

Figure 5. Data in figure 4(b) plotted in Arrhenius format.

Published data on the activation energy for electrical conduction in engineered cementitious composite are very limited; comparisons are, therefore, made with test data obtained from plain Portland cement mortar (water/cement ratio = 0.45), which was evaluated as 19.4 kJ/mol over the range 10–50°C [21], and Portland cement mortars containing various supplementary cementitious materials (with a constant water/binder ratio = 0.55), which were reported in the range 16–30 kJ/mol [22]. The activation energy obtained in this study could therefore be considered to lie within the range of the anticipated values.

4. Concluding remarks

An Arrhenius plot was established for the conduction process in an engineered cementitious composite, which represents an unresearched area in relation to the electrical response of cementitious systems that are reinforced with polymeric fibres and capable of exhibiting strain-hardening response in tension. Data are presented for ECC specimens undergoing continuous moist curing and, as such, results are directly relevant to ECC that remains saturated. It is shown that an Arrhenius relationship between temperature and bulk resistance does exist, highlighting the fact that electrical conduction in ECC involves a thermally activated process, as typically observed in other cementitious systems. As the self-monitoring version of ECC is now being developed and trialled in the field environment, the knowledge of activation energy will have direct practical significance (i.e. for removing the effect of natural temperature fluctuations) and work is continuing in this respect.

Acknowledgements

The Authors wish to acknowledge the financial support of the Engineering and Physical Sciences Research Council, United Kingdom (Grant EP/N028597/1), Kuraray Japan and GmbH for supplies of PVA fibres, and BASF UK for supplies of water reducers. DS also acknowledges the financial support provided by Heriot-Watt University. Thanks also go to Dr J Kim for assistance in the experimental work.

References

[1] Li V C 2008 Engineered cementitious composites (ECC)–material, structural, and durability performance *Concrete Construction Engineering Handbook* ed E G Nawy (Michigan: CRC Press) chapter 24 pp 1–48

[2] Kanda T, Tomoe S, Nagai S, Maruta M, Kanakubo T and Shimizu K 2006 Full scale processing investigation for ECC precast structural element *J. Asian Archit. Build. Eng.* **5** 2 333–40
[3] Liu H, Zhang Q, Gu C, Su H and Li V C 2016 Influence of micro-cracking on the permeability of engineered cementitious composites *Cem. Concr. Compos.* **72** 104–13.
[4] Kunieda M and Rokugo K 2006 Recent progress on HPFRCC in Japan *J. Adv. Concr. Tech.* **4** 1 19–33
[5] Lepech M D and Li V C 2006 Long term durability performance of engineered cementitious composite *Res. Build. Monum.* **12** 2 119–32
[6] Herbert E N and Li V C 2013 Self-healing of microcracks in engineered cementitious composites (ECC) under a natural environment *Materials* **6** 7 2831–45
[7] Yang E H, Li V C and Qiu J 2015 Early age cracking in a SHCC bridge deck link slab. *7th Congress on Forensic Engineering* (Reston: ASCE) pp 746-55.
[8] Suryanto B, Wilson S A, McCarter W J and Chrisp T M 2016 Self-healing performance of engineered cementitious composites under natural environmental exposure *Adv. Cem. Res.* **28** 4 211–20
[9] Rokugo K 2017 Applications of SHCC in Japan–tools and tips for promoting its use. *Proc. Int. Conf. on Strain-Hardening Cement-Based Composites* (Dordrecht: Springer) pp 671-80
[10] Ranade R, Zhang J, Lynch J P and Li V C 2014 Influence of micro-cracking on the composite resistivity of engineered cementitious composites *Cem. Concr. Res.* **58** 1–12
[11] Suryanto B, McCarter W J, Starrs G, Wilson S A and Traynor R M 2015 Smart cement composites for durable and intelligent infrastructure *Procedia Eng.* **125** 796–803
[12] Saraireh D, Walls S, Suryanto B, Starrs G and McCarter W J 2017 The influence of multiple micro-cracking on the electrical impedance of an engineered cementitious composite *Proc. Int. Conf. on Strain-Hardening Cement-Based Composites* (Dordrecht: Springer) pp 292-9
[13] BSI 2019 *BS EN 197-1:2011: Cement – Part 1: Composition, specifications and conformity criteria for common cements* (London: British Standards Institution)
[14] JSCE 2008 *Recommendations for Design and Construction of High Performance Fiber Reinforced Cement Composites with Multiple Fine Cracks (HPFRCC)* (Tokyo: Japan Society of Civil Engineers)
[15] Suryanto B, Takaoka H, McCater W J, Saraireh D and Taha H 2018 Impedance measurements on an engineered cementitious composite: a critical evaluation of testing protocols *Measurement* **129** 445–56
[16] Suryanto B, Tambusay A and Suprobo P 2017 Crack mapping on shear-critical reinforced concrete beams using an open source digital image correlation software *Civ. Eng. Dimens.* **19** 2 93–8
[17] Tambusay A, Suryanto B and Suprobo P 2020 Digital image correlation for cement-based materials and structural concrete testing *Civ. Eng. Dimens.* **22** 1 6–12
[18] Suryanto B, McCarter W J, Starrs G and Chrisp T M 2017 Characterization of fly-ash using electrochemical impedance spectroscopy *Procedia Eng.* **171** 705–14
[19] Whittington H W, McCarter J and Forde M C 1981 The conduction of electricity through concrete *Mag. Concr. Res.* **33** 114 48–60
[20] Suryanto B, McCarter W J, Starrs G and Ludford-Jones G V 2016 Electrochemical immittance spectroscopy applied to a hybrid PVA/steel fiber engineered cementitious composite *Mater. Des.* **105** 179–89
[21] McCarter W J 1995 Effects of temperature on conduction and polarization in Portland cement mortar *J. Am. Ceram. Soc.* **78** 2 411–5
[22] McCarter W J, Starrs G and Chrisp T M 2000 Electrical conductivity, diffusion, and permeability of Portland cement-based mortars *Cem. Concr. Res.* **30** 9 1395–400
[23] Chrisp T M, Starrs G, McCarter W J, Rouchotas E and Blewett J 2001 Temperature-conductivity relationships for concrete: an activation energy approach *J. Mater. Sci. Lett.* **20** 12 1085–7
[24] Suryanto B, McCarter W J, Starrs G and Jablonski M 2017 Assessing the performance of engineered cementitious composites under cyclic wetting and drying *Proc. Int. Conf. on Strain-Hardening Cement-Based Composites* (Dordrecht: Springer) pp 573-81

IOP Conf. Series: Materials Science and Engineering **930** (2020) 012054 doi:10.1088/1757-899X/930/1/012054

Effect of micro-cracking on the electrical and self-sensing properties of an engineered cementitious composite under tensile straining

D Saraireh[1], B Suryanto[1]* and A Tambusay[1,2]

[1] Institute for Infrastructure and Environment, School of Energy, Geoscience, Infrastructure and Society, Heriot-Watt University, Edinburgh, United Kingdom.
[2] Department of Civil Engineering, Faculty of Civil, Planning and Geo Engineering, Sepuluh Nopember Institute of Technology, Surabaya, East Java, Indonesia.

*Corresponding author's e-mail: B.Suryanto@hw.ac.uk

Abstract. The piezo-resistive response of a relatively mature engineered cementitious composite (ECC) under tensile straining is investigated and compared with previous studies. In this work, tensile tests were performed on four dog-bone shaped ECC samples and during the loading process, electrical impedance measurements were undertaken over the frequency range 100Hz–1MHz to identify the bulk resistance (hence accurate evaluation of resistivity). At the same time, digital images of the front face of the dog-bone samples were recorded throughout the entire loading process to enable detailed crack width analysis to be taken after testing and to monitor micro-crack formation during testing, using the digital image correlation. It is shown that tensile straining resulted in an overall increase in impedance, but retained a similar overall profile with a weakly developed spur evident at the low-frequency side of the impedance spectrum and a circular arc at the high-frequency side. It is also shown that the fractional change in resistivity increases nonlinearly with strain and is related to the nature of micro-crack formation. Published values for fractional change in resistivity and its relation with strain were found to be in a large scatter and in this study, attention is therefore focused on the crack width distribution during loading.

1. Introduction

Concrete cracking can have a detrimental effect on the durability and long-term performance of concrete structures and appropriate measures must therefore be taken to control cracking. Various methods have been employed in this regard, ranging from altering the concrete mix composition (i.e. limiting the cement content, use of chemical admixtures) to providing a well-distributed embedded reinforcement and/or a construction joint; these measures are generally taken to ensure that cracking will not affect the serviceability and durability of reinforced concrete. Given that control of cracking is of practical importance and that crack formation can greatly accelerate the deterioration process in reinforced concrete and reduce service life [1-3], it is of common practice to undertake periodic visual (on-site) inspections to identify potential defects. However, this is not always possible as some damage may not be readily visible, thereby remaining undetected until advanced stages which would be costly to repair. Furthermore, whilst visual inspections are simple to perform and allow direct evidence to be gathered, it can only give a qualitative measure and hence little insight into the integrity of a structure [4]. A number of automated, structural health monitoring systems have been developed in this regard [5, 6] and these systems generally require the installation of an extensive network of sensors.

Content from this work may be used under the terms of the Creative Commons Attribution 3.0 licence. Any further distribution of this work must maintain attribution to the author(s) and the title of the work, journal citation and DOI.
Published under licence by IOP Publishing Ltd

Recently, there has been a growing interest to use the concrete itself as a sensor due to its omnipresence in a structure [7]. At the same time, with increasing importance placed on producing durable concrete structures, there is also a growing interest in utilising fibre reinforcement as a means of more effective crack control. It is known that fibre additions into concrete can offer improvements in many of the mechanical properties of concrete such as enhanced toughness, reduced creep and shrinkage as well as improvements in the post-cracking behaviour. In this work, attention is given to a special type of fibre reinforced cement-based materials called the engineered cementitious composite (ECC) which is particularly known for its damage tolerant properties due to its high tensile strain capacity (in the order of a few percent) and superior crack width control (with crack width < 100μm commonly reported) [8-9]. As a cement-based material, ECC is shown to have a comparable electrical properties to a semi-conductor (10–10^5 Ωcm) [7]; the electrical properties of ECC could therefore be exploited to give a structure with self-monitoring capabilities with respect to internal damage and deformation under external loads.

With regard to damage sensing, work has predominantly focussed on the piezo-resistive properties (the resistivity change induced by mechanical strain) of ECC under tensile loading. To date, much of the work has studied these properties primarily through either d.c. or fixed-frequency a.c. resistivity measurements [7, 10-15]. In this paper, the a.c. electrical properties of ECC are investigated over a wide frequency range to identify the bulk electrical property more accurately. Detailed crack analyses through both visual (manual) observations and automated crack mapping technique (digital image correlation [16-18]) are also performed to provide supporting information in relation to the changes in electrical properties while under loading. The acquired electrical data are also compared with test data available in the literature [7,10-15] and a considerable scatter in published values is highlighted to call for the need for the development of a standardised test procedure.

2. Experimental Programme

2.1. Materials and mix proportion

The ECC mix used within the experimental program is presented in table 1. The binder comprised CEM I 52.5N Portland cement (PC) and fine fly-ash (Superpozz SV80), with a fly-ash/cement (FA/PC) ratios of 1.8. The water/binder (w/b) ratio was set as 0.28. Fine silica sand (RH110) with a mean grain size of 120μm was incorporated in the matrix at a sand/cement (S/C) ratio of 0.6 by mass. Polyvinyl alcohol (PVA) fibres were used at a dosage of 2% by volume. The fibres had an average length of 12mm and diameter of 39μm; the surface of which were coated with a thin layer of oiling agent (1.2% by weight). High-range water-reducing (HRWR) admixture was added at an dosage of 1% by cement weight.

Table 1. Summary of ECC mix and compressive strengths.

PC (kg/m^3)	FA (kg/m^3)	S (kg/m^3)	HRWR (kg/m^3)	PVA (kg/m^3)	w/b	Compressive strength (MPa)		
						28 days	90 days	180 days
454	818	273	4.54	2.6	0.28	40.3	53.1	65.5

Notes: PC: CEM I 52.5N; FA: fly ash; S: sand; HRWR: high-range water-reducing admixture; and PVA: polyvinyl alcohol.

2.2. Test specimens, fabrication and curing

A 10-litre Hobart planetary motion mixer was used to produce a total of 13 samples: four dog-bone shaped samples to determine the mechanical and electrical properties under tensile loading; and nine cubes to determine the compressive strength at 28th, 90th and 180th days after casting. Immediately after mixing, the fresh mix was placed into the corresponding moulds (Plexiglas moulds for the dog-bone samples and three-gang steel moulds for the cubes). All samples were then covered with cling film and allowed to cure for 24 h in a temperature controlled laboratory (21±2°C) before being demoulded and stored in a curing tank. After two weeks, two electrodes consisting of fine metal wire were wrapped securely at the opposing ends of the narrower portion of each dog-bone sample, as presented in figure

1. These wires were covered with several layers of silver-loaded conductive paint to seal any gaps and ensure intimate bonding with the sample. When the paint was fully dry, all dog-bone samples were returned to the curing tank until required for testing (180 days after casting).

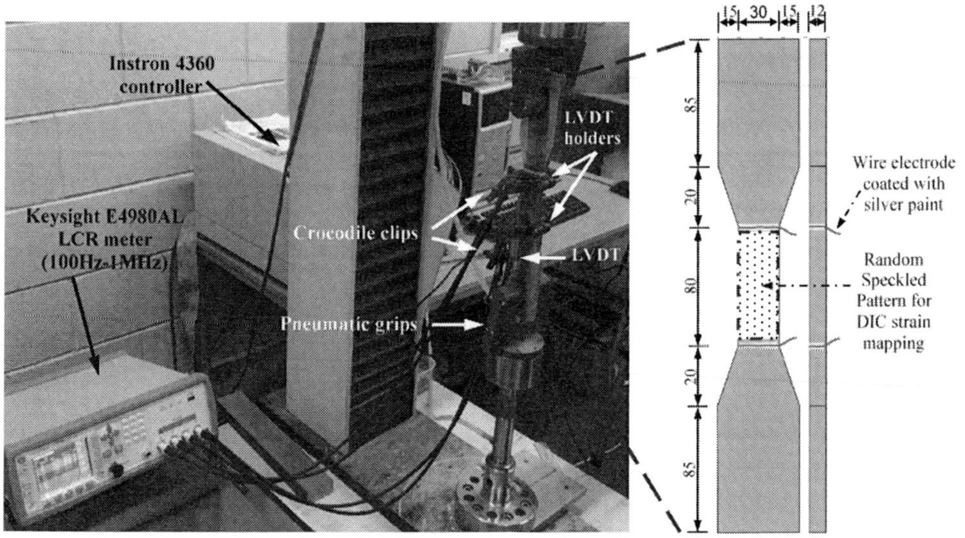

Figure 1. Dog-bone sample during tensile testing displaying the test setup and equipment used for testing. The dimensions of the dog-bone sample (in mm) are provided for clarity.

2.3. Testing equipment and procedures

A 100kN Instron machine was used to perform direct tensile tests. Each dog-bone sample was aligned in the machine and clamped at both ends using pneumatic grips. Two linear variable displacement transduscers (LVDTs) were used to measure the strain within the centre (narrower) region and these transducers were fixed at the opposing ends of this monitoring portion, one at each side of the sample, by means of lightweight plastic mounting blocks. Bluehill 2 software was used to manage the overall running of the experiment and log data from the load cell. Tensile stresses were determined by dividing the load cell readings with the cross-sectional area of the narrower section. In addition to this, a 16-bit 8-channel multifunction USB data acquisition system was used to log LVDT readings which were then converted to tensile strains using a calibration factor. These data were then used to plot the tensile stress-strain relation for each sample. All measurements were taken at a rate of 1Hz.

In addition to stress and strain measurements, electrical impedance measurements were undertaken during the loading process using an E4980AL precision LCR meter. This was done by connecting the pair of wire electrodes to four, individually screened, coaxial cables, with connection at the wire via an alligator clip. The LCR meter was controlled by a Desktop PC using LabVIEW virtual instrument. For each sample, electrical measurements were taken every three seconds at thirteen spot frequencies over the frequency range 100Hz–1MHz, covering 100Hz, 200Hz, 500Hz, 1kHz, 2kHz, 5kHz, 10kHz, 20kHz, 50kHz, 100kHz, 200kHz, 500kHz and 1MHz. These frequencies were chosen based on the time needed to perform a single full sweep measurement (slightly less than 3 seconds to complete).

To monitor micro-crack formation during testing, the digital image correlation technique was used. Prior to testing, the front surface of each dog-bone sample was painted white and then marked with random dotted patterns using a black marker (see figure 1). Once the sample was in position in the Instron machine, an 18.4MP Nikon J4 mirrorless digital camera was placed on a stable tripod in front of the sample. Digital images were then acquired at every 0.1mm extension to monitor the evolution of micro-cracks with increasing strain. This image acquisition was done wirelessly through a wireless mobile utility application installed on a mobile phone and this was specifically done to prevent camera movement which would render accuracy. Once the test was completed, the images were downloaded onto a Desktop PC and processed using a DIC software Ncorr [16-18] to follow the formation and propagation of micro-cracking during the entire loading process. In addition to this digital crack mapping, the crack pattern at various stages of loading was manually observed using the ImageJ software in order to measure individual crack widths, more accurately, and crack number.

3. Results and discussion

3.1. Complex impedance and bulk resistance

The typical complex impedance spectra for two dog-bone samples (DB2 and DB3) during tensile testing are presented in figures 2(a) and (b), respectively, as a Nyquist plot. Data markers represent the response acquired at the 13 spot frequencies (see section 2.3), with frequency increasing from right-to-left across each spectrum. In this figure, although measurements were obtained every 3 seconds, for clarity, only the response at 1% strain increment is presented (an example of a full impedance response over a 3-sec cycle during tensile straining is available in [19]). It is immediately apparent from figures 2(a) and (b) that during the entire loading process, the two samples displayed a typical impedance response with three distinct regions always evident:

(i) an arc on the high-frequency (left-hand) side of the spectrum;
(ii) a short intermediate plateau; and
(iii) a weakly developed spur at the low-frequency (right-hand) side of the spectrum.

The low-frequency spur is visible in the Nyquist diagrams presented due to polarization at the electrode/specimen interface and represents part of a much larger arc which would only become apparent at lower frequencies (viz, <<100Hz) and [20,21]. The arc represents the overall bulk impedance response of the sample. The bulk resistance of the test sample, R (Ω), which is related to ionic conduction processes [22], can be obtained from the intercept of the low-frequency end of the arc with the real axis.

With reference to figures 2(a) and (b), it is evident that tensile straining resulted in the entire impedance response being gradually displaced to the right, indicating an overall increase in impedance with increasing strain. This gradual increase in the impedance is due to the formation and widening of multiple micro-cracks during testing [19] (further presented and discussed in the section below).

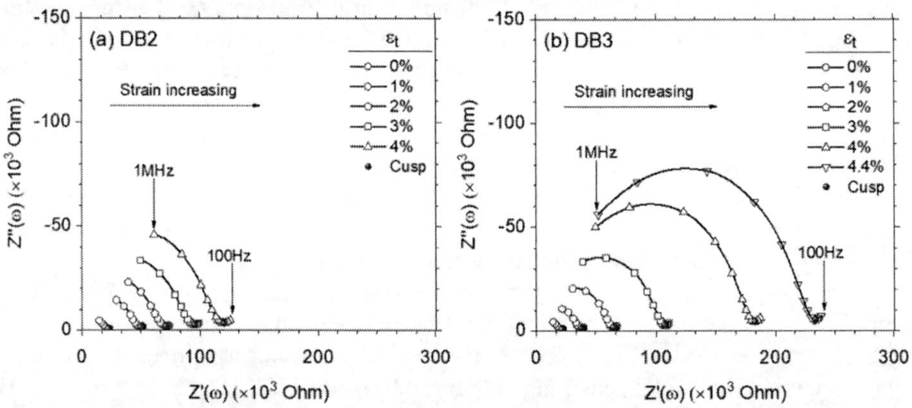

Figure 2. Typical complex impedance response during tensile loading for (a) DB2 and (b) DB3 samples. The cusp points are highlighted with circular (solid) data marker.

In this work, the minimum point within the intermediate region (the cusp point) is used to estimate the bulk resistance of the ECC samples due to its close location with the projected intercept of the low-frequency end of the arc with the real axis. In all samples, the frequency at which the cusp-point occurs was observed to occur at ~500Hz. Due to the amount of data collected, only the response at this spot frequency will be presented in the next section and discussed.

3.2. Fractional change in resistivity

The fractional change in resistivity (FCR) is a parameter that has been widely used in previous studies to highlight the self-monitoring feature of a material under loading, including various versions of ECC under tensile loading (see, for example, [7, 10-15]). The bulk resistivity, ρ (Ωm), can be determined by relating the bulk resistance obtained at the cusp point as discussed above to the electrode configuration and sample geometry through the relationship,

$$\rho = \frac{RA}{L} \tag{1}$$

where A is the cross-sectional area of the narrower (central) section of the dog-bone sample (approximately 0.03×0.012m^2); L is the measured distance between the two electrodes (approximately 0.08m); and A/L is the geometrical constant (/m). Since the electrodes used in this study did not cover the whole cross-section of the central region, current in regions close to the opposing ends of the central region would not flow across the full cross-sectional area and this may affect the value of geometry constant above. However, considering the relatively large length-to-depth ratio over the central section, this simple method of determination of geometrical constant could be considered sufficiently accurate. Alternatively, the geometrical constant can also be evaluated through a calibration procedure using the four-point method [23]. The FCR can be computed by normalising the change in resistivity after the start of testing, $\Delta\rho$, by the resistivity at zero strain (at the start of the test), ρ_{o}, through the following relationship,

$$\mathrm{FCR} = \frac{\Delta\rho}{\rho_{\mathrm{o}}} \tag{2}$$

Given that the change in length during testing is quite small (generally less than 5%), in this work, it is assumed that the geometric constant before and after deformation remains constant. This will have the effect of removing the need of calculating the resistivity; the bulk resistance can therefore be directly used in equation (2) in place of bulk resistivity. In addition to FCR, it is also possible to check the sensitivity of the material to mechanical strain and for this, the apparent gauge factor, GF, could be calculated through the following relationship,

$$\mathrm{GF} = \frac{\Delta\rho/\rho_{\mathrm{o}}}{\Delta\varepsilon_{\mathrm{t}}} \tag{3}$$

where $\Delta\varepsilon_{\mathrm{t}}$ is the change in tensile strain during testing.

Figure 3(a) displays the FCR at the cusp-point frequency (~500Hz) for all dog-bone samples, together with the tensile stress-strain response obtained from the same samples. It is evident that the FCR plots exhibit a nonlinear increase with strain, with the FCR value at failure of 19.7, 4.6, 9.7 and 17.2 for DB1, DB2, DB3 and DB4 samples, respectively. The increase in FCR is associated with the progressive development of micro-cracks during the strain-hardening phase. It is interesting to note that the FCR values for DB2 sample at 4% strain are comparable those reported earlier by [7, 15] who conducted similar tests on relatively young specimens (28 days of curing). However, the rest of the samples all attained much higher values of FCR (~11.7 (DB1), ~7.3 (DB3) and ~9.9 (DB4)) which would indicate higher sensitivity to tensile straining (or crack formation). This could be associated with the process of micro-crack development and the distribution of the micro-cracks during loading. It is anticipated that as the capillary pores within the ECC matrix break down (as being intercepted with a micro-crack), the crack space may be partially filled by pore-water thereby creating a water bridging conduction path across the micro-crack. With increasing time, however, the amount of free water within the ECC matrix could be expected to decrease due to the reduction in overall porosity resulting from on-going cement hydration and pozzolanic reaction. As a result, less water bridging effects could be expected in a mature ECC sample, as in the samples tested in this study. In addition to this aspect, it was observed that the crack widths were relatively large and this could also contribute to the enhancement in sensitivity.

IOP Conf. Series: Materials Science and Engineering **930** (2020) 012054 doi:10.1088/1757-899X/930/1/012054

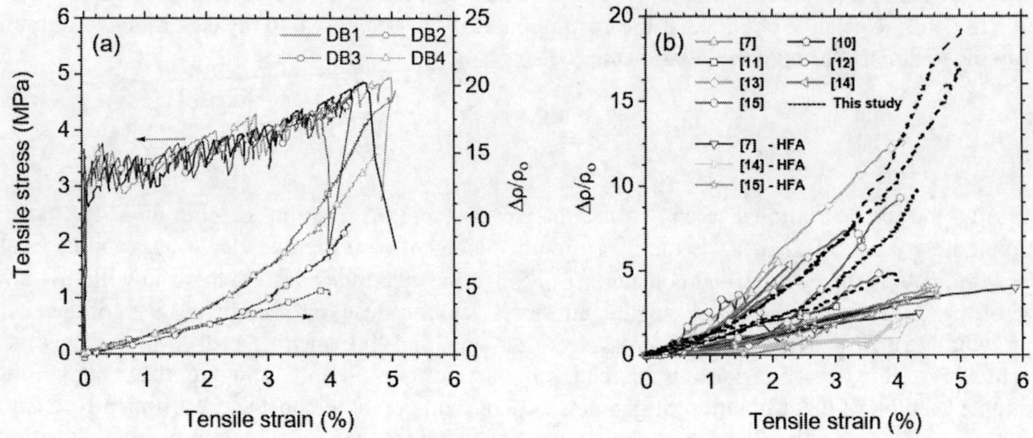

Figure 3. (a) The FCR for D1-D4 samples, together with the tensile stress-strain response; (b) comparison of the obtained FCR values with data reported in previous studies [7,10-15].

Figure 3(b) provides a summary of the FCR values of different versions of ECCs obtained from the literature [7,10-15]. It is apparent that there is clearly a wide scatter in published FCR values for similar ECCs and the values obtained in this study are within the range of the published values. With regard to the apparent slope of the FCR curve at 1% (i.e. the apparent GF over the initial 1% strain), this varies from only about 8 [15] and 15 [14] to as high as approximately 170 in [7]; 200 in [13]; and 260 in [12], with higher slope values indicating higher sensitivity. It is interesting to note that in cases with relatively low slope values, high volume of fly-ash (HFA) was used in the mix formulation and this type of ECC normally exhibits extremely small crack width, as reported in [7]. The large scatter in data could, therefore, be partly attributed to differences in crack width. It is also interesting to note that the FCR values obtained in this study are very similar to those reported by [15]. From figure 3(b), it is clear that there has been a few data published at strains levels of greater than approximately 4% and the data obtained in this paper could be useful in this regard. Given that the large variations in FCR values are likely related to the distribution and pattern of the micro-cracks, the number and width of the individual micro-crack during loading are presented and discussed below.

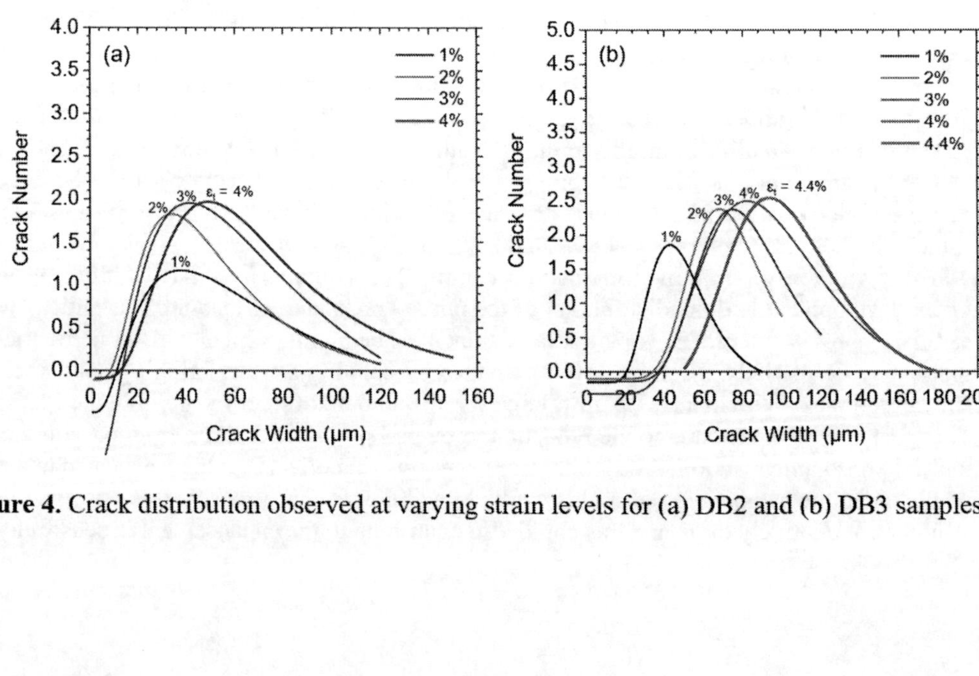

Figure 4. Crack distribution observed at varying strain levels for (a) DB2 and (b) DB3 samples.

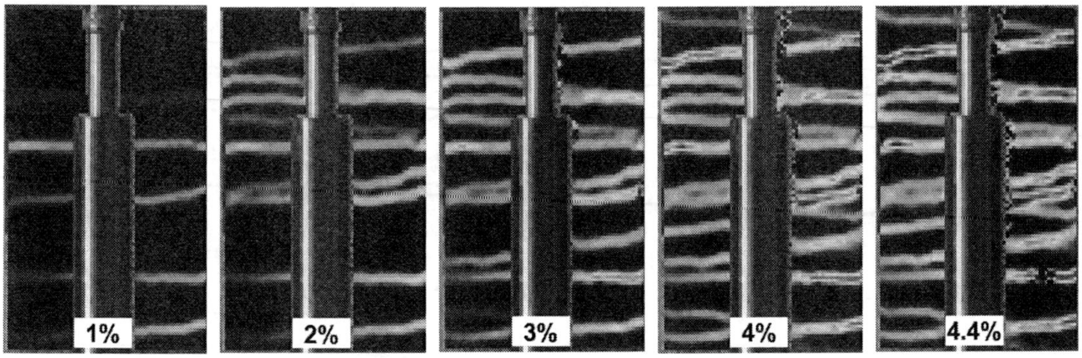

Figure 5. Progressive development of longitudinal strains obtained using the DIC for DB3 sample.

Figures 4(a) and (b) present the crack number-crack width relation, with the best-fit lognormal distribution curves presented for DB2 and DB3 samples, respectively. With reference to figure 4(a), it is evident that there is a progressive displacement of the curve upward, indicating an increase in crack number, and this occurs within the same crack width range (between 10μm and 120μm). At strain levels larger than 3%, however, the crack distribution curve starts to displace to the right while the peak value remains the same which is indicative of crack widening (without an increase in crack number). It is interesting to note that the crack width distribution of sample DB3 (see figure 4(b)) exhibits a different trend, displaying a continuous upward and progressive displacement to the right-hand-side up to 4% strain, when only progressive displacement to the right is observed. This would imply that there was a continuous increase in crack number and width of micro-cracks at strain levels less than 4%. Beyond that, only crack widening is operative. This observation agrees well with the DIC strain maps depicted in figure 5 which clearly display that there was indeed a progressive increase in crack number throughout the entire loading process, in particular between 1% and 2% strains.

4. Concluding remarks

1. The complex impedance response of cracked ECC during tensile loading displayed a classic impedance responses comprising an arc at the high-frequency end and a spur (only weakly developed) at the low-frequency end. Tensile straining was found to affect both the resistive and reactive components of the complex impedance, with both components shown to increase with increasing strain.
2. There is a nonlinear relationship between tensile strain and the fractional change in resistivity (FCR). The increase in FCR is shown to be gradual, due to the progressive micro-cracks formation within the ECC matrix.
3. Micro-cracks were found to exert an influence on the fractional change in resistivity, with the sensing sensitivity to tensile deformation increasing with increasing crack width.
4. There is a wide scatter in published FCR values for ECCs. Although the values obtained in this study are within the range of the published values, the authors want to call for the need for the development of a standardised test procedure to enable the assessment of the electrical and hence the self-sensing properties of ECC more accurately.

Overall, it has been shown that there is a close relationship between the electrical properties of ECC and crack width and, as such, the electrical properties of ECC could be further exploited as an alternative means of assessing the crack width in ECC (and hence in the quality control of an ECC mix). Although not considered as a variable in the present work, moisture could also be expected to exert a dominant influence on the measured resistivity and work is continuing in this respect.

Acknowledgements

The authors wish to acknowledge the financial support of the Engineering and Physical Sciences Research Council, United Kingdom (Grant EP/N028597/1), Kuraray Japan and Kuraray GmbH for supplies of PVA fibres, and BASF UK for supplies of water reducers. DS also acknowledges the financial support provided by Heriot-Watt University.

References

[1] Alyousif A, Lachemi M, Yıldırım G, Aras G H and Sahmaran M S 2015 Influence of cyclic frost deterioration on water sorptivity of microcracked cementitious composites *J. Mater. Civ. Eng.* 28 04015159.

[2] Kobayashi K, Le Ahn D and Rokugo K 2016 Effects of crack properties and water-cement ratio on the chloride proofing performance of cracked SHCC suffering from chloride attack *Cem. Concr. Compos.* **69** 18–27.

[3] Wang J, Basheer PM, Nanukuttan SV, Long AE and Bai Y 2016 Influence of service loading and the resulting micro-cracks on chloride resistance of concrete *Constr. Build. Mater.* **108** 56–66

[4] Hoult N, Fidler P, Hill P and Middleton C 2010 Long-term wireless structural health monitoring of the Ferriby road bridge *J. Bridge Eng.* **15** 2 153–9.

[5] Brownjohn J M W 2007 Structural health monitoring of civil infrastructure *Philos. Trans. R. Soc. Lond. A* **365** 1851 589–622.

[6] Abé M and Fujino Y (2017) Monitoring of long-span bridges in Japan. *P. I. Civil Eng.–Civil Eng.* **170** 3 135–44.

[7] Ranade R, Zhang J, Lynch J and Li V 2014 Influence of micro-cracking on the composite resistivity of Engineered Cementitious Composites *Cem. Concr. Res.* **58** 1–12.

[8] Li V C 2008 Engineered cementitious composites (ECC)–material, structural, and durability performance *Concrete Construction Engineering Handbook* ed E G Nawy (Michigan: CRC Press) chapter 24 pp 1–48.

[9] Suryanto B, Buckman J O, Thompson P, Bolbol M and McCarter W J 2016 Monitoring micro-crack healing in an engineered cementitious composite using the environmental scanning electron microscope *Mater. Charact.* **119** 175–85.

[10] Yıldırım G, Öztürk O, Al-Dahawi A, Ulu A A and Sahmaran M S 2020 Self-sensing capability of Engineered Cementitious Composites: Effects of aging and loading conditions *Constr. Build. Mater.* **231** 117132.

[11] Hou T C and Lynch J P 2005 Conductivity-based Strain Monitoring and Damage Characterization of Fiber Reinforced Cementitious Structural Components *In: Proc. of SPIE 12th Annual International Symposium on Smart Structures and Materials* pp 6-10.

[12] Lin V, Li M, Lynch J and Li V 2011 Mechanical and Electrical Characterisation of Self-sensing Carbon Black ECC *Proc. of SPIE* **7983** pp 1-12.

[13] Suryanto B, McCarter W J, Starrs G, Wilson S A and Traynor R M 2015 Smart cement composites for durable and intelligent infrastructure *Procedia Eng.* **125** 796–803.

[14] Huang Y, Li H and Qian S 2018 Self-sensing properties of engineered cementitious composites *Constr. Build. Mater.* **174** 253–62.

[15] Deng H and Li H 2018 Assessment of self-sensing capability of Carbon Black Engineered Cementitious Composites *Constr. Build. Mater.* **173** 1–9.

[16] Ncorr v1.2: Open source 2D-DIC Matlab software http://www.ncorr.com/

[17] Suryanto B, Tambusay A and Suprobo P 2017 Crack mapping on shear-critical reinforced concrete beams using an open source digital image correlation software *Civ. Eng. Dimens.* **19** 2 93–8

[18] Tambusay A, Suryanto B and Suprobo P 2020 Digital image correlation for cement-based materials and structural concrete testing *Civ. Eng. Dimens.* **22** 1 6–12.

[19] Saraireh D, Walls S, Suryanto B, Starrs G and McCarter W J 2017 The influence of multiple micro-cracking on the electrical impedance of an engineered cementitious composite *Proc. Int. Conf. on Strain-Hardening Cement-Based Composites* (Dordrecht: Springer) pp 292–9.

[20] McCarter W and Brousseau R 1990 The A.C. response of hardened cement paste *Cem. Concr. Res.* **20** 6 891–900.

[21] Suryanto B, McCarter W, Starrs G and Ludford-Jones G 2016 Electrochemical immittance spectroscopy applied to a hybrid PVA/steel fiber engineered cementitious composite *Mater. Des.* **105** 179–89.

[22] Whittington H W, McCarter J and Forde M C 1981 The conduction of electricity through concrete *Mag. Concr. Res.* 33 114 48–60.

[23] McCarter W J, Taha H M, Suryanto B and Starrs G 2015. Two-point concrete resistivity measurements: interfacial phenomena at the electrode–concrete contact zone. *Meas. Sci. Technol.* **26** 8 085007.

Nonlinear finite element analysis of reinforced concrete beam-column joints under reversed cyclic loading

A Tambusay[1,2], B Suryanto[1]* and P Suprobo[2]

[1] Institute for Infrastructure and Environment, School of Energy, Geoscience, Infrastructure and Society, Heriot-Watt University, Edinburgh, United Kingdom.
[2] Department of Civil Engineering, Faculty of Civil, Planning and Geo Engineering, Sepuluh Nopember Institute of Technology, Surabaya, East Java, Indonesia.

*Corresponding author's e-mail: B.Suryanto@hw.ac.uk

Abstract. This paper presents the application of three-dimensional nonlinear finite element analysis in studying the complex behaviour of reinforced concrete beam-column joints under reversed cyclic loading. Three different joint types were considered: an interior, an exterior, and a corner joint, each representing typical joint configurations in general moment resisting frame structures. These joints' test data were taken from the experimental investigations carried out by Shiohara and Kusuhara about 15 years ago and used here to showcase the practical application and value of an advanced numerical modelling. It is shown that the numerical models are able to accurately capture the full cyclic hysteretic response, progressive strength and stiffness degradation, cracking and damage evolution under load cycles, and failure mode. It is also shown that the numerical models provide a useful tool to characterise failure processes and behavioural mechanisms.

1. Introduction

Beam-column joint is widely considered as a critical element in reinforced concrete moment-resisting frame structures [1]. When subjected to earthquake-induced lateral forces, premature failure can occur in a poorly detailed joint element [1-5] and this failure can be very brittle as in concrete beams with little or no shear reinforcement [6-9]. Joint shear failure can also occur in overly reinforced concrete beams as flexural yielding cannot be achieved due to high ratio of longitudinal steel bars [10].

The key to ensuring the ductility and inelastic deformability of a beam-column joint depends primarily on the ability of the joint to sustain high shear stresses, particularly when flexural cracks have developed within the plastic hinge regions of the adjoining beams. In any case, it must be ensured that the columns remain essentially elastic throughout the load history to ensure global structural stability. This behavioural response is generally aimed for in the design as it will allow for inelastic large deformation to occur in the beams while preventing premature joint failure. A considerable amount of work [11-13] has been devoted since early 1960s to study the behaviour of beam-column joints under load reversals departing from the above concept and a number design recommendations have since been developed to ensure adequate connection behaviour in frame structures.

In the current building codes for beam-column joints [14, 15], a strength-based approach is generally adopted with joint shear strength typically being determined to meet the factored force demands imposed by adjoining members (i.e. beams and columns). By doing so, a joint can be expected to behave satisfactorily during load (shear) reversals. Provisions in a strong column-weak beam, anchorage length

Content from this work may be used under the terms of the Creative Commons Attribution 3.0 licence. Any further distribution of this work must maintain attribution to the author(s) and the title of the work, journal citation and DOI.
Published under licence by IOP Publishing Ltd

of longitudinal bars, and adequate shear reinforcements are also given to ensure the integrity of a structure in the post-yielding region. Although the ultimate stage limits are provided to protect against loss of lives and structural collapse, the formation of wide diagonal cracks in the joint can still occur [16].

Apart from the design point of view, the mechanics of a beam-column joint under load reversals have also presented a major challenge to finite element modellers, particularly those implementing the smeared approach as stresses could be highly localised. The difficulty is partly due to the highly complex behaviours and involvement of many interrelated factors, including the flexural actions from both beams and columns, joint shear response, bond-slip effects, and joint concrete confinement. To investigate these aspects, Shiohara and Kusuhara [17, 18] performed extensive experimental investigations on reinforced concrete beam-column joints, with the aim at providing reliable test data for model verifications. In their study, three loading types were considered to simulate typical boundary conditions in a structure i.e. (i) an interior, (ii) an exterior, and (iii) a corner beam-column joint. The results of their experiments have since been used as the subject for further studies and generally treated as a benchmark data for verifying the accuracy of nonlinear finite element procedures.

In this paper, the practical value and application of three-dimensional nonlinear finite element analysis is demonstrated through accurate simulations of the cyclic hysteretic responses of beam-column joints along with crack patterns. The beam-column joints tested by Shiohara and Kusuhara [17] were selected as a benchmark to testify the accuracy of the finite element analyses. A combination of plasticity and fracture model in conjunction with a smeared fixed crack approach and crack band model was adopted to this end.

2. Details of Shiohara and Kusuhara Beam-Column Joints

In 2006, Shiohara and Kusuhara [17] undertook a detailed experimental programme on six half-scale beam-column joints (hereinafter referred to as the SK beam-column joints). The primary objective of the test programme was to provide benchmark test data for the validation of their in-house mathematical models. Due to the high quality and comprehensive documentation of test results, their test data have been referred to by many researchers and used to support the corroboration in many software developments [19, 20].

In this paper, only one series (series A) of the three series of SK beam-column joints was analysed. In this series, there were three specimens (labelled A1, A2 and A3), each of which was tested under different loading patterns to cover possible types of beam-column joint in moment-resisting frame buildings. All specimens in this series are of critical form and were designed as per AIJ guidelines [21]. Loading type I was intended to simulate an interior joint and this was applied to specimen A1 to address significant joint core distress. Loading types II and III were intended to simulate an exterior and corner joint, respectively, and applied to specimens A2 and A3 (these are the specimens that sustained the least joint core distress). The joint shear capacity of specimen A1 was designed to be 10% higher than the joint shear demand. The strong column-weak beam concept was considered, with an overstrength factor of 1.25 to allow the beams to achieve their full flexural capacities prior to the columns.

The typical schematic representations of all beam-column joints geometry and reinforcement layout are displayed in figure 1(a), together with the schematic of the test setup in figures 1(b)-(d). All beam-column joints had a similar square section of 300 by 300 mm and were reinforced with steel bars of identical arrangement and properties (see table 1). The concrete used to cast these specimens had the mean compressive and tensile splitting strengths of 28.3 MPa and 2.67 MPa, respectively. The specimen details are summarised in table 2.

Table 1. Material properties of reinforcing bars in series A specimen [17].

Diameter (mm)	Grade	Young's Modulus (GPa)	Yield Strength (MPa)	Ultimate Strength (MPa)
13^{a}	SD390	176	456	582
13^{b}	SD390	176	357	493
6	SD295	151	326	488

Notes: a: steel bar used for beams; b: steel bar used for columns

Table 2. Details of SK beam-column joint specimens [17].

Description		A1	A2	A3
Model		Interior	Exterior	Corner
Loading type		I	II	II
Compressive strength of concrete		28.3 MPa		
Beams	Cross-section	300 × 300 mm		
	Span	2700 mm		
	Longitudinal bar	8D13 (top) and 8D13 (bottom)		
	Transverse bar	D6 at 50 mm		
Columns	Cross-section	300 × 300 mm		
	Height	1470 mm		
	Longitudinal bar	16D13		
	Transverse bar	D6 at 50 mm		
Joint	Transverse bar	D6 at 50 mm (3 NoS)		

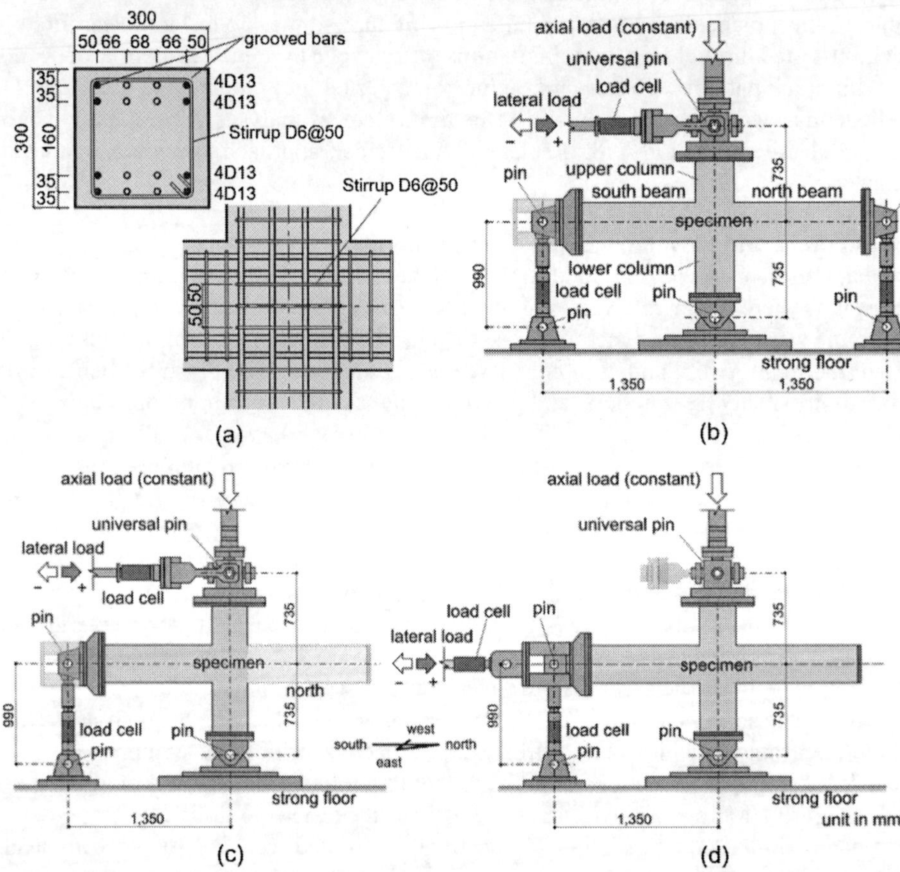

Figure 1. Schematic of series A of SK beam-column joints: (a) cross-section and bar arrangement; (b) loading type I; (c) loading type II; and (d) loading type III [17].

3. Finite Element Model

Three-dimensional nonlinear finite element analyses were performed using a specialist finite element software package ATENA Science developed exclusively by Červenka Consulting [22] for simulations of reinforced concrete structures [23, 24]. In this study, the accuracy of a smeared fixed crack approach to model the highly nonlinear cracked concrete behaviours experiencing bi-directional cracking [25, 26] resulting from reversed cyclic loads is tested. Figures 2(a) and (b) display the typical finite element meshes and bar arrangements used to represent SK beam-column joints, which was prepared in a pre-processor finite element software GiD. The concrete was modelled using 8-node hexahedral (brick) linear elements with a typical size of 25 mm (in columns, beams and joint region), thereby giving 12 elements across the overall depth or width. All plates were modelled using tetrahedral linear elements with larger unstructured mesh size as a means to expedite the runtime of analysis. Although different mesh sizes were used for the concrete and steel (end) plates, this would not affect the accuracy as there is a full compatibility between two mesh surfaces.

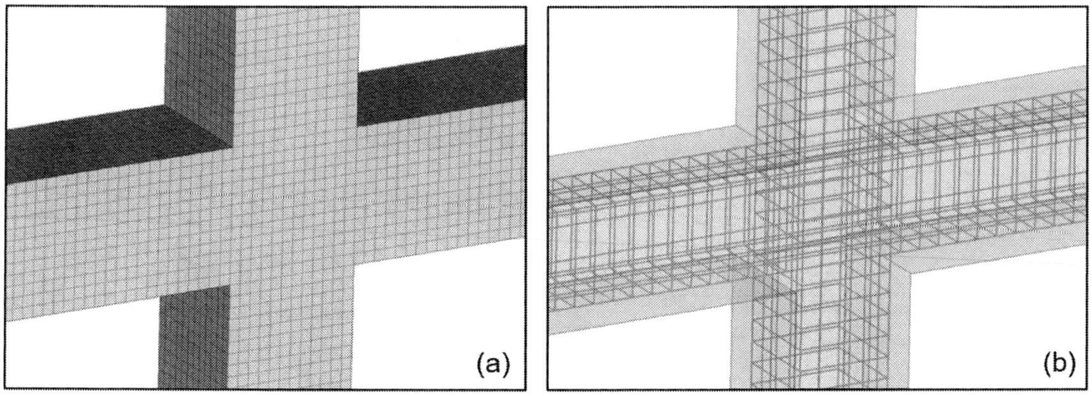

Figure 2. (a) Finite element mesh and (b) bar arrangement in ATENA Science.

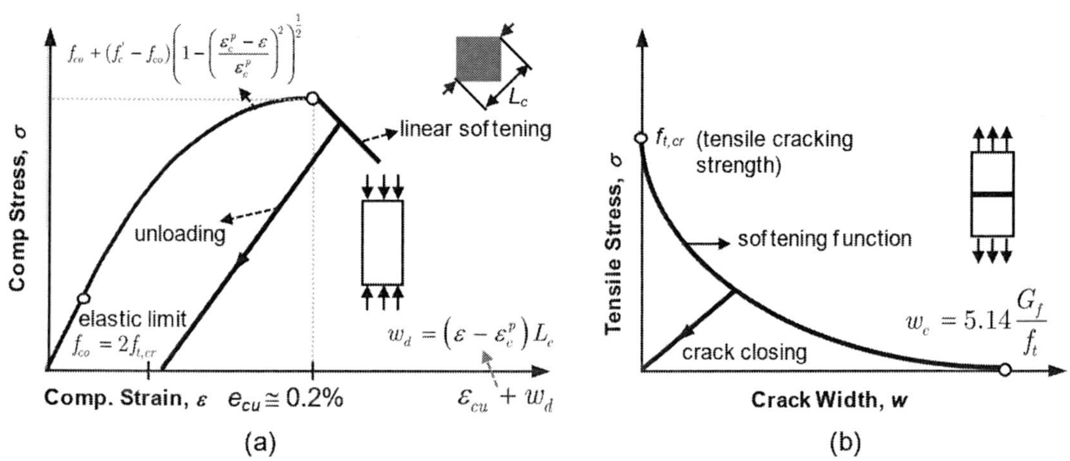

Figure 3. Concrete constitutive model: (a) compression and (b) tension.

The nonlinear concrete used in this study was the "Cementitious2" model which was formulated based on the CEB-FIP Model Code 1990 [27]. In this model, the response in compression is treated following the theory of plasticity, whereas the response in tension is formulated following the Rankine fracturing model for concrete cracking [22]. In the shear model, a constant shear factor coefficient (S_F), which defines a relationship between normal and shear (both modes II and III) crack stiffnesses, was used. Figures 3(a) and (b) show a summary of the constitutive laws adopted in this study. To model concrete behaviour under cyclic loading, the unloading factor parameter was activated to control the crack closure stiffness. In ATENA, this parameter can be set between 0 and 1, with 0 for unloading to the origin (default value for backward compatibility) and 1 for unloading parallel to the initial elastic stiffness. In this work, this factor was set to 0.2 and found to simulate residual displacement during unloading reasonably. Apart from this, the plastic flow was modified to a value of 0.5 to account for dilatancy resulting from the concrete volumetric expansion when undergoing compression failure.

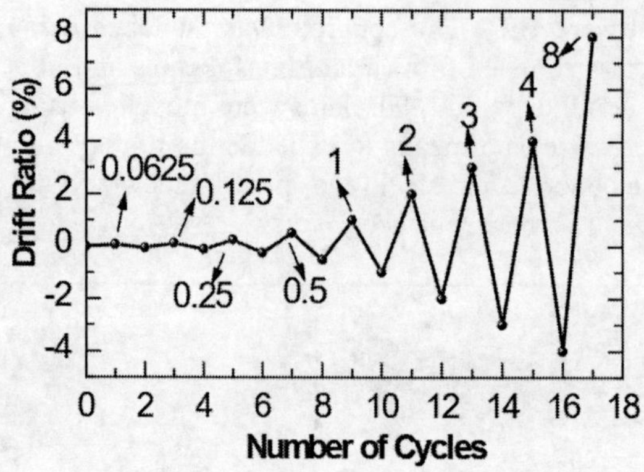

Figure 4. Loading history for reversed cyclic loading

Table 3. Summary of material parameters and finite element input parameters in ATENA.

No	Parameter	Value/Reference
Concrete constitutive model		
1.	Elastic modulus	CEB-FIP Model Code [27]
2.	Tensile strength	CEB-FIP Model Code [27]
3.	Smeared crack model	1 (fixed crack)
4.	Aggregate size	20 mm
5.	Unloading factor for cyclic loading	0.2
6.	Critical compressive displacement	0.5 mm
7.	Limit of comp. strength reduction due to cracking	0.8
8.	Eccentricity (defining the shape of failure surface)	0.52
9.	Plastic flow (defining dilatancy of plastic factor)	0.5
Reinforcement bar model		
10.	Stress-strain relationship	Bilinear with strain hardening
11.	Bond-slip model for cyclic (bar with memory bond)	CEB-FIP Model Code [27]
12.	Cyclic behaviour (Menegotto-Pinto)	R = 10; C1 = 0.925; and C2 = 0.15 [28]
Loading procedure and solution parameter		
13.	Loading procedure for axial load	Static (force-controlled)
14.	Loading procedure for cyclic load reversal	Quasi-static (displacement-controlled)
15.	Iteration method for cyclic	Modified Newton-Raphson
16.	Stiffness type	Elastic predictor with conditional break criteria
17.	Iteration limit	300
18.	Solver	Pardiso

All embedded steel bars were modelled in discrete representation using one-dimensional 2-noded linear truss elements. Elasto-plastic Manegotto-Pinto model [28] was used to accurately capture the cyclic behaviour of steel bars as it takes into account the Bauschinger's effect during unloading and reloading sequences. Bond-slip with memory bond was also considered following the nonlinear bond-slip formulation in the CEB-FIP Model Code 1990 [27].

In all models, the lower column was supported by a pin-joint. In the first loading interval, each beam-column joint was loaded with a constant axial load of 216 kN at the top of the upper column. In the second and subsequent intervals, cyclic lateral loads were then applied in the form of displacement increments, with a rate of 1.0 mm per step until reaching the drift ratio of 8% (see figure 4). For loading type I (specimen A1), the interior beam-column joint was supported at both sides of beam ends (with a pin and a roller support respectively) and the cyclic lateral loads were then applied at the top of the upper column. For loading type II (specimen A2 or an exterior beam-column joint), the lateral loads were applied in the same manner to that applied in loading type I – however, only one side of the beams was supported by a roller, accordingly, no internal stresses would develop in the other (dummy) beam. For loading type III (specimen A3 or a corner beam-column joint), the analysis was done by applying cyclic lateral loads at one of the beam ends. The support conditions were similar to that of loading type II.

In this study, the modified Newton-Raphson iterative solution parameter with elastic predictor was used to allow for consistent convergence at each of load steps. Input parameters in conditional break criteria were set higher than that of default values, typically ten times higher to provide more space when convergence difficulties are encountered. The convergence tolerance was set constant throughout intervals with a value of 1.0% for displacement, residual, and absolute residual error, whilst energy error was set at 0.1%. High iteration limits (e.g. 200-300) were found to be beneficial. The key input parameters used in the analyses are summarised in table 3.

4. Results and Discussion

4.1. Hysteretic response

Figures 5(a)-(c) display the load-drift responses for specimens A1 to A3, with a summary of the measured and predicted load capacities presented in table 4. As shown in the figures, the overall behaviours of the beam-column joints can be predicted reasonably well. The finite element models successfully exhibit comparable hysteretic shapes, particularly for specimens A2 and A3 which display similar response as observed in the experiment.

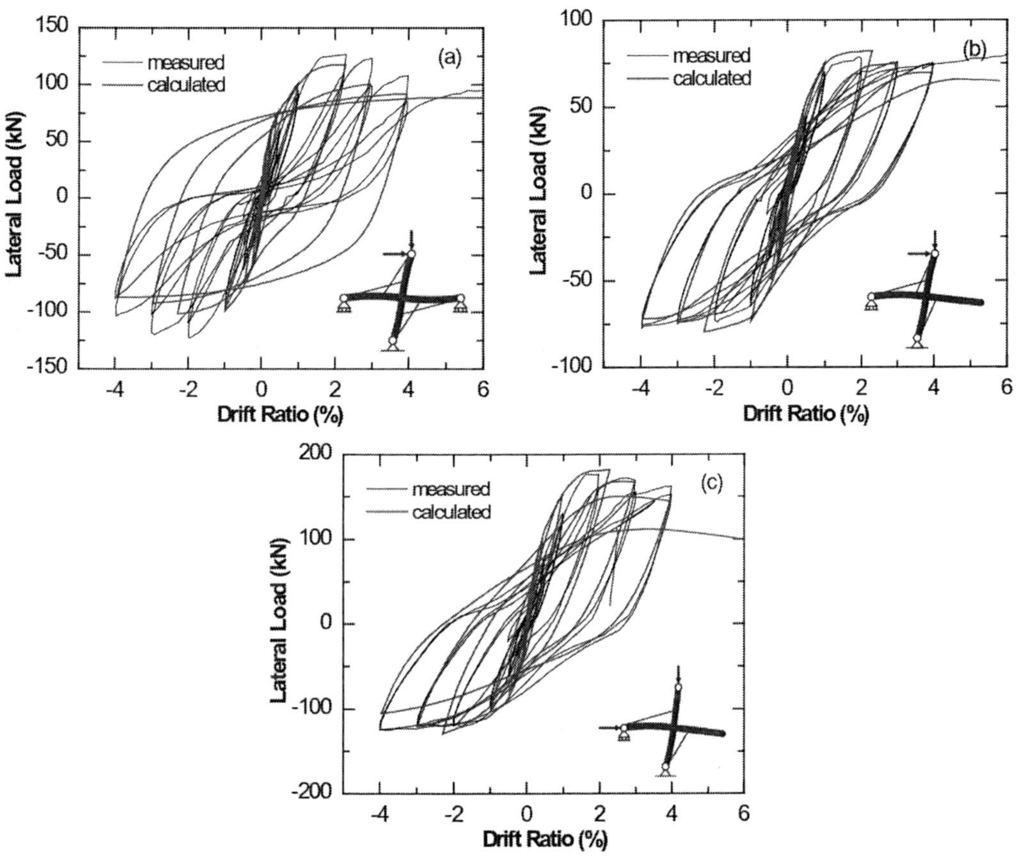

Figure 5. Loading history under reversed cyclic loading. Test data were taken from [17].

With regard to the response of interior beam-column joint (specimen A1), it is apparent from figure 5(a) that at low drift levels, the predicted hysteretic response displays accurate representation of strength and stiffness degradation. As the drift is increased further, however, the analysis starts to slightly underestimate the load capacity (particularly from the 5th cycle), but the agreement is still reasonable. Of interest to note is the overly high total energy dissipation in the analysis due to the more rounded shape of hysteretic loops during load cycles (compared to the highly pinched response). This could also be attributed to high complexity of joint distress incurred by the specimen, thereby making it difficult to predict with accuracy. As shown in table 4, the ratios of experimental-to-predicted load capacity for specimen A1 in the positive and negative loading direction are 1.08 and 1.16 respectively.

Table 4. Summary of experimental [17] and predicted load capacities.

Joint Specimen	Positive Loading Direction			Negative Loading Direction		
	$P_{u\text{-}Test}$ (kN)	$P_{u\text{-}Calc}$ (kN)	$P_{u\text{-}Test}/P_{u\text{-}Calc}$ (-)	$P_{u\text{-}Test}$ (kN)	$P_{u\text{-}Calc}$	$P_{u\text{-}Test}/P_{u\text{-}Calc}$ (-)
A1	126.6	117	1.08	-122.8	-106	1.16
A2	77.9	82	0.95	-77.1	-79	0.98
A3	176.4	182	0.97	-124.5	-129	0.97
		Mean	1.00		Mean	1.03
		CoV (%)	5.82		CoV (%)	9.46

Figures 5(b) and (c) compare the measured and predicted hysteretic responses for specimens A2 and A3 respectively. An excellent agreement is observed in terms of initial stiffness, load capacity, and the overall shape of the hysteretic loops. The strength and stiffness degradation can also be predicted with accuracy, although there is a slight variation of strength degradation during the 8th cycle which corresponds to the drift ratio of 4%. It is noteworthy that the models can better replicate the hysteretic shapes for specimens with less joint core distress. The ratios of experimental-to-predicted load capacity for specimens A2 and A3 in the positive and negative loading direction are 0.95/0.98 (A2) and 0.97/0.97 (A3) respectively which is indicative of marginal differences (consistently less than 5%), highlighting the excellent accuracy of finite element predictions.

4.2. Comparison of crack patterns

To facilitate further evidence of the accuracy of finite element models used in this study, figures 6-8 compare the observed and predicted crack patterns at selected drift ratio levels of 0.5%, 2%, and 4%. It is evident that, in general, the overall predicted crack patterns are in excellent agreement with the observed crack patterns in terms of the extent of accuracy of crack-alike development, location of crack, and mode of failure. The successful representation of crack angle (also principal strain profile) in all specimens under increasing loads is also appealing, highlighting once again successful appreciation of the highly nonlinear behaviours of the concrete by the finite element models.

With reference to the cracking behaviour of specimen A1, it is apparent that under load reversals, flexural cracks initially develop at the beam corners at the joint and then propagate diagonally within the joint core which is indicative of shear cracks. These existing shear cracks, during the increasing load levels, progress significantly within the joint, replicating more to a typical concrete strut action. The reason for this relates to the low overstrength factor of joint shear capacity (10%) as reported by Shiohara and Kusuhara [17]. At a drift ratio of 4%, shear cracks start to develop within the plastic hinge regions of both lower and upper columns. This is attributed to the marginal value of the overstrength factor in the strong column-weak beam design (i.e. 1.25). Therefore, it is not surprising that joint core exhibits heavy shear distress followed by diagonal concrete crushing, while the columns also experience significant damage. Furthermore, the analysis clearly shows that the longitudinal bars in the beams have reached their yield capacities, as evidenced by considerable flexural cracks formation along the beam length.

With respect to specimens A2 and A3, the failure modes are predicted to be flexure-shear in nature which is consistent with experimental observation. Significant flexural cracks manifest vastly in one side of the beam, following the formation of shear cracks in the joint core. In this specimen, damage is dominated primarily by flexural cracks with no existence of concrete crushing. It is interesting to note that in specimen A2 (exterior beam-column joint), concrete cracking is shown to develop primarily in the beam and within the joint core, whereas in specimen A3 (a corner or knee joint), concrete cracking is shown to also manifest in the lower column at locations within the plastic hinge region.

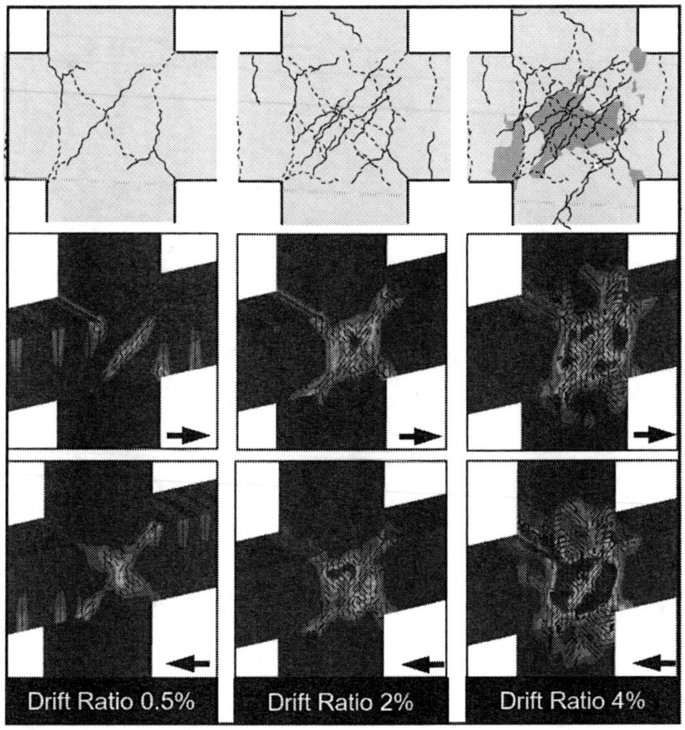

Figure 6. Computed and observed crack patterns and maximum principal strains of specimen A1.

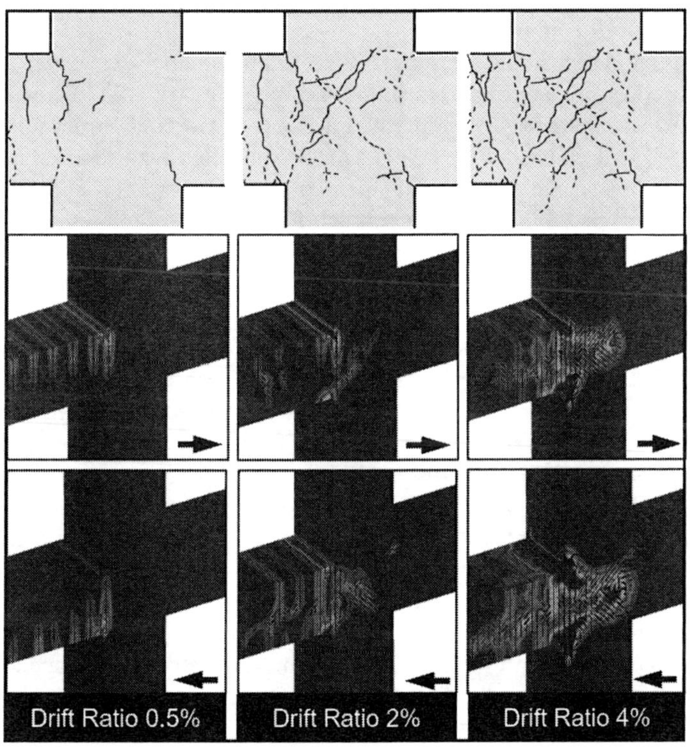

Figure 7. Computed and observed crack patterns and maximum principal strains of specimen A2.

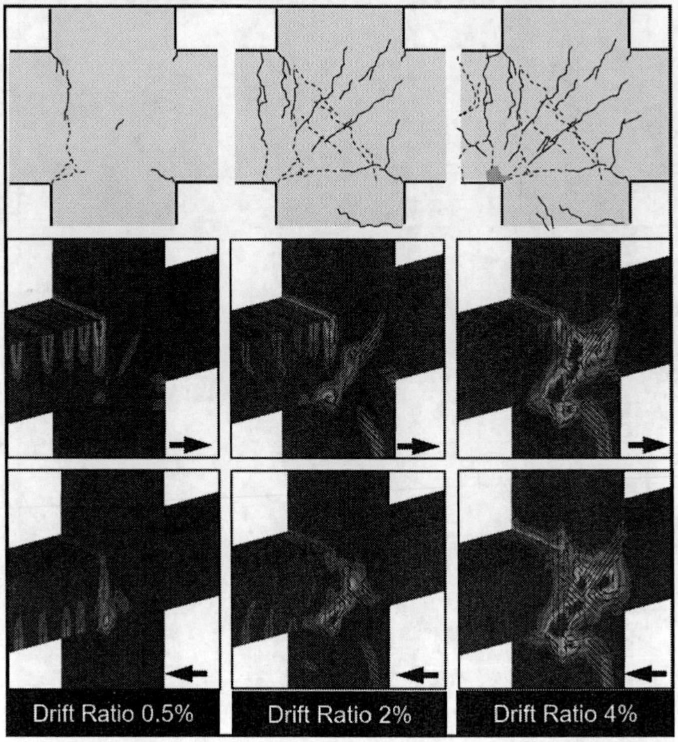

Figure 8. Computed and observed crack patterns and maximum principal strains of specimen A3.

5. Concluding remarks and conclusions

The application of three-dimensional nonlinear finite element analyses using ATENA Science is presented, with the models prepared using a pre-processor software GiD. Three beam-column joints tested by Shiohara and Kusuhara, representing the interior, exterior and corner joints, were thoroughly modelled and simulated under quasi-static displacement-controlled load reversals. The fracture-plastic model employing the classical path-dependent smeared crack formulation and crack band model was adopted to showcase the applicability of this approach for complex simulations of reinforced concrete under cyclic loading. This work also aims to respond to the need for establishing an appropriate modelling strategy under cyclic loading which still presents challenges to finite element modellers.

Based on the analytical work presented, some conclusions are drawn:

1. The use of ATENA Science is shown to provide good capabilities to simulate multifaceted nonlinear behaviour of beam-column joints under cyclic lateral loads. The nonlinear material models implemented in the software is shown to produce hysteretic loops and crack profiles which are in excellent agreement with experimental observations.
2. It is shown that the load capacity of the joints can be well predicted, with the predictions slightly underestimating the load capacity of the interior joint and slightly overestimating the load capacities of the exterior and corner joints. The computed crack patterns display good representation of the test data, with similar crack development demonstrated across the members. Accurate predictions of failure modes were also obtained.
3. The extent of damage in the analysis is shown to compare well with the actual damage in the test specimens. This is particularly true for the interior joint which is generally more critical and subjected to large shear stresses. The results demonstrate the suitability of reliable simulations to tackle challenges in predicting damage level under day-to-day design.

The analysis tool discussed in this paper can provide practising engineers and researchers with the ability to assess detailed behavioural response of structural elements subjected to unforeseen loading conditions such as earthquake. It could also be a useful tool in the assessment of structural elements with unusual detailing, particularly if constructed in seismic prone area.

Acknowledgements

This work was supported by an Institutional Links grant, ID 414707757, under the Newton Fund Institutional Links Grant and the Ministry of Research, Technology and Higher Education of the Republic of Indonesia partnership. The grant is funded by the UK Department for Business, Energy and Industrial Strategy and the Indonesian Ministry of Research, Technology and Higher Education (Grant No. 3/AMD/EI/KP.PTNBH/2020) and delivered by the British Council. The authors would also express their gratitude to Dr. Jan Červenka for his valuable discussions.

References

[1] Sharma A, Eligehausen R and Reddy G R 2011 A new model to simulate joint shear behavior of poorly detailed beam-column connections in RC structures under seismic loads–part 1: exterior joints *Eng. Struc.* **33** 3 1034–51
[2] Li B, Wu Y and Pan T C 2002 Seismic behavior of nonseismically detailed interior beam-wide column joints–part I: experimental results and observed behavior *ACI Struc. J.* **99** 2 791–802
[3] Li B, Wu Y and Pan T C 2003 Seismic behavior of nonseismically detailed interior beam-wide column joints–part II: theoretical comparisons and analytical studies *ACI Struc. J.* **100** 1 56–65
[4] De Risi M T, Ricci P, Verderame G M and Manfredi G 2016 Experimental assessment of unreinforced exterior beam-column joints with deformed bars *Eng. Struc.* **112** 215–32
[5] Hwang S-J, Lee H-J, Liao T-F, Wang K-C and Tsai H-H 2005 Role of hoops on shear strength of reinforced concrete beam-column joints *ACI Struc. J.* **102** 3 445–53
[6] Collins M P, Bentz E C and Sherwood E G 2008 Where is shear reinforcement required? Review of research results and design procedures *ACI Struc J.* **105** 5 590–600
[7] Suryanto B, Morgan R and Han A L 2016 Predicting the response of shear-critical reinforced concrete beams using response-2000 and SNI 2847:2013 *Civ. Eng. Dimens.* **18** 1 16–24
[8] Suryanto B, Tambusay A and Suprobo P 2017 Crack mapping on shear-critical reinforced concrete beams using an open source digital image correlation software *Civ. Eng. Dimens.* **19** 2 93–8
[9] Tambusay A, Suryanto B and Suprobo P 2018 Visualization of shear cracks in a reinforced concrete beam using the digital image correlation *Intr. J. Adv. Sci. Eng. Inf. Tech.* **8** 1 573–8
[10] Ghobarah A and Said A 2002 Shear strengthening of beam-column joints *Eng. Struc.* **24** 881–8
[11] Hanson N W and Connor H W 1967 Seismic resistance of reinforced concrete beam-column joints *J. Struc. Div.* **93** 5 533–60
[12] Hanson N W 1971 Seismic resistance of concrete frames with grade 60 reinforcement *J. Struc. Div.* **97** 6 1685–700
[13] Megget L M and Park R 1971 Reinforced concrete exterior beam-column joints under seismic loading *New Zaeland Eng.* **26** 11 341–53
[14] AIJ 2016 Standard for lateral load-carrying capacity calculation of reinforced concrete structures *Architectural Institute of Japan* (Tokyo: Japan)
[15] ACI 318-19 2019 Building code requirements for structural concrete and commentary *American Concrete Institute* (Michigan: United States of America)
[16] Parra-Montesinos G J, Peterfreund S W and Chao S H 2005 Highly damage-tolerant beam-column joints through use of high-performance fiber-reinforced cement composites *ACI Struc. J.* **102** 3 487–95
[17] Shiohara H and Kusuhara F 2006 Benchmark test for validation of mathematical models for non-linear and cyclic behavior of R/C beam-column joints *The University of Tokyo* (Tokyo: Japan).
[18] Kusuhara F and Shiohara H 2008 Test of R/C beam-column joint with variant boundary conditions and irregular details on anchorage of beam bars *Int. Proc. 14th World Conf. Earth. Eng.* (Beijing: China)

[19] Guner S and Vecchio F J 2010 Analysis of shear-critical reinforced concrete plane frame elements under cyclic loading *J. Struc. Eng.* **137** 8 834–43
[20] Pan Z, Guner S and Vecchio F J 2017 Modeling of interior beam-column joints for nonlinear analysis of reinforced concrete frames *Eng. Struc.* **142** 182–91
[21] AIJ 1999 Design guidelines for earthquake resistant reinforced concrete building based on inelastic displacement concept *Architectural Institute of Japan* (Tokyo: Japan)
[22] Červenka V, Jendele L and Červenka J 2018 Program documentation ATENA theory Czech Republic, Červenka Consulting
[23] Tambusay A and Suprobo P 2019 Predicting the flexural response of a reinforced concrete beam using the fracture-plastic model *J. Civ. Eng.* **34** 2 61–7
[24] Červenka J, Červenka V and Laserna S 2018 On crack band model in finite element analysis of concrete fracture in engineering practice *Eng. Fract. Mech.***197** 27–47
[25] Suryanto B, Nagai K and Maekawa S 2010 Modeling and analysis of shear-critical ECC members with anisotropy stress and strain fields *J. Adv. Conc. Techn.* **8** 2 239–58
[26] Suryanto B, Nagai K and Maekawa S 2010 Smeared-crack modeling of R/ECC membranes incorporating an explicit shear transfer model *J. Adv. Conc. Techn.* **8** 3 315–26
[27] CEB-FIP Model Code 1990 Comité Euro-International du Béton *Inf. Bullet.* **195**
[28] Menegotto M and Pinto P E 1973 Method of analysis of cyclically loaded RC plane frames including changes in geometry and non-elastic behavior of elements under combined normal force and bending *IABSE Symp.* (Lisbon: Portugal) pp 15–22

Experimental investigations on the durability performance of normal concrete and engineered cementitious composite

I Komara[1], P Suprobo[1*], D Iranata[1], A Tambusay[1], and W Sutrisno[1]

[1] Department of Civil Engineering, Faculty of Civil, Planning and Geo Engineering, Sepuluh Nopember Institute of Technology, Surabaya, East Java, Indonesia.

*Corresponding author's e-mail: priyo@ce.its.ac.id

Abstract. This paper presents the experimental investigations in studying the durability performance of normal concrete and engineered cementitious composite (ECC) under varying key influencing factors. To this end, direct measurements on sorptivity, water absorption and rapid chloride penetration were undertaken for both normal concrete and ECC; each of which was tested using 67-mm thick sliced cylinder specimens with a diameter of 100 mm. The results from this studies highlight that the inherent low permeability properties in ECC are shown to provide excellent durability performance, in particular to the improvement of sorptivity rate and water absorption which is twice higher than that of normal concrete, indicating the very dense capillary pores microstructure. It is also found that chloride concentration on the surface of ECC is twice lesser than normal concrete, owing to the fact marginal manifestation of the electrical conductance.

1. Introduction

Concrete cracking is known to have a knock-on effect on the durability performance and long-term service life of reinforced concrete structures [1, 2]. This phenomenon is primarily attributed to complex loading conditions that may arise from external loading under day-to-day service in addition to environmental exposures, creating even more complex physical and chemical changes in concrete properties [3-5]. This will cause not only localised distress and stiffness degradation of structural members but also poor resistance of penetration from aggressive agents (e.g. alkali-silica reaction and chloride transport) which will result in further significant deterioration [6, 7].

Concrete cracking due to chemical reactions (i.e. chloride) is regarded as one of the most severe and aggressive deteriorations in marine infrastructure, particularly to concrete members subjected to wetting and drying cycles which are normally located in the splash zone [8-10]. Serious concrete deteriorations have been reported to date and it was found that the typical cause was mainly due to cracking, allowing for chloride to entrance to the level of steel bars easily. If the amount of chloride ingress is beyond the critical threshold concentration, pitting corrosion may be induced and propagates immensely, thereby causing spalling of concrete [11-14]. Aside from cracking, the durability of concrete also relies primarily on the ability of water to penetrate the microstructure of uncracked concrete. Water acts as the medium for corrosive agents (e.g. chloride ingress) and is mainly occurred through capillary absorption (sorptivity) and diffusion [15-17]. If the uncracked concrete has high porosity as in case of normal concrete, ingress of chloride may initiate the corrosion process through depassivation of steel bars and continue to accelerate the corrosion product in steel bars or also known as rust.

Content from this work may be used under the terms of the Creative Commons Attribution 3.0 licence. Any further distribution of this work must maintain attribution to the author(s) and the title of the work, journal citation and DOI.
Published under licence by IOP Publishing Ltd

To avoid the premature deterioration, it is generally required to use a type of durable material in a sense that it is more relevant to the resistance of cracking in addition to being naturally less porous whilst in the uncracked state. With regard to this, there is a growing interest in the utilisation of cement-based composite as a means to offer durability improvement, such as enhanced pore microstructure, toughness and crack control. Engineered cementitious composite (ECC) is a special type of cementitious composite which is micro mechanically designed to exhibit high tensile strain capacity (in the order of a few percent) [17-20] and tight crack width control generally less than 0.1 mm [21, 22] (see figure 1). This material is regarded as having potential inherent durability and self-healing performance which thereby can be used to alleviate premature deterioration of marine infrastructure.

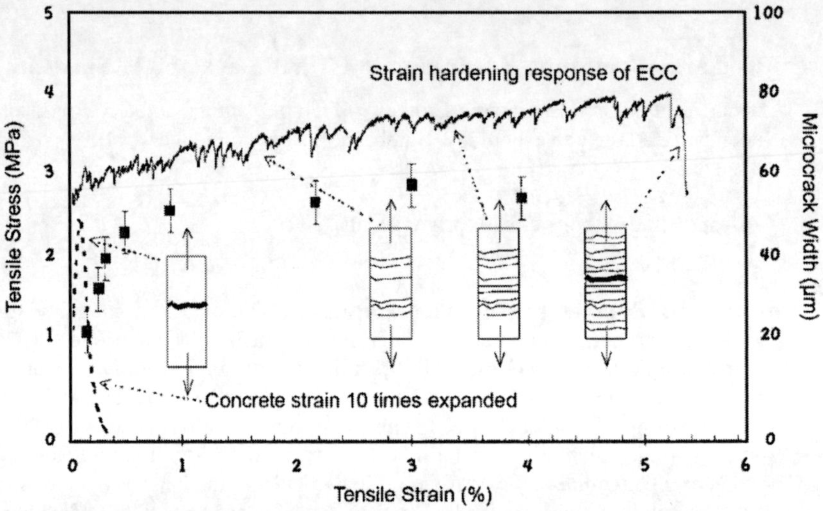

Figure 1. Uniaxial tensile stress-strain relationship and crack widths of ECC [20].

To confirm the suitability of ECC used in an aggressive environment, however, particularly under dry or partially saturated condition, research in the laboratory is necessary to perform. In this work, pilot study of experimental investigations on direct measurement of sorptivity, water absorption, and rapid chloride penetration of (uncracked) ECC specimens was undertaken; each of which represents the durability factor. Normal concrete specimens were also fabricated and measured in the same manner of ECCs to facilitate direct comparison on the extent of durability performance.

2. Experimental programme

2.1. Mixture proportion and fabrication

The mix proportions used to produce specimens for normal concrete and ECC are summarized in table 1 together with the results of 28-day mean compressive strengths. In normal concrete, the coarse aggregate from graded crushed granite was used and it had the maximum size of 10 mm. A well-graded fine aggregate with a size of less than 3 mm was also added into the mix. The binder for normal concrete comprised only of Portland cement referred to as CEMI–42.5R [23] with the water-to-cement ratio of 0.30 [24-26]. To ensure workability during mixing, a high-range water reducer (HRWR) superplasticiser was used with a dosage of 1.05 percent (by mass of cement). In ECC, a high volume of fine fly ash was added as a binder at a fly ash/cement ratio of 1.6/1 with the water-to-binder ratio of 0.28. The typical oxide analysis of Portland cement and fly ash are presented in tables 2 and 3 respectively. Unlike normal concrete, the sand used for ECC was silica sand in the form of powder with an average grain size of particles less than 0.1 mm. The fibre used for ECC was Kuralon RECS-15 polyvinyl alcohol (PVA) fibres, which were added into the matrix at a dosage of 2% by volume. The properties of PVA fibres are summarised in table 4.

Table 1. Mix proportions.

	CEM I (kg/m³)	10 mm (kg/m³)	Fine (<3mm) (kg/m³)	Silica Sand (kg/m³)	Fly ash (kg/m³)	Water (kg/m³)	HRWR (kg/m³)	PVA (kg/m³)	Compressive Strength f'_c (MPa)
NC	525	1054	764	-	-	158	5.5	-	47.2
ECC	465	-	-	390	744	338	3.25	26	46.2

Notes: NC is normal concrete; CEM I is Portland cement; 10 mm is coarse aggregate; HRWR is high-range water reducer; and PVA is polyvinyl alcohol.

Table 2. Oxide analysis of CEM I.

Compound	Content (%)
CaO	62.87
SiO_2	20.33
Al_2O_3	3.1
Fe_2O_3	4.81
MgO	0.1
SO_3	2.5
K_2O	0.45
TiO_2	0.43
V_2O_5	0.02
CuO	0.075
ZnO	0.027
SrO	0.053
ZrO_2	0.03
BaO	0.06

Table 3. Oxide analysis of fly ash.

Compound	Content (%)
CaO	9.32
SiO_2	43
Al_2O_3	12.8
Fe_2O_3	26.8
MgO	0.17
P_2O_5	0.81
K_2O	2.36
TiO_2	1.47
V_2O_5	0.074
Cr_2O_3	0.044
CuO	0.05
MoO_3	3
BaO	0.25

Table 4. Properties of Kuralon RECS-15 PVA fibre.

Type	Fibre diameter (μm)	Length (mm)	Specific gravity (g/cm³)	Tensile strength (MPa)	Elongation (%)	Young's modulus (GPa)
RECS-15	40	8	1.3	1600	6	41

100-litre tilting drum mixer and 30-litre Hobart planetary motion mixer were used to cast normal concrete and ECC specimens respectively, and it was done in a single batch mixing. In normal concrete, dry materials were first mixed following the addition of water and HRWR. In ECC, however, a special treatment was devoted with care to ensure the successful mixing and uniform fibres dispersion. At the first stage of mixing, all dry materials (i.e. fly ash, cement and silica sand) were inserted into the bucket and were manually mixed using a wooden scoop. The mixing was done gently to prevent the release (loss) of fly ash particles and hence trap them under the cement and silica sand particles. 80% of water was then added and mixed manually with these dry materials. After the water was roughly mixed, the bucket was then moved to mixer machine and a low-speed mixing was applied to create a uniform smooth paste. Mixing was continued at medium speed for at least five minutes where the superplasticiser was added slowly following the remaining 20% of water. The gradual inspection was performed to ensure that the bottom deposit disappeared and had mixed up consistently. It was also of importance to ensure that the viscosity level was achieved. Upon this stage, low-speed mixing was set and fibres were poured into the fresh matrix slowly. Mixing speed was changed to a high-speed rate for approximately five minutes to let the fibres dispersed uniformly. The final inspection was done by griping the fresh ECC with a hand for several times as a means of ensuring that there was no fibre ball occurred. Accordingly, the viscosity of fresh ECC was also evaluated using a marsh funnel cone.

The schematic representations of test specimens are displayed in figures 2(a) and (b). All test specimens were fabricated using a group of cylinder steel moulds with diameter and height of 100 by 200 mm, including cylinders used to define the compressive strengths. To confirm the quality of ECC mixtures, additional fabrication was done on a series of dog-bone shaped (coupon) with geometry and dimensions following the JSCE recommendation [27] for direct tensile tests. The example of tensile stress-strain response of an ECC coupon alongside the crack pattern obtained using a recently developed low-cost automated crack mapping technique [28-30] is shown in figure 3. Upon casting, all normal concrete/ECC specimens were covered with plastic sheet and allowed to harden for 24 hours. The moulds were then dismantled and the specimens were stored in a curing tank at a room temperature until testing (28 days).

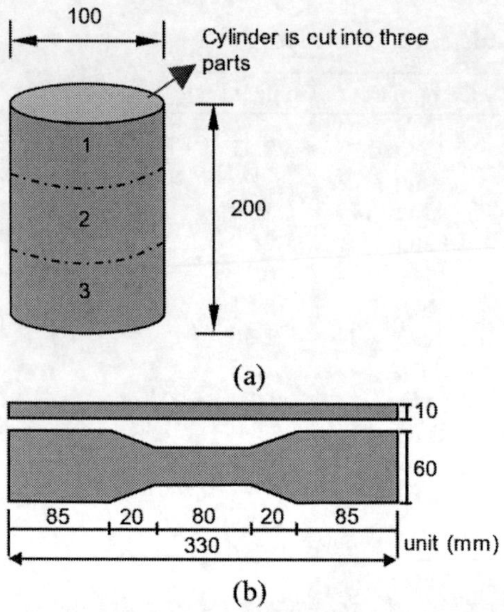

Figure 2. Test specimens: (a) 100×200-mm cylinder and (b) ECC coupon.

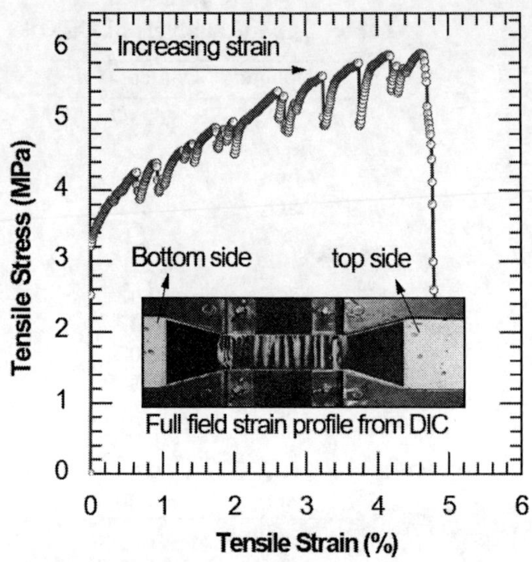

Figure 3. Uniaxial tensile stress-strain relationship and DIC crack pattern of ECC.

2.2. Sorptivity, water absorption and chloride penetration tests

The sorptivity test was undertaken following the requirement per ASTM C1585 [31]. It is aimed to determine the rate of absorption through the increase of mass resulting from the absorption of water. The schematic setup of sorptivity test is displayed in figure 4. Prior to testing, a 100×200 mm cylinder made of concrete/ECC, which was stored at a temperature of 50 ± 2°C for three days after curing, was sliced into three equal parts (67-mm thick); each of which was weighed for mass measurement. The surfaces of the sliced specimens (except for the bottom side) were coated with waterproof epoxy. This was done to allow for absorption penetrating only from the bottom side of the specimen. The top and circumference surfaces of the specimen was also wrapped with plastic sheet to prevent evaporation due to environmental exposure. Upon this, the specimens were transferred to the dedicated tank and placed onto the supports. The absorption procedure was then conducted at 23 ± 2°C with tap water until reaching 1 to 3-mm level of water measured from the bottom surface of the specimen. The test involved the measurement of mass of cylinder specimens at given time intervals, which, in the case of work presented, at 1, 5, 10, 20, 30, 60, 120, 180, 240, 300 and 360 minutes at the first cycle following the successive measurement at 30 hours later for the second cycle with time intervals of 30, 54, 102, 126, 150 and 174 hours, allowing for water absorption through capillary suction. The rate of absorption was measured through the change in mass (gram) at the time divided by the cross-sectional area of the specimen (mm2) and the density of water (g/mm^3).

The water absorption test was carried out in accordance with ASTM C642 [32] as a general means to obtain the percentage of absorption and voids in hardened concrete/ECC. In this test, the specimen was sliced in a similar manner with sorptivity specimens; however, it was dried inside the oven at a temperature of 100 to 110°C for at least 24 hours to determine the oven-dry mass of the specimen. The measurement was then taken after the removal from the oven at a temperature of 25°C. The oven-dry mass was compared with the initial mass of the specimen for control purposes until it reached the maximum difference value of 0.5%. Upon this procedure, the specimen was positioned in the container and immersed in the water at approximately 21°C for 174 hours until two successive values of mass at intervals of 24 hours showed an increase in mass of less than 0.5%, which hence define the saturated stage. It is worth noting that water absorption in this test could only take place in pores as a result of immersion process during testing. These pores could also be regarded as penetrable pores which are indicative of the degree of penetrability throughout the course of immersion until saturation.

The rapid chloride penetration test was undertaken based upon ASTM C1202-12 [33] to cover the evaluation of the electrical conductance of concrete and ECC for rapid indication of resistance due to penetration of chloride ions. The test involved 67-mm thick slice of 100-mm cross-sectional diameter of cylinders during 6-hour period. A potential difference of 60V in d.c. power supply was kept constant throughout testing; one of each was immersed in a 3% of sodium chloride (NaCl) solution, whilst the other in a 0.3N sodium hydroxide (NaOH) solution. The power supply cable connected to one side of specimen (immersed in NaCl) was treated as anode (-), whereas the other side which was immersed with NaOH was treated as cathode (+). The schematic setup of rapid chloride penetration test is displayed in figure. 5.

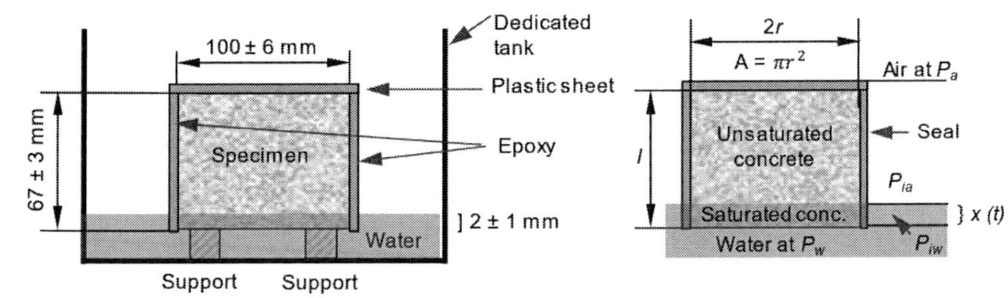

Figure 4. Schematic setup of sorptivity test.

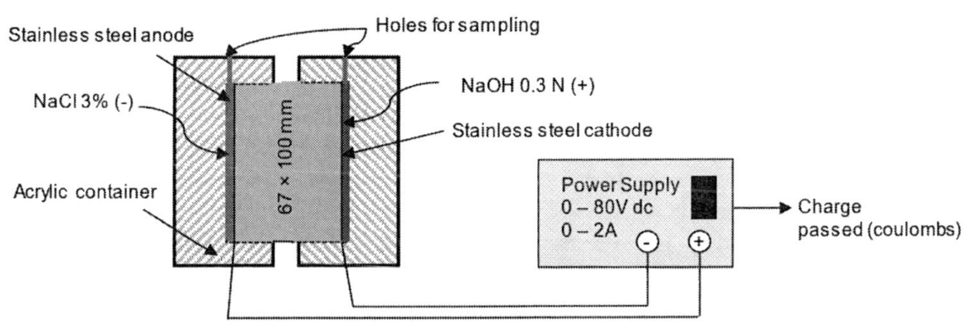

Figure 5. Schematic setup of rapid chloride penetration test.

3. Results and Discussion

3.1. Sorptivity

Figure 6 displays the relationship between the rate of capillary suction (sorptivity) and test time during the initial (first) and second cycle of mass measurement of three normal concrete and ECC specimens, with the detailed values of absorption summarised in tables 5 and 6 respectively. The cumulative mass of water absorption plotted against the test time is also depicted in figure 7.

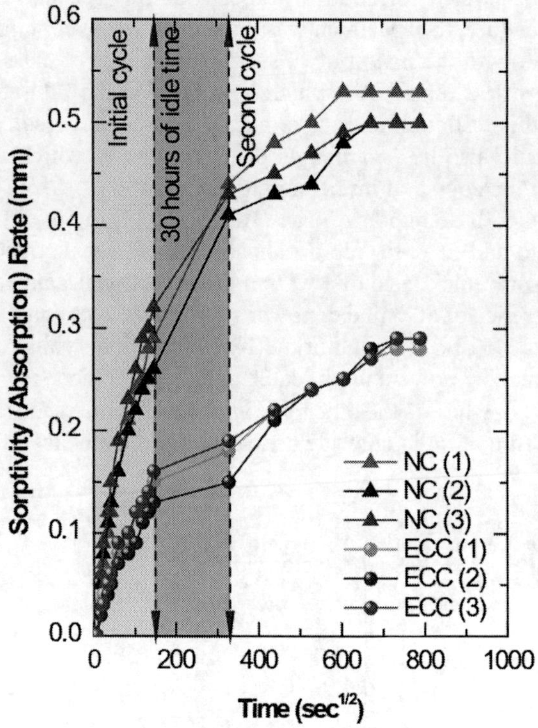

Figure 6. Relationship of sorptivity (absorption) rate and time.

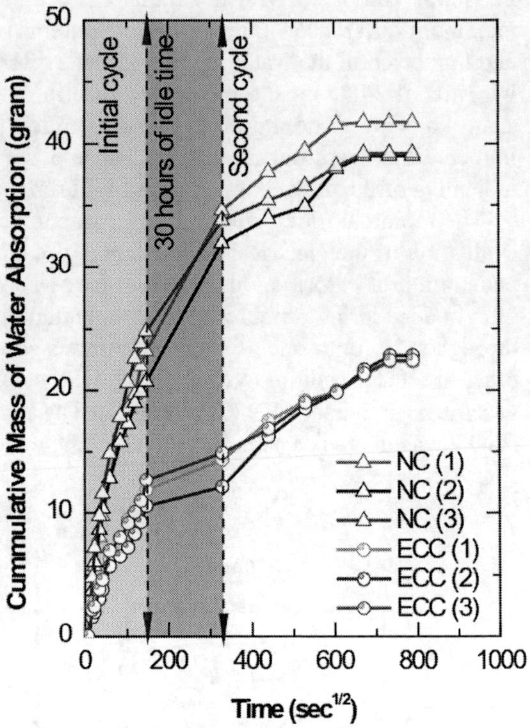

Figure 7. Relationship of cumulative water absorption mass and time.

With reference to figure 6, it is apparent that, in general, the rate of sorptivity is shown to increase as the test time is increased. During the first (initial) cycle of measurement, it can be seen that the typical rate of sorptivity in normal concrete specimens increases linearly with relatively high increment (steep line) up to 60 minutes. Dissimilar response is found in ECC specimens, in which the extent of increment is more insignificant. As the test time is continued to increase (i.e. circa 60 to 100 minutes), a noticeable spike of rate is evident in normal concrete specimens, whilst the ECCs still exhibit a minor linear-like increase. As the test time is further increased (beyond 100 minutes), the increase of sorptivity rate of normal concrete and ECC specimens becomes consistent up to the end of the second cycle measurement. With regard to the progression of water absorption measured in the experiment, the comparison obviously demonstrates that water penetrates faster in normal concrete as a result of inherent porosity in its microstructural properties. It is also evidenced from table 5 that the final sorptivity rate of normal concrete is 0.5 which is 40% (almost twice) higher than that of ECC (see table 6 for comparison). This certainly highlights the excellent durability performance of ECC specimens. It is worth mentioning that this manifestation is primarily attributed to the fact that ECC is structured with micro materials and tailored in such a way that it provides excellent resistance of absorption in addition to its strain hardening behaviour.

With regard to figure 7, it is evident that both normal concrete and ECC specimens exhibit comparable additional (cumulative) mass in the early stage of test time (i.e. up to 5 minutes). As the time is continued, water penetrates more drastically on normal concrete but still in linear (steep) increase, whilst in ECC, the response starts to deviate from linearity, creating a significant gap with normal concrete as the increase of mass in ECC is relatively trivial. The results obtained from water absorption tests indicate a reasonable trend as in case of sorptivity rate and it is evident that from the overall cumulative mass measured, ECC demonstrates the lower water absorption index. The average peak mass of three cylinder ECC specimens is at a value of 22 grams, whereas the normal concrete escalates considerably to a value of 42 grams which is approximately two times higher than ECC.

Table 5. Summary of sorptivity rate of normal concrete.

Test Time		Time	Mass			Δ Mass			Δ Mass/area/density of water, I		
Day	s	$s^{1/2}$	(g)			(g)			(mm)		
Specimen No.			1	2	3	1	2	3	1	2	3
	0	0	1243.7	1140.8	1197.7	0	0	0	0	0	0
	60	8	1243.7	1140.8	1197.7	0	0	0	0	0	0
	300	17	1247.3	1144.6	1202.5	3.6	3.7	4.8	0.05	0.05	0.06
	600	24	1249.4	1146.9	1204.9	5.7	6.1	7.2	0.07	0.08	0.09
	1200	35	1251.8	1149.5	1207.5	8.1	8.6	9.8	0.10	0.11	0.12
	1800	42	1253.5	1151.2	1209.3	9.8	10.4	11.6	0.12	0.13	0.15
	3600	60	1256.6	1153.7	1212.6	12.9	12.8	14.8	0.16	0.16	0.19
	7200	85	1260.3	1156.6	1215.5	16.6	15.8	17.8	0.21	0.20	0.23
	10800	104	1262.2	1158.1	1218.3	18.5	17.3	20.6	0.24	0.22	0.26
	14400	120	1263.3	1159.8	1220.3	19.7	18.9	22.5	0.25	0.24	0.29
	18000	134	1265.6	1160.6	1221.5	21.9	19.8	23.7	0.28	0.25	0.30
	21600	147	1266.8	1161.5	1222.5	23.1	20.7	24.8	0.29	0.26	0.32
1	108000	329	1278.2	1172.8	1231.5	34.6	31.9	33.8	0.44	0.41	0.43
2	194400	441	1281.4	1174.8	1233.2	37.7	33.9	35.4	0.48	0.43	0.45
3	280800	530	1283.1	1175.7	1234.4	39.4	34.9	36.6	0.50	0.44	0.47
4	367200	606	1285.2	1178.8	1235.9	41.5	37.9	38.2	0.53	0.48	0.49
5	453600	673	1285.4	1179.8	1236.9	41.7	39.0	39.2	0.53	0.50	0.50
6	540000	735	1285.4	1179.8	1236.9	41.7	39.0	39.2	0.53	0.50	0.50
7	626400	791	1285.4	1179.8	1236.9	41.7	39.0	39.2	0.53	0.50	0.50

Table 6. Summary of sorptivity rate of ECC.

Test Time		Time	Mass			Δ Mass			Δ Mass/area/density of water, I		
Day	s	$s^{1/2}$	(g)			(g)			(mm)		
Specimen No.			1	2	3	1	2	3	1	2	3
	0	0	966.3	1132.2	990.4	0	0	0	0	0	0
	60	8	966.3	1132.2	990.4	0	0	0	0	0	0
	300	17	968.7	1133.8	992.5	2.4	1.5	2.2	0.03	0.02	0.03
	600	24	969.0	1134.2	992.8	2.8	1.9	2.4	0.04	0.03	0.03
	1200	35	970.3	1135.3	994.2	4.0	3.0	3.8	0.05	0.04	0.05
	1800	42	970.9	1136.0	995.0	4.6	3.8	4.6	0.06	0.05	0.06
	3600	60	972.6	1137.9	997.3	6.3	5.7	6.9	0.08	0.07	0.09
	7200	85	973.4	1138.7	998.4	7.1	6.5	8.0	0.09	0.08	0.10
	10800	104	975.7	1139.4	999.5	9.5	7.2	9.1	0.12	0.09	0.12
	14400	120	976.2	1140.5	1000.5	9.9	8.3	10.1	0.13	0.11	0.13
	18000	134	977.3	1141.5	1001.4	11.0	9.3	11.0	0.14	0.12	0.14
	21600	147	978.1	1142.8	1003.1	11.9	10.6	12.7	0.15	0.13	0.16
1	108000	329	980.6	1144.3	1005.3	14.3	12.2	14.9	0.18	0.15	0.19
2	194400	441	983.8	1148.5	1007.3	17.5	16.2	16.9	0.22	0.21	0.22
3	280800	530	985.4	1150.8	1009.4	19.1	18.5	18.9	0.24	0.24	0.24
4	367200	606	986.2	1152.1	1010.2	19.9	19.8	19.7	0.25	0.25	0.25
5	453600	673	987.5	1153.9	1011.7	21.2	21.6	21.3	0.27	0.28	0.27
6	540000	735	988.6	1155.0	1012.8	22.3	22.8	22.4	0.28	0.29	0.29
7	626400	791	988.6	1155.0	1012.8	22.3	22.8	22.4	0.28	0.29	0.29

3.2. Water absorption

The comparison of the percentage of water absorption on three uncracked normal concrete and ECC specimens is presented in table 7. The mass of oven-dry specimen (labelled A) shown in the table was measured after the specimen was stored in the oven for 24-hour periods which also had 0.5% deviation with the initial mass (before stoved). The mass of immersed specimen (labelled B) was obtained after full immersion with water at testing time of 174 hours (or 7 days) which was measured directly after the removal of surface moisture with fabric material (i.e. towel). The results obtained generally indicate the presence of water ingress which is trapped inside the capillary structures of normal concrete and ECC. Of interest is the percentage of water absorption in ECC specimens which is significantly lesser than normal concrete, signifying once again the excellency of ECC in terms of durability performance. It is also evident in the table that the percentage obtained in ECC is twice lesser than the normal concrete which is indicative of low permeability properties.

Table 7. Summary of percentage of water absorption.

Sample	Dimension		Sample Weight (gram)			Percentage of absorption after immersion (%)
	Diameter (cm)	Height (cm)	Initial mass (g)	24-h oven-dry mass, *A* (g)	174-hr mass after immersion, *B* (g)	
NC1	10.07	6.635	1240.29	1230.18	1285.42	3.9
NC2	10.07	6.63	1136.17	1132.22	1179.83	3.6
NC3	10.07	6.7	1192.49	1190.13	1236.94	3.5
ECC1	10.02	6.4	962.55	961.32	988.59	2.0
ECC2	10.02	6.9	1129.5	1128.7	1155.03	1.4
ECC3	10.02	6.7	986.94	985.36	1012.83	2.0

Note: NC is normal concrete specimen

3.3. Rapid chloride penetration

The rapid chloride penetration test is widely known as one of the most important factors in defining the durability performance of concrete due to the ingress of corrosive (chloride) agent. The qualitative indications of chloride ion penetrability are determined based on the classification of charge passed (coulombs) as presented in table 8. The charge passed (coulombs), Q, is calculated based on the trapezoidal rule with the expression described in the following manner [33],

$$Q = 900(I_0 + 2I_{30} + 2I_{60} + \cdots + 2I_{300} + 2I_{330} + I_{360} \quad (1)$$

where I_0 is the current (amperes) immediately after voltage is applied and I_t is the current (amperes) at t (minutes) after the voltage is applied.

Table 8. Classification of chloride ion penetrability based on charge passed [33].

Charge passed (coulombs)	Chloride ion penetrability
> 4000	High
2000–4000	Moderate
1000–2000	Low
100–1000	Very Low
< 100	Negligible

Table 9 presents the records of current of normal concrete and ECC specimens during testing period at given time intervals, along with the summary of total charge passed (coulombs). The current (amperes) transmitted into the specimen is resulted from the electrical conductance through NaCl and NaOH solutions. These two solutions obviously have major key roles to play while migrating into the concrete/ECC as they serve as a medium to conduct electricity through ions. From the table presented, it is evident that the chloride penetration in normal concrete is considerably higher which is marked by the extent of charge passed at an average value of 2007 C, underlining the moderate level of chloride ion penetrability. On the contrary, ECC exhibits the marginal average value of charge passed (i.e. 981 C) which is indicative of very low level of chloride ion penetrability. The marginal manifestation of electrical conductance in ECC is found to be affected by the presence of polymeric admixtures, allowing for the improvement of durability. This finding is also aligned with the ability of ECC to provide slow water absorption and sorptivity as highlighted in the previous sections.

Table 9. Summary of charge passed (coulombs) recorded at 30-min interval.

	Current (amperes)						
	Specimen	Normal concrete			ECC		
	No.	1	2	3	1	2	3
Current (amperes)	0	0.05	0.05	0.04	0.04	0.04	0.04
	30	0.06	0.06	0.05	0.04	0.04	0.04
	60	0.06	0.07	0.06	0.04	0.04	0.04
	90	0.07	0.07	0.07	0.04	0.04	0.04
	120	0.08	0.08	0.08	0.04	0.04	0.04
	150	0.09	0.09	0.09	0.05	0.04	0.04
	180	0.09	0.09	0.1	0.05	0.04	0.05
	210	0.1	0.1	0.1	0.05	0.05	0.05
	240	0.11	0.11	0.11	0.05	0.05	0.05
	270	0.11	0.11	0.11	0.05	0.05	0.05
	300	0.12	0.12	0.12	0.05	0.05	0.05
	330	0.13	0.13	0.13	0.05	0.05	0.05
	360	0.14	0.13	0.14	0.05	0.05	0.05
	Q (C)	2007	2016	1998	999	963	981
	Mean (C)		2007			981	

Aside from the rapid chloride penetration, chloride diffusion (migration) coefficient is also considered important to be addressed herein as it provides the critical information of chloride ions speed penetrating in the concrete and ECC. Diffusion coefficient is determined by measuring the concentration of chloride ions on the surface of the cylinder specimen after the rapid chloride penetration and is calculated using these following expressions [34],

$$D = \frac{RTL}{zFE}\left(\frac{x_d - \propto \sqrt{x_d}}{t}\right) \tag{2a}$$

$$\propto = 2\sqrt{\frac{RT}{zFE}}\, erf^{-1}\left(1 - \left(\frac{2C_d}{C_0}\right)\right) \tag{2b}$$

$$E = \frac{U-2}{L} \tag{2c}$$

where D is the non-steady-state chloride diffusion coefficient (m^2/sec); R is gas constant (=8.314 (J/(°K mol))); T is the average (absolute) value of the initial and final temperature (K); z is the absolute value of chloride ion (=1); F is Faraday constant (=9.648×10^4 (J/(V mol))); E is the electric field (V/m); U is the absolute value of the applied voltage (V); L is the thickness of the specimen (m); x_d is the average value of the penetration depth (m); $\propto$ is laboratory constant (m$^{1/2}$) calculated based on equation 2(b); t is test duration (sec); erf^{-1} is the inverse of error function; C_d is chloride concentration at which the colour changes (mol/dm^3); and C_0 is chloride concentration in cathode solution (mol/dm^3).

Table 10. Summary of chloride diffusion coefficient.

	C_0 (%)	C_d (%)	T (K)	$\frac{RT}{zFE}$	x_d (m)	t (sec)	$\propto$ (-)	D (m^2/s)
NC	3	1.34	303	2.25×10^{-5}	0.01	21600	7.56×10^{-5}	1.04×10^{-11}
ECC	3	0.84	303	2.17×10^{-5}	0.01	21600	1.96×10^{-3}	9.85×10^{-12}

Note: NC is normal concrete specimen

Table 10 summarises the comparison of chloride diffusion coefficients (in the form of chloride ions speed, D (m2/sec)) between normal concrete and ECC specimens, which are calculated using equations 2(a)-(c). From the table presented, it is apparent that a slightly higher diffusion coefficient is found in normal concrete with a mean value of 1.04×10^{-11} m^2/sec which is 5.5% greater than that of ECC (i.e. 9.85×10^{-12} m^2/sec). Although the difference of diffusion coefficient is marginal, it is worth mentioning that the concentration of chloride (C_d) on the surface of normal concrete is found to be twice higher than ECC, which means that normal concrete suffers considerably during chloride ingress process.

4. Concluding remarks

The results of experimental investigation dealing with direct measurements of durability factors are presented to address the influence of sorptivity, water absorption, and rapid chloride penetration on (uncracked) normal concrete and ECC specimens. Systematic tests on these three parameters were thoroughly undertaken in the laboratory following the procedures and recommendations from the adopted codes. From the experimental results presented, it can be concluded that the sorptivity rate of ECC in the first and second cycle is shown to produce gradual increase which is in contrast with the response of normal concrete. A similar response is also found in the water absorption tests where ECC generally exhibits a considerably lesser percentage of water absorption when compared against normal concrete. In the rapid chloride penetration test, very low chloride ion penetrability level is also measured in ECC, whilst normal concrete exhibits moderate level which is indicative of high porosity inherent properties. The test results are now confirmed that ECC is shown to provide high durability performance, making it suitable to be utilised in aggressive environmental conditions such as for marine infrastructure.

Acknowledgements

The authors wish to acknowledge the support from Laboratory of Concrete, Advanced Materials and Computational Mechanics of Sepuluh Nopember Institute of Technology and PT. Wijaya Karya Beton Tbk. for supplies of Portland cement and fly ash. The first author wishes to acknowledge the financial support from Indonesia Endowment Fund for Education of the Ministry of Finance under LPDP scholarship programme. Thanks also to G.J.P Ghewa for his assistance in the experimental work.

References

[1] Li M and Li V C 2011 Cracking and healing of engineered cementitious composites under chloride environment *ACI Mater. J.* **108**, 333–40

[2] Castro J, Spragg R and Weiss J 2013 Crack-healing investigation in bituminous materials *J. Mater. Civ. Eng.* **25**, 864–70

[3] Kim J, McCarter W J, Suryanto B, Nanukuttan S, Basheer P A M and Chrisp T M 2016 Chloride ingress into marine exposed concrete: a comparison of empirical and physically-based models *Cem. Concr. Compos.* **72** 133–45

[4] Alaswad G, McCarter W J and Suryanto B 2018 Moisture movement within concrete exposed to simulated hot arid/semi-arid conditions *Proc. Inst. Civ. Eng. - Constr. Mater.* 1–15

[5] Maalej M, Chhoa C Y and Quek S T 2010 Effect of cracking, corrosion and repair on the frequency response of RC beams *Constr. Build. Mater.* **24**, 719–31

[6] Strauss A, Wendner R, Bergmeister K and Costa C 2013 Numerically and experimentally based reliability assessment of a concrete bridge subjected to chloride-induced deterioration *J. Inf. Sys.* **19** 2 166–175

[7] Sahmaran M, Li M and Li V C 2007 Transport properties of engineered cementitious composites under chloride transport properties of engineered cementitious composites under chloride exposure *ACI Mater. J.* **104** 6 604–11

[8] Cao C and Cheung M M S 2014 Non-uniform rust expansion for chloride-induced pitting corrosion in RC structures *Constr. Build. Mater.* **51** p. 75–81

[9] Berrocal C G, Lundgren K and Löfgren I 2016 Corrosion of steel bars embedded in fibre reinforced concrete under chloride attack: state of the art *Cem. Concr. Res.* **80** 69–85

[10] Maalej M, Ahmed S F U and Paramasivam P 2003 Corrosion durability and structural response of functionally-graded concrete beams *J. Adv. Concr. Technol.* **1**, 307–16

[11] Kim J, McCarter W J and Suryanto B 2018 Performance assessment of reinforced concrete after long-term exposure to a marine environment *Constr. Build. Mater.* **192** 569–83
[12] Yu K, Li L, Yu J, Wang Y, Ye J and Xu Q F 2018 Direct tensile properties of engineered cementitious composites: a review *Constr. Build. Mater.* **165** 346–62
[13] Kelham S, 1988 A water absorption test for concrete *Mag. Concr. Res.* **40** 106–10
[14] Booya E, Gorospe K, Ghaednia H and Das S 2019 Durability properties of engineered pulp fibre reinforced concretes made with and without supplementary cementitious materials *Compos. Part B Eng.* **172** 376–86
[15] Suryanto B, Saraireh D, Kim J, McCarter W J, Starrs G and Taha H M 2017 Imaging water ingress into concrete using electrical resistance tomography *Int. J. Adv. Eng. Sci. Appl. Math.* **9**, 2 109–18
[16] Li V C and Kanda T 1998 Engineered cementitious composites for structural applications *J. Mater. Civ. Eng.* **10** 66–69
[17] Suryanto B, Mccarter W J, Wilson S A and Chrisp T M 2016 Self-healing performance of engineered cementitious composites under natural environmental exposure *Adv. Cem. Res.* **28** 211–20
[18] Komara I, Tambusay A, Sutrisno W and Suprobo P 2019 Engineered Cementitious Composite as an innovative durable material: a review *ARPN J. Eng. Appl. Sci.* **14**, 822–33
[19] Tambusay A, Suprobo P, Faimun and Amiruddin A, 2017 Finite element analysis on the behavior of slab-column connections using PVA-ECC material *J. Tekn.* **79** 5 22–32
[20] Tambusay A 2017 Cyclic Behaviour of Slab-Column Connections using the Engineered Cementitious Composite *PhD Thesis Institut Teknologi Sepuluh Nopember, Surabaya*
[21] Suryanto, B, Wilson S A, Mccarter W J 2015 Self-healing of micro-cracks in engineered cementitious composites *Civ. Eng. Dimens.* **17** 187–93
[22] Suryanto B, McCarter W J, Starrs G, Wilson S A and Traynor R M 2015 Smart cement composites for durable and intelligent infrastructure *Proc. Eng.* **125**, 796–803
[23] DIN EN 197-1 Special cement composition and conformity evaluation 2000
[24] Mooy M, Tambusay A, Komara I, Sutrisno W, Faimun F and Suprobo P 2020 Evaluation of shear-critical reinforced concrete beam blended with fly ash *IOP Conf. Ser. Earth Environ. Sci.* **506** 012041
[25] Bastian M A, Tambusay A, Komara I, Sutrisno W, Irawan D and Suprobo P 2020 Enhancing the ductility of a reinforced concrete beam using engineered cementitious composite *IOP Conf. Ser. Earth Environ. Sci.* **506** 012044
[26] Oktaviani W N, Tambusay A, Komara I, Sutrisno W, Faimun F and Suprobo P 2020 Flexural behaviour of a reinforced concrete beam blended with fly ash as supplementary material *IOP Conf. Ser. Earth Environ. Sci.* **506** 012042
[27] JSCE 2008 Recommendations for design and construction of high performance fiber reinforced cement composite with multiple fine cracks, *Japan Society of Civil Engineers*
[28] Tambusay A, Suryanto B and Suprobo P 2020 Digital image correlation for cement-based materials and structural concrete testing *Civ. Eng. Dimens.* **22** 1 6–12
[29] Suryanto B, Tambusay A and Suprobo P 2017 Crack mapping on shear-critical reinforced concrete beams using an open source digital image correlation software *Civ. Eng. Dimens.* **19 2** 93–8
[30] Tambusay A, Suryanto B and Suprobo P 2018 Visualization of shear cracks in a reinforced concrete beam using the digital image correlation *Int. J. Adv. Sci. Eng. Inf. Technol.* **8** 2 573–78.
[31] ASTM C1585-13 2013 Standard test method for measurement of rate of absorption of water by hydraulic-cement concretes *ASTM Int.* (United States: West Conshohocken)
[32] ASTM C642-13 2013 Standard test method for density, absorption, and voids in hardened concrete *ASTM Int.* (United States: West Conshohocken)
[33] ASTM C1202-12 2012 Standard test method for electrical indication of concrete's ability to resist chloride ion penetration *ASTM Int.* (United States: West Conshohocken)
[34] Zych T 2014 Test method of concrete resistance to chloride ingress *Resist. Concr. to Chloride Ingress* **6-B** 117–39

IOP Conf. Series: Materials Science and Engineering **930** (2020) 012057 doi:10.1088/1757-899X/930/1/012057

Numerical investigation of HSC column under axial and flexural loading using 3D-NLFEA

P Pradnyanita[1], B Piscesa[1*], M M Attard[2], Faimun[1], and A K Samani[3]

[1] Civil Engineering Department, Institut Teknologi Sepuluh Nopember, 60111 Surabaya, Indonesia
[2]School of Civil and Environmental Engineering, The University of New South Wales (UNSW), Sydney NSW 2052, Australia
[3]B&G Consultant, Doha, Qatar

[*] Corresponding author: piscesa@ce.its.ac.id

Abstract. High strength concrete (HSC) has been widely used as material for Reinforced Concrete (RC) column in high rise building in the past two decades. However, ductility of HSC column was one of the main concerns in design. The ductility of HSC column is much lower compared to normal-strength concrete. To increase the ductility of HSC column, the use of adequate transverse reinforcement can be used. In this paper, numerical investigation of HSC column available in the literature is modelled using an inhouse finite element package called 3D-NLFEA. The modelled specimen consisted of HSC columns with dimension 305 m x 305 m x 1473 mm and is subjected to constant axial load and incremental displacement control which act as a shear load up to failure. From the comparisons with the available test results, it was found out that the prediction using 3D-NLFEA agrees well with the test results. This paper also discusses the yield point location of bars in the load-deflection curve, length of bar that yields, buckling of compression bar, and concrete which cracked or crushed.

1. Introduction

The use of high strength material, high-strength concrete (HSC) and bars, is widely applied for high rise building in the past two decades. The use of high strength material can reduce the structural element cost as a results of smaller member section for the same strength. By having smaller member section, the effective area of the floor increases [1]. However, HSC was found to be very brittle. For plain HSC column subjected to axial loads, sudden explosive failure most likely to occur [2] and snap-back behavior may present during the softening phase [3, 4]. Hence, in design, HSC column needs to be well confined to meet the minimum ductility demand for a Special Resisting Moment Frames (SRMF) column.

The new building codes such as ACI 318-19 [5] or SNI 2847-2019 [6] requires additional confinement equation applied for concrete with strength more than equal to 70 MPa or columns with axial load ratio (P_u/f_cA_g) higher than 0.3. As the axial load increases, the deformability of HSC RC column reduced (lower ductility) [7]. To meet with the minimum ductility requirements, confinement to the concrete core is required. One of the conservative way to provide confinement to the concrete core is to provide confinement rebars [8-12]. This confinement rebars around the concrete core produces confining pressure which is proportional to the core expansion [13, 14]. Experimental investigation which studies the effect of confinement bars for HSC column have been studied by many researchers

Content from this work may be used under the terms of the Creative Commons Attribution 3.0 licence. Any further distribution of this work must maintain attribution to the author(s) and the title of the work, journal citation and DOI.
Published under licence by IOP Publishing Ltd

[7, 8, 10, 15-26]. Although the responses from specific column were well studied, may other configuration still require extensive experimental validation which is extremely expensive. Furthermore, the specimen size is more likely to be limited due to the equipment availability. On the other hand, the use of numerical simulation provides more benefits and is extremely low cost.

In numerical simulation, the framework used to model the concrete constitutive model plays an important role to ensure the predicted response is sufficiently accurate. For that purpose, in this paper, the discussion is focused on the numerical simulation to investigate the load deflection curve of RC column made of HSC. There are two HSC RC columns specimen experimentally tested by Bayrak and Sheikh [8]. The RC columns are loaded under constant axial load and are loaded in the lateral direction as a shear load which will generate single curvature bending moment in column. The concrete constitutive model used is Piscesa et al. [11-14] plasticity-fracture model which was found to be in good agreement in predicting the cover spalling behaviour and load-deflection curve of RC column with HSC under concentric loading [14].

The numerical investigation is carried using an inhouse three-dimensional nonlinear finite element package called 3D-NLFEA [13]. 3D-NLFEA uses SALOME 9.30 platform [27] as the preprocessor and ParaView 5.8 [28, 29] as the postprocessor. There are two column specimens tested by [8] modelled and simulated using 3D-NLFEA. Details investigation on the behavior of RC column includes the yield point of bars, length of bar that yields, buckling of compression bar, and concrete region which cracked or crushing. It is expected that after the numerical investigation corresponds well with the available test result in the literature, extensive parametric study can be carried out to further verify the requirements of additional confining rebars clause stated in ACI 318-19 and SNI 2847-2019 which should be computed for HSC and axial load ratio greater than 0.3.

2. Reinforced concrete column model

There are two HSC RC column tested by Bayrak and Sheikh [8] being investigated. Specimens AS-5HT and AS-7HT. The axial load level ratio (P_u/f_cA_g) for the two columns are 0.45 which is higher than the codes limitation to enable the additional clause for confinement reinforcement. It should be noted that the confinement rebar for AS-5HT and AS-7HT are 1.08 and 0.8 times the ACI requirements (without additional clause for confinement), respectively.

2.1. Geometry details

Figure 1 shows the typical specimen geometry and rebars arrangement for columns AS-5HT and AS-7HT. Specimen AS-5HT and AS-7HT have the same geometry, longitudinal bar ratio, and concrete strength. The difference between those two are located on the confining bars arrangement. The cross-sectional shape is the same, but the use of bar type and diameter is different. The column cross-section is a square with a width of 305 mm and a height of 1473 mm. The cross-section of the stub is 508 mm x 762 mm and the height are 813 mm. The concrete cover for the column and the stub is set to 20 mm. The longitudinal bar ratio is 2.58% which consisted of eight 20M bars equally on each side. The concrete compressive strength for both specimens varies slightly. The concrete compressive strength for AS-5HT and AS-7HT are 102 and 101.8 MPa, respectively.

The reinforcing cage for the stub consisted of 10 M bars with 64 mm pitch spacing. The longitudinal bars are extended up to the interface between concrete and the steel loading plate to ensure the load are transferred effectively to the concrete. The confining bar arrangement for the column consisted of one square closed tie and one diamond shaped closed tie. For specimen AS-5HT, the confining bar size for the square closed tie is 10M and for the diamond shaped closed tie is 15M. The pitch spacing is 90 mm which gives the volumetric ratios of 4.02 % or equal to 1.08 times the ACI requirements for confinement. On the other hand, specimen AS-7HT have the same size for the confining bars which is 10M with the pitch spacing of 94 mm. This configuration gives the volumetric ratios of 2.72% or equal to 0.8 times the ACI requirements for confinement. To ensure the localized region occurred at the bottom of the column (when the load-deflection curve softens), a tighter pitch spacing near the loading plates are provided in the models.

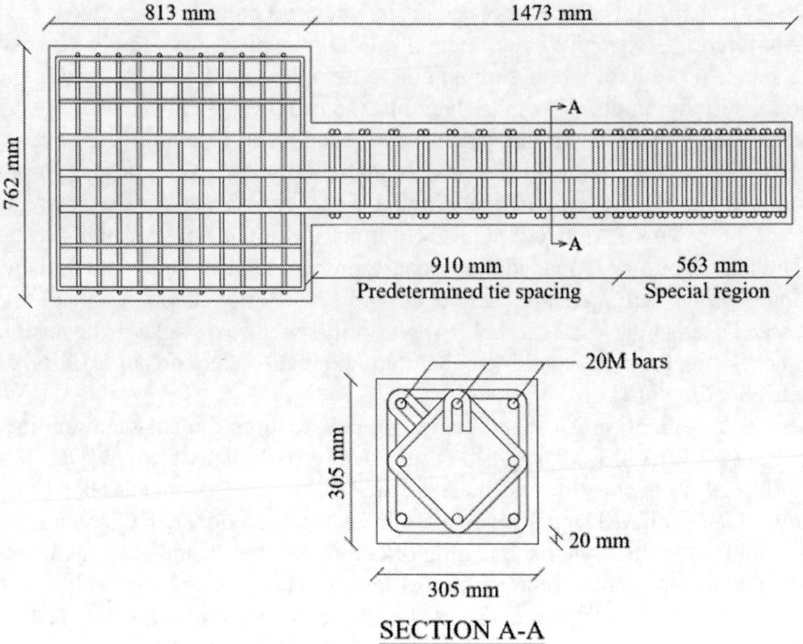

Figure 1. Specimen geometry detail

2.2. Boundary conditions and loading steps

Figure 2 shows the boundary conditions, constant axial load position, and displacement control position. There are two loading steps considered. The first loading step is force controlled. The axial load given is uniformly distributed on the steel loading plate (Point 1). The applied axial load ratio is 45% and is applied within 20 load steps. Hence, for each loading steps, the applied magnitude of the axial load ratio is 2.25%. The second load step is a full displacement control and is applied at the side steel loading plate (Point 2). The load is increased monotonically while axial load is held constant. The boundary condition at the stub (Point 3) is fixed in all three principal direction while in Point 1, is fixed in x and y directions.

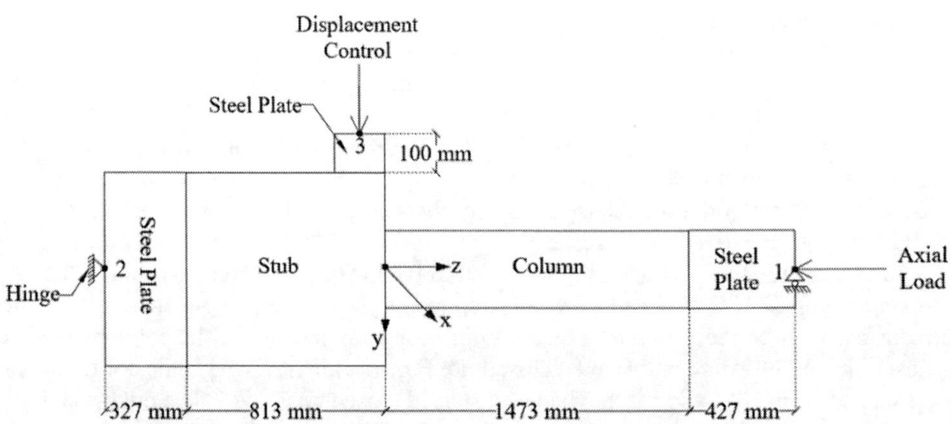

Figure 2. Loading step and boundary condition of specimen

2.3. Modelling procedure

The geometry model and the meshing procedure is prepared using SALOME 9.3.0 [30]. The mesh size is set to 20 mm. Figure 3 shows the 3D solid model of the column using hexahedral solid element. In figure 3a, the steel plate was shown as red coloured while the concrete was shown as grey coloured. In 3D-NLFEA, the hexahedral solid element is integrated using selective integration or BBar element method [31]. Figure 3b shows the steel rebar configurations. Inside 3D-NLFEA, the rebars element is modelled using embedded truss element inside the concrete element [32]. Perfect bond assumption is used for the interface between the rebar and concrete elements. The total solid and embedded truss elements are 54,596 and 25,958, respectively. Once the modelling in SALOME is finished, the data are extracted to the input file in 3D-NLFEA. Once the simulation completed, the output can be viewed using ParaView 5.8.0 [28, 29].

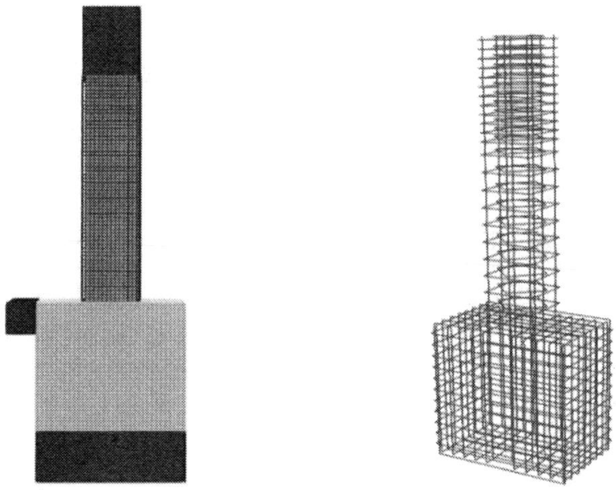

(a) (b)

Figure 3. Modelling of specimen: (a) Solid element; (b) Rebar element

3. Constitutive model of material

The constitutive model for the steel loading plates is based on J_2 metal plasticity model with elastic behaviour. The Young's modulus and the Poisson's ratio for the steel loading plate is 200 GPa and 0.3, respectively. Steel rebars are modelled using embedded truss element with perfect bond assumption. The material properties of reinforcement are detailed in Table 1. The stress-strain relationship of longitudinal bars is assigned to elastic-perfectly plastic model. For the confining rebar, stress-strain behaviour are obtained directly from the test result in [8] as shown Figure 4.

Table 1. Material properties of steel reinforcement

Type of Reinforcement	Longitudinal	Transverse	
Bar Size	20M	10M	15M
Nominal Diameter (mm)	19.5	11.3	16.0
f_y (MPa)	454	542	463
ε_y	0.0023	0.0026	0.0024
ε_{sh}	0.0067	0.0191	0.0207
f_u (MPa)	700	683	648
ε_u	0.1290	0.1550	0.1130

The constitutive model for concrete material under compression is based on plasticity model proposed by [12, 33]. The failure surface for the concrete material is based on the Menetrey and Willam [34] model but it is modified in [33] to improves both the prediction of peak and residual stress for concrete under confinement. One of the important aspects in the constitutive model to simulate the behaviour of HSC RC concrete is the ability of the constitutive model to capture the cover spalling behaviour in the load-deflection curve. In [12], a Rankine failure surface was introduced to include the fracture formulation into the model. This way, the model can distinguish between the tensile splitting cracks due to compression with a pure tensile crack due to tensile stress in concrete. In addition, the plastic dilation rate formulation was also extended to the tensile region to accelerate the dilation rate for concrete cover elements. This cover spalling behaviour is a compulsory concern in HSC column which cause premature drop of compressive load capacity of column [10].

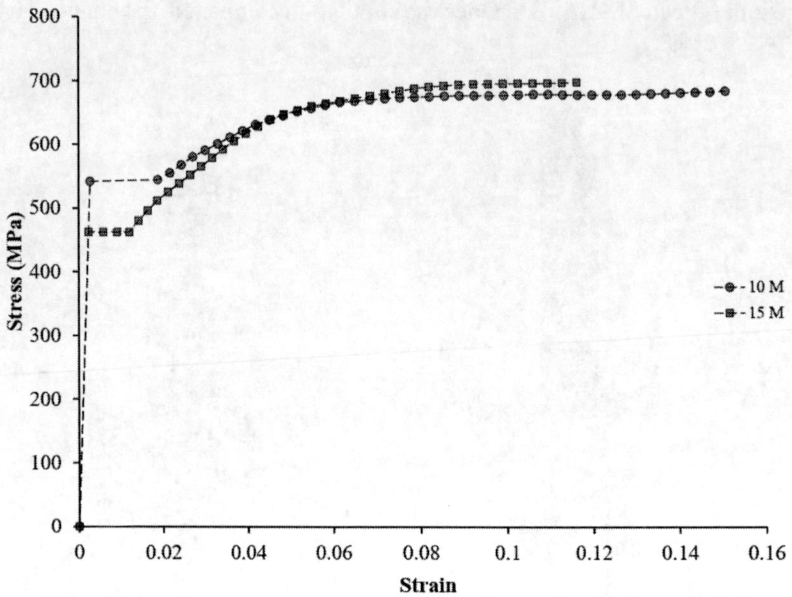

Figure 4. Stress strain relationship of transverse reinforcement [13]

4. Shear and tip deflection

Figure 5 shows the idealization of specimen under axial load and lateral displacement control**Error! Reference source not found.**. From 3D-NLFEA, the displacement in the z-direction at the end of stub (z_1), the displacement in the y direction at the column-stub interface (δ_1), and the reaction force (Q_L) applied at point 2 (see figure 3) can be obtained. The deformed angle, θ can be computed as follow:

$$\theta = \arcsin\left(\frac{z_1}{0.5d}\right) \tag{1}$$

where d is the stub width.

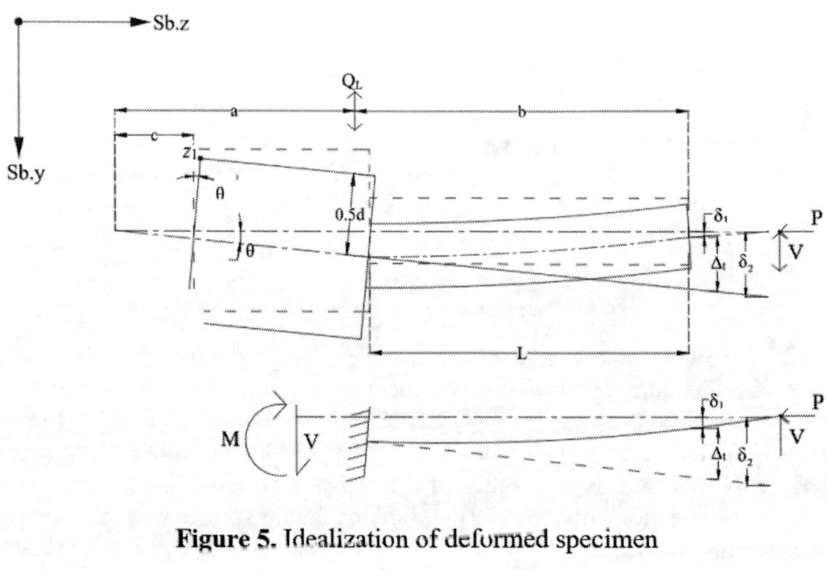

Figure 5. Idealization of deformed specimen

The deflection in y-direction at the top of the column (δ_2) which can be calculated by multiplying the length of column, L and the sinus of deformed angle, θ; i.e. $\delta_2 = L \sin \theta$. Figure 5 shows that the tip deflection, Δ_t is the difference between δ_2 and δ_1; i.e. $\Delta_t = \delta_2 - \delta_1$.

By using the reaction force of lateral displacement control, shear in column for each loading step can be formulated as:

$$V = Q_L \left(\frac{a}{a + b} \right) \tag{2}$$

In the above, *a* is the distance between the location of hinge and the applied lateral displacement and *b* is the distance between the location of the applied lateral displacement to the tip of the column. In addition, the bending moment can be computed:

$$M = P\delta + V\,(c + L) \tag{3}$$

where *c* is the distance between the end of stub to the hinge location.

5. Analysis results and discussions

This section presents the analysis results from 3D-NLFEA and are compared with the test result. In FE model using 3D-NLFEA, the lateral load applies to specimen monotonically using displacement control. To attain a proper comparison result, the backbone curve from the cyclic response [8] was used. Both backbone curve for positive and negative cycles were added into the comparison.

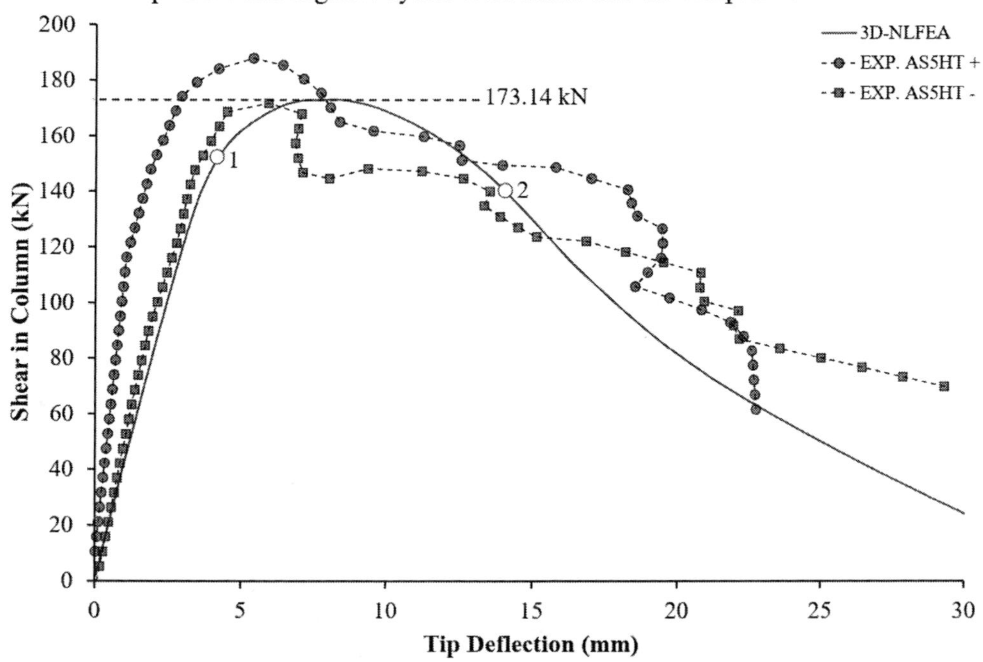

Figure 6. Shear in column versus tip deflection for AS-5HT

Figure 6 and Figure 7 show the shear force as functions of the tip deflection between the FE model and the test result [8] for specimen AS-5HT and AS-7HT, respectively. From figure 6, the 3D-NLFEA peak shear load prediction are within the backbone curves (prediction: 173.14 kN, max. value; 187.88 kN, min. value: 171.87 kN), the softening response can also be captured despite with a more conservative prediction than the test result. From figure 7, the predicted peak shear load prediction is slightly above the maximum back-bone curve of the test result (prediction: 170.58 kN, max. value; 166.97 kN, min. value: 150.5 kN). The softening response was also well capture with more conservative result than the experimental counterpart.

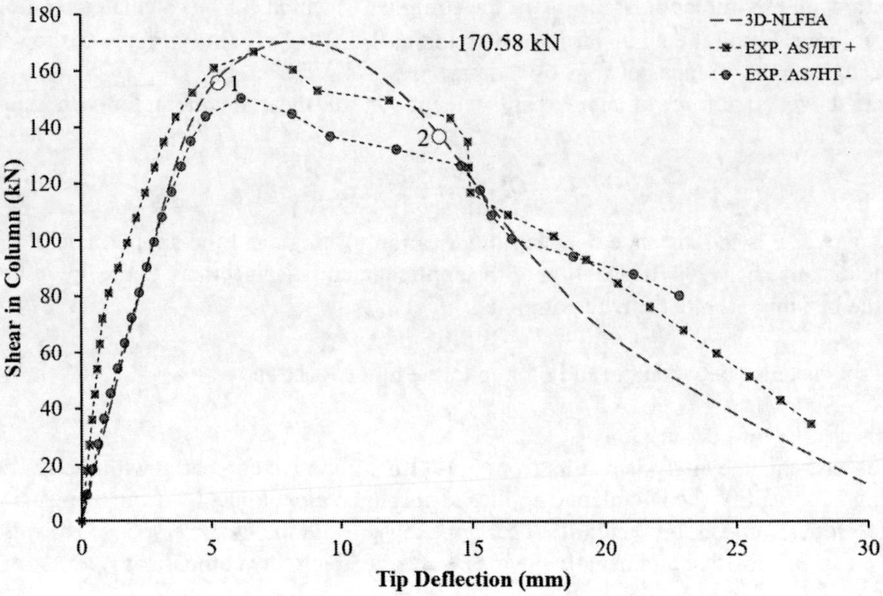

Figure 7. Shear in column versus tip deflection for AS-7HT

Figure 8 and figure 9 shows the deformed shape of the RC column along with the hardening and fracturing parameter. There are two points observation. The first point shows the condition before the peak shear load was reached while the second point shows the condition beyond the peak shear load (around 80% of the peak shear load).

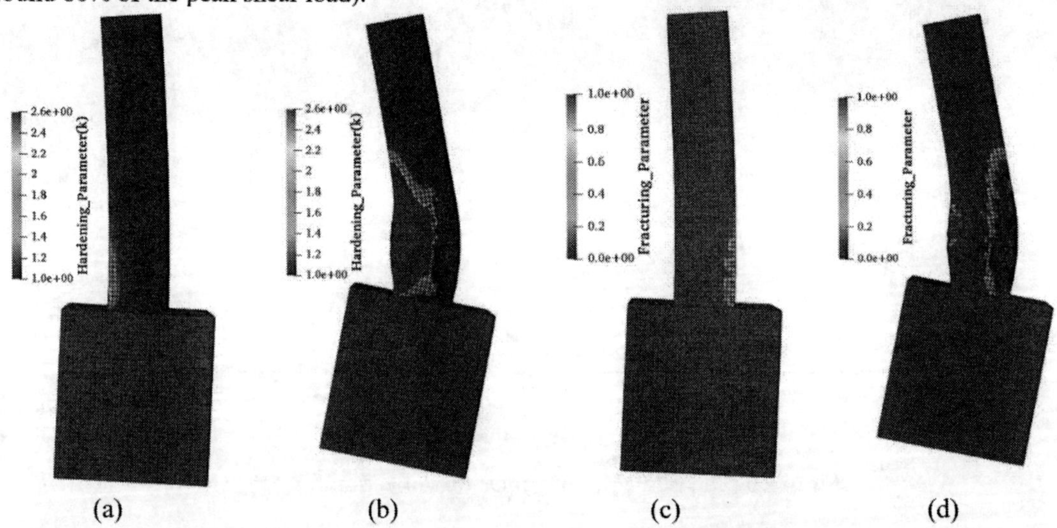

Figure 8. The deformed shapes of specimen AS-5HT with : (a) Hardening contour at point 1; (b) Hardening contour at point 2; (c) Fracturing contour at point 1; and (d) Fracturing contour at point 2

Figure 8a and Figure 9a shows the hardening parameter contour before the peak shear load. At this stage (point 1 in the load deflection curve), the damage in the concrete still localized at the extreme concrete fibre under compression. However, at the softening stage (point 2), the concrete that softens (showed as a hardening parameter value over than unity) are spread not only in the extreme concrete fibre under compression but also to the sides of the concrete cover. Furthermore, bulging appearances on the compression zone (see figure 8b and figure 9b) are clearly shown as a result of high dilation rate for concrete in the compression-tension region.

Figure 8c and figure 9c shows the fracturing parameter before the peak shear load. As shown in figure 8c and figure 9c, the fracturing region occurred in the tension region. It should be noted that the fracturing parameter is associated with the pure uniaxial tensile stress. At the softening phase, the fracturing region spreads in the tension region (figure 8d and figure 9d). The localised cracks can be shown by looking at the strain magnitude, but it is not shown in the discussion here.

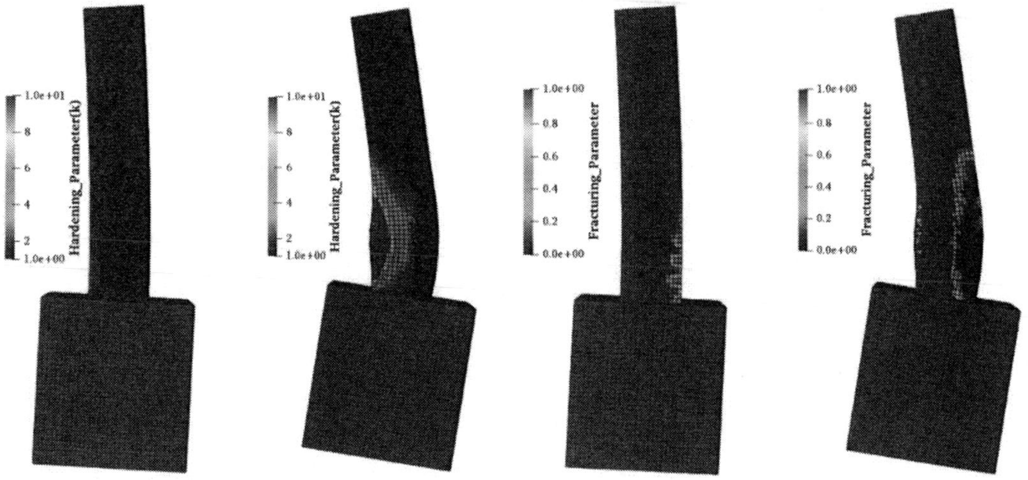

(a) (b) (c) (d)

Figure 9. The Deformed Shapes of Specimen AS-7HT with : (a) Hardening contour at point 1; (b) Hardening contour at point 2; (c) Fracturing contour at point 1; and (d) Fracturing contour at point

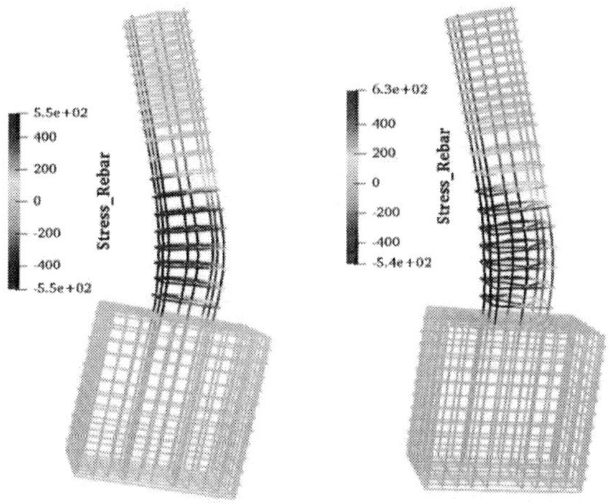

(a) (b)

Figure 10. Stress rebar contour of specimen: (a)AS-5HT; (b) AS-7HT

In [8], after the concrete cover spalls, the confining bar was found to be yielded. The failure of tested specimens was signed by buckling of longitudinal bars due to the absence of concrete covers and perimeter ties. Figure 10 shows the simulation of bars that are yield. At peak shear load, the tensile stress in the transverse bars are almost near it yield stress. There are variations in the confining bar stress as function of the column depth which caused by strain gradient. As shown in figure 10, in the tension region, the confining bar stress was found to be the lowest. The length of the plastic hinges, which is associated with the length of the longitudinal bar that are yield, are 470 mm for AS-5HT and 450 mm for AS-7HT. Furthermore, the buckling of longitudinal bars can be observed clearly in Figure 11 which present the condition in the end of simulation before all specimens fail. The absence of concrete cover and transverse reinforcement is affected to high compressive stress in longitudinal bars and the occurrence of buckling.

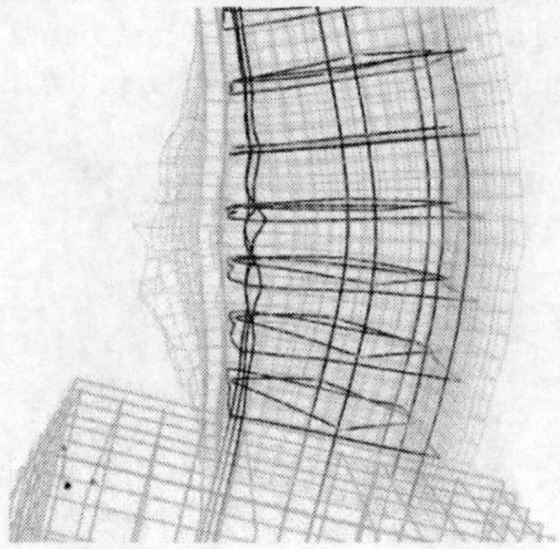

Figure 11. Buckling of longitudinal bars

Figure 11 shows buckling of the longitudinal bars. As shown in figure 11, the longitudinal bars that are in the extreme concrete fibre under compression were found to be curved which clearly indicates that the bar are buckled. The confining bars is also deformed outward which indicates that the concrete inside the ties are dilated outward. This buckling of bar phenomena is important and should be incorporated in the model by 2nd order effect analysis option being enabled.

6. Conclusions

This paper presents finite element analysis of high-strength RC column under constant axial load and shear force that generate a single curvature column. The numerical investigations are executed using the 3D-NLFEA program which incorporates the plasticity-fracture model [14]. Two specimens are adopted from the available literature [8]. The different of each specimen is emphasized on the volumetric ratio of the transverse reinforcement. The comparison of specimen behaviour from the experiment and 3D-NLFEA result are presented in the form of shear load versus the tip deflection diagram. The prediction using 3D-NLFEA agrees well with the test result. The peak shear prediction was within the range for AS-5HT and slightly higher for AS-7HT. The prediction of the load-deflection behaviour during the softening phase was found to be conservative compared to the test results.

The cover that spalls were shown clearly in the deformed shaped at the softening region. The hardening parameters propagates from the extreme concrete fibre in compression to the side concrete cover. By using the 2nd order analysis, the buckling of bars can also be captured, and the P-Delta effect behaviour of the whole RC column can also be considered. Further research should be carried out with varying input data such as higher concrete strength, different level of axial load, and the use of high-yield strength of the bars. Furthermore, simplified model such as fiber-based model [35] can be improved from parametric studies carried out in the future research.

References

[1] Gjørv O 2008 High strength concrete *Develop. Formulation Reinf. Concr.* 79-97
[2] Mendis P 2003 Design of high-strength concrete members: state-of-the-art *Progress in Struc. Eng. Mater.* **5** 1-15
[3] Samani A and Attard M A 2012 stress-strain model for uniaxial and confined concrete under compression *Eng. Struc.* **41** 335-49
[4] Jansen D C and Shah S P 1997 Effect of length on compressive strain softening of concrete *J. Eng. Mech.* **123** 25-35
[5] A. C. I. 2019 Committee, *Building Code Requirements Available for Public Review (ACI 318-19)*
[6] B. S. Nasional 2019 Persyaratan beton struktural untuk bangunan gedung (SNI 2847: 2019)

[7] Cusson D and Paultre P 1994 High-strength concrete columns confined by rectangular ties," *J. Struct. Eng.* **120** 783-804
[8] Bayrak O and Sheikh S A 1998 Confinement reinforcement design considerations for ductile HSC columns *J. Struct. Eng.* **124** 999-1010
[9] Foster S J 1999 Design and detailing of high strength concrete columns *Univ. New South Wales Sydney*
[10] Foster S J, Liu J and Sheikh S A 1998 Cover spalling in HSC columns loaded in concentric compression *J. Struct. Eng.* **124** 1431-37
[11] Piscesa B, Attard M M and Samani A K 2018 3D Finite element modeling of circular reinforced concrete columns confined with FRP using a plasticity based formulation *Compos. Struct.* **194** 478-93
[12] Piscesa B, Attard M M and Samani A K 2016 A lateral strain plasticity model for FRP confined concrete *Compos. Struct.* **158** 160-74
[13] Piscesa B 2018 Modeling confined concrete using plasticity formulation *PhD. Thesis School Civ. Environ. Eng.*
[14] Piscesa B, Attard M M, Prasetya D and Samani A K 2019 Modeling cover spalling behavior in high strength reinforced concrete columns using a plasticity-fracture model *Eng. Struct.* **196** 109336
[15] Yong Y, Nour M G and Nawy E G 1988 Behaviour of laterally confined high - strength concrete under axial loads *J. Struct. Eng.* **114** 332-51
[16] Sheikih S A and Khoury S S 1993 Confined concrete columns with Stubs *ACI Struct. J.*
[17] Sharma U K, Bhargava P and Kaushik S K 2005 Behavior of confined high strength concrete columns under axial compression *J. Adv. Concr. Technol.*
[18] Saatcioglu M and Razvi S R 1998 High-Strength concrete columns with square sections under concentric compression *J. Struct. Eng.* **124** 1438-47
[19] Razvi S R and Saatcioglu M 1994 Strength and deformability of confined high-strength concrete columns *ACI Struc. J.* **91** 678-87
[20] Razvi S, Saatcioglu, and Murat 1999 Confinement model for high - strength concrete *J. Struct.l Eng.* 281-89
[21] Paultre P, Légeron F, and Mongeau D 2001 Influence of concrete strength and transverse reinforcement yield strength on behavior of high-strength concrete columns *ACI Struct. J.*
[22] Paultre P, Eid R, Robles H I, and Bouaanani N 2009 Seismic performance of circular high-strength concrete columns *ACI Struct. J.*
[23] Pallewatta T M, Irawan P and Maekawa K 1995 Effectiveness of laterally arranged reinforcement on the confinement of core concrete *Doboku Gakkai Ronbunshu*
[24] Maekawa K and An X Shear failure and ductility of RC columns after yielding of main reinforcement
[25] Légeron F and Paultre P 2000 Behavior of high-strength concrete columns under cyclic flexure and constant axial load *ACI Struct. J.*
[26] Hwang S K and Yun H D 2004 Effects of transverse reinforcement on flexural behaviour of high-strength concrete columns *Eng. Struct.*
[27] CEA/DEN, EDF_R&D and OPEN_CASCADE 2017 SALOME - The open source integration platform for numerical simulation Ver. 8.3.0 ed
[28] Ahrens J, Geveci B, Law C, Hansen C, and Johnson C 2005 36-ParaView: An end-user tool for large-data visualization *The Visualization Handbook* 717
[29] Ayachit U 2015 The paraview guide: a parallel visualization application
[30] CEA/DEN, EDF_R&D and OPEN_CASCADE 2019 SALOME - The Open Source Integration Platform for Numerical Simulation Ver. 9.3.0
[31] Hughes T J 2012 The finite element method: linear static and dynamic finite element analysis *Courier Corp.*
[32] Ranjbaran A 1996 Mathematical formulation of embedded reinforcements in 3D brick elements *Communications num. methods eng.* **12** 897-903

[33] Piscesa B, Attard M M, Samani A K and Tangaramvong S 2017 Plasticity constitutive model for stress-strain relationship of confined concrete *ACI Mater. J.*

[34] Menetrey P and Willam K 1995 Triaxial failure criterion for concrete and its generalization *ACI Struct. J.* 92

[35] Piscesa B, Prasetya D, Irmawan M, and Alrasyid H 2019 Ductility evaluation of reinforced concrete column made of normal- to high-strength concrete under constant axial load level combined with flexural loading using nonlinear sectional fiber based model *J. Civ. Eng.* **34** 3-8

Effect of steel fiber volume fraction to the tensile splitting strength of concrete cylinder

M Maksum[1], H Alrasyid[1*], M Irmawan[1], and B Piscesa[1]

[1] Department of Civil Engineering, Faculty of Civil, Planning and Geo Engineering, Sepuluh Nopember Institute of Technology, Surabaya, East Java, Indonesia.

*Corresponding author's e-mail: harun@ce.its.ac.id

Abstract. Concrete is strong in compression but weak in tension. To improve the concrete strength and ductility under tension, embedded micro-cement reinforcement such as steel fiber can be used. In this paper, the effect of the volume fraction ratio to the tensile strength of concrete is investigated. The steel fiber volume fraction considered are 0%, 0.5%, 1%, 1.5%, and 2.0%. From the test result, a linear increased in concrete tensile strength as function of the steel fiber volume fraction was observed. The amount of the steel fiber volume fraction also affects the concrete workability. The fiber orientation in the concrete was also observed to capture its effect on tensile strength from the splitting cylinder test.

1. Introduction

The utilization of steel fiber in concrete can improve concrete mechanical properties, such as tensile strength, compressive strength, ductility, and toughness [1-4]. When a crack occurs in steel fiber reinforced concrete (SFRC), the tensile stress between cracks is transferred and resisted by steel fibers. At this stage, the steel fibers assign the tensile load at the crack location along its embedded length inside the concrete. This tensile load is transferred from fiber to concrete through shear stress that works on the steel fiber perimeter embedded in the concrete. The steel fiber reinforced concrete (SRFC) behavior can be affected by the properties of steel fiber, fiber orientation within the matrix, aspect ratio (length /diameter), and the fraction volume [5-8].

Concrete mix design plays an important role when preparing steel fiber reinforced concrete. The correct proportion from different materials must be well-designed. The physical and mechanical properties of composites depend on the volumetric fraction of the steel fiber, the steel fiber itslef, and the concrete matrix. After steel fibers are fully bonded with the concrete, the behavior of the final composite changes, particularly during the cracking phase. With the additional steel fiber, the tensile strength and ductility of the concrete can be improved. With the well-design concrete mix proportion, a goo quality of SFRC concrete can be achieved [9-12].

In this paper, one complete series of tensile splitting tests of SFRC concrete cylinder is reported. The tensile splitting test is preferred over the uniaxial direct tensile test due to the size of the steel fiber in which may not fit the formwork. The volumetric fraction of the steel fiber considered is 0%, 0.5%, 1%, 1.5%, and 2.0%. The Steel fiber trademark is Dramix with hooked ends was used in this study. In the discussions, the effect of the volumetric fiber content on the concrete slump, the fiber orientation factor, and the concrete tensile strength are presented in detail.

Content from this work may be used under the terms of the Creative Commons Attribution 3.0 licence. Any further distribution of this work must maintain attribution to the author(s) and the title of the work, journal citation and DOI.
Published under licence by IOP Publishing Ltd

2. Concrete mix design

Ordinary Portland Cement (OPC) used in this study was produced by PT Semen Indonesia (Persero) Tbk. The OPC is considered as Type I cement and is comply with SNI 15-7064-2004 [13]. Coarse aggregate is with compliance to SNI 1969-2008 [14]. The coarse aggregate is from PT. WiKa Beton (Persero) Tbk. and has nominal size of 10 mm. The fine aggregate which consisted of Lumajang sand is prepared and fall into the Zone 2 category according to SNI 2834:2000. The water cement ratio is set to 0.52 to ensure the workability of the fresh concrete is sufficiently high.

The mix design is prepared using the DOE method. The steel fibers are additional material and therefore share the same portion with the coarse aggregates. The concrete strength is designed to have 30 MPa strength. The factors that influence the mechanical behavior of SRFC, such as volumetric fraction, water ratio, and aggregate quality, are considered during the mix design process. The concrete was poured into a cylinder steel formwork with 100 mm diameter and 200 mm height.

Table 1. Mix proportions of concrete 1 m^3.

Material	Specimens				
	S1-0%	S1–0.5%	S1–1.0%	S1–1.5%	S1–2.0%
OPC (kg/m^3)	619	619	619	619	619
Coarse Aggregate(kg/m^3)	1128	1128	1128	1128	1128
Gravel Sand (kg/m^3)	923.5	923.5	923.5	923.5	923.5
Water (ltr/m^3)	288	288	288	288	288
Fiber Steel (gr)	-	194	388	583	777
Slump (cm)	10 ± 2	10 ± 2	10 ± 2	10 ± 2	10 ± 2

V_f : Volumetric fraction; OPC: Ordinary Portland Cement

The steel fiber used in this research is Dramix steel fiber which was produced through a cold drawn process with a curve at the end that will provide optimal binding. Steel fiber with the type of Dramix 65/60 BG was used. These fibers have a length of 60 mm and a diameter of 0.9 mm, with an aspect ratio of $l_d/d_f = 60$. They are bonded together into a bundle of 30 fibers with water-soluble glue to perform better in fresh concrete. Thus, increasing the workability of the mixture and can reduce the initial tensile strength in concrete. The main characteristics of steel fiber are presented in Table 2.

Table 2. Properties of steel fiber$^+$.

Geometry and Properties	Values
Diameter (mm)	0.9
Length (mm)	60
Aspect Ratio (l_f/d_f)	65
Young's Modulus (N/mm^2)	200000
Tensile Strength (N/mm^2)	1600

3. Experimental program

Five concrete cylinders were prepared for the test. The experimental test was performed at Laboratory of Concrete, Advanced Material and Computational Mechanics, Civil Engineering Department, Institut Teknologi Sepuluh Nopember. Slump test, fiber orientation test and splitting test were conducted to observe the workability and performance of SRFC.

3.1. Slump test

The purpose of this test is to determine the concrete workability. The slump test is carried out using Abrams Cone. The Abrams Cone had a base diameter of 200 mm, a width of 100 mm, and a height of 300 mm. A rod compacter with a diameter of 160 mm and a length of 600 mm was used to compact the concrete. The cones are placed on a flat and wet surface. To prepare the concrete for slump test, there are three stages involved. In each of these stages, the layer is compacted 25 times by rod compactor. At

IOP Conf. Series: Materials Science and Engineering **930** (2020) 012058 doi:10.1088/1757-899X/930/1/012058

the end of the third stage, leveling the concrete surface with rod and all fresh concrete that falls around the molds should be cleaned. The mold is lifted slowly and upright. Concrete then declined. The cone is then placed in an upside-down position next to the test specimen as a reference height to measure the concrete slump. The slump was measured using a ruler from the top of the slumped concrete to the height of the Abrams Cone. This test is carried out within 2.5 minutes.

3.2. Fiber orientation

Fiber orientation is often used to estimate the fiber volumetric fraction in the specimen. The method works by manually counting the number of fibers that crosses the failure plane after the tensile splitting test was performed [15]. The number of fibers is the total amount of withdrawal fibers that appear on both sides. The orientation factor is estimated using the equation below [16].

$$\alpha = \frac{N}{N_{th}} \tag{1}$$

where α is the fiber orientation, N is the ratio number of fibers counted, and N_{th} is the cross section and theoretical number. Among all methods available (image analysis, non-destructive testing) to check the fiber orientation, this method was found to be the simplest [17-20].

3.3. Splitting test

After curing for 28 days, the splitting test was conducted for each cylinder to investigate the tensile strength of SFRC. Figure 1 shows the splitting tensile test procedure based on [21]. Two parallel steel plates are used to uniformly distribute the stress at the top and bottom sides of the concrete cylinder. The load-bearing strip should not be very narrow to avoid stress concentration at the concrete surface. The tensile strength of concrete (f_{st}) can be determined by:

$$f_{st} = \frac{2P}{\pi l D} \tag{2}$$

where P is the applied compressive load to split a concrete cylinder, L is the length of the specimen and D is the diameter of the specimen. Universal Testing Machine (UTM) is used to apply the compressive loads (P) at rates of 0.5/mm^2/sec.

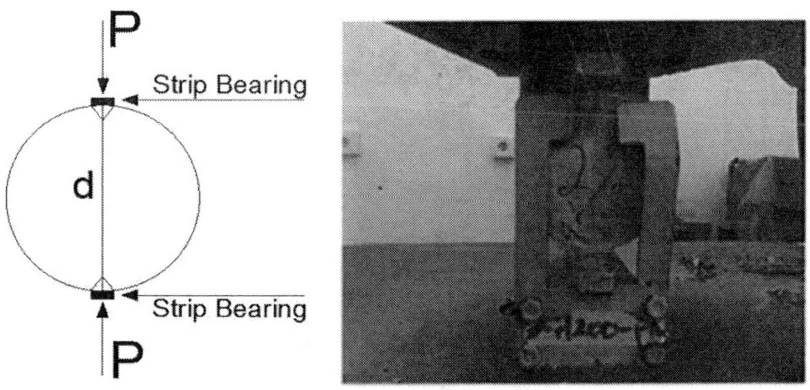

(a) (b)

Figure 1. Splitting test (a) geometry test; (b) experimental equipment

4. Results and discussion

In this part, the test result from five specimens are presented. The discussion is focused on the effect of steel fiber volumetric ratio to the concrete slump test, the fiber orientation, and the concrete tensile strength. The specimen with notation S1-0.0% is the control specimen without steel fiber.

IOP Conf. Series: Materials Science and Engineering **930** (2020) 012058 doi:10.1088/1757-899X/930/1/012058

4.1. Slump test

The value of concrete slump was designed for 100 ± 20 mm. The slump of the concrete is measured from the top of the slumped concrete to the top level of the Abrams Cone (see Figure 3). From the measurement it was found out that as the steel fiber volumetric fraction increases, the concrete slump value is decreases. This shows that the presence of the steel fiber reduces the workability of concrete. It should be noted that in this experimental program, no admixture was used in the mixture proportion.

Table 3. Slump Test.

Specimen	V_f (%)	Slump Test (cm)
S1 – 0.0 %	0.0	12
S1 – 0.5 %	0.5	10
S1 – 1.0 %	1.0	10
S1 – 1.5 %	1.5	9
S1 – 2.0 %	2.0	8

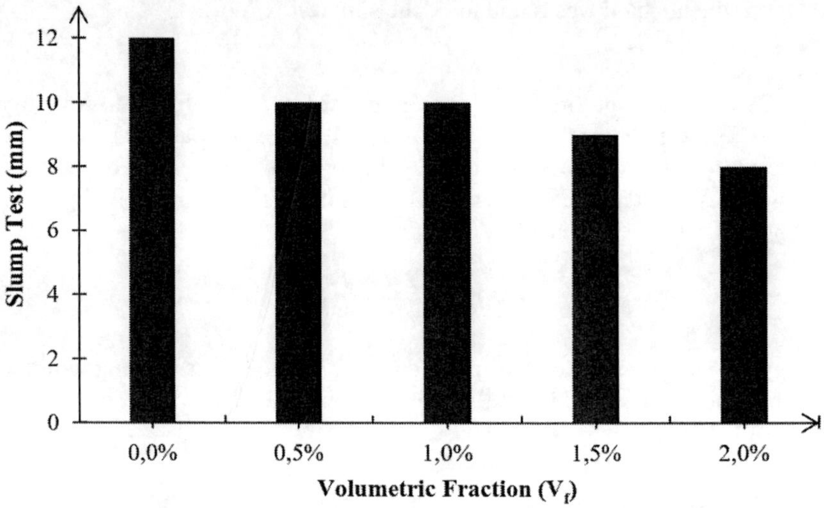

Figure 2. Slump test of SFRC concrete as function of V_f

Figure 3. Slump measurement

4.2. Fiber orientation

Like conventional reinforcement, fibers are most efficient when aligned parallel to the direction of tensile stress, and least efficient when they are perpendicular to the direction of tensile stress [22]. Table 4 shows the fiber orientation value for all of the specimens performed using manual counting of the fibers that cross the failure plane. Figure 4 shows the fiber orientation as the function of the volumetric fraction of the steel fibers. Figure 5 shows the cross-section of the fiber failure plane that used to manually counted the number of steel fiber that cross the failure plane. Some researcher noted that the fiber orientation was dependent on the concrete workability [23, 24]

Table 4. Fiber orientation of the tested specimens

Specimen	Volumetric fraction	Fiber Orientation (α)
S1 – 0.0 %	0.0 %	-
S1 – 0.5 %	0.5 %	0.486
S1 – 1.0 %	1.0 %	0.864
S1 – 1.5 %	1.5 %	1.440
S1 – 2.0 %	2.0 %	1.960

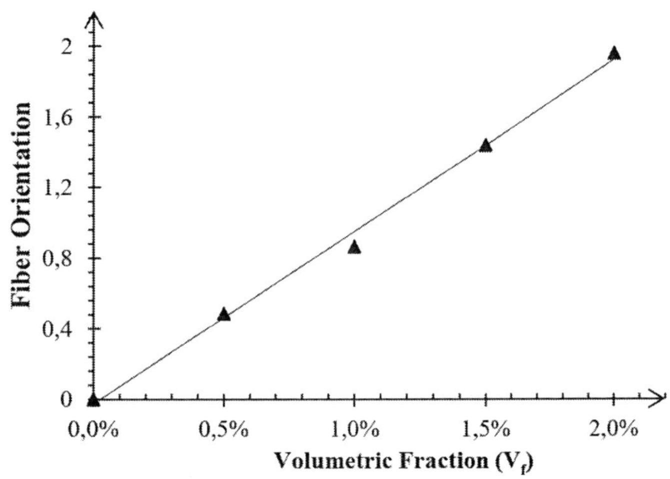

Figure 4. Fiber orientation of the specimen as function V_f

Figure 5. Cross-section of the concrete after tensile splitting test

4.3. Concrete tensile splitting test

Table 5 show the result of the concrete tensile splitting strength. As shown in Table 5, as the fiber volumetric fraction increases, the concrete tensile splitting strength also increases. Figure 6 shows the graph of the concrete tensile splitting strength as functions of the fiber volumetric fraction. As shown in Figure 6, linear increase in the concrete tensile splitting strength was observed from the test result. Figure 6 also shows the linear regression line for forecasting purposes. Despite of significant increase of concrete tensile strength due to the steel fiber, the improvement of the concrete compressive strength was found to be less significant [25-27].

Table 5. Concrete tensile splitting strength

Specimen	Volumetric Fraction (V_f)	Tensile strength (f'_{st})
S1 – 0.0 %	0.0 %	2.93
S1 – 0.5 %	0.5 %	4.13
S1 – 1.0 %	1.0 %	5.75
S1 – 1.5 %	1.5 %	6.34
S1 – 2.0 %	2.0 %	7.68

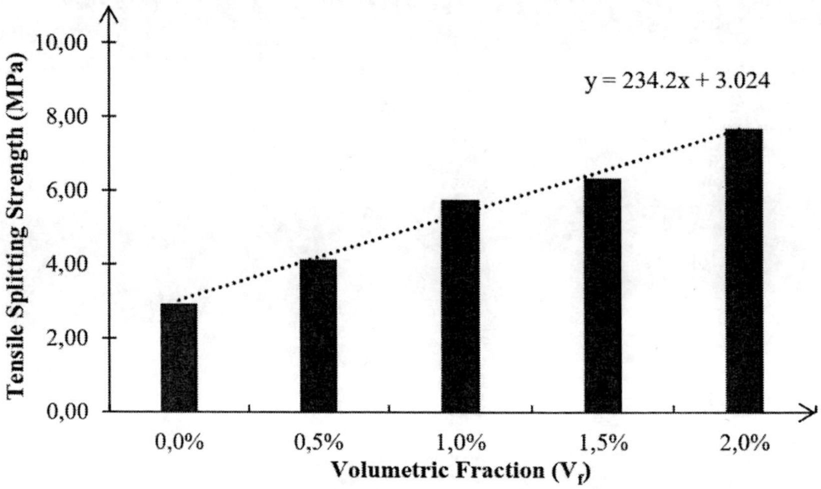

Figure 6. Concrete tensile splitting strength as function of V_f

5. Conclusions

This paper presents experimental test of concrete cylinder to evaluates the workability, fiber orientation and the concrete tensile splitting strength of the SFRC. From the experimental test, it was found out that the additional steel fiber affects the workability of concrete. The slump test value of SFRC concrete was found to be lower than normal concrete without the steel fiber. As the steel fiber content increases, the slumps decreases. In this experimental program, there are no admixture used in the mix proportion. From the investigation of the fiber orientation that measured at the cross-section of the failure plane showed that the result agrees well with the fiber volumetric fraction of the tested specimen. From the tensile splitting test of concrete, it was found out that the concrete tensile splitting strength is function of the volumetric fraction of the steel fiber and it shows a linear relation ship. Hence, linear regression analysis should be sufficient for forecasting purposes.

Acknowledgments

This research was supported by Civil Engineering Department of Institut Teknologi Sepuluh Nopember through Penelitan Dana Department Institut Teknologi Sepuluh Nopember (Grant no 1709/PKS/ITS/2020). The authors would also express their gratitude to PT WIKA BETON (Persero)

IOP Conf. Series: Materials Science and Engineering **930** (2020) 012058 doi:10.1088/1757-899X/930/1/012058

References

[1] Amin A and S J Foster 2015 Predicting the flexural response of steel fibre reinforced concrete prisms using a sectional model *Cem. Concr. Compos.* **67** 2016 p 1-11
[2] Dahake A G and K S Charkha 2016 Effect of steel fibers on strength of concrete. *J. Eng. Sci. Manage Educ.* **9** 1 p 45-51
[3] Sahin, Y and F Koksal 2010 The influences of matrix and steel fibre tensile strengths on the fracture energy of high-strength concrete *Constr. Build. Mater.* **25** p 1801-6
[4] Zhang S, C Zhang and L Liao 2018 Investigation on the relationship between the steel fibre distribution and the post-cracking behaviour of SFRC *Constr. Build. Mater.* **200** p 539-50
[5] Li B et al 2018 Effects of fiber type, volume fraction and aspect ratio on the flexural and acoustic emission behaviors of steel fiber reinforced concrete. *Constr. Build. Mater.* **181** 2018 p 474-86
[6] Yazıcı S, G Inan and V Tabak 2006 Effect of aspect ratio and volume fraction of steel fiber on the mechanical properties of SFRC *Constr. Build. Mater.* **21** 2007 p 1250-3
[7] Lee S J, D Y Yoo and D Y Moon 2019 Effects of hooked-end steel fiber geometry and volume fraction on the flexural behavior of concrete pedestrian decks *Appl. Sc.* **9** p 1241
[8] Soufeiani L et al 2016 Influences of the volume fraction and shape of steel fibers on fiber reinforced concrete subjected to dynamic loading – A review. *Eng. Struct.* **124** 2016 p 405-17
[9] Barros J A O and J A Figueras 1999 Flexural behaviour of SFRC: testing and modelling *J. Mater. Civ. Eng.* **11** p 15853
[10] Chang D I et al 1991 A study on the fracture energy of steel fiber reinforced concrete structures with initial cracks *Trans. SMiRT* **11** p 139-44
[11] Jo B W, Y H Shon and Y J Kim 2001 The evalution of elastic modulus for steel fiber reinforced concrete. *Russ. J. Nondestruct.* **37** p 152-61
[12] Koksal F, A Ilki, and M A Tasdemir 2012 optimum mix design of steel-fibre-reinforced concrete plates *Arab. J. Sci. Eng.* **38** p 2971-83
[13] Badan Standar Nasional 2004 *SNI 15-7064-2004 in Semen portland komposit* (Badan Standard Nasional)
[14] Badan Standard Nasional 2008 *SNI 1969:2008 in Cara uji berat jenis dan penyerapan air agregat kasar* (Badan Standard Nasional)
[15] Martinie L and N Roussel 2011 Simple tools for fiber orientation prediction in industrial practice *Cem. Concr. Res.* **41** p 993-1000
[16] Soroushian P and C D Lee 1990 Distribution and orientation of fibers in steel fiber reinforced concrete. *ACI. Mater. J.* **87** 5 p 433-9
[17] Aveston D and A Kelly 1973 Theory of multiple fracture of fibrous composites *J. Mater. Sci.* 1973 **8** p 352-62
[18] Ferrara L and A Meda 2006 Relationships between fibre distribution, workability and the mechanical properties of SFRC applied to precast roof elements *Mater. Struct.* **39** p 411-20
[19] Ferrara L and N Ozyurt 2011 High mechanical performance of fibre reinforced cementitious composites: the role of ''casting-flow induced'' fibre orientation *Mater. Struct.* **44** p 109-28
[20] Robins P J, S A Austin and P A Jones 2003 Spatial distribution of steel fibres in sprayed and cast concrete *Mag. Concrete Res.* **55** p 225–35
[21] ASTM C496/C496M - 11 2011 in Standard Test Method for Splitting Tensile Strength of Cylindrical Concrete Specimens *ASTM International* p PA 19428-2959
[22] Shah S P and B V Rangan 1971 Fiber Reinforced Concrete Properties *ACI Mater. J.* 1971 **68** 2 p 126-35
[23] Boulekbache B et al 2012 Influence of yield stress and compressive strength on direct shear behaviour of steel fibre-reinforced concrete *Constr. Build. Mater.* **27** p 6-14
[24] Boulekbache B et al 2015 Failure mechanism of fibre reinforced concrete under splitting test using digital image correlation *Mater. Struct.* **8** p 2713-26
[25] Clayton N 1990 The cylinder splitting test as a method for determining the compressive strength of concrete *Mag. Concrete Res.* **42** p 51-5

IOP Conf. Series: Materials Science and Engineering **930** (2020) 012058 doi:10.1088/1757-899X/930/1/012058

[26] Lin Z and L Wood 2003 Concrete Uniaxial Tensile Strength and Cylinder Splitting Test *J. Struct. Eng.* **129** p 692-8
[27] Rocco C et al 1999 Size effect and boundary conditions in the brazilian test: theoretical analysis *Mater. Struct.* **32** p 437-44

IOP Conf. Series: Materials Science and Engineering **930** (2020) 012059 doi:10.1088/1757-899X/930/1/012059

Railway capacity analysis using Indonesian method and UIC code 405 method

H Widyastuti[1*] and W S Budhi[1]

[1] Department of Civil Engineering, Faculty of Civil, Planning and Geo Engineering, Sepuluh Nopember Institute of Technology, Surabaya, East Java, Indonesia.

*Corresponding authors e-mail: hera@ce.its.ac.id

Abstract. The government has been constructing double track at Surabaya – Madiun segment to improve railway operations performance. Accordingly, after the double track has been constructed in this segment, it is necessary to analyze existing railway capacity to determine the effect of development as a cost-effective solution. Nowadays, Indonesia is using the Indonesian method as a capacity analysis method and only provides railway capacity analysis using fixed block. On the other hand, another method such as the UIC method has different variables. Therefore in this study, the railway capacity analysis uses two analytic methods that are Indonesian and UIC code 405. The analysis result of both methods shows an improvement in railway capacity because of the double-track construction. These two methods have different results due to each variable. Moreover, one of the UIC variables is additional time due to maintenance differentiated based on its utilization. Utilization in the UIC method is differenced based on its operational conditions by peak and non-peak hours or normal conditions that may also affect the capacity. It also happens in MRT Jakarta operating systems in which the headway is divided into peak and non-peak hours. Accordingly, this study could be a preliminary study for policymakers to analyze capacity in the next near future condition.

1. Introduction

Railways Strategic Plan of Transportation Ministry concerning Railway Sector of 2015-2019 [1] mentions delay data of passenger and freight trains have tended to fluctuate, and in 2014, both increased. Freight train is the most frequently delayed, the arrival delay time of which reached out 140 minutes in 2014. While, the longest arrival delay time of the passenger train happened in 2010 which was 50 minutes. Delay data of passenger and freight trains have tended to fluctuate, and in 2014 both increased. According to Sogin [2], it was caused by the different characteristics of passenger and freight trains, e.g. power, weight, braking performance, speed, priority, and on-time arrival. Therefore, delay indicates the existing railway capacity problem. In this case, capacity is represented by indicators related to infrastructure with traffic data, while the timetable is the main input in determining capacity consumption and delay indication [3].

Content from this work may be used under the terms of the Creative Commons Attribution 3.0 licence. Any further distribution of this work must maintain attribution to the author(s) and the title of the work, journal citation and DOI.
Published under licence by IOP Publishing Ltd

Railway capacity analysis is one of the cost-effective solutions because of the ability to manage railway capacity more effectively. This analysis is aimed to determine the maximum number of trains that can be operated on certain infrastructure and time intervals by observing existing operational conditions [4]. Way to offset the flow of passenger and cargo mobility is by increasing connectivity on the rail network [5]. Connectivity enhancement can be done by constructing new infrastructure, using existing infrastructure, and also reactivating old infrastructure [6]. A new infrastructure project such as improving the track from single to double track is also one of solutions to increase the capacity. In 2019, Surabaya – Madiun segment was double-track constructed, where Jombang – Madiun segment has begun operating [7]. Concerning this condition, it is necessary to analyze how the capacity will be affected when the double track is in operation.

Nowadays, Indonesia uses the Indonesian method, it only provides railway capacity analysis for fixed blocks [8], it uses 3 variables as input in the capacity calculation which is the total time per day (1440 minutes), headway, and multiplier factor. Another method in railway capacity analysis namely UIC Code 405 method provides a formula which has different variables. There are 5 variables as input in the capacity calculation of the UIC method, which are reference time, average minimum headway, additional time due to the number of intermediate blocks, and additional time due to maintenance which is differenced by its utilization. Considering these conditions, it is necessary to analyze using both methods. And also, it could be a preliminary study on the railway capacity method in the next near future condition.

2. Methodology

2.1. Experimental objective

Surabaya – Madiun railway segment is currently under construction from single track become double track. The length of this segment is 152,828 km and consists of 23 stations, starting from Surabaya Gubeng Station to Madiun Station [9]. In this case, DAOP (Daerah Operasional) 8 region scope authority is Surabaya Gubeng Station to Mojokerto Station. While, DAOP (Daerah Operasional) 7 region scope authority is Curah Malang Station to Madiun Station. For detailed locations are shown in figure 1 below:

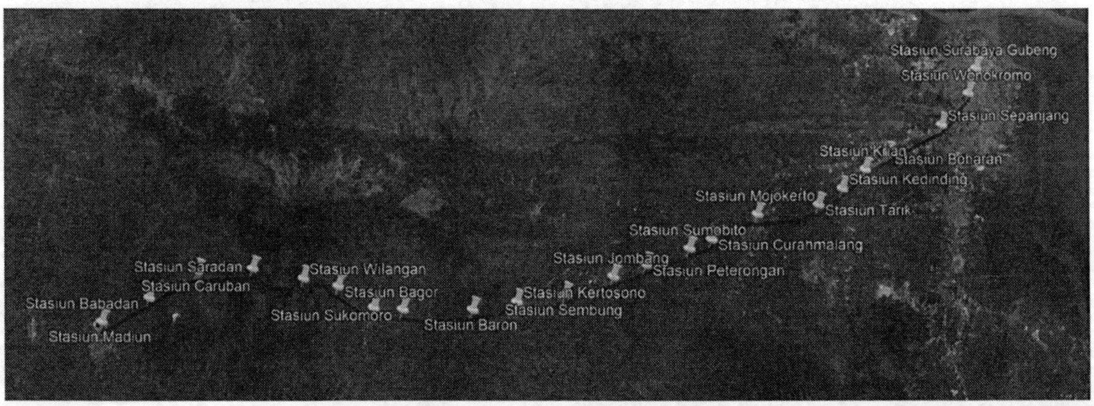

Figure 1. Location of double track project.

2.2. Data sources

The data used in this study are only secondary data. Secondary data are obtained from related institutions, including from PT.KAI DAOP 8 Surabaya and DAOP 7 Madiun. Data used in this study are GAPEKA (Grafik Perjalanan Kereta Api), track system, signaling and block system, signal service time, rolling stock data, and train speed. The detail of data function can be seen in table 1.

Table 1. Types of data and function.

Data Types	Function
Grafik Perjalanan Kereta Api (GAPEKA)	To count the number of operating trains, train departure and arrival time at a station. Furthermore, to find out the railway segment, signaling and block system, track system, etc.
Signaling and block system	Signaling and block system needed in railway capacity data processing and its location will determine the block section length in the UIC method.
Track system	Track system either the single track or double track becomes an important variable in railway capacity analysis
Signal service time	Signal service time used for headway analysis, for both single and double track.
Rolling stock data	The data referred to types of trains and train succession. Furthermore, the train type will determine its power, resistance, and also acceleration and deceleration. These data are used in the UIC method
Travel time	Travel time is used to determine operation speed
Train speed	The speed used is the maximum speed and operation speed. The maximum speed is obtained from the applicable GAPEKA. Whereas, the operation speed is average speed of trains when passing a certain segment.

2.3. Methods of railway capacity analysis

Supriadi [8] mentioned that railway capacity is a railway ability to accommodate train operations in the period of 1440 minutes [8]. While, UIC (1996) stated railway capacity can be acknowledged by three main parameters: infrastructure, operating systems, and service quality. Railway capacity is not static, but dynamic and bound by components forming a railway system that makes capacity depend on the specific train mix and operating conditions [3, 10]. Therefore, railway capacity will be varied along with changes in infrastructure and operating conditions. Infrastructure parameters, traffic parameters, and operating parameters are several factors that affect railway capacity [3, 11, 12]. Railway capacity is categorized into 4 types i.e. theoretical capacity, practical capacity, used capacity, and available capacity [4]. Theoretical capacity is the number of operating trains during a certain time interval; practical capacity is the actual traffic volume condition which is calculated using realistic assumptions; used capacity is the actual traffic volume occuring on the existing rail network; and available capacity is the difference between practical capacity and used capacity [4]. Furthermore, the explanation of the Indonesian method and the UIC code 405 method is as follows:

1. Indonesian Method

Supriadi [8] stated the formula used by Indonesian railways is differenced by track systems (single track and double track). This method is an analytical method in the form of mathematical equations [8], [13].

$$K = \frac{1440}{H} x\eta \tag{1}$$

$$K = \frac{1440}{\frac{1}{2}H} x\eta \tag{2}$$

$$H = t_{A-B} + t_p + C \tag{3}$$

Where:

K : capacity on the calculated track section

1440: total time for 24 hours (24x60 minutes)

H : average headway

η : The multiplier factor after deducting time factor for maintenance and time due to the operating pattern of train travel, 60% for single track and 70% for double track
tp : travel time on a track section
t_{A-B} : travel time from the sight point of the outer signal to the station
C : signal and block service time

Equation (1) is a formulation that applies to a single-track system. Equation (2) is a formulation used for the double-track system. Meanwhile, equation (3) is the headway calculation.

Headway calculation is a step before analyzing railway capacity. Equation (3) is the main formula to calculate the headway of Indonesian method. However, the headway calculation of Indonesian method is more simplified. It is divided based on the track system and block system with the value of several variables having been determined. The Indonesian method only provides formulas for fixed block system. As in the Surabaya - Madiun railway segment, the signaling system used is an electrical signal system with closed automatic block system. The Indonesian electrical signaling scheme can be seen in figure 2.

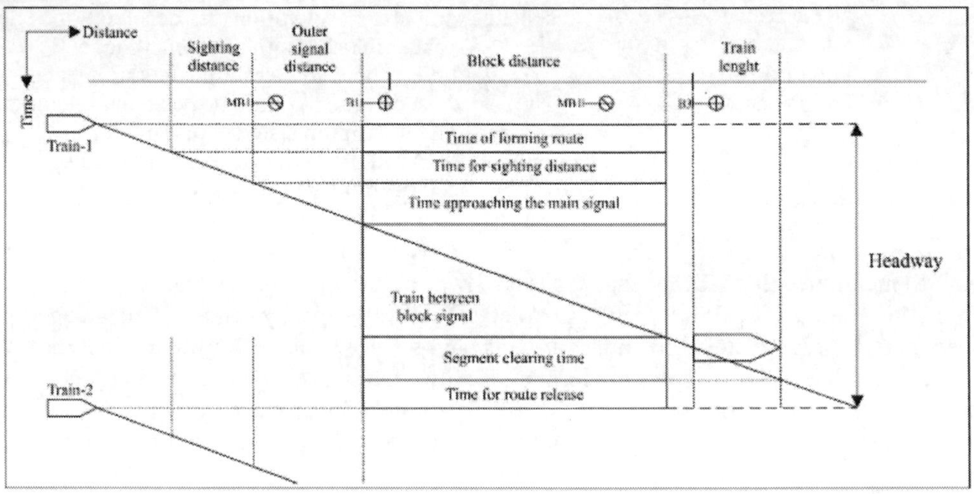

Figure 2. Electric signaling headway scheme [8].

2. UIC code 405 Method

International Union of Railways (UIC) in a leaflet code 405 proposed analytical method for calculating railway capacity between stations sequentially. This method calculates the proportion of operating trains to calculate the headway [14–16]:

$$P = \frac{T}{t_{fm} + t_r + t_{zu}} \tag{4}$$

Where:

P : capacity (per day, per hour, etc according to the T used)
T : reference time (usually 24 hours for daily capacity or 1440 minutes)
tfm : average minimum headway
tr : additional time due to maintenance
  tr = 0,67 x tfm if utilization 0,6 (applicable for normal operation of the system)
  tr = 0,33 x tfm if utilization 0,75 (applicable during peak hours)
tzu : additional time based on the number of intermediate blocks in the track and calculated using the formula tzu = 0.25 x a
a : number of intermediate block sections

Average minimum headway for each track is calculated using the average minimum headway between two consecutive trains of the same category [17]:

$$t_{fm} = \frac{\alpha_L . t_{fm_L} + \alpha_R . t_{fm_R} + \alpha_M . t_{fm_M}}{\alpha_L + \alpha_R + \alpha_M} \tag{5}$$

This procedure considers three different types of trains
L = intercity train
R = local/regional train
M = freight train
Where α_L, α_R, α_M factors in the formula above represent the percentage of each category of total trains.

Similar to the Indonesian Method, this method also requires headway to analyze railway capacity. F.Rotoli [17] pointed out that headway at a single-track system will be affected by travel time, acceleration and deceleration time, and signal service time. On thc other hand, at a double-track system, the headway will be affected by the block length, train length, speed and signal service time.

a. Headway for single track
Determine travel times with constant operating speed and acceleration and deceleration time using Equation 6 – Equation 8 [17]:

$$t_{V_{L,R,M}} = \frac{l_b}{V_{L,R,M}} \tag{6}$$

$$t_{\alpha_{L,R,M}} = \frac{V_{L,R,M}}{\alpha} \tag{7}$$

$$t_{d_{L,R,M}} = \frac{V_{L,R,M}}{d} \tag{8}$$

Where:
$t_{V\,L,R,M}$ = travel time with a constant operating speed
$V_{L,R,M}$ = operating speed of each train category
l_b = length of railway segment
a = acceleration
d = deceleration

The travel time is the result of the average train operating speed of each category divided by the length of the railway segment. Therefore, the minimum average headway can be calculated using Equation 9 [17]:

$$t_{fm_{L,R,M}} = t_{V_{L,R,M}} + t_a + t_d + t_p \tag{9}$$

Where:
t_a and t_d = time of acceleration and deceleration (the value of acceleration and deceleration is indicated by *a* and *d*)
t_p = signal service time
$t_{fm_{L,R,M}}$ = average minimum headway

In this case, it is assumed in single track that the railway segment between two stations can only be occupied by one train per time based on the direction of the track. Furthermore, signal service time value (t_p) is assumed as representing additional time from the preparation of the electro-mechanical block routing.

b. Headway for double track

The average minimum headway for double track is calculated using Equation 10 [17]:

$$t_{fm_{L,R,M}} = \frac{l_{b1} + l_{b2} + L}{V_{L,R,M}} + t_s \tag{10}$$

Where:

l_b = length of block section
L = length of train succession
$V_{L,R,M}$ = operating speed considered for each category
t_s = signal service time

In this case, to calculate the double-track headway, it is assumed that the railway is equipped with a three-aspect automatic block signaling system. Where the minimum distance between two trains consecutively is based on the first block that functions to guarantee the train braking distance, added with the second block that functions to guarantee sighting distance (the operating train has to find the approaching signal location clearly) and additional distance to see signals and switch, and additional train length from the block system [17]. The blocking time calculation scheme for double track can be seen in figure 3.

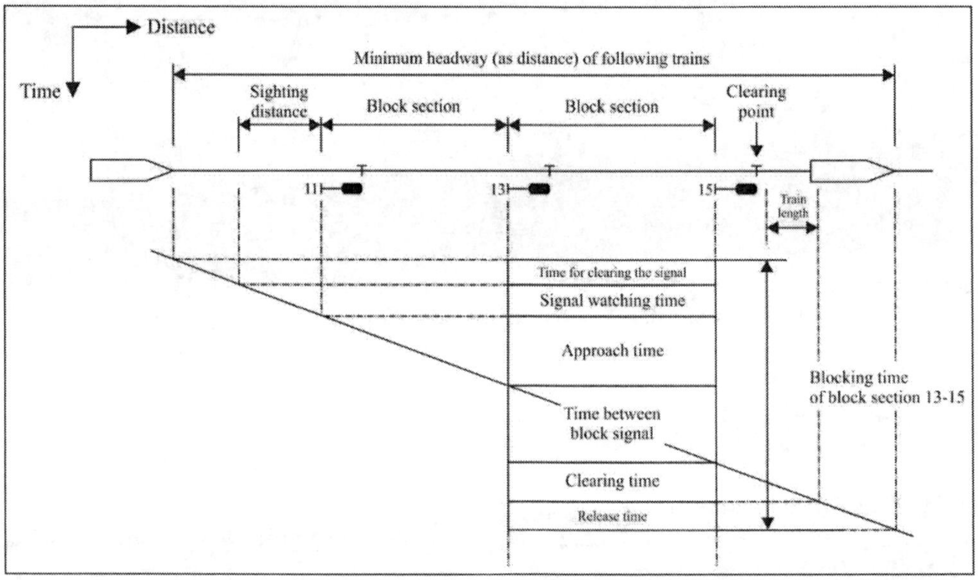

Figure 3. Blocking time calculation scheme for double track [17].

3. Result and discussion

In this study, the railway capacity analysis application of Indonesian method and UIC code 405 method in the Surabaya – Madiun railway segment was carried out. The analysis was performed on the existing conditions with 2 assumptions: full single-track condition which reflects GAPEKA 2017 and 50% double-track condition which reflects GAPEKA 2019. The Surabaya – Madiun segment has different number of operating trains. It is caused by several local trains that only operate in several lines in Surabaya – Madiun segment. The Surabaya – Madiun segment is divided into 22 segments for GAPEKA 2017 and 21 segments for GAPEKA 2019, which is caused by Wilangan Station which turns into an intermediate block. The railway capacities analyzed in this study are the theoretical capacity and practical capacity. Railway capacity analysis result of the Indonesian method and UIC Code 405 method can be seen in figure 4 – figure 7.

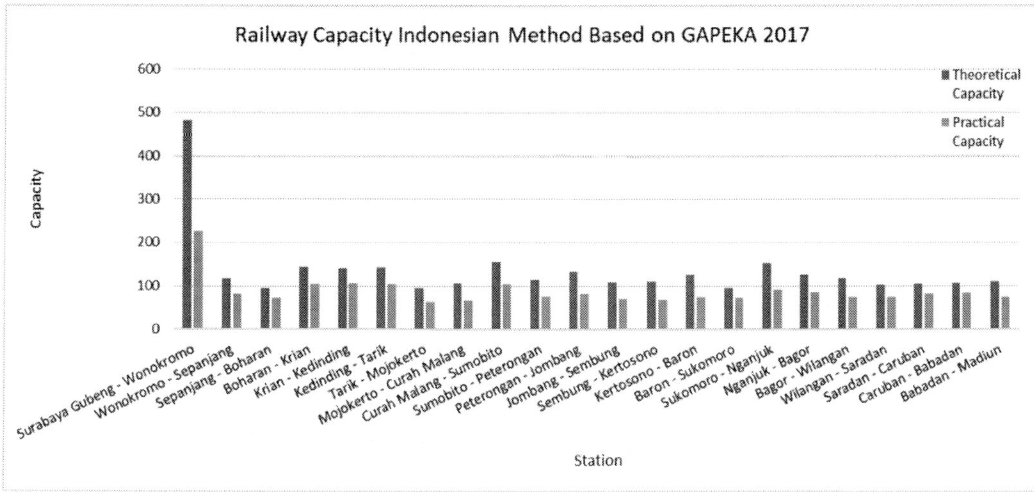

Figure 4. Railway capacity of the Indonesian method based on GAPEKA 2017.

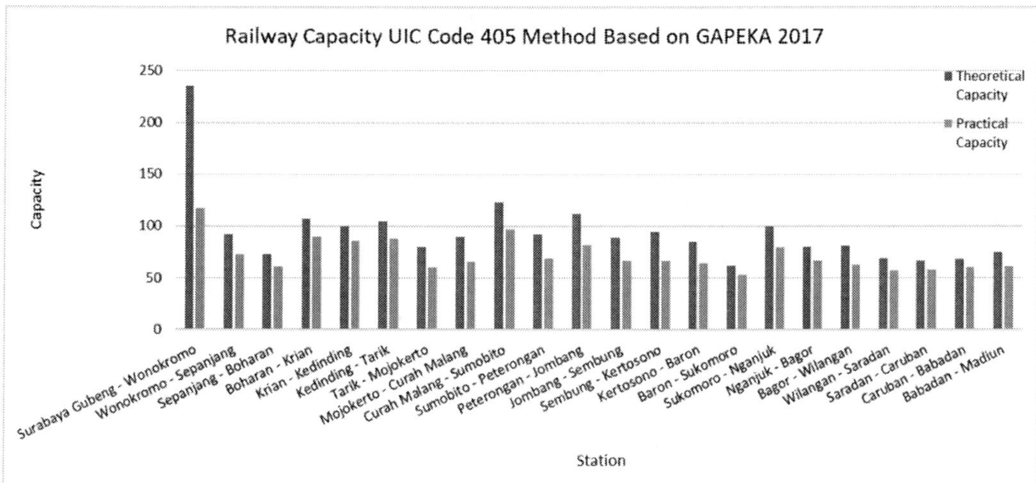

Figure 5. Railway capacity of the UIC Code 405 method based on GAPEKA 2017.

Based on figure 4 and figure 5, the largest theoretical capacity of Indonesian method and UIC method is located in the Surabaya Gubeng Station – Wonokromo Station segment. The average theoretical capacity based on GAPEKA 2017 is 136 trains/day for the Indonesian method and 95 trains/day for the UIC code 405 method. The largest practical capacity of Indonesian method and UIC method is also located in the Surabaya Gubeng Station – Wonokromo Station segment. The average practical capacity based on GAPEKA 2017 is 89 trains/day for the Indonesian method and 73 trains/day for the UIC code 405 method.

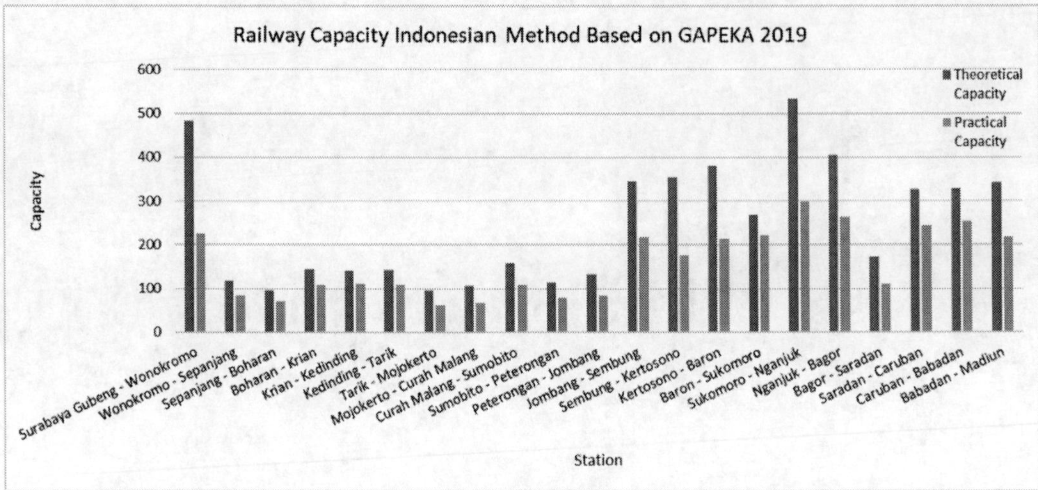

Figure 6. Railway capacity of the Indonesian method based on GAPEKA 2019.

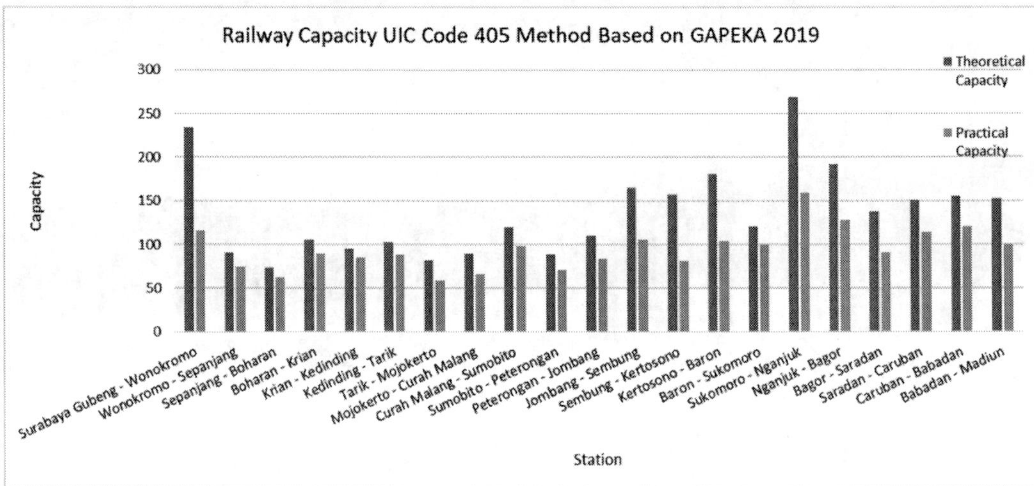

Figure 7. Railway capacity of the UIC Code 405 method based on GAPEKA 2019.

Based on figure 6 and figure 7, the largest theoretical capacity of Indonesian method and UIC method is located in the Sukomoro Station – Nganjuk Station segment. The average theoretical capacity based on GAPEKA 2019 is 247 trains/day for the Indonesian method and 137 trains/day for the UIC code 405 method. The largest practical capacity of Indonesian method and UIC method is also located in the Sukomoro Station – Nganjuk Station segment. The average practical capacity based on GAPEKA 2019 is 159 trains/day for the Indonesian method and 96 trains/day for the UIC code 405 method. The railway capacity increment is due to the double-track operation and the replacement of the signaling and block system. Moreover, it is affected by the Wilangan station that turns into an intermediate block.

4. Conclusion

The results of railway capacity analysis using these two methods indicate an improvement in railway capacity because of double-track construction in Surabaya - Madiun segment, although the railway capacity value of Indonesian Method and UIC method is different. This is caused by the different variables in each method. Besides, these two different methods will have different methodologies and complexity levels in carrying out railway capacity analysis [15], even though the headway value of both methods is quite similar. One of the different variables is additional time due to maintenance which is differentiated based on its utilization. Utilization in the UIC method is differenced based on its operational conditions by peak and non-peak hours that may also affect the capacity. It also happens in MRT Jakarta operating systems in which the headway is divided into peak and non-peak hours or normal conditions. In addition, MRT Jakarta and LRT Jabodebek have already used a moving block system [18]. This study implies that UIC Method can be a relevant comparison method to the existing conditions, and to generalize the results of these two methods, more case studies are needed. However, it can be used as a preliminary study recommendation for policymakers to analyze capacity in the next near future condition. Furthermore, regardless the method that will be applied in the next near future, safety must be considered more, especially at railway level crossings which are still many in Indonesia, taking into account that the crossings may create several serious potential conflicts for collisions between road vehicles and trains [19]. In addition, these conflicts will also affect the railway operating system performance.

Acknowledge
The authors would like to thank the Indonesian Railway Company (PT KAI (Persero)) for supporting us in the development of this study.

References
[1] Direktorat Jenderal Perkeretaapian, "Rencana Strategis Kementerian Perhubungan Bidang Perkeretaapian Tahun 2015-2019," Jakarta, 2019.
[2] S. Sogin, Y. C. Lai, C. Dick, and C. Barkan, "Comparison of capacity of single- and double-track rail lines," *Transp. Res. Rec.*, vol. 2374, no. 2374, pp. 111–118, 2013, doi: 10.3141/2374-13.
[3] A. Lindfeldt, *Railway capacity analysis - Methods for simulation and evaluation of timetables, delays and infrastructure*. 2015.
[4] M. Abril, F. Barber, L. Ingolotti, M. A. Salido, P. Tormos, and A. Lova, "An assessment of railway capacity," *Transp. Res. Part E Logist. Transp. Rev.*, vol. 44, no. 5, pp. 774–806, 2008, doi: 10.1016/j.tre.2007.04.001.
[5] Y. Wiarco, S. Malkhamah, and I. Muthohar, "Tinjauan Atas Faktor-Faktor Yang Memengaruhi Kapasitas Jalur Kereta Api," in *The 17th FSTPT International Symposium, Jember University*, 2014, pp. 1291–1298.
[6] R. Ravitharan, A. Labrooy, H. Widyastuti, and W. K. Chiu, "Rail Infrastructure in Port City - Surabaya, Indonesia," in *Procedia Engineering*, 2017, vol. 188, pp. 486–492, doi: 10.1016/j.proeng.2017.04.512.
[7] PT. Kereta Api Indonesia, *Grafik Perjalanan Kereta Api (GAPEKA)*. Bandung, 2019.
[8] U. Supriadi, *Kapasitas Lintas dan Permasalahannya*. Bandung, 2008.
[9] PT. Kereta Api Indonesia, *Grafik Perjalanan Kereta Api (GAPEKA)*. Bandung, 2017.
[10] U. Supriadi, *Perencanaan Perjalanan Kereta Api dan Pelaksanaannya*. Bandung, 2008.
[11] A. Landex, "Methods to estimate railway capacity and passenger delays," *Tech. Univ. Denmark*, p. 218, 2008.
[12] E. Kontaxi and S. Ricci, "Railway capacity analysis: Methodological framework and harmonization perspectives," in *Proceedings of the 12th World Conference on Transport Research*, 2010, pp. 1–21.
[13] R. Prihatanto, A. Wicaksono, and L. Djakfar, "Evaluasi Kapasitas Lintas Jalur Ganda Kereta Api Segmen Bojonegoro – Surabaya Pasarturi," in *19th International Symposium of FSTPT*, 2016, vol. 10, no. October, pp. 1445–1453.
[14] F. Rotoli, S. Ricci, E. Navajas Cawood, and G. Malavasi, "Capacity versus punctuality assessment procedures and accessibility measures for rail networks," *Ing. Ferrov.*, vol. 12, no. January 2016, pp. 1011–1040, 2015.
[15] F. Rotoli and S. Ricci, "Methodological Review for Capacity and Punctuality Assessment

Procedures," *Int. J. Railw. Res.*, vol. 1, no. 2, pp. 31–47, 2015.

[16] T. Dimitrios and G. Charalaipous, *Relationship And Management Of Important Transportation, Space Submission Significant Line – Application (ΣΧΕΔΙΑΣΜΟΣ ΚΑΙ ΔΙΑΧΕΙΡΙΣΗ ΣΙΔΗΡΟΔΡΟΜΙΚΩΝ ΜΕΤΑΦΟΡΩΝ, ΥΠΟΛΟΓΙΣΜΟΣ ΧΩΡΗΤΙΚΟΤΗΤΑΣ ΣΙΔΗΡΟΔΡΟΜΙΚΗΣ ΓΡΑΜΜΗΣ - ΕΦΑΡΜΟΓΗ)*. Thessaloniki: Aristotle University Of Thessaloniki, 2004.

[17] F. Rotoli, G. Malavasi, and S. Ricci, "Complex railway systems: capacity and utilisation of interconnected networks," *Eur. Transp. Res. Rev.*, vol. 8, no. 4, 2016, doi: 10.1007/s12544-016-0216-6.

[18] Direktorat Jenderal Perkeretaapian, "FIXED BLOCK DAN MOVING BLOCK," *http://djka.dephub.go.id/*, 2019. .

[19] H. Widyastuti, A. Utami, and Z. M. Dzulfiqar, "Model of queuing in the railway level crossing (case study: Imam Bonjol railway level crossing in Blitar)," in *IOP Conference Series: Materials Science and Engineering*, 2019, vol. 650, no. 1, pp. 0–9, doi: 10.1088/1757-899X/650/1/012053.

IOP Conf. Series: Materials Science and Engineering **930** (2020) 012060 doi:10.1088/1757-899X/930/1/012060

Intercity train dwell time estimation by using robust regression method: a study case on Surabaya-Yogyakarta line

P Dewi[1] and H Widyastuti[2*]
[1] The Indonesian Railways Polytechnic, Madiun 63129, East Java, Indonesia
[2] Department of Civil Engineering, Sepuluh Nopember Institute of Technology, Surabaya 60111, East Java, Indonesia

*Corresponding author's e - mail: hera@ce.its.ac.id

Abstract. Dwell time is important an issue to optimise not only for train operation efficiency but also punctuality. This paper proposes a model based on a field survey of intercity trains along the Surabaya–Yogyakarta line by using a robust regression analysis of 138 observations from 13 trainsets operating on the line. Assuming no differences in door or platform width, a model has been formulated to include the number of people who are alighting and boarding, the type of platform and the number of train doors on one side. There were three robust estimators applied to the model. With a 2.037% mean absolute percentage error, the robust regression using the M estimator formed the dwell time model for this study. The results show that, between the dummy variable for platform type using a binary number (0 or 1), dummy variables for high-level platforms and mid-level platforms have opposing tendencies. High-level platform use reduces dwell time by 2.014 sec based on continuous variables; in contrast, dwell time increases by 9.353 sec based on continuous variables by using mid-level platforms. Furthermore, people who are alighting need less time than those boarding. Dwell time will increase by 1.222 sec per alighting passenger and 1.314 per boarding passenger, while it will decrease by 2.727 sec per door on a one-sided trainset. Moreover, dwell time estimates with the model have reduced trainset travel time.

1. Introduction

Dwell time is an essential issue in train scheduling. Indonesian railway scheduling is charged to the Ministry of Transportation as the infrastructure owner and to PT Kereta Api Indonesia (Persero) as the intercity train operator. The Surabaya–Yogyakarta line is serviced by 13 trainsets; however, there are only two trainsets that service this line's main business. The rest trainset has further terminal station far off Yogyakarta and Surabaya. However, passenger demand on this line is high. PT Kereta Api Indonesia transported 2,478,064 people in 2017, 2,894,819 people in 2018 and 2,093,638 people from January to September 2019 on the Surabaya–Yogyakarta line with their 13 trainsets. Dwell time and stop stations for each trainset vary by classes (economy, executive and mixed class). The longest total dwell time along the line is 1 hour and 8 minutes with 13 stops, and the shortest is 15 minutes with four stops based on the 2019 Indonesian Train Schedule.

The Surabaya–Yogyakarta line has upgraded to a double track recently. It serves 45 stations in three provinces. For passenger purposes, there are only 19 stations where loading and/or unloading activities for the intercity trainset occur. In order to make dwell times for passenger flows in stations run effectively and fairly without dwell time classifications, mathematical models were established to estimate the effective dwell time for each serviced station.

Content from this work may be used under the terms of the Creative Commons Attribution 3.0 licence. Any further distribution of this work must maintain attribution to the author(s) and the title of the work, journal citation and DOI.
Published under licence by IOP Publishing Ltd

2. Literature review

Railway scheduling determines short stops in stations for loading and/or unloading passengers, train operation crossing and overtaking. There are no standard provisions for assigning dwell times in stations. Many researchers have studied dwell time using varied methods and approaches. A dwell time model including the number of alighting, boarding and through passengers using multiple linear regression was established in the Massachusetts Bay Transportation Authority (MTBA) Red Line. The model resulted in a linear liaison between the number of alighting and boarding passengers, but there was nonlinear relation in vehicle congestion. [1]

Another study was conducted in Mass Rapid Transit Swiss that developed a multiple linear regression model including the number of alighting and boarding passengers at critical doors, delays, morning peak hours, train type and rainy days. The study indicated that passenger boarding (1.2 sec/person) takes more time than alighting (0.9 sec/person) [2]. A study applied to London Underground Ltd (LUL) proposed a linear method to estimate stop times by including critical door factors, the number of boarding passengers, the number of alighting passengers, the number of train doors, the number of train seats, the number of through passengers and the door widths. However, it resulted in a nonlinear liaison for dwell time and number of alighting and boarding passengers [3].

A multi-agent simulation was established in Japan and found agent (passenger) decisions to alight and board were influenced by the environment and other agents. The environment consists of the platform and the train. The platform attributes were its length and width, stair locations, elevators, walls, pillars and where the train stops. The train attributes were the number of cars; car length and width; door locations; number of doors; door width; and seat location, width and length. For the agents, calculations of walking speed and direction towards the nearest door based on the environment were completed [4].

Multiple linear regression was chosen to model Istanbul Turkey Tramways' dwell time on weekends and weekdays. There were some variables that have important effects on the dwell time that were considered in the models, which included the number of alighting and boarding passengers, platform occupancy, wagon occupancy, peak hours and off-peak hours on weekdays and weekends and time spending for operation by the driver. High platform occupancy had a greater influence on tramway dwell time than high wagon occupancy on weekdays, but this was the opposite on weekends. Among the continuous variables, in both the weekday and weekend models, the time spent by the driver had the highest effect on dwell time. The number of passengers who were alighting in the weekday model and the number of passengers who were boarding in the weekend model had the lowest impact on dwell time [5]. A linear regression also resulted in a statistically significant liaison in a Taiwan Railway Administration (TRA) study. Through a field survey with digital applications for surveyors on each platform, the study resulted in average time of 1.64 sec/person for boarding and 1.38 sec/person for alighting [6].

Regression is one of the most commonly used statistical techniques. Out of many possible regression techniques, the ordinary least squares (OLS) method has been generally adopted because of its ease of computation. However, the OLS estimation of regression weights in a multiple regression is affected by outliers, non-normality, multicollinearity and missing data [7]. Outliers are observations that appear inconsistent with the rest of the data. Often, such influential points remain hidden to the user, because they do not always appear in least squares residual plots. To remedy this problem, new statistical techniques have been developed that are not so easily affected by outliers. These are robust methods, such as the Huber M Estimation, MM Estimation and S Estimation. [8]

In order to find the best regression model, mean absolute percentage error (MAPE) can be used as a benchmark. The best MAPE model is equivalent to a weighted mean absolute error (MAE) regression. Its universal consistency in empirical risk minimisation remains possible using MAPE instead of MAE [9].

3. Data collection

The data used in this paper were obtained from on-board observations of the Surabaya–Yogyakarta line on 13 trainsets for each station stop. The dwell time in this paper is the time between when the trainset fully stopped on the station platform and when the on-board operator (usually the on-board janitor) closed the door. The surveyor would stand in the middle of the trainset (in front of the dining car) to observe passenger flows and measured dwell time using a stopwatch. The platform types were observed directly for each stop and written in an observation form (as below). High-level platforms (1000 mm from the head rail) were defined as having no step differences between the platform height and the train floor; mid-level platforms (430 mm from the head rail) were defined as having a small step between the platform and train floor, which less requires stairs to accommodate passenger alighting and boarding; and the low-level platforms (180 mm from the head rail) were described as having a huge step between the platform and the train floor, which require stairs to ease passenger flow. The number of passengers who were alighting and boarding was informed by on-board officers and was updated during the observation day by the passenger section of Daerah Operasi (DAOP) 8 Surabaya, a regional office of PT Kereta Api Indonesia who manage train operations for certain line on the Surabaya–Yogyakarta route. The observation days were conducted from 3 February to 13 February 2020. The observations were gathered from Surabaya to Yogyakarta on 13 trainsets (one run), comprising 138 station observations using the observation form in figure 1 below.

Trainset	: /						
One-sided doors number	:						
Route	: Surabaya-(Yogyakarta / Lempuyangan)* *Strikethrough One from The Bracket						
Observation Date	:						
Station	**Type of Platform**	**Boarding**	**Alighting**	**Time**	**Arrival**	**Departure**	**Note**
(1)	*(2)*	*(3)*	*(4)*	*(5)*	*(6)*	*(7)*	*(8)*
SURABAYA GUBENG							1 None 2 Ops
WONOKROMO							1 None 2 Ops
. . .							
LEMPUYANGAN							1 None 2 Ops
YOGYAKARTA							1 None 2 Ops

Figure 1. Observation form example

To complete the observation form in figure 1, before the observation started, the surveyor had to write the trainset identity, one-sided doors number and observation date and strikethrough one route choice (trainset origin–destination). The following columns were written during the observation period with the help of an on-board officer for passenger number data. To determine the platform type, the surveyor had to assume platform types based on the vertical distance between the car floor and the platform, where some stations used portable stairs to help passengers up onto mid-level or low-level platforms. The time column was filled in based on the duration of passenger activities on platform, which the surveyor stopped counting when no more alighting and boarding activities took place. The arrival column was filled in with the time the trainset spent fully stopped on the platform, and the departure column was filled in with the time the trainset started to run to depart the station.

There are 19 stations where passenger loading and/or unloading occurs. The types of platform and serviced stations are presented in table 1.

Table 1. Station and platform types

No	Station	Platform Type
1	Yogyakarta	High-Level
2	Lempuyangan	High-Level
3	Klaten	Mid-Level
4	Purwosari	Mid-Level
5	Solo Balapan	High-Level
6	Sragen	Mid-Level
7	Walikukun	Low-Level
8	Ngawi	Mid-Level
9	Magetan	Mid-Level
10	Madiun	High-Level
11	Caruban	Mid-Level
12	Nganjuk	Mid-Level
13	Kertosono	Mid-Level

No	Station	Platform Type
14	Jombang	Mid-Level
15	Mojokerto	Mid-Level
16	Krian	Low-Level
17	Sepanjang	Low-Level
18	Wonokromo	Mid-Level
19	Surabaya Gubeng	High-Level

Based on table 1, there are five stations that use high-level platforms, 11 stations that use mid-level platforms and only three stations that used low-level platforms for passengers to board and alight. Furthermore, one predicting variable for dwell time was the number of doors on one side of the trainset, which were opened during passenger boarding and alighting. The number of doors varied depending on the trainset; the lowest number was 12 on Sritanjung, and the highest was 18 on a number of trainsets. Passenger flow in each station was divided into two steps. The first step was passenger alighting, and the next step was passenger boarding, during the transition period in which passengers who intended to board had to wait until there were no more alighting passengers on the platform area. Then, there was mostly no friction between alighting and boarding passengers on the platform area.

4. Data analysis

The modelling was approached by a linear liaison assumption between the independent variables, including number of doors, number of alighting passengers, number of boarding passengers and type of platform for each stop station. For the platform type, a dummy variable was generated. The dummy variable using a binary number (1 or 0) for each platform type quantified the originally qualitative platform data. Although there were three types of platform, the dummy variable included only two. In this case, mid-level and high-level platforms were chosen because of their frequency in the observation data. OLS was the first method conducted followed by the classic assumption test and outlier identification. The violation of the classic assumption and the presence of outliers transformed the OLS regression into a robust regression with M, S and MM estimators. To choose the best model, MAPE was calculated.

4.1. Ordinary least square model

An OLS is introduced to build the first estimation model. The independent variables are fitted by running a regression analysis and collinearity test. Furthermore, the classic assumptions for OLS were conducted, which consisted of a linearity test, a normality test, a heteroscedasticity test and a multicollinearity test. The OLS model is sensitive to outliers, a small number of which could affect its results. To investigate outliers, Cook's distance, DfFIT and DfBETA were applied. The general OLS model is:

$$Y_i = \beta_0 + \beta_1 X_{1i} + \beta_2 X_{2i} + \cdots + \beta_p X_{pi} + \varepsilon_i \tag{1}$$

1. Linearity test
   The linearity test is conducted to confirm that the mean of the data group is placed in a straight line.

2. Normality test
   The normality tests are supplementary to the graphical assessment of normality. One of main tests for the assessment of normality is the Shapiro–Wilk test, which compares the scores in the sample to a normally distributed set of scores with the same mean and standard deviation. The null hypothesis is that the sample distribution is normal.

3. Heteroscedasticity test
   Heteroscedasticity is the unequal variance of the data along the regression line (i.e., when the spread of the data points is not constant). Heteroscedasticity has serious consequences for the OLS estimator. Although the OLS estimator remains unbiased, the estimated standard error (SE) is wrong.

Because of this, confidence intervals and hypotheses tests cannot be relied upon. In addition, the OLS estimator is no longer Best Linear Unbiased Estimation (BLUE).

4. Multicollinearity test
   Multicollinearity occurs when the independent variables in a regression model are correlated. This correlation is a problem because independent variables should be independent. If the degree of correlation between the variables is high enough, it can cause problems when fitting the model and interpreting the results. Variance inflation factors (VIFs) measure inflation in the variances of the parameter estimates due to collinearities that exist among the predictors. It is a measure of how much the variance of the estimated regression coefficient β_k is inflated by the existence of a correlation among the predictor variables in the model. A VIF of 1 means that there is no correlation among the kth predictor and the remaining predictor variables, and the variance of β_k is hence not inflated. The general rule is that VIFs exceeding 4 warrant further investigation, while VIFs exceeding 10 are signs of serious multicollinearity requiring correction. The VIF is calculated as:

$$VIF = \frac{1}{1 - R_k^2} = \frac{1}{Tolerance} \quad (2)$$

4.2. Outliers

A common problem in regression analysis is the presence of outliers. As defined by Barnett and Lewis (1994), outliers are observations that appear inconsistent with the rest of data. Outliers may occur as a result of unusual, but explainable, events, such as faulty measurements, the incorrect recording of data, the failure of a measurement instrument and so on.

1. Leverage point
   A data point has high leverage if it has extreme predictor *X* values. With a single predictor, an extreme *X* value is simply one that is particularly high or low. With multiple predictors, extreme *X* values may be particularly high or low for one or more predictors or unusual combinations of predictor values.

$$hii = X_i^T (X^T X)^{-1} X_i; i: 1,2, \ldots., n \quad (3)$$

$$2\overline{h_{ii}} = \frac{2 \sum_{i=1}^{n} h_{ii}}{n} = \frac{2k}{n} = \frac{2(p-1)}{n} \quad (4)$$

   An observation data-i would be defined as a leverage on *X*, if $hii > 2\overline{h_{ii}}$.

2. Cook's distance
   Cook's distance is the scaled change in fitted values, which is useful for identifying outliers in the *X* values (observations for predictor variables). Cook's distance shows the influence of each observation on the fitted response values. An observation with a Cook's distance greater than three times the mean Cook's distance might be an outlier.

$$(Cook's\ Distance)i = \left[\frac{R_{standar}^2}{2}\right] \times \left[\frac{h_{ii}}{1 - h_{ii}}\right] \quad (5)$$

3. Difference in fits (DfFITS)
   The DfFITS is a scaled measure of the change in the predicted value for the ith observation and is calculated by deleting the ith observation with the formula given as:

$$(DfFITS)_i = \frac{\hat{y}_i - \hat{y}_{i-1}}{s_{i-1}^2 - \sqrt{h_{ii}}} \quad (6)$$

Data-i would be defined as an outlier if:

$$|DfFITSi| > 1 \qquad \text{for} \leq 30 \tag{7}$$

$$|DfFITSi| > 2\left(\frac{p}{n}\right)^{0.5} \qquad \text{for } n > 30 \tag{8}$$

As the data for this research have important value that could not be deleted at any cost, a robust regression method that provides an alternative to the least squares regression and that works with less restrictive assumptions was applied. Specifically, it provides much better regression coefficient estimates when outliers were present in the data. Outliers violate the assumption of normally distributed residuals in least squares regression. They tend to distort the least squares coefficients by having more influence than they deserve.

4.3. Robust regression model

Robust regression is a regression method that is used when the distribution of residuals is not normal or there are outliers that affect the model. This method is an important tool for analysing the data affected by outliers so that the resulting models are stout against outliers. The robust estimation method is a form of regression analysis designed to circumvent the limitations of traditional parametric and non-parametric methods. Robust estimation methods are designed to not be overly affected by outliers. Under these conditions, a robust regression that is resistant to the influence of outliers is the best method [10]. Therefore, we introduce a comparison of robust regression methods to determine the best method.

1. M estimation

   The most common general method of robust regression is the M estimation method, introduced by Huber (1964). The M estimation method is a generalisation to a maximum likelihood estimation in the context of location models. This is nearly as efficient as an OLS regression. Rather than minimising the sum of squared errors as the objective, the M estimation method principle is minimising the residual function. The M estimation objective function is:

$$\sum_{i=1}^{n} \rho(e_i) = \sum_{i=1}^{n} \rho(Y_i - X_i^t b) \tag{9}$$

   With the Huber weight function:

$$P(e) = \begin{cases} \frac{e^2}{2} & ; |e| \leq k \\ k|e| - \frac{k^2}{2} & ; |e| > k \end{cases} \tag{10}$$

   The general function for the M estimator is:

$$\hat{\beta}_M = (x'wX)^{-1}(x'wX) \tag{11}$$

2. S estimation

   The regression estimators associated with M scales are the S estimators proposed by Yohai (1987). The S estimation method is based on the residual scale of the M estimation method. S estimators are a generalisation of Least Median Square (LMS) and Least Trimmed Squares (LTS). They have the same asymptotic properties corresponding to M estimators and handle 50% of the outliers appearing in the data. The weakness of the M estimation method is the lack of consideration of the data distribution and that it is not a function of the overall data because it only uses the median as the weighted value. S estimators refer to the fact that this estimator essentially is based on the minimisation of a (robust) scale M estimator. This method uses the residual standard deviation to overcome the weaknesses of median [11]. The S estimator is defined by:

$$\hat{\beta}_S = \min \hat{s}\,(e(\beta)) \tag{12}$$

With $\hat{s}(e)$ being obtained from the M estimation residual scalar *s(e)*, which is the solution from $\frac{1}{n}\sum_i \rho\left(\frac{e_i(\beta)}{\hat{s}(e(\beta))}\right) = \delta$.

3. Modified M estimation

The modified M (MM) estimation method is a special type of M estimation method developed by Yohai (1987). MM estimation is a combination of the high breakdown value estimation method and the efficient estimation method. Yohai's MM estimator was the first estimation with a high breakdown point and high efficiency under normal error. The MM estimator regression is defined as:

$$\hat{\beta}_{MM} = \frac{min}{\beta}\sum_{i=1}^{n}\rho\left(\frac{e_i}{\hat{\sigma}}\right) = \frac{min}{\beta}\sum_{i=1}^{n}\rho\left(\frac{y_i - \sum_{j=0}^{k}x_{ij}\beta_j}{\hat{\sigma}}\right) \quad (13)$$

4.4. Model determination

In order to choose the best model, Mean Absolute Percentage Error (MAPE) was applied to get the best estimation model with the lowest MAPE value:

$$MAPE = \sum_{t=1}^{n}\left|\frac{y_i - \hat{y}_i}{\hat{y}_i}\right| \times 100\% \quad (14)$$

5. Result and discussion

5.1. Data description

The on-board observation for 13 trainset stop stations resulted in 138 data point to further analyse The descriptive statistics of the observation results show that the number of doors is between 12 and 18 unit per trainset, excluding the wagon, dining and generator cars. The number of boarding and alighting passengers varied for each stop station; there were no passengers alighting or boarding in some cases. The summary of the observation results is shown in table 2 below.

Table 2. Variable summary

Variable	Obs	Mean	Std.Dev.	Min	Max
Door	138	15.899	2.069	12	18
Boarding	138	41.13	70.37	0	386
Alighting	138	32.833	62.037	0	493
Dwell time	138	132.797	133.431	7	665

Based on table 2, the standard deviation of the continuous variables was high; therefore, outliers and leverages were assumed to be present. The highest number of boarding passengers was 386, while there was no boarding activity for some stop stations.

Table 3. Dummy variable summarize

Dummy Variables	Freq.	Percent	Cum.
Low-Level Platform	3	2.17	2.17
Mid-Level Platform	87	63.04	65.22
High-Level Platform	48	34.78	100.00
Total	138	100.00	

As shown in table 3, varied platforms were used to service passengers at each stop station. Mid-level platform use dominated with more than 50% of cases, while low-level platforms were rarely used. For the statistical analysis, the type of platform was defined as a dummy variable of 0 if the serviced station used another type platform and 1 if the station used the specific platform type.

Table 4. Correlation matrix among variables

Variables	(1)	(2)	(3)	(4)	(5)	(6)
(1) Time	1.000					
(2) Door Number	-0.089	1.000				
(3) Boarding	0.801*	0.027	1.000			
(4) Alighting	0.619*	-0.106	0.184*	1.000		
(5) Mid-Level Platform	-0.583*	-0.067	-0.497*	-0.484*	1.000	
(6) High-level Platform	0.623*	0.065	0.528*	0.514*	-0.954*	1.000

* shows significance at the 0.05 level

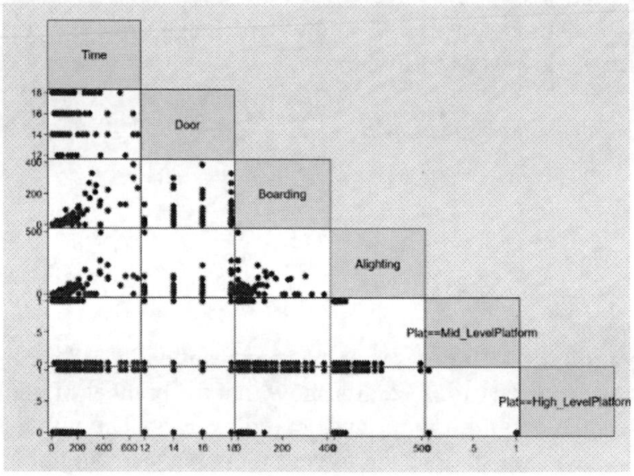

Figure 2. Matrix of variables Scatter-plot

Based on table 4 and figure 2, among the six independent variables, four had statistically significant connections to dwell time. However, the multiple linear regression analysis result was not satisfactory (see table 7). The studentized residual of the multiple linear regression model deviated from the linear assumption, while its adjusted R2 was 0.87. The normality test for the studentized residual with the Shapiro–Wilk test resulted in 0.00000, which forced the rejection of the null hypothesis (that the residual is normally distributed). The result was supported by a kernel density estimation, as seen in figure 3.

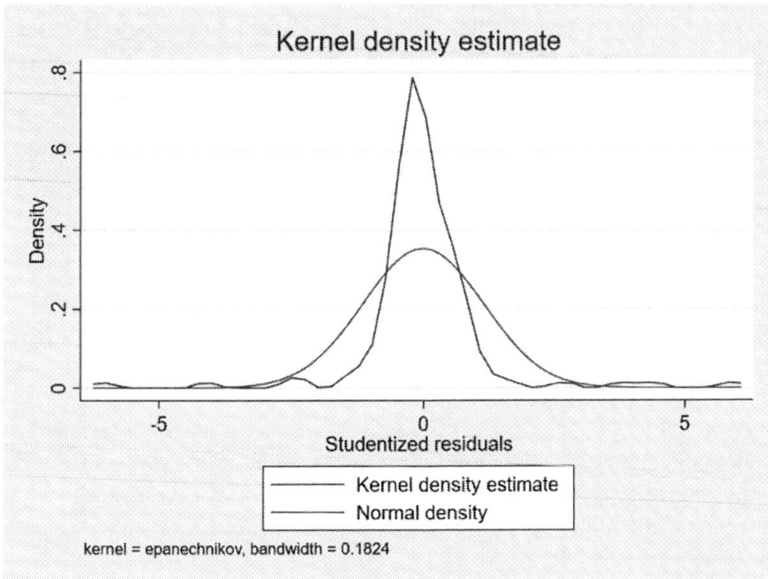

Figure 3. Kernel density estimate

In figure 3, the kernel density estimation based on observation data has a similar shape and pattern to the normal density curve; however, the peak point of the observation data exceeds the normal density. This indicates that the observations contain extreme data, which potentially influence the regression model's accuracy.

In terms of multicollinearity, the VIF values for continuous variables were below 5, while the dummy variable VIFs were greater than 10. For this case, the multicollinearity test can be ignored for the dummy variable because it was sourced from three types of platform that interact with each other. In addition, the heteroscedasticity test using the Breusch–Pagan/Cook–Weisberg test did not have a satisfactory result at valued 0.0000, which forced the rejection of H0 (constant variance).

5.2. Outlier identification

In order to identify outliers and leverage points, the Cook's distance and DfFITS methods indicated the same data points as outliers with the highest exceeding values from the limit. The limitation for Cook's distance is $4/n$, and the boundary value of DfFITS for detecting outliers is $2\sqrt{p/n}$, where p is number of parameters and n is number of data points. The top ten data points that most violated the value based on Cook's distance and DfFITS are listed in table 5 and table 6 below.

Table 5. The result of identifying outlier by Cook's distance

	ID	Door	Boarding	Alight~g	Mid_Le~m	High_L~m	Time	d
7.	s7	18	0	269	0	1	373	.0485313
15.	s15	18	269	0	0	1	290	.1038137
67.	s67	12	0	493	0	1	379	4.376512
81.	s81	12	129	143	0	1	572	.1788606
82.	s82	18	130	68	0	1	532	.1442872
103.	s103	16	386	0	0	1	633	.1074935
111.	s111	16	171	126	0	1	295	.0401431
121.	s121	18	331	0	0	1	311	.4390887
130.	s130	14	91	133	0	1	434	.0400041
138.	s138	14	236	125	0	1	665	.1849594

Table 6. The result of identifying outlier by estimate DfFITS

	ID	Door	Boarding	Alight~g	Mid_Le~m	High_L~m	Time	dfit
7.	s7	18	0	269	0	1	373	.5411851
15.	s15	18	269	0	0	1	290	-.8038641
67.	s67	12	0	493	0	1	379	-5.778908
81.	s81	12	129	143	0	1	572	1.105113
82.	s82	18	130	68	0	1	532	1.043153
103.	s103	16	386	0	0	1	633	.8078269
111.	s111	16	171	126	0	1	295	-.4992299
121.	s121	18	331	0	0	1	311	-1.717546
130.	s130	14	91	133	0	1	434	.5010158
138.	s138	14	236	125	0	1	665	1.105176

Comparing both table 5 and table 6, outlier points were detected in s7, s15, s67, s81, s82, s103, s111, s112, s130 and s138. The outlier data contain important data that represent real situations in the field where there were no passengers boarding but hundreds of passengers alighting. In addition, to uncover more influencing data, the leverage points and normalised squared residual were calculated.

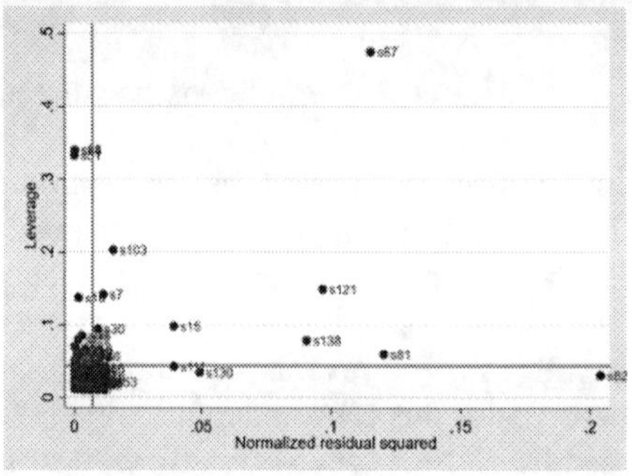

Figure 4. Outlier and leverage identification

The lines on figure 4 show the average values of leverage and the (normalised) residuals squared. Points above the horizontal line have higher-than-average leverage, and points to the right of the vertical line have larger-than-average residuals. As the chart above shows, there are many data point that have unbalanced values of leverage and normalised residual squared. For instance, s82 has larger-than-average residuals while s81 has larger-than-average residuals and leverage. The data points exceeding the average greatly affect the slope of the regression line. Based on table 5, table 6 and figure 4, the observation data have many outliers and influencing data. Next, robust regression methods were conducted to get the best estimation model.

5.3. Robust regression result

Given the deviation from the OLS assumptions, robust regressions were developed using M estimation with Huber's weight, S estimation and MM estimation. The comparison between them and OLS is presented in table 7.

Table 7. Comparison model

VARIABLES	OLS DwellTime	M-Estimator DwellTime	S Estimator DwellTime	MM Estimator DwellTime
Door Number	-3.685*	-2.747**	-2.727***	-2.122**
	(2.021)	(1.218)	(1.020)	(0.971)
Boarding	1.349***	1.314***	1.567***	1.514***
	(0.0694)	(0.144)	(0.0251)	(0.204)
Alighting	1.031***	1.222***	1.194***	1.267***
	(0.0790)	(0.131)	(0.0484)	(0.0812)
Mid-Level Platform	11.08	9.353**	1.779	6.699
	(28.30)	(4.638)	(5.014)	(5.492)
High-Level Platform	12.07	-2.014	-7.912	-7.912
	(30.06)	(7.833)	(5.829)	(8.510)
Constant	90.88**	76.06***	74.48***	65.05***
	(42.65)	(20.15)	(16.77)	(15.76)
Observations	138	138	138	138
R-squared	0.874			

Standard errors in parentheses
*** p<0.01, ** p<0.05, * p<0.1

As table 7 shows, the four models show that mid-level platforms increase dwell time while high-level platform reduce it, except the high-level platform on OLS model. Among the robust estimators, the M estimator had four independent variables with a less than 0.05% probability error. The comparison of R-square values between the OLS and robust models is inaccurate because R-square in robust regressions is different from R-square in OLS.

5.4. Model determination

The determination between the models in table 7 is based on MAPE calculation result as below:

Table 8. MAPE comparison between model

No	Method	MAPE
1	Ordinary Least Square	3.368%
2	Robust Regression S Estimation	3.507%
3	Robust Regression M Estimation	2.037%
4	Robust Regression MM Estimation	2.352%

Based on table 8 above, the M estimation had the lowest MAPE value, indicating the lowest error of all the models. Here, a robust regression model using M estimation for dwell time was taken. The model is written as:

$$Dwell\ Time = -2.727\ Door\ number + 1.314\ Boarding + 1.222\ Alighting + 9.353\ MidLevel\ Platform - 2.014\ HighLevel\ Platform + 76.06$$

Based on the model, alighting passengers need less time than boarding, with 1.222 sec added to dwell time per alighting passenger and 1.314 sec per boarding passenger. By using high-level platforms on the stop stations, dwell time will be reduced by 2.014 sec based on continuous variables, and using mid-level platforms will increase dwell time by 9.353 sec based on continuous variables.

6. Applied model

The model application reduced dwell time for each station per trainset. For instance, the dwell time model applied to the Sancaka and Argo Wilis data resulted in more than 40% in time savings compared to the actual dwell time based on passenger volume on the day of observation.

From table 9 above, the total actual dwell time for Argo Wilis was 29 minutes, while the model resulted in 18 minutes of total dwell time. For Sancaka, the model resulted in 23 minutes of total dwell time, whereas in the actual operation, total dwell time for Sancaka was 35 minutes. Furthermore, with dwell time modelling, total travel time per trainset will be reduced.

Table 9. Argo Wilis and Sancaka saving time from dwell time

Data	Trainset	STATION	Platform	Boarding (person)	Alighting (person)	Door (unit)	Dwelltime in Field	Dwelltime With Model in second	Roundup Dwelltime With Model in Minute	Saving Time	Saving per Trainset
s1	ARGO WILIS/1/SGU-GMR	YOGYAKARTA	High	0	117	16	00:13	173,068	00:02	00:11	
s2	ARGO WILIS/1/SGU-GMR	SOLO BALAPAN	High	20	26	16	00:07	88,146	00:02	00:05	
s3	ARGO WILIS/1/SGU-GMR	MADIUN	High	79	62	16	00:05	209,664	00:04	00:01	00:19
s4	ARGO WILIS/1/SGU-GMR	KERTOSONO	Mid	7	0	16	00:02	50,659	00:01	00:01	
s5	ARGO WILIS/1/SGU-GMR	JOMBANG	Mid	7	8	16	00:02	60,435	00:01	00:01	
s8	SANCAKA/179/SGU-YK	KLATEN	Mid	4	12	18	00:02	55,887	00:01	00:01	
s9	SANCAKA/179/SGU-YK	SOLO BALAPAN	High	19	74	18	00:05	139,994	00:03	00:02	
s10	SANCAKA/179/SGU-YK	MADIUN	High	42	17	18	00:16	100,562	00:02	00:14	
s11	SANCAKA/179/SGU-YK	NGANJUK	Mid	7	4	18	00:02	50,053	00:01	00:01	00:23
s12	SANCAKA/179/SGU-YK	KERTOSONO	Mid	8	2	18	00:03	48,923	00:01	00:02	
s13	SANCAKA/179/SGU-YK	JOMBANG	Mid	15	18	18	00:03	77,673	00:02	00:01	
s14	SANCAKA/179/SGU-YK	MOJOKERTO	Mid	39	7	18	00:04	95,767	00:02	00:02	

7. Conclusions

The dwell time study on the Surabaya–Yogyakarta trainset estimation using statistical approaches has resulted in a model with robust regression methods due to the violation of the OLS assumptions of heteroscedasticity and normality and the presence of outliers. In the comparison of OLS and the three robust regression estimators, the robust regression with the Huber M estimator had best results with minimal error. According to the model, alighting passengers need less time than boarding passengers. The model's applications to passenger volume, number of doors and platform type caused the trainset travel time to run effectively.

Acknowledgements

We would like to acknowledge the fruitful discussion and help during our observations from Asad Triadi and team from DAOP 8 Surabaya PT Kereta Api Indonesia. The author would also thank to the vice president of passenger marketing at PT Kereta Api Indonesia for providing passenger data on the Surabaya–Yogyakarta line.

References

[1] Puong A 2000 Dwell Time Model and Analysis for MBTA Red Line *Massachusetts Institute of Technology Research Memo* 1-10

[2] Gysin K 2018 *An Investigation of the Influences on Train Dwell Time* (Zurich: Swiss Federal Institute of Technology, ETH Zurich) 601-700

[3] Harris N G 2006 Train Boarding and Alighting Rates at High Passenger Loads *J. of Adv. Trans.* **40** 249–63

[4] Yamamura A, Koresawa M, Inagi T, Tomii N 2013 *Proc. of World Conf. on Transportation Research (WCTR)* (Rio de Janeiro: World Conference on Transport Research Society) **13**

[5] Amirnazmiafshar E 2019 Tramway Dwell Time Estimation and Analysis by using Multiple Linear Regression Method European *J. Eng. Sci. Tech.* **2** 36-48

[6] Jong J-C and Chang E-F 2011 *Proc. of World Cong. on Railway Research (WCRR)* (Lilie: The Rail Safety and Standards Board(UIC)) **9**

[7] Ho K and NaugherJ R 2000 Outlier Lies: An Illustrative Example of Identifying Outliers and Applying Robust Models *Mult. Linear Regres. Viewpoints* **26** 2–6

[8] Berk R A 1990 A Primer on Robust Regression *Modern Methods of Data Analysis* Eds J Fox and J S Long (Newbury Park: Sage Publications) chapter 7 292–323

IOP Conf. Series: Materials Science and Engineering **930** (2020) 012060 doi:10.1088/1757-899X/930/1/012060

[9] Myttenaere A D, Golden B, Grand B L, Rossi F 2016 *J. Adv.in artific. neural net., mach. learning and comp. intel.: Selected papers from the 23rd Eur. Symp. on Artific. Neural Networks (ESANN 2015)* **192** 38-48

[10] Draper N R and Smith H 1998 *Applied Regression Analysis* (Canada: John Wiley and Sons, Inc) Third Edition 567-584

[11] Susanti Y and Pratiwi H 2014 M-estimation, S-estimation, and MM- estimation in robust regression *Int. J. Pure Applied Math.* **91** 349-360

IOP Conf. Series: Materials Science and Engineering **930** (2020) 012061 doi:10.1088/1757-899X/930/1/012061

Design of the strong and durable of flexible pavement on road segment with the heavy traffic and overloads

Sutoyo[1*]

[1] Road and Bridge Engineering, Road and Bridge Infrastructure Office, East Java Provincial Government, Jalan Gayung Kebonsari 167 Surabaya, Indonesia.

[*]Corresponding author's e-mail: sutoyo.207012@mhs.its.ac.id

Abstract. The arterial road sections are generally heavy vehicles that tend to exceed the standard load so that the existing pavement surface is not able to survive according to the design life despite using modified asphalt. This has been proven by the occurrence of a heavy damage in a relatively short time on most primary arterial roads in Indonesia. In this paper, the method of implementation and criteria for the construction of pavement layers with a certain thickness has been prepared according to the planning standard for the volume of excessive vehicle loads and taking into account extreme weather conditions in Indonesia. There are three (3) main things in this paper that become an important factor in determining the structure of a solid and durable; they are the sand as layer controlling deflection (stabilizer), the weight of the Vibratory roller more than 25 tons to achieve maximum of compaction energy for base course, and a single-layer of AC-Base as a surface layer as anticipation of rutting and bleeding. Single overlays in each layer are rigidly constraining overall construction, while the sand layer can neutralize deflection due to overload and force due to the vertical force of subgrade, so that these three factors are energy deposits in the pavement construction that will work when an increase in load exceeds the standard.

1. Introduction

The primary arterial road in the north coast area of Java is a road with a super heavy load, so that the average pavement age of less than 3 years has experienced quite severe damages. The damage begins with rutting, bleeding down, and waves and grain spills occur until it becomes a fairly large hole and increases very quickly, especially in the rainy season. Damage repair only lasts a maximum of 1 year so that the surface conditions are relatively uneven because there are so many fillings of holes and waves. The average annual maintenance cost for the north coast area of Java road per kilometer is around 80 million rupiah (4 lanes), but with the allocation of these funds, the road surface conditions are still uncomfortable and tend to be damaged.

AC-WC and AC-BC are not able to serve repetition of heavy loads that especially tend to be excessive, so that the carrying capacity of the surface layer is smaller than the compression force load of the vehicle with tire pressure wheel. From lecture module book of Indrasurya, it is stated that the minimum marshal stability value (in kg) is 10 vehicle tire pressure (psi), so that if the truck tire pressure is 160 psi then the minimum stability value is 1600 kg. If the stability value of the Marshall AC-WC and AC-BC Modification is only 1000 kg, it is very reasonable that it is not able to support the vehicle load with tire pressure up to 160 psi. Therefore in this paper, we will discuss in detail about the placement

Content from this work may be used under the terms of the Creative Commons Attribution 3.0 licence. Any further distribution of this work must maintain attribution to the author(s) and the title of the work, journal citation and DOI.
Published under licence by IOP Publishing Ltd

of AC-Base as a surface layer in an effort to support repetition of heavy loads that tends to be excessive to minimize the occurrence of rutting and bleeding.

In addition to AC-Base as a surface layer used as a material for discussion, the overlay method, especially in relation to a single AC-Base overlay, is also the material of analysis in this paper because there is a significant difference between a single stretch and layer by layer when viewed in terms of the robustness of the pavement layer structure. Analysis of the occurrence of failures for laying on a layer by layer mixture system is also discussed in this paper, so that the selection of a single overlay against the AC-Base layer to achieve a sturdy and durable surface layer structure can be achieved to the maximum.

The foundation layer was chosen by CTB as the right material because CTB is not affected by the presence of high ground water during the rainy season, as well as it has a relatively higher coefficient value than the grained aggregate foundation layer so as to minimize the thickness of the upper foundation layer. Another thing that is important to use CTB as the upper foundation layer is that the CTB is spread with optimum moisture content conditions, so that it is easier to achieve maximum density at all points of the base course layer. This is very beneficial because so far the optimum water content in the base course layer (aggregate A) and sub-base course (Aggregate B) has never achieved optimum water content in all parts of the thickness of the layer including the surface layer.

Condition of water content is often considered trivial, whereas the maximum density is achieved when the optimum moisture content is in the range of +1 to -3% of the value of the optimum moisture content. The grained foundation aggregate that is sent in the field is definitely not clear with what the water content is, if it has been piled in the field and left for days, sometimes it gets rained on and sometimes sunburned so that it gets farther from the optimum water content conditions. At the time of distribution and compaction, the average water content in each pile is also not tested so that an amount of water can be added based on the field density value determined in the JMF. What happens is that the field only adds water without any calculation, resulting in varying degrees of density at a number of surface locations, and eventually the surface becomes a longitudinal and transverse direction waves.

The existence of waves on the road surface cannot be considered trivial because these conditions are the triggers for more severe damages, the presence of waves can cause shock loads that are up to 2-2.5 times the weight of the existing loads. This is very dangerous especially when overloaded by heavy vehicle loads, so as if the location of the valley (concave) receives multiple overloads again. This is evident in the road section that is traversed by heavy and heavy traffic if there are symptoms of deflection and small waves, then in a relatively short time it will turn into a large puddle during the rainy season. On the basis of this compaction method, providing optimum water content is a matter that must become a separate consideration.

Compaction energy produced by the weight of the compactor is very influential on the robustness of the construction of the aggregate foundation layer, the greater the weight of the compactor the greater the effect on the thickness of the layer. This should become the main consideration when carrying out compaction, if a lightweight compactor is used then the thickness of the overlay layer is small enough, so that to achieve the thickness of the plan, several layers are required. The weakness of the structure of the foundation layer construction which is composed of several thin layers is not as solid as a single layer according to the design thickness. Analysis of weaknesses in the arrangement of layer thicknesses composed of several layers will be discussed further in the analysis chapter.

The sand layer deliberately installed in the design of flexible pavement construction structures is intended to control the existence of excessive load which is expected to cause deflection that exceeds the permit deflection of the rigidity (modulus) of the existing design construction structures. Besides, sand is also needed as a stabilizer when the subgrade which has a very low CBR is submerged in water during the rainy season, where conditions are very weak, then some of the sand thickness will function as a form of subgrade repair, so that the structural conditions remain stable. Similarly happens during the compaction of the foundation layer with a compactor that is heavy enough, so that the sand can be used as a supporting layer so that the subgrade layer is not completely disturbed by the compaction energy that occurs.

2. Intent and purpose

This paper aims to solve the problem of flexural road damage due to overload and tropical climate that can soften hot asphalt mixes during the day, as well as subgrade conditions that are disturbed by the level of rigidity of the pavement construction layers which are designed according to the technical requirements and technical planning criteria for road pavement thickness.

The main goal of this paper is to create a design system that is sturdy and durable on flexible pavement based on the characteristics of the material, the stabilizer layer, the method of implementation and weight of the compactor for the foundation layer, and the selection of the most appropriate type of surface layer.

3. Literature study

A sturdy and durable road in serving traffic volume during the life of the plan is highly desirable, moreover it can serve up to more than 40 years of age (perpetual pavement) where during the service period only minor routine maintenance is carried out, but the value of surface conditions is quite good, a maximum IRI value of 6. In Indonesia, it seems that this is impossible because many factors can cause damage in a relatively short time, including road users not obeying regulations (overloading), tropical climate so that the heat on the road surface is more than 60 ° C, construction material, construction method, capacity and condition of compactor, and relatively high rainfall. These things should become the variables taken into account in the comprehensive flexible pavement construction arrangement. Meanwhile, standard design only considers 3 main variables, namely subgrade (CBR), traffic data (ESAL accumulation) according to design life of pavement and drainage system.

Calculation of traffic data must be examined carefully, especially the problem of excess vehicle planning as a result of incorrect volume prediction during the service period (planning year), incorrect setting growth factors, not paying attention to land use (spatial mapping), and transportation industry changing configuration axis so that the amount of load increases, but the number of wheels remains, or modifies the type of tire so that it is able to accept greater pressure and lift capacity. This must be understood by planners so that in calculating the volume of heavy traffic, they must take into account these things and convert them into ESAL values. The accumulation of ESAL values resulting from the conversion is expected to approach the actual conditions in the field during the service life according to the planning results.

The occurrence of rutting is caused by 2 important things, namely the characteristics of asphalt as a binding material and the aggregate quality as the results of the aggregate characteristics test. The characteristics of asphalt include softening point and viscosity, where asphalt is expected to have high elasticity at low temperatures so it is not easy to break / crack, and high viscosity so that when the temperature in the field is high and the mixture is not easily softened / weakened. In other words, the asphalt softening point is high, but the penetration is also still high so that the mixture is neither brittle nor softened. Meanwhile, the quality of the aggregate is mainly related to the coarse cubical shape of the coarse grains and has a maximum abrasion value of 30% for each grain. To get the maximum mechanical bonding, it takes more than 50% of the coarse aggregate fraction in the mixture. Mechanical ties are bonds that lock together/ bind (interlock) so that they are eternal. Therefore, if the conditions in the field obtained asphalt mixture with a more dominant type of mechanical bonding, it is certain to be very resistant to repetition of heavy vehicle loads with excessive loads so that rutting and bleeding do not occur.

To obtain the characteristics of asphalt that is able to provide high elasticity and high viscosity values, it must be modified with synthetic elastomeric material and rubber (natural and in the form of crum rubber). If the mixture uses the modified asphalt, then there is an appropriate temperature calibration for the implementation of mixing, spreading and compaction. The composition of the modified modification asphalt must be known in advance to meet all the characteristics, especially the portion of added ingredients (additives and rubber / latex). After obtaining the correct modification of asphalt, a mixture of aggregate and asphalt was tested with proportions that meet the characteristics of the mixture in the laboratory. Modification of asphalt viscosity testing is needed to find the right temperature at the time of mixing, spreading and compaction. Meanwhile, elastic recovery test is needed to get the modified asphalt elasticity value so that it can be used to control the occurrence of cracks.

The SHRP researchers determined that rutting is a controlled tensile stress to the phenomenon of repetitive traffic loads, commonly referred to as workloads. Some of the workloads are accommodated by the strength of the pavement surface elasticity, while others are stored in the pavement in the form of permanent deformation and heat. To minimize the occurrence of permanent deformation, it must minimize the stored work energy when there is a repetitive traffic load. The mathematical formula for stored energy due to repetitive traffic loads is as follows:

$$W_c = \pi \text{ x } (\sigma°)2 \text{ x } \frac{1}{G * / \sin\delta} \quad (1)$$

Where :
Wc = stored workload every repetition of traffic loads
σ° = tensile stress during traffic load repetition
G* = complex modulus
δ = space angle

From the formula above, it appears that the rutting value can be reduced or asphalt material can be made more resistant to rutting by increasing the modulus of the complex, G * or making asphalt more rigid and reducing the value of δ, or making asphalt more elastic. This concept has been developed by the Superpave method.

Modified asphalt used as a binder in the mixture in the 2018 Bina Marga technical specifications must meet the technical requirements as Type 2 asphalt namely PG-70 and PG-76 asphalt. However, implementation control in the field is not fully able to meet these requirements because not all laboratories have testing equipment for the modified asphalt modification criteria, specifically the complex shear modulus (G *) and the space angle (δ) on dynamic shear strength testing (Dynamic Shear Rheometer, DSR).

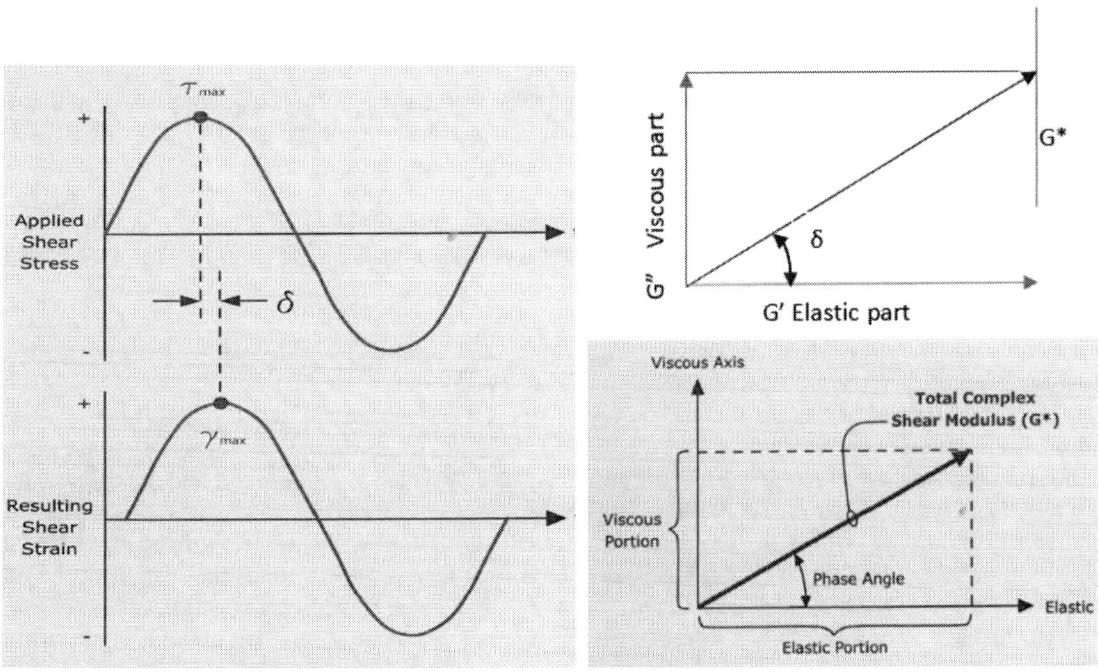

Figure 1. DSR testing to determine the values of G "and δ

The formula for calculating variables in the DSR measurement is as follows:

$$G^* = \frac{\tau_{max}}{\Upsilon_{max}} \quad (2)$$

$$\tau_{max} = \frac{2T}{\pi r^2} \quad (3)$$

$$\Upsilon_{max} = \frac{\Theta r}{h} \quad (4)$$

Where :

G* : complex modulus
δ : space angle (time difference)
τ max : maximum shear stress
Υmax : maximum shear strain
Q : maximum rotating voltage
r : the radius of the test specimen 4 mm or 12.5 mm
Θ : deflection (rotation angle)
h : height of test specimens is 1 mm or 2 mm

Material is called viscoelastic when the value is: 0 <δ <90 ºC

Meanwhile, if the proportion of mixed grading has been obtained where the coarse grain fraction is more than 50% and cubical, then the type of asphalt modification is no longer needed because the interlock ability between coarse grains is much greater than the strength of the modified asphalt bond, meaning that the asphalt function as material binders can be minimized, and mechanical ties of this kind if they occur evenly in asphalt mixtures will become very strong and durable mixes due to the effects of weather and water cannot work, so they can be said to be an eternal mixture or referred to as perpetual pavements that can last up to more than 40 years.

What is meant by the coarse aggregate fraction in this paper is not merely retained in sieve no.4 (4.75 mm), but the coarse fraction which is the body of the mixture (interlock dominant fraction), where the AC-WC mixture is 5 - 10 crush-stone (medium), and AC-BC is 10 - 15 crush-stone (medium tends to be rough). In accordance with the grading requirements in the specification that combined grading for each type of surface layer, the maximum grain size is set only in the range of 0-10%. In AC-Base mixture, the aggregate, which is a mixed body, is 10-20 grain size around 38-48 %, so that the aggregate fraction as the main support of mechanical bonding is able to increase the Marshall Stability value of AC-Base of more than 2250 kg when using modification asphalt, whereas if using non modification asphalt, Marshall values can be more than 1800 kg, with this value alone is sufficient to support heavy traffic and overloads.

4. Problems and alternative solutions

As discussed earlier, the problem of damage includes several causes, ranging from the supporting layers (if any), the grained aggregate foundation layers, the CTB, and the surface layers, all of which are mutually supportive and integrated as a whole. For example, if the strength of the support layer is not uniform because there are large grains (of more than 10 cm) which are spread out and compacted, then in less than 2 years the surface conditions must have waves with a peak point at the location where the large rock is located. Waves on the surface of the asphalt mixture will cause a shock load, so that the location of the break (concave) will receive an additional force of 2-2.5 times the weight of the load carried when heavy vehicles pass. Then within that relatively short time, it will become an increasingly severe damage if not promptly given a leveling layer at the location of the sinkhole or the trough (concave section).

In general, activities in the field do not use compactors in accordance with the weight requested in the requirements, although existing compactors can be converted with thinner layer thickness and number of passes. Compaction of aggregate foundation layers made with a small compactor (small energy) will be very vulnerable to damage because it must be done layer by layer, which often occurs in the field not done with a thickness that is appropriate to the weight of the compactor. For example, a

30 cm thick bed uses only a 10 ton vibro roller weight, so that the bottom of the bed does not get enough energy, so parts that do not get enough compaction energy will be a source of weakening. Weakening in question is the occurrence of deflection and sink in when heavy vehicles pass, if the deflection that occurs has exceeded the permit limit, cracks will occur and if filled with rain water, the damage will be more severe quickly.

If the thickness of the aggregate foundation is compacted layer by layer in accordance with the weight of the compactor, and there is a load that exceeds the compaction energy (heavy load exceeds the weight capacity of the compactor) there will be a deflection between layers so that each layer will work independently, and this is a weak form. Thick overlays composed of several layers will remain sturdy if the load that passes does not exceed the permit capacity of each layer. If there is one layer that collapses as a result of overload, then the collapsed part becomes weak and will trigger the collapse of the other layers, so that it will collapse completely (total layers). Meanwhile, almost 50% of heavy vehicles in Indonesia tends to carry loads with heavy loads exceeding heavy load capacities.

In general, the condition of subgrade CBR values along the road segment is uneven, and this condition will provide different support for the flexible pavement construction layers. Differences in carrying capacity of the subgrade layer along the road can cause longitudinal and transverse direction waves if the total stiffness of the entire layer is unable to support the repetition of heavy loads that passes, therefore a stabilizer layer is needed that is able to control the forms of subgrade movement / pressure when experiencing volume changes due to the presence of ground water levels during the rainy season. This paper chooses a layer of 20 cm thick sand as a subgrade pressure stabilizer layer and as an overload controller when the rigidity of the flexible pavement layer construction is exceeded by the overload path. Sand is a single gradation that can function very effectively to divide or distribute the load acting on the layer evenly, so that the subgrade layer is not disturbed. Besides, conversely, the subgrade pressure at some locations will be neutralized by the sand layer so that it is not reflected to the surface layer.

5. Implementation of experimental activities in the field

The activity is carried out on roads that have the criteria of soft land and land that tends to be prone to high shrinkage, with heavy vehicle reps that are quite dense and tend to be overloaded, so the implementation steps are as follows:

1. Excavating the existing as deep as 95 cm to dispose of existing material because it is assumed the foundation layer has experienced mixing with clay pumped up to the surface layer, so that all excavated material will be used as the fill around outside the road shoulder.
2. The width of the excavation for half the body of the road is 4.50 meters (1 line + 1 meter shoulder), with the aim that the structure of the shoulder is the same as the pavement structure, the location of the road should be bordered by markers.
3. Spread a layer of sand zone 2 or 3 criteria as thick as 20 cm above the surface of the subgrade that has been dug and flattened, doused with water until saturated, allowed to stand for at least 4 hours so that the water seeps into the subgrade layer .
4. In addition to compacting sand, water that seeps into the subgrade layer will make the subgrade the weakest condition and represent the real conditions in the field.
5. Non-woven geotextile extend just above the surface of the sand layer that serves as an interface layer (separator), in case the pumping does not mix with layers of foundation.
6. Carry out a single layered underlay foundation (class B aggregate) with optimum moisture content as thick as 30 cm and compacted with vibro roller liner with a minimum static weight of 25 tons, with the number of compaction equipment trajectories according to field trials (JMF).
7. Carry out a single 30 cm CTB overlay, which is compacted with a vibro roller with a static weight of at least 25 tons, with the number of trajectories in accordance with the trial (JMF).
8. For curing purposes of the CTB layer which has fulfilled the technical requirements, namely flatness, density and slope of the surface according to the technical plan drawings are carried out overlaying the prime coat.

9. Particularly on heavy traffic roads, at minimum of 4 hours after the overlay of the prime coat can be sprinkled with sand or rock ash so that traffic can be opened, as well as functioning as proof-rolling to determine the degree of density visually.
10. Carry out tack coat as much as 0.2-0.5 liters / m2 before exposure to the surface layer.
11. Carry out a single AC-Base overlay 12-15 cm thick with a tool that is capable of the thickness with the results according to the plan.
12. In an effort to monitor surface subsidence, initial elevation measurements are made at several locations tied to the BM that has been provided previously, so that monitoring of the decline in whole or in part can be monitored with details and can be known periodically (every year).
13. The trial location was conducted on the primary collector road of the Bojonegoro-Nganjuk section Km 58 + 450 to 58 + 675 with the road conditions as attached.

6. Analysis of determination of research layer arrangement thickness

The composition of the pavement construction layers that are the object of research is as shown in figure 2 below. With a total thickness of 95 cm, consisting of 20 cm tidal sand layer, bottom foundation layer (class B aggregate) 30 cm, upper foundation layer, CTB as thick as 30 cm and surface layer with a single AC-Base single layer 15 cm thick.

1. A 20 cm thick sand layer is used as a stabilizer layer, which is to anticipate the vertical movement of the subgrade force due to its development. The sand layer also functions as a layer of controlling heavy traffic loads that tend to be excessive because sand is a material that is easy to adjust the shape of the surface direction of the load acting on it and spread it out to all the interacting grains, so that the load pressure becomes very small along with the thick layers, in other words, sand is able to neutralize several types of work load.
2. The foundation layer that is spread with a single layer using a heavy compaction tool will produce a high degree of density and save excess compaction energy in the form of rigidity of the pavement layer structure which is very high, so that it will be able to support the overloads without exceeding the permit deflection of the existing pavement structure.
3. Surface structure designed with interlock-based bonding mechanism between cubical coarse grains will be able to survive indefinitely as long as coarse aggregate grains do not weather. Therefore, the mechanical granular bonding system based on the mechanical (interlocking) can minimize the function of asphalt as a binder, so that the surface layer is not affected by aging factors of the binder, including modified asphalt material which is not needed to create flexible pavement structures that are resistant to rutting and bleeding.
4. The accumulation of the three strengths of the pavement structure as analyzed above can be referred to as the excessive energy stored in a flexible pavement system in the form of stiffness, and will work when the load exceeds the plan so as to be able to provide services according to the age of the plan with small maintenance costs.
5. The composition of the layers referred to above is shown in figure 2.

7. Resume specification for each materials

7.1. Sand material specifications

1. Weight of loose contents of at least 1.2 tons / m^3
2. Sand equivalent value of at least 50%
3. Pass the sieve No. 200 of maximum 10%

Table 1. Gradation of sand type

Filter Size	Percentage of Weight that Passes the Filter			
	Zone I (Coarse) Gradation	Zone II Gradation (rather Rough)	Zone III Gradation (Fine)	Zone IV Gradation (Fine)
9,60 mm	100	100	100	100
4.80 mm	90-100	90-100	90-100	95-100

IOP Conf. Series: Materials Science and Engineering **930** (2020) 012061 doi:10.1088/1757-899X/930/1/012061

2.40 mm	60-95	75-100	85-100	95-100
1.20 mm	30-70	55-90	75-100	90-100
0.60 mm	15-34	35-59	60-79	80-100
0.30 mm	5-20	8-30	12-40	15-50
0.15 mm	0-10	0-10	0-10	0-15

7.2. Geotextile material specification

Table 2. Geotextile requirement

Performance	Test Methode	Unit	Requirement
Permittivity	SNI 08-6511-2001 (ASTM D4491/ D4491M-17)	second $^{-1}$	0,02
Apparent Opening Size (AOS)	SNI 08-4418-1997 (ASTM D4751-16)	mm	0,60 (maximum average of roll value)
Ultraviolet Stability (residual strength)	ASTM D4355/ D4355M-14(2018)	%	50% after being exposed for 500 hours

7.3. Specification of lower base layer and CTB material

1. Bottom foundation material is Class B Aggregate
2. CTB material is Class A and Cement Aggregate (3-5) %

Table 3. Technical requirements of layered foundation

Properties	Aggregate Foundation Layer			Drainage Layer
	Class A	Class B	Class S	
Course Aggregate Abrasion (SNI 2417 2008)	0 - 40%	0 - 40%	0 - 40%	0 - 40%
Angularity, retained No.4 sieve size (SNI 7619, 2012)	95/90 [1)]	55/50 [2)]	55/50 [2)]	80/75 [3)]
Liquid limit, (SNI 1967, 2008)	0 - 25	0 - 35	0 - 35	-
Plasticity Index (SNI 1966, 2008)	0 - 6	4 - 10	4 - 15	-
The result of the multiplication between the plasticity index and % of material passing the sieve size no. 200	Max. 25%	-	-	-
Clay lumps and friable materials (SNI 414, 2015)	0 - 5%	0 - 5%	0 - 5%	0 - 5%
Soaked CBR (SNI 1744,2012)	Min. 90%	Min. 60%	Min. 50%	-
The comparison of % sieve passing no. 200 and no. 40	Max 2/3	Max 2/3	-	-
Coefficient of uniformity $C_v = D_{60} / D_{10}$	-	-	-	> 3,5

Note :

1) : coarse aggregate which has 95% 1 or more divisions and 90% 2 or more divisions

2) : coarse aggregate which has 55% 1 or more divisions and 50% 2 or more divisions

3) : coarse aggregate which has 80% 1 or more divisions and 75% 2 or more divisions

7.4. Cement requirement

1. On the market in zaq (40/50) kg.
2. OPC / PPC / PCC type

IOP Conf. Series: Materials Science and Engineering **930** (2020) 012061 doi:10.1088/1757-899X/930/1/012061

7.5. Asphalt mixture

1. AC-Base Gradation as shown in table 4
2. Thickness of asphalt covering 10 -12 microns
3. Has a mixed performance like table 5

Table 4. Arrangement gradation of asphalt surface layer mixture

Sieve size		% Of Weight Passed to Total Aggregate							
		Stone Matrix Asphalt (SMA)			Hot Rolled Sheets (HRS)		Asphalt Concrete (AC)		
ASTM	(mm)	Thin	Fine	Coarse	Wearing	Base	Wearing	Binder	Base
1 1/2"	37,5								100
1"	25			100				100	90-100
3/4 "	19		100	90-100	100	100	100	90-100	76-90
1/2"	12,5	100	90-100	50-88	90-100	90-100	90-100	75-90	60-78
3/8"	9,5	70-95	50-80	25-60	75-85	65-90	77-90	66-82	52-71
No.4	4,75	30-50	20-35	20-28			53-69	46-64	35-54
No.8	2,36	20-30	16-24	16-24	50-72	35-55	33-53	30-49	23-41
No.16	1.18	14-21					21-40	18-38	13-30
No.30	0,600	12-18			25-60	15-35	14-30	12-28	10-22
No.50	0,300	10-15					9-22	7-20	6-15
No.100	0,150						5-13	5-13	4-10
No.200	0,075	8-12	8-11	8-11	6-10	2-9	4-9	4-8	3-7

Table 5. Properties of hot asphalt mixtures

Mixture Properties		**Asphalt Concrete Layer**		
		Wearing	**Binder**	**Base**
Number of blows	Times	75		112
The ratio of particles passing through sieve no. 200 with effective asphalt content	Min	0,6		
	Max	1,2		
Volume of Void in mixture (%)	Min	3,0		
	Max	5,0		
Volume of Void in Mineral Aggregate (VMA) (%)	Min	15	14	13
Volume of Void Filled with Asphalt (VFA) (%)	Min	65	65	65
Marshall Stability, (kg)	Min	800		1800
Flow, (mm)	Min	2		3
	Max	4		6
Marshall stability remaining after soaking for 24 hours at 60 °C (%)	Min		90	
Volume of Void in in mixture in refusal density condition, (%)	Min		2	

8. Conclusion

1. The use of sand layers (Zones 2 & 3) as stabilizer layers to control the pressure of subgrade development and the influence of forces from overload repetition.
2. The use of vibratory roller with a static weight of at least 25 tons to support the implementation of single layer foundation exposure with the effect of maximum compaction energy, so that compaction energy stored in the foundation layer is not exceeded by compaction energy from overload repetition.
3. Placement of 15 cm thick AC-Base in a single layer as a surface layer to anticipate rutting and bleeding due to repetition of heavy loads that tend to be excessive.

4. The accumulation of the three strengths of the pavement structure can be referred to as the excess energy stored in a flexible pavement system in the form of stiffness, and will work when the load exceeds the plan so as to be able to provide services according to the design life with small maintenance costs.

Acknowledgements
I would like to express my gratitude to the Head of the Public Works Department of Bina Marga in East Java Province for facilitating this research in the form of costs and facilities for material testing laboratories in supporting the implementation of this research.

To all laboratory staff which I cannot mention one by one who were very enthusiastic in carrying out this research.

Special thanks for Prof. Indrasurya, and also my group discussion partner who have helped in the completion of this paper. Thank you so much for everything.

References
[1] General Specifications of the Director General of Highways 2018 *Ministry of Public Works and Human Settlement for Road and Bridge Works* 501-668
[2] Revision-1 General Specifications of the Director General of Highways 2018 *Ministry of Public Works and Human Settlement for Road and Bridge Works* 501-669
[3] Lanham and Maryland 1996 Hot mix asphalt materials, mixture design, and construction *NAPA Educ. Found.*
[4] McGennis, Shuler R B S and Bahia H U 1994 Background of superpave asphalt binder test methods *FHWA report* FHWA-SA-94-069
[5] Indrasurya 2008 Lecture module book *Institute Sepuluh Nopember Technology*

Assessing the impact of the incorporation of aramid and polyolefin to hot and warm asphaltic mixture using dry and wet process: A Review

C G Daniel[1*]

[1] Department of Civil Engineering, Pelita Harapan University, Tangerang, 15811, Indonesia.

[*]Corresponding author's e-mail: christian.daniel@uph.edu

Abstract. The utilisation of polymer has become a prominent trend in the civil engineering industry, one of them is in the production of asphalt for pavement structure. There have been currently a vast range of polymer product variation used in the asphalt production, among which are polyolefin (polyethylene, polypropylene, PET) and aramid. This paper attempts to assess the benefit of the incorporation of the polymer to produce both hot and warm mix asphalt, both by the wet process to develop a modified bitumen, also dry process to form a fibre reinforced mixture. The results indicate that the addition of up to 10% of polyolefin increase the softening point and dynamic stiffness of the bitumen as well as decrease the penetration grade of bitumen by means of wet process. Nevertheless, an extra binding agent is required to prevent the separation between the polymer and bitumen. Furthermore, the inclusion of the polymers using the dry process has been discovered to promote improvements in the mechanical performances of both hot and warm mix asphalt regarding the resistance against rutting, fatigue life, stiffness, tensile and compressive strength, and fracture energy given a wide range of working temperature, as well as reduce binder drain down. These outcomes concluded that the use of polymer as an additive in bituminous mix can combat the failure mechanism in the pavement structures.

1. Introduction

The amount of road infrastructure in Indonesia is not well-spread throughout the country. According to the report by National Statistical Bureau of Indonesia in 2017, the major part of non-highway asphaltic road infrastructure centralised in both Java and Sumatera, leaving a substantial need to improve such situation to ensure the availability of good quality pavements in the entire nation (1) (2). Furthermore, the estimated growth of available vehicle in Indonesia has outgrown the development of the road itself, establishing the necessary need to erect the new infrastructure (3). A road construction, however, may only last so much within its serviceability period before being degraded and needs to be replaced or completely demolished to create a new structure. Such occurrence has been a regular practice in the civil engineering field, thus this raises an issue with the environmental sector regarding the employment of finite resources, such as bitumen. Moreover, the repeated process may come in a considerable expense, ultimately if not designed or constructed properly. Last but not least, the asphalt concrete mix production standard needs an enormous amount of energy to produce the necessary heat (recorded at 4-5 Giga Joule/Ton asphalt production), and thus, generating undesired carbon dioxide (CO_2) pollution (recorded at 0.3ton/ton asphalt production) (4). Therefore, the engineers and scientists have tried to invent advanced materials to enhance the performance and durability of asphaltic material as well as develop a modified production technique to produce a higher quality pavement infrastructure. One of the

Content from this work may be used under the terms of the Creative Commons Attribution 3.0 licence. Any further distribution of this work must maintain attribution to the author(s) and the title of the work, journal citation and DOI.
Published under licence by IOP Publishing Ltd

available advanced material to date is the polymer substance modifier, while a warm asphalt mix approach is designed to produce the bituminous mixture in a lower temperature to output a substantial reduction in fuel used in the production as well as the greenhouse gas emission by around 25% (5)(6)(7). This research focuses on extracting the information regarding the influence of both aramid and polyolefin (polyethylene and polypropylene) to the mechanical performance of asphalt concrete (AC) mixture both in the typical hot mix (HMA) and warm mix asphalt (WMA), where the incorporation can be classified into two ways: blended with bitumen to create a polymer modified bitumen, called wet method; and incorporate the polymer directly into the mixture, called dry method. The outcomes from both methods are discussed in this paper.

2. Fibre as modifier material in pavement engineering field

2.1. The history of fibre-reinforced asphalt concrete

Fibre as reinforcement material has been widely used in road engineering discipline, it has been started in the early 1920s by the use of asbestos fibre. Asbestos fibre was commonly used as the reinforcing component until the 1970s, when its application was banned following its declaration as a hazardous material (8)(9). Since then, numerous reports have recorded the use of various types of fibre in bituminous mixtures: namely *cellulose*, reportedly being able to promote binder stability in an open-graded mix, with however, no effect to tensile strength; *mineral*, such as slag, basalt, asbestos, steel and carbon; and *synthetic polymer* fibre, among them are polyester, polypropylene, polyethylene and aramid (9). The general premises of the benefits brought by the incorporation of fibre to the asphaltic mixture are to increase stiffness, tensile strength, fatigue resistance, rutting resistance and abrasion resistance, which lead into higher durability as well as life service that result in lower life cycle costs (8)(9). In addition, steel fibre is reported to carry a limited potential of healing capacity to the asphaltic mixture through the process of induction heating (10). Also, the polymer product can either be used as reinforcing material or to alter the characteristics of the binder by directly mixing them, forming a polymer modified bitumen (11). On the other hand, these admixtures could also bring several disadvantages to the mixture. One example is asbestos, which brings an issue to health. Moreover, steel fibre could be exposed to corrosion, which will surely reduce the performance of the bituminous mix. Additionally, polymer fibre tends to absorb moisture, which will later degrade the strength of the asphaltic mixture and introduce moisture damage to it (12)(13). Lastly, the additional fibre will also need more compaction time to avoid the mix to harden and too difficult to be compacted; this is particularly important in an open graded mixture (14). Regardless, the potential brought by the inclusion is proven to be beneficial that such application is encouraged, and the use of fibre-reinforced material remains a common practice to date.

2.2. History and properties of aramid and polyolefin fibre

Aramid and polyolefin are both the main clusters of synthetic polymer fibre; both of them share a long history and application throughout the century. The history of aramid fibre could be traced back to 1938 when nylon was the first commercialised, followed by the introduction of Nomex fibre by Dupont Co. in 1962 (15). Since then, aramid can be associated with the term "Kevlar" under the brand of Dupont. The production process of aramid can be described in the following way (16)(17). Firstly, the condensation between two monomers called 1,4-Para-Phenylenediane (PPD), and Terephthaloyl Dichloride (TDC) occurs to form Poly Para-phenylene terephthalamide (PPD-T), whose molecular structure is described in figure 1. PPD-T is then washed to eradicate the acidic substance. Afterwards, the dry PPD-T is extruded from a spinneret inside a robust acidic liquid, in order to form the orientation and structure of the fibre by means of coagulation and crystallisation process. The final stage is to dry the fibre filament, which already develops its physical tensile strength. The molecular structure of aramid contributes to the physical properties, where the aromatic ring structure generates the high thermal stability, and the para configuration forms the stiffness and strength of the fibre (18). Altogether, aramid fibre has the tensile strength and modulus of approximately 4 GPa and 180 GPa, respectively. Moreover, the physical attributes of aramid fibre can be maintained to be about 80% of them at the temperature of 200°C, making it insensitive to temperature change. However, aramid is prone to plastic deformation when subjected to compressive loading, suffering under the effect known as "kink bands" (18)(15).

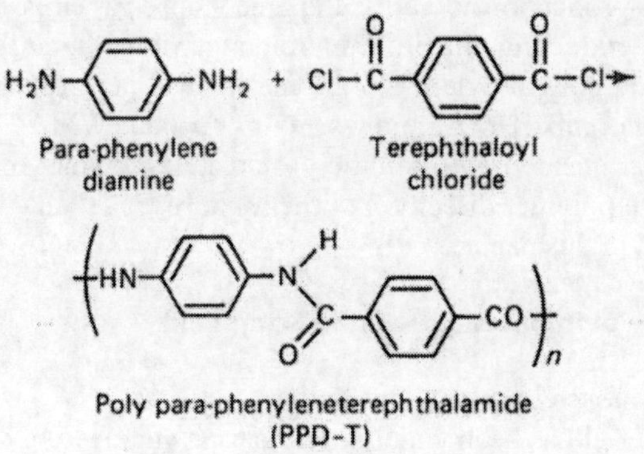

Figure 1. Molecular structure of aramid fibre (18)

Nowadays, aramid is widely used in the aerospace industry. For instance, the combination of aramid and carbon fibre is commonly used as structural parts in the commercial aircraft and choppers. Such application exploits the capacity of aramid to absorb energy as well as its high tensile strength and stiffness, which are found out to be 2-4 times and 25-30% larger than that of steel and glass fibre, respectively. It is also commonly used as conveyor belt due to its high fatigue resistance as well as excellent adhesion with rubber. More than that, Kevlar products can also be used as safety suits and gloves for the firemen and safety operatives. Lastly, aramid serves as a component of bulletproof armour for being an energy-absorbent material (15)(19).

Meanwhile, the polyolefin is a class of a synthetic polymer thermoplastic type which can be further classified into two subcategories, such as polyethylene (PE) and polypropylene (PP). The term "polyethylene" was firstly invented by a French chemist named Pierre Eugene Marcellin Berthelot in 1869, though the first discoveries were remarked by Pechmann in the 1890s and later by Staudinger in 1920, during which it was introduced through the concept of high molar mass macromolecules obtained from polymerisation process which created a covalent bond between small monomers. Polyethylene has two different sub-clusters, namely Low-Density Polyethylene (LDPE) and High-Density Polyethylene (HDPE), both of which come from a similar process with only a small variation in terms of the reaction pressure. LDPE was firstly found in 1933 by Fawcett and Gibson as an unintentional, waxy solid, by-product of a reaction between ethylene and benzaldehyde on a massive pressure (more than 1000 atm). HDPE was discovered two decades later by Philips along with the discovery of nickel and chromium oxide/metal oxide on silica/alumina as catalysts to produce polyethylene under significantly lower pressure. On another occasion, Ziegler (1953) developed another way to construct polyethylene by using triethyl aluminium as the catalyst to create a low molecular weight polyethylene. Later on, Kaminsky developed a new method to produce PE utilising compound named methyl aluminoxane (MAO) as an activator for metallocene catalysts, which is approximately 100 times more reactive than that of the Ziegler's catalyst (20)(21). The documents show that the production of HDPE as plastic products has already exceeded 50 million tons in 2018 (22). In terms of physical and mechanical properties, LDPE and HDPE exhibit distinctive performances. Regardless of the density, which ranges from 0.92-0.94 for LDPE and 0.94-0.97 for HDPE, the differences become more evident regarding the melt temperature, tensile strength, and tensile modulus. LDPE will melt under the heat of 115-122°C, whereas HDPE will melt in 135-140°C (23)(24). Moreover, the tensile strength of HDPE doubles the strength of LDPE, which ranges from 17-34 MPa. Lastly, the tensile modulus of LDPE ranges from 140-210 MPa, whereas the stiffness of HDPE is approximately five times higher than LDPE, ranges from 550-1250 MPa (25) (26).

On the other hand, the development of polypropylene follows a similar direction to that of polyethylene since both of them come from the identical monomer, from which their molecular structures are basically the asymmetric chain of α-olefins with general formula CH2=CHx (x represents an alkyl group) (27). The melting point and tensile strength of PP is slightly higher than PE, while the stiffness is almost the same as HDPE (28)(29). PP currently serves as the second most popular ingredient of almost every plastic-based product that exists these days, with the estimated value on global market at 155.57 billion dollars in 2026, with the projected production volume to 88 million metric tons at the same timeframe (30).

Another one of the most remarkable derivatives of polyolefin is polyethylene terephthalate (PET), also known under the brand Dacron (31) by Dupont, where 30% of the fabricated product is utilised as plastic bottles and other plastic-related products (32). PET is formed from the extrusion of synthesis product of ethylene glycol and terephthalic acid. It has a higher melting temperature than the other ethylene derivatives (up to 275°C) (33) with the stiffness up to 4 GPa and strength of 80 MPa (34). All of these establish the significant role of PET as one of the principal components of lightweight materials.

3. The application of aramid and polyolefin fibre in bituminous mixture

3.1. Wet vs dry technique

Many synthetic polymer compounds can be used to enhance the characteristic and performance of a bituminous mixture, among them are aramid and polyolefin products. In particular, polyolefin can be poured directly into the heated bitumen to form a polymer-modified bitumen (PMB). Polyolefin, along with other types of polymer such as Styrene-Butadiene-Styrene (SBS), EVA, etc. are typically associated with this method of incorporation. The technique used is called the wet method. Meanwhile, aramid can also be used through direct integration to the mixture of bitumen, aggregate and filler, in a process called dry method. The differences between both techniques are illustrated in figure 2 (35). In the wet process, the polymer is blended with the neat bitumen to form a polymer modified binder (step a-b). The modified bitumen is then mixed together with aggregates to form the AC mixture (step d). Meanwhile, the neat binder is directly mixed with aggregates as well as the polymer (step c-e) in the dry process (36)(37).

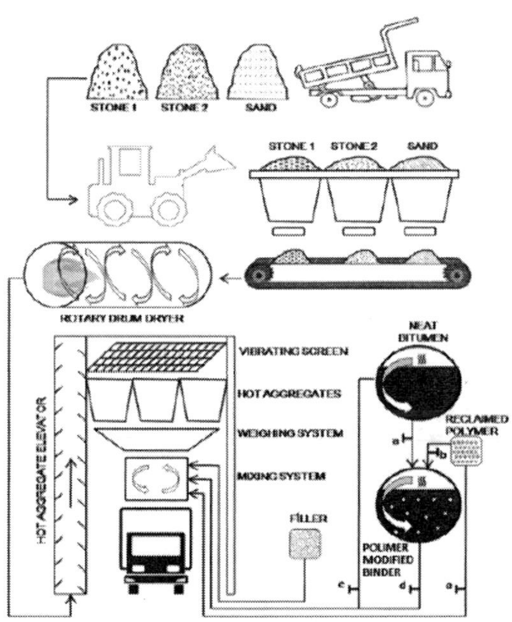

Figure 2. Description of wet and dry process (35)

However, the wet process suffers from a significant issue of segregation that occurs when the polymer-bitumen mixture is not stirred continuously, where the olefin molecules and bitumen do not form a stable reaction; hence olefin molecules will come up on top due to their lower density (38). The segregation phenomena are linked to asphaltene, which is the main component of bitumen whose density is high, and form is stiff and waxy solid (39). figure 3 depicts exactly this mechanism, which takes place even at a high temperature (40). The mixture earns more stability as the temperature declines, indicating that a polymer modified bitumen can be categorised as a thermodynamically unstable product. Stability can only be promoted by the kinetic process of constant stirring (41)(42).

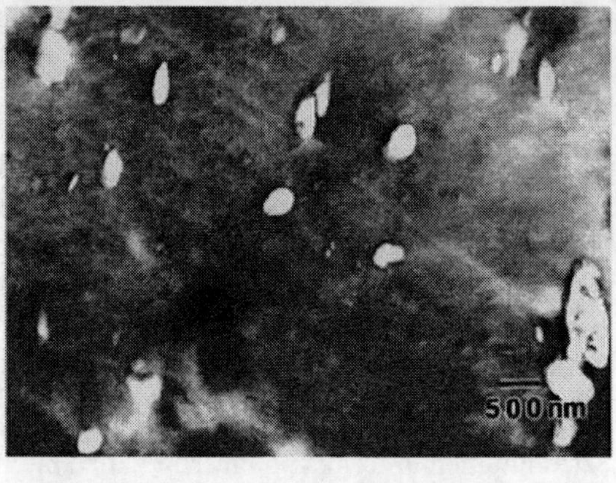

Figure 3. SEM Figure of the separation between polymer molecule and bitumen (40)

3.2. Chloride penetration test

The inclusion of polyolefin using the wet method has been found to improve the rheological and mechanical performances of the bitumen. These effects include penetration and softening point, which can be summarised in figure 4 (42)(43)(44)(45).

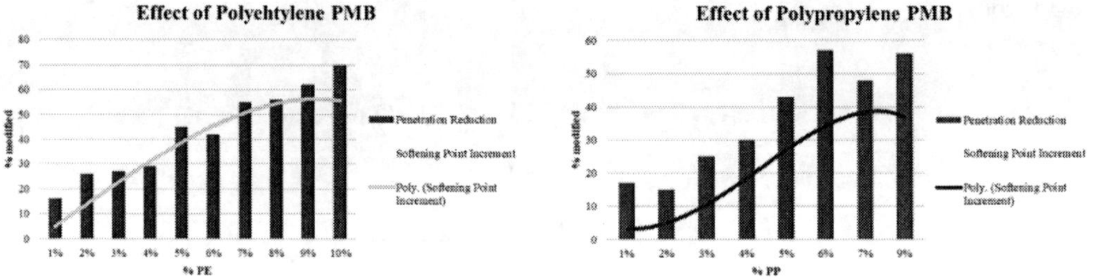

Figure 4. Effect of (left) PE and (right) PP to penetration and softening point of modified bitumen (42)(43)(44)(45)

It is apparent from the charts that the softening point of the PMB has increased while the penetration grade is reduced by increasing the dosage of polyolefin. Furthermore, another research has found that the inclusion of polyolefin could enhance the complex shear modulus (G*) of the bitumen (35)(44). Conclusively, the addition of polymer product will increase the stiffness, which supports the general knowledge that a PMB holds a significantly higher stiffness than the standard bitumen product. Consequently, the finding also implies that the addition of polyolefin product will yield a higher resistance to permanent deformation (rutting). Thus, the PMB product is suitable to be applied to a high traffic load in high-temperature regions (46)(47). On the other hand, the segregation and instability issues can be solved by means of a chemical solvent named Maleic Anhydride (MA) which enables the polymer molecules to be dispersed uniformly and form a high stability mixture (43), captured by figure 5.

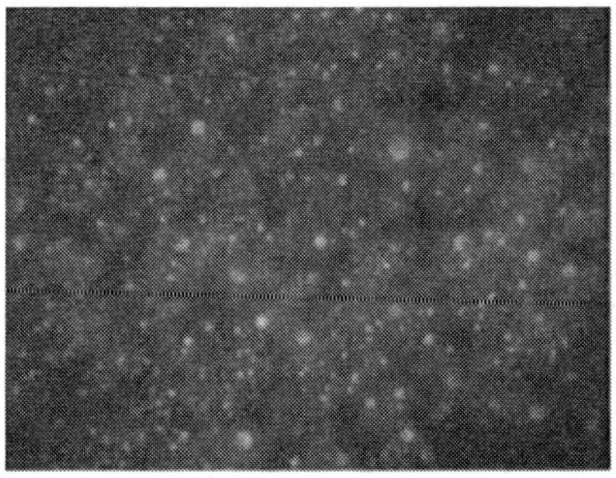

Figure 5. Microscopic image of molecular structure of PMB modified with maleic anhydride (43)

Additionally, PET is also occasionally used as a bitumen modifier, which the application was reported to increase the dynamic shear modulus (G*), and consequently, resistance against rutting for a stone mastic asphalt (SMA) mixture, as shown in figure 6 (48). Further application of PET PMB could also promote the significant use of recycled asphalt mixture with the enhanced results in terms of softening point, tensile strength and moisture susceptibility, with a lower value of penetration, indicating a stiffer material (figure 7) (49)(50). It is interesting that the enhancement is not necessarily in the linear proportion with the increase of polymer dosage; rather, the benefit declines after a specific dosage of approximately 6%. That being shown, it can be concluded that after the critical dosage, the presence of the polymer becomes unused, and even, disturb the bonding inside the mixture. Moreover, the same issue regarding segregation and instability also need to be taken into account for the results. Therefore, the aforementioned treatment with a dispersal promotor substance is recommended.

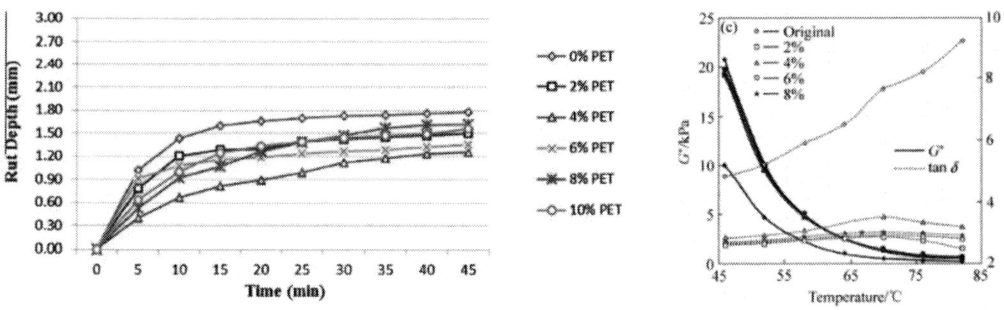

Figure 6. (Left) Rut depth and (Right) Dynamic modulus of PET-modified PMB (48)

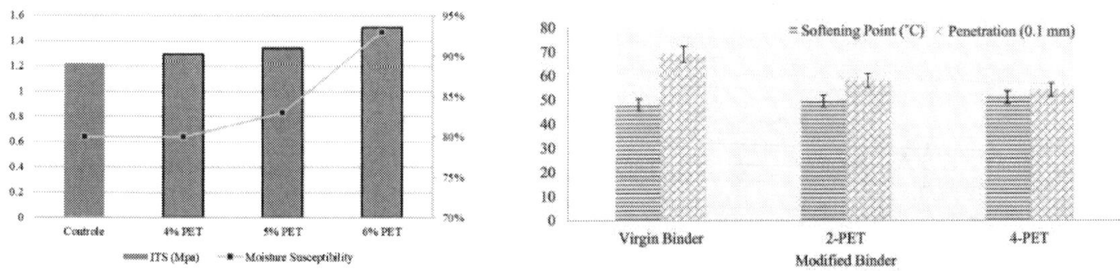

Figure 7. (Left) Tensile strength and moisture susceptibility and (Right) Softening point and penetration test of PET-modified PMB (49)(50)

From the gathered results, it can be concluded that the incorporation of polyolefin, be it PE, PP, or PET, has improved each aspect of the mechanical properties of the bitumen. However, the proper dosage is a consideration, and as suggested from the outcomes, the optimum effect to the results is evident until approximately the dosage of 6%, beyond which the presence of the substance becomes negligible and does not help to improve the performance furthermore. This amount of dosage can then be established as the critical dosage. Aside from it, more researches need to be conducted to establish the general dosage for the application in various types of asphaltic mixtures.

The other possible application is the dry method. Aramid can be used with this practice as a fibre reinforced material, which may also be applied in combination with polyolefin. Such combination has become widely popular nowadays, with quite extensive research has been performed to examine the effect on the characteristic of asphaltic mixture. The application of aramid fibre will enhance the mechanical performance physically following its capability to withstand its work even when mixed at a high temperature. Thus, the reinforcing mechanisms are mainly in the form of adhesion and friction between aramid and asphalt matrix so that load transfer between fibre and matrix is enabled, which is comparable to the mechanism of steel bar in a reinforced concrete structure (51)(52). Such mechanism is clearly illustrated in figure 8, where the filament is displaced after a specimen is subjected to tensile loading, leaving a so-called micro-crack due to the loss of adhesion between aramid and asphalt matrix.

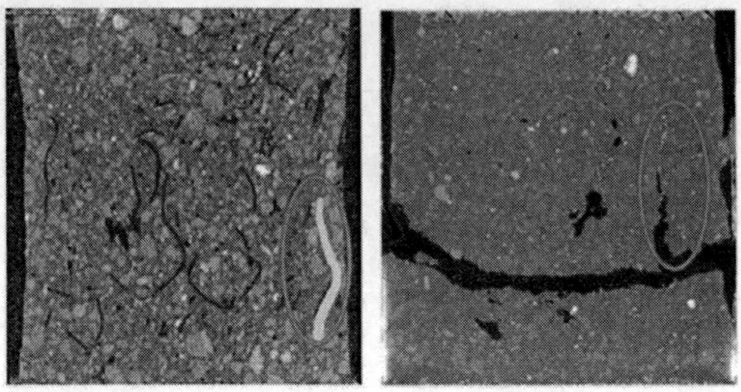

Figure 8. Bonding mechanism of fibre reinforcement (Left) before and (Right) after cracking (51)

The load transfer mechanism can be explained by observing the pull-out test mechanism and result, illustrated in figure 9, where the examinations were performed at room temperature. The graph could also indicate the impact of having a different loading speed on the performance and mechanism. In combination with the research of Park (52)(53), the full picture of the mechanism can be explained in the following manner. Firstly, at the high temperature, the bitumen matrix has begun to transform into more viscous state, and therefore, the presence of interface bonding and load transfer mechanism between the fibre and mortar matrix becomes critical to ensure that the matrix failure, in which the mortar will become sufficiently deformed, will occur in the end. On the other hand, at relatively low temperature, or high loading speed, the bitumen remains in its stiff/elastic state, and as a result, fibre-matrix adhesion has not gained the full potential and becomes prone to the interface failure. These could also then magnify the benefit brought by the fibre at high temperature.

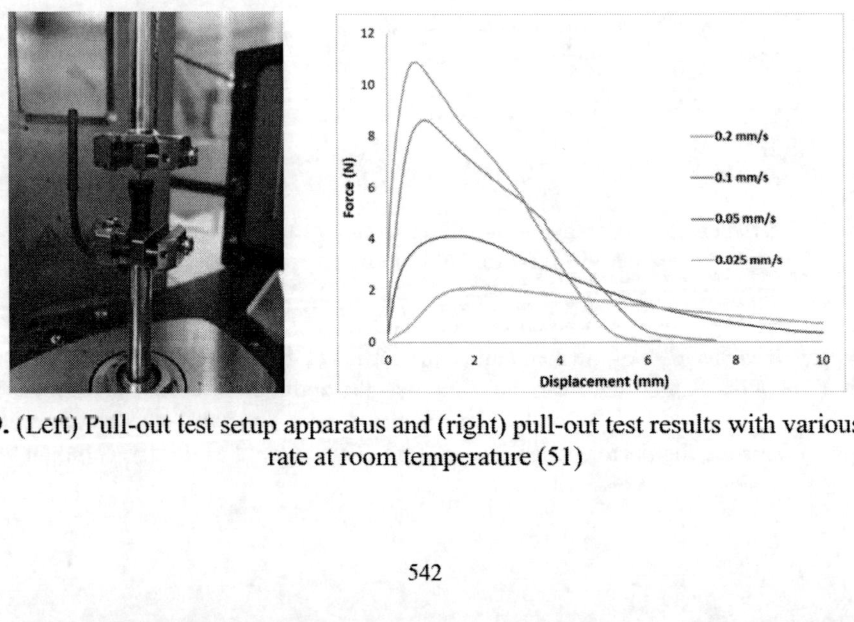

Figure 9. (Left) Pull-out test setup apparatus and (right) pull-out test results with various loading rate at room temperature (51)

The research about the dry method application has been extensively executed. As for the implementation of polyolefin only, investigations conducted using Marshall stability and indirect tensile tests to check the effect on the performance of asphalt concrete mix. Several conclusions that can be drawn from the researches are the increase of fibre dosage up to 1% of total weight of mixture will improve the Marshall stability and fatigue life of bituminous mixture by 58% and 27% respectively, whereas flow is reduced by 142%. Furthermore, the addition of more dosage of the fibre minimises the amount of permanent deformation and air voids, as well as raises elastic modulus of the dense asphalt mix. These phenomena indicate a raise in stiffness of the mix (54)(55)(56).

Meanwhile, there are several contributions of aramid-polyolefin to the mechanical performance of bituminous mix as highlighted in this paper. Firstly, the inclusion in a hot mix asphalt will establish an extended secondary stage as well as a lower creep coefficient from triaxial test that indicates a higher resistance to permanent deformation, both tested with the hot and warm mix asphalt with 0.05% dosage of fibre mixture weight (12)(57)(58). Another paper reveals the effect of fibre length and dosage to the resistance of rutting, both of which are illustrated in figure 10. A lower creep coefficient indicates the less deformation impacted by the cyclic loading subjected to asphaltic mortar specimens. It is apparent that the increase of fibre dosage and length brings a beneficial effect to the mixture resistance. Additionally, the coefficient of the long fibre with a dosage of 0.1% is somehow equivalent to that of the short fibre with 0.5% dosage (51). The results indicate that the mixture becomes stiff enough to withstand more impact without suffering any permanent deformation, which in turn could cause inconvenience should the condition implemented to the real pavement structure.

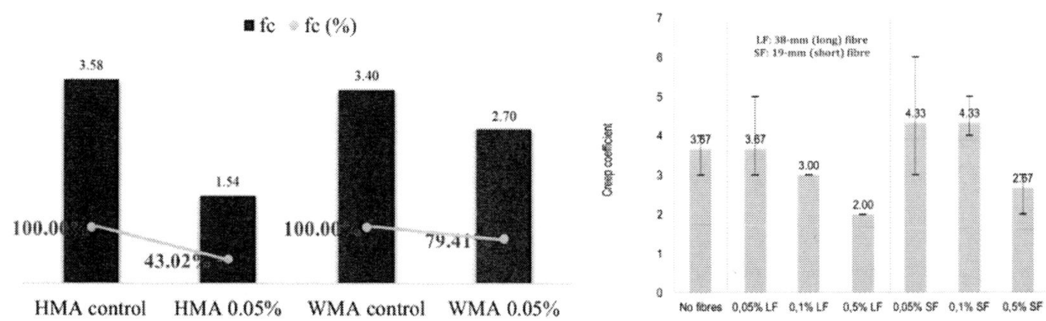

Figure 10. Creep coefficient as a result of cyclic (left) triaxial (12)(57) and (right) uniaxial loading test (64)

Moreover, both stiffness and fatigue life of the modified mixture with 0.05% dosage of fibre, hot and warm mix, show inclination trends as indicated by the higher value of dynamic modulus depicted in figure 11. Notice that the increase of both parameters for the hot mix mainly occur at the low strain/load level, indicating that the contribution of both fibres is more obvious in a road with high traffic speed. Conversely, the stiffness of warm mix asphalt specimens is not significantly affected by the existence of the fibre, whilst the fatigue life of the specimens is in contrast with the result of hot mix asphalt, indicating a possible different direction of development for a road with the warm mix asphalt applied to it (figure 12). Last but not least, the effect of temperature is also shown herein where the difference in stiffness due to fibre modification mainly appears at the high temperature (59)(60).

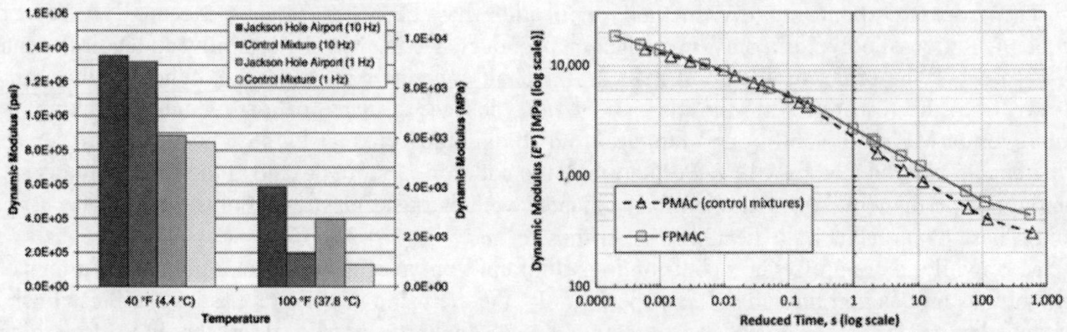

Figure 11. (Left) Dynamic modulus of FRAC at two temperatures and frequencies and (Right) dynamic modulus at various frequencies (59)(60)

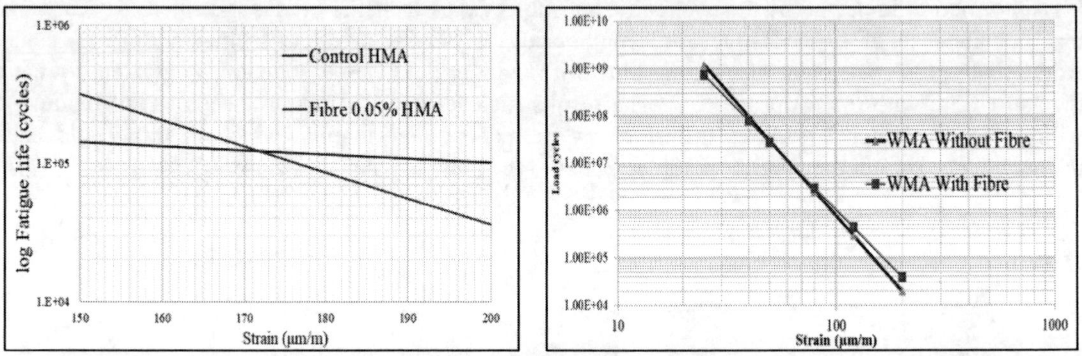

Figure 12. Fatigue life results of (left) hot mix and (right) warm mix fibre reinforced asphalt (12)(57)

Furthermore, tensile strength and energy exhibits the benefit of the fibre additives, both in hot and warm asphalt mixture, ultimately at the high temperature as shown in figure 13 (left), for indirect tensile test and figure 13 (right) for direct tension test results. The results, linked with the outcome of stiffness, yield a conclusion about the better exhibition of fibre-reinforced asphalt mixture at high temperature, where bitumen becomes more viscous, hence allow the fibre to react as a bonding agent to promote more adhesion between aggregate and the matrix and generates a mixture with higher capacity to undergo extra deformation before cracking (51)(61).

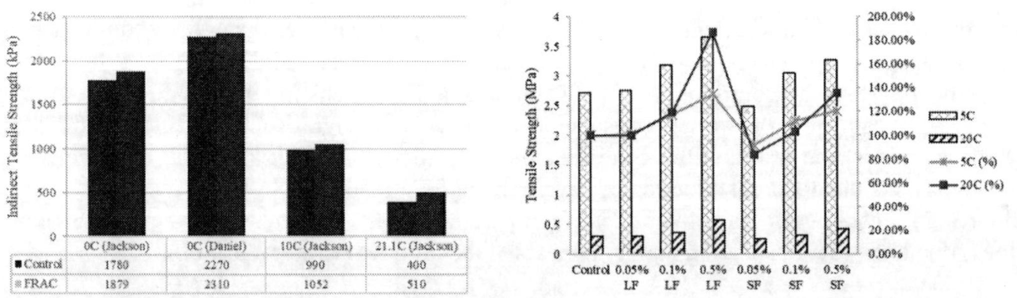

Figure 13. (Left) Tensile strength from Indirect tensile test (12) (57) and (Right) from direct tension test (51)

Another critical finding is obtained by drawing a comparison between the strength and fatigue life of the asphalt mortar specimen filled with a 19-mm long fibre and 38-mm fibre, where the latter generates better outcomes than the former, as shown in figure 14 by performing semi-circular bending and direct tension test to obtain toughness (left) and fatigue life (right). The higher phenomena can be attributed to the more significant bonding length, and thus, larger interface area possible between the fibre and the mortar matrix. Both the bonding capacity/adhesion and the frictional interface area contribute to the final mechanical performance of the bituminous mixture. Furthermore, it is also apparent that higher values of tensile strength, toughness or fracture energy and fatigue life can be achieved by adding more fibre to the equation. However, there is also a critical value in which the optimum condition between mechanical performance and the efficiency, beyond which the additional amount of fibre will become less usable and the improvement seems to be negligible (62)(63).

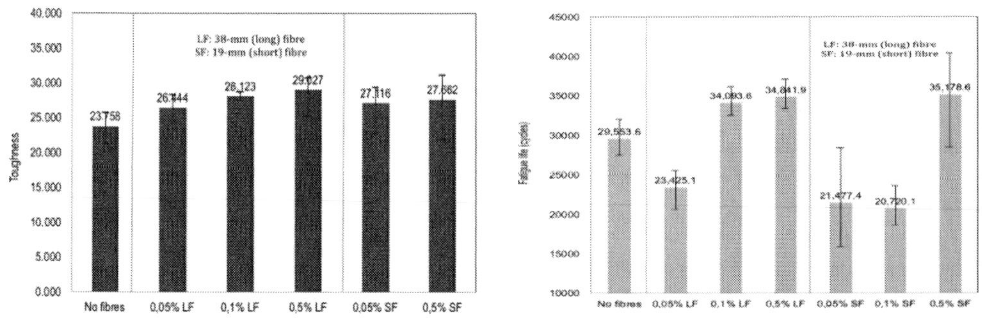

Figure 14. (Left) Toughness and (Right) Fatigue life of fibre-reinforced asphalt mortar mix with various fibre dosages (51)(62)(63)

The case presented herein concludes the amount of 0.1% of aramid-polyolefin fibre combination to become the critical dosage (64). Thus, the fibre amount, as well as fibre length, are of paramount parameters to achieve the optimum performance of an asphalt mixture. All the combined knowledge could bring the conclusion that the length, or dimension, of the fibre is more critical to the performance than merely the dosage. This could also be understood from the point of view given in figure 8, which the load transfer mechanism will not be sufficient once the amount of fibre exceeds the needed dosage, and instead, will get clumped and become unnecessary inside the mixture. In contrast, the correct dosage and orientation as well as the length of fibre will ensure the bonding to occur well, and the load transfer mechanism could be performed swiftly. On the other hand, the fibre-reinforced hot asphalt mixture exhibits a less component loss than the control mixes as shown through the Cantabro test, indicating higher resistance to stone loss or ravelling (65). Further assessments also find the positive impact brought by the fibre to the identical mix configuration by mitigating the drainage of the specimens as well as increasing the tensile energy by almost 50% in the dry condition. However, as the moisture begins to partake, the improvement is reduced greatly by almost 40% of the initial finding, confirming the susceptibility of the fibre to moisture (66)(67).

Lastly, another most well-known polymer that is normally employed to the modified asphalt concrete (AC) mix using the dry method is PET, which comes normally in the form of recycled plastic products as a partial substitution of aggregates. Till nowadays, the substitution with the dosage up to 1% one-fourth of the total aggregate can improve fatigue life about twice the control mix (68), whereas the dosage up to one-fourth of the total aggregate decrease rut depth until approximately half of it (69). Furthermore, the replacement could also improve Marshall stability, amplifies the compressive strength by almost 10%, as well as reduces the flow or drainage of the mixture in a direct proportion to the dosage increase. Such applications and positive impacts can be linked to the high natural melting point of PET, which enables to maintain the strength and integrity within the blending temperature (70)(71).

4. Conclusion

The application of polymer product as an admixture in bituminous mixtures has become a regular practice nowadays. There are currently various polymers available to be used, among them are aramid and polyolefin, with two methods of application: the wet process, wherein the polymer is blended with the heated bitumen and stirred to form a uniform Polymer Modified Bitumen; and dry method, wherein the polymer is blended directly with the aggregates and bitumen to create a so-called fibre-reinforced asphalt mixture. While aramid is traditionally linked to the dry process due to its capability to withstand the mixing heat, polyolefin can also be used to create the PMB product. Nevertheless, both can also be combined to create a fibre-reinforced asphalt concrete mix, whether hot or warm mixture. The gathered outcomes regarding the application of polyolefin in the wet method have suggested the critical dosage of 6% mixture to achieve the optimum mechanical performance of the mix in terms of, stiffness, penetration and softening point of bitumen, as well as tensile strength and resistance of rutting, whereas the excessive dosage does not guarantee further improvement to the results. However, suitability between bitumen and polymer is already a well-known issue, and one of the possible solutions is to add Maleic Anhydride as the dispersal promoter to avoid the segregation and separation of components. Meanwhile, polyolefin can also be employed using the dry method, with similar effects to the mechanical properties have been discovered where the dosage of up to 1% of mixture weight is taken into account. However, the more popular choice is to combine aramid and polyolefin to the asphaltic mixture using the dry method, which is backed up by numerous researches. First of all, the reinforcement effect can be divided into three mechanisms, which is also load- and temperature- dependent: interfacial bonding/friction between the physical form of aramid and asphalt matrix, adhesion brought by the polyolefin, and load transfer between fibre-matrix; at which the benefit is more apparent at high temperature and low loading speed due to the viscous nature of bitumen that is activated in such situations. The inclusion of aramid-polyolefin combination has several effects on the mechanical performance of both hot and warm bituminous mixture. Firstly, the fibre reinforcement increases the resistance to rutting by approximately 57% and 21% for hot and warm mix asphalt, respectively. The increased amount and length of fibre also increase improvement. Secondly, the fibre will enhance the stiffness and fatigue life of both types of mixture, mainly when subjected to low-frequency loading that is translated to high-speed traffic for the hot mix. In contrast, the same improvement for a warm mix is found within a high strain-level loading. Such inconsistency is worth exploring more in the future. Moreover, tensile strength and toughness of both hot and warm mix asphalt are also improved by the inclusion of the fibre. However, the benefit lasts only until a precise dosage of fibre, after which the states improve quite insignificantly, and thus, nearly negligible. It is also found that the higher length of the fibre contributes to the better performance of the mix. Conclusively, the specific combination of aramid-polyolefin fibre with 0.1% dosage of mix weight and a longer dimension works best to improve the characteristics of the bituminous mixture within the high working temperature using the three working mechanisms. This could be attributed to the network formed between the fibre and mortar matrix, and the interface bonding between them that boost the performance of the mixture. Furthermore, the fibre reinforcement will also react positively when injected to the open graded / Stone mastic asphalt to reduce the potential of raveling. Lastly, PET could contribute to the increase of fatigue life and lower rut depth by employed as partial aggregate substitution. This synthesis is able to sum up the benefits brought by the three most polymer substances to the mechanical and physical characteristics of the bituminous mixture. There is a consideration that the enhanced moisture susceptibility brought by the inclusion of the polymer products due to their capability to absorb the water.

References

[1] Badan Pusat Statistik. No Title. 2019.
[2] Katadata.co.id. 2017 No Title
[3] INDONESIAN COMMERCIAL NEWSLETTER (ICN). 2012 The development of road infrastructure *INDONESIAN COMMERCIAL NEWSLETTER*
[4] Yang R, Ozer H, Kang S and Al-Qadi I L 2014 Environmental impacts of producing asphalt mixtures with varying degrees of recycled asphalt materials *Int. Symposium on Pavement LCA*
[5] Zaumanis M 2011 Asphalt is going green: Overview of warm mix asphalt technologies and research results from all over the world *Lambert Aca. Pub.*
[6] Rubio MC, Martínez G, Baena L, Moreno F. 2012 Warm mix asphalt: An *overview J. Clean. Prod.* **24** 76–84

[7] Mazumder M, Sriraman V, Kim H H and Lee S J 2016 Quantifying the environmental burdens of the hot mix asphalt (HMA) pavements and the production of warm mix asphalt (WMA) *Int J. Pavement Res. Technol.* **9** 3 190–201
[8] Serfass J P and Samanos J 1996 Fiber-modified asphalt concrete characteristics, applications and behavior *Asph. Paving Technol.* **65** 193–230.
[9] Mcdaniel R S 2015 Fiber Additives in Asphalt Mixtures *NCHRP SYNTHESIS* 475
[10] Apostolidis P 2017 Experimental and numerical investigation of induction heating in asphalt mixes *Delft Univ. Technol.*
[11] Venturini L, Monti F. 2020 Graphene-enhanced recycled asphalt pavements. in: springer, editor. proceedings of the 5th international symposium on asphalt pavements & environment (APE) *Springer* 44–54
[12] Delft T 2016 DIBEC *Dutch Fibre Trading Delft*
[13] Aramid Fiber [Internet]. Greece: Fibremax Ltd. [cited 2020 Mar 30]. Available from: http://www.aramid.eu/advantages---disadvantages.html
[14] National Asphalt Pavement Association (NAPA) - US Department of Transportation Federal Highway Administration. Designing and Constructing SMA Mixtures — State-of-the-practice. Lanham, Maryland; 2002. Available from: https://www.google.com/url?sa=t&rct=j&q=&esrc=s&source=web&cd=&ved=2ahUKEwiktrr d0qTrAhXIcn0KHQSPATcQFjABegQIBBAB&url=https%3A%2F%2Fwww.il-asphalt.org%2Ffiles%2F5114%2F4890%2F3754%2FNAPA-QIP122-DesigningAndConstructingSMAMixtures.pdf&usg=AOvVaw3aC5WL2iN86iwuJdDk9K2r
[15] Jassal M and Ghosh S 2002 Aramid fibres - An overview *Indian J. Fibre Text Res.* 290–306
[16] Aramid 1994 Fiber of poly para-phenylene terephthalamide from the Netherlands Int. Trade Comission (Washington, D.C: U.S)
[17] Zwaag V D S 2009 Structure and properties of aramid fibres. In: Handbook of textile fibre structure - fundamentals and manufactured polymer fibres *Woodhead Publish. Limited* 394–412
[18] Chang K 2001 Aramid fibers 41–5
[19] Dupont (TM) 2019 No Title *Dupont*
[20] Hutley T J and Ouederni M 2016 Polyolefins—The History and Economic Impact. *In: AlMa'adeed MA, Krupa I,* ed. Polyolefin compounds and materials - fundamentals and industrial applications *Springer Cham* 13–50
[21] Zhong X, Zhao X, Qian Y and Zou Y 2018 Polyethylene plastic production process *Insight – Mater. Sci.* **1** 1 1–8
[22] Garside M 2018 Global production volume of HDPE resin 2016-2022 (in million metric tons) *Statista*
[23] Jun L, Yuxia Z and Yuzhen Z 2008 The research of GMA-g-LDPE modified Qinhuangdao bitumen *Constr. Build Mater.* **22** 6 1067–73
[24] Al-Hadidy A I and Yi-qiu T 2009 Effect of polyethylene on life of flexible pavements *Constr. Build Mater.* **23** 3 1456–64
[25] Sauter D W, Taoufik M and Boisson C 2017 Polyolefins, a success story polymers (Basel) 185
[26] Su B, Zhou Y G and Wu H H 2017 Influence of mechanical properties of polypropylene/low-density polyethylene nanocomposites: Compatibility and crystallization *Nanomater Nanotechnol* **7** 1–11
[27] Mather R R 2009 The structure of polyolefin fibres in: Handbook of textile fibre structure fundamentals and manufactured polymer *Fibres Woodhead Pubish. Lim.* 276-304
[28] Hindle C Polypropylene (PP) *British Plastics Fed.*
[29] Hinsley N 2016 Polypropylene- Is it different from Polyethylene? *Global Plastic Sheeting*
[30] Reports & data 2019 Polypropylene market to reach usd 155.57 billion by 2026 | reports and data *Global News Wire* 1
[31] Petresin.org [Internet]. New York: PET Resin Association [updated 2015; cited 2020 Feb 20]. Available from: http://www.petresin.org/news_introtoPET.asp

[32] Poulikakos L D, Poulikakos L D, Papadaskalopoulou C, Hofko B, Gschösser F, Falchetto A C, et al. 2017 Harvesting the unexplored potential of European waste materials for road construction. *Resour. Conserv. Recy.* 32–44
[33] Horvath T, Kalman M, Szabo T, Roman K, Zsoldos G and Kollar M, et al. 2018 The mechanical properties of polyethylene-terephthalate (PET) and polylactic-acid (PDLLA and PLLA), the influence of material structure on forming. *11th Hungarian Conf. Mater. Sci.*
[34] AZO Materials 2003 AZoM *AZO Materials*
[35] Brasileiro L, Moreno-Navarro F, Tauste-Martínez R, Matos J, Rubio-Gámez M and del C 2019 Reclaimed polymers as asphalt binder modifiers for more sustainable roads: A review. *Sustain.* **11** 3 1–20
[36] Modarres A and Hamedi H 2014 Effect of waste plastic bottles on the stiffness and fatigue properties of modified asphalt mixes *Mater. Des.* **61** 8–15
[37] Earnest MD 2015 Performance characteristics of polyethylene terephthalate (PET) modified asphalt 95
[38] Polacco G, Berlincioni S, Biondi D, Stastna J and Zanzotto L 2005 Asphalt modification with different polyethylene-based polymers. *Eur Polym. J.* 2831–44
[39] Hofko B, Eberhardsteiner L, Fussl J, Grothe H, Handle F and Hospodka M, et al. 2015 Impact of maltene and asphaltene fraction on mechanical behavior and microstructure of bitumen *Mater. Struct.* 1–13
[40] Ho C H, Shan J, Wang F, Chen Y and Almonnieay A 2015 Performance of fiber-reinforced polymer modifed asphalt two-year review in northern arizona *J. Transp. Res. Board* 138–49
[41] Taki Z N M, Abed A H and Al-mosawe H 2019 Based on strength ratio and fracture energy parameters 1
[42] Attaelmanan M, Feng C P and Ai A H 2011 Laboratory evaluation of HMA with high density polyethylene as a modifier *Constr. Build Mater.* **25** 5 2764–70.
[43] Vargas M A, Vargas M A, Sánchez-Sólis A and Manero O 2012 Asphalt/polyethylene blends: Rheological properties, microstructure and viscosity modeling *Constr. Build Mater.* **45** 243–50
[44] Roman C, Cuadri A A, Liashenko I, García-Morales M 2016 Partal P. Linear and non-linear viscoelastic behavior of SBS and LDPE modified bituminous mastics *Constr. Build. Mater.* **123** 464–72
[45] Fernandes S, Costa L, Silva H and Oliveira J 2017 Effect of incorporating different waste materials in bitumen *Cienc e Technol dos. Mater.* **29** 1 204–9
[46] Polacco G, Stastna J, Biondi D and Zanzotto L 2006 Relation between polymer architecture and nonlinear viscoelastic behavior of modified asphalts *Curr. Opin. Colloid Interface Sci.* 230–45
[47] Yu R, Fang C, Liu P, Liu X and Li Y 2015 Storage stability and rheological properties of asphalt modified with waste packing polyethylene and organic montmorillonite *Appl. Clay Sci.* 1–7
[48] Zheng C, Shao-peng W, Zu-huang Z and Jie-sheng L 2008 Experimental evaluation on high temperature rheological properties of various fiber modified asphalt binders **15** 135–9
[49] Ahmadinia E, Zargar M, Karim M R, Abdelaziz M and Ahmadinia E 2012 Performance evaluation of utilization of waste Polyethylene Terephthalate (PET) in stone mastic asphalt *Constr. Build. Mater.* **36** 984–9
[50] Sreeram A, Leng Z, Padhan R K and Qu X 2018 Eco-friendly paving materials using waste PET and reclaimed asphalt pavement *HKIE Trans Hong Kong Inst. Eng.* **25** 4 237–47
[51] Apostolidis P, Liu X, Daniel G C, Erkens S and Scarpas T 2019 Effect of synthetic fibres on fracture performance of asphalt mortar *Road Mater. Pavement Des.* **0** 0 1–14
[52] Park P, El-Tawil S and Naaman A E 2017 Pull-out behavior of straight steel fibers from asphalt binder *Constr. Build. Mater.* **144** 125–37
[53] Wheat M 2007 This study was supported by the US Department of Transportation, Office of the Secretary *Grant* DTOS59-06-G-00026
[54] Tapkin S 2008 The effect of polypropylene fibers on asphalt performance *Build Environ.* **43** 6 1065–71
[55] Bayat R and Talatahari S. 2016 Influence of polypropylene length on stability and flow of fiberreinforced asphalt mixtures *Civ. Eng. J.* **2** 10 538–45

[56] Ahmed A I 2012 laboratory investigation into the impact of polypropylene fiber content on temperature susceptibility of dense graded mixtures **5** 4 424–38

[57] Kaloush K, Waleed A. Zeiada, Biligiri K P, Rodezno M C and Reed J Evaluation of fiber-reinforced asphalt mixtures using advanced material characterization tests

[58] Mello L G, Kaloush K E and Biligiri K P 2008 Evaluation of fiber reinforcement in hot mix asphalt using advanced material characterization tests *First Pan. Am. Geosynth. Conf. Exhib. 2-5 March 2008, Cancun, Mex.* 0–7

[59] Ho C H, Shan J, Wang F, Chen Y and Almonnieay A 2016 Performance of fiber-reinforced polymermodified asphalt: Two-year review in Northern Arizona *Transp. Res. Rec.* **2575** 138–49

[60] Stempihar J J, Souliman M I and Kaloush K E 2012 Fiber-reinforced asphalt concrete as sustainable paving material for airfields *Transp. Res. Rec.* **2266** 60–8

[61] Daniel C G failure performance of synthetic fibre-reinforced warm-mix asphalt FAILURE PERFORMANCE OF SYNTHETIC FIBRE

[62] Apostolidis P, Liu X, Daniel C G and Erkens S 2019 fracture performance of synthetic fibre-reinforced asphalt mortar fracture performance of synthetic fibre-reinforced asphalt mortar section of pavement *Engineering Faculty of Civil Engineering and Geosciences Delft University of Technology Department of C*

[63] Daniel C G, Liu X, Apostolidis P, Erkens S and Scarpas A 2019 Impact of synthetic fibres on asphalt concrete mix *Bitum. Mix Pavements VII* 2019 0–0

[64] Daniel C G 2020 Analysis of the effect of using fiber aramid-polyolefin on the strength, stiffness, and durability of warm-mix asphalt *J. Tek. Sipil ITB* **27** 1 9–16

[65] Takaikaew T, Tepsriha P, Horpibulsuk S, Hoy M, Kaloush K E and Arulrajah A 2018 Performance of fiber-reinforced asphalt concretes with various asphalt binders in Thailand *J. Mater. Civ. Eng.* **30** 8

[66] Slebi-Acevedo C J, Pascual-Muñoz P, Lastra-González P and Castro-Fresno D 2019 Multi-response optimization of porous asphalt mixtures reinforced with aramid and polyolefin fibers employing the CRITIC-TOPSIS based on Taguchi methodology *Materials* (Basel). **12** 22

[67] Slebi-Acevedo C J, Lastra-González P, Indacoechea-Vega I and Castro-Fresno D 2020 Laboratory assessment of porous asphalt mixtures reinforced with synthetic fibers *Constr. Build Mater.* **234** 117224

[68] Baghaee Moghaddam T, Soltani M and Karim M R Evaluation of permanent deformation characteristics of unmodified and Polyethylene Terephthalate modified asphalt mixtures using dynamic creep test *Mater. Des.* 2014 **53** 317–24

[69] Rahman W M N W A and Wahab A F A 2013 Green pavement using recycled Polyethylene Terephthalate (PET) as partial fine aggregate replacement in modified asphalt *Procedia Eng.* 124–8

[70] Saiyari D M, Contreras L R Z, Cabanag K L N, De Alday H A, L. P J P and Manio M V M 2016 Effects of Temperature in Mechanical Properties of Stone Mastic Asphalt (SMA) Mixture with Waste Polyethylene *Int. Conf. Environ. Qua.l Concern, Control Conserv.*

[71] Ahmad A F, Razali A R, Razelan I S M 2017 Utilization of polyethylene terephthalate (PET) in asphalt pavement: A review. *IOP Conf. Ser. Mater. Sci. Eng.* **203** 1

Demand analysis at Tanah Grogot Airport East Kalimantan

A B K Suharso[1*] and E Ahyudanari[1]

[1] Department of Civil Engineering, Faculty of Civil, Planning, and Geo Engineering, Institut Teknologi Sepuluh Nopember, Sukolilo 60111, Surabaya, Indonesia.

[*]Corresponding author's e-mail: akbarsuharso.18031@mhs.its.ac.id

Abstract. Tanah Grogot Airport is an airport proposed by the Paser Regency Regional Government to facilitate the people's movement. During this time, the people of Paser Regency have to travel long distances with long travel time between regions. The airport's presence is expected to be able to have shorter travel time between regions because the transportation access between regions is already equipped with air transportation. If there are people or foreign tourists who want to travel quickly, they can use air transportation at this airport. Thus, it is necessary to analyze whether Tanah Grogot Airport has sufficient demand, which will affect the continuity of the airport's operations. The analysis in this research considers the following variables, i.e., the land and air accessibility based on travel time, vehicle operating costs, the flight network, market share value, and analysis of demand from the existing airports to Tanah Grogot Airport. The method uses stated preference and binary logistic regression. The outcome of this research is the estimation of demand at Tanah Grogot Airport East Kalimantan.

1. Introduction

Along with the increasing economic growth in Indonesia, the need for safe, comfortable, and inexpensive transportation is higher to support the mobility of the people who are also increasing day by day. Several choices of transportation modes, land, sea, and air transportation are also increasingly improving facilities and performance to provide the best service for passengers. Air transportation as a mode of transportation widely used by the public because it can connect long distances quickly is also improving in terms of safety and comfort for its users. One particular concern is the airport as the main supporting facility for the public to use air transportation modes, namely airplanes. In Indonesia, there are 298 airports, of which PT Angkasa Pura I and PT Angkasa Pura II manages 13 airports each. The private sector, the TNI, Udata Airport Operational Unit, and Regional UPT/Regional Government manage the remain airports.

Tanah Grogot Airport is an airport proposed by the Paser Regency Regional Government to facilitate the movement of the people to and from Paser Regency. During this time, the Paser Regency community must travel to Balikpapan City if necessary and use the mode of aircraft to go to other cities. The Indonesia government accelerates the development of 20 airports spread across the regencies/cities in Kalimantan, including airports in the border region. Figure 1 shows the location of airports in East Kalimantan.

Content from this work may be used under the terms of the Creative Commons Attribution 3.0 licence. Any further distribution of this work must maintain attribution to the author(s) and the title of the work, journal citation and DOI.
Published under licence by IOP Publishing Ltd

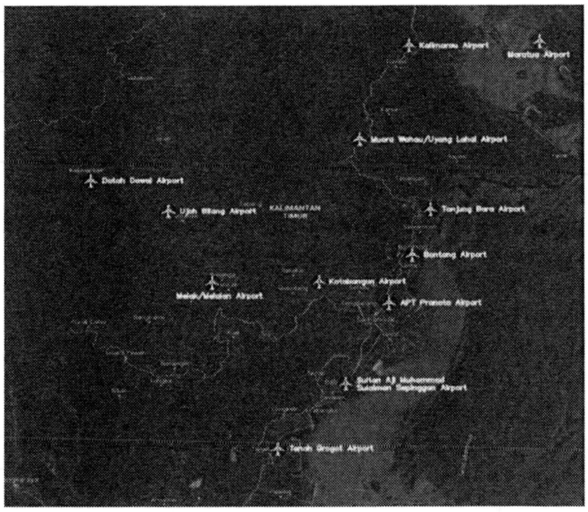

Figure 1. Location of airports in East Kalimantan.

There are several airports in Kalimantan that will compete to get the same passengers due to the close area of services that will be shared with other airports around. The airline is likely to assign flights to each airport by considering each airport's characteristics and the demand patterns. The decision of the airline will lead to a competitive relationship between these airports in Kalimantan. Therefore, the construction of Tanah Grogot Airport also needs to consider the air travel requests from neighboring cities and at the nearest airport.

Besides, all this time, the people of Paser Regency have to take a long time to travel between regions. To make the trip, the people of Paser Regency can only use land transportation such as cars, motorbikes, and buses or can also use sea transportation such as ferries and speed boats. Moreover, if they have to travel outside the island, such as Java Island, the people of Paser Regency are required to go to the nearest airport, requiring a long travel time. For example, to go to Balikpapan is around 5 hours by driving, and the travel time to go to Banjarmasin is around 8 hours. The presence of this airport will help to support the faster travel time between regions. The short travel time will be applicable for transportation access between regions due to air transportation availability. However, in preparing the development of an airport, one thing that needs to be available in the airport's demand. Therefore, this study aims to determine the level of demand at Tanah Grogot Airport, Paser Regency, East Kalimantan.

2. Study Area Data

An area can have more than one airport, especially in the metropolitan area, which is influenced by higher demand. Productive activities that occur between two or more adjacent airports in a region raise a form of Multiple Airports Regions (MAR). Therefore, the concept of Multiple Airport Regions (MAR) is a practical starting point for air transportation research. The MARs concept was introduced in the 1990s: MARs are based on groups of two or more major commercial airports in the metropolitan area. MARs in large-scale studios are often determined by spatial distance metrics to estimate airport catchment areas, ranging from 50 km to 250 km [1].

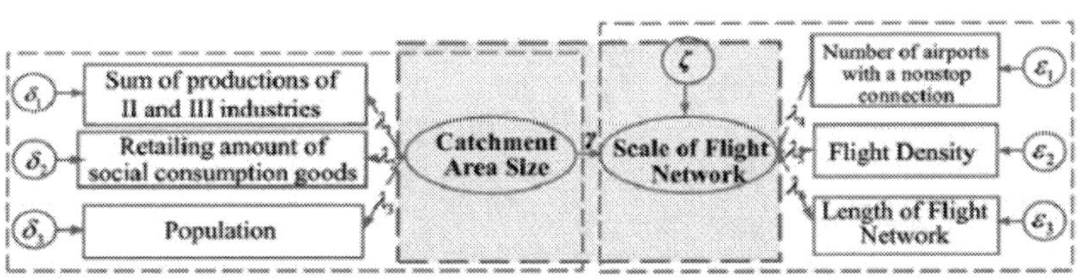

Figure 2. Concept of development catchment area between airports (source : Yang et al., 2016).

Figure 2 shows the model's concept that explains the factor of land use development around the airport. The land use development is a determining factor on the land side because it forms a catchment area and the flight network [2].

The market share measured the coverage area, taking into account factors such as access to the airport, airline ticket costs, fees access time to the airport, and time charges in the air. The market share determines the pattern of airport selection by passengers and the competition that occurs between airports in the region. However, within MARs, air passengers can indeed choose between different airports not only based on the location of the airport that is in the vicinity but also a series of other airport service level (LOS) attributes. In general, airline ticket prices are the most important factor and are called the LOS attributes of major airports [3].

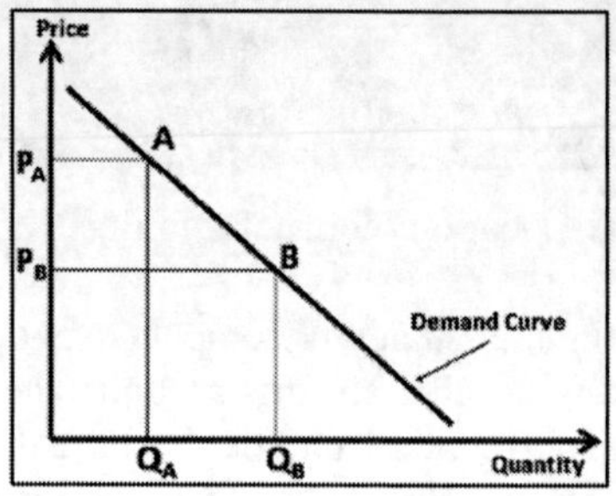

Figure 3. Demand curve (source : Raisová and Ďurčová, 2014).

Demand also depends on the type of customer, type of trip, destination, competitors, flight network, market behavior, and price & income elasticity in specific markets. The demand curve can be seen in figure 3 [4]. The purpose of estimating demand is to meet passenger needs, create appropriate sales strategies, maximize profits, and offer optimal ticket pricing structures in a reservation system. Accessibility has a close relationship with demand. One of the essential factors in demand growth is accessibility. If the value of accessibility has increased, then demand will also increase [5].

3. Methods

Several things need to be reviewed in the demand analysis at Tanah Grogot Airport. The methodology aims to determine the market demand for the airport and solve the problems that need to be done in several work stages. These stages include the problem of identification, literature study, and data collection. An analysis of the land and air accessibility is carried out, as well as an analysis of demand shifts from the existing airports to Tanah Grogot Airport.

3.1 Analysis of Land and Air Accessibility

Land accessibility will be calculated based on the closest distance to the road network from the origin point to the airport. This land accessibility is illustrated by travel time between the origin *i* and the departure airport *j* [6]. Therefore, the distance will be calculated, as well as the travel time from the passenger's area of origin based on the airport service radius in East Kalimantan. Furthermore, airport air accessibility will be analyzed using a formula developed by Yang, et al (2016). The formula is as follows:

$$Ac_x^{air} = Num_x^{0.949(19.1)} \; x \; Freq_x^{0.851(21.4)} \; x \; NAirRoute_{in\ use}^{0.218(4.5)} \tag{1}$$

Where,

Ac_x^{air} = Air Accessibility x
Num_x = Number of airports with direct connections x
$Freq_x$ = Airport flight density x

$NAirRoute_{in\ use}$ = Airport flight network length x

After land accessibility and air accessibility are obtained, the next step is to calculate the accessibility of the regional airport air transportation ($Access_x$) to the existing airports and after the addition of Tanah Grogot Airport with the following formula.

$$Access_x = Ac_x^{air}\ x\ Pop_x\ x\ \sum_{i=1}^{m} \frac{1}{(Ac_{ix}^{land})^{\beta}\ x\ Pop_{ix}} \tag{2}$$

Where,
$Access_x$ = Accessibility of regional air transportation x
i = Traffic zone code (origin)
j = Airport code
Ac_{ix}^{land} = Land accessibility
Pop_x = Population in the airport catchment area
Pop_{ij} = Population in zone i in the airport cathment area j
m = Number of traffic zones
β = Effect of land transportation resistance on flight accessibility, use value 1

Henceforth, the calculation follows the formula as follows if there are multiple airports in an area.

$$Access = \sum_{x=1}^{n} \frac{Pop_x}{Pop}\ x\ Access_x \tag{3}$$

Where,
n = Number of airports
Pop = Total regional population

3.2 Analysis of Demand Movements from Existing Airport to Tanah Grogot Airport
The data used in this study are primary data obtained through direct interview surveys in the field with respondents. Determination of the Number of sampling that will be used in this analysis is only a portion of the population taken. It is used to determine the nature and characteristics of a population. The number of samples will be calculated using the Slovin method with the following formula [7].

$$n = \frac{N}{1+Ne^2} \tag{4}$$

Where,
n : minimum number of samples
N : regional population
e : error margin (the maximum error that can be tolerated)

In this study, the analysis technique used is a logistic regression model using the binary logit model. This model can be used to determine the percentage of passengers who are willing to move from an existing airport and can also be used to determine the factors that influence the respondents to move. The model used in binary logistic regression is as follows:

$$\text{Log}\ (P / 1 - p) = \beta_0 + \beta_1 X_1 + \beta_2 X_2 + \ldots. + \beta_n X_n \tag{5}$$

Where p is the possibility that Y = 1, and X_1, X_2, X_n are independent variables, and β is the regression coefficient.

The Aircraft Operating Cost method (AOC) will be used to determine the flight ticket rates. AOC is the value of aircraft operating costs per passenger and can also be interpreted as a plane ticket fare per passenger. The calculation of AOC is based on the revision of KM 26 of 2010.

4. Results and Discussion

4.1 Land and Air Accessibility

This analysis reviewed Tanah Grogot Airport and several other airports, including Sepinggan International Airport Balikpapan, Kalimarau Airport Berau Regency, APT Pranoto Samarinda Airport, Melalan Airport West Kutai Regency, Batulicin Bersujud Airport Tanah Bumbu Regency, and Warukin Airport Tabalong Regency. These airports were chosen for review because the airport area has a short distance from Tanah Grogot Airport, which means the potential of overlapping catchment areas. The overlapping catchment areas will give rise to potential market share values (market share scores) with Tanah Grogot Airport. The depiction of the radius of service from Tanah Grogot Airport can be seen in figure 4, adjusting to PM 39 of 2019 in the Kalimantan Island region of 60 km.

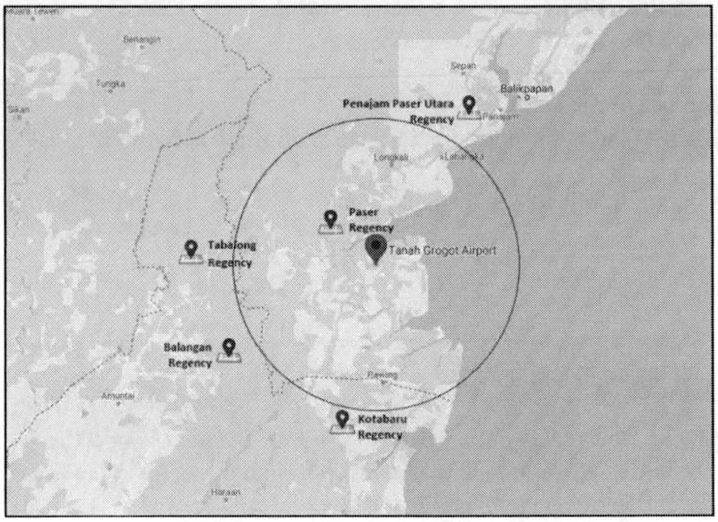

Figure 4. Tanah Grogot service radius.

Examples of travel time data from several sub-districts to the Tanah Grogot Airport can be seen in table 1. The calculation of travel time in table 1 can also be called land accessibility. This analysis aims to determine travel time from the area of origin to the existing departure airport and planning. The travel time results can also be called the value of land accessibility from which this result data will then be used for further analysis, namely, air accessibility.

Table 1. Travel time.

Regency	Sub-Regency	Travel Time (hour)	Distance (km)	Velocity (km/hour)
Penajam Paser Utara	Babulu	2,47	105,0	42,6
	Penajam	3,88	164,0	42,2
	Sepaku	5,27	199,0	37,8
	Waru	3,47	143,0	41,3
Paser	Batu Engau	1,50	57,1	38,1
	Batu Sopang	1,55	58,8	37,9
	Kuaro	1,27	46,2	36,5
	Long Ikis	1,97	75,9	38,6
	Long Kali	5,27	224,0	42,5
	Muara Komam	4,00	143,0	35,8
	Muara Samu	2,17	76,4	35,3
	Paser Belengkong	1,08	26,7	24,6

Regency	Sub-Regency	Travel Time (hour)	Distance (km)	Velocity (km/hour)
	Tanah Grogot	1,42	39,3	27,7
	Tanjung Harapan	4,17	115,0	27,6

Airport air accessibility is the accessibility of an airport in terms of its flight network. The variables needed include the number of airports that can be reached directly (Num_x), frequency ($Freq_x$) and length of the flight network ($NAirRoute_{in\ use}$). The calculations and results can be seen in table 2.

Table 2. Value of air accessibility.

Airport	Code	Num_x	$Freq_x$ (weeks)	NAirRoute (10,000 km)	Air Accesibility
APT Pranoto	AAP	10	127	126,34852	1019,38
Kalimarau	BEJ	3	69	24,72143	157,02
Sultan Aji Muhammad Sulaiman Sepinggan	BPN	18	526	389.121,79	16685,66
Melalan	MLK	2	18	3.073,51	63,14
Tanah Grogot	TNG	5	76	14.281,28	624,75

After land accessibility and air accessibility are obtained, the next step is to calculate the accessibility of the regional airport air transportation ($Access_x$) to the existing airports and after the addition of Tanah Grogot Airport which can be seen in table 3 and table 4.

Table 3. Regional air transport accessibility at existing airports.

Airport	Total Population Catchment Area	Accessibility of Air Transportation	(Popx/Pop)*Access
APT Pranoto (AAP)	1.377.493	431.922,62	117.565,44
Kalimarau (BEJ)	703.839	13.356,24	1.857,56
Sultan Aji Muhammad Sulaiman (BPN)	1.951.050	1.463.228,54	564.111,66
Melalan (MLK)	1.028.376	2.370,35	481,67
Total	5.060.758	1.910.877,74	684.016,32

Table 4. Regional air transport accessibility with the addition of Tanah Grogot Airport.

Airport	Total Population Catchment Area	Accessibility of Air Transportation	(Popx/Pop)*Access
APT Pranoto (AAP)	1.377.493	431.922,62	115.720,17
Kalimarau (BEJ)	703.839	13.356,24	1.828,40
Sultan Aji Muhammad Sulaiman (BPN)	1.951.050	1.463.228,54	555.257,55
Melalan (MLK)	1.028.376	2.370,35	474,11
Tanah Grogot	80.699	4.004.544,17	62.853,99
Total	5.141.457	5.915.422	736.134,21
		Addition (%)	7,62%

In table 3, the value of the accessibility of regional air transportation in East Kalimantan Province to the existing airports is 684,016.32. On the other hand, in table 4 with the addition of Tanah Grogot Airport, the accessibility value of regional air transportation increased by 7.62% with a value of 736,134.21.

4.2 Analysis of Demand Movements from Existing Airport to Tanah Grogot Airport

The methodology was previously explained to determine the Number of samples needed for conducting surveys in the field using the formula of Slovin theory. The population of Paser Regency in 2019 is 279,975 based on data from BPS (Statistics Indonesia) and the margin error value used is 10%. Calculation of the Number of samples is as follows:

$$n = \frac{N}{1+Ne^2}$$

$$n = \frac{279975}{1+(279975*(10\%)^2)}$$

$$n = 99{,}9643$$

$$n = 100 \text{ samples}$$

The interview survey was conducted directly in Paser Regency for 3 days. Respondents who were the targets of conducting the survey were the respondents who relatively frequently traveled in and out of the district of Paser where these respondents would later become potential passengers of Tanah Grogot Airport. The interview succeeded in getting as many as 103 respondents who had been surveyed from the conditions in the field.

Before analyzing binary logistic regression, a Chi-Square test (bivariant selection) will be conducted to select variables that can be entered into the multivariant stage of binary logistic regression. In the Chi-Square test, a selection is made between only two variables in the survey results, the dependent variable (WTM) and the independent variable (age, gender, education, employment, income, origin, destination, purpose, frequency, vehicle, travel time, travel costs, and ticket costs). The WTM variable is a willingness to move.

Table 5. Chi-Square Test results.

Omnibust Tests of Model Coefficient

Information	Chi-square	df	P value	Result
WTM * Age	0,694	1	0,405	Not a Candidate
WTM * Gender	2,224	1	0,136	Candidate
WTM * Education	5,12	1	0,024	Candidate
WTM * Employment	0,576	1	0,448	Not a Candidate
WTM * Income	3,535	1	0,06	Candidate
WTM * Origin	13,716	1	0	Candidate
WTM * Destination	0,002	1	0,962	Not a Candidate
WTM * Purpose	0,036	1	0,85	Not a Candidate
WTM * Frekuency	2,712	1	0,1	Candidate
WTM * Vehicle	9,378	1	0,002	Candidate
WTM * Travel Time	3,441	1	0,064	Candidate
WTM * Travel Cost	22,044	1	0	Candidate
WTM * Ticket Cost	0,001	1	0,993	Not a Candidate

From table 5, it can be seen that eight variables become variable candidates to be multivariant variables in the binary logistic regression stage (Pvalue <0.25), namely gender, education, income, origin, frequency, vehicle, time, and travel costs. Furthermore, binary logistic regression analysis of the multivariant variables from the previous result is obtained. The control of the results of this analysis is to pay attention to the aspects of Sig (Pvalue) <0.1. If it does not meet, it must be re-analyzed by eliminating one variable with the most significant Sig (Pvalue) value. After the re-analysis, the next thing to control is to pay attention to changes in Oddratio / Exp (B) [ΔExp (B)] <10%. Suppose some variables do not meet these requirements. In that case, the eliminated variable can be re-entered or replaced by eliminating another variable with the next largest Sig (Pvalue) value. This multivariant analysis process is carried out repeatedly until it is obtained a suitable variable that meets the (significant) requirements. The results of all binary logistic regression tests were analyzed 9 times. The selected multivariant results can be seen in table 6.

Table 6. Binary logistics multivariant regression results.

Variables in the Equation

		B	S.E.	Wald	df	Sig.	Exp(B)	95% C.I.for EXP(B)	
								Lower	Upper
Step 1[a]	Income	,377	,330	1,304	1	,253	1,458	,763	2,784
	Origin	1,055	,522	4,077	1	,043	2,872	1,031	7,996
	Vehicle	,407	,183	4,976	1	,026	1,503	1,051	2,149
	Travel Cost	1,017	,325	9,768	1	,002	2,766	1,461	5,234
	Constant	-5,270	1,328	15,748	1	,000	,005		

[a] Variable(s) entered on step 1: Income, Origin, Vehicle, Travel Cost.

Even though there is one variable whose value is sig (Pvalue)>0.1, it can still be used for further calculation. This variable is called the confounding variable. From table 6, the following equation is obtained.

$$\text{Log}\,(P / 1 - p) = \beta_0 + \beta_1 X_1 + \beta_2 X_2 + \ldots + \beta n X n$$

$$Y = -5{,}270 + 0{,}377\, X_1 + 1{,}055\, X_2 + 0{,}407\, X_3 + 1{,}017\, X_4$$

Where;
Y = Logit (p/1-p)
X_1 = Income
X_2 = Origin
X_3 = Vehicle
X_4 = Travel Cost

Table 7. Result of probability.

Number of Responden	Income (X_1)	Orgigin (X_2)	Vehicle (X_3)	Travel Cost (X_4)	Logit (p/1-p)	Probability (p)
1	2	1	1	2	-1,020	0,2689
2	1	1	4	1	-1,192	0,2315
3	1	1	2	2	-0,989	0,2689
∫						
100	1	2	3	5	3,523	0,9707
101	1	2	4	2	0,879	0,7109
102	1	1	4	1	-1,192	0,2315
103	1	2	4	1	-0,138	0,4750
			Probability of WTM from existing airport =			0,6741
					=	67,41%

From the results in table 7, the probability of willingness to move from the existing airports to Tanah Grogot Airport is 67.41% (average of all probabilities). The calculation of Aircraft Operating Cost (AOC) is based on the revision of KM 26 of 2010 regarding the analysis and evaluation of the formulation mechanism of the calculation and determination of passenger fares for economy class. The aircraft studied is the ATR72 aircraft, which is following the planned aircraft for Tanah Grogot Airport. The calculation of AOC includes direct and indirect costs, as can be seen in table 8.

Table 8. Aircraft operating costs.

Number	Aircraft Operating Cost Components	%	Range Flight 1 Hours (Rp/hour)
(1)	(2)	(3)	(4)
I	Fixed Direct Operating Costs		
1	Depreciation of aircraft	18,7	9.100.000
2	Insurance premium	7,57	3.683.333
3	A fixed salary for a airplane crew	4,1	1.993.333
4	A fixed salary for a technician	2,67	1.300.000
5	Training Cost	0,56	270.833
	Sub-Total	33,6	16.347.499
II	Variable Direct Operating Costs		
1	Use of fuel (avtur)	21,54	545.600
2	Use of Lubricants	0,56	10.481.419
3	Flight Crew Allowances	2,49	2.426.278
4	Airplane overhaul & maintenance	13,63	13.265.417
5	Airplane services	0,63	447.036
6	Ground handling	3,67	1.784.444
7	Catering	2,35	1.141.233
	Sub-Total	44,88	30.091.427
	Total Direct Operating Costs	78,48	46.438.926
III	Indirect Operating Costs		
1	General and organization	4,24	2.062.500
2	Marketing & sales (agent commission)	8,19	3.985.648
	Total indirect operating costs	12,43	6.048.148
	Total operating costs before profit	90,91	52.487.074
	Profit margin 10%	9,09	5.248.707
	Total Aircraft Operating Costs	100,00	57.735.781

Table 8 shows the operational cost value of the ATR72 type aircraft in a 1-hour flight operation with a total of Rp. 57,732,781 per hour. The flight route was chosen based on the survey results of the destination's origin, with the most destinations: Tanah Grogot Airport, the city regency of Paser - Balikpapan Sepinggan Airport. The distance between the two airports is 99.24 km, with a travel time of 23.26 minutes. The calculation of operational costs for one flight in determining ticket prices is as follows:

$$\frac{Aircraft\ Operating\ Cost\ (route)}{(Paser - Balikpapan)} = \frac{AOC}{1\ jam}\ x\ travel\ time$$

$$= \frac{Rp.57.735.781}{1\ x\ 60\ minutes}\ x\ 23{,}26\ minutes$$

$$= Rp.22.382.238$$

$$AOC\ per\ unit\ of\ output = \frac{Aircraft\ Operating\ Cost\ (route)}{Number\ of\ aircraft\ seats}$$

$$AOC\ per\ unit\ of\ output = \frac{Rp.22.382.238}{52}$$

$$AOC\ per\ unit\ of\ output = Rp.430.428\ per\ passenger$$

From the calculation that has been done, the value of AOC per unit of output is Rp. 430.428, - per passenger. The AOC value per unit of output is the value of aircraft operating costs per passenger or can also be interpreted as a plane ticket rate per passenger.

5. Conclusion

The development of Tanah Grogot Airport has an influence on the accessibility of regional air transportation to the surrounding airports. Before the operations of this airport, the value of regional air transportation accessibility in East Kalimantan Province was 684,016.32. With the Tanah Grogot Airport, the accessibility value of regional air transportation has increased by 7.62% to 736,134.21. This increasing value means that the development of Tanah Grogot Airport can increase regional accessibility in terms of land and air accessibility in the East Kalimantan region. One of the essential factors in demand growth is accessibility. If the value of accessibility has increased, then demand will also increase.

From the interview survey results that have been analyzed by binary logistic regression statistics, the results are obtained in the form of a large percentage of the probability of transfer of demand from the existing airports to Tanah Grogot Airport, which amounted to 67.41%. Variables that affect, among others, namely the variable level of income, the variable origin of the trip, the variable vehicle commonly used, and the variable cost of travel.

Reference

[1] Hansen M and Li A 2017 Multiple airport regions based on inter-airport temporal distances *Transp. Res. Part E Logist. Transp. Rev.* **101** 84–98
[2] Yang Z, Yu S and Notteboom T 2016 Airport location in multiple airport regions (MARs): The role of land and airside accessibility *J. Transp. Geogr.* **52** 98–110
[3] Lieshout R 2014 Measuring the size of an airport's catchment area *Transp. Geogr.* (Elsevier) **25** 27–34
[4] Raisová M and Ďurčová J 2014 Economic growth-supply and demand perspective *Proc. Econ. Financ.* **15** 184–91
[5] Sivrikaya O 2013 Demand Forecasting for Domestic Air Transportation in Turkey *The Open Transp. J.* **7** 20-26
[6] Mackie P J, Jara-Díaz S and Fowkes A S 2001 The value of travel time savings in evaluation *Transp. Res. Part E Logist. Transp. Rev.* **37** 91–106
[7] Taherdoost, H 2016 Sampling methods in research methodology: how to choose a sampling technique for a research *IJARM.* **5** 18–27

IOP Conf. Series: Materials Science and Engineering **930** (2020) 012064 doi:10.1088/1757-899X/930/1/012064

Analysis the use of steel slag as a replacement of natural aggregate in the asphalt concrete binder course (AC-BC) mixture

M Kasaf[1*] and C A Prastyanto[2]

[1]Civil Engineering Department, Sepuluh Nopember Institute of Technologi, Surabaya, Indonesia
[2]Lecture of Civil Engineering Department, Sepuluh Nopember Institute of Technology, Surabaya, Indonesia

*Corresponding author's e-mail: michelkasaf.18031@mhs.its.ac.id

Abstract. Aggregates are the largest component of the flexible pavement structure. The quality of the road is largely determined by the nature and type of aggregates used. Currently, the development of road infrastructure in Indonesia continues to experience increasing amount of natural aggregate use, so this will cause the availability of natural aggregate raw materials to continue to thin out. For that reason, an effort should be made to find alternatives to the use of natural aggregates, one of which is the utilization of steel slag. In this study, it will be examined the characteristics of the steel slag aggregates by referring to the 2018 Bina Marga specifications and the Marshall characteristic test method. This study used steel slag addition with variations of 0%, 50%, 80% and 100%. The results obtained in this study are that the value characteristics of the slag aggregate meet the standard specifications. The use of steel slag at 80% variation in coarse aggregate produces the best stability value which is equal to 2320,19 Kg. Overall, the use of steel slag can be an alternative to the use of natural aggregates. Moreover, steel slag can also be an alternative to an effort to reduce waste.

1. Introduction

Material is an important aspect that greatly influences the quality of road pavement construction, including aggregate type, soil carrying capacity, and asphalt. The selection of the appropriate type of aggregate for use in pavement construction is influenced by several factors, namely gradation, strength, grain shape, surface texture, viscosity to asphalt and cleanliness, and chemical properties. The material used must produce a pavement mixture that is able to withstand strong and durable loads [1]. The biggest component in the construction of pavement is an aggregate. In the classification of processing, the aggregate is divided into two, namely natural aggregate and artificial aggregate. Natural aggregates are aggregates that are produced from rock exploration in nature, whereas artificial aggregates are aggregates obtained from physical or chemical processes of a material so that their properties resemble aggregates. Flexural hardness is very dependent on the form of aggregate and gradation used. Therefore, the shape of the aggregate also affects the skid-resistance of pavement. Aggregates must have properties that are capable of adhering to asphalt [2].

Infrastructure development in Indonesia, particularly road infrastructure, continues to increase in the use of natural aggregates, so this will cause the availability of natural aggregate raw materials to run low. Road improvement by adding additional layers continuously will cause the thickness of the

Content from this work may be used under the terms of the Creative Commons Attribution 3.0 licence. Any further distribution of this work must maintain attribution to the author(s) and the title of the work, journal citation and DOI.
Published under licence by IOP Publishing Ltd

pavement layer to get thicker and the required materials to be reduced [3]. For this reason, an effort should be made to find alternatives to the use of natural aggregates, one of which is the utilization of steel slag.

Steel slag is an irregularly shaped cubical rock. This rock is formed from minerals that are used as steel purification from a high kitchen, and then the steel slag is channelled and accommodated in a slag pot. To form splinters, the spread steel slag is sprayed with water. A sudden change in temperature makes the slag steel break and form an aggregate. The steel slag which has a rough shape and pits (porous) will provide a better bond. Steel slag is also more resistant to chemical reactions and temperature changes because the metal has been excreted through high combustion. When viewed from its potential, steel slag is an alternative material that has a high degree of stability, resistant to chemical reactions, and resistant to extreme temperature changes [4].

Steel slag is a good type of aggregates with excellent durability and surface specifications as features of friction and durability. Steel slag has a high effect on road resistance and plays an important role to prevent aggregates from stripping [5]. Porosity in steel slag can increase bonding strength in mixtures. Blends containing steel slag show short-term and long-term Marshall stabilities, and the use of steel slag in the mixture can increase the aging index of asphalt mixture [6]. Steel slag has a number of advantages with high engineering properties, and economically, the steel slag may be cheaper if utilized in urban roads, but it would be expensive for rural roads due to the transportation charges [7].

Besides the aggregates, the quality of asphalt used also largely determines the quality of the mixture. Asphalt is a brownish black material that is thermoplastic, so it will soften and melt when heated to a certain temperature. Asphalt freezes again if the temperature drops. Asphalt used in road construction has important physical properties, including: consistency, long-lasting or weathering resistance due to weather, degree of hardening, and resistance to water [8]. Asphalt with low penetration is used in areas with hot weather or high volume traffic, whereas asphalt cement with high penetration is used for areas with cold weather or low volume traffic. In Indonesia, asphalt cement is generally used with 60/70 penetration [9].

For all above reasons, this study is intended to explore the properties of steel slag aggregates for road pavement mixture. The aggregate test follows the 2018 Bina Marga specification standards. This research was conducted to see the characteristics of the mixture with Marshall parameters. In this study, four variations of slag mixture are made, namely 0%, 50%, 80% and 100%. In variations 2 and 3, steel slag aggregates are only used in coarse aggregates and combined with natural aggregates, whereas in variation 4 steel slag aggregates will be used in coarse and fine aggregates.

2. Method

In this study, the method is the characteristics of steel slag material used as aggregate for road pavement mixtures were examined of properties. Checks are carried out on coarse aggregates and fine aggregates, while material checks are carried out to determine the characteristics of the material used as a mixture that meets the requirements specified in the specification.

(a) (b) (c)

Figure 1. (a) Aggregate Filter; (b) Los Angales Machine; (c) Ball Mill Machine.

Tests conducted include abrasion value with *los angles* machine, specific gravity, absorption and affinity aggregate of asphalt. The testing is carried out on coarse and fine aggregates. Because aggregate slags are chunks, they must be grounded using a ball mill machine.

Examination of characteristic values is guided by the 2018 Bina Marga specification standards. Requirements for specifications can be seen in table 1 below:

Table 1. Requirements for aggregate characteristics of Bina Marga Specifications (2018).

Test	Requirements value for aggregate characteristic
Bulk Spesific Gravity, (gr/cm^3)	-
Absorption, (%)	Maks. 3
Abrasion, (%)	Maks. 40
Affinity Aggregate of Ashpalt, (%)	Min. 95

In the previous studies, the use of slag steel aggregates in asphalt mixes was 10%, 20%, and 30%. The results obtained by adding 30% slag aggregate still meet the requirements [10]. The results of this analysis provide the still small use of steel slag in the mixture, so it needs to be developed with the use of more steel slag aggregates to reduce the use of massive natural aggregates.

The prepared aggregate is then mixed with asphalt using the Marshall Method following the 2018 Bina Marga Specifications. Asphalt mixture variations in this study are:

a) Variation 1, 100% Natural Aggregate (NA).
b) Variation 2, 50% Slag Agregate (SA) and 50% Natural Aggregate (NA).
c) Variation 3, 80% Slag Agregate (SA) and 20% Natural Aggregate (NA).
d) Variation 4, 100% Slag Agregate (SA).

In variations 2 and 3, steel slag aggregates are only used in coarse aggregates and combined with natural aggregates, whereas in variation 4 steel slag aggregates will be used in coarse and fine aggregates.

This study uses continuous gradations following the 2018 Bina Marga Specifications. Gradations used can be seen in figure 2.

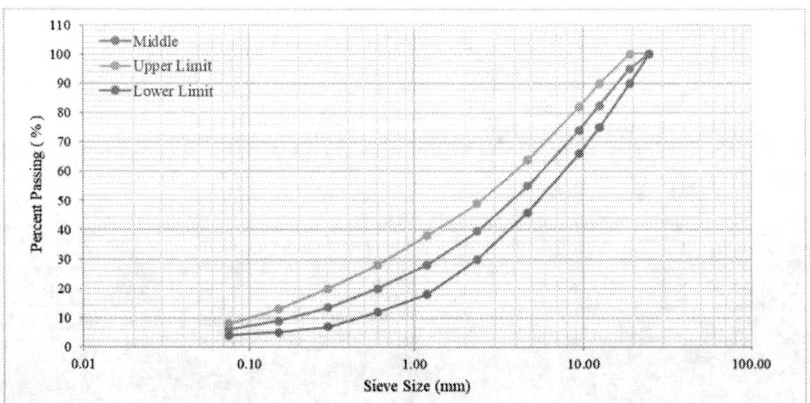

Figure 2. Continuous graded distribution.

2.1 Marshall Test

The Marshall test is an important step in determining the characteristics of an asphalt mixture. The Marshall Design method for porous asphalt mixes is the most widely used as mixed-design method for experimental laboratory studies at present [11]. Asphalt mixture characteristics which are Marshall parameters are density, stability, flow, void in mixture (VIM), void in mineral aggregates (VMA), void filled with asphalt (VFA), and Marshall Quotient. Requirements for the Marshall Characteristics of 2018 Bina Marga Specifications can be seen in table 2 below:

Table 2. Bina Marga spesifications (2018) of asphalt concrete binder course of flexible pavement.

Properties	Requirements Value for Marshall characteristic
Density (g/cm^3)	-
Stability (Kg)	Min. 800
Flow (mm)	2 - 4
VIM (%)	3 - 5
VMA (%)	Min. 14
VFA (%)	Min. 65
MQ (Kg/mm)	Min. 250

3. Result and Discussion

3.1 Steel Slag Aggregate Characteritics

To find out the aggregate characteristic values used in asphalt mixtures, several tests were conducted. This is needed to see the feasibility of the aggregates used in the mixtures. Tests carried out include specific gravity, aggregate wear, absorption and adhesion to asphalt. In table 3, it is prensented the result of testing the steel slag aggregate characteristics.

Table 3. Result of steel slag characteristics.

Test	Spesification	Result
Coarse Aggregate		
Bulk Spesific Gravity, (gr/cm3)	-	5.405
Absorption, (%)	Maks. 3	1
Abrasion, (%)	Maks. 40	20.60
Affinity Aggregate of Ashpalt, (%)	Min. 95	97
Fine Aggregate		
Bulk Spesific Gravity, (gr/cm3)	-	3.378
Absorption, (%)	min. 95	2.229

Based on table 3, it can be seen that the value of slag aggregate characteristics obtained meets the specifications. Therefore, the aggregates can be used in an asphalt mixture.

Figure 3. (a) Asphalt Penetration 60/70; (b) Natural Aggregate; (c) Steel Slag Aggregate

3.2 Characteristics of Asphalt Mixture

After checking the characteristics of the material used, the next step is to plan the asphalt mixture using the Marshall method.

3.2.1 The Determination of Optimum Asphalt Content

This test will produce optimum asphalt content in each mixture variation including VIM, VMA, VFA, stability, flow and MQ values. Optimum asphalt content values are obtained from the average asphalt content that meets all the properties of the mixture. The optimum asphalt content (KAO) in each variation can be seen in figure 4.

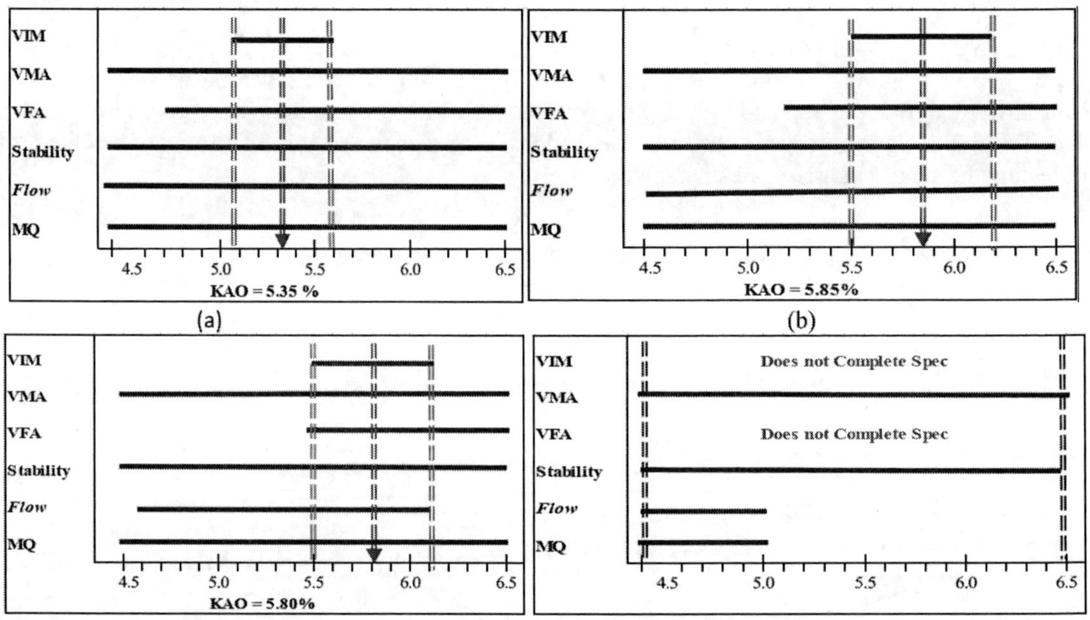

Figure 4. (a) KAO 100% Natural Aggregate; (b) KAO 50% Slag Aggregate; (c) KAO 80% Slag Aggregate; (d) KAO 100% Slag Aggregate

In figure 4, it can be seen the optimum asphalt content values in Variations 1, 2, and 3 have met the specified specifications. However, in Variation 4, the optimum asphalt content cannot be determined because the VIM and VFA values cannot meet the 2018 Bina Marga specifications.

3.3 Marshall Test Optimum Asphalt Content

Marshall testing on optimum asphalt content (KAO) was carried out to determine the best Marshall characteristics in each mixture variation. Marshall KAO testing is only on variations that meet the 2018 Bina Marga specifications. From the results of this test, it will be obtained a mixture of steel slag aggregate variations that are suitable for use in the Asphalt Concrete Binder Course flexible pavement. Following are the Marshall testing results for each variation using optimum asphalt content.

Table 4. Result of marshall characteristic on optimum asphalt content (100% natural aggregate).

Marshall Characteristic	Results	Spesification
Optimum Asphalt Content (%)	5.35	-
Density (g/cm3)	2.37	-
Flow (mm)	3.85	2 - 4
VIM (%)	3.10	3 - 5
VMA (%)	17.62	≥ 14
VFA (%)	82.42	≥ 65
Stability (Kg)	1245.21	≥800
MQ (Kg/mm)	323.66	≥ 250

Table 5. Result of marshall characteristic on optimum asphalt content (50% slag aggregate).

Marshall Characteristic	Results	Spesification
Optimum Asphalt Content (%)	5.85	-
Density (g/cm3)	2.70	-
Flow (mm)	3.22	2 - 4
VIM (%)	4.06	3 - 5
VMA (%)	21.63	≥ 14
VFA (%)	81.26	≥ 65
Stability (Kg)	1937.06	≥800
MQ (Kg/mm)	603.38	≥ 250

Table 6. Result of marshall characteristic on optimum asphalt content (80% slag aggregate).

Marshall Characteristic	Results	Spesification
Optimum Asphalt Content (%)	5.80	-
Density (g/cm3)	2.99	-
Flow (mm)	2.92	2 - 4
VIM (%)	4.76	3 - 5
VMA (%)	23.16	≥ 14
VFA (%)	79.47	≥ 65
Stability (Kg)	2320.19	≥ 800
MQ (Kg/mm)	800.66	≥ 250

Based on tables 4, 5 and 6, it can be seen the value of the characteristics of Marshall was produced at each of optimum asphalt contents. The effect of steel slag gives different results on each mixture variation, and the testing results using optimum asphalt content show that all variations of the mixtures have met the specifications set by Bina Marga 2018.

3.4 Marshall Test Analysis
After Marshall test results have been obtained on each variation of the mixture, an analysis is carried out to see the effects of the slag steel aggregates on the characteristics of asphalt mixtures.

3.4.1 The Effect of Slag Aggregate on Density Value
In figure 5, it can be seen the comparison of density values for each mixture variation. The density value continues to increase with the increasing levels of steel slag in the mixture. The highest density is found in the addition of 80% steel slag mixture of 2.99 g/cm^3. This is due to the specific gravity of the large steel slag, so more steel slag content is used, the density value will increase.

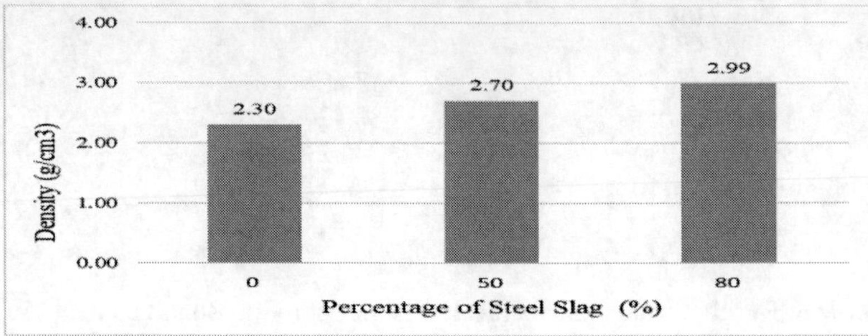

Figure 5. The effect of slag aggregate on density value.

3.4.2 The Effect of Slag Aggregate on VIM Value
In figure 6, it can be seen that the value of Void in Mix (VIM) on each variation of the mixture has met the specified requirements. VIM value continues to increase with the increasing levels of steel slag in the mixture. This is because the steel slag is porous, causing more cavity volumes in the mixture. A good VIM value is needed to provide enough space for the compaction process caused by traffic loads.

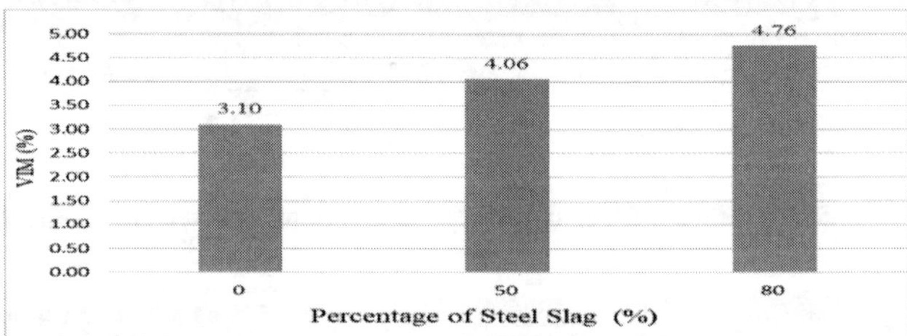

Figure 6. The effect of slag aggregate on VIM value.

3.4.3 The Effect of Slag Aggregate on VMA Value
In figure 7, it can be concluded that the effect of adding steel slag to the mixture gives higher Void in Mineral Aggregate (VMA) values than that of the natural aggregates. VMA value continues to increase with increasing levels of steel slag. This is due to the nature of the steel slag which has a cavity (porous) so that the fine aggregate fraction in the mixture does not fill the empty cavities between the aggregates. VMA values that are close to the minimum limit indicate that the mixture has more durability, while VMA values that are too high will result in the mixture experiencing a large deformation. However in this case, the VMA value in each mixture variation still meets the specifications set by the 2018 Bina Marga of ≥14%.

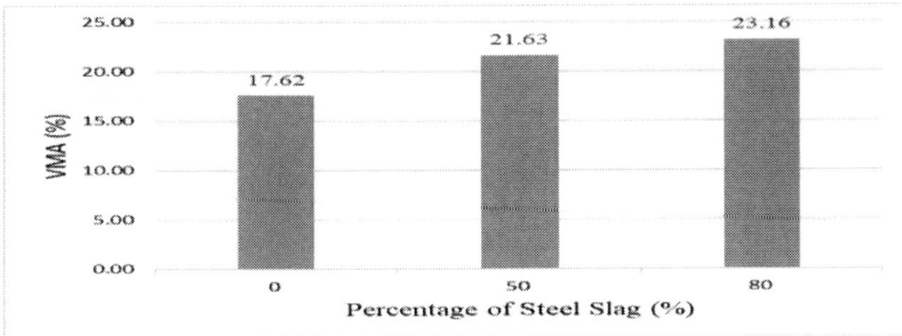

Figure 7. The effect of slag aggregate on VMA value.

3.4.4 *The Effect of Slag Aggregate on VFA Value*

Figure 8 shows the comparison of VFA values in each variation of the mixture which has fulfilled the 2018 Bina Marga minimum limit of ≥65%. The use of steel slag aggregates in the mixture tends to decrease in the VFA value. This decrease is due to the nature of the slag steel which has many cavities, so that smaller particles fill the cavities in the aggregate and cause the aggregate not covered by asphalt to the maximum.

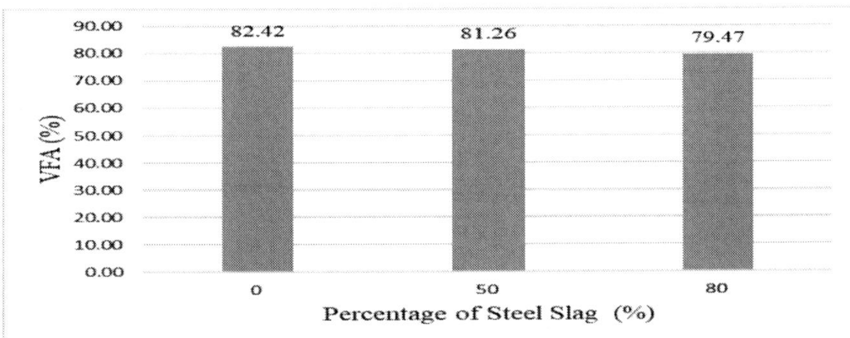

Figure 8. The effect of slag aggregate on void filled with asphalt (VFA)

3.4.5 *The Effect of Slag Aggregate on Stability Value*

The use of steel slag as a mixture of pavement tends to increase the value of stability. It can be seen in figure 9 that the highest stability is found in an 80% steel slag mixture of 2320.19 kg, and it can be seen that the stability values in each variation still meet the minimum limit of ≥800 kg. The physical properties of hollow and irregularly shaped steel slag aggregates cause good interlocking. Therefore, the addition of higher levels of steel slag will result in a road more able to withstand deformation due to traffic loads that work on it without undergoing a change in shape or tendency to be rigid.

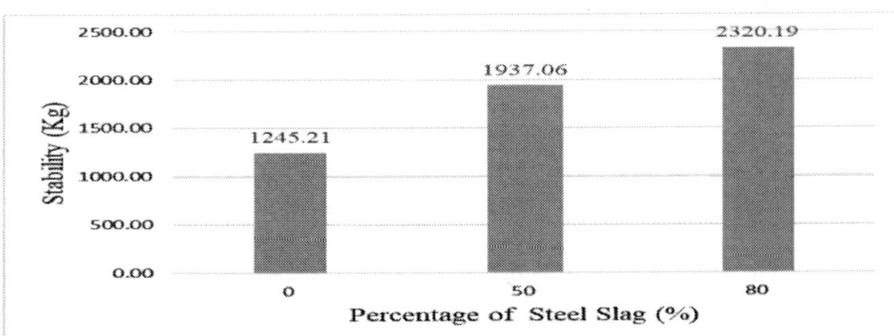

Figure 9. The effect of slag aggregate on stability value.

3.4.6 The Effect of Slag Aggregate on Flow Value

In figure 10, it can be seen that the flow value in each variation still meets the specified requirements. The flow value decreases with increasing levels of steel slag in the mixture. Mixtures that have low flow values and high stability tend to be stiff, brittle and prone to cracking, while mixtures that have high flow values and low stability tend to be plastic and easily deformed due to traffic loads on them.

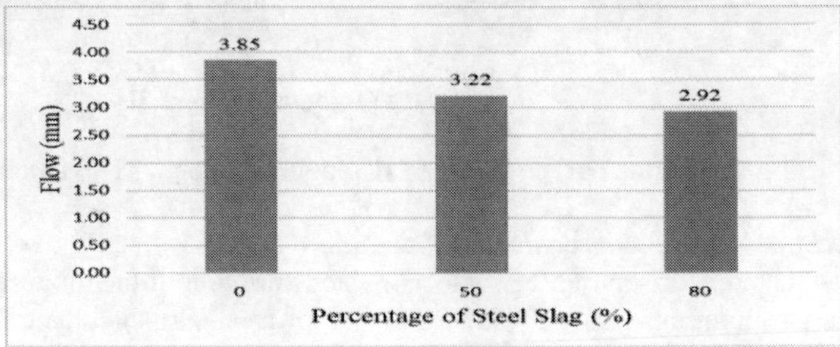

Figure 10. The effect of slag aggregate slag on flow value.

3.4.7 The Effect of Slag Aggregate on Marshall Quotient

In figure 11, it can be seen that all variations of the mixture meet the specified requirements. MQ values that are too large indicate a mixture which tends to be rigid so that it is not easily deformed, and vice versa. The highest MQ value is found in the variation of 80% steel slag mixture that is equal to 800.66 kg / mm and the lowest MQ value is found in a mixture of 100% natural aggregate, which is 323.66 kg / mm.

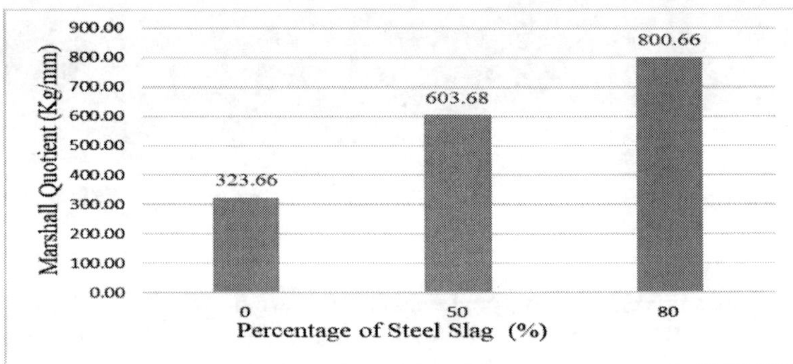

Figure 11. The effect of slag aggregate on marshall quotient.

4. Conclusions

Based on the research that has been carried out, the tests for the values of round steel slag aggregate characteristics have fulfilled the specified specifications so that they can be used on asphalt concrete binder course (AC-BC). The use of steel slag aggregate gives high density values due to the specific gravity of the large steel slag. In addition, low abrasion value in steel slag aggregates can increase the stability value in the mixture. This increased stability is also caused by the characteristics of hollow steel slag and irregular shape which cause good bonding interlocking. The highest stability value was found in the addition of 80% slag steel in the amount of 2320.19 kg.

The optimum asphalt content in each variation of the mixture produces different values. KAO value in variation 1 is 5.35%, in variation 2 is 5.85%, in variation 3 is 5.80%. Meanwhile, in variation 4 the optimum asphalt content cannot be identified because VIM and VFA values do not meet specifications of Bina Marga 2018. The resulting characteristic values for all variations with optimum asphalt content meet the specified specifications. The addition of steel slag aggregates in the mixture results in a higher VIM, VMA and Marshall Quotient value.

Based on the results of this study, the use of slag aggregates in asphalt mixtures can be an alternative to the use of natural aggregates. If the use of slag aggregates can be applied, this will help protect the nature from damages caused by the continuing use of natural aggregates. The use of slag aggregate can also become an effort to reduce waste that can damage the environment.

References

[1] Sukirman S 2003 *Beton Aspal Campuran Panas* (Jakarta: Granit)
[2] Sengoz B, Onsori A and Topal A 2014 Effect of aggregate shape on the surface properties of flexible pavement *KSCE J. Civ. Eng.* **5** 1364–71
[3] Balitbang 2012 Recycling (Teknolgi Daur Ulang Perkerasan Jalan) litbang.pu.go.id/ recycling teknologi-daur-ulang-perkerasan jalan
[4] Djakfar L H B and Zaika Y 2013 Evaluation of use of steel slag material and phyrophylite materials in the porous asphalt mix **18** 1524-29
[5] Wang and George C 2018 The Utilization of slag in civil infrastructure construction July 1–274
[6] Masoudi, Sajjad, Abtahi S M and Goli A 2017 Evaluation of electric arc furnace steel slag coarse aggregate in warm mix asphalt subjected to long-term aging *Constr. Build. Mat.* **135** 260–6
[7] Ali, Zulfiqar, Putra Jaya R, Moetaz M El-sergany and Yaacob H 2015 Steel slag as a road construction material **4** January 33–8
[8] Asphalt Institute 1993 Mix design methods for asphalt concrete and other hot-mix types *Manual Series No. 6* Sixth Edition (Lexington USA)
[9] Sukirman S 1999 Perkerasan lentur jalan raya (Bandung: Nova)
[10] Bethary R T, Subagio B S, Rahman H and Suaryana N 2019 Aging effect condition on hot asphalt mixtures marshall (AC-BC) performance by using slag *IOP Conf. Series:Materials Science and Engineering* vol 673 1 p 1–7
[11] Xu B, Chen J, Zhou C and Wang W 2016 Study on marshall design parameters of porous asphalt mixture using limestone as coarse aggregate *Constr. Build. Mater.* **124** 846–54
[12] Kementerian Pekerjaan Umum dan Perumahan Rakyat; Direktorat Jenderal Bina Marga 2018 General specifications of Bina Marga 2018 for road works and bridges September

IOP Conf. Series: Materials Science and Engineering **930** (2020) 012065 doi:10.1088/1757-899X/930/1/012065

Development of risk-based standardized Work Breakdown Structure (WBS) to improve time performance on high-speed railway construction project

D B Jati[1*] and Y Latief[1]

[1]Civil Engineering Department, Universitas Indonesia, Depok, Indonesia

[*]Corresponding author's e-mail: didik.bintoro@ui.ac.id

Abstract. In Indonesia plans and studies for High Speed Railway (HSR) are not a new idea as they have been conducted before 2010. However, the plan to build HSR was just recently announced in 2015. Specifically, for the high-speed railway construction project in Indonesia, there is no standards that contains details of work activity packages and their methods for carrying out work in the field. This resulted in every railway-based construction project still using several methods from both overseas and Indonesian domestic experiences. Therefore, it is now necessary to standardize both the methods and work activities and suggested to be implemented for any highspeed railway project so that the work breakdown structure (WBS) can be standardized for any future rail and rail-based construction project. This can also minimize any potential risks and ensure the project is completed according to the targets in terms of quality, cost, and time. The methodology used in this study are to identify risks at construction project with a WBS matrix (observation and questionnaire). Analysis data in this study used statistical analysis. The result of this study is 15 risks selected from the 3 highest risks and 1 risk representing each category. The 3 (three) highest are Delay in issuing designs by the planner / designer, Delay in land acquisition by the project owner, Work activity has stopped due to weather factors. Risk representing each category are The subcontractor's productivity does not match planning (Work packages), Delay in issuing designs by the planner / designer (Method/ design), Work activity has stopped due to weather factors (Activity), Delay from suppliers in delivering material (Material resources), The planned productivity of equipment is not as needed (Equipment resources) and Human productivity is lower than demand (Human resources).

1. Introduction

The integration of urban and rural areas of our country constantly, especially improving regional requirements of transport capacity, objectively promoted the steady development of the domestic rail transit construction. In Indonesia plans and studies for High Speed Rail (HSR) are not recent as they have been conducted since before 2010. However, the plan to build HSR was just recently announced in 2015. The first HSR construction to connected Jakarta with Bandung. The distance around 150 km.

The High-Speed Railway Project has a very complicated construction on its work that requires environmental problems (hydrology, geotechnical, and seismic), train schedules and safety, as well as the infrastructure needed so that it requires a good management system in its implementation [1]. Therefore, to guarantee construction work to run smoothly and in accordance with company goals, a system / structure is needed that describes each method and package of activities to be carried out in the field such as WBS (Work Breakdown Structure).

Content from this work may be used under the terms of the Creative Commons Attribution 3.0 licence. Any further distribution of this work must maintain attribution to the author(s) and the title of the work, journal citation and DOI.
Published under licence by IOP Publishing Ltd

At present, there are still many projects that carry out construction work without making a formal WBS first, even though WBS plays an important role in the management of the project itself from the planning stage to the implementation stage [2]. Specifically, for the high-speed railroad construction project in Indonesia, there is no standard that contains details of work activity packages and their methods for carrying out work in the field. This resulted in every railway-based construction project still using several methods from outside Indonesia. Therefore, it is now necessary to uniform methods and work activities that must exist and must be implemented for the highspeed railway project so that the WBS of the project can be standardized and used in every rail and rail-based construction project work going forward. This can also minimize any risks occurred and ensure the project is completed according to targets both in terms of quality, cost, and time.

2. Research Objectives

The aim of this research is to create WBS standardized for work on high-speed railway construction projects, identify risk factors that come from resources, activities, and work packages that affect the time performance of high-speed railway construction projects, and develop risk-based WBS standards for planning project management that can improve time performance on high-speed railway construction project.

3. Literature Review

3.1. Work Breakdown Structure for Railway Project

The Project Management Body of Knowledge (PMBOK) defines the work breakdown structure (WBS) as a hierarchical decomposition of the total scope of work to be carried out by the project team to accomplish the project objectives and create the required deliverables. WBS is a deliverable-oriented breakdown of a project into smaller components. Many management software such as resource management, scheduling, and financing can be used in creating WBS diagram. WBS will also breaks the project into units according to specific requirements and these units will be as project of a series of manageable work. All the units will be consolidated as a whole process of construction project by a project manager [3].

Railway project management mostly use bottom-up cost estimate method, based on bottom up method can related with work breakdown structure and project scheduled plans, etc. that make to know time performance in each sub-contractor in each packages work. The work breakdown structure can arrange the schedule construction to be on time. By means of analyzing plentiful railway project construction cases, there are two types of WBS, the one is similar to the organizational chart with the contents representing activities and the other is like to the tasks list. In top-down type WBS, the upper layer is the biggest task and lower layers are smaller one or more detailed specific tasks, that will ensure the biggest task finished [4].

The basis for the development of WBS of railway project were various bill of quantity of railway construction projects and adopt some regulations: the Undang-Undang No. 23 Tahun 2007 about Perkeretaapian, Peraturan Pemerintah No. 61 tahun 2016 about Lalu Lintas Angkutan Kereta Api, Peraturan Pemerintah No. 6 tahun 2017 about Penyelenggaraan Perkeretaapian, and Peraturan Menteri Perhubungan No. 69 tahun 2019 about Standar Spesifikasi Teknis Kereta Api Kecepatan Tinggi.

3.2. Risk Management

In a construction project, risk management will focus on the identification of any factors, and the determination of their occurrence probability and impact during the construction phase. The problems, which often occur in terms of the risk analysis are the methodology of risk analysis for a project [5], and risk classification [6]. In many literatures, a risk is treated as a possibility of incurring of a loss. While in risk management it should be paid more attention to three general process: risk identification, risk assessment and risk response/reaction.

3.3. Process of the Work Breakdown Structure

The Work Breakdown Structure steps are (7):

IOP Conf. Series: Materials Science and Engineering **930** (2020) 012065 doi:10.1088/1757-899X/930/1/012065

- Define the major project tasks, and then decompose the tasks into sub tasks in construction project.
- Identify each deliverable to the level of detail with the preparation of the budget and cost then schedule project.
- Work packages, make sure all work packages can be detected and allocated. The result quantified performance of the work packages can be tested it related for inspection in site.

3.4. WBS Dictionary
The description of the WBS dictionary are WBS codes, levels, job responsibilities, description of work packages, activities and resources. WBS dictionary format consists of work packages, activities and resources which are interconnected.

3.5. WBS Checklist
WBS Checklist is a part of assessing the level of accuracy of the WBS with reference to detailed activities in each section in the WBS and based on a description of each activity in the Project in the form of a form as an assessment guide for each section in the WBS

4. Methodology
This research was carried out in qualitative approach to achieves the research objectives. The method in this study are surveys and interviews that performed by structured and programed questionnaire for railway infrastructure construction professional with an experience of 10 years or more.
The research flow model shown in figure 1 below.

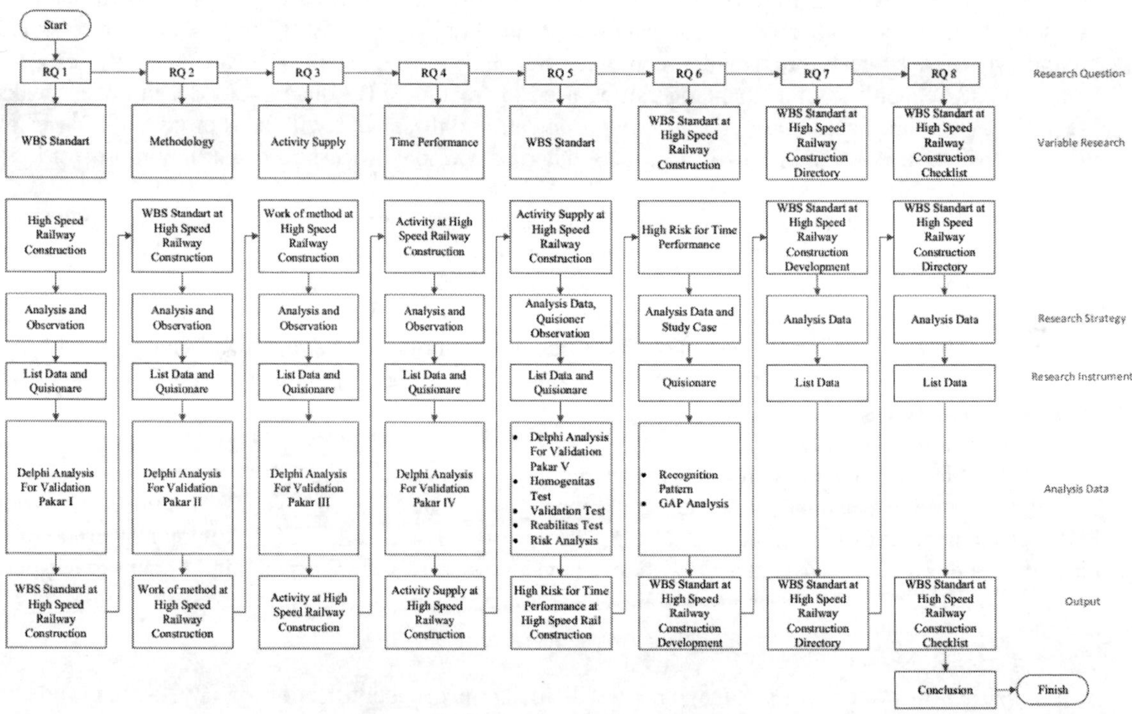

Figure 1. The research flow diagram.

5. Results and Discussion
Based on review of literatures related to railway infrastructure construction and 7 (seven) previous similar project, it was found that railway construction work can be categorized into 5 (five) infrastructure: Civil Infrastructure, Track Infrastructure, Depo Station Infrastructure, Passenger Station Infrastructure, and Signal and Telecommunication Infrastructure. Civil contains preparatory land, bridge, culvert and tunnel. Track contains ballast, sleepers, and rail. Signal and telecommunication

contain interlocking system, interlocking power supply, signal indicator and marker, mechanism, track circuit, axle counter, intermediate block, level crossing modification, communication, and electrical power supply. Depo station contains structure, track, rail, signal and communication. Passenger station contains structure, architecture, interior, mechanical, electrical, and landscape.

Each category has its own WBS that typically shown in figure 2. Figure 2 shows the resources and activities of special equipment in the WBS of the Passenger Station.

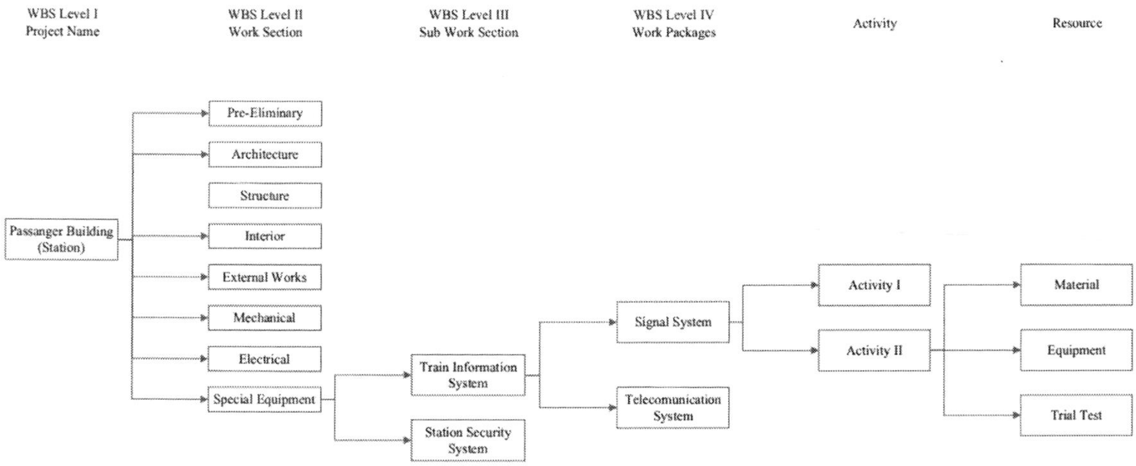

Figure 2. Work Breakdown Structure diagram.

The structure of the WBS consists of section 1 (L I), which is the identity of the project, section 2 (L II) is section of work, part of architectural work, structural work, electrical work, mechanical work, preparatory work and special work. In section 3 (L III) it is a sub-part of the work, and in sections 4, 5 and 6 it is a Package of Work (L IV), Activities (L V) and Resources (L VI) for each of the work elements contained in the High-Speed Railway Project.

The high-speed rail project is very complex in its implementation, so it needs a special section, for example a security system or a station that has special specifications in the implementation of its construction.

In identifying the risks obtained from several previous research studies, expert validation is then analyzed and the impact and causes of each of the risk factors have been validated with the following formula:

$$R = P \times I \quad (1)$$

Where, R = risk factor,
P = probability of risk,
I = impact of risk.

IMPACT		Very Low	Low	Med	High	Very High
PROBABILITY		1	2	3	4	5
Very Frequent	5	5	10	15	20	25
Frequent	4	4	8	12	16	20
Somewhat frequent	3	3	6	9	12	15
Occasional	2	2	4	6	8	10
Rare	1	1	2	3	4	5

Figure 3. Risk level matrix.

As for the levels of each category, there are high, medium, and low risks:

1. High Risk, when the risks residing in the cells marked in red, are the most critical risks and should be treated on a high priority basis.
2. Medium Risk, when the risk is in one of the cells marked in yellow, which has implications and can affect time performance, is immediately responded to at the project level.
3. Low Risk, if the risk that enters the green cell, has a slight impact on cost performance and can be responded directly by engineers or related parties.

The next method is to assess and determine potential risks and sources of associated risks. Potential risks will be seen at level 4 and level 5 which contains a package of work and activities. After that, for control points based on time performance can be seen the priority scale and time risk control. The result is as shown in Figure 4

RISK IDENTIVICATION, RISK ASSESSMENT, PRIORITY SCALE, SAFETY RISK CONTROL AND RESPONSIBILITY

Company Name :
Activity :
Location :
Date :

No	WBS Level 4 Work Package	WBS Level 5 Activity	Risk Identification	Risk Assessment			Control Point	
				Frequency	Severity	Risk Level	Priority Scale	Time Risk Control

Adding work packages and activity

Risk control point based time performance

Figure 4. Risk response planning.

WBS Dictionary in this research was made by the form as in figure 5. Figure 5 is an example of the WBS dictionary for one of railway infrastructure component: passenger station.

WBS DICTIONARY HIGH SPEEN RAILWAY INFRASTRUCTURE CONSTRUCTION
PASSANGER BULDING (STATION)
TRAIN INFORMATION

2	Work Section	: Special Equipment
2.3	Sub Work Section	: Train Information System
2.3.3	Work Package	: Signal System
	Pesponsible Person	:
	Description	: operating facility that serves to provide train arrival and departure signals at the station
	Deliverable	: Time arrival and depature train
	Reference	: Contract Document Expert validation BoQ

Code	Activity	Sumber Daya			
		Manpower	Material	Equipment	Test
2.3.3.1	Starting Signal	Worker	Signal Tool	Tool for building the system	-
		Specialist ME and IT	System coding	Computer	trial test instrument

Figure 5. WBS dictionary.

The WBS checklist is created the format as in figure 6. Figure 6 is one of the WBS checklists for the components of the railway infrastructure: Passenger Station.

CHECKLIST WBS SINGNAL AND COMMUNICATION
PASSENGER BUILDING (STATION)

WBS Level 1	:	Project Name
WBS Level 2	:	Work Section
WBS Level 3	:	Sub Work Section
WBS Level 4	:	Work Package

LEVEL WBS	CODE	DESCRIPTION
1	1	Station
2	1.1	Signal and Telecommunication
3	1.2	Interlocking System
4	1.2.1	Electronic Interlocking Sytem

Figure 6. WBS Checklist.

5.1. Validation

Validation is carried out on the overall results of the research, including WBS development result, identification high risk related for railway construction, corrective and preventive based on cause and impact high risk variable. Selected experts and professionals are chosen in the validation phase. The experts have professional experience in railway infrastructure construction project. Experts have various positions ranging from division manager with 30 year of experience, general manager with 28 years of experience, project manager with 25 years of experience, construction manager with 16 year of experience and engineering managers with 16 years of experience. Experts in this study are specialist for railway infrastructure construction. The experts have been involved at Light Rail Transit, Mass Rapid Transit, Commuter Line, and Inter Regional Train. Shown by the following table 1.

Table 1. Profile of expert

No.	Expert	Position	Experience	Education
1	Pakar 1	Division Manager	30 years	S2
2	Pakar 2	General Manager	28 years	S2
3	Pakar 3	Project Manager	25 years	S2
4	Pakar 4	Construction Manager	16 years	S1
5	Pakar 5	Engineering Manager	16 years	S1

5.2. Survey for Risk Variables

The survey was carried out in order to obtain the dominant risk factors/variables in Jakarta-Bandung high speed railway construction project. The respondents are professional civil engineers with work experience in the railway construction projects for at least 3 years. Questionnaire form was sent to 40 respondents and 30 was returned. At the end we have 61 risk factors/variables from respondents that could be used for this research analysis. The respondent profile in this survey is shown in table 2 below.

Table 2. Profile of respondent

No	Description	Total
1	**Position**	
	Manager Divisi	2
	General Manager	15
	Project Manager	5
	Construction Manager	6
	Engineering Manager	2
2	**Institution**	
	(PT. A)	15
	(PT. B)	2
	(PT. C)	3
	(PT. D)	4
	(PT. E)	3
	(PT. F)	3

No	Description	Total
3	**Work Experience**	
	≤ 15 years	8
	16 - 20 years	10
	21 - 25 years	6
	≥ 26 years	6
4	**Education**	
	Bachelor degree	10
	Master degree	20

Based on the expert's validation all these 61 risk variables were valid and related to time performance. The fifteen dominant risk data shown in table 3 as follows.

Table 3. Higher risk level identification.

No		Risk Variables	Rank
1	X24	Delay in issuing designs by the planner / designer	1
2	X25	Delay in land acquisition by the project owner	2
3	X31	Work activity has stopped due to weather factors	3
4	X42	The planned productivity of equipment is not as needed	4
5	X45	Equipment specifications are not according to plan	5
6	X53	Labor productivity is lower than demand	6
7	X33	Lack of cash flow so that payments to subcontracts are hampered	7
8	X20	Extreme weather changes	8
9	X37	Delay from suppliers (third parties) in sending materials	9
10	X10	The subcontractor's productivity does not match planning	10
11	X11	There is a lack of design in the design planning team	11
12	X13	Change request due to differences in planning with actual (Incompatible scope)	12
13	X16	There was an error in the results of the land investigation in the Feasibility Study	13
14	X12	Many design changes, drawings, specifications and scope of the work package during the project	14
15	X17	Underground material conditions are different from the conditions set forth in the contract documents	15

6. Conclusion

The results of the standardized WBS for High Speed Railway infrastructure construction development process was categorized based on part of the rail infrastructure, while rail infrastructure construction work categorized into four parts: rail infrastructure containing station for depo, station for passenger, track, signal and telecommunication. Station for depo contain rail, load lifter, mechanical and electrical for train. Station for passenger contain passenger's waiting room, train information, and check in system. Track contain ballast, rail and sleepers. Signal and telecommunication contain information about arrival and departure of train, signal traffic, and gate system. Each of part high-speed infrastructures divided becomes 6 (six) parts. There are preparatory, civil, track, depo station, station for passenger, signal and telecommunication. Civil contains earth works, bridge, culvert and tunnel. Track contains ballast, sleepers, and rail. Depo station contains structure, track, rail, signal and communication. Passenger station contains structure, architecture, interior, mechanical, electrical, and landscape. Signal and telecommunication contain interlocking system, interlocking power supply, signal indicator and marker, mechanism, track circuit, axle counter, intermediate block, level crossing modification, communication, and electrical power supply.

The research results 15 dominant risks that selected from the 3 highest risks and 1 risk representing each category. The 3 (three) highest risks are the delay in issuing designs by the designer, delay in land acquisition by the project owner, and work activity has stopped due to weather factors. Risk representing each category are:

1. Work packages, the subcontractor's productivity does not match planning (X10)
2. Method/ design, Delay in issuing designs by the planner / designer (X24)
3. Activity, Work activity has stopped due to weather factors (X31)

4. Material resources, Delay from suppliers in delivering material (X37)
5. Equipment resources, the planned productivity of equipment is not as needed (X42)
6. Human resources, Human productivity is lower than demand (X53)

The risk response become the input of risk groupings as follows: activity at the WBS concerned, additional activities, additional management construction, additional requirements on construction and impacted the coefficient of WBS. Time schedule arrangement can be performed based on the standardized WBS and it can be managed to the level of resources.

References

[1] Limon T and Crozet Y 2017 Risk analysis and high speed rail projects in France: introducing economic slowdown into appraisal methodologies *Transportation Research Proc.* Vol 25 pp 2824–42
[2] Nguyen L D, Ogunlana S O and Lan D T X 2004 A study on project success factors in large construction projects in Vietnam *J. Eng. Constr. Manag.* (Emerald) **11** 6 404-13
[3] Dziadosz A 2010 Przegląd wybranych metod wspomagających analizę ryzyka przedsięwzięć budowlanych *Przegląd Budowlany* **7-8** 76- 7
[4] Railway 23rd Bureau 2010 Project management system of China railway 23th Bureau [R] (China Ministry of Railways)
[5] Kuchta D Skorupka 2012 Project risk management taking into consideration the influence of various risk levels based on linguistic approach *World Scientific Proc. Series on Computer Engineering And Information Science 7; Uncertainty Modeling in Knowledge Engineering And Decision Making – Proc. of the 10th Int. FLINS Conf. 7* pp 1093-8
[6] Kapliński 2010 Klasyfikacja ryzyka w zarządzaniu według Andreasa Klinkego i Ortwina Renna *Przegląd Organizacji* 5 13-7
[7] Su Lei 2012 WBS-Based risk identification for the whole process of the real estate project and countermeasures *National Conference on Information Technology and Computer Science, November 2012*
[8] Undang-Undang No. 23 Tahun 2007 tentang Perkeretaapian (Jakarta: Kementerian Perhubungan)
[9] Peraturan Pemerintah No. 61 tahun 2016 tentang Lalu Lintas Angkutan Kereta Api (Jakarta: Kementerian Perhubungan)
[10] Peraturan Pemerintah No. 6 tahun 2017 tentang Penyelenggaraan Perkeretaapian (Jakarta: Kementerian Perhubungan)
[11] Peraturan Menteri Perhubungan No. 69 tahun 2019 tentang Standar Spesifikasi Teknis Kereta Api Kecepatan Tinggi (Jakarta: Kementerian Perhubungan)

IOP Conf. Series: Materials Science and Engineering **930** (2020) 012066 doi:10.1088/1757-899X/930/1/012066

Competition measures of TOD point to Central Business Districts in Surabaya using travel time approach (a case study: Joyoboyo Terminal)

A A Zahra[1] and E Ahyudanari[1*]

[1] Departement of Civil Engineering, Institut Teknologi Sepuluh Nopember (ITS) , Raya ITS Street, Surabaya 60111

*Corresponding author e-mail: ervina@ce.its.ac.id

Abstract. Several areas are planned to be transit-oriented development in Surabaya, one of which is Joyoboyo. Joyoboyo Terminal will be developed as a transit point in Joyoboyo TOD. At present, Joyoboyo Terminal functions as a park and ride building. The accessibility of the Joyoboyo terminal is measured by total travel time to Surabaya CBD. In this study, we used Google maps and Moovit to obtain the total travel time by private vehicles and public transport to 7 Surabaya CBDs. The results of the travel time survey show better accessibility using a private vehicle (car) than public transport (bus/city transport). The travel time difference was about 4 to 20 minutes. This research also provided an accessibility map from Joyoboyo TOD to Surabaya CBD

1. Introduction

People need the development of facilities that integrates their activities with the required movements. This integration is the concepts of Transit-Oriented Development (TOD). The concept of TOD is attempted to be applied in Surabaya. This effort is shown from the proximity of the city center and transit points [1]. In 2014, Surabaya Government with the World Bank, Hansen, City Form Lab, and several instances conducted research related to city corridor development in Surabaya. Joyoboyo transit is one of the terminal that will be developed following the TOD concept. The previous research recommended five potential locations along the Surabaya Mass Rapid Transit (SMART) corridor to be a TOD area, one of which is the Joyoboyo area [2]. The Surabaya Mass Rapid Transit is planned in the north-south corridor by *Suro* tram and east-west corridor by *Boyo* rail, and both corridors meet in the Joyoboyo area. Thus, this research took place in Joyoboyo Terminal. Future development of Joyoboyo planned by the Transportation Authority of Surabaya and the World Bank can be seen in figure 1.

The development of TOD requires support from the use of public transport for daily commuting. The Surabaya government need to restrict the usage of private vehicles by implementing traffic demand management (TDM). Some countries have successfully reduced the number of vehicles entering the city with the TDM concept, such as Cambridge, Singapore, Perth, Beijing, Dublin, Greater Vancouver Region, London, Trondheim, Southern California, Portland, and Auckland [3]. Providing park and ride buildings is one of the TDM implementations by the Surabaya Government.

Content from this work may be used under the terms of the Creative Commons Attribution 3.0 licence. Any further distribution of this work must maintain attribution to the author(s) and the title of the work, journal citation and DOI.
Published under licence by IOP Publishing Ltd

IOP Conf. Series: Materials Science and Engineering **930** (2020) 012066 doi:10.1088/1757-899X/930/1/012066

Currently, public transportation in Surabaya merely depends on the city transport and bus that has not been upgraded yet into the public transport system, such as bus rapid transit system. The Bus Rapid Transit (BRT) system has been successful in shifting private vehicle usage to public transport in several countries [4–6]. One of the facilities to accommodate the transfer of private vehicle mode to public transport is the availability of park and ride. However, the availability of park and ride will not optimally function if there is no public transport system pass through its location.

In Europe, the park and ride are built to prevent traffic congestion and high parking demand in the city center [7]. The location of the parking lot must be adjacent to the transit corridor, with a maximum 15-minute headway of public transport [8]. Surabaya has already had seven parks and ride spots along the Suroboyo Bus route, which is managed by the government. One of those parking and ride buildings is located in Joyoboyo Terminal. Therefore, this research would determine the accessibility of transit point in Joyoboyo TOD.

Accessibility is a measure of comfort or convenience about the interaction of land use with one another [9]. If the land use is far to one another and transportation network between the land uses are in bad condition, then the accessibility is low. High internal TOD accessibility can be characterized by the 'safe, uninterrupted, and interconnected' access to pedestrians and cyclists within the block, so all the blocks are connected. Arterial roads still dominate the access to the CBD in the Surabaya area because the city railroad is across along the outer city. It is then assumed that there is a need to do analysis related to the accessibility from the terminal to the CBD. The accessibility of this research focuses on competitive measures in travel time.

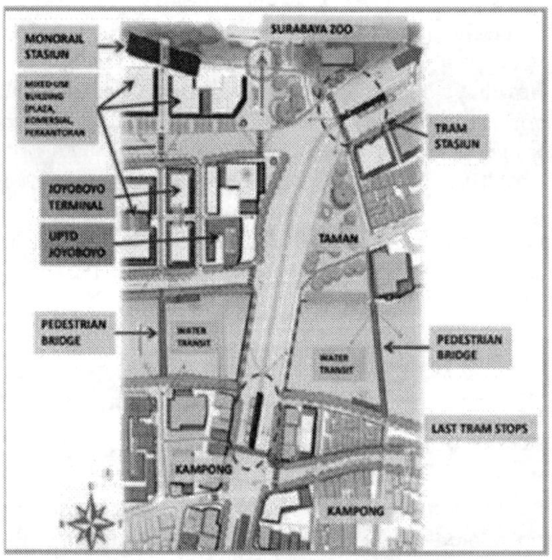

Figure 1. Joyoboyo TOD Development [2].

2. Study Area

From previous research [10], Joyoboyo TOD is divided into five blocks, in which delimitation of each block can be seen in figure 2. Those five blocks are located in Wonokromo Subdistrict, consisting of Darmo, Jagir, Ngagel, Ngagel Rejo, Sawunggaling, and Wonokromo Districts. The area of Joyoboyo TOD is180.54 ha and dominated by residential areas. The Joyoboyo Terminal, as a transit point, is located in Block 1 as presented in figure 2.

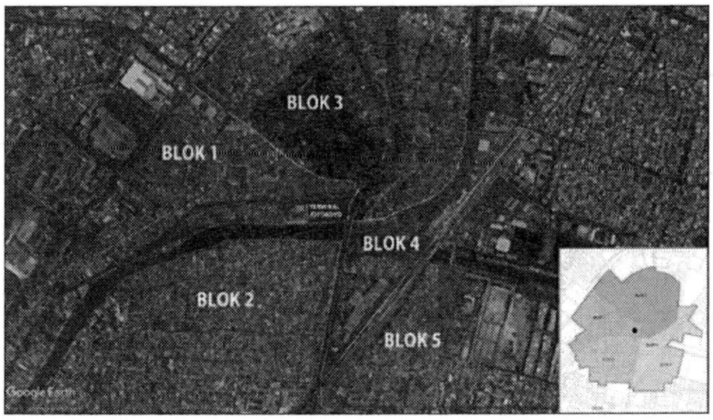

Figure 2. Joyoboyo TOD Delimitation [10].

3. Methodology

Indicators to determine accessibility can be measured by travel distance, travel cost, and travel time [10]. This research used travel time as a parameter of accessibility. The travel time data from Joyoboyo TOD to Surabaya CBD was obtained from a one-week travel time survey for private vehicles with Google maps. For public transportation travel time data was obtained from Moovit per 30 minutes of departure. The methodology used in this study can be seen in figure 3.

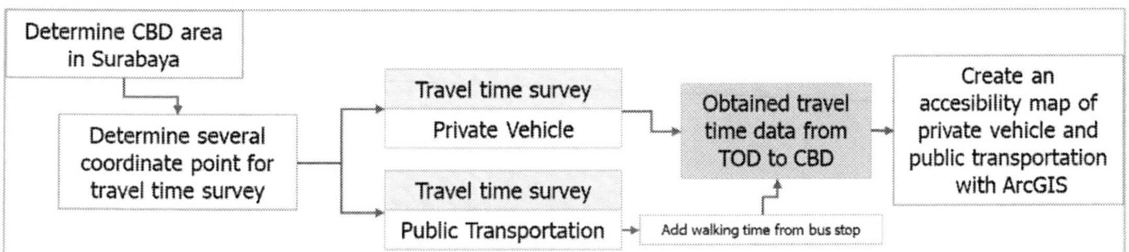

Figure 3. Research Methodology.

The travel time obtained from the survey was classified into six classes Level of Services [11], as shown in table 1. Those classifications of LOS were added to GIS to create an accessibility map.

Table 1. Classification of LOS [11]

Level of Services (LOS)	Speed (km/jam) Road Class I	Road Class II	Road Class III	Map Colour Indicator
A	≥ 56	≥ 48	≥ 40	
B	≥ 45	≥ 38	≥ 31	
C	≥ 35	≥ 29	≥ 21	
D	≥ 28	≥ 23	≥ 15	
E	≥ 21	≥ 16	≥ 11	
F	< 21	< 16	< 11	

4. Walking Speed

This research needed to adopt the walking speed to determine walking access time. Walking speed can be calculated by the distance traveled divided by the time taken to walk. It is influenced by environmental factors, traffic conditions, and pedestrian characteristics. People will walk slower in dense traffic areas with uncomfortable and unsafe pedestrian facilities. However, it tends to be faster if people walk on wide sidewalks equipped with roadside protection barriers, even in dense traffic areas.

IOP Conf. Series: Materials Science and Engineering **930** (2020) 012066 doi:10.1088/1757-899X/930/1/012066

The average people walking speed is 0.83 m/s [12]. Manual on Uniform Traffic Control Devices (MUTCD) states that the people walking speed when crossing a signal crossing is 1.22 m/s [13]. Meanwhile, the Institute of Transportation Engineers surveyed in Florida at a pedestrian-dense location and found the people walking speed was 0.76 m/s.

In Indonesia, from a pedestrian survey in Bandung found the people walking speed was 1.08 m/s [14] and was 1.4 m/s while disaster evacuation simulation in Padang [15]. Nusdwinuringtyas (2018) did a research using 6 Minute Walk TES (6MWT) method in healthy Indonesian ethnic aged 18-50 years and found the average walking speed of male was 1.61 m/s and female was 1.56 m/s. This research used 0.83 m/s as walking speed for analysis of total travel time by public transport.

5. Central Business District

CBD can be interpreted as a part of the city that has the highest density from diverse activities [17]. At present, CBD is not just a place for a commercial and retail center, but also residential and entertainment facilities, e.g. malls. Thus, people can work, get entertained, eat, and sleep in one area. Several characteristics of CBD are shown in table 2.

Table 2. Characteristics of CBD.

Researcher	High Buildings	Office Cluster	Financial District	multi-story	At Main Road	High Land Value	Not Densely Housing
[18]	✓	✓		✓		✓	✓
[19]	✓	✓		✓		✓	
[20]	✓	✓	✓	✓	✓	✓	
[21]	✓	✓				✓	✓
[22]	✓	✓		✓	✓	✓	✓
[17]		✓	✓	✓			
[23]	✓	✓	✓				

Table 2 presents that all sources stated that CBD has the characteristics of high office building concentration. Besides, all sources also stated that CBD is in the area of many multi-story buildings that have differentiation of functions, except [17]. Yu (2015) argues that a CBD area can be rated from the density of population activity, so it does not have to consist of high buildings. CBD is in an area that has a high land value and not dense residential areas [17,18,20–22]. It is different from which states that the CBD area is in a densely populated area.

In the CBD, there is a "multi-story" building that provides various trade and services characterized by the presence of supermarkets or malls [17,18,20,22]. Battino, Borruso, & Donato (2012), Yu (2015) and Yogi (2017) state that the CBD area is marked by the presence of a banking headquarters (financial district). About the traffic condition, Battino, Borruso, & Donato (2012) and Pissourios (2014) state that the CBD is located at the main arterial road and has high traffic density. Thus, it can be concluded that CBD area has the following characteristics:

1. CBD consists of high office building concentration.
2. CBD is in an area of many high-rise buildings that have differentiation of functions.
3. CBD is located across a high land value zone.
4. Within the CBD, there is a multi-story building that provides trade and services.

5.1. CBD in Surabaya

The development of the CBD area in Surabaya was discussed in the master plan for spatial planning (*Rencana Induk Tata Ruang dan Wilayah)* in 1973 [24]. At that time, the city government had permitted the construction of a Segi Delapan CBD area in West Surabaya. According to Johan Silas, the Segi Delapan area had indeed been projected to become Surabaya's business center since 1973.

In the 2016-2021, the document of Surabaya City Mid-Term Development Plan (*Rencana Pengembangan Jangka Menengah Daerah Kota Surabaya*) states that the CBD area in Surabaya is located at Jl. Mayjend Sungkono, Jl. Kutisari, Jl. Mulyosari, Jl. Ngagel-Pucang, Jl. Kertajaya, Jl. Ahmad Yani, West Outer Ring Road, West Inner Ring Road, Jl. Kalianak, Jl. Dharmahusada, Jl. HR Muhammad, UP VIII Hamlet Pakis, Pakuwon Trade Center, East Coast and Puncak CBD Wiyung. Meanwhile, the document of Surabaya City Corridor Development Report by the World Bank in 2014 states that Surabaya CBD stretches between Keputran and Genteng (Tunjungan Area).

Surabaya City is divided into 12 Development Units (DU), and then DUs were assessed as having proximity with four main characteristics of CBD. It obtained 7 CBD areas in Surabaya located in Tunjungan; Genteng; Segi Delapan; Yono Suwoyo; Sungkono; Wiyung and MERR. Those 7 CBDs are located across the high land value zone in Surabaya as can be seen in figure 4.

Figure 4. Location of Surabaya CBD.

6. Travel Time Survey

6.1. Travel Time by Private Vehicle

The data needed for Google Maps travel survey is 'origin' and 'destination' point. Then, it will show several alternative routes that can be selected. The first step is selecting 44 road segments to be surveyed as can be seen in figure 5. Those 44 selected roads are the route passed from Joyoboyo TOD to 7 CBD points in Surabaya. The coordinates of 44 road sections obtained from Google Earth seen in table 3.

Figure 5. 44 Road segments to be surveyed.

After a one-week survey, the private vehicle travel time data of 44 segments can be seen in figure 6. Figure 6 shows that several road segments have a relatively extended travel time,x while other streets have a peak hour every day.

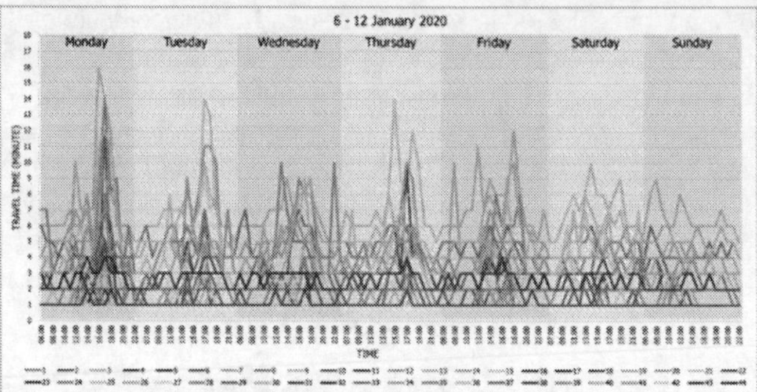

Figure 6. Private vehicle travel time data graphs in one week.

The graph in figure 6 shows the travel time of each segment. The total travel time was obtained by summing up the travel time of the roads that have passed; the result can be seen in table 3. The total travel time was then classified according to the LOS classification in table 1.

Table 3. The coordinates points and total travel time of private vehicle.

No.	Start point	Endpoint	Street Name	Total Travel Time	LOS
1	-7.29823870784, 112.737781844	-7.29639805772, 112.739124018	Jl. Darmo	00:01:02	D
2	-7.29639805772, 112.739124018	-7.28992506819, 112.738881701	Jl. Darmo	00:03:18	D
3	-7.28992506819, 112.738881701	-7.27749267544, 112.741215289	Jl. Darmo	00:07:22	D
4	-7.27749267544, 112.741215289	-7.27314036875, 112.742049686	Jl. Urip Sumoharjo	00:08:23	D
5	-7.27314036875, 112.742049686	-7.26976035728, 112.74162918	Jl. Basuki Rahmat	00:09:25	D
6	-7.27314036875, 112.742049686	-7.262512946, 112.740559958	Jl. Basuki Rahmat	00:12:15	C
7	-7.26976035728, 112.74162918	-7.2667242045, 112.741137571	Jl. Basuki Rahmat	00:10:27	D
8	-7.2667242045, 112.741137571	-7.26280847192, 112.741303595	Jl. Basuki Rahmat	00:11:31	C
9	-7.262512946, 112.740559958	-7.25860688033, 112.73347313	Jl Embong Malang	00:14:41	C
10	-7.25860688033, 112.73347313	-7.25600014063, 112.734178477	Jl. Blauran	00:15:43	C
11	-7.25600014063, 112.734178477	-7.25586283001, 112.736984927	Jl. Praban	00:17:14	C
12	-7.25586283001, 112.736984927	-7.26280847192, 112.741303595	Jl. Tunjungan	00:19:31	C
13	-7.26280847192, 112.741303595	-7.26442907377, 112.745456975	Jl. Gubernur Suryo	00:12:36	C
14	-7.26442907377, 112.745456975	-7.26775139628, 112.744419424	Jl. Pang. Sudirman	00:13:38	C
15	-7.26775139628, 112.744419424	-7.27301235816, 112.74242045	Jl. Pang. Sudirman	00:12:46	C
16	-7.2653412481, 112.7410235871	-7.26667422189, 112.744732191	Jl. Embong Wungu	00:11:27	C
17	-7.2678900218, 112.74128694	-7.26881111447, 112.743943599	Jl. Embong Gayam	00:11:27	C
18	-7.28992506819, 112.738881701	-7.29129271186, 112.742970984	Jl. Bengawan	00:05:24	D
19	-7.29129271186, 112.742970984	-7.27700692881, 112.745261067	Jl. Ngagel	00:09:59	E
20	-7.27700692881, 112.745261067	-7.27628007308, 112.746172825	Jl. Sulawesi	00:10:59	E
21	-7.27628007308, 112.746172825	-7.27453273475, 112.746763704	Jl. Gubeng	00:11:59	E
22	-7.27628007308, 112.746172825	-7.26864104369, 112.750534955	Jl. Gubeng	00:14:03	E
23	-7.26864104369, 112.750534955	-7.26635152369, 112.75104222	Jl. Sumatera	00:15:03	D
24	-7.26635152369, 112.75104222	-7.26454518641, 112.745828673	Jl. Pemuda	00:16:42	D
25	-7.29639805772, 112.739124018	-7.2924814325, 112.737171023	Jl. Diponegoro	00:02:05	E
26	-7.2924814325, 112.737171023	-7.29564309765, 112.733234991	Jl. Ciliwung	00:03:16	C
27	-7.29564309765, 112.733234991	-7.29255107038, 112.72844178	Jl. Adityawarman	00:05:37	C
28	-7.29255107038, 112.72844178	-7.29186310027, 112.719166271	Jl. M. Sungkono	00:07:35	C
29	-7.29186310027, 112.719166271	-7.28844525977, 112.707236172	Jl. M. Sungkono	00:11:00	C
30	-7.28844525977, 112.707236172	-7.28142241207, 112.684788282	Jl. HR Muhammad	00:18:00	C

No.	Start point	Endpoint	Street Name	Total Travel Time	LOS
31	-7.28142241207, 112.684788282	-7.28494337739, 112.681387476	Jl. Yono Suwoyo	00:19:24	C
32	-7.28142241207, 112.684788282	-7.28891783742, 112.677607966	Jl. Yono Suwoyo	00:22:30	C
33	-7.28142241207, 112.684788282	-7.27693721619, 112.689733904	Jl. Yono Suwoyo	00:22:38	C
34	-7.29938485441, 112.735993254	-7.30038353421, 112.730219591	Jl. Joyoboyo	00:01:03	B
35	-7.30038353421, 112.730219591	-7.30727765921, 112.719650771	Jl. Gunungsari	00:05:41	C
36	-7.30727765921, 112.719650771	-7.31289448492, 112.709665377	Jl. Gunungsari	00:09:44	C
37	-7.31289448492, 112.709665377	-7.31349174903, 112.702368469	Jl. Wiyung	00:11:45	C
38	-7.27453273475, 112.746763704	-7.27497384318, 112.749608707	Jl. Bali	00:13:01	D
39	-7.27497384318, 112.749608707	-7.27707317614, 112.749182194	Jl. Biliton	00:14:03	D
40	-7.27707317614, 112.749182194	-7.27959029509, 112.764585678	Jl Kertajaya	00:20:44	D
41	-7.27959029509, 112.764585678	-7.28052167191, 112.780725472	Jl Kertajaya	00:24:39	C
42	-7.28052167191, 112.780725472	-7.27620708026, 112.781170256	Jl. Ir Soekarno	00:25:40	E
43	-7.28052167191, 112.780725472	-7.2848607513, 112.780950866	Jl. Ir Soekarno	00:25:42	E
44	-7.28052167191, 112.780725472	-7.28938265529, 112.780669905	Jl. Ir Soekarno	00:28:16	E

Table 3 shows the average travel time of private vehicle (car) from Joyoboyo TOD to Tunjungan CBD located in Jl. Basuki Rahmat (number 6) was 12 minutes 15 seconds; The travel time is in the range of 1-28 minutes. The LOS classification in table 3 was added to GIS to create the accessibility map by private vehicle, as shown in Figure 7.

6.2. Travel Time by Public Transport

Those 44 points were surveyed with Google Maps for a private vehicle, 19 points of which were surveyed with Moovit as can be seen in table 4. The origin point entered was Joyoboyo Terminal and the destination points entered were those 19 points. Public transport travel time data was taken per 30 minutes from 06:00 - 22:00 in Moovit. This software shows three recommendation lines that can be selected and show the time for bus/city transport arrival. Thus, the travel time data from Moovit was the accumulation of travel time itself and waiting time for bus/city transport arrival. The locations of 19 destination points, selected bus/city transport line, and travel time to the CBD can be seen in table 4.

Table 4. The coordinates points, selected line and travel time of public transport.

No.	Endpoint coordinate	Line	Travel Time	LOS
1	-7.27749267544, 112.741215289	R1R2	00:20:00	F
2	-7.27314036875, 112.742049686	R1R2	00:24:00	F
3	-7.262512946, 112.740559958	R1R2	00:27:00	F
4	-7.25586283001, 112.736984927	V → R3R4	00:34:00	F
5	-7.26442907377, 112.745456975	V	00:13:00	C
6	-7.26775139628, 112.744419424	V/ P2 / PAC 2 / DA / BUS C / P1	00:33:00	F
7	-7.27301235816, 112.74242045	V/ R1R2 / P2 / PAC 2 / DA/ BUS C/ P1	00:35:00	F
8	-7.29255107038, 112.72844178	TV	00:06:00	D
9	-7.29186310027, 112.719166271	TV	00:10:00	D
10	-7.28142241207, 112.684788282	TV	00:19:00	C
11	-7.28891783742, 112.677607966	TV	00:21:00	C
12	-7.27693721619, 112.689733904	TV → DKB	00:17:00	E
13	-7.29129271186, 112.742970984	F	00:05:00	D
14	-7.27453273475, 112.746763704	F	00:11:00	E
15	-7.26454518641, 112.745828673	F→N	00:22:00	D
16	-7.27620708026, 112.781170256	S→WK	00:30:00	D
17	-7.28938265529, 112.780669905	S	00:27:00	C
18	-7.30727765921, 112.719650771	DKM	00:06:00	C
19	-7.31349174903, 112.702368469	JM	00:14:00	D

IOP Conf. Series: Materials Science and Engineering **930** (2020) 012066 doi:10.1088/1757-899X/930/1/012066

Using public transport from the Joyoboyo TOD to Tunjungan CBD (number 3) requires 27 minutes. Other CBDs areas are taken at various times, from 6 minutes to 37 minutes. Those LOS classification in table 3 were added to GIS to create the accessibility map by public transport, as shown in figure **7**.

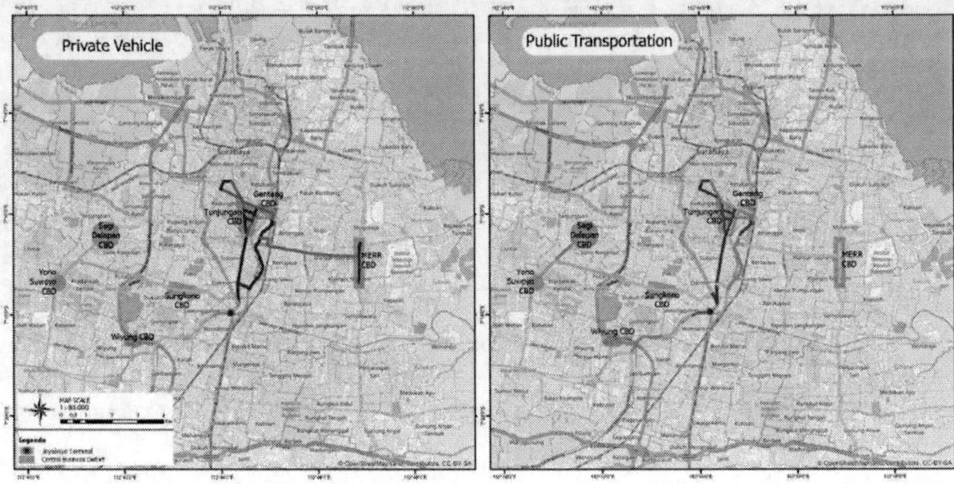

Figure 7. Accessibility map of private vehicle and public transportation.

6.3. Total Travel Time

The advantages offered by private vehicles is 'door to door service'. While the travel time data of public transit in table 4 above have not been added with walking time from the bus stop to CBD points. The walking distance was calculated using ArcGIS. For example, in the Segi Delapan CBD, it needs to walk from the bus stop to the CBD point, as shown in figure 8.

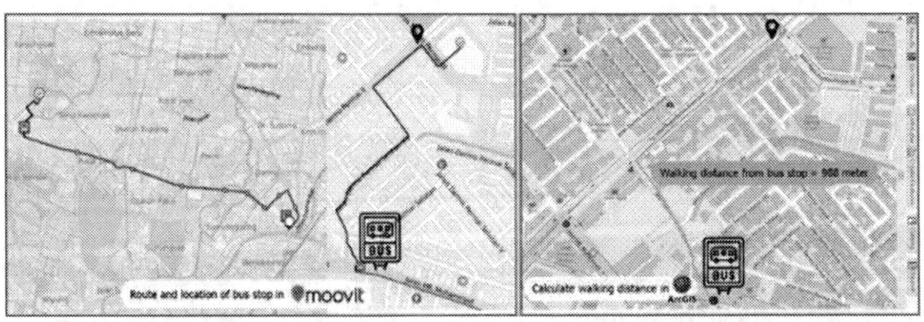

Figure 8. Walking distance from bus stop to segi delapan CBD.

Figure 8 shows the walking distance from the bus stop to Segi Delapan CBD (coordinate number 33, see table 4) is 988 meters. The walking time equals to the distance from the bus stop was divided by walking speed (0,83 m/s). Thus, the total travel time by public transport to Segi Delapan CBD can be calculated as follow:

$$\text{Walking Time} = \frac{Distance\ from\ bus\ stop\ (meter)}{Walking\ speed\ (0{,}83\ m\ /s)}$$

$$= \frac{988\ meter}{0{,}83\ m\ /s}$$

$$= 1190 \text{ second} \approx 19 \text{ minute } 50 \text{ second}$$

$$\text{Total Travel time} = (\text{Waiting time} + \text{estimated bus / angkot travel time}) + \text{walking time}$$

$$= 17 \text{ minute} + 19 \text{ minute } 50 \text{ second}$$

$$= 36 \text{ minute } 50 \text{ second}$$

The walking distance and walking time in another 6 CBDs were calculated in the same step as above, and the result can be seen in table 5. Thus, it obtained the whole total travel time using public vehicles and private vehicles as can be seen in table 6 and figure 9.

Table 5. Walking time from bus stop to CBD points.

Central Business District	Distance from Bus Stop (meter)	Walking Time
Tunjungan	283	00:05:41
Genteng	264	00:05:18
Sungkono	188	00:03:47
Segi Delapan	988	00:19:50
Yono Suwoyo	297	00:05:58
Wiyung	669	00:13:26
MERR	175	00:03:31

Table 6. Total travel time by public transport and private vehicle.

Central Business District	Public Transport Waiting + Travel Time	Walking Time	Total Travel Time by Public Transport	Total Travel Time by Private Vehicle
Tunjungan	00:27:00	00:05:41	00:32:41	00:12:15
Genteng	00:22:00	00:05:18	00:27:18	00:16:42
Sungkono	00:10:00	00:03:47	00:13:47	00:07:35
Segi Delapan	00:17:00	00:19:50	00:36:50	00:22:38
Yono Suwoyo	00:21:00	00:05:58	00:26:58	00:22:30
Wiyung	00:14:00	00:13:26	00:27:26	00:11:45
MERR	00:30:00	00:03:31	00:33:31	00:25:40

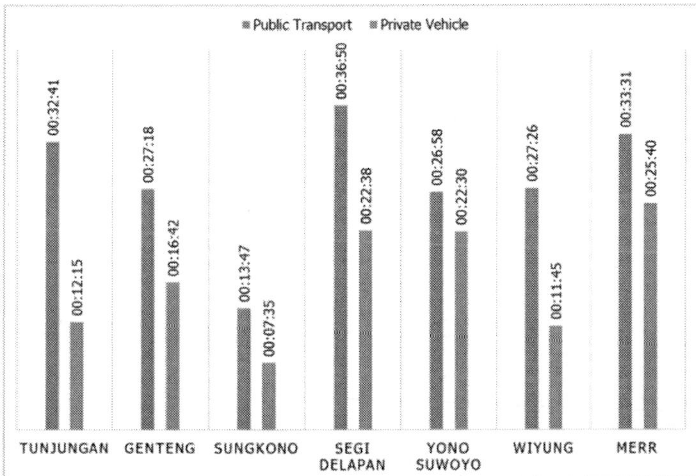

Figure 9. Total travel time from Joyoboyo TOD to Surabaya CBDs.

From the table 6 and figure 9 shows the total travel time of taking public transport (city transport/bus) from Joyoboyo TOD to Tunjungan CBD, in which endpoint coordinate number 3, was 32 minutes 41 seconds; Sungkono CBD (number 9) was 13 minutes 47 seconds; Yono Suwoyo CBD (number 11) was 26 minutes 58 seconds; Segi Delapan CBD (number 12) was 36 minutes 50 seconds; Genteng CBD (number 15) was27 minutes 18 seconds; MERR CBD (number 16) was 33 minutes 31 seconds; and Wiyung CBD (number 19) was 27 minutes 26 seconds. Table 6 also shows that the private vehicle had faster travel time to those 7 CBD points.

7. Conclusion

This research was designed to calculate the travel time from Joyoboyo Terminal, a TOD transit point, to Surabaya CBD. The results of the travel time survey show that private vehicles had better accessibility to Surabaya CBD. The difference travel time by public transport and a private vehicle in Tunjungan CBD was about 20 minutes 26 seconds; in Genteng CBD was about 10 minutes 36 seconds; Sungkono CBD was about 6 minutes 12 seconds; Segi Delapan CBD was about 14 minutes 12 seconds; Yono Suwoyo CBD was about 4 minutes 28 seconds; Wiyung CBD was about 15 minutes 41 seconds, and MERR CBD was about 7 minutes 51 seconds. This difference occurred because public transport needed extra time for waiting for bus/city transport arrival and walking time from the last bus stop. It is recommended for future research to find several scenarios to develop better accessibility from Joyoboyo Terminal, especially for public transport mode. Therefore, it can bring the success of 'ride' in the park and ride concept as well as 'transit' in transit-oriented development concept itself..

References

[1] Handayani K D M E and Ariastita P G 2016 Sustainability of transportation in Surabaya through the development of transit oriented development areas *J. Tataloka* **16** 108–15
[2] Hansen & SUTD Lab 2014 *Surabaya Urban Corridor Development Program* (Surabaya)
[3] Syabri I 2011 The Influence of Railway Station on Residential Property Values-Spatial Hedonic Approach The Case of Serpong's Railway Station *J. Teor. dan Terap. Bid. Rekayasa Sipil* **18** 291–300
[4] Robert C 2006 *Public Transport and Sustainable Urbanism : Global Lesson* (California: California Digital Library)
[5] Pradipta E G, Suroso and Suharini E 2014 Efektivitas BRT (Bus Rapid Transit) Trans Semarang Sebagai Moda Transportasi di Kota Semarang *Geo-Image* **3** 1–4
[6] Singh Y J, Lukman A, He P, Flacke J, Zuidgeest M and Maarseveen M van 2015 *Planning for Transit Oriented Development (TOD) Using a TOD Index* (Washington D.C.: Transport Research Board)
[7] Milosavljevic N and Simicevic J 2019 *Sustainable Parking Management* (Serbia: Elsevier)
[8] Spillar R J 1997 *Park and Ride Planning and Design Guidelines* (New York: Parsons Brinckerhoff Inc.)
[9] Black J A 1979 Urban Acessibility and Transport Policy *Transp. Commun. Bull. Asia Pacific*
[10] Nadyla A and Nurlaela S 2018 Pengukuran Tingkat Keseimbangan Node dan Place di Kawasan Transit Oriented Development (TOD) Terminal Joyoboyo, Surabaya *J. Transp.* **1** 100–5
[11] Nahdalina and Tamin O Z 1998 Analisis Dampak Lalu Lintas (Andall) *J. Perenc. Wil. dan Kota* **5** 16–25
[12] ITDP 2017 *TOD Standard* vol 3 (Jakarta: ITDP)
[13] Knoblauch R L, Pietrucha M T and Nitzburg M 1996 Field studies of pedestrian walking speed and start-up time *Transp. Res. Rec.* **1538** 27–38
[14] Auhari A 2010 *Pedestrian facilities planning on Dr. Setiabudhi street , Bandung* (Pasundan University)
[15] Yosritzal, Kemal B M, Purnawan and Putra H 2018 An observation of the walking speed of evacuees during a simulated tsunami evacuation in Padang, Indonesia *IOP Conf. Ser. Earth Environ. Sci.* **140** 1–8
[16] Nusdwinuringtyas N 2018 Six Minute Walking Distance Cut-off Point in Indonesian (Mongoloid) Population *Med. J. Indones.* **68** 389–94
[17] Yu W, Ai T and Shao S 2015 The analysis and delimitation of Central Business District using network kernel density estimation *J. Transp. Geogr.* **45** 32–47
[18] Murphy R E and Vance J E 1954 Delimiting the CBD *Econ. Geogr.* **30** 189–222
[19] Sun Y 2011 Development and characteristics of central business district under the philosophy of health *Procedia Eng.* **21** 258–66
[20] Battino S, Borruso G and Donato C 2012 Analyzing the Central Business District: The case of

IOP Conf. Series: Materials Science and Engineering **930** (2020) 012066 doi:10.1088/1757-899X/930/1/012066

Sassari in the Sardinia Island *Lect. Notes Comput. Sci. (including Subser. Lect. Notes Artif. Intell. Lect. Notes Bioinformatics)* **7334 LNCS** 624–39
[21] Taubenbock H, Klotz M, Wurm M, Schmieder J, Wagner B and Esch T 2013 Delimiting central business districts - A physical approach using remote sensing *Joint Urban Remote Sensing Event 2013* (Sao Paulo: IEEE) pp 17–20
[22] Pissourios I A 2014 A historical overview and critical analysis of town centre delimitation methodologies *Bull. Geogr.* **25** 155–65
[23] Yogi A 2017 *Central Business District (CBD) - Explained with Examples*
[24] Erawan A 2015 Segi Delapan Surabaya Barat, CBD Baru Kota Pahlawan *Rumah.com*

Valuing subjective cost of motorcyclists used willingness-to-pay in Surabaya

H Widyastuti[1*], A Utanaka[2]

[1]Department of Civil Engineering, Institut Teknologi Sepuluh Nopember, Surabaya, Indonesia
[2]Department of Civil Engineering, Universitas Sebelas Maret, Surakarta, Indonesia

[*]Corresponding author's e-mail: hera@ce.its.ac.id

Abstract. Motorcycles have increased for the last few decades and could increase more. Besides, the number of accidents is mostly dominated by motorcycles. It is caused by the motorcycles which stand on two wheels and are unstable while riding. Considering this problem, it is needed to know how the motorcyclists are willing to protect themselves from an accident. WTP method has been selected as this method which is commonly used in developed countries and recommended to be used in developing countries to calculate accident costs. This study used data obtained from the stated preference survey such as age, income, number of children, and willingness-to-pay choices. The equation model is obtained for each severity level by analyzing the data using SPSS software. The binary logit equation models show that an increase in age and income will improve the probability for motorcyclists to protect themselves from an accident, however when the number of children increases, it will reduce the probability. WTP value of motorcyclists on the risk reduction of serious injuries is 126,500,000 IDR, while WTP value of motorcyclists on the risk reduction of slight injuries is 13,000,000 IDR.
Keywords: accident, motorcycle, willingness-to-pay

1. Introduction

For the last few decades, the number of motorcycles has increased significantly in Indonesia. In 2018, the number of the motorcycles in Indonesia has reached 120,101,047 and could increase more [1–4], and even could exceed the population number [2]. Unfortunately, the increasing number of vehicles is usually followed with numbers of accidents, which also happened in Surabaya, which are dominated by motorcycles [5–8]. Moreover, it has been proved that motorcyclists have a high risk of having a traffic accident since the motorcycles stand on two wheels only [9–12]. Furthermore, the World Health Organization has informed that the accident is included in the top 10 causes of death in the world [13, 14]. Several factors such as human factor, vehicle factor, and environmental factor are factors that influence the occurrence of traffic accidents [15], [16]. Besides, one of the causes of motorcycle accidents is the lack of driving skills and awareness about traffic safety and its regulations [11].

Content from this work may be used under the terms of the Creative Commons Attribution 3.0 licence. Any further distribution of this work must maintain attribution to the author(s) and the title of the work, journal citation and DOI.
Published under licence by IOP Publishing Ltd

Concerning about the highest number of motorcyclists, the subjective cost which is expressing pain, grief and suffering as the result an accident is needed to be carried out to know how the motorcyclists are willing to protect themselves from an accident. Subjective costs are costs needed to cover the suffering and bereavement experienced by accident victims [10]. Willingness-to-Pay (WTP) method has been selected as this method which has been commonly used in developed countries and it is recommended to be used in developing countries to calculate accident costs [17, 18].

2. Method

Data used in this study are primary data obtained from stated preference survey. A survey was conducted to citizens of Surabaya as respondents. Several locations that were surveyed were the city park, food court, and university. The respondents were given some willingness-to-pay choices by hypothetical situations [19]. In this case, these choices are how much they will spend their money to reduce the accident probability.

2.1. Binary logit model

From the stated preference survey, there were obtained some data such as age, income, number of children, and willingness-to-pay choices which will be analyzed. Data were analyzed using SPPS (Statistical Package for the Social Sciences) software. Furthermore, the binary logit equation model has been obtained to express people's preferences on the reduction of accident probabilities. The general equation form of willingness-to-pay binary logit model can be seen in Equation 1. The significant level limits of the variables for the model used in this study are 0,1 where variables that have significance less than 0,1 still affect the equation model [20].

The utility function can be formulated as follows [18]:

$$U = V + \varepsilon = \beta' x + \varepsilon \tag{1}$$

U = the utility of willingness to pay the amount for severity reduction
V = the systematic (deterministic) component of the utility of willingness to pay the amount for severity reduction
ε = the random (disturbance or error) component of the utility of willingness to pay the amount for severity reduction
x = the vector of attributes that is related to the willingness to pay the amount for severity reduction
β' = the vector of unknown parameters

The probability of the binary model choice is estimated based:

$$p = \frac{exp^{logit(p)}}{1+exp^{logit(p)}} \tag{2}$$

In other word

$$p = \frac{probability}{1+probability} \tag{3}$$

2.2. Willingness-to-pay (WTP) value

There are two stages to determine the subjective cost of an accident: the calculation of the average value of the willingness-to-pay and cost estimation. The equation used is as follows [21]:
Stage 1 : Average value calculation of willingness-to-pay

$$\bar{W} = \sum_{n=1}^{K}\left[\sum_{i}^{N} P_n(i)W(i)\right]/K \tag{4}$$

Where :
$\bar{W}$ = individual willingness-to-pay average value
n = willingness-to-pay choices
K = total number of choices
N = individual total number in the database
i = each individual
$P_n(i)$ = alternative probability i chosen by individuals n
$W(i)$ = willingness-to-pay choices (yes or no)

Stage 2 : Value estimation of accident costs

$$VOCC = \frac{\bar{W}}{\beta} \tag{5}$$

Where :
VOCC = subjective cost as a result of an accident
$\bar{W}$ = average willingness-to-pay per individual to avoid accidents
β = changes in statistical risk

2.3. Willingness-to-pay (WTP) choices

The WTP choices in the form of reducing the risk probability for serious injuries and slight injuries are provided to be chosen by respondents. The choices provided are reductions of the accident risk probability by 25% and 50% for each severity level. The choice in the form of a percentage is equipped by the amount of money that must be spent by the respondents. The detail of WTP choices can be seen in table 1.

Table 1. Willingness-to-pay choices of the accident risk probability reduction for each severity level

Severity level	Additional cost		
	0% reduction	25% reduction	50% reduction
Serious injury	-	7,200 IDR	14,400 IDR
Slight injury	-	5,600 IDR	11,200 IDR

3. Result and Discussion

3.1. Respondent characteristics

There are variations in respondents based on the stated preference survey results. The characteristics grouping can be seen in figure 1 – figure 3. Figure 1 shows the lowest percentage is age category of more than 44 years old by 2.0% and the highest percentage is age category of 20 – 24 years old by 37.5%. Figure 2 shows the lowest percentage is income category of more than 6,250,000 IDR by 1.2% and the highest percentage is income category of 2,500,000 IDR – 3,750,000 IDR by 30.9%. Figure 3 shows the lowest percentage is number of children category of "have no children" by 61.3% and the highest percentage is number of children category of "have 4 children" by 37.5%.

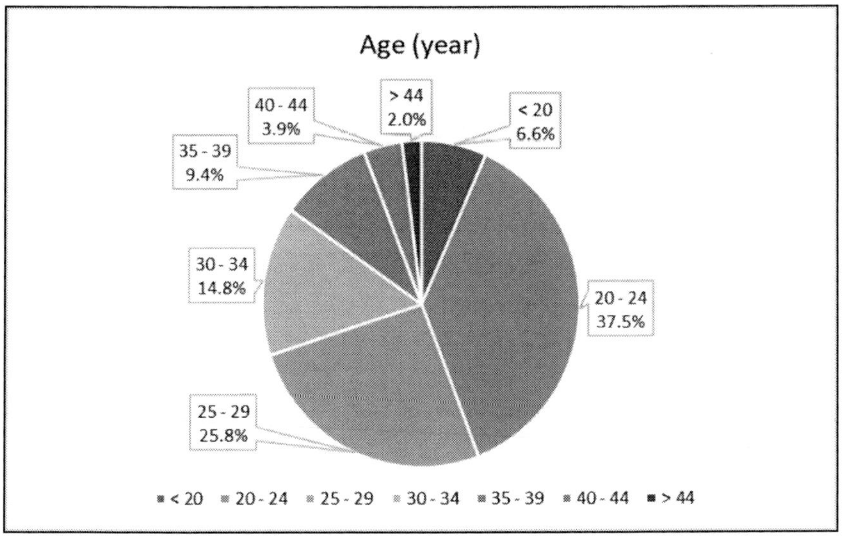

Figure 1. Respondent characteristics based on age

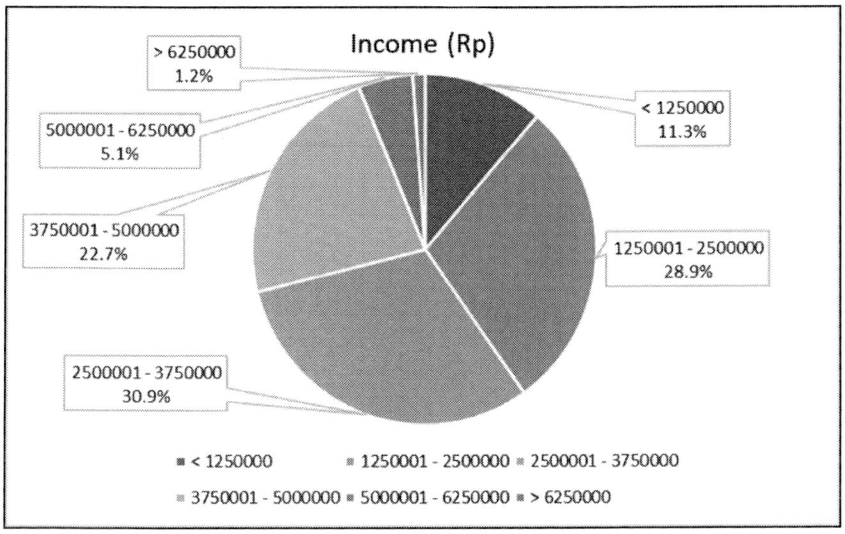

Figure 2. Respondent characteristics based on income

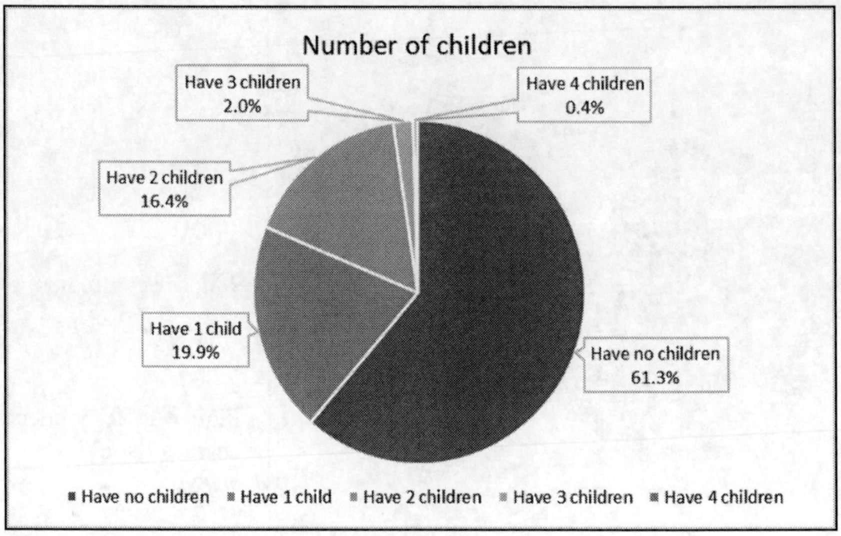

Figure 3. Respondent characteristics based on number of children

3.2. Binary logit model analysis

There are some data obtained from the stated preference survey, but the data that affect the binary logit model are age variable, income variable, and number of children variable from each respondent as in previous studies [8, 18, 22–25]. The three variables mentioned earlier (age, income, and number of children) were analyzed for the choice of WTP using the binary logit model by SPSS software. There are two binary logit models that have been analyzed for each severity category, which are reducing accidents probability by 50% and 75%. The results of the binary logit model analysis using SPSS software can be seen in table 2 – table 3.

Table 2. Results of SPSS output binary logit model for reducing the risk of serious injuries

	Reduction by 25%			Reduction by 50%		
Variable	Parameter	p-value	Exp (β)	Parameter	p-value	Exp (β)
Age	0.142	0.006	1.153	0.099	0.036	1.104
Income	0.393	0.073	1.481	0.376	0.066	1.457
Number of children	-1.429	0.000	0.240	-1.186	0.000	0.305
Constant	-2.307	0.046	0.100	-1.455	0.167	0.233

Table 3. Results of SPSS output binary logit model for reducing the risk of slight injuries

	Reduction by 25%			Reduction by 50%		
Variable	Parameter	p-value	Exp (β)	Parameter	p-value	Exp (β)
Age	0.121	0.019	1.129	0.091	0.047	1.095
Income	0.383	0.075	1.457	0.332	0.096	1.393
Number of children	-1.125	0.001	0.325	-1.152	0.000	0.316
Constant	-1.938	0.091	0.144	-1.192	0.245	0.304

Table 2 and table 3 show the SPSS output of binary logit biner analysis. Every variable has p-value less than 0,1 in which the variable is still significant to the equation model. The binary logit model equation for each willingness-to-pay choice can be written as follows:

- Binary logit equation model for reducing the risk of serious injuries by 25%

$$\text{Logit(p)}=\ln\left(\frac{p}{1-p}\right)= -2.307 + 0.142\,Age + 0.393\,Income - 1.429\,Number\ of\ children$$

- Binary logit equation model for reducing the risk of serious injuries by 50%

$$\text{Logit(p)}=\ln\left(\frac{p}{1-p}\right)= -1.455 + 0.099\,Age + 0.376\,Income - 1.186\,Number\ of\ children$$

- Binary logit equation model for reducing the risk of slight injuries by 25%

$$\text{Logit(p)}=\ln\left(\frac{p}{1-p}\right)= -1.938 + 0.121\,Age + 0.383\,Income - 1.211\,Number\ of\ children$$

- Binary logit equation model for reducing the risk of slight injuries by 50%

$$\text{Logit(p)}=\ln\left(\frac{p}{1-p}\right)= -1.938 + 0.121\,Age + 0.383\,Income - 1.211\,Number\ of\ children$$

From the 4 equation models that have been obtained, they show the similarity. It can be interpreted that every increase in age and income will increase willingness-to-pay log probability, but if the number of children increases it will reduce willingness-to-pay log probability. To prove the equation, 3 scenarios of variable change were performed. For scenario 1, the age variable is increased, but the income variable and number of children variable are constant. For scenario 2, the income variable is increased, but the age variable and number of children variable are constant. And for scenario 3, the number of children variable is increased, but the age variable and income variable are constant. The analysis result can be seen in table 4 – table 7.

Table 4. Probability scenario for the risk reduction of serious injuries in traffic accidents by 25%

Scenario	Age	Income	Number of Children	Logit p	Probability	Probability (%)
	20	3,000,000 IDR	1	0.283	0.570	57.0%
1	30	3,000,000 IDR	1	1.703	0.846	84.6%
	40	3,000,000 IDR	1	3.123	0.958	95.8%
	20	3,000,000 IDR	1	0.283	0.570	57.0%
2	20	4,000,000 IDR	1	0.676	0.663	66.3%
	20	5,000,000 IDR	1	1.069	0.744	74.4%
	40	3,000,000 IDR	1	3.123	0.958	95.8%
3	40	3,000,000 IDR	2	1.694	0.845	84.5%
	40	3,000,000 IDR	3	0.265	0.566	56.6%

Table 5. Probability scenario for the risk reduction of serious injuries in traffic accidents by 50%

Scenario	Age	Income	Number of Children	Logit p	Probability	Probability (%)
	20	3,000,000 IDR	1	0.467	0.615	61.5%
1	30	3,000,000 IDR	1	1.457	0.811	81.1%
	40	3,000,000 IDR	1	2.447	0.920	92.0%
	20	3,000,000 IDR	1	0.467	0.615	61.5%
2	20	4,000,000 IDR	1	0.843	0.699	69.9%
	20	5,000,000 IDR	1	1.219	0.772	77.2%
3	40	3,000,000 IDR	1	2.447	0.920	92.0%

	40	3,000,000 IDR	2	1.261	0.779	77.9%
	40	3,000,000 IDR	3	0.075	0.519	51.9%

Table 6. Probability scenario for the risk reduction of slight injuries in traffic accidents by 25%

Scenario	Age	Income	Number of Children	Logit p	Probability	Probability (%)
	20	3,000,000 IDR	1	0.506	0.624	62.4%
1	30	3,000,000 IDR	1	1.716	0.848	84.8%
	40	3,000,000 IDR	1	2.926	0.949	94.9%
	20	3,000,000 IDR	1	0.506	0.624	62.4%
2	20	4,000,000 IDR	1	0.889	0.709	70.9%
	20	5,000,000 IDR	1	1.272	0.781	78.1%
	40	3,000,000 IDR	1	2.926	0.949	94.9%
3	40	3,000,000 IDR	2	1.801	0.858	85.8%
	40	3,000,000 IDR	3	0.676	0.663	66.3%

Table 7. Probability scenario for the risk reduction of slight injuries in traffic accidents by 50%

Scenario	Age	Income	Number of Children	Logit p	Probability	Probability (%)
	20	3,000,000 IDR	1	0.472	0.616	61.6%
1	30	3,000,000 IDR	1	1.382	0.799	79.9%
	40	3,000,000 IDR	1	2.292	0.908	90.8%
	20	3,000,000 IDR	1	0.472	0.616	61.6%
2	20	4,000,000 IDR	1	0.804	0.691	69.1%
	20	5,000,000 IDR	1	1.136	0.757	75.7%
	40	3,000,000 IDR	1	2.292	0.908	90.8%
3	40	3,000,000 IDR	2	1.140	0.758	75.8%
	40	3,000,000 IDR	3	-0.012	0.497	49.7%

Based on table 4 – table 7, scenario 1 proves that with age increment, but income and number of children are constant, the probability of reducing a traffic accident risk will increase. Scenario 2 proves that with income increment, but age and number of children are constant, the probability of reducing a traffic accident risk will increase. Meanwhile, scenario 3 proves that with the number of children increment, but age and income are constant, the probability of reducing a traffic accident risk will decrease.

3.3. Willingness-to-pay (WTP) value analysis

Willingness-to-pay (WTP) value of the risk reduction of serious injuries by 25% is 121,120,670 IDR and willingness-to-pay (WTP) value of the risk reduction of serious injuries by 50% is 132,165,695 IDR. Hence, the average willingness-to-pay (WTP) value of the serious injuries risk reduction is 126,143,182 IDR or can be rounded to 126,500,000 IDR

Whereas, willingness-to-pay (WTP) value of the risk reduction of slight injuries by 25% is 13,069,431 IDR and willingness-to-pay (WTP) value of the risk reduction of serious injuries by 50% is 12,472,838 IDR. Hence, the average willingness-to-pay (WTP) value of the serious injuries risk reduction is 12,771,134 IDR or can be rounded to 13,000,000 IDR.

4. Conclusion

Variables that affect the WTP choices are age, variable, and number of children. Variables in each equation have a significance level less than 0,1 so that it is still acceptable and has an effect on the equation model. The four equation models that have been obtained show the similarity. It can be interpreted that increases in age and income will improve the probability for motorcyclists to protect themselves from an accident, however, when the number of children increases, it will reduce the probability. WTP value of motorcyclists on the risk reduction of serious injuries is 126,500,000 IDR, while WTP value of motorcyclists on the risk reduction of slight injuries is 13,000,000 IDR.

References

[1] Badan Pusat Statistik 2018 *Perkembangan Jumlah Kendaraan Bermotor Menurut Jenis, 1949-2018*. https://www.bps.go.id/linkTableDinamis/view/id/1133 (accessed Jul. 13, 2020).

[2] Kresnanto N C 2019 Model Pertumbuhan Sepeda Motor Berdasarkan Produk Dosmetik Regional Bruto (PRDB) Perkapita (Studi Kasus Pulau Jawa) *Media Komun. Tek. Sipil* **25** 1 107. doi: 10.14710/mkts.v25i1.18585.

[3] Pangestu F, Widodo A W and Rahayudi B 2018 Prediksi Jumlah Kendaraan Bermotor di Indonesia Menggunakan Metode Average-Based Fuzzy Time Series Models *J. Pengemb. Teknol. Inf. dan Ilmu Komput.* **2** 9 2923–92.

[4] Priyambodo 2018 Analisis Korelasi Jumlah Kendaraan dan Pengaruhnya Terhadap PDRB di Provinsi Jawa Timur *War. Penelit. Perhub.* **30** 59–65.

[5] Polrestabes Kota Surabaya 2017 *Data Kecelakaan Lalu Lintas Kota Surabaya.*

[6] Badan Pusat Statistik Kota Surabaya 2019 *Kota Surabaya Dalam Angka 2019* (Surabaya: Badan Pusat Statistik Kota Surabaya).

[7] Machsus, Prayogo I, Chomaedhi, Hayati D W and Utanaka A 2017 Road safety analysis on Achmad Yani frontage road Surabaya *IOP Conference Series: Materials Science and Engineering* **267** 1 1–11. doi: 10.1088/1757-899X/267/1/012027.

[8] Utanaka A and Widyastuti H 2019 Traffic Accident Cost Analysis Using Willingness-to-pay Method in Surabaya *Advances in Engineering Research* **186** Apte 2018 103–6.

[9] Rahmawati N and Widyanti A 2017 Comparison between motorcyclist' violation behavior and accidents in urban and rural area in Indonesia: A comparative study *IOP Conference Series: Materials Science and Engineering* **277** 1 012026. doi: 10.1088/1757-899X/277/1/012026.

[10] Widyastuti H and Mulley C 2005 The casualty cost of slight motorcycle injury in Surabaya, Indonesia *Transp. Commun. Bull. Asia Pacific* 75 57–73. [Online]. Available: http://www.unescap.org/sites/default/files/bulletin74_Article-4.pdf.

[11] Li Y *et al.* 2008 Motorcycle accidents in China *Chinese J. Traumatol. - English Ed.* **11** 4 243–6. doi: 10.1016/S1008-1275(08)60050-4.

[12] Mccarthy M G, Walter L K, Hutchins R, Tong R and Keigan M 2017 *Comparative Analysis Of Motorcycle Accident Data From OTS and MAIDS.*

[13] World Health Organization 2016 *The top 10 causes of death.* http://www.who.int/news-room/fact-sheets/detail/the-top-10-causes-of-death.

[14] World Health Organization 2019 *World Health Statistics Overview 2019* **23** 3 (Switzerland: WHO).

[15] Mohanty M and Gupta S 2015 Factors affecting road crash modeling *J. Transp. Lit.* **9** 2 15–9. doi: 10.1590/2238-1031.jtl.v9n2a3.

[16] Afolabi J A and Kolawole T G 2017 Road Traffic Crashes In Nigeria: Causes And Consequences *Int. J. Shipp. Transp. Logist.* **17** April 40–9.

[17] Widyastuti H and Mulley C 2005 Evaluation of Casualty Cost of Motorcyclist'S Slight Injury in Indonesia *J. East. Asia Soc. Transp. Stud.* **6** 3497–507. doi: 10.11175/easts.6.3497.

[18] Widyastuti H 2012 Valuing Motorcycle Casualties in Developing Countries using Willingness-to-Pay Method : Stated-Preference Discrete Choice Modelling Approach (University of Newcastle upon Tyne, Newcastle, England).

[19] Le H, Van Geldermalsen T, Lim W L and Murphy P 2014 Deriving accident costs using willingness-to-pay approaches - A case study for Singapore *ATRF 2011 - 34th Australasian Transport Research Forum* September 16.

[20] Figueiredo Filho D B *et al.* 2013 When is statistical significance not significant? *Brazilian Polit. Sci. Rev.* **7** 1 31–55. doi: 10.1590/s1981-38212013000100002.

[21] Widyastuti H, Dissanayake D and Bell M 2011 Value of Slight Injury Reduction (Case Study :

Surabaya - Indonesia), *Proceedings of the Eastern Asia Society for Transportation Studies, Vol.8, 2011* **8** 1–17.

[22] Ghee C, Silcock D, Astrop A and Jacobs G 1997 *Socio-economic aspects of road accidents in developing countries* (Crowthorne).

[23] Silcock B R and Transport Research Laboratory 2003 *Guidelines for Estimating the Cost of Road Crashes in Developing Countries*. doi: Project R7780.

[24] T R L 1995 *Overseas Administration Development*.

[25] Widyastuti H, Mulley C and Dissanayke D 2007 Binary choice model to value motorcyclist's slight injury cost in Surabaya *Proceedings of the Eastern Asia Society for Transportation Studies* **6**.

Student satisfaction with the performance of Surabaya school buses

M C P Lincoln[1] and H Widyastuti[2]

[1]Department of Civil Engineering, Institut Teknologi Sepuluh Nopember, Surabaya, Indonesia
[2]Lecturer of Civil Engineering Department, Sepuluh Nopember Institut of Technology, Surabaya, Indonesia

*Corresponding author's e-mail: mardhika.18031@mhs.its.ac.id

Abstract. School transportation facilities in the city of Surabaya are increasingly being upgraded to serve the needs of education. In 2019, there have been 7 School Bus units operating in several points in South, East, North and West Surabaya. The addition of this mode is intended to fulfill and facilitate the journey of the citizens of Surabaya to school; in addition, to reduce the number of students using private vehicles to school. However, to find out the interests of school bus users, it is necessary to know the performance of school bus services and passenger satisfaction. This study aims to measure the performance and service of school buses, as well as the level of student satisfaction. Data collection in this study uses several methods, namely questionnaire survey, observations and interviews. The data were analyzed using two methods, namely descriptive analysis with standard public service calculations and the Importance Performance Analysis method. Based on the results of the analysis, the performance of the Surabaya School Bus was categorized as "Good". At the level of satisfaction, there are 5 attributes considered important, 7 attributes have good performance, 1 attribute is not prioritized, and 3 attributes have good performance but are not considered important.
Keywords: performance, satisfaction, Importance Performance Analysis

1. Introduction

Transportation problems are still a trending issue in several big cities like Surabaya. The reason is that the increasing population growth will affect the demand for transportation modes as a means of mobility, so that it will also affect the density of motorized vehicles. The increasing number of motorized vehicles in urban areas has the potential to cause various traffic problems such as accidents and congestion on various roads. The Surabaya City Government continues to improve services to ensure the availability of safe and comfortable transportation for the community to reduce these problems. Providing public transportation services can be a solution to unravel congestion and reduce accident rates. However, the provision of public transportation services will not have a significant impact without being supported by the optimization of community use. Thus, public interest in the use of public transportation must be encouraged. The large number of modes of transportation needs of school children in various cities such as in Surabaya has caused an increase in the density of motorized vehicles at certain times. Therefore, it needs government efforts to provide public transportation services specifically for school children such as school buses. However, the quality of school bus service quality must always be improved to provide facilities for students to use school buses as one of the expectations of students who are not old enough to use motorbikes. The implementation of the free school bus transportation service must pay attention to several indicators based on Minister of Transportation Regulation No. 29 of 2015 consisting of Security, Safety, Comfort, Equality, and Regularity. Passenger satisfaction is one important component in transportation services, so that the school bus service organized by the government can run according

Content from this work may be used under the terms of the Creative Commons Attribution 3.0 licence. Any further distribution of this work must maintain attribution to the author(s) and the title of the work, journal citation and DOI.
Published under licence by IOP Publishing Ltd

to the expectations and needs of the community.

Actually, public transportation services in the city of Surabaya have increased. In addition to the city bus which has been operating for a long time, the Surabaya city government has added public transportation facilities in the form of 20 units of Suroboyo Bus and 10 units of school bus. Improving this transportation service can gradually unravel congestion and reduce the number of accidents in the city of Surabaya. However, it must be balanced with public interest in the use of public transportation. The rise and pull of land use, especially in educational facilities, is one of the activities that often causes traffic jams. Based on data released by the Surabaya City Transportation Agency in 2016, the use of private vehicles as a mode of transportation to schools in the city of Surabaya is 88%, while 12% of students uses public transportation modes.

Users of private transportation mode go to school for junior high school students more than public transportation. However, the results of his research revealed the willingness of respondents to use school buses as a mode of transportation at Wijaya Kusuma road area if school buses were provided is as much as 56.9% of 4,195 students [1]. These data can also be strengthened by the results of research which state that 71% of the respondents surveyed at Jl. Governor Suryo and Jl. Prof. Dr. Moestopo is also willing to use the school buses as a mode of transportation to school, and some of those who use the school bus are satisfied with the service [2]. That is, there are still many students who are willing to use the school bus if it is available on the route they are traveling on.

The delivery of free school bus services in Surabaya by the city government is one form of good governance because school transportation is a very important community need that must be met. However, in terms of operation and management of the school bus, the government as the manager must pay attention to the quality of service and passenger satisfaction because the choice of transportation mode can be known from how good the quality of the service is, to the increasing user satisfaction.

2. Study Area Data

School buses in Europe are seen as one of the safest modes of transportation because they are specifically designed to protect students from road hazards. Statistics show that students who travel by private vehicles are seven times more likely to be involved in a traffic accident than those who use bus transportation [3]. In addition, the National Highway Traffic Safety Administration (NHTSA) compares the number of deaths of children aged five to eight years in the United States caused by private cars and school buses, and 87 times school buses are safer than private cars. In Greece, school bus transportation is divided into three categories, namely private school buses that serve elementary school students (door-to-door service), private school buses that serve middle school students (pickup as parents wish), and public school buses that serve students primary and secondary schools (stopping points are determined by schools and in collaboration with transport providers) [4]. There are special rules that regulate and supervise the school bus transportation system such as the maximum speed, the provision of seat belts, the presence of officers on the bus, and others [5].

Unlike the public school transportation system, that of the private school is more organized. It is proven that school students in the private sectors are served inter-bridge transportation from home, while students in public schools must go to certain places or stops to wait for the school bus pickup. In addition, all private primary schools provide the services of school bus officers who are responsible for student safety. This service does not exist in public primary schools [6].

There are five dimensions of quality of public transport services namely, Tangibility, Reliability, Responsiveness, Assurance, Empathy [7].

3. Methodology

This research data collection method uses a questionnaire survey, observation and interview. The data collected from the results of the questionnaire survey will be analyzed using two methods. First, the performance of school bus services will be analyzed using descriptive analysis with standard public service calculations which include load factor, speed, headway, travel time, service time, frequency, number of vehicles operating, waiting time, and end of trip. Second, Surabaya School Bus Passenger satisfaction will be analyzed using the Importance Performance Analysis (IPA) Technique method to determine the effect between passenger satisfaction and expectations of the school bus. This IPA method has two stages of analysis, i.e., looking for conformity prices and Cartesian charts.

3.1 Performance Analysis of School Bus Services

School bus service performance was analyzed using descriptive analysis with Public Transport Service Standard parameters. School bus service performance data obtained from observations and interviews are first analyzed by load factor, travel time, travel speed, and service frequency. From the load factor analysis, travel time, travel speed, and service frequency, it can be seen that the analysis value is lower, equal, or higher than the standard indicators of public transport services that have been determined by the Director General of Land Transportation in table 1.

Table 1. Standard Indicators for Public Transport Services

Number	Parameter	Unit	Assessment Standards		
			Bad (1)	**Average (2)**	**Good (3)**
1	Load factor busy time	%	>100	80 – 100	<80
2	Load factor busy outside Rush Hour	%	>100	70 – 100	<70
3	Speed Traveling	KM / hour	<5	5 – 10	>10
4	Headway	Minute	>15	10 -15	<10
5	Travel Time	Minute / km	>12	6 – 12	<6
6	Service Time	Hour	<13	13 – 15	>15
7	Frequency	vehicle / hour	<4	4 -6	>6
8	Number of Vehicles	%	<82	82 – 100	100
9	Waiting Time	Minute	>30	20-30	<20
10	Start and end of trip		05.00-18.00	05.00-19.00	05.00-20.00

Source: Director General of Land Transportation 2002

All assessments are added up and the quality of service is assessed using table 2 below:

Table 2. Performance Standards for Public Transport Services

Criteria	Total Value
Good	18 – 24
Average	12 – 17,9
Bad	< 12

Source: Director General of Land Transportation 2002

3.1.1 Load Factor

Load factor is the ratio between the number of passengers and the capacity of public transport vehicles with the following formula:

$$\text{LF max} = \text{LF}_{\text{max}} = \frac{\textit{Number of passengers transported}}{\textit{Passenger seating capacity}} \text{ x } 100\ \% \qquad (1)$$

The load factor calculation is done by calculating the capacity of passengers transported in each segment from the starting point of departure to the end. The calculation of each segment is carried out to anticipate changes in the number of passengers up and down at each stop. Then, an average load factor is calculated to determine the load factor from the starting point to the end.

3.1.2 Travel speed

Urban public transport travel speed is the ratio of operating distance and the travel time required by the transport to carry out its service operations. The equations used in measuring travel speed are:

$$V = \frac{S}{T} \tag{2}$$

Where:
V = Travel speed of public transport (km / hour)
S = Distance of public transport routes (km)
T = Travel time for public transport (minutes)

3.1.3 Travel time

Travel time is used to measure the travel time of a public transportation per kilometer traveled. Travel time can be calculated using the equation:

$$\text{Travel Time} = \frac{\textit{Travel Length}}{\textit{Route Length}} \tag{3}$$

Frequency is the number of vehicles operating within 1 hour. Frequency calculation uses the following formula:

$$f = \frac{N}{60} \tag{4}$$

f = frequency (number of vehicles per minute)
N = number of vehicles (unit)

3.2 Analysis of passenger satisfaction with school bus services

Satisfaction questionnaire is designed based on the satisfaction dimension which consists of reliability, responsiveness, assurance, empathy, and tangibility. The survey in this study used the revealed preference method. This method is a real expression based on the experience that has been felt by respondents and passenger expectations. This satisfaction analysis is to find out the level of satisfaction of performance, the level of importance or expectations of passengers, and the level of correspondence between the level of satisfaction and importance. Satisfaction and importance are analyzed using the Importance Performance Analysis (IPA), where there are two stages in working on this method, namely looking for conformity prices and Cartesian charts. The formula used is as described below.

3.2.1 Looking for Price Conformity (with IPA analysis)

$$Tki = \frac{Xi}{Yi} \text{ x } 100\% \tag{5}$$

Where :
Tki = Level of fitness of respondents.
Xi = School bus performance assessment scores
Yi = Scores for the importance of school bus users

The horizontal axis (X) will be filled by the performance level score, while the vertical axis (Y) will be filled by the importance level score. In simplifying the formula, each attribute that affects customer satisfaction can be determined by the formula:

$$\overline{X_i} = \frac{\sum Xi}{n} \tag{6}$$

$$\overline{Y_i} = \frac{\sum Yi}{n} \tag{7}$$

Where :
$\overline{X_i}$ = The average score of each variable i at the level of performance.
$\overline{Y_i}$ = The average score of each variable i at the level of importance.
$\sum Xi$ = The total score of each variable i at the level of implementation of the respondents
$\sum Yi$ = The total score of each variable i at the level of importance of all respondents
N = Total Respondents

3.2.2. Cartesian Diagram

$$\overline{\overline{X}} = \frac{\sum_{i=1}^{n} \overline{X_i}}{k} \tag{8}$$

$$\overline{\overline{Y}} = \frac{\sum_{i=1}^{n} \overline{Y_i}}{k} \tag{9}$$

Where :
$\overline{\overline{X}}$ = Average of the total weighted implementation level.
$\overline{\overline{Y}}$ = Average of the total weighted importance level.
K = Number of variables specified

4. Results and Discussion

4.1 School Bus Service Performance Analysis
The seven school buses that operate serving students at the four departure points have a capacity of 35 passengers with 25 seats, 10 stands. This load factor analysis is to measure the performance of school bus services based on the 2002 Director General of Land Transportation provision regarding public transport service standard indicators [8]. Load factor calculation is carried out by analyzing seven school buses at four points. This calculation is based on the results of counting each segment on each route to find out the number of load factors for each segment because there is a change in the amount of ups and downs from the start point of departure to the end. Therefore, to find out the number of load factors on each route, the average load factor of each segment is calculated. The results of the load factor analysis of seven school buses produce the following:

Table 3. Total recap of average load factor

Number	Bus Code	Route	Load Factor Recap	
			Depart	Return
1	Bus 02	Transportation Department - SMKN 5	0,95	0,82
2	Bus 10	Transportation Department - SMPN 1	0,87	0,81
3	Bus 03	Rungkut sub district office - SMPN 1	0,69	0,56
4	Bus 07	Rungkut sub district office - SMPN 1	0,79	0,50
5	Bus 05	Tandes sub district office - SMPN 1	0,80	0,72
6	Bus 09	Tandes sub district office - SMPN 1	0,92	0,84
7	Bus 08	Romokalisari Flat House - Barata Jaya	0,64	0,25
		Average	**0,81**	**0,64**

Based on the standard indicators of public transport services established by the Director General of Land Transportation in 2002 which explains that the assessment standard is said to be "insufficient" if the load factor at peak hours and outside rush hour is above 100%, the assessment standard is said to be "medium" if the load factor at rush hour between 80% - 100% and outside rush hour between 80% - 100%, and the standard assessment is said to be "good" if the load factor at rush hour is below 80% and outside rush hour is below 70%. By looking at the details of the recapitulation of the load factor analysis above, it can be concluded that there is a load factor that is in the "good" category, namely Bus 03 and 07 of the Kec. Rungkut - SMPN 1, Bus 05 route Kec. Tandes - SMPN 1, and Bus 08 route Rusun Romokalisari - SDN Baratajaya. Meanwhile, Bus 02 and 10 of the Dishub Office route - SMKN 5, Bus 10 of the Dishub Office route - SMPN 1, and Bus 09 Office route Kec. Tandes - SMPN 1 is in the "medium" category because the value of the load factor reaches above 80%. The average load factor of all school buses is in the category of "moderate" with a load factor of 81% at the hours of departure, and the category of "good" at the home hours with a load factor of 64%.

4.2 Travel Speed Analysis
From secondary data obtained from the Surabaya City Transportation Department school bus, it is known that bus travel speed is as in table 4 below:

Tabel 4. Travel Speed

Number	Route	Bus Number	Distance (Km)	Travel Time (Hours)	Speed (Km / hr)
1	Transportation Department	02	18.65	0.633	29
2	Transportation Department	10	21.55	0.65	33
3	Rungkut SubDistrict Office	03	19.3	0.76	25
4	Rungkut SubDistrict Office	07	19.5	0.67	29
5	Tandes Sub-District Office	05	16	0.57	28
6	Tandes Sub-District Office	09	14.9	0.48	31
7	Romokalisari Flat House	08	63.55	1.57	40
		Average			**31**

The analysis results from the survey data above show the average speed of the bus is 31 km / hour, and from the table, it can be seen that the highest speed occurs in Bus no. 08 route from Romokalisari Flat to SDN Baratajaya with an average speed of 40 km / hour. The departure route of the Romkalisari Flat starting point is the longest route, the length of which is 63.55 km. Meanwhile, the lowest speed is 25 km / h on Bus no. 03 route from Rungkut District to SMAN 1 Surabaya with a route length of 19.3 km.

4.3 Headway Analysis
Headway time is the departure interval between one bus and the next bus which is calculated in time units at a certain point on each route. Headway is one thing that affects the service level. Headway arrangement results in passenger transportation. A headway that is too low will result in a capacity that exceeds demand because the bus arrival rate will be greater than the rate of the arrival of passengers. Meanwhile, the headway that is too high can result in too long waiting times for passengers. However, in the Surabaya School Bus study, there was no Headway because all buses depart at the same time at 05.30 WIB, except Romokalisari buses depart at 05.00 WIB.

4.4 Travel Time
Travel time from the initial segment to the final segment, the time required to travel including the time of lowering and raising passenger units used minutes / km. Therefore, the amount of travel time is as

follows:

Table 5. Total recap of average travel time

Number	Bus Code	Route	Recap of Average Travel Time	
			Depart	**Return**
1	Bus 02	Transportation Department - SMKN 5	2,42	2,08
2	Bus 10	Transportation Department - SMPN 1	2,62	2,32
3	Bus 03	Rungkut SubDistrict Office - SMPN 1	2,68	2,14
4	Bus 07	Rungkut SubDistrict Office - SMPN 1	2,51	1,93
5	Bus 05	Tandes Sub-District Office - SMPN 1	2,76	2,85
6	Bus 09	Tandes Sub-District Office - SMPN 1	2,41	2,21
7	Bus 08	Romokalisari Flat House – SDN Baratajaya	2,09	1,80
		Average	**2,50**	**2,19**

The results of the analysis of the travel time from the table above, the average travel time is of 2.50 minutes / km with the fastest time value seen on bus No. 08 Romokalisari flat departure route with a route length of 63.55 km and a trip length of 1 hour 57 minutes, so that it is obtained travel time of 2.09 minutes / km, while the lowest travel time can be seen on bus no.05 with the original departure route from Tandes District with a route length of 15.99 km, 34 minutes of travel time so that the bus travel time is 2.76. On the return bus, the average travel time is 2.19 minutes / km with the fastest time value seen on the bus no. 08 Romokalisari route, 42 minutes travel time and 1.8 minutes / km travel time, with a 25 km return route length. Meanwhile, the lowest bus on the return journey can be seen on bus No. 05 with the initial departure route from Tandes District with a travel time of 24 minutes, a route length of 12.81 km and a travel time of 2.85 minutes. From table 5, it can be seen as a whole that the average travel time when departing is greater than the return trip time.

4.5 Service time

Long school bus service times in operation per day indicate a poor performance. The standard commonly used for branch and local routes is not less than 15 hours per day for good categories and for categories less than 13 hours. Surabaya school bus services start at 05:30 until 16:30 so that the school bus service time is 12 hours.

4.6 Frequency of service

Frequency is the number of vehicles operating within 1 hour.

Table 6. Vehicle Frequency

Number	Route	Bus Number	Number of Buses	Minute	Frequency
1	Menanggal	02 and 10	2	60	0,03
2	Kecamatan Rungkut	03 and 07	2	60	0,03
3	Kecamatan Tandes	05 and 09	2	60	0,03
4	Rusun Romokalisari	08	1	60	0,02
		Average			0,03

Based on the table above, the average frequency count results of the number of schools that pass every hour are 0.03.

4.7 Analysis of Passenger Satisfaction with Surabaya Free School Bus Services

Measurement of satisfaction and service quality are used to study the perceptions of each customer towards the desired service quality, find out the needs and expectations, improve service quality in accordance with expectations, and improve service quality [9].

Analysis of passenger satisfaction with the services of Surabaya school buses refers to five dimensions of service quality namely, reliability, responsiveness, assurance, empathy and tangibilitys [10]. The survey results based on the five dimensions are analyzed using Importance Performance Analysis (IPA) and illustrated with Cartesian diagrams to measure the level of satisfaction of someone on the performance of the school buses by comparing the level of expectations with the performance of the school bus manager.

Previously, it was known the level of conformity or the results of the comparison of satisfaction and expectation scores to find out whether the performance of the Surabaya school buses was in line with passenger expectations or not. In this level of conformity analysis, the letter Y represents expectations and the letter X represents satisfaction. Following are the results of the suitability calculation.

Table 7. Results of Calculation of Conformity Level

Criteria	Satisfaction (X)	Hope (Y)	Xi	Yi	Tki
Punctuality	4,25	4,44	901	941	96%
Suitability of Operating Hours with Needs	4,02	4,52	853	959	89%
Distance from home to bus stop	4,05	4,45	858	944	91%
Travel time from home to bus stop	3,94	4,39	835	930	90%
Driver behavior	4,17	4,08	883	866	102%
Travel time to school	4,15	4,47	880	947	93%
Passenger Comfort	3,88	4,46	822	946	87%
Ease of getting a seat	3,72	4,47	788	947	83%
Waiting time for the next bus	4,09	4,32	868	916	95%
Order	4,24	4,56	898	967	93%
Driver's courtesy / friendliness	4,29	4,26	909	904	101%
Security	4,10	4,47	870	947	92%
Driver alertness	3,84	3,91	814	829	98%
Cleanliness	4,21	4,46	892	945	94%
Bench condition	3,78	4,58	802	970	83%
First Aid Kit	3,42	4,64	726	984	74%
Safety equipment	3,81	4,63	808	982	82%
Total	**67,96**	**75,11**	**14.407**	**15.924**	**90%**
Average	**4,00**	**4,42**	**847**	**937**	**90%**

Source: Calculation results

Based on the results of the calculation of the suitability level above, there are 10 criteria that have a score of expectation or level of importance higher than the satisfaction of school bus performance, that is, P3K completeness with an expectation score of 4.64% and a respondents' suitability level of 74%, completeness of safety equipment 4.63 % and the respondents' suitability level of 82%, the next bus waiting time is 4.32% and the respondents' suitability level is 95%, the distance of the house to the bus stop is 4.45% and the respondents' conformance level is 91%, the driver's readiness is 3.91% and the respondents' compliance level is 99% , 4.56% Order and 93% respondents' conformity level, 4.44% Accuracy and 96% respondents' correspondence level, Operating Hours Compliance with Needs

4.52%, and 89% respondents' conformity level, Travel time to school 4.47% and the suitability level of respondents 93%, Security 4.47% and the level of conformity of respondents 92%, Cleanliness 4.46% and the suitability level of respondents 94%. Travel time from home to bus stop is 4.39% and respondent's compliance level is 90%, Passenger Comfort is 4.46% and respondent's compliance level is 87%, Accessibility is 4.47% and respondent's compliance level is 83%, Cleanliness is 4.46 % and the level of conformity of respondents 94%, Bench conditions 4.58% and the level of conformity of respondents 83%. Meanwhile, there are 2 criteria that have a higher performance satisfaction score than expectations, namely, driver behavior 4.17% and the respondents' suitability level 102%, driver's courtesy / friendliness 4.29% and the respondents' suitability level 101%. However, overall, the average expectation value is higher than the performance with a satisfaction score of 4.00% compared to 4.42% expectation and the respondents' conformity level 90%. However, the percentage of suitability between the performance and passengers of the Surabaya school buses can be said to be good because it is approaching 100% with a gap of 10%.

Then, the level of passenger satisfaction with the performance of Surabaya school buses can be seen in figure 1. The following are the results of the Importance Performance Analysis (IPA) analysis of the level of passenger satisfaction with school bus services using the Cartesian diagram in the SPSS 23.0 Program. In the Cartesian diagram, the quadrant boundary is on the X axis = 4.00 and the Y axis = 4.42, so the following results are obtained:

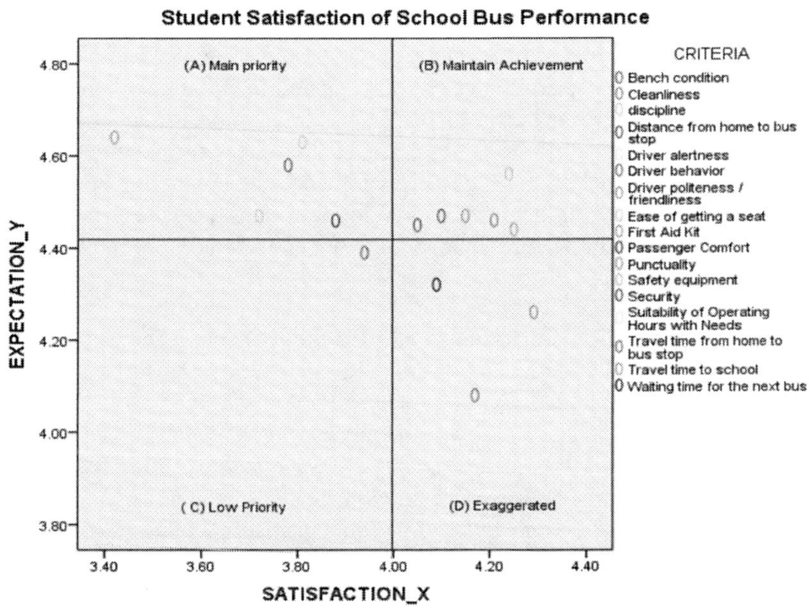

Figure 1. Quadrant Analysis of School Bus Passengers with SPSS 23.0

Based on the Cartesian diagram above, it can be seen that there are 5 variables in quadrant A. Thus, these variables are among the top priorities that must be considered by Surabaya school bus managers because they have high importance. The 5 variables are passenger comfort, ease of getting a seat, bench condition, first aid kit, safety equipment. On the other hand, awareness of B, there are 7 variables namely, timeliness, conformity of operating hours with needs, distance of the house to the bus stop, travel time to school, order, safety, and cleanliness. Quadrant B can be interpreted that all the variables in the quadrant have a good performance, and the respondents are satisfied. Therefore, it must be maintained because it matches the expectations or interests of Surabaya School Bus passengers.
In addition, there are 2 variables included in quadrant C and 3 variables in quadrant D. Two variables included in quadrant C are travel time from home to bus stop and driver alertness. Driver alacrity

variable is not prioritized by the respondents. However, the travel time from home to this stop variable is in quadrant C which is adjacent to quadrant A with an expected score of 4.39%, higher than the satisfaction value of 3.94%. Furthermore, the variables included in quadrant D are driver behavior, waiting time for the next bus, and driver courtesy / friendliness. These variables in Quadrant D have a good level of satisfaction or performance, but are considered not important by the respondents.

5. Conclusions

Research on Student Satisfaction Analysis of School Bus Performance produces two analyzes and discussions on School Bus performance and passenger satisfaction with school buses. The results of this study can be concluded that there are 2 categories of load factors on Surabaya School Buses if referring to the standard indicators of public transport services established by the Director General of Land Transportation in 2002. The "good" category is on Bus 03 and 07 of Kec. Rungkut - SMPN 1, Bus 05 route of Kec. Tandes - SMPN 1, and Bus 08 route of Rusun Romokalisari - SDN Baratajaya. Meanwhile, Bus 02 and 10 of the Dishub Office route - SMKN 5, Bus 10 of the Dishub Office route - SMPN 1, and Bus 09 route of Kec. Tandes - SMPN 1 are in the "medium" category, however, the average load factor of all school buses is in the "medium" category. Surabaya school bus speed reaches > 10 km / hour, the number of vehicles operating is of 70%, and waiting time is <20 minutes, so that it belongs to the category of "good". Meanwhile, the average service time, vehicle frequency, and the beginning of the trip are included in the category of "less". There are 5 attributes of satisfaction that are considered important by the respondents to be improved, namely: passenger comfort, ease of getting a seat, bench conditions, first aid kit, and safety equipment completeness. Seven (7) attributes that are considered to have good performances and are maintained, namely: timeliness, operational suitability with needs, distance from home to bus stops, travel time to school, order, safety, and cleanliness. The driver's alertness attribute is not prioritized by the respondents. Attribute travel time from home to the bus stop is normal because the average is close and not too far from the house. Finally, the attributes of the driver's behavior, the next bus waiting time, and the driver's courtesy / friendliness have a good level of satisfaction or performance, but are not considered important by the respondents.

References

[1] Sambodja R S, Widyastuti H dan Herijanto W 2015 Studi Demand and Supply Bus sekolah untuk Siswa Menengah Pertama dan Atas di area Jalan Wijaya Kusuma Surabaya *Prosiding Seminar Nasional Aplikasi Teknologi Prasarana Wilayah (*Surabaya: 11 Juni 2015S).

[2] Sandy B D A dan Widyastuti H 2016 *Evaluasi kinerja, tingkat kepuasan penumpang, demand bus sekolah untuk siswa sekolah menengah atas di Kota Surabaya area Gubernur Suryo dan Jl. Prof. Dr. Moestopo* (Surabaya: Institut Teknologi Sepuluh Nopember).

[3] European Commission 2004 *Road safety in school transport study – Final report* (Brussels) retrieved January 2016 from http://ec.europa.eu transport/roadsafety_library/publications/rsst_final_report_v1.3.pdf

[4] Sakellariou A, Kotoula K M, Morfoulaki M and Mintsis G 2017 Identification of quality indexes in school bus transportation system *Transportation Research Procedia*. https://doi.org/10.1016/j.trpro.2017.05.110.

[5] Wang J T J, Li J S, Chiu W T, Chen S H, Tsai S D, Yu W Y, Liao C C dan Choy C S 2009 Characteristics of bicycle-related head injuries among school-aged children in Taipei area *Surgical Neurology*. https://doi.org/10.1016/j.surneu.2009.07.035.

[6] Sakellariou A, Kotoula K M, Morfoulaki M and Mintsis G 2017 Identification of quality indexes in school bus transportation system *Transportation Research Procedia*. https://doi.org/10.1016/j.trpro.2017.05.110.

[7] Lupiyoadi, Rambat dan Hamdani A 2006 *Manajemen Pemasaran Jasa* (Jakarta: Salemba Empat).

[8] Petunjuk Teknis Penyelenggaraan Angkutan Penumpang Perkotaan 2002 Direktorat Jendral Perhubungan Darat (Republik Indonesia).

IOP Conf. Series: Materials Science and Engineering **930** (2020) 012068 doi:10.1088/1757-899X/930/1/012068

[9] Kotler P, Hamlin M A, Rein I dan H D H 2002 *Marketing Asian Places* (Singapore: John Wiley & Sons).
[10] Lupiyoadi, Rambat dan Hamdani A 2006 *Manajemen Pemasaran Jasa* (Jakarta: Salemba Empat).

IOP Conf. Series: Materials Science and Engineering **930** (2020) 012069 doi:10.1088/1757-899X/930/1/012069

Public transportation in Jabodetabek: performance satisfaction analysis

E N Jannah[1], A N H Ibrahim[1], and M N Borhan[1*]

[1] Department of Civil Engineering, Faculty of Engineering & Built Environment, Universiti Kebangsaan Malaysia, 43600 UKM Bangi, Selangor, Malaysia.

[*]Corresponding author's e-mail: mnazri_borhan@ukm.edu.my

Abstract. A chronic traffic congestion in JABODETABEK or Greater Jakarta has become a problem that needs to be addressed by the government. The existing public transportation system is associated with poor vehicle condition, poor infrastructure for non-motorized modes, poor customer service and uncomfortable traveling conditions. This study focused on Kereta Rel Listrik (KRL) in JABODETABEK. This study also aims to evaluate the level of passengers' satisfaction towards KRL performance and to identify what needs to be improved on public transportation to attract private transportation users into using public transportation while maintaining its existing passengers. The research was conducted using quantitative method. Questionnaires were distributed to 390 passengers to gather empirical data. Data analysis used to examine the effect of satisfaction on the performance of public transportation is Structural Equation Modelling (SEM) using AMOS 22.0. Result shows that service quality has positive relationship towards perceived value and satisfaction, but insignificantly negative to behavioural intention. On the other hand, perceived value is significantly positive towards behavioural intention, but it has a negative relationship with satisfaction. The satisfaction towards behavioural intention has a positive relationship. The findings in this study may help to improve the performance of existing Kereta Rel Listrik in JABODETABEK.

1. Introduction

In this era, with constantly increasing population and world economy growth, the demands on transportation are also on the rise. The dependency on vehicles by individuals led to its extensive use. People who mostly stay in cities and prefer to drive their own vehicles relegating it to be the main factor of urban congestion. The space requirements for the automobile, which are much greater than any other transportation mode, aggravated the traffic congestion problem, especially in medium-sized and big cities [1, 2].

Public transportation is a sustainable, reasonable alternative to private car use [3, 4]. It is necessary to summarize information concerning what drives customer satisfaction and dissatisfaction on public transportation, so that a more marketable and attractive public transportation can be designed based on it. This study focused on JABODETABEK public transportation where the number of private vehicles is too high. It is widely known that public transportation is essentially popular amongst children, adolescents, and elderly. Less obvious is the reliance of the city for the existing capacity of public transportation are limited only for those who do not have private vehicles; people are slowly starting to realize that it is important to seriously find ways of reducing the use of private vehicles and that will involve some opting into public transportation.

Content from this work may be used under the terms of the Creative Commons Attribution 3.0 licence. Any further distribution of this work must maintain attribution to the author(s) and the title of the work, journal citation and DOI.
Published under licence by IOP Publishing Ltd

Utilization of public transportation systems in Jakarta is the main key solution in reducing the heavy traffic congestion that requires the need to understand the public's satisfaction in relation to the public transportation itself. Previous works have reported the crucial of understanding the passengers' perceptions on the quality of service provided. Service quality reflects the passenger's perception of public transportation performance and it depends to a great extent on the operating decisions made by a service providers, particularly on decisions of how long and how often it is to be provided, where transit service should be provided, and what kind of services the passengers needed [5]. Understanding the passenger's satisfaction may be effective to meet the passenger needs, retain the existing public transportation passengers, while attracting more people to switch from private transportation into public transport. In addition, according to Cunningham et al. [6], the accurately assessing passenger's perception towards public transportation service quality is the vital indicator for service quality improvement. Thus, this study was conducted to explore the passenger's perception towards public transportation which focused on Kereta Rel Listrik services in order to formulate the effective strategies to increase Kereta Rel Listrik service quality in JABODETABEK as well as enhance the passenger's satisfaction towards the service.

2. Literature review and hypotheses formulation

2.1. Service quality (SQ)
Service quality (SQ) is essentially a set of properties of services that meets passenger's expectation and as flaws are eliminated, enhances the passenger's satisfaction [7, 8]. Therefore, when a passenger becomes satisfied with public transportation service, he/she perceives the service as being of good or high quality and vice versa. A high level of service quality lead to increase in the number of loyal passengers, creating a favourable image of the service providers that may lead to the reduction in traffic congestion. Lai and Chen [9] was reported the service quality of public transportation have ability to influence the perceived value, passenger's satisfaction and passenger's behavioural intention towards the service provided. Thus, the following hypotheses are suggested:

- Hypothesis 1: Service quality positively influences the passenger's perceived value
- Hypothesis 2: Service quality positively influences the passenger's behavioural intentions
- Hypothesis 3: Service quality positively influences the passenger's satisfaction

2.2. Perceived value (PV)
The value of a service perceived by passengers is a valuation carried out by comparing the benefits of a service, or relationship against costs or sacrifices incurred [10]. Consumer value is a comparison of the benefits of a product or service received to costs incurred by consumers [11]. If the benefits for consumers are higher than the costs incurred by consumers, it will turn into a high consumer value. If the value received by consumers is continuously given it will ensure a competitive advantage for the company, therefore it is important that companies ensure high consumer satisfaction [10]. The next two formulated hypotheses are:

- Hypothesis 4: Perceived value positively influences the passenger's behavioural intentions
- Hypothesis 5: Perceived value positively influences the passenger's satisfaction

2.3 Passenger's satisfaction (PS)
Satisfaction is predetermined by how the expectations of the passengers are met [12]. It is related to passengers' needs. The level to which these needs are fulfilled define the enjoyment or disappointment from discrepancy [13]. Oliver [14] highlighted that satisfaction is the passenger's fulfilment response. It is a consideration that a service feature, or the service itself, provided (or is providing) a pleasurable level of consumption-related fulfilment, including levels of under-or over fulfilment". Therefore, the last hypothesis is:

- Hypothesis 6: Satisfaction positively influences the passenger's behavioural intentions

2.4 Behavioural intention (BI)
Behavioural intention as an act to represent customers' loyalty [15]. Therefore, behavioural intention includes repurchases and recommendations [16]. Measurement of behavioural intention can be the best way to predict future buying behaviour [15, 17]. Zeithaml et al. [16] stated that consequence arising from perceptions of service quality in the form of intention to behave by individual consumers is a signal of success or failure of the company to retain its customer.

2.5 Model framework
Based on comprehensive literature review, the current study proposed a model framework to explore the relationship between four constructs, namely service quality (SQ), perceived value (PV), passenger's satisfaction (PS), and behavioural intention (BI), as displayed in figure 1.

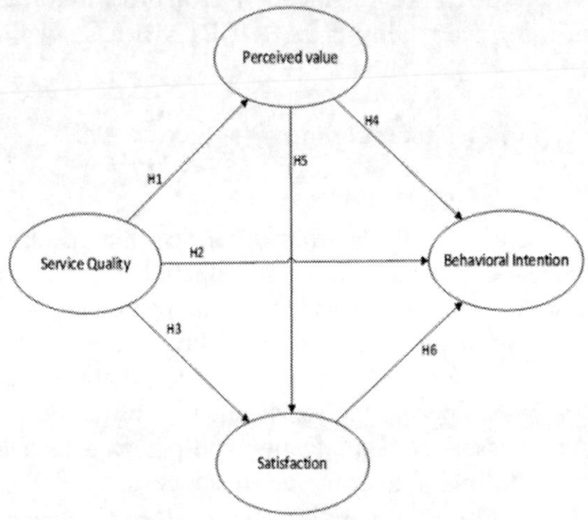

Figure 1. The model framework

3. Research methodology
The questionnaire that used in this study was adopted by Lai and Chen [9] and was modified based on the detailed service in JABODETABEK and back translated to Indonesian language. The data is collected from 390 KRL passengers in JABODETABEK. The total of sample size in this study was decided based on recommended by Krejcie & Morgan [18]. Random sampling method is employed when selecting the respondents. The valid responses are analysed using the Statistical Package for Social Sciences Software (SPSS) 22.0 and Analysis of Moment Structure (AMOS) version 22.

4. Results and discussion
Based on the respondents' information obtained, majority of respondents participated in this research are female with a total of 240 or in a percentage of 61.5%, while male respondents are 150 which is equivalent to 38.5%. From the data obtained, age group are categorised into five as shown in the table 1.

4.1 Reliability and convergent validity
Confirmatory Factor Analysis (CFA) was used to evaluate the measurement model in this study. It was also utilized to assess dimension, convergent validity and construct reliability for overall model. The result obtained shows that the factor loading for all items ranging between 0.59 – 0.93, which is exceed the set threshold value of 0.5 as recommended by Hair et al. [19] as shown in table 2.

Table 1. Respondents characteristics

Groups	Frequency	Percentage (%)
Gender		
Male	150	61.5
Female	240	38.5
Age		
18 - 25	175	44.9
26 - 30	70	17.9
31 - 35	52	13.3
36 - 40	33	8.5
> 40	60	15.4
Occupation		
Government employee	82	21
Private employee	105	26.9
BUMN employee	68	17.4
Entrepreneur	58	14.9
Not working	8	2.1
Student	76	19.5
Trips		
Everyday	63	16.2
Working days only (Monday to Friday)	105	26.9
Every week	67	17.2
Once a month	79	20.3
Every 2 - 3 months	81	20.8
Everyday	63	16.2
Route(s)		
Bogor/Depok - Manggarai - Jakarta Kota	112	24
Bogor/Depok - Tanah Abang - Pasar Senen	85	18.2
Bekasi - Jatinegara - Manggarai - Jakarta Kota	74	15.8
Rangkasbitung - Tanah Abang	55	11.8
Tangerang - Duri	87	18.6
Tanjung Priok - Jakarta Kota	24	5.1
Other routes	30	6.4

Based on table 2, it can be concluded that behavioural intention construct has higher square root of AVE compared to other correlation of constructs and it shows a good sign of convergent validity. Other square roots of AVEs are not as high as other correlation of constructs. All loading values from the constructs are above 0.7, where it is considered accepted as good validity [20].

Table 2. Reliability and convergent validity results

Constructs	Indicator	Factor Loading	AVE	CR
Service Quality	SQ1	0.79	0.64	0.88
	SQ2	0.78		
	SQ3	0.75		
	SQ4	0.80		
	SQ5	0.75		
	SQ6	0.75		
	SQ7	0.78		
	SQ8	0.75		
	SQ9	0.78		
	SQ10	0.79		
Perceived Value	PV1	0.86	0.74	0.86
	PV2	0.90		
	PV3	0.89		
Passenger's Satisfaction	PS1	0.90	0.63	0.87
	PS2	0.86		
	PS3	0.59		
	PS4	0.75		
	PS5	0.85		
Behavioural Intention	BI1	0.92	0.81	0.90
	BI2	0.93		
	BI3	0.85		

4.2 Goodness of fit indices

The goodness-of-fit of the measurement models explained how good the item in examining the intended constructs [21]. The Goodness of fit indices that assess the measurement model encompass the normed chi-square test, p-value, Goodness of Fit (GFI), Adjusted of Goodness of Fit (AGFI), Comparative Fit Index (CFI), Tucker Lewis Index (TLI), Incremental Fit Index (IFI), Root Mean Square Error of Approximation (RMSEA), and Normed Fit Index (NFI). According to range of acceptable value by Shah and Goldstein [22], the values of goodness of fit indices obtained in this study are in the acceptable range and considered fit as shown in table 3.

Table 3. Goodness of fit indices results

Goodness of Fit Indices	Results	Range of acceptable value	Remarks
Chi-square	1129.58	(0.0, 1270.0)	Acceptable
P-value	0.00	(0.0, 0.94)	Acceptable
Chi-square/df	3.16	(0.02, 4.80)	Acceptable
GFI	0.82	(0.75, 0.99)	Acceptable
AGFI	0.76	(0.63, 0.97)	Acceptable
CFI	0.93	(0.88, 1.00)	Acceptable
TLI	0.91	(0.73, 1.07)	Acceptable
IFI	0.93	(0.88, 0.98)	Acceptable
RMSEA	0.07	(0.00, 0.13)	Acceptable
NFI	0.90	(0.72, 0.99)	Acceptable

4.3 Evaluation of structural model

The relationship between each construct for the proposed model in this study is shown in table 4. The service quality of public transportation in JABODETABEK had a significant and direct effect on perceived value ($\beta = 0.884$, $p < 0.001$) and passenger's satisfaction ($\beta = 0.982$, $p < 0.001$). In addition, the service quality did not show any significant effect on behavioural intention ($\beta = -0.251$, $p = 0.547$). Thus, the hypothesis 1 and hypothesis 3 were supported and hypothesis 2 was rejected. In addition, the perceived value had a significant effect on behavioural intention ($\beta = 0.530$, $p < 0.001$), while the effect of perceived value and passenger's satisfaction was not significant ($\beta = -0.040$, $p = 0.191$). Therefore, hypothesis 4 was supported while hypothesis 5 was rejected. Furthermore, passenger's satisfaction reported had a significant influence on behaviour intention ($\beta = 0.621$, $p < 0.001$) towards public transportation (KRL) in JABODETABEK. This proved that hypothesis 6 is supported. The summary of the interrelationship among the constructs is summarized in table 4.

Table 4. Summary of hypothesis testing model

Path	β	*p*-value	Hypotheses result
SQ > PV	0.884	***	H1-Supported
SQ > BI	-0.251	0.547	H2-Rejected
SQ > PS	0.982	***	H3-Supported
PV > BI	0.530	***	H4-Supported
PV > PS	-0.040	0.191	H5-Rejected
PS > BI	0.621	***	H6-Supported

5. Conclusions
From a questionnaire that was distributed to 390 respondents who are the passengers of Kereta Rel Listrik (KRL) in JABODETABEK, the results were then analysed using SEM analysis and used to measure the satisfaction on KRL performance. From this study, it can be concluded that:

- Overall, KRL passengers in JABODETABEK feel dissatisfied with the services provided and the performance of KRL. Nevertheless, the involvement of passengers in public transportation, especially KRL and the behavioural intention of passengers to continue using KRL remains high. This is due to the high needs for transportation to travel from one place to another. In addition, the high level of behavioural intention by passengers is also influenced by several factors, such as fares, income levels, public transportation routes, frequency of public transportation for the routes itself, the quality of services provided are not much different than other, have no option but to use other transportation, and so on.
- There are some things that need to be improved in public transportation to attract private transportation users into public transportation. Based on the results obtained, factors that influence the intention of people in using KRL are punctuality, comfortable waiting area, staff's behaviour, fares, cleanliness, and so on. By using KRL, passengers can reach many places when in Jakarta. Availability of information is very helpful for passengers when traveling.

Acknowledgements
The authors would like to express their gratitude to Universiti Kebangsaan Malaysia (UKM) for the financial support for this work (GUP-2019-015).

References
[1] Vuchic V 2017 *Urban Transit: Operations, Planning, and Economics* (New Jersey: John Wiley and Sons) pp 1-672
[2] Borhan M N, Ibrahim A N H, Miskeen M A A, Rahmat R A O K and Alhodairi A M 2017 Predicting car drivers' intention to use low cost airlines for intercity travel in Libya *J. of Air Transp. Manag.* **65** 88–98
[3] Holmgren J 2007 Meta-analysis of public transport demand *Transp. Res. Part A: Policy and Pract.* **41A** 10 1021–35
[4] Borhan M N, Ibrahim A N H and Miskeen M A A 2019 Extending the theory of planned behaviour to predict the intention to take the new high-speed rail for intercity travel in Libya: Assessment of the influence of novelty seeking, trust and external influence *Transp. Res. Part A: Policy and Prac.* **130** 373–84
[5] Irtema H I M, Ismail A, Borhan M N, Das A M and Alshetwi A B Z 2018 Case study of the behavioural intention of public transportation passengers in Kuala Lumpur *Case Studies on Transp. Policy* **6** 462-74
[6] Cunningham LF, Young CE and Moonkyu L 2002 Cross-cultural perspectives of service quality and risk in air Transportation *J. of Air Transp.* **7** 1 3-26
[7] Ibrahim A N H, Borhan M N, Md. Yusoff N I and Ismail A 2020 Rail-based public transport service quality and user satisfaction: A literature review *Promet-Traffic and Transp.* **32** 423–35
[8] Zeithaml V A, Bitner M J and Gremler D D 2010 Services marketing strategy *Wiley Intl. Ency. of Mktg.* 1-11
[9] Dw Lai WT and Chen CF 2011 Behavioral intentions of public transit passengers—the roles of service quality, perceived value, satisfaction and involvement *Transp. Policy* **18** 2 318– 25
[10] Roig JCF, García JS and Tena MAM 2009 Perceived value and customer loyalty in financial services *The Svcs. Industries J.* **29** 775-89
[11] Heskett JL, Jones TO, Loveman GW, Sasser WE and Schlesinger LA 1994 Putting the service-profit chain to work *Harvard Bus. Rev.* **72** 4 164-74
[12] Kotler P and Armstrong G 2014 *Principle of marketing* (London: Pearson) pp 1-695
[13] Hill N, Roche G and Allen R 2007 *Customer satisfaction: the customer experience through the customer's eyes* (London: Cogent) pp 1-314.
[14] Oliver R 2010 Customer satisfaction *Wiley Intl. Ency. of Mktg.* 1-5

[15] Petrick JF 2004 The roles of quality, value, and satisfaction in predicting cruise passengers' behavioral intentions *J. of Travel Res.* **42** 397- 407
[16] Zeithaml AV and Bitner MJ 1996 *Ser. Mktg.* (New York: McGraw-Hill) pp 1-700
[17] Sumaedi S, Mahatma IG, Bakti Y and Yarmen M 2012 The empirical study of public transport passengers' behavioral intentions: the roles of service quality, perceived sacrifice, perceived value, and satisfaction (Case study: Paratransit passengers in Jakarta, Indonesia) *Intl. J. for Traffic and Trans. Engr.* **2** 1 83–97
[18] Krejcie RV and Morgan DW 1970 Determining sample size for research activities *Ed. and Psych. Meas.* **30** 607-10
[19] Hair JF, Jr. Black WC, Babin BJ, Anderson RE and Tatham RL 2006 Multivariate data analysis (6th ed.) *Upper Saddle River* (New Jersey: Prentice Hall) pp 1-928
[20] Ghozali I 2004 *Aplikasi Analisis Multivariate Dengan Program SPSS* (Semarang: Badan Penerbitan Universitas Diponegoro)
[21] Hulland JS 1999 Use of partial least squares (PLS) in strategic management research: a review of four recent studies *Strat. Manag. J.* **20** 4 195–204
[22] Shah R and Goldstein SM 2006 use of structural equation modelling in operations management research: looking back and forward *J. of Ops. Manag.* **24** 2 148–69

The selection of the International Port Design Standards applicable for Indonesia

I W Dyah[1*] and Fuddoly[1]

[1] Department of Civil Engineering, Institut Teknologi Sepuluh Nopember (ITS), ITS Campus, Keputih, Sukolilo, Surabaya, East Java, Indonesia 60111

*Corresponding author's email: dyah@ce.its.ac.id

Abstract. As of today, almost all the engineer, experts, and student in Indonesia face confusion and edginess concerning designing of Port Infrastructure, mainly due to the absence of a design Standards legally valid for domestic purpose. On the other hand, several standards or guide lines or codes are available and internationally recognized and applied in many places for many years solely without unaffiliated standards. 3 (three) International standards are selected for evaluation to be thoroughly adapted to the Design Standards of Indonesia are British Standards (BS 6349), the German Standards (EAU 2004), and the Standards Design Criteria for Port and Harbor in Japan (OCDI, 2009). The criteria applied during evaluation and selection process based on the observation output from about 300 selected projects within Indonesia. Finally, the British Standards proved to give the best matched for the condition of Indonesia.

1. Introduction

Indonesia controls 1,905,000 km^2 area with two-thirds consist of sea water and the number of islands reach about 16,000. Variations of islands size and spread of port functions for each points considered exceptional by any standards. Here, Port variation in all cases wide open to be grouped or make a simple categorization. It reflects a high the demand for new port as well as maintaining the available port over the coast of Indonesia

Some drawbacks often found in the field of port infrastructure in Indonesia are:

a) The design used mixed or unknown Standards of design without legal consequences.
b) Procurement process implements incorrect procedures.
c) The life of the structure is very short.
d) The construction considered made of low workmanship as well as the materials.
e) Ignorance to the maintenance works

The design phase considered as source and lead to further distortions. Pertinent to the availability of valid Standard related to government regulation considered urgent. The implementation and connection to the legal basis from any digressions requires immediate actions.

As of today, lack of discussion and research concerning the standard applicable particularly to Indonesia. The comparison study on the port design standard using International code of practice for jetty or other port structure in Indonesia are few or almost none.

There are several standards for Port Design and planning valid internationally. The designer, engineer and university students in Indonesia most frequently apply 3 (three) standards of Port Design specifically the British Standards, the German Standards, and the Japan Standards. Therefore these standards selected and recommended as the best candidate to be the Indonesian Standards. The selection will be

Content from this work may be used under the terms of the Creative Commons Attribution 3.0 licence. Any further distribution of this work must maintain attribution to the author(s) and the title of the work, journal citation and DOI.
Published under licence by IOP Publishing Ltd

based on some key criteria. And each criteria is selected based on the observation and research output from more than 300 port project locations scattered all over Indonesia covering various function of terminal as well as type of structure particularly jetty, breakwater, dredging and reclamations.

The criteria will be weighted based on level of urgency and the extent of the impact to the output of the design or to the finished structure. The score of each criteria split in to four (4) due to the relatively broad and general overview given to each criteria. The worst situation or insufficient receives score 1, moderate (sufficient but not complete) earn score of 2, Good or complete will earn score of 3, and the Best means complete and easy to follow will earn 4. Recommendation will be withdrawn considering the highest weight-score from each standards.

2. Purpose and benefit of the research

The purpose of the research are to give idea on the necessity of Indonesia Standard for Port Design to be prepared as soon as possible and accelerate the implementation with legal basis elaborate government regulations or ministerial regulation or published as special book of Indonesian Standards for Port.

The benefit will more in effect to the implementation of the use of the Standards. We can expect for the design to be more accountable, to open more opportunity to the involvement of other expertise, and especially in the field of education will direct the course material straight to the Standards.

3. The Selection Criteria

As previously mentioned, there will be three (3) standards to be evaluated here are:

1) "British Standards - Maritime Structures", BS 6349 part 1 to 8 (abbreviated as BS 6349).
2) "Recommendations of the Committee for Waterfront Structures, Harbours and Waterways", EAU 2004 (abbreviated as EAU 2004).
3) "Technical Standards for Port and Harbor Facilities in Japan", OCDI, 2009 (abbreviated as OCDI, 2009).

The criteria to be applied for the selection of the standards is chosen based on the most significant requirement for designing of Port facilities in Indonesia especially considering the weakness or the problem appear in Indonesia. Priority scale put on to weight the criteria as follows:

1) The overall coverage and detail guidance

The overall coverage and detail guidance considered essential for any standards to give clear guidelines to the users. It shall consist of the various possibility of type of structure and the future change. The extensive interrelation to other standards or guide lines bring positive merit. Certain detail of directions considered advantageous to most user in Indonesia especially to reduce disputes.

The weight for this case is 25.

2) The design procedures and the load patterns

This criteria regarded as deliberately crucial to be adopted in the Standards of Port Design in Indonesia. It is due to the fact that mostly during design the port facilities, that designer faced with the dilemma to follow or to freely apply certain design Standards without legal basis. The liability to any options

As well as the load patterns, several experts and designer meet the conflict to other standards applicable widely in design structure in Indonesia. Confusion of the designer often occur concerning The confusing application of several different standards lead to several problems concerning inefficiency of the structure, over or under design, the load pattern shall be adopted in strictly in accordance to the Standards individually.

The weight for this case is 20.

3) The construction life time and the safety factors

As the most susceptible elements found on the structure in the sea environment, the age of the port structure in average reach about 12 years. The actual design life time mostly unknown but

about 20% stated for 30 years. The reality of the structure shows difference to the expected life span from the feasibility study.
The weight for this case is 20.

4) The formula to calculate the all structure components
The confusion and uncertainty as well as the topic of discussion and criticism occur to about 90% of the design. And the case of failure or subject of findings of audit institutions.
Most cases deal with problem the use of Standards applicable of other structures state suitability only for certain type of structure (including building and bridge) other than port or sea water environment.
Some standards did not covering all part of the structure in the terminal or other coastal structures.
The weight for this case is 20.

5) The Monitoring and inspections
The monitoring or supervision and the inspection phase determine the quality, time and cost of construction in Indonesia. 60% of the port projects face with detention problems. Tracing the root of this obstacle shows due to unavailability of explicit and literally exact phrase as well as inconsistency of the monitoring and inspections documents.
Evaluation shall examining the detail and overage of the guideline for monitoring and inspection to be performed during construction phase.
The weight for this case is 15.

4. The Evaluation of the Standards

Each criteria rated and evaluated concerning the suitability of the standards to the environment and to the common practice of the society of designer in Indonesia.
Each criteria will be juxtaposed to each Standards, 15 points of review has been thoroughly examined to meet the best score.
Evaluation mainly focused to the suitability of the standards on the environment and to the common practice of the society of designer here are as follows:

The overall coverage and detail guidance

a) BS 6349
This Standards regularly overviewed and revised adapting the latest information and development in area of knowledge. This is a multi-part document divided into the following parts:
1. Part 1 Maritime structures. General criteria
   Part 1-1 Maritime works. General. Code of practice for planning and design for operations
   Part 1-2 Maritime works. General. Code of practice for assessment of actions
   Part 1-3 Maritime works. General. Code of practice for geotechnical design
   Part 1-4 Maritime works. General. Code of practice for materials
2. Part 2 Maritime structures. Design of quay walls, jetties and dolphins
3. Part 3 Maritime structures. Design of dry docks, locks, slipways and shipbuilding berths, ship lifts and dock and lock gates
4. Part 4 Maritime structures. Design of fendering and mooring systems
5. Part 5 Maritime structures. Code of practice for dredging and land reclamation
6. Part 6 Maritime structures. Design of inshore moorings and floating structures
7. Part 7 Maritime structures. Guide to the design and construction of breakwaters
8. Part 8 Maritime structures. Code of practice for the design of Ro-Ro ramps, linkspans and walkways

The scope and coverage of this Standards comprise of 90% of the type of structure to protect coast and port operation.

The score given to this standards is 4.

b) EAU, 2004
This Standard rarely overviewed and revised. And the document comes in one thick book. The latest version published in 2005 as 8th edition to EAU, 2004.
Table of contents of the document shows the coverage of the guidance to the demand on the port infrastructure as follows:
1. Statical calculations
2. Subsoil
3. Active and passive earth pressures
4. Overall stability, foundation failure and sliding
5. Water levels, water pressure, drainage
6. Ship dimensions and loads on waterfront structures
7. Configuration of cross-section and equipment for waterfront structures
8. Earthwork and dredging
9. Sheet piling structures
10. Anchor piles and anchors
11. Waterfront structures, quays and superstructures of concrete
12. Piled structures
13. Embankments
14. Dolphins
15. Experience with waterfront structures
16. Monitoring and inspection of waterfront structures in seaports

The scope and coverage of this Standards comprise of 80% of the type of structure to protect coast and port operation with document systematic referring to the basic knowledge mix with structure function. This systematic less familiar to most designer in Indonesia with the tendency to approach based on package of works to be tendered such as Jetty works, Dredging works, etcetera.

The score given to this standards is 3.

c) OCDI, 2009
This standard provides a single thick document of 1028 pages, with systematic as follows:
Part I: General
Part II: Actions and Material Strength Requirements
Part III: Facilities
Each part give comment or explain all matter in detail covering almost 85% port facilities. The reference adopting various countries design Standards.
The score given to this standards is 3.

The design procedures and the load patterns

a) BS 6349
The most common type of port structure in Indonesia is Jetty. In this design standards, introduced as suspended deck structures various configuration of pile foundation, see **Figure** *1*and **Figure** *2*. The bridge between berth and land side introduced as trestle which is commonly term for such part.
The type of structure as in **Figure** *1* bring to flexible type of structure to allow rather substantial deformation when extensive horizontal force working on. On the other hand **Figure** *2* gives rigid structure to resist large horizontal force.

Both type of jetty or berth structure extensively adopted in Indonesia reach about 70% of total type of structure as observed.

The design procedure in systematical order is not available but sequence of discussion in sub chapter lead to design procedure. The sub chapter started with selection concerning type of structure, type of pile foundation, type of upper structure, design method, loadings, design of sub structure (pile) and structure tolerances.
In sub chapter 8.10.1, shows configuration principle for the pile to withstand the upper loading to cause on pile deformation. Limited value of deformation recommended to adjoin the capacity of the elastic deformations of the pile.
The load working on the structure classified as:

i. Self weight and imposed load
ii. Horizontal actions
iii. Environmental actions

The load scale determined in conjunction to other British Standard. Load combination inter related to load safety factor calculated based on probability of event. Generally loading grouped into normal situation, extreme situations, and temporary load. Detail factor linked to the load is tabulated incorporate almost the entire possibility of load type and the occurrence of load to take place.
It is concluded that guidance on load is thorough and complete.

The score given to this standards is 4.

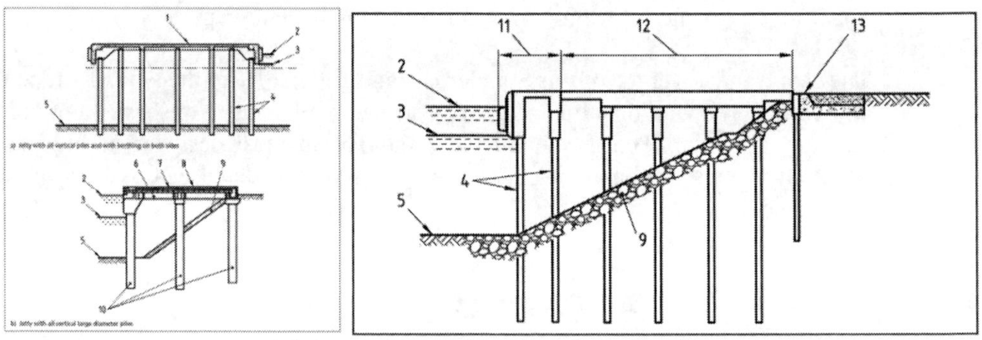

Figure 1. Jetty with all vertical piles [4].

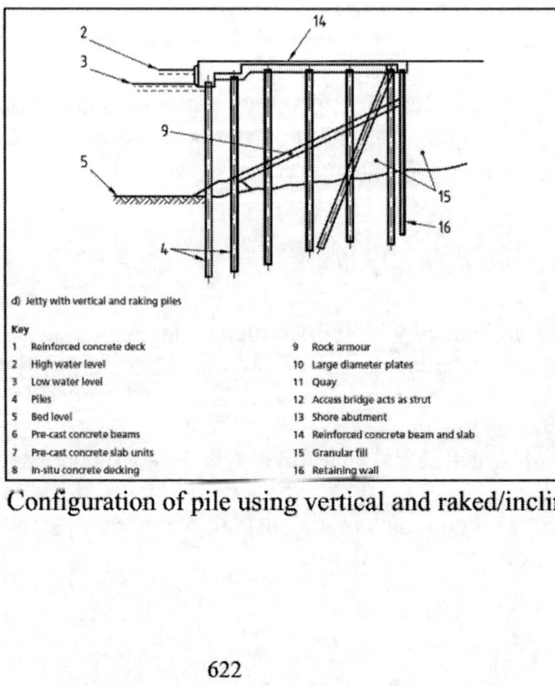

Figure 2. Configuration of pile using vertical and raked/incline pile [4].

b) EAU, 2004

The calculation procedure step by step is not available for overall structure. In the case of Jetty introduced as *General Pile-founded Structures (R 157)*. The design procedure available of each part of the structure such pile foundation, beam and slab.

EAU, 2004 discuss quite extensive about structure of Free-standing, high pile-founded structures, Statical System and Calculation, including Structural Advice. Clear recommendation concerning the pile position can not be found but simply that joint to each beam required pile support.

Complete guidance found related to general guide to determine beam for crane position, longitudinal beam as well as the lateral beam or portal including position of beam for fender and bollard. The suggested typical structure of pier or jetty as in **Figure 3**.

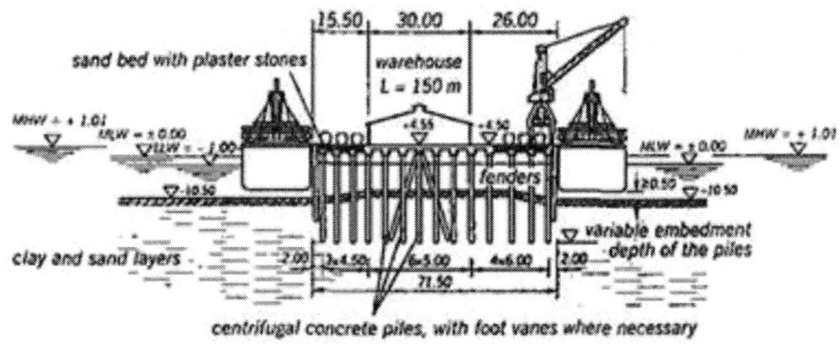

Figure 3. Example of port pier system using reinforced cement concrete or RCC and concrete piles [11].

Safety factor partially applicable for load categorized into three (3) cases including:

- Load case 1 for load due to active soil
- Load case 2 for load due to scouring, water pressure difference, water suction passing through vessel, active pressure of extra ordinary soil load, bollard pull and fender pressure, and temporary load during construction phase.
- Load case 3 considered equal to load case 2 with additional load for the possibility of drainage failure, unexpected wave forces and possible over load precious unpredicted.

Multiplier factor to load cases categorized as partial safety factor. Another multiplication applied the load factor for the overall structure with relatively low number.

The score given to this standards is 3.

c) OCDI, 2009

OCDI, 2009, introduced the term of jetty similar to breakwater. The Jetty as common practice in Indonesia grouped as Open-type wharves, see "Part VIII Mooring Facilities". 2 type of structure for berthing a vessel are the open-type wharves on vertical piles and the open-type wharves with coupled raking piles, see **Figure *4***.

Compared to 80% of jetty structure configuration found in the 400 locations, rather obvious difference can be directly recognized especially the use of low vertical wall or gravity wall along the shore and the absence of trestle.

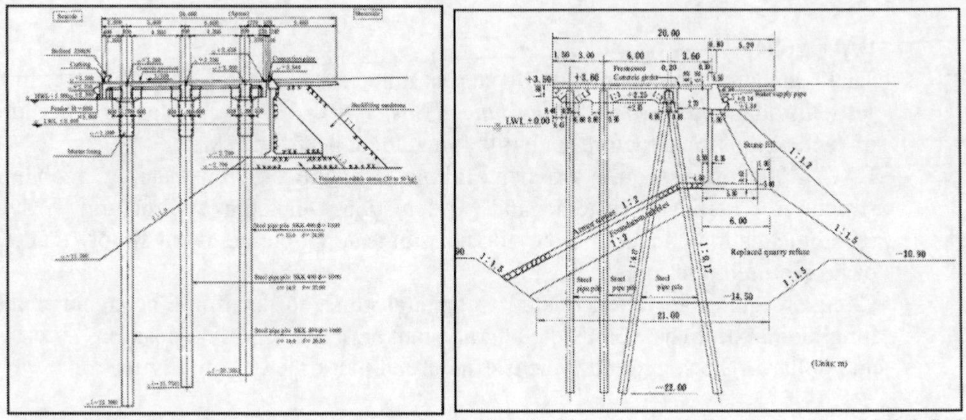

Figure 4. Open-type wharves on vertical piles (left) and coupled raking piles (right) [12].

Design procedure and the flow chart for *open-type wharves on vertical* available in detail explaining step by step the process including the loop or control as required at eertain step.

The load is grouped into vertical and horizontal load and described in detail. The load variety considered sufficient and covers the entire possible loads. However, it is found that regarding the load due to cargo stacked on the berth, as common practice in Indonesia, ignored.

The load factor as part of safety factor separated into 3 types are permanent load, variable load, and accidental load. The scale factor presented in a table and simply straight appointed as a set of load combination.

The score given to this standards is 3.

The construction life time and the safety factors

The determination of the structure age or life span shall take into account the feasibility of the project and the benefit generated by the purpose or the role of the port to the industry. The stretch especially in the reduction regime of service time most probable bring to loss of profit. The use of Standard or guidelines expected to help maintain the design life time or age of the structure.
Safety factor (F) principally to obtain the Resistance capacity (R) of the material as part of the structure higher than the forces working on the material or the Solicitation part (S). The fraction of R to S is the safety factor. By exaggeration of each side both R and S still has to lead $R/S \geq F$. When R and S each applying the safety factor it may lead to extra safe condition.

a) BS 6349
Recommendation of construction life time is 50 years with consequences in several parts as follows:
   i. Factor of the running time to the safety of the structure considering fatigue loading, corrosion, marine growth and reduction to soil capacity.
   ii. Probability level applied in boundary number for the design and in determining the return period or recurrence interval.
   iii. Economic feasibility of the project and the future development.

Practical recommendations available in this Standard concerning the vehicle or equipment load shall adjust to the latest technology progress during the life time of the construction. The wave forces, tide and current shall apply maximum magnitude during the life time.
Safety factor related to the life time has been applied at the calculation of factor of fatigue, durability including the equation preferably to be used, one example shown in BS 6349 Part 1 sub chapter 39 and section 7.

Conclusion concerning the life time of the structure has been comprehensively discussed in BS 6349. The application of life time also applied in the material requirement. The use of this Standard estimated to maintain the structure aged as designed.

The theory behind safety factor introduced in BS 6349 considered unclear. The safety factor mixed in some parts with the life time of structure. The safety factor clearly stated in the load combinations.

The safety factor explain in part 1 and 2 considered relatively detail. The safety refers to the action measure both during construction as well as operational such as fire fighting, road access in emergency stage, emergency response and the safety equipment including the requirement of *speed-of-approach monitoring aids*. The deformation limit given with limit value.

The *safety factor* considered less thorough and need more focus to the structure. The safety factor applicable for load combination, and the structure life time considered complete.

The score given to this standards is 3.

b) EAU, 2004

The construction life time applied to the pile concerning corrosion prevention as long as 60 years. For RCC in conjunction to the density of the concrete, type of cement, thickness of concrete decking suggested for 50 years. The structure continuously monitored yield to service life computed starting from the date of operation.

Average service life of the structure at several buildings discussed in R 46, it shows that port structure shorter in life time compare to buildings and better maintenance, and revitalization in regular basis including deepening and spare parts change.

The principle to determine safety factor is based on the stochastic approach to determine the probability of the Resistance (R) value and the Solicitation (S) value.

The structure failure as the result of exceeded value of the limit state or known as Limit State 1 (LS 1), in the form of failure at the foundation or the building. The exceedance to the limit state of the serviceability known as Limit State 2 (LS 2) in the form of over in deformation.

F taken based on experience and adopt the Erope Code. Partial safety factor still applicable specific structure such as quaywall or wharf.

The safety factor for the Bearing Capacity applicable by assuming each pile as Individual pile. And partial safety factor for loads is separated in 3 Load Cases as LC 1, LC 2, and LC 3 using *Ultimate Limit State* (ULS).

The application of load factor in load combination depends on safety level, load combination and the load cases. The F for individual pile in LC 1, LC 2, LC 3 each yield to 1.4, 1.2, and 1.1. For jetty, the load cases suggested only of LC 2.

Deformation and fatigue of the structure give less impact.

The safety factor application considered quite clear.

The score given to this standards is 4.

c) OCDI, 2009

In Part II, the structure life time suggested to be determined considering functional operability of the facilities, economic value, functional role to the society, and physical characteristic. Encounter probability is elaborated to compares with life time and the recurrence interval. The material selection determined apply the structural life time. The life time also connected to the earthquake calculation (see 2.2 as in OCDI, 2009) and damage level of the structure (see 5.3 in OCDI, 2009).

Safety to the structure based on empirical approach. The boundary value considered as the best approach at the limit state sparated into ultimate limit state, serviceability limit state, and the fatigue limit state.
Partial safety factor as in the material factor, load factor, structural analysis factorthe number factor), and the structural factor.
The deformation problems more for the deformation at the foundation not at the state limit of the structure.
The safety factor has been thoroughly examined but unclear about the basic assumption.

The score given to this standards is 3.

The formula to calculate the structure components

The approach to the equation or formula applicable to the design help significantly to reduced the design error.

a) BS 6349
The Standards provide complete equation and formula connected to different series of Standard. The equation for technical calculation in conjunction to the bearing capacity (for example) mentioned in separate Standards. Equation related to the life time of construction is available in the same document of BS6349. Most equation related to the port design presented in separate standards.

The score given to this standards is 4.

b) EAU, 2004
The formula and equation available in one book, especially to the basic computation and the safety factors. Technical equation on bearing capacity and deep foundation is available in the separate document. The limit value for each part did not clearly stated.

The score given to this standards is 3.

c) OCDI, 2009
The formula available at limited application but covering extended topics. For example bearing capacity calculation limited to certain type of soil but extensively cover detail of corrosion prevention.
The equation for waterways layout available in complete.

The score given to this standards is 4.

The Monitoring and inspections

a) BS 6349
The *supervision* discussed in one with the structure as well as part of the structure for example point 52 and 59. No specific quantity concerning the volume or time to be supervised. *Monitoring* mean at the supervision during construction to certain part of the project for example the foundation issue such as land slides.
Time and time interval of supervision and monitoring not specifically determined.

The score given to this standards is 3.

b) EAU, 2004
This Standards include the construction methods and inspection during construction phase, including the monitoring procedure as in R 193.
Dimension of each part, loading and safety factor focused on the wharf structure as in 6.2.1, R 118, R 209, R 86.
Inspection during construction to check the entire movement of the structure especially:
- Alignment
- Leveling
- And deflections.

Monitoring proceed during construction for fabrication component of the structure such as in 8.1.8.4 and in R 195.
Some pint in inspection and monitoring work explained in complete including the record, report and inspection interval.

The score given to this standards is 4.

c) OCDI, 2009
Monitoring recommended for some parts of the structure. For example the pile joints shall be in lien with inspection works. Monitoring suggested for the entire reclamation works. The inspection suggested periodically done during maintenance phase. No guideline concerning supervision to the entire works, inspection periodically applicable for lighting and other small facility.

The score given to this standards is 4.

5. Conclusion and Recommendation
Conclusion to the three Standard as presented in **Table *1*** shows that the British Standard gives the best conformity for Indonesian Standard.

Table 1. The Comparison between the Standards.

Criteria	BS 6349	EAU, 2004	OCDI, 2009
The overall coverage and detail guidance	25 x 4	25 x 3	25 x 3
The design procedures and the load patterns	20 x 4	20 x 3	20 x 3
The construction life time and the safety factors	20 x 3	20 x 4	20 x 3
The formula to calculate the all structure components	20 x 4	20 x 3	20 x 4
The Monitoring and inspections	15 x 3	15 x 4	15 x 4
Total	365	335	335

Recommendation concerning further detail shall be focused to British Standard.

References
[1] Comin C and De Souza R 2017 *Revista IBRACON de Estruturas e Materiais. 10. 626-638. 10.1590/s1983-41952017000300005*
[2] Roubos A, Peters D J, Groenewegen L and Steenbergen R 2018 Partial safety factors for berthing velocity and loads on marine structures *J. Mar. Struct.* **58** 73-91 https://doi.org/10.1016/j.marstruc.2017.11.003
[3] BS 6349-1 2016 *Maritime works: General – Code of Practice for Assessment of Actions* (London, United Kingdom) ISBN 978-0-580-76229-1

[4] BS 6349-2 2019 *Maritime Works. Code of Practice for the Design of Quay Walls, Jetties and Dolphins* (London, United Kingdom) ISBN 978 0 580 98478 5
[5] BS 6349-3 2013 *Maritime Works: Code of Practice for the Design of Shipyards and Sea Locks* (London, United Kingdom) ISBN 978 0 580 78084 4
[6] BS 6349-4 2014 *Maritime Works: Code of Practice for Design of Fendering and Mooring Systems* (London, United Kingdom) ISBN 978-0-580-66969-9
[7] BS 6349-5 2016 *Maritime Works: Code of Practice Dredging and Reclamation* (London, United Kingdom) ISBN 978-0-580-66969-9
[8] BS 6349-6 1989 *Maritime Works: Code of Practice Design of Inshore Moorings and Floating Structures* (London, United Kingdom) ISBN 0 580 16761 5
[9] BS 6349-7 1991 *Maritime Works: Guide to the Design and Construction of Breakwaters* (London, United Kingdom) ISBN 978-0-580-66969-9
[10] BS 6349-8 2007 Code of practice for the design of Ro-Ro ramps, linkspans and walkways *Maritime structures* (London, United Kingdom)
[11] Recommendations of the Committee for Waterfront Structures, Harbours and Waterways EAU 2004 (abbreviated as EAU 2004)
[12] OCDI 2009 Technical Standards and Commentaries for Port and Harbour Facilities in Japan (Tokyo, Japan: The Overseas Coastal Area Development Institute)
[13] Nagao T, Watabe Y, Kikuchi K Honjo Y 2009 Recent revision of Japanese Technical Standard for Port and Harbour Facilities based on a performance based design concept *Geotechnical Risk and Safety* (London, United Kingdom)
[14] PIANC 2002 *Guideline for design of fender Systems* (Brussels, Belgium) ISBN-2-87223- 125-0
[15] PIANC 2014 *Design of Lock Gates for Ship Collision* (Brussels, Belgium) ISBN 978-2- 87223-215-4
[16] Dyah Iriani and Fuddoly 2017 *Perencanaan Pelabuhan* (Jogjakarta: CV.Absolute Media) ISBN 978-602-1083-86-4

2D numerical modeling of the Jeneberang River Flood due to the overflow of the Bili-Bili Dam.

Aslam[1*], and U Lasminto[1]

[1]Department of Civil Engineering, Institut Teknologi Sepuluh Nopember

*Corresponding author's e-mail: aslam.18031@mhs.its.ac.id

Abstract High rainfall at the end of January 2019 in South Sulawesi, especially in the upstream area of the Jeneberang watershed, caused an increase in the height and volume of water in the Bili-Bili dam reservoir. The elevation of the water level was recorded on January 22, 2019 at 18:00 WITA reaching +101.87 m ASL above the normal water level elevation of +99.50 m ASL. In accordance with the spillway door operating pattern, if the water level exceeds the normal water level elevation limit, the dam floodgates must be opened to prevent dam collapse. To find out the impact caused, it is necessary to numerically model the flow velocity, depth, and inundation area in the downstream area of the Jeneberang river caused by overflow of discharge from the Bili-Bili Dam spillway. The data used are the flow hydrograph data from the outflow of 10-hour dam outflow as an upstream boundary condition and the rating of the rubber dam rating curve downstream of the Jeneberang river as a downstream boundary condition. The 2D research area focuses on 2 sub-districts directly adjacent to the Jeneberang river, namely Pallangga and Somba Opu sub-districts. From the results of flood simulations using the 2D HEC-RAS numerical model, the flow velocity is between 0.57m / s to 4.84 m / s, the depth is between 1.70 m and 7.38 m, and the inundation area is 1.7 km^2. These results were then calibrated with the results of observations and measurements in the field of the former high flood inundation in the building and the results of interviews with the affected communities which found that the simulation results correlated well at the seven points of measurement in the field using RMSE which is 0.37.

1. Introduction

The background of this research is the occurrence of high rainfall at the end of January 2019 in South Sulawesi, especially in the upstream area of the Jeneberang watershed causing an increase in the height and volume of dam reservoir water. The elevation of the water level was recorded on January 22, 2019 at 18.00 WITA reaching +101.87 m ASL above the normal water level elevation of +99.50 m ASL (Sisda BBWS Jeneberang Pompengan, 2019). In accordance with the spillway door operating pattern, if the water level exceeds the normal water level elevation limit, the dam floodgates must be opened to prevent dam collapse (BPSDM 2017). The highest flow outflow of the Bili-Bili dam outflow spillway was 1,426 m3 / d on January 22, 2019 from 18.00 to 21.00 WITA, causing an increase in the volume of the Jeneberang river downstream which then overflowed and caused floods in several Gowa regency areas bordering directly with the Clear river. The flood disaster is the first event that hits the riverbanks in the Palangga and Somba Opu sub-districts within the last 10 years.

Observing these data, this research needs to be done to analyze the area, flow velocity and flood inundation based on the outflow discharge data of the Bili-Bili dam in the area along the Jeneberang river to be used as an early anticipation of flood hazards, and then determine the affected areas as one of the efforts to mitigate floods in the preparedness stage.

To be able to do an assessment of the element of preparedness, then we need a mathematical calculation that can describe the behavior of flooding in the area around the Jeneberang river. This mathematical calculation can also be said as a flood modeling to see the amount of discharge over time through natural event approaches through hydrodynamic model. The modeling results obtained are then

Content from this work may be used under the terms of the Creative Commons Attribution 3.0 licence. Any further distribution of this work must maintain attribution to the author(s) and the title of the work, journal citation and DOI.
Published under licence by IOP Publishing Ltd

IOP Conf. Series: Materials Science and Engineering **930** (2020) 012071 doi:10.1088/1757-899X/930/1/012071

used as a virtual flood control effort schematically to determine the results before the application in the field; one of the mathematical methods used in flood simulations is a two-dimensional numeric simulation (2D). With 2D simulations approaching the actual situation on the ground, it is expected that the flood disaster can be anticipated earlier.

The purpose of this study is to analyze how much influence the flow outflow of the Bili-Bili dam from January 17 to 27 January 2019 on floods in the downstream watershed that is directly adjacent to the Jeneberang river through 2D numerical modeling using the HEC-RAS application when compared with inundation floodwaters data, area and flow velocity based on field observation data and efforts to support flood disaster mitigation.

2. Literature review

Flooding is an event where the overflow of river water exceeds a riverbed or a pool of water that occurs in areas that are low and cannot be drained, (SNI 2415: 2016). Flooding is a disaster caused by high rainfall, topography of a watershed, inadequate cross section of a river, and changes in land use. High rainfall and watershed topography are natural conditions that cannot be changed so that river capacity has an important role in flowing rainwater to prevent flooding (Wasim 2018). Based on the occurrence process, the possibility of a river channel or channel which is not able to accommodate normal discharge capacity, in extreme conditions suddenly can cause river water to overflow and enter areas that have lower topography.

2.1. Flood Mapping

To reduce the risk of flood events, a proper and integrated flood control system planning is needed. Kodoatie (2013) explains that in planning a flood control system in an area, there needs to be an evaluation and analysis that take into account the area or the extent of the flood inundation. The characteristics of floods in the form of the area / scope of inundation, inundation depth and flow velocity need to be mapped so that it can become a very basic and important foundation in flood control planning. Flood map is also a necessity in planning an area by developers to determine flood insurance and insurance from buildings, houses, offices and so on. In addition, the government also needs flood maps to support infrastructure plan, transportation facilities, and other plans, including flood emergency preparation and response (Al Amin 2018).

Flood mapping is a series of analyses conducted to produce flood maps that at least contain the scope or boundaries of inundation, flood risk zones, and depth of inundation. There are four methods that can be used for flood studies and mapping as explained in the National Research Council (2009), namely: 1) detailed studies, 2) limited detailed studies 3) approach studies, and 4) redline. The detailed study method requires river measurement data and / or rain-runoff models in hydrological analysis, while hydraulic analysis requires flow modeling (steady or unsteady) with detailed river building survey data. The limited detailed study method requires river measurement data for hydrological analysis, whereas for hydraulic analysis a steady flow modeling is required without detailed information about the river structure. The method of study approach is done by determining or estimating inundation boundaries without technical studies, for example by reading topographic maps and field surveys. The fourth method, namely redelineation, is to re-create flood maps based on the historical elevation of floodwaters on a new topographic map (Al Amin 2018).

The most commonly used methods in flood mapping are limited to the approach study and redesign method, such as loading flood map information on the depth and area of flood inundation areas carried out by Siregar (2016), Suleman (2017), Utomo (2019), and Seniarwan (2013). Meanwhile, to produce complete information about the characteristics of flood inundation which include discharge, flow velocity, inundation depth and extent of flood, it is necessary to model and simulate hydrodynamics as has been done by Alif (2018), Al Amin et.al (2018) and Sulaiman (2017).

According to Goel (2005), one method to help flood management is done by numerical simulations. Numerical simulations of flood estimates cannot immediately get good results because inundation depends on the topography and it changes with time (dynamic). This makes flood prediction more complicated and requires a long process. Flood information with a visual display helps better planning (Suleman 2017).

2.2. Flood Simulation Modeling

At present, mathematical modeling in solving a flood problem can be simulated using 1-dimensional (1D), 2-dimensional (2D) hydraulic modeling or a combination of 1-and 2-dimensional (1D / 2D) modeling. 1D Modeling is a model that has one direction of flow that is the direction of flow along the main channel, whereas 2D Modeling is a model that has two directions of flow, namely the direction of flow along the main channel and the direction of flow in the area around the flow.

The modeling of flood flow propagation is carried out based on the numerical settlement of the St. Venant equation using 1D calculations for along the river flow and 2D calculations if there is inundation at the surface of the land due to overflow from the river. The calculation process that occurs between rivers and land is done by 1D and 2D interconnection approaches.

2D flow calculations will use three equations as follows: 2D continuity equation, 2D momentum equation of the x-axis direction, and 2D momentum equation of the y-axis direction. The 2D continuity equation is as follows:

$$\frac{\partial \zeta}{\partial t}+\frac{\partial (uh)}{\partial x}+\frac{\partial (vh)}{\partial y}=0 \quad (1)$$

Information:

u : velocity of the x-axis direction
v : velocity of the y-axis direction
V : velocity refer to $V=\sqrt{v^2+u^2}$
ζ : water level above reference
h : depth $\zeta+d$
d : water level bottom reference (m)

Flow equation of x-axis 2D flow is as follows:

$$\frac{\partial u}{\partial t}+u\frac{\partial u}{\partial x}+v\frac{\partial u}{\partial y}+g\frac{\partial \zeta}{\partial x}+g\frac{u|V|}{C^2h}+au|u|=0 \quad (2)$$

Information:

C : Chezy coefficient
a : wall roughness coefficient

Flow equation of x-axis 2D flow is as follows:

$$\frac{\partial v}{\partial t}+u\frac{\partial v}{\partial x}+v\frac{\partial v}{\partial y}+g\frac{\partial \zeta}{\partial x}+g\frac{u|V|}{C^2h}+av|v|=0 \quad (3)$$

HEC-RAS can be used to simulate flood inundation of 1D or 2D flow or a combination of 1D and 2D using the Saint Venant or Diffusion Wave equation in both permanent and non-permanent flow (steady and unsteady one-dimensional flow model). Research that uses the application of HEC-RAS 5.0 is Sulaeman (2017), Mamuaya (2018) and Alsorigussa (2018).

2.3. Calibrating Model

Validation is needed so that the accuracy of the results of modeling done can be known. Parameters that can be used to carry out the validation process are the elevation of the surface of the former floodwaters in the field with the results of the analysis on the HEC-RAS program.

Validation is done using the root mean square error (RMSE) method. RMSE is calculated by squaring the error divided by the amount of data, then rooted. If the RMSE value is close to 0, then the model is considered good or can represent the existing conditions in the field.

$$RMSE = \sqrt{\frac{1}{n}\sum_{i=1}^{i}(Hobs - Hsim)^2} \quad (4)$$

3. Methods of the research

3.1. Research sites
This research takes a case study in the Jeneberang Sub-watershed in the western part after the Bili-Bili dam, where the outflow discharge of the dam is channeled through the spillway to the Jeneberang river along ± 30 km which then empties into the Makassar Strait (Figure 1).

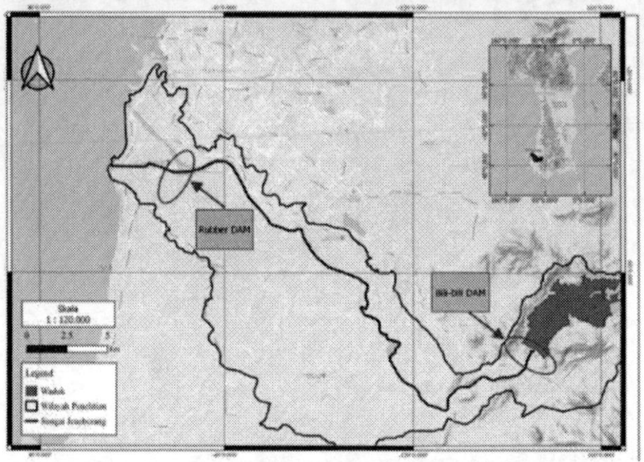

Figure 1 Research sites in the Jeneberang Sub-watershed downstream

3.2. Data collection
In this study some secondary data are used, namely the Bili-Bili dam outflow discharge data as input data on the upstream boundary condition and the relationship data between the discharge and river water level (rating curve) of rubber dam (BBWS Pompengan Jeneberang). Measurement of flood inundation height is carried out by measuring the former inundation and interviewing affected communities.

3.3. Hydraulic Model
The hydrodynamic model is used to calculate the elevation of flood water levels. With topographic data input in the form of digital elevation model (DEM) and a base map in the Mapper RAS menu, it can be simulated and modeled on characteristics, flood inundation area, flow velocity, and flood depth using HEC-RAS 5.0. In this study, the outflow spillway hour flow rate data of the Bili-Bili dam during the high rainfall events on January 17 to January 27, 2019 became the upstream boundary conditions and the rating curve of the rubber dam in the downstream area of the Jeneberang river was made as downstream boundary conditions.

The components of the geometry model used in this study are features of river geometry parameters that are connected schematically, with various hydraulic elements in the river which are the physical representation of the river. River geometry data used in this study consist of cross sections, banks, and flow paths. These four-geometry data are configured in the RAS Mapper menu integrated in the HEC-RAS program.

For the flow, this research used unsteady flow. The boundary conditions in unsteady flow can be divided into 2, namely upstream boundary conditions and downstream boundary conditions. The upstream boundary conditions are applied as flow hydrographs with connection to the time (Flow Hydrograph). Meanwhile, the downstream boundary conditions are needed at the downstream end, which is not connected to the reach or other storage area.

3.4. Calibration of Simulation Results

Calibration of simulation results on the HEC-RAS program are conducted using the flood events on the study site that occurred on January 21 to January 23, 2019 through observation. The calibration data are collected based on the interviews with affected residents and the measurement of inundation height in the field from the former flood that is still visible in the walls of people's houses. The research flowchart can be seen in Figure 2

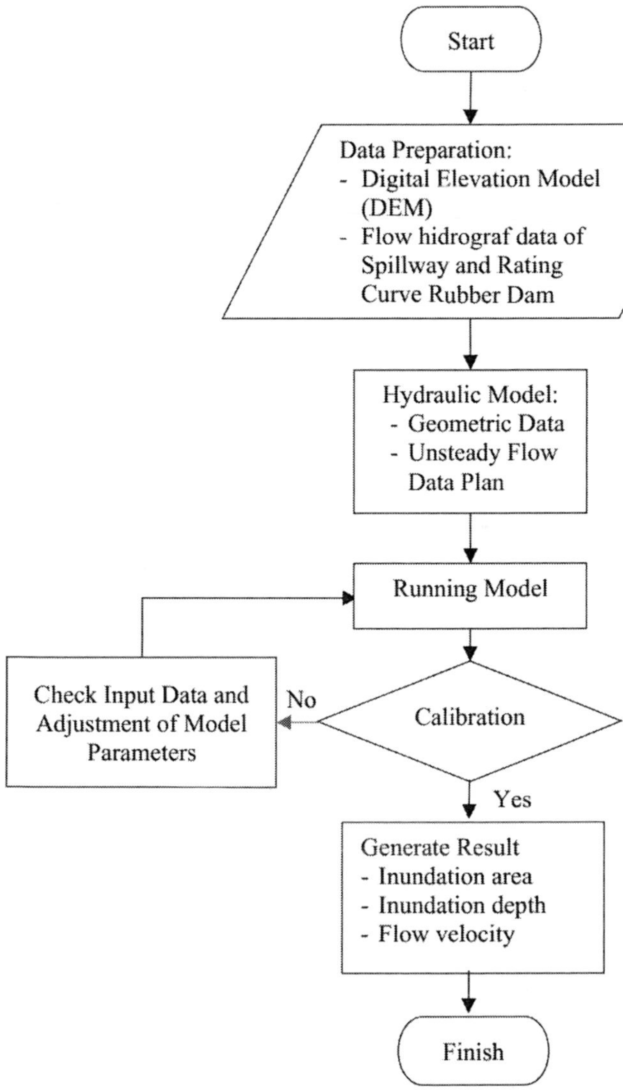

Figure 2 Research Flowchart

4. Results and discussion

4.1. DEM Data Processing

DEM data are used as terrain conditions at the study site affected by the overflow of the Jeneberang river processed using the help of a GIS-based program. Based on the DEMNAS map data sheet, it is known that the research location is divided into several map sheets (Figure 3). These map sheets are then combined into 1 sheet according to the area to be analyzed using the RAS-Mapper program in the HEC-RAS program sub-menu.

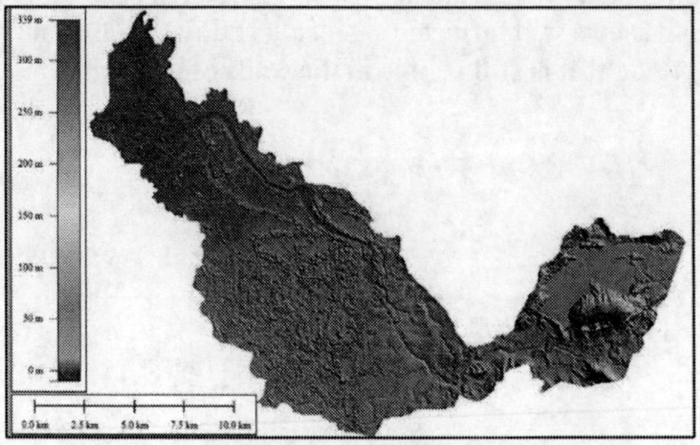

Figure 3 DEM map on research location
(source: modelling result)

The DEM data are used to get the catchment area delineation from the research location using a GIS based program. Digital elevation model data used are sourced from the DEMNAS website of the Geospatial Information Agency (http://tides.big.go.id/DEMNAS/) in the Geotiff32bit float format with a resolution of 0.27-arcsecond (8.1 mx 8.1 m) and a scale of 1: 25,000 using the EGM2008 datum with the geographic system coordinate projections, which is converted using the 50S zone's WGS84 UTM projection.

The hydrograph data on spillway discharge at the Bili-bili dam on January 17 to January 27, 2019 (11 days) are the upstream boundary conditions, whereas the rating curve of the rubber dam data located at downstream of Jeneberang river are the downstream boundary conditions.

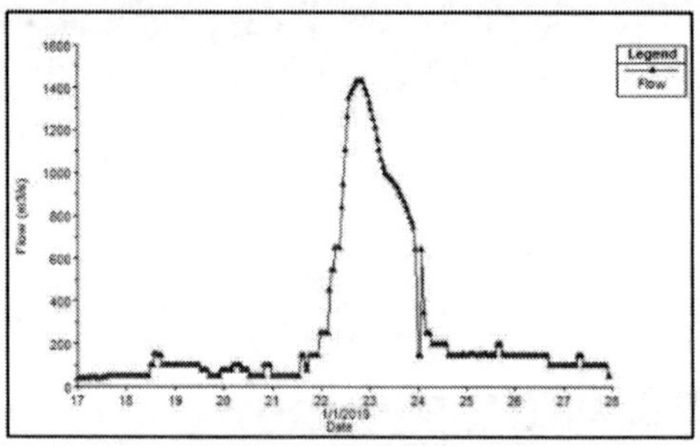

Figure 4 Graphic of Hydrograph on spillway of Bili-Bili dam

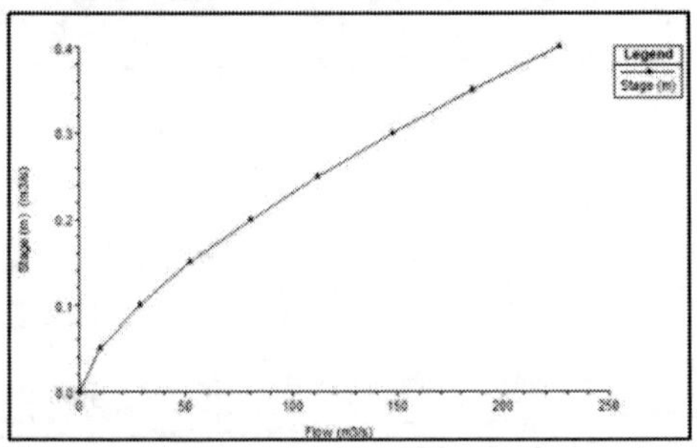

Figure 5 Graphic of rating curve on the rubber dam

(Source: Balai Besar Wilayah Sungai Pompengan Jeneberang)

In Figure 4, it can be seen that there was an increase in the flow of discharge through the outflow of the Bili-Bili dam, starting on January 22, 2019 with the recorded debit from 245 m3/s at 03.00 WITA increasing to 545 m3/s at 05.00 WITA until its peak was 1,426 m3/s at 20:00 WITA. Then it gradually dropped and returned to normal on January 24, 2019 to 246 m3/s at 04.00 WITA. Then in downstream, stage rating curve for water level is depicted on the vertical axis while discharge is the horizontal axis on the cross section (Figure 5).

4.2. Hydraulic Modeling

The results of the DEM map delineation are used as an input for terrain model in the RAS Mapper menu in the HEC-RAS 5.0 program. Following is the imitation of Jeneberang river geometry in HEC-RAS 5.0 including river coverage, cross sections, lateral structures and 2D areas that include inundation areas

(2D areas). Geometry model of the Jeneberang river on HEC-RAS in Figure 6 shows that the upstream boundary condition is the hydrograph of spillway from the Bili-Bili dam, while the downstream boundary condition is the rating curve of the rubber weir. Along the river there is a lateral structure which is a direct runoff connection from the Jeneberang river. The distance between cross sections (cross section) is ± 100 m with a total of 270 cross sections (Figure 6).

The flow scenario is unsteady flow. The upstream boundary condition data can be seen at figure 4, whereas the downstream boundary conditions can be seen at figure 5.

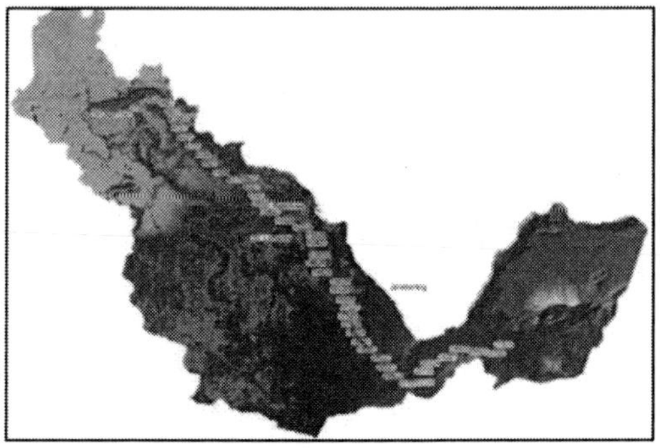

Figure 6 Geometry model of Jeneberang river

4.3. Result

After running simulation using HEC-RAS, it can be seen that in the side view of the left and right overbanks profile there is an increase in the river water level through the riverbanks (figure 7 and 8). The slope of the river from the Bili-bili dam at RS 2070 tends to be critical up to RS 9228 and the influence of the narrowing of the river crossing width at RS 9327 to RS 8326 (figure 9) can be a cause of the overflow of Jeneberang river if flowing is 1,426 m3/d suddenly.

The simulation results of the profile of the water surface on this longitudinal surface are then used as the boundary conditions for runoff flow simulations in the 2D model and the determination of the lateral structures as a medium for connecting 1D and 2D areas. The lateral structure is made following the topographic profile of the river flow boundary that directly intersects the 2D area.

The cross-section profiles of the simulation results will be analyzed as a reference in making 2D areas. If the cross section of the river is not able to withstand the flow of the river, the river will run off into the riverbanks and cause flooding. From 270 cross-section profiles, there are 4 cross-section profiles that overflow exceeding the carrying capacity, namely at RS 6625, RS 9327, RS 8126, and RS 17824 (figures 10 to 12).

The 2D flow area on the HEC-RAS is modeled on the area behind the embankment or overbank flow by connecting the river to the 2D area using a lateral structure. In mesh creation in 2D areas, it is implemented in the RAS Mapper sub-menu in the HEC-RAS application. In this model, a 10x10m mesh was used with the segmentation of the stagnation area on the right and left sides of the river in accordance with the side view of banks profile (Figure 7 and 8) of 13 lateral structures and many cells on the mesh 207793 cells on the left side of the river. Whereas, in the right stream there are 6 lateral structures and 80311 cells.

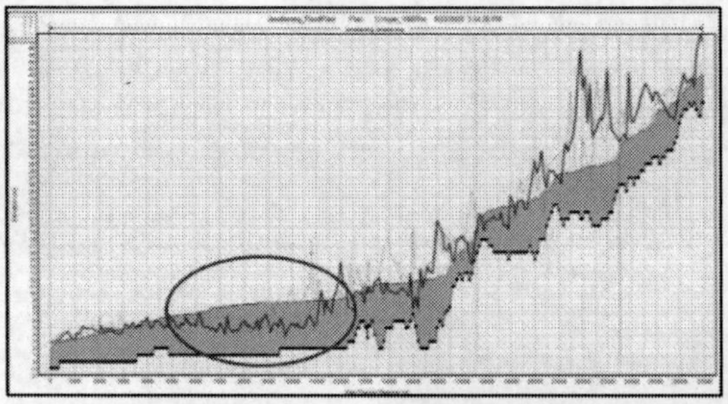

Figure 7 Left over banks profile

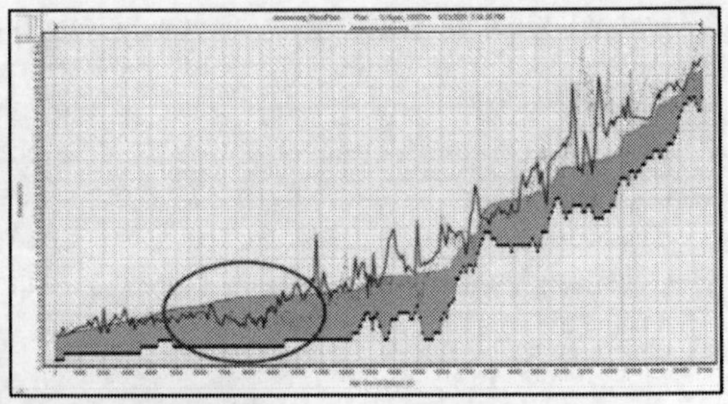

Figure 8 Right over banks profile

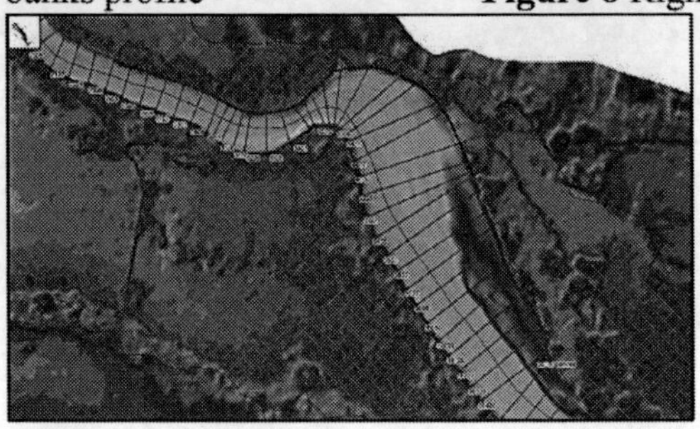

Figure 9 River reach section 9327 to 8326
(Source : Simulation result)

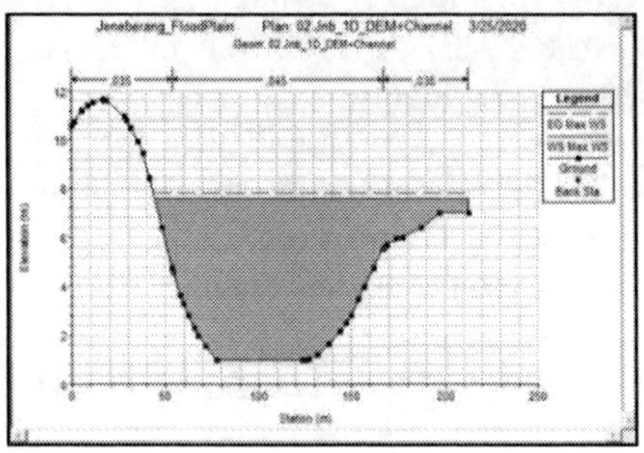

Figure 10 Water Surface Elevation RS 6625

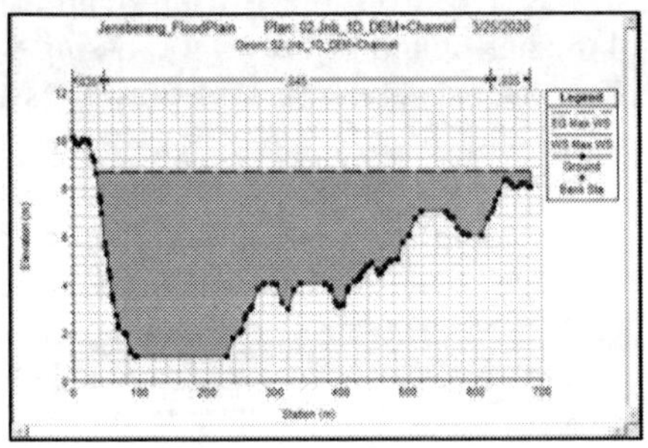

Figure 11 Water Surface Elevation RS 9327

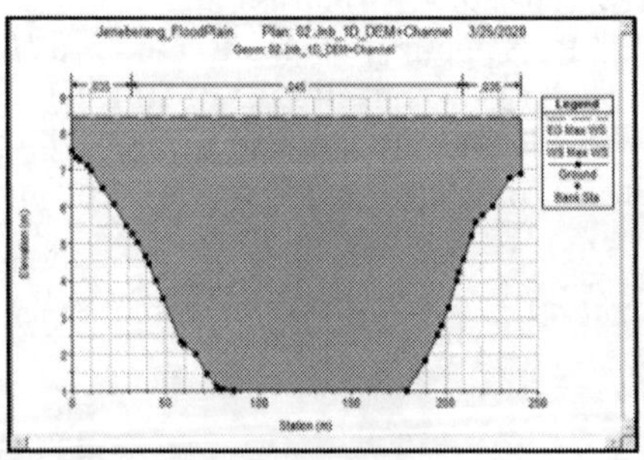

Figure 12 Water Surface Elevation RS 8126

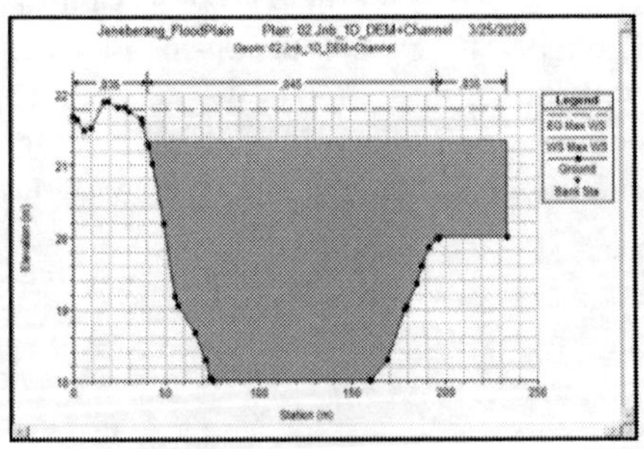

Figure 13 Water Surface Elevation RS 17824

(Source : Simulation result)

IOP Conf. Series: Materials Science and Engineering **930** (2020) 012071 doi:10.1088/1757-899X/930/1/012071

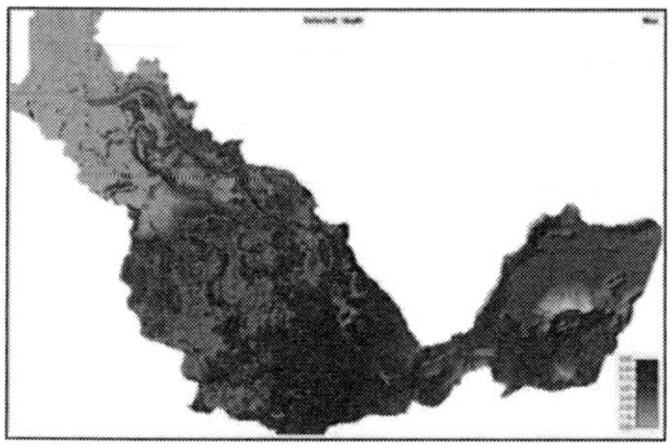

Figure 14 Result simulation of Depth 2D Jeneberang river

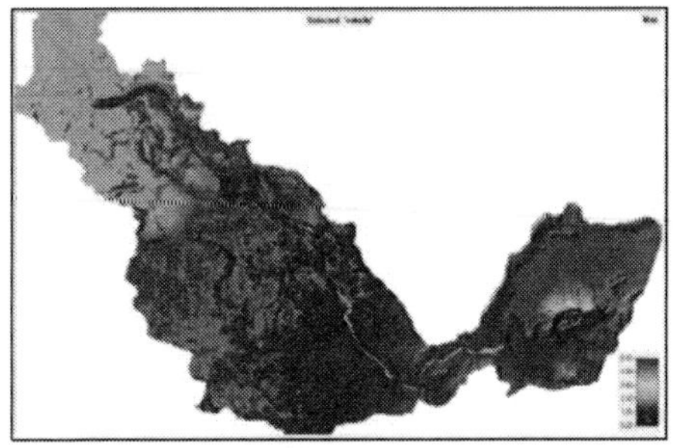

Figure 15 Result simulation of Velocity 2D Jeneberang river

(Source : Simulation result)

From the results of field observations, 7 measurement points were determined based on the former high inundation within two sub-districts. The 7 measurement points can be seen in figure 15.

Figure 16 Measurement point limit inundation field observations

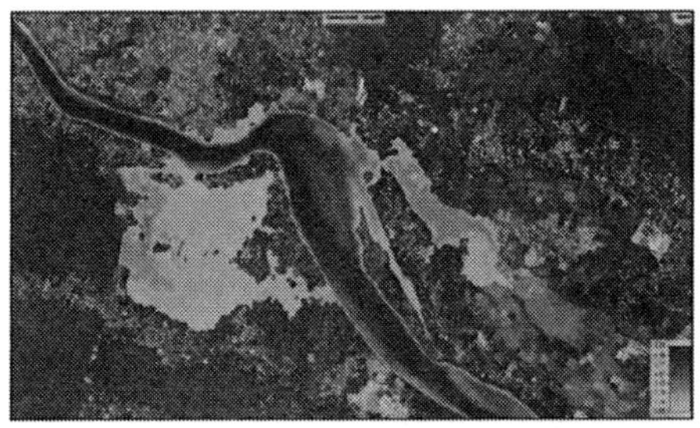

Figure 17 Area and inundation height of simulation results

(Source : Simulation result)

The results of the simulation are known that the flood inundation area that occurs is 1.7 km2 with the flow velocity at the maximum river discharge conditions between 0.57 to 4.84 m3 / s and the depth varies between 1.70 to 7.38 m. The overflow of the Jeneberang river passes through the right bank first and enters the settlement area in Somba Opu sub-district at RS 9027 on January 22, 2019 at 09.54 WITA with a speed of 0.068 m3 / d until the peak on January 23, 2019 at 02.17 Eastern Indonesian Time and receds at 22.31 Eastern Indonesian Time. As for the Pallangga sub-district, the river flow overflowed at RS 8126 on January 22, 2019 at 13.08 WITA with a speed of 0.054 m3 / until the peak on January 23, 2019 at 05.24 WITA and then slowly receded at 04.37 WITA. To test the reliability of the simulation model, then measurements were taken of flood inundation in the field, measurements were taken at 7 points of flood on the former flood on the walls of people's houses.

From Graphic 1 below, it appears that the simulation results correlate quite well with the results of the field observation using RMSE, which is 0.37. From the RMSE calculation to test the model, an RMSE value of 0.37 is obtained where the data will be considered more accurate if it approaches 0, then based on the test of the RMSE data the HEC-RAS modeling results are considered to be feasible, so the modeling results can be used as information on flood disaster mitigation efforts in the preparedness stage.

The results of the field survey are based on observations of traces or tracks of flooding and interviews with local communities that are highly biased. For better results, it is recommended to use river discharge and also carry out hydrological modeling to obtain truly accurate data.

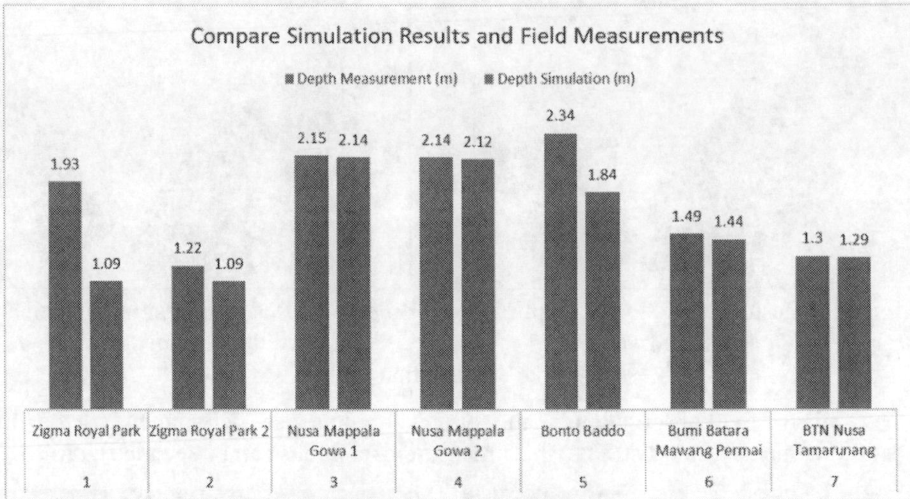

Graphic 1 Comparison of inundation height from field observations and simulation results (Source : Simulation result)

5. Conclusion

Based on the analysis and discussion above, several conclusions can be drawn, namely:

1. Discharge from the outflow of the Bili-Bili dam on January 17, 2019 to January 27, 2019 can cause the overflow of the Jeneberang river downstream along the riverbank on River Sta (RS) 6625 to 9327, 6928 to 8126 and 17824 to 19225, and affect by narrowing of the river crossing width on the River Station.
2. In the downstream Jeneberang watershed case study, 2-dimensional numerical modeling can be used to analyze the Jeneberang river runoff that has runoff because the cross section of the river is not able to hold the flow of the river flow to the area directly adjacent to the river.
3. Based on data from flood inundation height measurement results in the field, the post-assessment simulation model correlates reasonably well with the results of the HEC-RAS simulation using RMSE, which is 0.37. It is known that the area of flood inundation is 1.7 km2 with the speed at maximum river discharge conditions between 0.57 m3/s to 4.84 m3/s and depths between 1.70m to 7.38m.
4. The HEC-RAS 5.0 program can be used for mapping flood inundation, so it is expected to become a standard in planning flood control systems in the preparedness stage as an effort to mitigate disasters in the downstream Jeneberang watershed.

References

[1] Al Amin M B, Ulfah L Haki H and Sarino 2018 Simulasi karakteristik genangan banjir menggunakan HEC-RAS 5 *Studi Kasus Subsistem Sekanak di Kota Palembang* (Jurnal Penelitian dan Kajian Bidang Teknik Sipil: Cantilever) **7** 2

[2] Alif C J and Andawayanti U 2018 Aplikasi HEC-RAS untuk analisa dan penanganan banjir di sungai mujur, tempeh lor, kabupaten lumajang (Universitas Brawijaya)

[3] Alsorigussa P, Pandulu G D and Sulistyani K F 2019 Evaluasi kapasitas sungai haruyan hilir kabupaten hulu sungai tengah provinsi kalimantan selatan dengan menggunakan HEC-RAS 5.0 *Sem. Nas. Tek. Ind. Ling. Infra.* (Prosiding SENTIKUIN) **2**

[4] Brunner and Gary W 2014 Combined 1D and 2D modeling with HEC-RAS *Eng. Hyd. Eng. Cen.* (US Army Corps: Davis CA)

[5] Brunner and Gary W 2016 HEC-RAS River analysis system 2D modeling user's manual *Eng. Hyd. Eng. Cen.* (US Army Corps: Davis CA)

IOP Conf. Series: Materials Science and Engineering **930** (2020) 012071 doi:10.1088/1757-899X/930/1/012071

[6] Mamuaya F L, Sumarauw J S and Tangkudung H 2018 Analisis kapasitas penampang sungai roong tondano terhadap berbagai kala ulang banjir *Jur. Sip. Stat.* **7** 2

[7] Marhendi T, Wardana P N Nurhadi S Bramanti I A P 2017 Alternatif pengendalian banjir kali juana berbasis model HEC-RAS *Jurn. Ilm. Din. Rek.* **13** 1 p 37-42

[8] Pusat Pendidikan dan Pelatihan Sumber Daya Air dan Konstruksi (BPSDM) 2017 *Mod. Oper. Wad.* (Bandung: Jawa Barat) **8**

[9] Sisda 2019 Pola operasi pintu spillway balai besar wilayah sungai pompengan jeneberang *Bahan Informasi Bendungan Bili-bili Kabupaten Gowa* (Makassar: Sulawesi Selatan).

[10] Siregar R I and Indrawan I 2016 Studi komparasi permodelan hidrologi dan permodelan hidrolika dalam memprediksi banjir *Konf. Nas. Tek. Sip.* (Universitas Atma Jaya: Yogyakarta) **10**

[11] Seniarwan, Baskoro D T and Gandasasmita K 2013 Model spasial genangan banjir *Studi Kasus Wilayah Sungai Mangottong, Kabupaten Sinjai, Provinsi Sulawesi Selatan* (Majalah Ilmiah Globe) **15** 1

[12] Sulaeman A, Ery S and Sumiadi S 2017 Analisis genangan banjir akibat luapan bengawan solo untuk mendukung peta risiko bencana banjir di kabupaten bojonegoro *Jur. Tek. Peng.* **8** 2 p 146-57

[13] Utomo et.al 2019 Analisis banjir dan pemetaan kawasan terdampak banjir di kelurahan laweyan, kota surakarta *Mat. Tek. Sip.* **7** 3

[14] Wasim Z E and Wardhana P N 2018 Flood mapping of juwana river with unsteady flow simulation using hec-georas program

IOP Conf. Series: Materials Science and Engineering **930** (2020) 012072 doi:10.1088/1757-899X/930/1/012072

Numerical rainfall-runoff model of Cimanuk Watershed before and after the operation of Jatigede Reservoir

R M Widayati[1*], and U Lasminto[1]

[1] Department of Civil Engineering, Faculty of Civil, Planning and Geo Engineering, Sepuluh Nopember Institute of Technology, Surabaya, East Java, Indonesia

*Corresponding author's e-mail: ramadani.18031@mhs.its.ac.id

Abstract. Cimanuk watershed is one of the watersheds in West Java that has critical land. Critical land resulted in a Cimanuk watershed becomes a flood and drought. The management of Cimanuk Watershed water availability is done by regulating water supply from the Cimanuk River with the contruction of Jatigede Reservoir. The watershed hydrological process is the relationship between inputs in the form of rainfall, process, and output in the form of runoff. The process of converting rainfall into runoff is very complex, therefore, simplifying the rainfall-runoff model approach is needed. Hydrologic Engineering Center-hydrologic Modeling System (HEC-HMS) is one of the models for simulating rainfall into runoff. The purpose of the study is to simulate the availability of water surface at Cimanuk Hulu watershed using HEC-HMS hydrological modeling system based on before and after the operation of Jatigede reservoir. The results showed that the calibration of the model produced an NSE value of 0,640. On the other hand, the best validation results was in 2018 with NSE of 0,576 with the peak discharge value before the operation of the reservoir was of 456,8 m^3/s, and after modified with the presence of Jatigede Reservoir, peak discharge became 131,9 m^3/s.

1. Introduction

Cimanuk watershed is one of the watersheds located in West Java within the working area of the Balai Besar Cimanuk-Cisanggarung. Geographically, Cimanuk Watershed is at $7{,}45° - 6{,}42°\ LS\ and\ 107{,}6° - 108{,}8°\ BT$, while administratively, it is located in four districts, including Garut Regency, Sumedang Regency, Majalengka Regency, and Indramayu Regency. The total area of Cimanuk watershed is 3.216,14 km^2 with Cimanuk river becomes the main river having a length of 258 km [1]. The Cimanuk river is an upstream in Garut Regency and a downstream in Indramayu Regency.

Based on information from the Cimanuk-Citanduy BPDAS in 2003, Cimanuk watershed has a critical land reaching 178.794 Ha on forest and non-forest area. The biggest critical land is in Garut Regency, reaching 90,045 hectares [2]. This critical land has caused worsening hydro-orological condition in the watershed. This can be seen from the very large discharge fluctuation in the Rentang Weir, $Q_{maks} = 1.004\ m^3/s$ and $Q_{min} = 4\ m^3/s$ [1].

The impact of the critical land in Cimanuk watershed is frequent flooding, drought, erosion, landslides, and sedimentation. The occurrence of flood in the Cimanuk watershed is a routine event that occurs both upstream and downstream of the watershed. The management of Cimanuk watershed water availability is carried out by the construction of Jatigede Reservoir. This reservoir is a multifunctional

Content from this work may be used under the terms of the Creative Commons Attribution 3.0 licence. Any further distribution of this work must maintain attribution to the author(s) and the title of the work, journal citation and DOI.
Published under licence by IOP Publishing Ltd

reservoir that aims to control flood, raw water supply, irrigation, and hydropower in Cimanuk Watershed.

Efforts to conserve Cimanuk watershed have been carried out by several researchers as well as policies and evaluations from the BBWS Cimanuk-Cisanggarung. Several studies have been carried out such as the research conducted by Santoso on the study of determining the priotity sequence of the management of critical land in Cimanuk Hulu watershed using delphi method, resulting in the conclusion of increasing criteria for determining critical land priority [3]. Furthermore, regarding the geomorphological characteristics, Cimanuk watershed is composed of easily destroyed rocks which can increase the rate of erosion and cause siltation of Jatigede Reservoir [4]. In addition, a study has been conducted on the Cimanuk Hulu flood analysis caused by land use in a poorly precise area, high rainfall intensity, and population density in flood prone areas [5]. Research conducted by Pradwipa regarding the utilization of Jatigede reservoir as a surface water resource was carried out based on the history of Cimanuk river discharge which was on the basis on the upstream recording of the Jatigede reservoir on the use of the reservoir as the priority for irrigation and hydropower [6].

The watershed hydrology process is the relationship between input/inflow in the form of rainfall, process, and output in the form of runoff. The process of transforming rainfall into runoff is a very complex [7]. To simplify such a very complex process, a rainfall-runoff model approach is needed [8]. The rainfall-runoff model can predict the daily and monthly runoff value of a watershed based on the rainfall, evaporation, infiltration and watershed characteristic parameters. Hydrologic Engineering Center-hydrologic Modeling System (HEC-HMS) is one of the rainfall-runoff models developed by U. S Army Corps of Engineering. HEC-HMS simulates watershed hydrological processes by simplifying systems in nature based on watershed characteristics.

Based on research related to the condition of the Cimanuk watershed, it is necessary to analyze the use of Jatigede reservoir on the conservation of surface water resources in Cimanuk watershed. Therefore, this research used rainfall-runoff model on the availability of Cimanuk watershed in the conditions before and after the operation of Jatigede reservoir using the HEC-HMS rainfall-runoff model.

2. Research Method

2.1. Study Area

The location of this research is the Cimanuk Hulu watershed in the Cimanuk-Cisanggarung BBWS working area. The Cimanuk Hulu watershed which is the watershed of the Jatigede Reservoir in this study has an area of $1.457{,}26\ km^2$ and a river length of $105{,}1\ km$. The Cimanuk Hulu Watershed is dominated by a rather steep to steep area with a slope of > 15% [9]. The watershed is in the administrative area of Garut Regency and Sumedang Regency. Here, the Jatigede Reservoir becomes a multifunctional reservoir in optimizing the use of surface water in Cimanuk watershed. Jatigede Reservoir is located in Cijeungjing village, Jatigede District, Sumedang Regency. As a multifunctional reservoir, Jatigede reservoir is used for irrigation, hydropower, raw water, and flood control.

2.2. Research Data and Processing

2.2.1. Research Data

The data used in this research are hydrological, climatological, spatial data, profile of Jatigede Reservoir, Jatigede Reservoir capacity arch, and the profile of Cimanuk River. Hydrological data in the form of Cipasang post daily discharge data were obtained from BBWS Cimanuk-Cisanggarung. Daily rainfall data used in this study consisted of sta rainfall data from Sta. Jatigede, Sta. Bayongbong, Sta. Cikajang, and Sta. Leuwengitiis. The rainfall data were obtained from BBWS Cimanuk-Cisanggarung. Climatology data were obtained from BMKG online data that can be downloaded through the official website of BMKG (http://dataonline.bmkg.go.id/). Climatological data obtained were in the form of daily wind speed data, daily sun radiation, daily temperature data, and daily humidity data. Spatial data used were in the form of Digital Elevation Model (DEM) maps, land use map, and soil types maps.

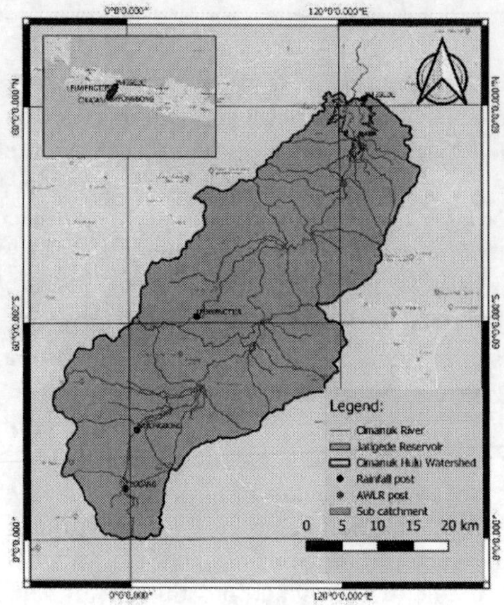

Figure 1. Location of study area

2.2.2. *Processing*

The framework in this study begins with the collection of secondary data. Then, the secondary data is processed. Secondary data processing such as rainfall data, discharge data, and climatological data that are time series which are done by filling in the missing data first. To fill in the missing data, the normal ratio method is used, and the result is tested for the data consistency that has been corrected. After completing the time series data, then the data are processed data according to the data requirements entered in the HEC-HMS model. The rainfall data of each rainfall station are calculated to be the regional average using Thiessen method, while the climatology data are used to calculate the evapotranspiration value using Panman modification method.

After conducting secondary data processing, the model development is carried out using HEC-HMS 4.2.1. The structure of the development of the HEC-HMS model consists of six components in which there are various methods. Calculation that can be performed using HEC-HMS includes the precipitation model, runoff volume model, direct runoff model, baselow model, search model, and water control model [10]. This study will use the calculation of the precipitation model using specified hyetograph, evapotranspiration by manual calculation using the Panman modification method, calculation of surface flow using SCS UH, calculation of baseflow using constant monthly, and the routing calculation on the channel using kinematic wave.

Development of the HEC-HMS model begins with the creation of basin model. This model consists of the created watershed elements such as rivers, junctions, reservoirs, sinks, sub-basins that are arranged to describe the physical representation of the watershed [11]. In the basin model, there are parameters that describe the watershed characteristics. The parameters used in the study include of Manning coefficien, Curve number (CN), LagTime, Initial abstraction, impervious, canopy, surface, and baseflow. Furthermore, it is also carried out the making time series data components containing discharge data, rainfall data, and evapotranspiration data needed in this study. The latter makes a paired data component in the form of storage discharge which is used when simulating the operation of Jatigede Reservoir on the model.

After building the model, it is ready in run and calibrates the parameters that affect the characteristics of the watershed in the field. After the calibration, modifications are made by adding reservoir to the model. Adding the reservoir is done by duplicating the Basin Model during the calibration, then replacing Junction 12 elements into a reservoir. The results of the model before and after the operation of the reservoir were analyzed for the availability of surface water at Cimanuk Hulu Watershed.

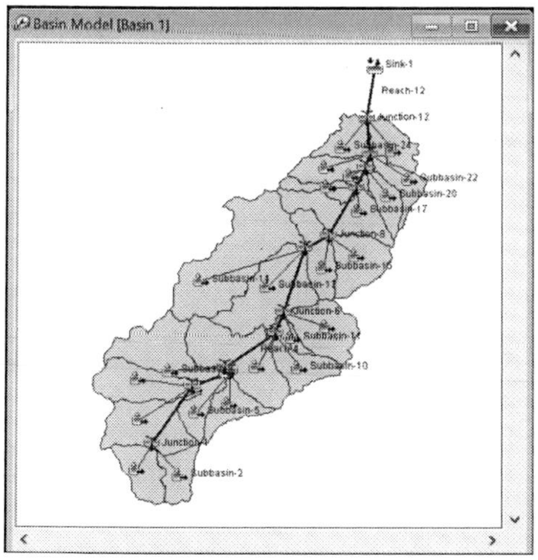

Figure 2. Basin model before the operation of Jatigede Reservoir
Source: Analysis result

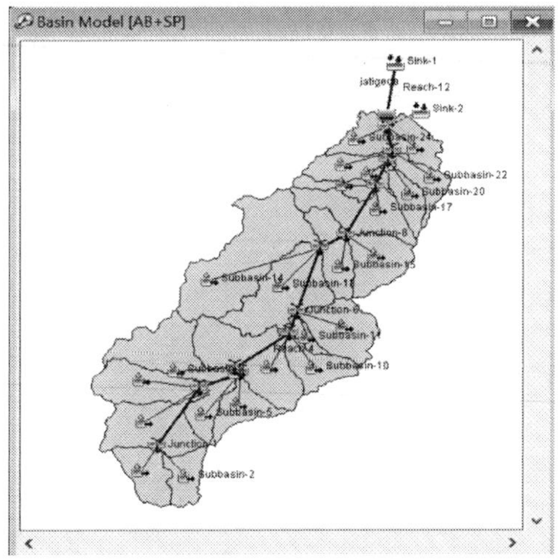

Figure 3. Basin Model after the operation of Jatigede Reservoir
Source: Analysis result

2.3. Calibration and Validation Model

Calibration is the process of selecting parameter combinations [12]. In principle, the model is able to provide responses that match expectations, but which models are able to present the conditions in nature. The output of the model is expected with data from field observation. In this study, the calibration was performed on the data discharge model for the observation of the discharge. The calibration model is carried out by trial and error. In this study, the calibration model was carried out on July 1, 2011 – December 31, 2014 on the recording of daily discharge in Cipasang post.

Validation is the process of evaluating the model to get a picture of the level of uncertainty in predicting the hydrological process. The data are used for the validation period using data outside of the calibration period data [12]. In this study, the data used for validation are data from Cipasang post discharge recording in 2016, 2017, and 2018.

To test the reliability of the model, this study used the Nash Sutcliffe Efficiency (NSE) test. The value of efficiency coefficient can range from infinite minus (low model) to 1 (perfect model) [13]. The NSE value criterion is said to be good if the NSE value is more than 0,75; If the NSE value is between 0,36 to 0,75 then the model meets, and if the NSE value is less than 0,36 then the model does not meet [14].

Table 1. Weight of the rain stations

Station	Area	Thiesen coefficient
Bayongbong	295,885	0,203
Cikajang	129,567	0,089
Jatigede	343,766	0,236
Leuwingitiis	688,046	0,472
Amount	1.457,263	1

Source: Analysis result

3. Results and Discussion

3.1. Secondary Data Analysis

The daily rainfall data from the four rain stations are calculated as the average rainfall of the regions using the polygon Thiessen method. The weight Thiessen at each rainfall station can be seen in table 1.

Spatial data in this study were used to obtain the delineation of the catchment area of the study location, topography, and rivers. Delineation of catchment area is obtained by watershed boundaries and sub- catchment division to be simulated. From the area of each sub-catchment, it will be used to calculate the Curve Number (CN), Initial abstraction, lagtime, canopy, surface, and impervious. The resulted topographic map will be used for the calculation of the Jatigede Reservoir rating curve. The value of curve number of each sub-catchment is obtained from the map overlay of land use, map of soil type, and boundaries of each sub-catchment using GIS tools.

The calculation of the rating curve of Jatigede Reservoir is based on the data planning data that has been done by the Jatigede Reservoir development planner. Data obtained from the planning that has been made include data on the effective width of the Jatigede Reservoir spillway and the maximum discharge that exits in the spillway when the four doors are open. In addition, the curved graph of the planned reservoir capacity can be used as an elevation reference used for rating curve calculation. The rating curve calculation can be carried out using the formula (1).

$$Q = C \times L \times H^{3/2} \quad (1)$$

where, Q is the discharge that passes through the overflow (m^3/s) ; C is the coefficient of runoff, L is the effective width (m) and H is the water pressure height above the lighthouse weir (m)

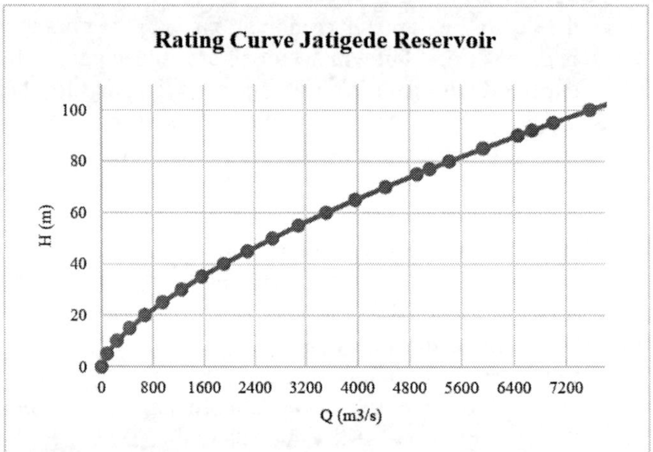

Figure 4. Rating curve of Jatigede Reservoir
Source: Analysis result

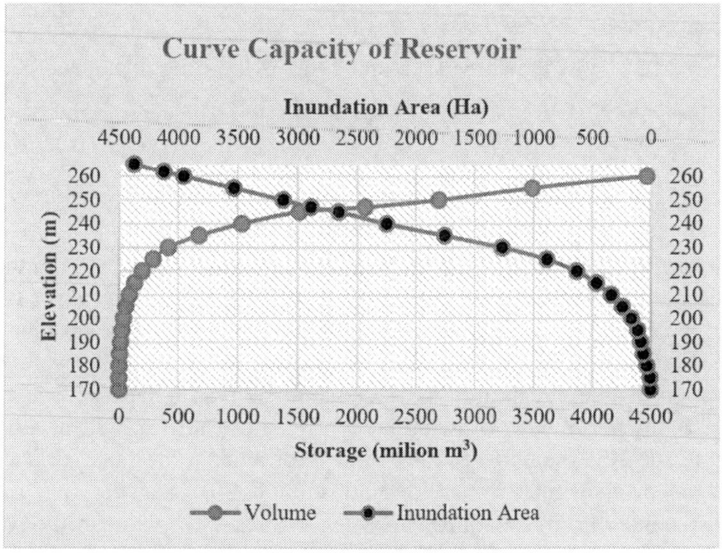

Figure 5. Curve capacity of the reservoirs
Source: BBWS Cimanuk-Cisanggarung

The calculation of Curve Number can be carried out using formula (2) [15].

$$CN_{composit} = \frac{A_i CN_i + A_{i+1} CN_{i+1} + \ldots + A_n CN_n}{\sum_{i=1}^{n} A_i} \quad (2)$$

where, A is the area of each type of land use (km^2) and CN is the curve number coefficient based on soil type and land use.

From the results the CN value is then used to calculate the initial abstraction of each sub-catchment. Initial abstraction is calculated using the following formula [15]

$$Ia = 0{,}2\,S \quad (3)$$

where, Ia is the initial loss of abstraction and S is the maximum storage capability.

$$S = \frac{25400 - 254CN}{CN} \quad (4)$$

where, S is Parameter retention and CN is the curve number.

Daily climatology data are used to calculate the daily evapotranspiration value of Cimanuk Hulu watershed. In this study, Penman Modification method is used to calculate the value of evapotranspiration using formula (5) as follows:

$$Etp = c \times \{W \times Rn + (1 - W) \times f(u) \times (ea - ed\} \quad (5)$$

where, Etp is evapotranspiration (mm), c is a condition adjustment factor due to weather at days and nights, W is a factor that affects the solar radiation, $f(u)$ is a function of wind speed in comparison, Rn is the radiation in the ratio of evaporation or solar radiation net (mm/day), ed is the real vapor pressure (mbar), and ea is a saturated vapor pressure (mbar).

Cimanuk River profile is used for lagtime calculation and river routing data in modeling using HEC-HMS. Calculation of the lagtime of each sub-catchment uses the following formula (7) [16]

$$Tc = \frac{L^{0{,}8}[S+1]^{0{,}7}}{1140 y^{0{,}5}} \quad (6)$$

$$Tlag = 0{,}6\,Tc \quad (7)$$

where, Tc is the concentration time (hours), L is the length of the main river (ft), S is the maximum retention $\left(S = \frac{1000}{CN} - 10\right)$, y is the slope (%), and $Tlag$ is the time difference between the center of mass of excess rainfall and peak of the hydrograph unit (hours).

3.2. Calibration

The results of the running model before calibration produce an NSE value of $-3{,}868$. To produce the optimum NSE, a trial error is carried out on the constituent parameters of the watershed. This study uses baseflow parameter with a minimum discharge value recorded during the simulation period. The resulting model discharge is far above the observed discharge, then a trial error is carried out on the baseflow to obtain an NSE value of $-2{,}665$. With NSE $-2{,}665$ the model discharge still exceeds the observation discharge, so it needs to increase the water loss in the watershed. By conducting a trial error on the initial abstraction value, it can produce NSE value $-2{,}642$. Then, the trial error of CN is up to the NSE value of $-0{,}337$. Trial error of manning coefficient results in NSE of $-0{,}309$. Trial error of max storage is up to the NSE value to $0{,}100$. Trial error against the max surface of the NSE value becomes $0{,}376$. Trial error against impervious becomes $0{,}440$ and the last trial error of lagtime parameters gets NSE value of 0,640 with RMSE 30,3. With an NSE value of 0,640 the model that has been prepared is reliable enough to present the Cimanuk Hulu Watershed.

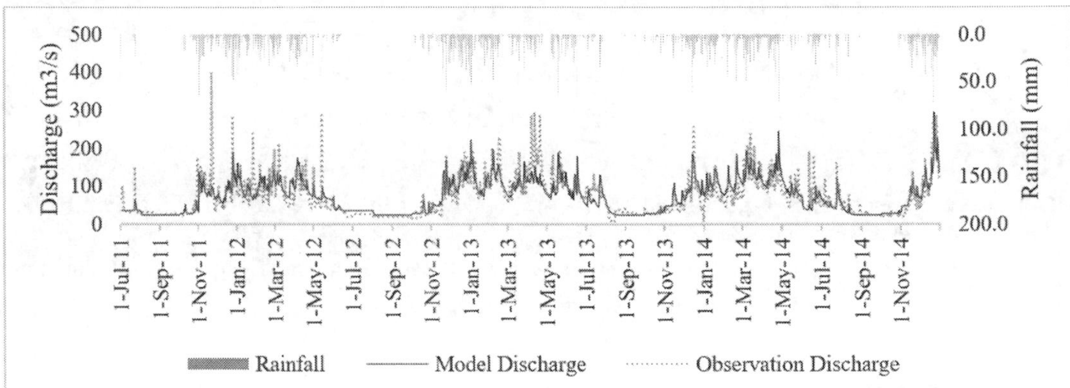

Figure 6. The relationship between precipitation data, discharge observation and discharge model.
Source: Analysis result

Based on the graph of simulation results using HEC-HMS, the model discharge can approach the observation discharge trand, but only a few moments can the model follow the peak and low discharge. The model tends to follow the rainfall data trand because the model simulates the watershed based on input of a rainfall process in the watershed form evapotranspiration infiltration, direct flow and baseflow based on the characteristics of watershed without any addition and taking of water in the watershed. Therefore, the observation discharge that shows high discharge during the rainfall is moderate and the discharge which is reduced during the rainfall does not occur due to the addition and taking of water occurring in the watershed.

3.3. Model validation

The model validation period is carried out using data in the years of 2016, 2017, and 2018. The parameter value of the watershed characteristic uses the optimum parameter value of the calibration period results. From the validation result to the model, it is obtained the value of Nash and RMSE as presented in table 2.

Table 2. Validation results table

Years	Nash	RMSE
2016	-0,104	65,9
2017	0,472	40,4
2018	0,576	35,3

Source: Analysis result

Based on the results of the validation of the model, the year of 2018 has the best result with Nash value approaching one (1) and the RMSE value approaching 0.

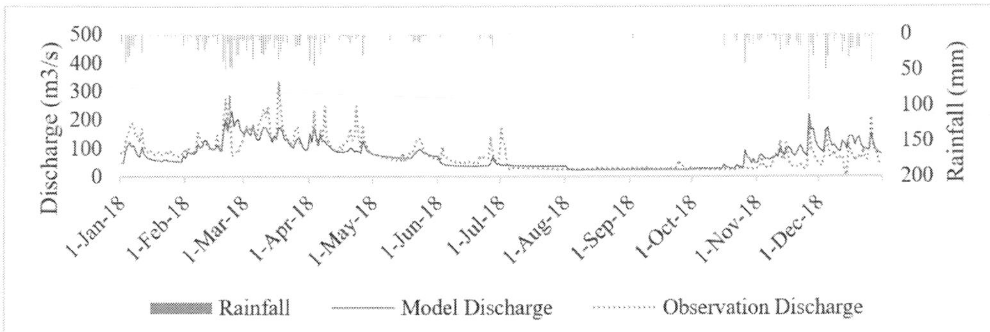

Figure 7. The relationship between rainfall, observation discharge and model discharge in 2018
Source: Analysis result

3.4. Results modeling before and after the operation of Jatigede Reservoir

3.4.1. Before the operation of the reservoir

Before the operation of the Jatigede reservoir, water flowing from the upstream will flow naturally towards the downstream. It can be seen from the inflow and outflow graphs presented in Figure 8, showing that inflow and ouflow are equally. Thus, the surface water cannot be utilized maximally. When abundant water flows and is wasted to the sea, it can even cause flood, otherwise, a shortage of water will be experienced during the dry months (seasons). Therefore, there is a need for management of the use of surface water, one of which is the existence of reservoir. Reservoirs can hold surface water when water surplus occurs and can be utilized during water deficit.

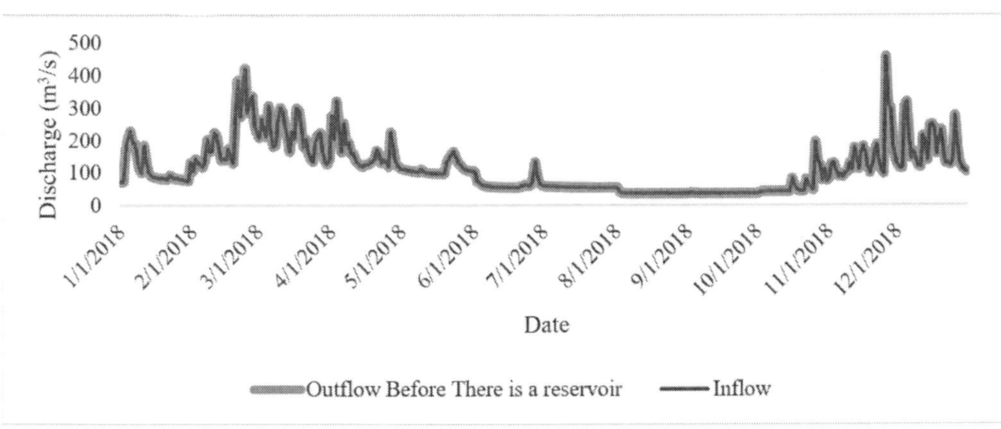

Figure 8. The relationship between Inflow and Outflow before the operation of the reservoir
Source: Research result

3.4.2. After the operation of the reservoir reservoir

Jatigede Reservoir is a multifunctional reservoir for flood control, raw water supply, irrigation, and hydropower. Jatigede Reservoir has three outlets and a spillway with four-door radial type operations. The spillway is in the body of the right bank of the cliff with a light elevation of + 247 m. Irrigation outlets are in the overflow body with inlet elevation of + 221 m. Divertion tunnel as a raw water outlet is under the spillway with inlet elevation of + 164 M. The last hydropower inlet apron is at elevation of + 221 m to the right of the weir pile and hydropower outlet is downstream of the Jatigede Reservoir.

Simulation of the operation of Jatigede reservoir using HEC-HMS was carried out using the outflow structure method with an initial condition of spillway elevation (+247 m). Resevoir operation simulation using HEC-HMS by adding outlets for water use in accordance with reservoir sunctions cannot be done optimally. Intake of water using an outlet cannot be controlled. Water in the reservoir will continue to flow through the lowest outlet. The operation of the door control door can only be done on the spillway. Therefore, to be able to take water in accordance with needs is done by modifing reservoir inflow. Before the water enters the reservoir, the water will be deflected using diversion element assuming the water for collection at the outlet has been taken first. The rest of the diverting water through the diversion will be fed into the reservoir and will overflow through the spillway. Water collection for the needs of each outlet and operation of the spillway door opening are carried out constantly during the simulation period.

Simulation results show the operation of the Jatigede reservoir can reduce peak discharge by 28,87 % and can control the surface water of Cimanuk watershed, so that it can be put into good use. Jatigede reservoir can fulfill the raw water need of 3,5 m^3/s and hydropower at 59,63 m^3/s. From the results of the operation of reservoir for raw water and hydropower, Jatigede reservoir can still filfull the irrigation need of 8,9 m^3/s. Water that has been used to turn turbines in the hydropower plant will be channeled back to Cimanuk river, so that it can be reused as a supply irrigation water at the downstream of Cimanuk watershed with an area of 90.000 ha.

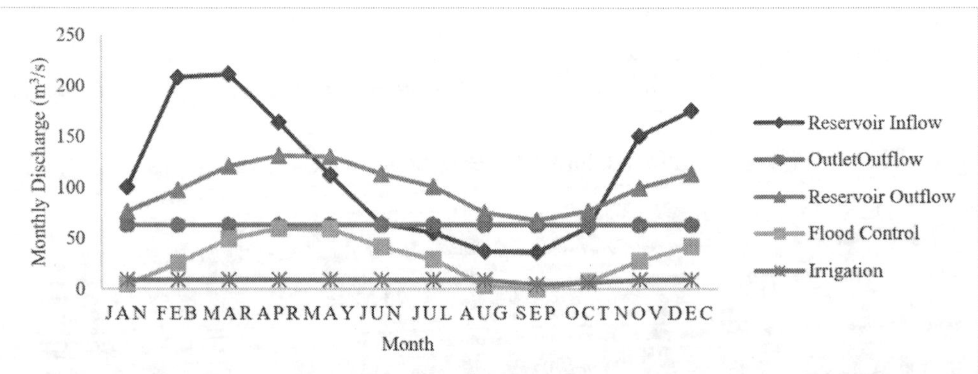

Figure 9. Jatigede Reservoir operating pattern in 2018
Source: Research Results

Table 3. Changes of peak discharge

Elements	Peak Discharges	
	Before	**After**
Junction 12/Jatigede	456,8 M3/S	131,9 M3/S
Reach 12	450,2 M3/S	131,9 M3/S

Source: Research Results

Based on the peak discharge value in the condition before the operation of the reservoir, it is obtained a value of 450,2 m^3/s in the downstream of Jatigede Reservoir. On the other hand, after the operation of the reservoir, the value has dropped to 131,9 m^3/s. Therefore, Jatigede Reservoir can reduce flood discharge by storing water in the reservoir.

3.5. Conclusion

Based on research that has been done, it can be concluded that the results of the Cimanuk Hulu watershed calibration produce an NSE value of 0,640 and RMSE value of 30,3. With the results of the NSE model, it is quite reliable in presenting the condition of Cimanuk Hulu watershed. Meanwhile, the best validation result is in 2018 with NSE value of 0,576 with the peak discharge value before the operation of the reservoir is of 456,8 m^3/s, and after the operation of the reservoir becomes 131,9 m^3/s. The availability of surface water in the Cimanuk watershed can be utilized properly after the operation of Jatigede Reservoir. This can be seen from the results of the operation of the reservoir which can fulfill the needs of raw water and hydropower in accordance with the purpose of dam development. Besides, Jatigede Reservoir can still provide irrigation water of 8,9 m^3/s.

Reference

[1] Sukardi S, Warsito B and Kisworo H 2013 *Pengelolaan Sungai di Indonesia*
[2] Anonim 2010 Pengelolaan sumber daya air wilayah sungai Cimanuk Cisanggarung
[3] Santosa R A B 2003 *Kajian Penentuan Urutan Prioritas Penanganan Lahan Kritis Studi Kasus Sub SUb DAS Cimanuk Hulu* (Bandung: Institut Teknologi Bandung)
[4] Sulaksana N, Sukiyah E, Sjafrudin A and Haryanto E . 2013 Karakteristik Geomorfologi DAS Cimanuk bagian hulu dan Implikasinya terhadap Intensitas Erosi Serta Pendangkalan Waduk Jatigede *Bionatura J. Ilmu-ilmu Hayati dan Fis.* **15** 100–6
[5] Savitri E and Pramono I B 2017 Analisis Banjir Cimanuk Hulu 2016 *J. Penelit. Pengelolaan Drh. Aliran Sungai* **1** 97–110
[6] Pradwipa D P, Jayadi R and Istiarto 2015 Kajian Pemanfaatan Sumberdaya Air Waduk Serbaguna Jatigede, Jawa Barat
[7] Munajad R and Suprayogi S 2012 Kajian Hujan–Aliran Menggunakan Model HEC–HMS Di Sub Daerah Aliran Sungai Wuryantoro Wonogiri, Jawa Tengah *J. Geogr.* **2** 150–7
[8] Harsoyo B 2010 Review Modeling Hidrologi Das Di Indonesia *J. Sains Teknol. Modif. Cuaca* **11** 41–7
[9] Fakhrudin M, Wibowo H, Ridwansyah I, Daruati D, Setiawan F and Sutrisno N 2013 Karakteristik Fisik DAS Cimanuk Sebagai Dasar Konservasi Waduk Jatigede *Pros. Pemaparan Has. Penelit. Puslit Geoteknologi LIPI* 353–63
[10] Anonim 2015 Hydrologic Modeling System HEC-HMS Applications Guide
[11] Prayudi M, Handayani Y L and Sujatmoko B 2017 Analisis Sensitivitas Parameter Kalibrasi dalam HEC-HMS *Jom FTEKNIK* **4**
[12] Indarto, Adriyani I and Novita E 2008 Kalibrasi Model IHACRES pada Dua DAS Identik *Din. Tek. SIPIL* **8** 89–100
[13] Allen R V, Rusnam, Arlius F and Herdianto R 2019 Analisis Perubahan Penggunaan Lahan Daerah Aliran Sungai (DAS) Air Dingin dan Dampaknya Terhadap Aliran Permukaan *J. Tek. Pertan. Lampung* **8** 153–233
[14] Liew M W Van and Garbrecht J 2003 Hydrologic Simulation of the Little Washita River Experimental Watershed Using SWAT *J. Am. Water Resour. Assoc.* 413–26
[15] Feldman A D 2000 Hydrologic Modeling System HEC-HMS Technical Reference Manual *Hydrogeol. J. Hydrol. Model. Syst. Tech. Ref. Man.*
[16] Anonim 2010 Time of Concentration *National Engineering Handbook* pp 1–15

Diversion canal to decrease flooding at Kemuning river, Sampang district

G Idfi[1*], T Rahayuningsih[1], and N Suryoputro[1]

[1] Department of Civil Engineering, Faculty of Engineering, Universitas Negeri Malang, Malang, Jawa Timur, Indonesia.

*Corresponding authors: gilang.idfi.ft@um.ac.id

Abstract. Flooding occurs when rain runoff water cannot be accommodated by a river, due to the flow of water that exceeds its drainage capacity. Flooding can occur anywhere, one of which occurs in the Kemuning river, Sampang Regency. The existing ag factor of the river profile that is experiencing siltation is also a cause of flooding, especially in the rainy season. This study aims to reduce flood discharge in the Kemuning river by planning a diversion channel as a solution using HEC HMS and HEC RAS modeling. Diversion channel is a divider discharge channel or shortcut to reduce flood discharge in rivers that flow directly into the sea. In this study three alternative solutions are planned, alternative 1 is planned to use a rectangular cross-section, alternative 2 is planned to use a trapezoidal cross-section and alternative 3 is planned to use a trapezoidal cross-section and adding and raising existing river embankments. The results of the analysis and planning showed that the flood discharge plan for the 25 year return period in the HEC HMS modeling was 781.9 m^3/sec. The results of the HEC RAS modeling with flood hydrograph input during the 25-year return period, Kemuning river conditions overtopping at all stations as high as 1-4 m. From the trial and error results in the HEC RAS modeling, the maximum capacity of the Kemuning river is 90 m^3 / second. Based on the maximum capacity of the Kemuning river, alternative diversion channels 1 and alternative 2 are planned to accommodate flood discharge of 681.9 m^3 / sec. An alternative diversion channel 3 is planned to only discharge flood discharges of 300 m^3 / sec by adding and raising embankments on the Kemuning river. All alternative diversion channel planning can reduce flood discharge by a percentage of 88.5% for alternatives 1 and 2, 38.4% for alternative 3. Based on the maximum capacity of the Kemuning river, alternative diversion channels 1 and alternative 2 are planned to accommodate flood discharge of 681.9 m^3 / sec. An alternative diversion channel 3 is planned to only discharge flood discharges of 300 m^3 / sec by adding and raising embankments on the Kemuning river. All alternative diversion channel planning can reduce flood discharge by a percentage of 88.5% for alternatives 1 and 2, 38.4% for alternative 3. Based on the maximum capacity of the Kemuning river, alternative diversion channels 1 and alternative 2 are planned to accommodate flood discharge of 681.9 m^3 / sec. An alternative diversion channel 3 is planned to only discharge flood discharges of 300 m^3 / sec by adding and raising embankments on the Kemuning river. All alternative diversion channel planning can reduce flood discharge by a percentage of 88.5% for alternatives 1 and 2, 38.4% for alternative 3.

Content from this work may be used under the terms of the Creative Commons Attribution 3.0 licence. Any further distribution of this work must maintain attribution to the author(s) and the title of the work, journal citation and DOI.
Published under licence by IOP Publishing Ltd

1. Introduction

Flooding can occur anywhere, especially in areas that are crossed or have river flow [2]. Floods that often occur in the Sampang Regency due to the overflow of the Kemuning river caused many losses. The overflow of the Kemuning river resulted in delays in transportation between regions around the location, one of which was the interruption of transportation activities between The Bangkalan Regency and Pamekasan Regency [3]. Because one of the locations of the flood was on the provincial road, it killed off-road user activity. Besides, floods that often hit with depths ranging from 0.4 to 2 m also killed local community activities which indirectly had an impact on the economy[9]. Many alternatives can be applied to overcome or at least reduce flood discharge in the Sampang Regency. One of the studies was conducted by the Department of public works, represented by the Pamekasan Irrigation Area Coordinator who planned a project entitled flood control and drainage project to control the Kemuning river flood.

The writer has another alternative, which is to plan the development of a diversion channel. Diversion channel is a divider debit channel or shortcut to reduce flood discharge on the river that can be directed directly to the sea or back to the river[1]. With the diversion channel, when the Kemuning river receives flood discharge from upstream and at the same time high rainfall is expected to reduce the flood discharge at the Kemuning river. Flood discharges that were originally only accommodated by the Kemuning river can be reduced by dividing flood discharges into the diversion channel. the reduction of flood discharge in the Kemuning river flow is directly proportional to the risk of inundation in The Sampang Regency because the flood discharge in the river can be reduced and does not overflow to undue places.

2. Research method

2.1 Research location

The Kemuning river is in Sampang Regency, Madura, East Java Province, and is located between 7°10 "-7°20" south latitude and 113°13'18 "-113°23'74" east longitude [3]. The Kemuning watershed has an area of 345 km2 with an elevation upstream +200 m and elevation downstream + 4 m above sea level. This river has a length of 58 km and empties into the Madura Strait. The Kemuning River is the estuary of several fairly large tributaries, such as the Kelokat river, the Serpang river, and the Gunung Maddah river. The upstream part of the Kemuning river is a barren, arid and dry hilly area. These areas are the Kedungdung sub-district, the Robatal sub-district, and the Omben sub-district. Also, the position of Sampang Regency is in the basin area. That is because, in the east, west, and north are higher land.

2.2 Research design

The initial concept of this plan is to reduce flood discharge in the Kemuning river when receiving peak discharge by planning a diversion channel. The method used in this planning is data analysis. Data used in this plan include population surveys of inundation, daily rainfall data, land use data, watershed area data (area topography), and river cross-section data[10]. This thesis uses two alternative diversion channel sections, namely quadrilateral, and trapezoid. As well as three alternative solutions planned namely rectangular cross-section channel diversion for The first alternative , trapezoidal cross channel diversion for the second alternative, and trapezoidal cross channel diversion with addition and height of embankments on the Kemuning river for the third alternative.

2.3 Data analysis

After the necessary data is collected then a hydrological analysis is carried out to obtain the value of the flood discharge plan which is used to determine a sufficient cross-section and can accommodate the flood discharge. In the hydrological analysis, there are several calculations before finally getting the flood discharge plan value [5]. The steps taken are an analysis of regional average rainfall, calculating the height of the planned rainfall, conducting a distribution test and drawing conclusions, calculating the curve number, calculating the time lag, calculating the flood discharge plan based on a 25-year return period (Q25) [8]. In the work of flood discharge, this plan uses the HEC-HMS SCS Model aids program [11][12].

After calculating the flood discharge plan is obtained, then the next step is to calculate the river cross-section capacity[6]. In this hydraulic analysis using HEC-RAS aids in the process[13]. The steps taken are to calculate the capacity of existing river conditions using HEC-RAS modelling[7]. Data needed for the modelling assistance program are river geometric data to illustrate river shape, river cross-section data to illustrate river elevation and state, manning coefficient, river flow data, and boundary conditions [4]. Then compare the river capacity with the discharge plan, plan the Diversion Channel, calculate the river capacity with the Diversion Channel using HEC-RAS modelling [15]

3. Result and discussion

3.1 Rainfall data analysis and Watershed

Rainfall data obtained from the Sampang District Public Works Irrigation Agency thirteen rain stations namely the Ketapang rain gauge station, Banyuates, Sokobanah, Sampang, Torjun, Jrengik, Omben, Camplong, Kedungdung, Robatal, Labuhan, Karang Penang, and Tambelang. The Kemuning watershed area of 358,165 km2 which can be seen in Figure 1. The data used is the maximum daily rainfall at each station for 24 years recorded from 1992 to 2015. From these data then performed calculations to get the average. The value of the coefficient of Thiessen Polygon is presented in the table 1. The source of rainfall data analysis is presented of figure 2 and table 2.

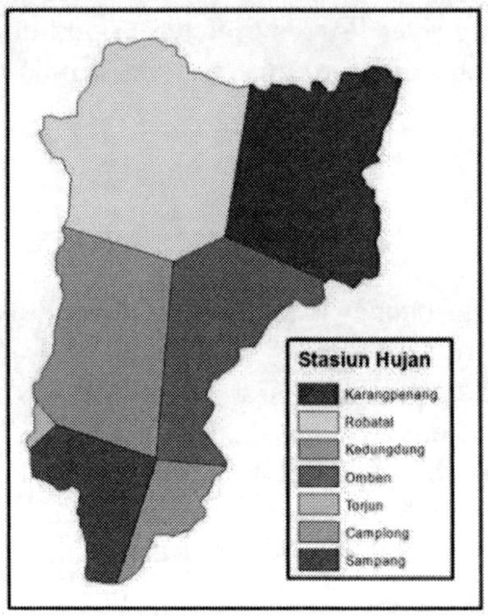

Figure 1. The Kemuning watershed.

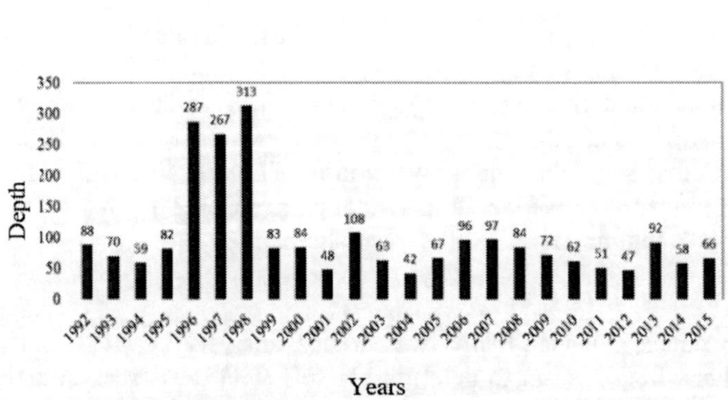

Figure 2. Rainfall data.

Table 1. Thiessen coefficient.

Station	Large (km^2)	Thiessen Coefficient
Sta. Robatal	98,244	0.274
Sta. Penang Reef	82,003	0.229
Sta. Lacquer	32,419	0.091
Sta. Torjun	1,845	0.005
Sta. Omben	56,826	0.159
Sta. Camplong	16,630	0.046
Sta. Kedungdung	70,198	0.196
amount	358,165	1

Table 2. The rainfall return period

Rainy Return Period	(mm)
2	73.444
5	121.200
10	171.361
25	265.514
50	365.624
100	501.315

3.2 Flood peak discharge of 25-years return period

From the results of running the HEC HMS program each time it will produce each hydrograph unit. Each hydrograph unit is presented in the figure 3.

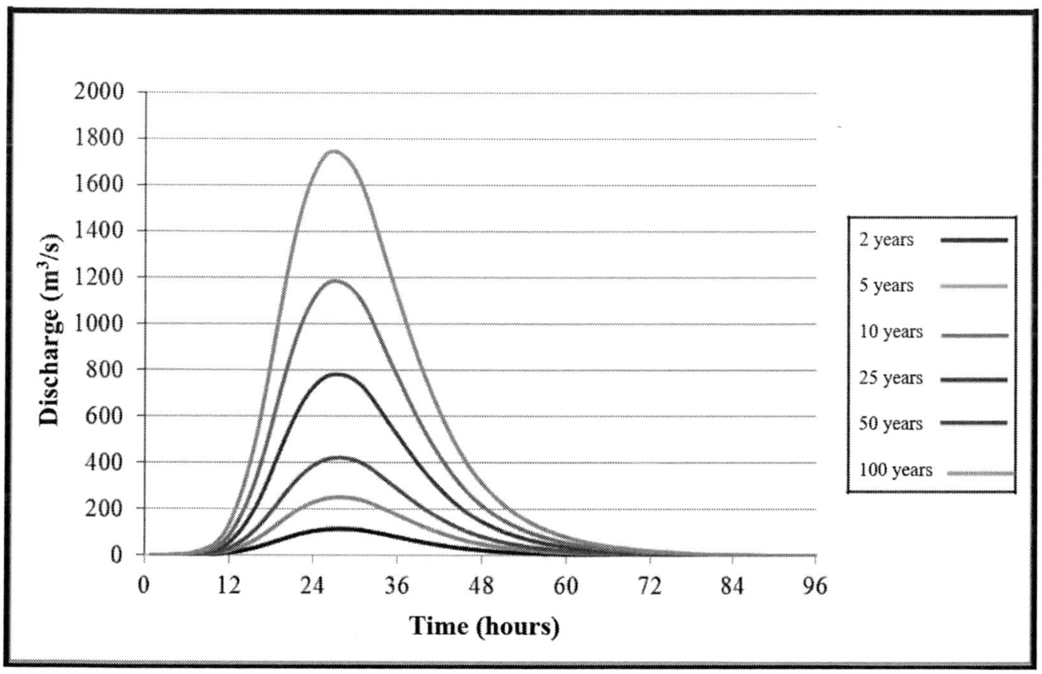

Figure 3. The hydrograph flood return period.

IOP Conf. Series: Materials Science and Engineering **930** (2020) 012073 doi:10.1088/1757-899X/930/1/012073

In this diversion channel planning, the return period flood discharge used is 25 years, this is because the kemuning river is categorized as a natural channel or main channel with a high risk. Peak discharge for the 25-year return period is 781.9 m3 / s.

3.3 The Flood water surface elevation using 25-years return period

From the results of the HEC RAS output obtained the ability of the river to accommodate the discharge that occurs [9]. From the HEC RAS output, it appears that the capacity of the river is not able to accommodate the planned flood discharge during the 25 year return period. This is indicated by the elevation of water in the river higher than the elevation of the right and left embankments so that the water overflows to the right and left sides of the river [14]. There are output cross-sections as shown in Figure 4 and Figure 5 for longitudinal sections.

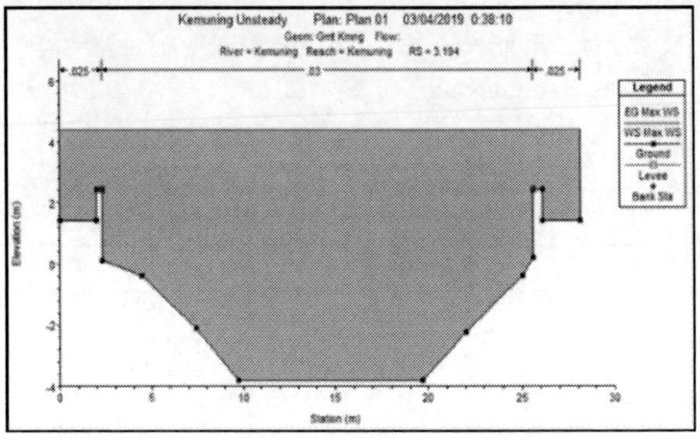

Figure 4. Cross section of the Kemuning river.

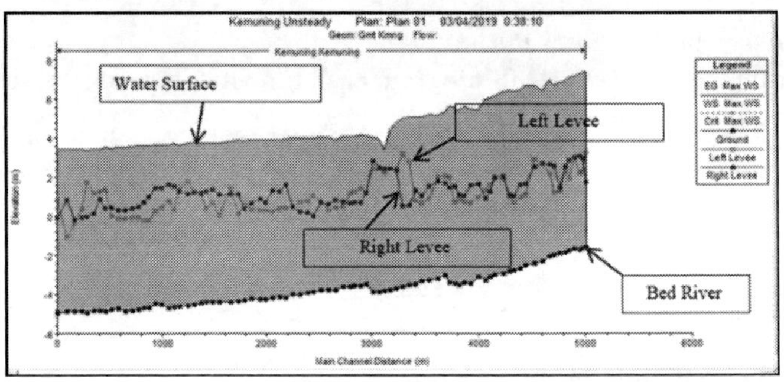

Figure 5. Long Section of the Kemuning river.

3.4 Design of diversion canal

To overcome the wider flood, diversion channel planning was carried out with several alternatives to reduce flood discharge in the Kemuning river. That condition is presented in figure 6. Alternative figure design of canal is presented of 7, 8, 9.

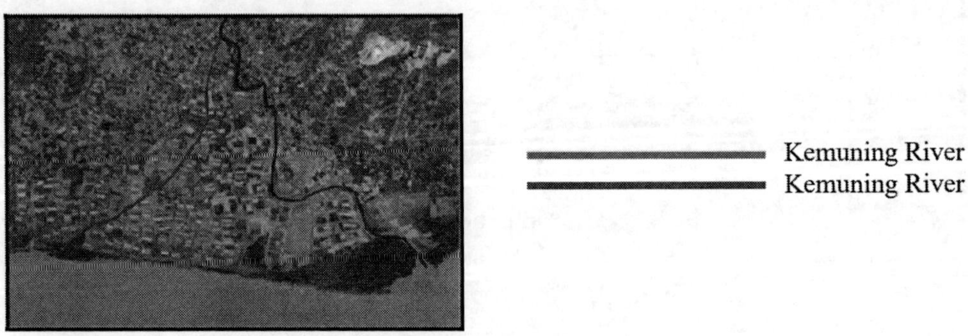

Figure 6. Location of diversion canal.

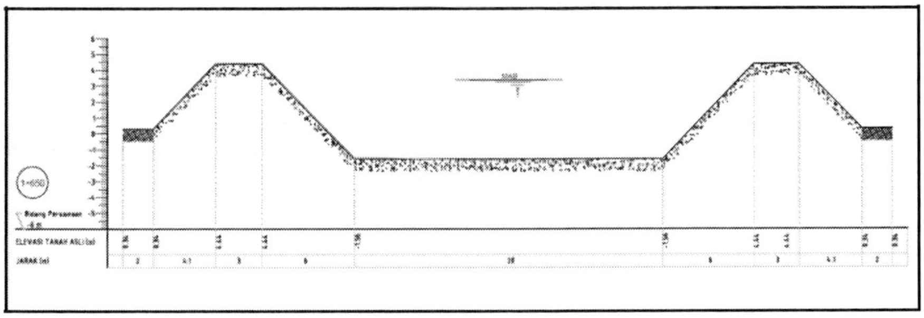

Figure 7. Alternative diversion channel plans 1.

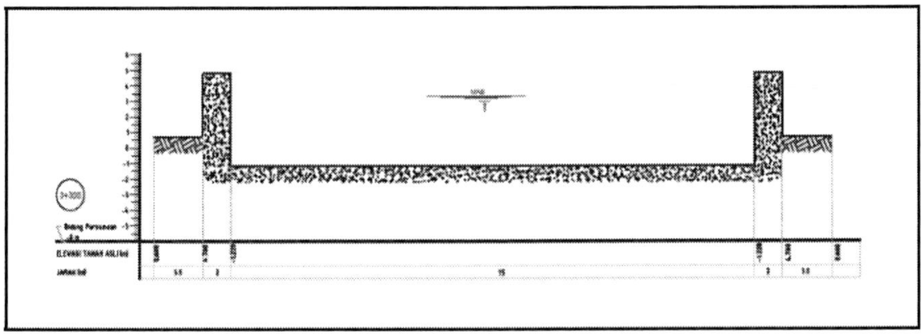

Figure 8. Alternative diversion channel plans 2.

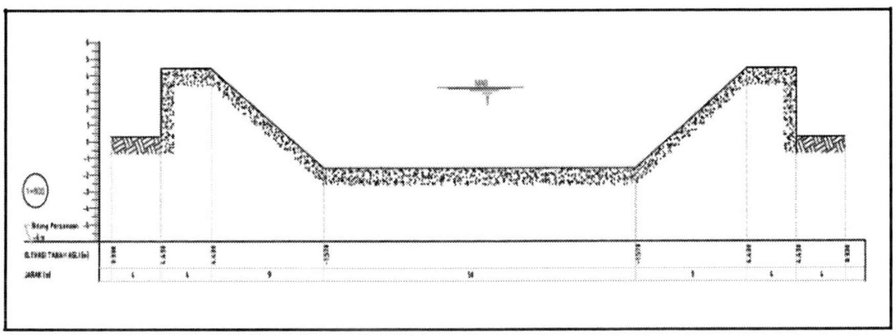

Figure 9. Alternative diversion channel plans 3.

3.5 Comparation of the result of design

All alternative diversion channel plans can reduce flood discharge by a percentage of 88.5% for alternatives 1 and 2 by decreasing the elevation of flood water surface in the downstream of the Kemuning river +3.46 m to +2.21 m, and 38.4% for alternative 3 with a decrease in the elevation of the flood water surface at the downstream of the Kemuning river +3.46 m to +2.07 m. the result is presented of table 3

Table 3. The result of design.

Alternative	Flood Discharge	Discharge of Canal	Existing Water Surface	Design of Water Surface	Reduction of Peak Discharge
	m^3/sec	m^3/sec	m	m	%
1	781.9	681.9	3.46	2.21	88.5
2	781.9	681.9	3.46	2.21	88.5
3	781.9	300	3.46	2.07	38.4

4. Conclusion

The results of the discharge on the 25 year return plan are 781.9 m^3/sec. The results of HEC RAS modeling with flood hydrograph input when the 25-year return (Q_{25}) Kemuning river exisiting conditions overtopping at all stations as high as 1-4 m. The first and the second diversion channel alternatives are planned to accommodate flood discharge (Q) 681.9 m^3/sec. The third alternative is planned to only flow flood discharge (Q) 300 m^3/sec by adding and raising embankments on the Kemuning river. The addition and elevation of the Kemuning river embankment an the third alternative planning is planned with height adjusting water level elevation with free board (w) 0.8 m and width 1 m. All alternative diversion channel plans are able to reduce flood discharge by a percentage of 88.5% for alternatives 1 and 2 by decreasing the elevation of floodwater levels in the downstream of the Kemuning river +3.46 m to +2.21 m, and 38.4% for alternative 3 with a decrease in the elevation of the flood water level at the downstream of the Kemuning river +3.46 m to +2.07 m.

References

[1] Harto S 1993 *Analisis Hidrologi (Hydrological Analysis)* (Jakarta: Gramedia Pustaka)
[2] Limantara L M 2010 *Hidrologi Praktis (Practical Hydrology)* (Bandung: Lubuk Agung)
[3] Cahyono AT 2011 *Planning for Flood Control in Sampang City* (Surabaya: Institut Teknologi Sepuluh Nopember)
[4] Chow VT 1959 *Open Channel Hydraulics (Nens Rosalina Translation)* (Jakarta: Erlangga Publisher)
[5] Xu C 2002 *Hydrologic Models* (Uppsala: Uppsala University Department of Earth Sciences Hydrology)
[6] Suripin 2004 *Sistem Drainase yang Berkelanjutan (Sustainable Drainage System)* (Yogyakarta: Andi Offset)
[7] Idfi G 2017 *Simulation Of River Pond Scenario On Reducing The Discharge Peak Of Ngotok River* (Surabaya: Sepuluh November Institute of Technology)
[8] Jayadi R 2015 *Instructions on how to use the HEC - HMS Model Package* (Yogyakarta: Gadjah Mada University)
[9] Soemarto C D 1999 *Hidrologi Teknik* (Jakarta: Erlangga)
[10] Soewarno 1995 *Hidrologi Aplikasi Metode Statistik untuk Analisa Data* (Bandung : Nova)
[11] USACE 1998 *Hydrologic Modelling System HEC HMS Technical Reference*
[12] USACE 2000 *Hydrologic Modelling System HEC HMS Technical Reference*
[13] USACE 2000 *User's Manual HEC-RAS*
[14] Nash J E and Sutcliffe J V 1970 River Flow Forecasting Through Conceptual Models *J. Hydrol* **101** 282-90
[15] SNI 03-2415-1991 2004 *Tata Cara Pengendalian Debit Banjir (Procedures for Control of Flood)* (Jakarta: National Standardization Agency)

The real operational cost for managing Semarang river polder drainage system

F Hartawan[1] and S I Wahyudi[1*]

[1] Department of Civil Engineering, Sultan Agung Islamic University, Semarang, Central Java, Indonesia.

*Corresponding authors: wahyudi@unissula.ac.id

Abstract: To resolve tidal flood and flooding problem, the Semarang city government has made various efforts such as building reservoirs in the upstream area, normalizing the rivers, developing polder drainage systems, and preventing sea water intrusion by building sea dike and river embankments. The method is quite effective to prevent the occurrence of rob (tidal flood), but water from the upstream area cannot flow by gravity to the sea. Water flows from upstream through the canals into a retention pond, and then pumping it into the sea. The Semarang River drainage polder system consists of some rivers (Semarang River, Asin River, Baru River), Retention Pool and Pump Station at the mouth of the Semarang River. Sustainability of the polder system requires an effective system of operation and maintenance. This study aims to determine the real need cost for a good operation of the Polder Drainage System. The research method used is quantitative description analysis. Data collection techniques using interviews, observation, and document review. The result show that budget component consists of fuel consumption, electricity, lubricants, spare parts, services, and inspection step. The difference in the amount of financing each year is due to fluctuations in rainfall intensity per month, the combination power supply by electricity and generator.

1. Introduction

Semarang city is a metropolitan city with population of ±1.7 million. As a metropolitan city located in a coastal area, the city of Semarang has a drainage problem that is identical to several Central Java north coast areas, namely tidal flood (rob) and flood. Floods mainly occur in the rainy season, due to large discharges exceeding the cross-sectional capacity of drainage system [1]. This is caused by sedimentation and waste water from the upstream watershed or its sub-watershed. Besides sedimentation, the decline in function and capacity of rivers and urban drainage. There is also caused by illegal buildings on the banks or even river bodies or channels, which reduce the function of the overflow capacity of the river [2].

Other problems affecting the drainage system in the coastal area of Semarang City are the phenomenon of tidal flood (flood due to tides), saltwater intrusion in the lower Semarang City, land subsidence that is local impacts of the ongoing process of land consolidation in the coastal area where the coast is generally composed of alluvial compressive layers [3].

Content from this work may be used under the terms of the Creative Commons Attribution 3.0 licence. Any further distribution of this work must maintain attribution to the author(s) and the title of the work, journal citation and DOI.
Published under licence by IOP Publishing Ltd

To deal with the problem of tidal flood and flooding that occurred in the city of Semarang, the Semarang city government has made various efforts such as building reservoirs and reservoirs in the upstream area as water conservation efforts. They work to normalize the rivers, to enlarge the capacity of the reservoir and to facilitate the water flow, and to develop some polder systems drainages. To prevent sea water entering the land, they build sea dike, gate and river embankments. The construction of sea dike that have been carried out in the city of Semarang. The drainage is for the ditch that was built aiming at facilities for the basic needs of planning or designing an urban infrastructure area [4]. Drainage development is one of the city's infrastructures based on environmentally friendly concepts [5].

Drainage design is based on the philosophy that water flows as fast as possible and inundates the service area to a minimum. To increase good water balance (usage and availability), a philosophical drainage design is needed not only safe from inundation but also at the same time for water conservation [6]. The Implementation of urban drainage systems explained that the operational activities for polder systems include planning, implementation, monitoring and evaluation activities [7]. Operational infrastructure and facilities for polder drainage systems are carried out to optimally regulate the flow of water and sediment management including the activities of opening and closing water gates, cleaning trash in the trash rack, turning the water pump off, and regulating water flow.

The polder system is a water management subsystem which is expected to be democratic and independent that is developed and operated by and for the community in terms of flood control in residential areas [8]. The most important elements in a polder system are the management organization, a system of governance based on community participation that is democratic and independent, and water management structures that are designed, operated and maintained by the community.

With the operations management polder drainage system can be obtained an overview of the actual field conditions and financing items. The polder drainage need a guideline for operational management systems and can be used as a reference for other systems [9]. Therefore, the procedure for preparing the real needs management operational of the polder drainage system must be prepared in accordance with local conditions based on applicable legal principles in order to contribute to the achievement of the overall drainage system management objectives with an example of the Semarang River system.

2. Data and method

The research method is based on a quantitative research approach that is by collecting data in the form by questionnaire and choice questions which are then analyzed quantitatively (inference). This research approach is also integrated qualitatively with data collection in the form of sentences, descriptions, through interviews or go directly to the field and supported by photo documentation. Field observations were made in the Semarang river polder drainage pump station. In this research, we collected all the data needed to be analyzed quantitatively and qualitatively, to be able to formulate and arrange scientific conclusions. In this study the research instruments were the researchers themselves (human instruments), primary and secondary data [10]. The human instrument serves to determine the focus of the study, select informants as data sources, conduct data collection, assess data quality, analyze data, interpret data, and make conclusions [11].

3. Results and discussion

3.1 Semarang river drainage system

The Semarang Central Drainage System has low land in the north and hilly areas in the south. The drainage system area has a catchment area of 2,307 ha and is bordered by the Java Sea to the north, the West Flood Canal to the west, the East Flood Canal to the east and Mount Ungaran to the south. Primary channels in the Central Semarang Drainage System include the West Canal Flood, Bulu channel,

Semarang River, Baru River and Banger River. There is cross drain (belt channel) that holds water from the upstream area and flows it to the West Flood Canal and East Flood Canal.

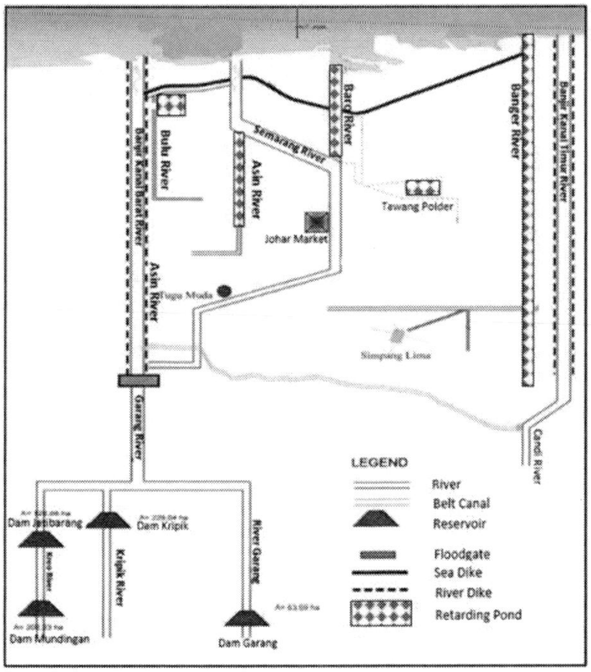

Figure 1. Schematic of Central Semarang drainage system and discharge distribution.

The Central Semarang drainage system is divided into 7 (seven) sub-systems, namely Bulu River Sub-System, Asin River Sub-System, Semarang River Sub-System, Baru River Sub-System, Bandarharjo River Sub-System, Simpang Lima River Sub-System, and Banger River Sub System [12].

Table 1. Sub-System area in Central Semarang.

No	Sub System	Sub System Area (ha)
1	Bulu River	93,57
2	Asin River	281,35
3	Semarang River	576,28
4	Baru River	185,55
5	Bandarharjo River	302,07
6	Simpanglima River	340,3
7	Banger River	523,79

Semarang River has 576.28 ha of catchment area. Currently Semarang River is also the estuary of Asin River, Baru River and Bandarharjo River. Based on the previous survey, inundation in the Semarang River sub-system is about 30% of total area

3.2 Semarang river sub-system infrastructure components

3.2.1 *Pump and generators.* The runoff from some of the canals is pumped through retention basin and outflows into the Java Sea. The drainage system is planned for a 5 year flood return period [13]. Number of Pumps installed Pumps with a capacity of 5 m3 / sec are 6 units; Pumps with a capacity of 2.5 m3 / sec are 2 units. The small capacity pumps are used during daily operations in the dry season.

Figure 2. Semarang river pump and generators.

Pump inlet elevation is at- 3.0 m sea level, this elevation is chosen so that the pump still can be operated when the water level is at it lowest. While the maximum elevation that is maintained is +0.5 m. From the research results it can be known that the pump resources were using electricity and generators. For generators there are the capacity of 2,250 kVa (2 units) for large generators, 1,000 kVa (1 unit) and 180 kVa (1 unit) for small generators.

3.2.2 Retention pond and trash rack. This retention pond construction is located at the mouth of Semarang River with an area of 68,000 m2 (6.8 ha), water storage capacity of 238,000 m3, elevation base is at EL = -3.5 m, embankment elevation is +4.00 m. Planning sea level tide level with consideration of land subsidence at +3.5 m, Free board of 0.5 m, embankment around the 1250 meters of retention pond. At present observation, the retention pond capacity has decreased by 30% from the initial capacity.

Figure 3. Retention pond and trash rack.

The trash rack construction plays an important role because it affects the pump performance system. The trash rack construction consists of upstream trash rack with 51 meters' length and 10 meters' height, a total of 11 units of garbage catchers. The drive of this trash rack system can be automatically or manually. The trash rack structure can be seen in figure 3. It is the structure under the bridge.

There is floodgate that border from sea water. The floodgate will be functioned or opened if there is an emergency situation due to extreme rainfall intensity. Until now the floodgate at Semarang River Drainage has never been used since the construction built, because the sea level has higher elevation than the Semarang river area water level.

3.3 Operational cost analysis

Operations and Maintenance (O&M) activities are two different activities, but they cannot be separated because each other influences one another. In terms of engineering maintenance can be defined as the art of maintaining equipment, buildings, and other related facilities, for providing service to be conducive condition [14]. The operation of a drainage system requires not only physical operation of various components, but its operation in an emergency and on-call condition. Operational financing can be divided into constant and dynamic categories.

3.3.1 Constant Operational Costs. Results of the study obtained data on the operation of the Semarang River Pump Station which is constant every month, including replacement of parts, service and inspection (monitoring) activities. The following data are for the last 3 (three) years operation from 2017 to 2019.

1) Replacement Parts Financing
The cost of replacing parts at the Semarang River Pump Station from 2017 to 2019 every month per year is relatively increasing. The amount of costs required for 2017 was IDR. 12,000,000.00, for 2018 was IDR. 24,000,000.00, and for 2019 was IDR. 36,000,000.00. From these results it can be concluded that there is a trend of increasing for replacement parts financing each year.
2) Service Financing
Service financing at Semarang River Pump Station from 2017 to 2019 every month per year is constant. The recapitulation of service costs required for 2017 and for 2018 are IDR. 24,000,000.00 while service fees for 2019 are IDR. 36,000,000.00.
3) Inspection / Monitoring Activities Financing
The financing of Semarang River Station Pump inspection activities from 2017 to 2019 every month per year are increasing, because the cost of each activity per month uses a lump sum calculation. The recapitulation of costs required for 2017 is IDR. 24,000,000.00, for 2018 is IDR. 30,000,000.00, and for 2019 is IDR. 36,000,000.00. From these results it can be concluded that each year there is an increasing trend of financing needs for inspection activities.

3.3.2 Dynamic Operational Financing Analysis. In operating of the Semarang River Polder Pump System, there is a large amount of funds each month, namely operational fuel financing, operational financing for electricity usage, and operational financing for oil / lubricant used.

1) Fuel Operational Financing
The trend of fuel operational cost needs from 2017 to 2019 has the same pattern, namely in the rainy season and then the trend decreases in the dry season. From the data obtained above, it can be concluded that the highest operational costs occur in the rainy season and the lowest operational costs occur in the dry season. They need low cost on June to August about IDR. 40,000,000.00, and peak cost in December during the season rain about IDR. 200,000,000.00. In the rainy season the pump works 24 hours nonstop so it requires a large operational cost. Whereas in the dry season the pump works only when needed, so the operational costs are relatively low.
2) Electricity Operational Financing
In 2017 the highest electricity operating costs occurred in December amounting to IDR. 138,450,855.00 and the low electricity operating costs occurred in February, March, July, August, September, October, November with the same expenses as IDR. 114,772,924.00. In 2018 the highest electricity operating costs occurred in March amounting to IDR. 227,755,807.00 and the low electricity operating costs occurred in June, July, August, September, October, November, December with the same cost of IDR. 114,772,924.00. In 2019 the highest electricity operating costs occurred

in April amounting to IDR. 238,476,519.00 and the lowest electricity operating costs incurred in February of IDR. 114,113,786.00.

From the data obtained above, it can be concluded that the operational costs of electricity do not have much influence with the seasons, because each month the demand for electricity is relatively the same. The energy needs beyond that are supported by the operation of generators that use fuel as discussed previously.

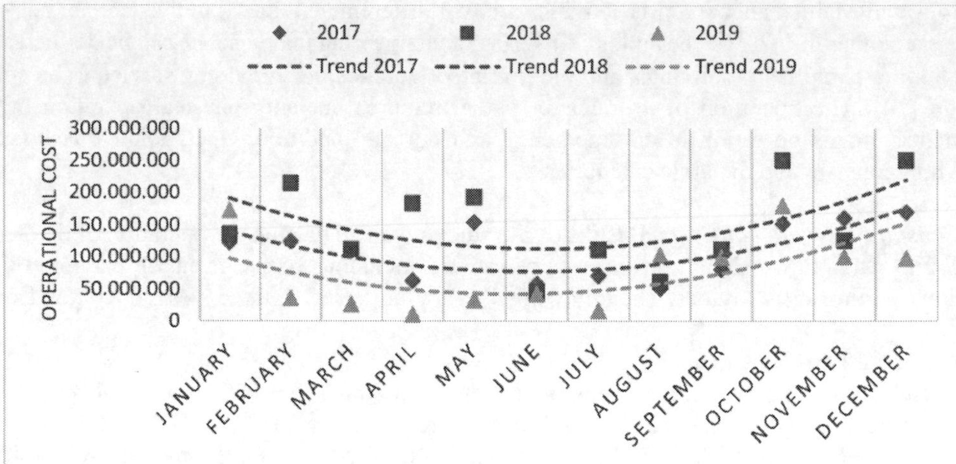

Figure 4. Graphic of fuel operational financing.

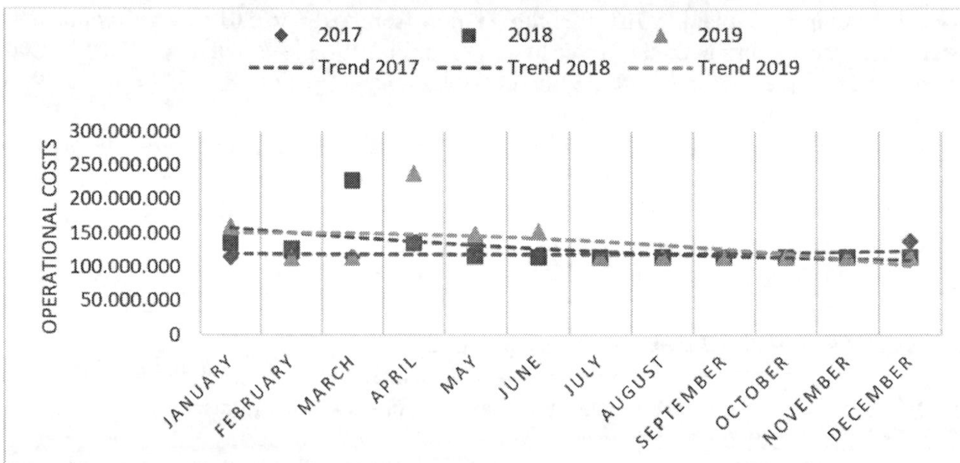

Figure 5. Graph of electricity operational financing.

3) Oil / Lubricant Operating Financing

From the graph in figure 6, it is obtained that in 2017 the highest operational costs for oil / lubricant use occurred in January, February, March, April with the same cost of IDR. 82,624,000.00 and the lowest oil / lubricant operating costs incurred in October, November, December with the same expenses of IDR. 62,848,000.00. In 2018 the highest operational costs for oil / lubricant use occurred in January, February, March with the same costs of IDR. 57,037,500.00 and the lowest operating costs for oil / lubricant use occurred in August, September, October, November, December with the same cost of IDR. 48,945,000.00. In 2019 the highest operational costs for oil / lubricant use occurred in January, February, March, April, May, June with the same costs of IDR. 52,710,000.00 and the lowest operating costs for oil / lubricant use occurred in December at a cost of IDR. 26,390,000.00. The trend of operational costs for the use of oil / lubricant in 2017 from the beginning of the year to the end of the year has relatively decreased, the trend of the need for operational costs for the use of oil / lubricant in 2018 from the beginning of the year to the end of the year has decreased but is relatively flat, whereas in 2019 since the beginning of the year has decreased until the end is similar to the 2017 trend.

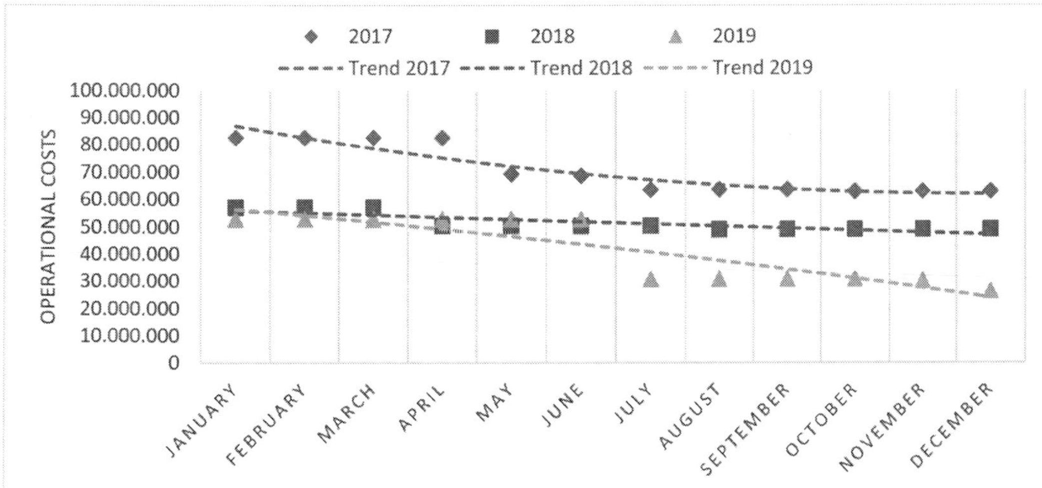

Figure 6. Chart of oil / lubricant operating financing.

3.3.3 Recapitulation of Real Operating Costs. The Operation and Maintenance (O&M) activities are two different activities but cannot be separated, because they influence each other [11]. The results showed Operation cost of Semarang River Polder drainage system in 2017 of IDR. 3,66 billion; in 2018 IDR. 4,03 billion; in 2019 IDR. 3,13 billion with cost components consisting of fuel operational financing, electricity operational financing, oil / lubricant operational financing, parts replacement financing, service service financing, and financing for inspection / monitoring activities. The results showed that the difference in the amount of financing each year was different due to the fluctuations of rainfall intensity per month.

4. Conclusion

Based on the existing conditions of the Semarang River Polder Drainage Pump Station, as well as from the results of the analysis conducted, several conclusions can be drawn as follows:

1. The pump is still in good condition where overall 8 pump units with a capacity of 35 m3 / sec. The operation of the Semarang River Polder Drainage System Pump Station is carried out by working steps by checking the availability, supplying of fuel to the generator set, checking the electrical panel and pump start panel, operating the pump.
2. The results of analysis of the constant cost for Operational process. The cost of replacing parts for 2017 is IDR. 12,000,000.00; in 2018 IDR. 24,000,000.00 and in 2019 IDR. 36,000,000.00. The need for services financing increases every year due to an increase in labor costs and along with the decline in equipment function, service activities will increase. The need for financing inspection activities has increased due to an increase in honorarium.
3. The results of the dynamic cost analysis. The fuel operational costs for 2017 amounting to IDR. 1,314,695,621.00 for 2018 IDR. 1,794,200,340.00 and for 2019 IDR. 915,965,130.00. The need for the highest operational fuel costs occurs in the rainy season and the lowest operational costs occur in the dry season. The electricity operating costs in 2017 of IDR. 1,442,881,262.00, in 2018 IDR. 1,546,093,090.00 and in 2019 IDR. 1,618,653,366.00. The electricity operational costs is not too influential with the seasons, because every month electricity needs are relatively the same. The

operational costs for the use of oil / lubricant in 2017 is IDR. 847,360,000.00; in 2018 IDR. 616,817,500.00 and in 2019 amounting to IDR. 496,160,000.00.

4. The Real operation cost of the Semarang River Drainage System Pump Station in 2017 is IDR. 3,664,936,883.00, in 2018 is IDR. 4,035,110,930.00 and in 2019 is IDR.3,138,778,496.00. The difference in the amount of financing each year is due to fluctuations of rainfall intensity per month and per year.

Acknowledgment
The authors are grateful for our colleagues from Magister of Civil Eng., Unissula, Semarang. Our thanks are addressed to the Directorate of Research and Society Service (DRPM), Ministry of Research, Technology and Higher Education Indonesia and "Bappeda" of Semarang City.

References

[1] Boogaard F C, Groningen H., Restemeyer B, Groningen H, and Venvik G 2019 *46th IAH Congress-Malaga* (Spain)
[2] Williams S J 2013 Sea-Level Rise Implications for Coastal Regions *J. Coast. Res.* **63**
[3] Adi H P and Wahyudi S I 2018 Tidal Flood Handling through Community Participation in Drainage Management System (A case study of the first water board in Indonesia) *Int. J. Integr. Eng. Spec. Issue 2018 Civ. Environ. Eng.* **10** 19–23
[4] Rudiarto I, Handayani W, and Setyono J S A 2018 Regional Perspective on Urbanization and Climate-Related Disasters In The Northern Coastal Region Of Central Java, Indonesia *Land* **7**
[5] Capodag/io A G, Ghilardi P, and Boguniewicz-Zab/ocka J 2016 New paradigms in urban water management for conservation and sustainability *Water Pract. Technol.* vol 11 pp 176–86
[6] Suripin 2004 *Sustainable Urban Drainage System* (Yogjakarta: Andi Offset)
[7] Wahyudi S I, Heikoop R, Adi H P, Overgaauw T, Schipper B, and Persoon R 2017 Emergency scenarios in the banger polder, semarang city: A case study to identify different emergency scenarios *Water Pract. Technol.* **12**
[8] 2013 *Tata Cara Perencanaan, Pelaksanaan, Operasi Dan Pemeliharaan Sistem Pompa* (Jenderal Cipta Karya Kementerian Pekerjaan Umum)
[9] Damayanti L D 2017 Perencanaan Sistem Drainase Wilayah Tawang Sari dan Tawang Mas Semarang Barat *Karya Tek. Sipil* vol 6 pp. 194–203
[10] Arns A, Dangendorf S, Jensen J, Talke S, Bender J, and Pattiaratchi C 2017 Sea-level rise induced amplification of coastal protection design heights *Sci. Rep* **7**
[11] Vedhuis J A E T and Clemens F H L R 2009 Uncertainty in risk analysis of urban pluvial flooding: a case study *Water Pract. Technol.* **4**
[12] Bappeda 2017 *Master Plan of Drainage* (Semarang City: Government)
[13] Wahyudi S I, Adi H P, Santoso E, and Heikoop R 2017 Simulating on water storage and pump capacity of " Kencing " river polder system in Kudus regency, Central Java, Indonesia *AIP Conf. Proc.* vol. 1818, pp. 1–11
[14] Spencer T *et al.* 2016 Global coastal wetland change under sea-level rise and related stresses: The DIVA Wetland Change Model *Glob. Planet. Change* **139**

IOP Conf. Series: Materials Science and Engineering **930** (2020) 012075 doi:10.1088/1757-899X/930/1/012075

The comparative study of peak discharge at Ngotok watershed by using the method of SCS, Snyder and Nakayasu for flood control needs

G Idfi[1*], A Yulistyorini[1], and N Suryoputro[1]

[1] Department of Civil Engineering, Faculty of Engineering, Universitas Negeri Malang, Malang, Jawa Timur, Indonesia.

[*]Corresponding authors e-mail: gilang.idfi.ft@um.ac.id

Abstract. Flood problems often hit the Ngotok River watershed every year. The problem of flooding is generally caused by changes in land use and a decrease in river functions in the Ngotok River watershed area as well as frequent back water from rivers which empties into Brantas River when experiencing maximum discharge. For that reason, a comparative study of peak flood discharge needs to be done because there is no research on flood control. So that research is conducted on the analysis of synthetic unit hydrograph models. The synthetic unit hydrograph method used is SCS, Snyder, and Nakayasu. Rainfall data used is data from 1998-2016 from 14 rain stations in the Ngotok River watershed. The Thiessen polygon method is used to determine the amount of rain scattered in the Ngotok River watershed. The average rainfall in the Ngotok River watershed in the period 1998-2016 was 97.05 mm. In the modeling phase, the sub-catchment division of the watershed is carried out by dividing it into 5 sub-catchments. The modeling results using the SCS, Snyder, and Nakayasu methods show the amount of discharge for the return period of 2 years, 5 years, 10 years, 20 years, 25 years, 50 years, 100 years, and 200 years which varies. AWLR data approaching the modeling results are 2014 data. The hydrograph calibration results for the SCS method with a 25 year return period are 0.88, for the Snyder method with a 25 year return period of 0.74, and the Nakayasu method with a 25 year return period of 0.43. Thus the SCS hydrograph approaches the existing hydrograph model based on the AWLR data that exists and is following the observations during the population survey.

1. Introduction

Floods in the Ngotok watershed occur almost every year during the rainy season [7]. The change in land use and the decline in river functions in the Ngotok watershed causes frequent flooding. From the impact and the absence of research on flood control, it is necessary to research the analysis of synthetic unit hydrograph models [21]. The synthetic unit hydrograph method used is SCS, Snyder, and Nakayasu [1]. By using these three synthetic unit hydrograph methods, we can see the comparison of the peak flood discharge values that occur in the Ngotok watershed by reading the graph of the hydrograph of each method [16]. The use of SCS, Snyder, and Nakayasu methods aims to determine which method is close to or following the characteristics of the Ngotok watershed [20].

Idfi defines Ngotok River watershed is administratively located in Mojokerto Regency, Mojokerto City, and Jombang Regency. Ngotok River is one of the rivers that functions as a collector of several

Content from this work may be used under the terms of the Creative Commons Attribution 3.0 licence. Any further distribution of this work must maintain attribution to the author(s) and the title of the work, journal citation and DOI.
Published under licence by IOP Publishing Ltd

IOP Conf. Series: Materials Science and Engineering **930** (2020) 012075 doi:10.1088/1757-899X/930/1/012075

rivers such as Tembelang River, Sambong River, Jombang Kulon River, Jombang Wetan River, Bening River, Sewedang River, Gunting River, Balong River and Brangkal River [7]. The length of Ngotok River is approximately 26.50 km with watershed coverage covering an area of ± 722 km^2.

The aim of this research is expected to help the government in efforts to control flooding. With this study, it can be seen the peak flood discharge in the Ngotok River watershed. While for researches, the results of the research are expected to add a contribution to researchers and can be a reference for further research [17].

2. Research method

2.1 Research location

The study was conducted in the the study location is the Ngotok River watershed (Figure 1) that passes through Mojokerto Regency, Mojokerto City, and Jombang Regency [7].

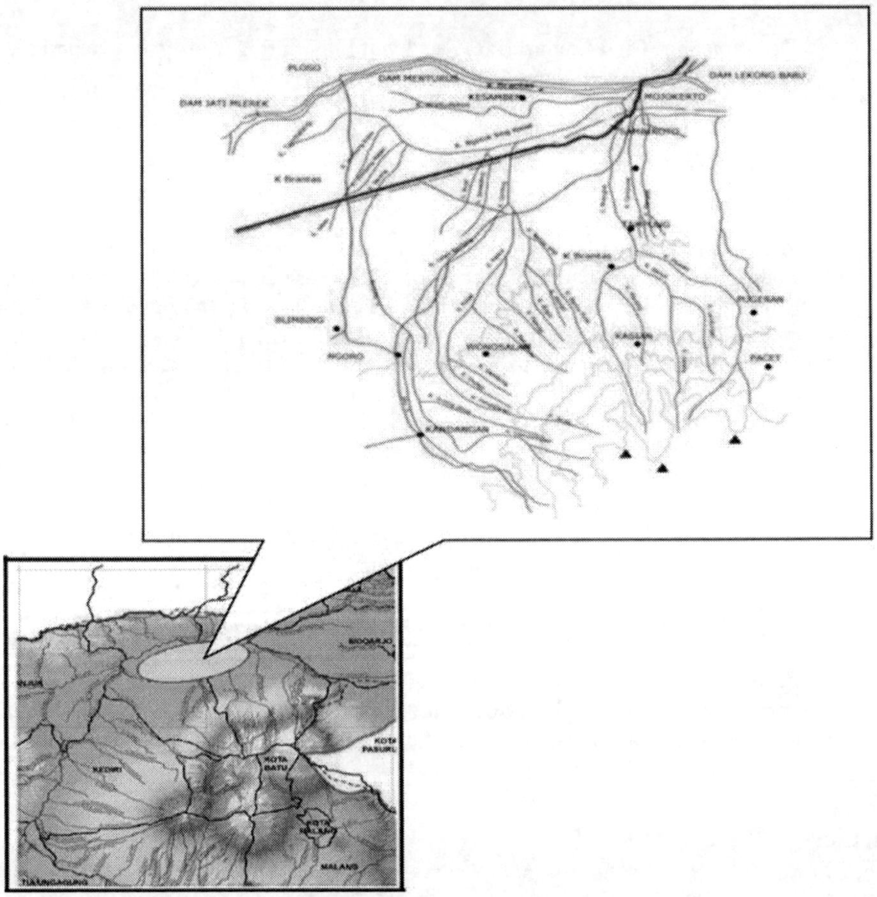

Figure 1. Research location.

2.2 Rainfall data analysis

The rainfall data obtained is point rain data, so that it can be used for further calculations, the point rain must be changed to regional rainfall by making Thiesen Polygon [2]. The equation of Thiessen Polygon is :

$$P = \frac{P_1 + P_2 + P_3 + \ldots + P_n}{n} \quad (1)$$

Where :
P : Daily Maximum Rainfall (mm)
$P_1, P_2 ... P_n$: Daily Maximum Rainfall at Rainfall point gauge (mm)
n : point gauge

In order to obtain the best frequency distribution, the existing data were analyzed first with 4 kinds of probability distribution methods, Gumbel, Normal, Log-Normal, and Pearson Log Type III [18][5]. Test probability distribution by using two methods, namely, Chi-square and Smirnov-Kolmogorof methods [16][4]. The distribution of sub-catchment watersheds into 5 sub-catchments is carried out before the modeling phase. Then perform data modeling rain with SCS, Snyder, and Nakayasu methods using software HEC-HMS 4.0 and Microsoft Excel 2007 [19].

2.3 Modeling input parameter
Before doing modeling with HEC-HMS 4.0 software, there are several parameters that must be calculated first [3]. The calculated parameters affect the output results from the modeling of the SCS and Snyder methods [11][12]. The parameters calculated are curve number (CN), impervious area, initial abstraction, watershed slope, river slope, and time lag for each sub-watershed [10][6]. The following are the formulas used for calculating the modeling input parameters. The equation of CN is

$$CN(I) = \frac{4.2CN(II)}{10 - 0.058CN(II)} \tag{2}$$

2.4 Calibration model
Model calibration can be done by comparing the hydrograph of the model with the observed hydrograph. According to Nash [13], as for the method for determining the model calibration criteria for the results of observations in the field as follows:

$$EI = 1 - \frac{\sum_{i=1}^{n}(Q_O - Q_s)^2}{\sum_{i=1}^{n}(Q_O - Q_a)^2} \tag{3}$$

Where :
EI = efficiency index
Qo = measurement discharge (observation)
Qs = Discharge simulation
Qa = average discharge measurement
The value of EI according to the Nash method is divided into three categories: low accuracy level if $EI \leq 0.50$, moderate accuracy level if $0.50 < EI < 0.70$, and high accuracy level if $EI \geq 0$ [9].

3. Result and discussion

3.1 Rainfall data analysis and sub catchment area
The rainfall data used is rainfall data from 1998-2016. From the calculation results, the average rainfall is 97.05 mm. The amount for each year is presented in the table 1.

Rainfall frequency analysis was carried out using the Log Pearson Type III method. From the results of the calculation of rainfall design with the Pearson Type III Log method obtained the value of the amount of rain based on the return period of rain [14][8]. The sub-catchment of the watershed is divided into 5 parts based on the topography and its constituent rivers. The value of the amount of rain based on the return period of rain and subcatchment area is presented in the table below.

Table 1. Average rainfall of the ngotok river watershed.

Year	Rainfall (mm)	Year	Rainfall (mm)
1998	84.34	2008	76.33
1999	85.52	2009	85.34
2000	119.82	2010	100.15
2001	98.58	2011	101.82
2002	90.16	2012	109.26
2003	105.31	2013	115.43
2004	120.39	2014	85.26
2005	80.73	2015	87.61
2006	82.02	2016	119.94
2007	95.94		

Table 2. Rainfall plan calculation.

Return Period (year)	Xt (mm)
2	95.440
5	108.518
10	116.476
20	124.330
25	125.964
50	132.706
100	139.220
200	145.593

Table 3. Division of watershed sub catchment.

Sub Catchment	Name of River	Area (km^2)
1	Tembelang Sambong Jombang Kulon Jombang Wetan	127.854
2	Bening Trawasan	112.954
3	Sebani Drainage System Gunting Balong	241.159
4	Panemon Prajurit Kulon Drainage System	44.850
5	Brangkal	195.182

3.2 Calculation of model input parameter

The result of model input parameter is presented in the table below.

Table 4. Curve number (CN).

Sub	CN
1	80.75
2	79.67
3	79.44
4	80.49
5	79.04

Table 5. Impervious area and initial abstraction.

Sub	Impervious (%)	Initial Abstraction
1	10.10	0.477
2	8.83	0.511
3	8.55	0.518
4	14.45	0.485
5	6.77	0.530

Table 6. River and watershed slope.

Sub	River Slope (%)	Watershed Slope (%)
1	6.59	4.50
2	5.83	3.96
3	5.08	3.82
4	6.26	4.30
5	5.33	4.64

Table 7. Time lag.

Sub	Watershed Time Lag		River Time Lag	
	Hour	Minute	Hour	Minute
1	8.872	532.338	15.811	948.679
2	8.532	511.892	17.457	1047.390
3	11.198	671.894	26.794	1607.615
4	4.667	280.009	13.758	825.508
5	9.477	568.602	24.002	1440.141

3.3 Rainfall run off modelling result

3.3.1 SCS method. From the computational results, the final results are in the form of a graph of the hydrograph model and peak flood discharge data that occur based on certain returns in each sub watershed

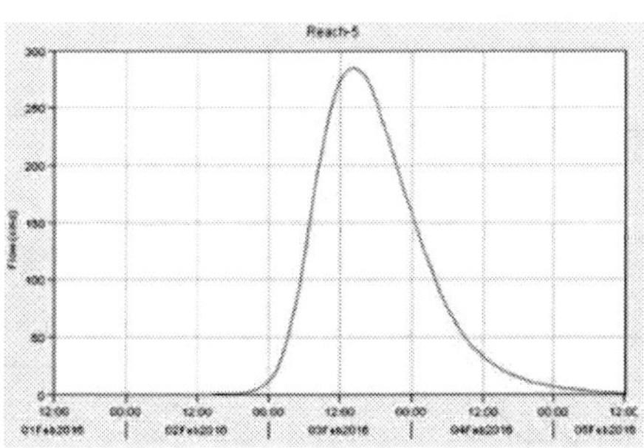

Figure 2. SCS Hydrograph chart return period 25 years of sub watershed 3.

Table 8. Recapitulation of SCS method modeling results.

Return Period (year)	Flood Duration (hour)	Sub 1 Peak Discharge (m^3/sec)	Flood Duration (hour)	Sub 2 Peak Discharge (m^3/sec)	Flood Duration (hour)	Sub 3 Peak Discharge (m^3/sec)	Flood Duration (hour)	Sub 4 Peak Discharge (m^3/sec)	Flood Duration (hour)	Sub 5 Peak Discharge (m^3/sec)
2	64	124.2	62	108.1	68	196.4	44	63.9	67	172.1
5	64	147.6	62	129.9	68	233.9	44	75.9	68	205.6
10	64	162.1	63	142.8	68	257.2	44	83.3	68	226.5
20	64	176.6	63	155.8	68	280.4	45	90.7	68	247.3
25	64	179.7	63	158.5	68	285.3	45	92.3	68	251.7
50	65	192.2	63	169.7	68	305.5	45	98.7	68	269.8
100	66	204.5	63	180.7	69	325.2	45	105.0	68	287.4
200	66	216.5	63	191.5	69	344.5	45	111.1	68	304.8

3.3.2 Snyder method. From the computational results, the final results are in the form of a graph of the hydrograph model and peak flood discharge data that occur based on certain returns in each sub watershed

Table 9. Recapitulation of Snyder method modeling results.

Return Period (year)	Flood Duration (hour)	Sub 1 Peak Discharge (m^3/sec)	Flood Duration (hour)	Sub 2 Peak Discharge (m^3/sec)	Flood Duration (hour)	Sub 3 Peak Discharge (m^3/sec)	Flood Duration (hour)	Sub 4 Peak Discharge (m^3/sec)	Flood Duration (hour)	Sub 5 Peak Discharge (m^3/sec)
2	40	141.9	40	123.3	44	223.4	32	74.8	42	194.3
5	40	168.5	40	146.7	44	266.0	32	88.7	42	231.9
10	40	185.0	40	161.3	44	292.5	32	97.3	42	255.2
20	40	201.5	40	176.0	44	318.9	32	105.9	42	278.5
25	40	205.0	40	179.1	44	324.4	32	107.7	42	283.3
50	40	219.4	40	191.8	44	347.3	32	115.1	42	303.6
100	40	233.3	40	204.2	44	369.7	32	122.4	42	323.3
200	40	247.1	40	216.3	44	391.7	32	129.5	42	342.7

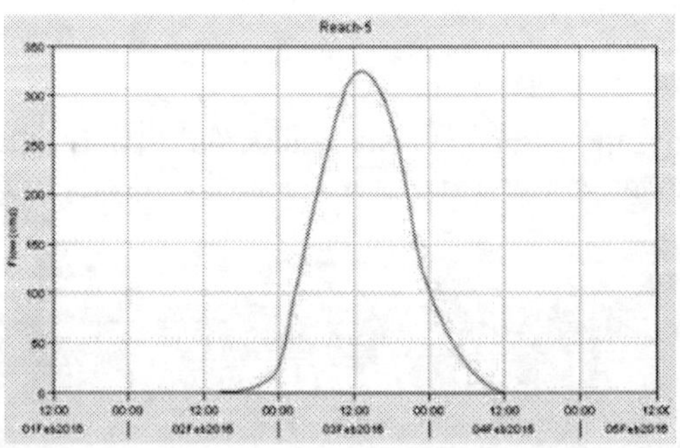

Figure 3. Snyder hydrograph chart return period 25 years of sub watershed 3.

3.3.3 Nakayasu method. From the computational results, the final results are in the form of a graph of the hydrograph model and peak flood discharge data that occur based on certain returns in each sub watershed.

Table 10. Recapitulation of Nakayasu method modeling results.

Return Period (year)	Flood Duration (hour)	Sub 1 Peak Discharge (m^3/sec)	Flood Duration (hour)	Sub 2 Peak Discharge (m^3/sec)	Flood Duration (hour)	Sub 3 Peak Discharge (m^3/sec)	Flood Duration (hour)	Sub 4 Peak Discharge (m^3/sec)	Flood Duration (hour)	Sub 5 Peak Discharge (m^3/sec)
2	86	194.9	88	168.6	96	255.1	68	74.9	96	226.9
5	88	239.1	90	206.8	96	313.2	69	92.0	96	278.4
10	89	266.6	91	230.5	96	349.1	70	102.5	96	310.4
20	90	293.9	92	254.2	96	385.1	71	113.0	96	342.3
25	90	299.7	92	259.1	96	392.6	71	115.3	96	348.9
50	91	323.5	93	279.7	96	423.9	71	124.4	96	376.7
100	91	346.6	93	299.7	96	454.3	72	133.3	96	403.7
200	92	369.5	94	319.5	96	484.3	72	142.1	97	430.3

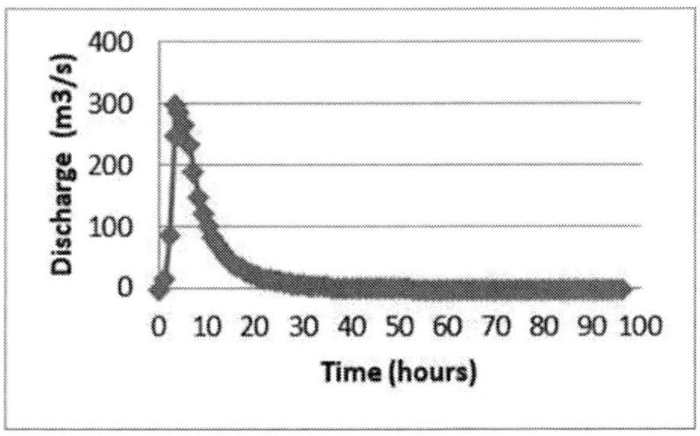

Figure 4. Nakayasu hydrograph chart return period 25 years of sub watershed 3.

3.3.4 Calibration model. The calibration process relies on AWLR (Automatic Water Level Recorder) data located in the Gunting River (Sub Watershed 3). The results of the hydrograph graph of the AWLR data are compared with the results of the modeling hydrograph . In addition to using AWLR data, calibration also considers the survey results of residents living in the Ngotok River watershed area. The model used is a model hydrograph in the sub basins Guntin River with a return period of 25 years and data AWLR used was 2014. From the calculation using the formula Nash shows a model with a 25 year SCS method, the calibration value is 0.88. The Snyder method give the calibration value is 0.74 and the Nakayasu method give the calibration value is 0.43. the result of the comparation hydrograph is presented in figure below.

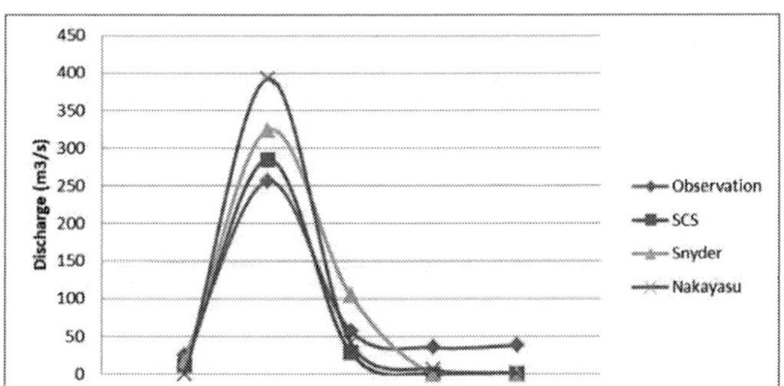

Figure 5. The comparation of hydrograph chart return period 25 years of sub watershed 3.

4. Conclusion

Based on the formulation of the problem, the results, and discussion that the hydrograph model on Ngotok watershed has been described previously, it can be simulated using the SCS, Snyder, and Nakayasu method. The hydrograph model that occurs when compared or calibrated with the field hydrograph model is quite significant. For the SCS method with a 25 year return period, the suitability of the calibration has a high level of accuracy of 0.88. For the Snyder method with a 25 year return period, the suitability of the calibration has a high level of accuracy of 0.74. Whereas for the Nakayasu method with a 25 year return period, the suitability of the calibration has a low level of accuracy of 0.43. Models that approach existing or existing conditions based on existing calibration data are models with the SCS method with a 25 year return period. The SCS method for 25 years return produces a peak discharge of 285.30 m^3/s and a peak existing discharge of 255.99 m^3/s.

References

[1] Harto S 1993 *Analisis Hidrologi (Hydrological Analysis)* (Jakarta: Gramedia Pustaka)
[2] Limantara L M 2010 *Hidrologi Praktis (Practical Hydrology)* (Bandung: Lubuk Agung)
[3] Azizah N and Affandy 2011 *Pemodelan Hujan-Debit Menggunakan Model Hec-Hms Di Das Sampean Baru* (Surabaya : Institut Teknologi Sepuluh Nopember)
[4] Viessman W, Knapp J W, Lewis G L and Harbaugh T E 1977 *Introduction to Hydrology* (Newyork: Harper and Row)
[5] Xu C 2002 *Hydrologic Models* (Uppsala: Uppsala University Department of Earth Sciences Hydrology)
[6] Suripin 2004 *Sistem Drainase yang Berkelanjutan (Sustainable Drainage System)* (Yogyakarta: Andi Offset)
[7] Idfi G 2017 Perbandingan Model Aliran Banjir Unsteady Flow dan Steady Flow Pada Sungai Ngotok Ring Kanal Jurnal Bangunan **22** 31-40
[8] Jain S K 1993 *Sciences -Journal- des Sciences Hydrologiques* **38** 431-41
[9] Xu C 2002 *Hydrologic Models* (Uppsala: Uppsala University Department of Earth Sciences Hydrology)
[10] USACE 1998 *Hydrologic Modelling System HEC HMS Technical Reference*
[11] USACE 2000 *Hydrologic Modelling System HEC HMS Technical Reference*
[12] USACE 2000 U ’ M H C-RAS
[13] Nash J E, Sutcliffe J V 1970 River Flow Forecasting Through Conceptual Models *J.Hydrol* **101** 282-290
[14] SNI 03-2415-1991 2004 *Tata Cara Pengendalian Debit Banjir (Procedures for Control of Flood)* (Jakarta: National Standardization Agency)
[15] Stewart M K 2015 *Hydrol. Earth Syst. Sci.* **19** 2587–2603
[16] Zhang R, Li Q, Chow T L, Li S and Danielescu S 2013 *Hydrol. Process* **27** 2659–2665
[17] Nathan R J and McMahon T A 1990 *Water Resources Research* **26** 1465–73
[18] Neitsch S.L, Arnold J.G, Kiniry J.R, Williams J.R 2005 *Soil and Water Assesment Tool Theoritical Documentation Temple* (Texas : Grassland, Soil and Water Research Laboratory USDA Agricultural Research Service)
[19] Sebastian L 2008 *Pendekatan Pencegahan dan Penanggulangan Banjir* (Palembang: Universitas Sriwidjaja)
[20] Soemarto C.D 1999 *Hidrologi Teknik* (Jakarta : Penerbit Erlangga)
[21] Soewarno 1995 *Hidrologi Aplikasi Metode Statistik untuk Analisa Data* (Bandung : Nova China)

IOP Conf. Series: Materials Science and Engineering **930** (2020) 012076 doi:10.1088/1757-899X/930/1/012076

Numerical simulation of the water level of the channel and the Kalidami boezem by pump operation

B Setiawan[1] **and U Lasminto**[1*]

[1] Department of Civil Engineering, Faculty of Civil, Planning and Geo Engineering, Sepuluh Nopember Institute of Technology, Surabaya, East Java, Indonesia.

*Corresponding authors: umboro@ce.it.ac.id

Abstract. Analysis of water level in a reservoir such as Boezem is needed. The purpose of this research is to find out whether the procedures implemented in managing the Boezem are appropriate to deal with various situations. To do this, a simulation is needed so that the movement of water in the channel to the Boezem can be analyzed. This research was conducted at Kalidami Boezem in the city of Surabaya which has an area of 20,936.69 m2 with an average depth of Boezem of 3.8 m. The simulation is carried out using the Storm Water Management Model (SWMM 5.1) application with rain data on February 25, 2020 and a 5-year return period rainfall data. The simulation results show that the model constructed has been calibrated based on the RMSE method and produced an error value of 0.11. Because the model has been well calibrated, it will be possible to do next simulation to see how the Channel and Boezem are performing to deal with various scenarios that might be encountered later on. Based on the results of hydraulic simulation using a 5-year return period rainfall data, it can be seen that within 45 minutes, Kalidami Boezem has been filled with water.

1. Preliminary

Kalidami Boezem is an estuary of various channels in the city of Surabaya. Boezem has an area of 2.09 Ha and a water storage capacity of 61,971 m^3. The function of Boezem is as a pond to collect rainwater and waste water coming from various sewers before being discharged to the sea. Because it is located at the estuary of various channels in the city of Surabaya, it can be seen that the elevation of the land at the Kalidami Boezem location is relatively low. In the upper part of Boezem, there is a primary channel, the Kalidami channel which is one of the important channels in the city of Surabaya. This channel has a length of 44,65.27 m with different river widths between 11-35 m. The position of the channel is parallel to Jl. Kertajaya, Surabaya City.

Based on Local Government Regulation No. 12 of 2014 concerning Urban Planning of Surabaya in 2014-2034, it can be seen that Boezem and Kalidami Channel have a vital function for the City of Surabaya. As a flood controller, the Kalidami Channel functions to drain wastewater and rain water from the catchment area, various secondary channels and tertiary channels around the Kalidami channel. Waste water and rainwater that enter this channel will be directed to the sea in the eastern part of Surabaya by passing Kalidami Boezem. Because the Kalidami Channel and Boezem have very important functions, especially to prevent flood in the eastern part of the city of Surabaya, an analysis of the water flow that enters the channel and Boezem becomes very necessary. With this research, it is expected that information will be obtained related to the movement of water flow in the Kalidami Channel and

Content from this work may be used under the terms of the Creative Commons Attribution 3.0 licence. Any further distribution of this work must maintain attribution to the author(s) and the title of the work, journal citation and DOI.
Published under licence by IOP Publishing Ltd

IOP Conf. Series: Materials Science and Engineering **930** (2020) 012076 doi:10.1088/1757-899X/930/1/012076

Boezem. In addition, information on the performance of water pumps that are operated to deal with various conditions such as the 5-year return period rainfall that might occur at the location will also be obtained.

1.1 Research purpose
The purpose of this study was to conduct a water level simulation at the Kalidami Boezem using hydrological and hydraulic modeling tools, namely SWMM 5.1. This research was conducted to determine whether the model made was in accordance with the real conditions based on the recording of the Boezem water level on February 25, 2020. Then, the resulted model will be analyzed using the 5-year return period rainfall data obtained based on the results of rain data processing from several rain stations that affect the catchment area around the Kalidami Channel and Boezem.

2. Literature review

2.1 Surabaya city drainage system
Development of the Surabaya City drainage system has been carried out by dividing its service area through the establishment of several regional sections. Based on Article 36 of Local Government Regulation No. 12 of 2014, it can be seen that Surabaya City has divided its service area into 5 Regional Areas which include:

1. Region of Genteng
2. Region of Gubeng
3. Region of Jambangan
4. Region of Wiyung, and
5. Region of Tandes

Each regional area is divided into three parts, namely primary system, secondary system, and tertiary system. In this case, Kalidami Channel is primary channel that is connected with several secondary channels, including (Surabaya City Government Regulation: 2014):

1. Pucang Anom Timur
2. Pucang Adi
3. Kertajaya X
4. Unair
5. Srikana
6. Karang Menjangan
7. Dharmahusada Indah
8. Wisma Permai II
9. Mleto
10. Raya ITS/Sawahan
11. ITS Barat
12. Bhaskara.

2.2 Flood
The human need for water cannot be denied. Some regions in Indonesia experience water scarcity, especially during dry season. However, when rainy season comes, these areas are often flooded due to the enormous volume of water caused by rain. The large number of land changes and the reduction in infiltration sites are often a source of flood. This often happens in big cities where infiltration sites are usually much reduced so that the areas become very dependent on reliable drainage to drain the large volume of water in the areas.

Flood is a condition where the discharge/volume of water flowing in a river and drainage channel is above its drainage capacity (Rosyidie, 2013). In Indonesia, known as a tropical country, the rainy season usually occurs in January-February. With this, there will come a sense of worry both in the community and government. Rainfall in this period is usually higher than the other months (Rosyidie, 2013).

2.3 Modeling using SWMM 5.1

SWMM 5.1 is an application developed by U.S. Environmental Protection Agency (EPA). SWMM is a model used to simulate dynamic rainfall that can cause runoff which can then be measured in terms of quality and quantity, especially for urban areas (SWMM User's Manual Version 5.1: 2015). Modeling with SWMM is carried out based on various hydrological processes, such as rainfall with time variations, evaporation on the surface of the water, rainfall in the reservoir area and infiltration of rainfall that enter the water unsaturated soil layer, taking into account the presence of runoff and the existing drainage system (Suprapto, 2018). Thus, it can be seen that this application has advantages because it can do both modeling of hydrological and hydraulic elements as well in one application.

2.4 Thiessen polygon

Each rain station located around the study site will have a different effect on the paint area. The Thiessen Polygon is one method for determining which rain stations will affect water runoff on the surface of a catchment area. The working principle of this method is to assign weight based on the area of the sub-catchment area to the associated rain station.

Weighted average for each rain station is determined by the area of the affected area based on the polygons formed (illustrated by the presence of axis lines on the connecting lines between two rain stations that are close together) (Ningsih, 2012). The method of calculating the average rainfall using the Thiessen Polygon method is by using the following calculation formula:

$$d = \frac{A_1 d_1 + A_2 d_2 + A_3 d_3 + \dots A_n d_n}{A} = \frac{\sum A_i d_i}{A} \quad (1)$$

Where:

A : Area (km^2)

d : Value of average rainfall

$d_1, d_2, d_3 \dots d_n$: Rainfall value at the rain station 1,2,3… n

$A_1, A_2, A_3 \dots A_n$: The area covered by the rain station 1,2,3… n.

2.5 Root mean square error (RMSE)

One important thing in determining whether a model that is made suitable for use as a reference for analysis is to first measure the error rate. According to Astari (2018), RMSE is the average value of the sum of the squares of the error results of the analysis. A fairly high level of sensitivity is an advantage of the RMSE method.

To calculate the error value of the water level model on Kalidami Boezem, then it can use this RMSE method which is calculated with the following calculation formula:

$$RMSE = \sqrt{\frac{1}{n}\sum_{i=1}^{n} e_i^2} \quad (2)$$

Where n is the number of samples used to measure the error value of the observation water level data and the water level value generated from the model (e_i, $i=1,2,3 \dots n$).

2.6 Gumbel distribution

In hydrology, there are several types of distribution that are very well known among researchers, including the Normal Distribution, Log-Normal Distribution, Log-Pearson III Distribution, and Gumbel Distribution. To be able to use them in a study, it is necessary to test each of these distribution models. The testing mechanism is done by using the goodness of fit test method, namely by doing the chi-square test and the Smirnov-Kolmogorov test.

In a previous study, according to Sari (2017) based on the Chi-Square and Smirnov-Kolmogorov test results, it is known that one type of distribution can be used to analyze the performance of the

Kalidami Channel, which is the Gumbel Distribution. The calculation of the rain return period using the Gumbel Distribution method can use the following formula:

$$X_T = \overline{X} + s\square K \quad (3)$$

Where:

X_T : Estimated value of rainfall that is expected to occur with a T-year return period.

$\bar{X}$: The average value of the variate

s : Standard deviation of variate values

K : The frequency factor is a function of the return period and the type of frequency distribution

where the probability factor K is expressed by the following equation:

$$K = \frac{Y_{Tr} - Y_n}{S_n} \quad (4)$$

Where:

Y_n : reduced mean with n amount of data

S_n : reduced standard deviation of n data

Y_{T_r} : reduced variate

Where the value of Y_{T_r}d can be calculated with the following equation:

$$Y_{Tr} = -\ln\left(-\ln\frac{Tr-1}{Tr}\right) \quad (5)$$

Where T_r is return period.

2.7 Novelty

This study has an update from previous studies because in previous studies there were two pump houses named Kalidami Boezem Pump House and Kalidami Screw which had 5 sluice gates and 5 pumps, while the conditions in this study were only one pump house namely Pump House Kalidami Boezem. Then, this study is also different from previous studies because it has measured the Boezem dimensions directly so that there are some differences in data for example the Boezem depth value. In addition, this study also uses rain data from the rain station located at ITS which has fairly detailed rain data that is per 5 minutes and has been integrated with recording water levels directly so that the model can be calibrated.

3. Research methods

3.1 Research sites

Kalidami Boezem is located in the District of Mulyorejo, Surabaya City, East Java Province at coordinates 7 ° 16'26.8 "S 112 ° 48'14.4" E or -7.274111, 112.804000. As shown in the map image above, the location of this research study is close to sea waters located at the east of Surabaya.

3.2 Spatial analysis

Catchment area data used in this study were obtained from the Dinas Pekerjaan Umum Bina Marga dan Pematusan Kota Surabaya based on the Surabaya Drainage Master Plan (SDMP) in 2018-2038. This study uses a number of sub-catchment areas that are approximately 4 m above sea level, and will provide runoff water into the Kalidami Channel and Boezem.

At the Kalidami Channel and Boezem locations there are several rain stations scattered around the catchment area. Based on the Thiessen Polygon concept and supported by the GIS (Geographic Information System) technology, it will be known that there are two rain stations that can affect the Kalidami Channel and Boezem namely the Gubeng and Keputih Rain Stations.

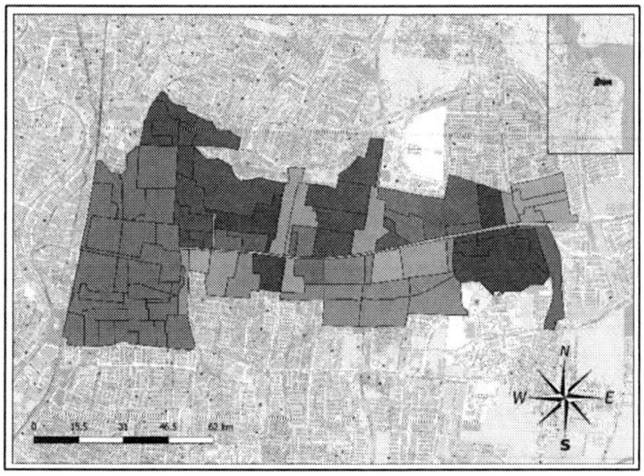

Figure 1. Kalidami boezem location in the City of Surabaya (processed data using gis based on SDMP 2018-2038).

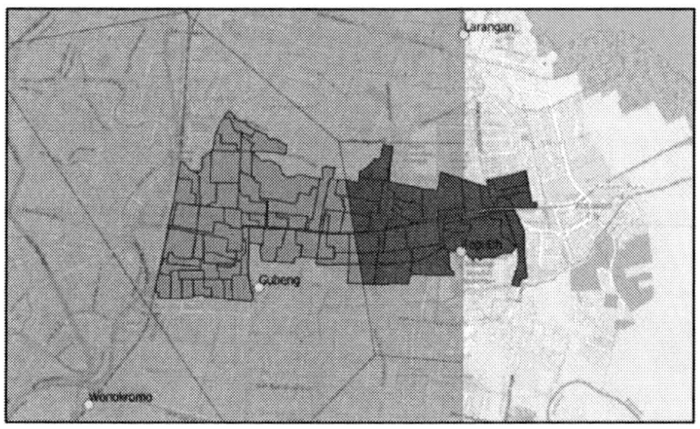

Figure 2. Thiessen polygon on the Kalidami channel (processed data using GIS).

3.3 Hydrological and hydraulic analysis

Kalidami Channel and Boezem Modeling are done by combining hydrological and hydraulic data obtained during the study. These data are as follows:

3.3.1 Rain data. Rain data are obtained by the Rain Station of the Sepuluh Nopember Institute of Technology which publishes weather information on page *https://wunderground.com/dashboard/pws/ISURABAY4*. These data are fairly good and can be used for simulating water pumps in the Kalidami Channel and Boezem because they have a time step per 10 minutes. These data will then be processed into hourly rain data so that they can be inputted into the SWMM 5.1 application. The return period rainfall data used in this study are to use rainfall data with a return period of 5-year in accordance with the rules of hydrological planning in the Ministry of Public Works Regulation No. 12 of 2014 concerning the Implementation of the Urban Drainage System. Following are the 5-year return period rainfall data of 122.99 mm with duration of 4 hours which will be used for the analysis of the Kalidami Boezem and Channel hydraulics. Calculations made to produce rain data for this return period are calculated using the Gumbel Distribution.

IOP Conf. Series: Materials Science and Engineering **930** (2020) 012076 doi:10.1088/1757-899X/930/1/012076

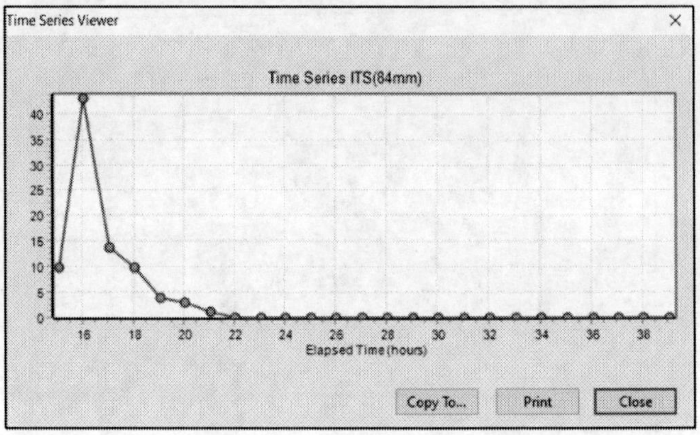

Figure 3. Rainfall data for february 25, 2020 (SWMM 5.1).

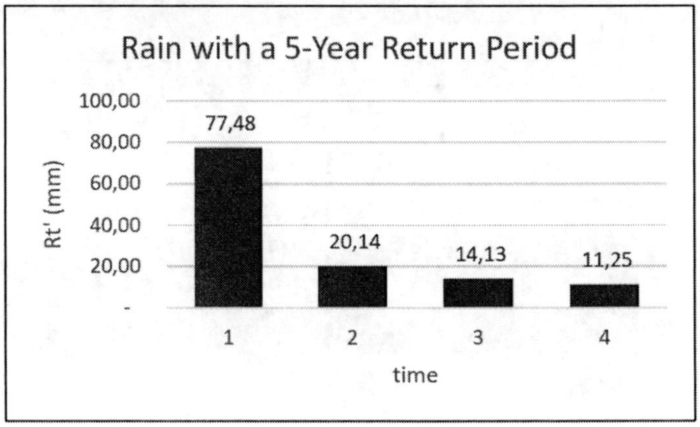

Figure 4. Rain with a 5-year return period rainfall (processed rainfall data of Gubeng and Keputih rain stations from ITS).

3.3.2 Catchment area data. These data include the size of the catchment area, slope, impervious area and others. The size of the catchment area used in this study is 982.81 Ha with a total sub-catchment area of 33 units.

Figure 5. Catchment area (SWMM 5.1).

The area of each sub-catchment area in the image above is as follows:

Table 1. Sub-catchment area.

Sub-cathment	Area (Ha)	Sub-cathment	Area (Ha)
1	270.28	18	65.05
2	4.02	19	18.85
3	7.06	20	18.25
4	121.45	21	7.00
5	6.33	22	12.64
6	6.51	23	24.46
7	15.26	24	20.71
8	7.40	25	90.42
9	14.40	26	33.54
10	22.33	27	4.22
11	15.78	28	13.41
12	13.27	29	5.41
13	12.28	30	71.00
14	2.15	31	20.32
15	7.46	32	4.07
16	9.30	33	29.92
17	8.25	Total	982.81

3.3.3 Cross section and geometry boezem data. These data include cross section, Kalidami boezem geometry, slope and others. The following is an example of a cross section and the geometry of the Kalidami Boezem schematically in the SWMM 5.1 application.

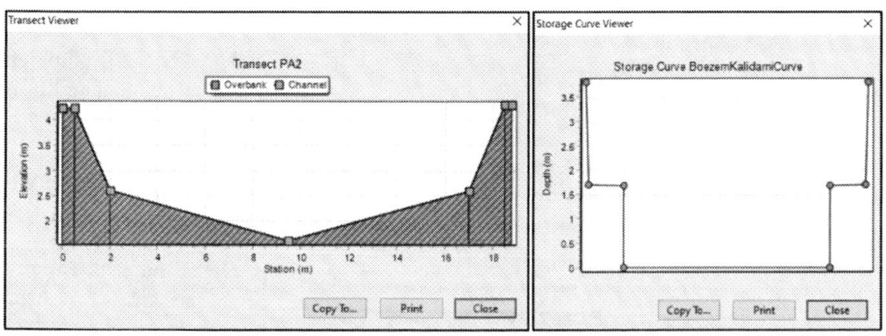

Figure 6. Cross section and boezem geometry (SWMM 5.1).

3.3.4 Water pump data. The water pump located in Kalidami Boezem has a very important function. This is due to the frequent occurrence of tidal conditions every day. Therefore, to continue being able to drain wastewater coming from the center of Surabaya, a water gate and water pump are needed at Kalidami Boezem.At the pump house located at the Kalidami Boezem site, there are 7 flood pumps with capacity of 3 m^3/s. The following is the pump curve from the pump.

3.3.5 Calibration of Simulation Results. The modeling results obtained using the SWMM 5.1 application will be matched with the data recording of the water level that occurred on February 25, 2020. Then, the error rate of the model formed will be measured using the RMSE method. The data used are obtained directly through field observation at Kalidami Boezem on that date.

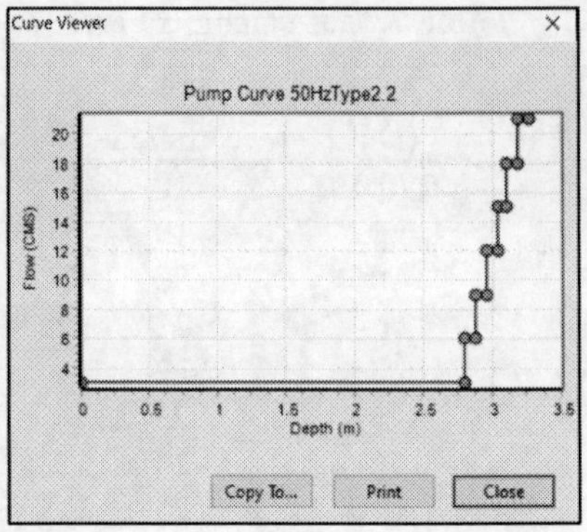

Figure 7. Pump curve (SWMM 5.1).

3.3.6 Flowchart. In general, the stages of conducting research from beginning to end can be seen in the following flowchart:

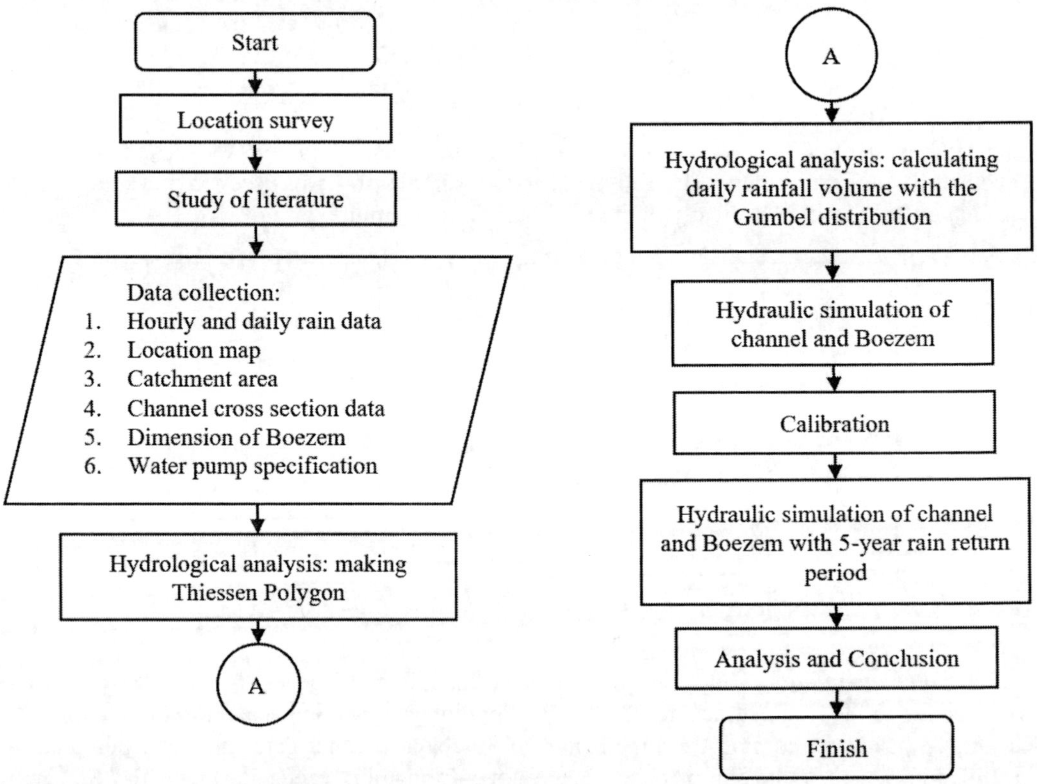

Figure 8. Flowchart research methodology.

4. Results and discussion

4.1 Kalidami channel

The Kalidami channel is a primary channel located in Surabaya that functions to flow water from the center of Surabaya to the sea in the eastern part of Surabaya. In the figure above, the initial condition of the Kalidami channel has an average depth of 0.3 m originating from the domestic disposal of the area around the channel.

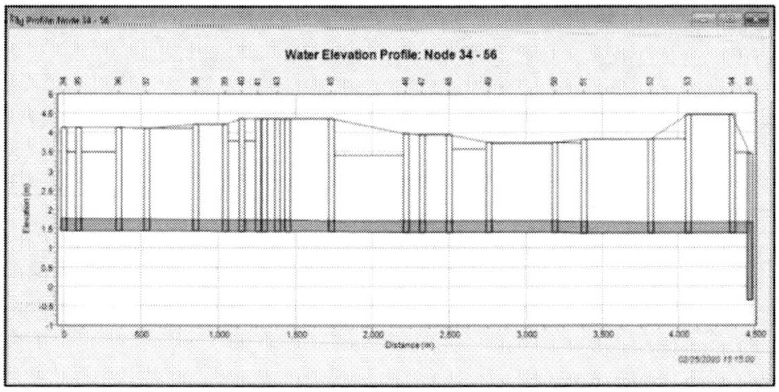

Figure 9. Initial conditions of the Kalidami channel (SWMM simulation).

This initial condition will be intervened by an increase in water level caused by the falling rain in the catchment area. The following is the results of the Kalidami channel simulation on February 25, 2020 at 21:00 where on that day there was a fairly high rainfall of 84.6 mm starting at 15:04 with the highest peak intensity at 16:04 Western Indonesia Time.

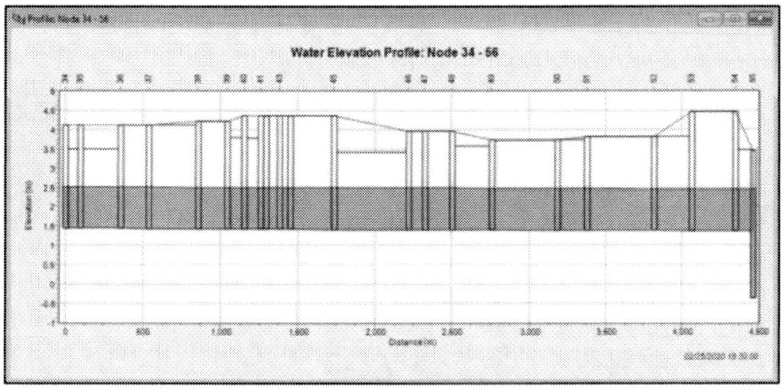

Figure 10. Condition of Kalidami channel after the simulation (SWMM simulation).

Based on the results of the simulation it can be seen that for rainfall condition below 100 mm which is equal to 84.6 mm, the Kalidami Channel and Boezem are sufficiently filled with water. The highest increase in water level occurred at 18:30 WIB since the rain began.

4.2 Model calibration results

By entering the hydrological and hydraulic data that have been obtained into the SWMM 5.1 application, we will get the results of a simulation of the rise in the water level of Kalidami Boezem. Then, based on the results obtained, it can be calculated how deviating the results of the simulation are when compared with observation of the water level of Kalidami Boezem on February 25, 2020.

The following is a comparison chart between the depth values based on observation on February 25, 2020 compared with data on the results of simulated water level generated by the SWMM 5.1 application.

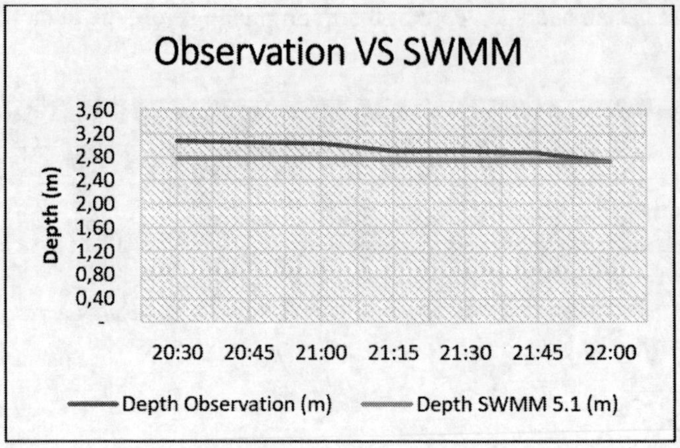

Figure 11. Comparison of water depth between the observation and SWMM simulation (observation and simulation results).

Based on the values obtained, the calibration results using the RMSE method will produce a value of 0.11. The RMSE value is good enough because it is below 1 and close to 0. Because the Kalidami Channel model has been calibrated, various scenarios in the resulting model can be carried out such as by replacing 5-year return period rainfall data to see performance pumps and water level profiles on the channel.

4.3 Simulation with 5-year return period rainfall data

The following is the simulation result of the Kalidami Channel system using 5-year return period rainfall data in SWMM 5.1 application.

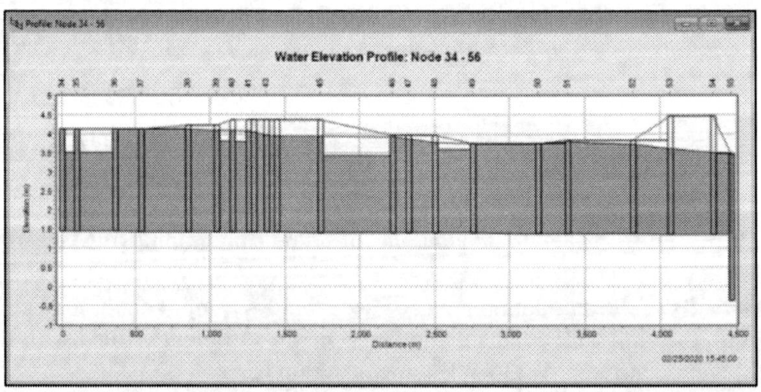

Figure 12. Results of simulation using input of 5-year return period rainfall data (SWMM simulation).

Based on the simulation results with the assumption of rain at 15:00 which occurred for 4 hours and the same water pump settings as in the previous discussion, it can be seen that the water level at Kalidami Boezem reached its peak at 15:45. Then, it was discovered that the water level at the Boezem began to decrease gradually at 19:30.

5. Conclusion

Based on the analysis and discussion above, several conclusions can be written as follows:

1. The water level of Kalidami Boezem on February 25, 2020 increased due to quite high rainfall which is 84.6 mm.

2. The error value generated from the water level data in the model compared to the field observation data measured using the RMSE method gives a pretty good value that is equal to 0.11. It shows that the resulting model has been able to adequately describe the real conditions that occur in Kalidami Boezem.

3. Based on the simulation results using a 5-year return period rainfall, it can be seen that the Kalidami Channel has a capacity that is not large enough to hold the incoming water runoff. It can be seen that only in the span of 45 minutes, the Kalidami Channel has been flooded.

References

[1] EPA United States Environmental Protection Agency 2015 *Storm Water Management Model User's Manual Version 5.* EPA/600/R-14/413b
[2] Pemerintah Kota Surabaya 2014 *Peraturan Daerah Kota Surabaya Nomor 12 Tahun 2014 Tentang Rencana Tata Ruang Wilayah Kota Surabaya Tahun 2014-2034*
[3] Rosyidie A 2013 Banjir: Fakta dan Dampaknya, Serta Penngaruh dari Perubahan Guna Lahan *J. Perencanaan Wilayah dan Kota* **24**
[4] Suprapto M, Yusuf M A, Prilbista A S, 2018 Analisis Sistem Drainase untuk Penanganan Genangan di Kecamatan Magetan Bagian Utara *e-Jurnal MATRIKS TEKNIK SIPIL* **231**
[5] Ningsih D H U 2012 Metode Thiessen Polygon untuk Ramalan Sebaran Curah Hujan Periode Tertentu pada Wilayah yang Tidak Memiliki Data Curah Hujan *J. Teknologi Informasi DINAMIK* **17**
[6] Astari K F, Hendri A, and Fauzi M 2018 Analisis Pasang Surut Perairan Dumai Menggunakan Metode Admiralty *Jom FTEKNIK* **5**
[7] Upomo T C and Kusumawardani R 2016 Pemilihan Distribusi Probabilitas pada Analisa Hujan Dengan Metode Goodness Of Fit Test *J. Teknik Sipil & Perencanaan* **18**
[8] Basuki, Winarsih I, and Adhyani N L 2009 Analisis Periode Ulang Hujan Maksimum dengan Berbagai Metode *J. Agromet* **23** 76-92
[9] Kementerian Pekerjaan Umum 2014 *Peraturan Menteri Pekerjaan Umum Nomor 12/PRT/M/2014 Tentang Penyelenggaraan Sistem Drainase Perkotaan* (Indonesia: Kemen. PUPR)

The benefits of river normalization of Guntur weir upstream to irrigation area service in Demak Regency Central Java Indonesia

A Y Imawan [1], S I Wahyudi [2*] and N R Wahyudi[3]

[1]Master of Civil Engineering Student, Faculty of Engineering, Universitas Islam Sultan Agung, Semarang, Indonesia
[2]Professor of Civil Engineering Department, Faculty of Engineering, Universitas Islam Sultan Agung, Semarang, Indonesia
[3]Master of Ocean Engineering Student, Faculty of Ocean Engineering, Institut Teknologi Sepuluh Nopember, Surabaya, Indonesia

*Corresponding authors: wahyudi@unissula.ac.id

Abstract. Guntur weir is located in Demak Regency, Central Java Province, Indonesia. The damage occurred in the upstream of Guntur Weir, mainly due to sedimentation. The cross-sectional area of the river becomes smaller so that when heavy rainfall, river water overflows and breaks the levees. These problems also disrupt water supply to irrigation channels where sedimentation covers the entrance of the Guntur Intake Weir. The purposes of this study are to determine the condition of the Guntur weir before and after the normalization work, the benefits of the Jragung river normalization work. This study was performed by analyzing hydrological data to find out the water balance, flood discharge, and the benefits analysis of the Jragung River normalization work. Based on the analysis using the HEC RAS application at Guntur Weir, an increase of Guntur Weir storage from 1002.5 m^3 to 88280.525 m^3. Flood discharge calculation using the HEC RAS application shows normalization work has benefits to reduce the potential flooding in the upstream of the Guntur Weir and also increase the ability of Guntur Weir to supply water needs for the Guntur Irrigation Area. The result used to adapt the operation Guntur Weir gate.

Introduction

The Guntur Irrigation had an area of 2,036 Ha in the last data and become 1,944 Ha area after updating to the current condition map. Guntur weir is located at coordinates 6° 43' 26" - 7 ° 09' 43" latitude South and 110° 27' 58" - 110° 48' 47.40" East Longitude, Demak Regency, Jawa Tengah, Indonesia. All Guntur irrigation areas are in Demak Regency [1]. Number of villages which entered the Guntur Irrigation Area are 12 villages consisting of 2 regions sub-districts namely Guntur sub-district with 5 villages and Karangtengah sub-district with 7 villages. According to Irrigation history, in 1912-1914 The Dutch East Indies Government built the Guntur Weir. Then In 1974 Indonesia government, by the Jratun Seluna Project,

Content from this work may be used under the terms of the Creative Commons Attribution 3.0 licence. Any further distribution of this work must maintain attribution to the author(s) and the title of the work, journal citation and DOI.
Published under licence by IOP Publishing Ltd

Guntur Weir was rehabilitated by increasing the crest height ± 0.90 m and replacing the water gate . Then in 1982, the 10 gates were replaced with wooden sliding doors which were driven by electricity [2]. In 2008 by the Pemali Juana, Guntur Weir was rehabilitated again by increasing the height of the crest ± 0.48 m and replacing the sliding door which was originally from wood becomes steel door. Damage occurred in the Jragung river mainly due to sedimentation in the upstream of the Guntur Weir and the lack of awareness of the people who live along the river in disposing of the residual agricultural waste [3]. These problems also hampered the flow of water into the irrigation channel so that sediment covered the right and left Intake gates and extended to the Primary and Secondary channel. In 2018 the Jragung River normalization work was carried out by the Ministry of Public Work to overcome the problems in the Jragung River.

The purposes of this study are to determine the rainfall intensity, to analyze water needs, to know the volume of Guntur weir storage before and after implementing normalization work, an d then to know its benefits. After normalizing the Jragung River, it expected to be more effective function for Guntur Irrigation Area.

Research methods

The research method is a step of work to be carried out to achieve a maximum, efficient, and effective result [4]. The stages of the data collecting and analysis of the benefits of normalization in the Jragung River are as follows :

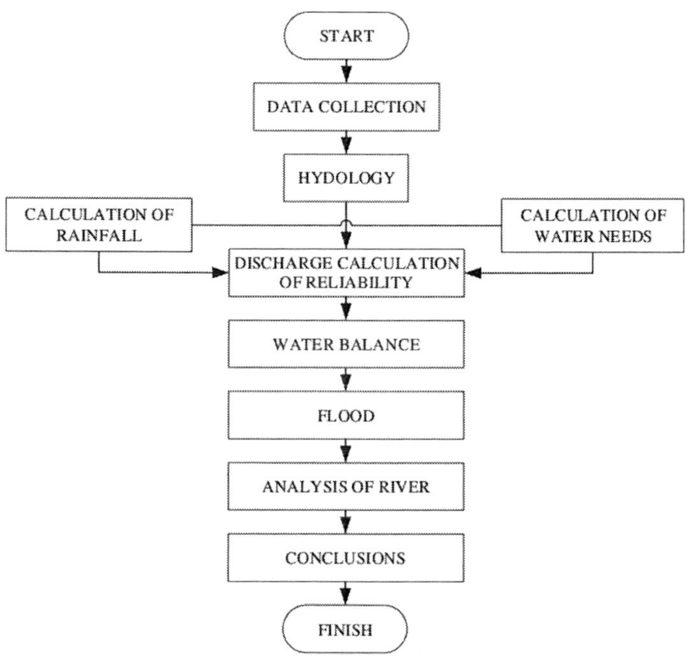

Figure 1. Research stages.

Result and discussion

The hydrological analysis phase aims to analyze the water balance to determine the water ability of the Guntur Weir to supply the Irrigation Area. The analysis stages are to analyze the rainfall plan, calculation of rainfall intensity, analysis of flood discharge plans, analysis of the discharge and water requirements. One method used in the rainfall analysis plan is the Thiessen method [5]. The Thiessen polygon is required for calculation and is shown in Figure 2 below:

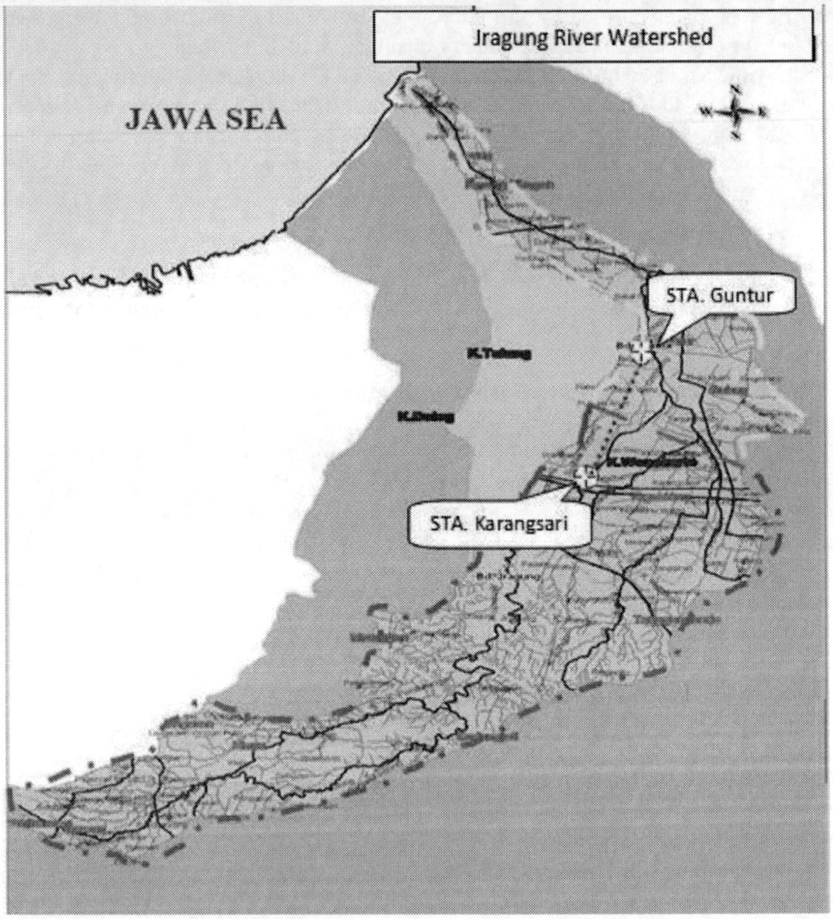

Figure 2. Polygon Thiessen Guntur Weir.

From the Thiessen polygon drawings, the Thiessen coefficient value can be calculated for each region. The results of the calculation of the Thiessen coefficient for each region are shown in table 1 below :

Table 1. Factors of rain stations in the Jragung River.

Rainfall station	Area (Ha)	Percentage (%)
Sta Karangsari	1.069,28	54,99

Sta Guntur	875,22	45,01
Total	1.944,50	100,00

From the above table, the percentage comparison results are obtained between the area of the polygon of the rain station and the total area of the River Basin [6]. After obtaining the Thiessen coefficient for each region, it can be calculated the average monthly rainfall. The results of the calculation of average monthly rainfall are shown in table 2 below:

Table 2. Average monthly rainfall.

Station	Polygon thicsson %	Description	Jan	Feb	Mar	Apr	Mei	Jun	Jul	Aug	Sep	Oct	Nov	Des
Sta. Karangsari	54.99	X	461.89	423.67	296.17	192.11	78.83	48.00	31.56	13.28	40.00	159.44	227.06	322.83
		Sd	223.53	183.10	134.36	95.42	73.87	55.34	38.89	25.13	63.72	112.36	87.81	92.51
		R5 mounthly	273.68	269.50	183.04	111.77	16.63	1.40	0.00	0.00	0.00	64.83	153.12	244.94
		R5 daily	8.83	9.62	5.90	3.73	0.54	0.05	0.00	0.00	0.00	2.09	5.10	7.90
Sta. Guntur	45.01	X	365.06	362.33	260.61	212.33	96.67	75.89	36.39	16.33	44.83	167.61	235.33	308.17
		Sd	173.76	198.15	85.84	104.81	53.54	75.06	38.89	27.17	77.35	122.92	109.34	100.73
		R5 mounthly	218.75	195.49	188.34	124.08	51.59	12.68	3.64	0.00	0.00	64.11	143.27	223.35
		R5 daily	7.06	6.98	6.08	4.14	1.66	0.42	0.12	0.00	0.00	2.07	4.78	7.20
Result	100.00	X	418.30	396.06	280.16	201.21	86.86	60.55	33.73	14.65	42.18	163.12	230.78	316.23
		Sd	201.13	189.88	112.52	99.65	64.72	64.22	38.89	26.05	69.85	117.12	97,50	96.21
		R5 mounthly	248.96	236.19	185.42	117.31	32.26	6.48	1.64	0.00	0.00	64.51	148.68	235.22
		R5 daily	8.03	8.44	5.98	3.91	1.04	0.22	0.05	0.00	0.00	2.08	4.96	7.59

From the calculation of average monthly rainfall using the Thiessen Polygon method above, it is necessary to analyze the distribution using statistical methods [7]. There are 4 distributions in the calculation of rainfall statistical parameters namely Normal, Gumbel, Log Normal and Log Pearson III. Table 2 present the statistical parameters and selected by some distribution model in table 3.

According to the table 3, the formula selected is the Normal Distribution. The next step is to do the distribution test with the Chi-Square and Smirnov-Kolmogorov test to find out whether the distribution is acceptable or not. From the Chi-Square and Smirnov-Kolmogorov test results obtained that the distribution of data can be accepted. The calculation of flood debit plan is calculated using the FSR Jawa Sumatra method. Calculation of flood debit plan can be seen in table 4 below:

IOP Conf. Series: Materials Science and Engineering **930** (2020) 012077 doi:10.1088/1757-899X/930/1/012077

Table 3. Selection of distribution types.

Number	Distribution	Requirements	Results	Explanation
1	Normal	Cs ~ 0 ± 0,3	0.3194	Fulfill
		Ck ~ 3,00	1.8116	Fulfill
2	Gumbel	Cs = 1,14	0.3194	Not Fulfill
		Ck = 5,4	1.8116	Not Fulfill
3	Log Normal	$Cs = Cv^3 + 3\ Cv = 0{,}21$	-0.0033	Not Fulfill
		$Ck=Cv^8+6Cv^6+15Cv^4+16Cv^2+3 = 3.05$	0.1272	Not Fulfill
4	Log *Pearson* Tipe III	$Cs \neq 0$	-0.0033	Fulfill
		$Ck = 1{,}5Cs^2+3 = 3$	0.1272	Not Fulfill

Table 4. Calculation of designed flood discharge using the FSR Jawa Sumatra method.

T years	GF	A km^2	V	PBAR mm	ARF	APBAR mm	H km	L km	MSL	SIMS	MAF m^3/sec	Qt m^3/sec
5	1.275	264.22	0.95	203.95	1.12	228.86	0.83	71.67	64.50	0.013	329.54	420
10	1.550	264.22	0.95	203.95	1.12	228.86	0.83	71.67	64.50	0.013	329.54	511
20	1.860	264.22	0.95	203.95	1.12	228.86	0.83	71.67	64.50	0.013	329.54	613
50	2.325	264.22	0.95	203.95	1.12	228.86	0.83	71.67	64.50	0.013	329.54	766
100	2.750	264.22	0.95	203.95	1.12	228.86	0.83	71.67	64.50	0.013	329.54	906

From the results of the analysis using the FSR Jawa Sumatra method as the table above, the flood discharge results with a return period of ± 50 years is equal to 766 m^3 / sec. After the flood discharge plan is obtained, the next step is the analysis of water requirements, in this analysis the calculation uses the Penman method to calculate potential evapotraspiration, which will be used later to calculate water needs. After that the calculation of effective rainfall is carried out which will also be used to calculate water needs [8]. The results of the calculation of water demand are obtained from the analysis of cropping patterns. The next calculation is the calculation of the mainstay debit which will later be entered into the water balance. Water balance is obtained by comparing water availability and water demand [9]. If there is a surplus, it means that the water demand is smaller than the water availability. and if the deficit means that the water demand is greater than the water availability.

Table 5. Calculation of water balance in the Jragung catchment area.

Description	Unite	October		November		December		January		February		March	
		I	II	I	II	I	II	I	II	I	II	I	II
Discharge Needs For Irrigation Water	m^3/dt	0.94	1.45	2.74	2.65	1.91	2.11	2.12	1.63	0.69	1.30	2.53	2.32
Jragung Discharge	m^3/dt	0.66	1.52	2.44	4.57	5.60	6.62	7.78	10.14	9.99	9.62	10.24	9.78
Reliability	%	70.50	100.00	88.83	100.00	100.00	100.00	100.00	100.00	100.00	100.00	100.00	100.00
Description		April		May		June		July		August		September	
		I	II	I	II	I	II	I	II	I	II	I	II
Discharge Needs For Irrigation Water	m^3/dt	2.41	2.53	2.92	2.43	1.28	1.68	2.03	1.07	1.20	1.58	0.00	0.00

Jragung Discharge	m^3/dt	9.55	8.49	5.25	3.63	2.88	1.28	1.14	1.05	0.55	0.51	0.39	0.60
Reliability	%	100.00	100.00	100.00	100.00	100.00	76.32	55.98	97.87	45.82	32.18	100.00	100.00

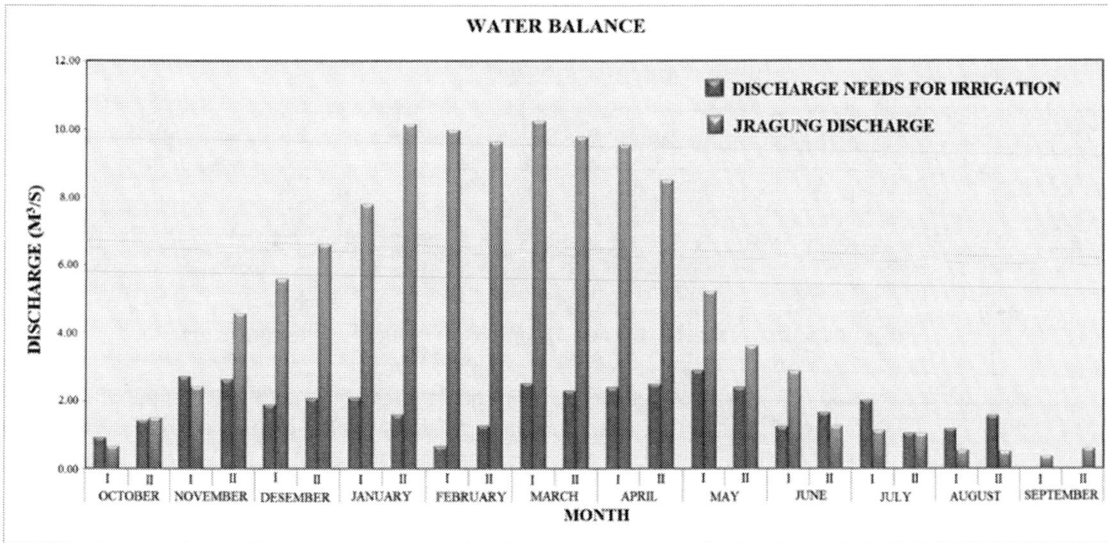

Figure 3. Water balance graph.

From the calculation of the above water balance table, it was found that the largest water availability deficit occurred in the August II period where the mainstay discharge only met water needs of 32.18% and there was a deficit of 67.82%. The next calculation is to calculate the volume of the Guntur Weir. The area of each cross can be seen in the image below:

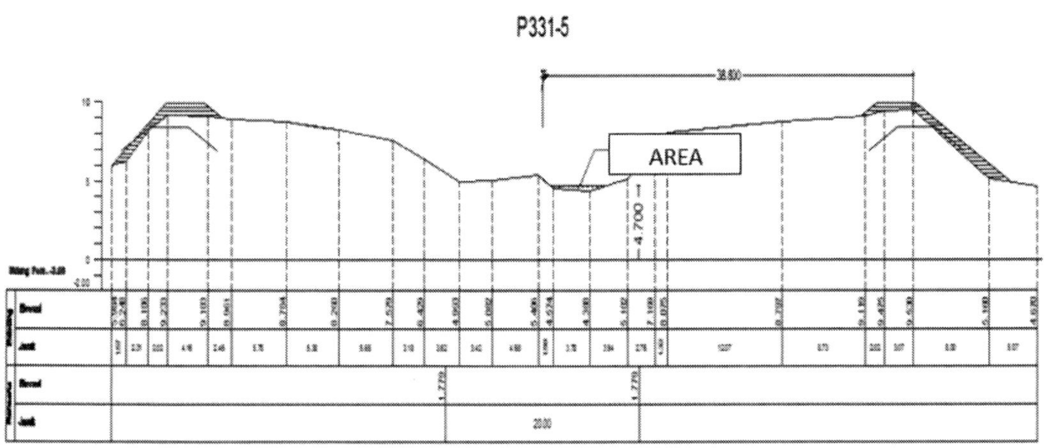

Figure 4. Cross section of River Basin area conditions, before normalization.

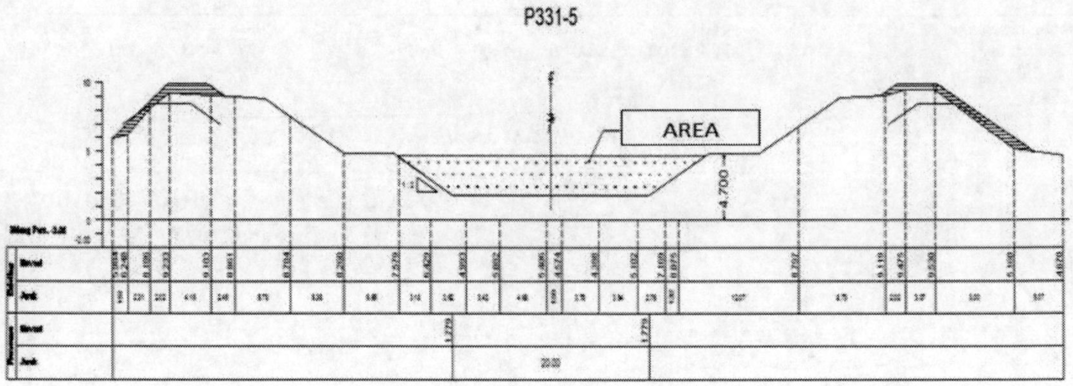

Figure 5. Cross section of river basin area conditions, after normalization.

After the basic elevation and water level elevation from the Jragung River are known in the upstream of the Guntur Weir when the conditions before and after the normalization, the volume of the reservoirs can be calculated. The first step is to make a number of pieces upstream of the Guntur Weir to calculate the cross-sectional area [10]. The figure 6 present HEC-RAS simulation after river normalization.

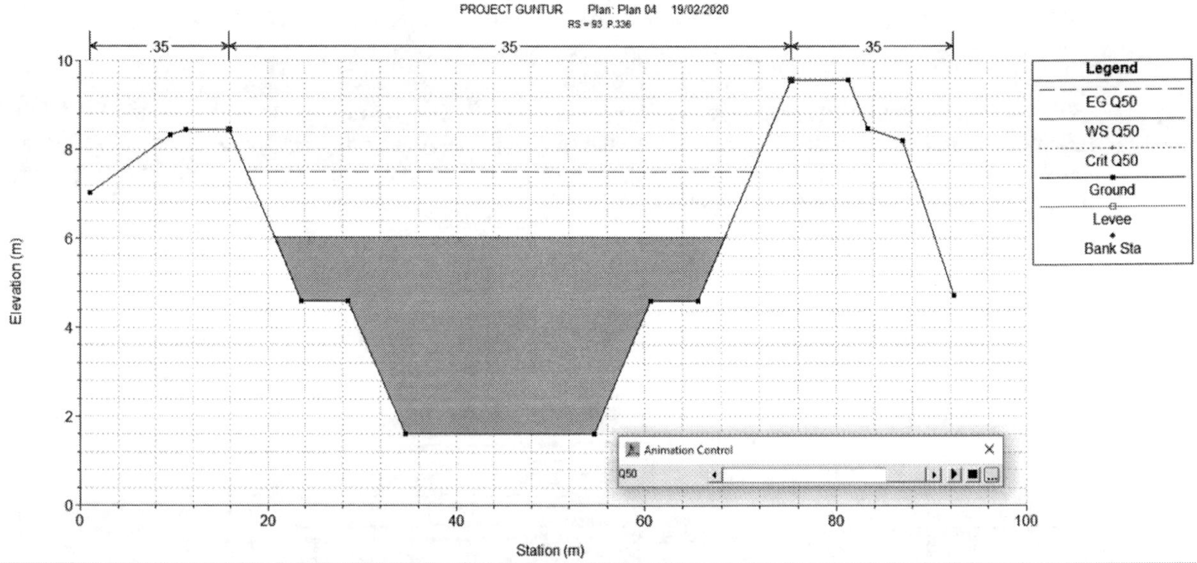

Figure 6. HEC-RAS Simulation result after river normalization in one river section.

The next step is to calculate the cross-sectional calculation can be used to calculate the volume from the upstream part of the Guntur Weir. For the calculation of the cross-sectional area and volume of the Guntur Weir can be seen in the table below:

Table 6. Cross-sectional area calculation table.

No	STA	Area (m^2)	
1	P349	Before Normalization	2.90
		After Normalization	98.07
2	P340	Before Normalization	0
		After Normalization	86.95
3	P 331	Before Normalization	1.11
		After Normalization	73.11

Table 7. Table calculation volume of Guntur Weir.

Distance		Volume (m^3)	
STA	Long (m)	Before Normalization	After Normalization
P349 - P340	250	725	24517.5
P340 - P349	250	0	21737.5
P340 - P331	250	0	21737.5
P331 - P340	250	277.5	20288.025
Total	1000	1002.5	88280.525

The difference volume in Weir Reservoir between before and after Normalization is 87278.03 m3. Base on the analysis result can be used to supply to irrigate Guntur Irrigation Area [11]. The recapitulation of the service ability of Guntur Weir can be seen in the table below:

Table 8. Calculation of capability of Guntur Weir service.

No	Condition	Storage Volume (m^3)	Irrigation Need (m^3/sec)	Service-ability (seconds)	Service-ability (hours)
1	Before Normalization	1002.5	1.07	936.92	0.260
2	After Normalization	88280.525	1.07	82505.16	22.92

According to the table 8, it was found that the increase in service ability of the Guntur Weir after the normalization work was carried out increased from only 0.2 hours to 22.92 hours' time services.

Conclusion

Base on the technical and benefit analysis of river normalization of Guntur Weir, It can be concluded on some following poins:

1) Before the normalization, the storage volume of the Guntur Weir is 1002.5 m^3. Then after normalization the storage volume of Guntur Weir is 88280,525 m^3. So the difference storage volume of Guntur weir before and after normalization is 87278.03 m^3.
2) The normalization work provides benefits for Guntur weir for increasing the availability of water for irrigation during the dry season. The ability after normalization, the Guntur Irrigation Area has increased 22.92 hours
3) River normalization increases flood protection and aesthetic value of the Jragung River because after the normalization process a long of the river becomes wider and looks more beautiful than before.
4) The simulation result can be used to adapt the manual of Guntur weir gate operation

Acknowledgment
The authors are grateful for our colleagues from Magister of Civil Eng., Unissula, Semarang. Our thanks are addressed to the Directorate of Research and Society Service (DRPM), Ministry of Research, Technology and Higher Education Indonesia and BBWS Pemali Juwana.

References

[1] Tallar R Y and Suen J 2015 Identification of waterbody status in Indonesia by using predictive index assessment tool *Int. Soil Water Conserv. Res.* vol. 3 pp. 224–238

[2] Wahyudi S I, Heikoop R, Adi H P, Overgaauw T, Schipper B, and Persoon B 2017 Emergency scenarios in the banger polder, semarang city: A case study to identify different emergency scenarios *Water Pract. Technol.* **12**

[3] Pitanggi G T, Lestari I T, Darsono S, and Salamun 2017 Normalisasi Sungai Dolok Semarang – Demak, Jawa Tengah *J. Karya Tek. Sipil* **6** 367–76

[4] Adi H P and Wahyudi S I 2018 Tidal Flood Handling through Community Participation in Drainage Management System (A case study of the first water board in Indonesia) *Int. J. Integr. Eng. Spec. Issue Civ. Environ. Eng.* **10** 19–23

[5] Triatmodjo B 2008 *Hidrologi Terapan* (Yogjakarta: Beta Offset)

[6] Remondi F, Burlando P, and Vollmer D 2016 Exploring the hydrological impact of increasing urbanisation on a tropical river catchment of the metropolitan Jakarta, Indonesia *Sustain. Cities Soc.*, vol. 20

[7] Setiono, Hadiani R, Erlangga, Edo, and Solichin 2016 Rainfall Simulation at Bah Bolon Watershed with Backpropagation Artificial Neural Network Based on Rainfall Data Using Scilab *Appl. Mech. Mater.* **845** 10–7

[8] Sriyana I, Wahyudi S I, Heikoop R, Kusmiyati F 2019 Performance of Sumur Gendong (Sedrainpond) Technology on Harvesting Water and Sediment from Agricultural Field toward Effective Rainday A Case Study on Jragung Irrigation Area, Jawa Tengah, Indonesia. *IRECE* **10** 41-7

[9] Boogaard F C, Restemeyer B, and Venvik G 2019 Knowledge exchange on Climate Adaptation with Nature-based solutions and Best Management Practices for Sustainable (ground) water management in Resilient Cities in *46th IAH Congress-Malaga, Spain*

[10] Sa'ud L 2008 Prediksi Sedimentasi Kali Mas Surabaya *J. Apl. Tek. Sipil* **4** 20

[11] Sherly M A, Karmakar S, Chan T, and Rau C 2016 Design Rainfall Framework Using Multivariate Parametric-Nonparametric Approach *J. Hydrol. Eng.* **21**

Hydrological analysis of moveable weir planning for tidal flood handling in Cilacap, Central Java

M Taufiq[1], H P Adi[2*] and S I Wahyudi[2]

[1] Thesis Student in Department of Civil Engineering, Sultan Agung Islamic University, Jl. Raya Kaligawe Km. 4, Kaligawe, Semarang, Indonesia
[2] Department of Civil Engineering, Sultan Agung Islamic University, Jl. Raya Kaligawe Km. 4, Kaligawe, Semarang, Indonesia

*Corresponding author's e-mail: henni@unissula.ac.id

Abstract. Global warming resulting in climate change has become a new disaster in the world. One of the changes in climate is the rise of sea water to land surface, which causes tidal flooding. One area that experienced tidal flooding was Kawunganten District, Cilacap, Central Java Province. The area is traversed by the Parit River with the river estuary located in Ujungmanik Village. The problem of tidal flooding in Kawunganten District can be overcome by constructing a moveable weir in the estuary of the Parit River. In planning on the moveable weir, a hydrological analysis is needed to determine the flood discharge plan. Rainfall data used as a basis for calculation of flood discharge at the study site is rainfall data from 3 observation stations namely Cilacap Station, Majenang Station and Ujungbarang Station with the recording period from 2009 to 2018. Calculation of rainfall plan is done using the Pearson Log Method III with a value of T100 = 165.463 mm. The results of the calculation of the flood discharge plan with the haspers method Q100 = 280.989 m3/sec, while the weduwen method Q100 = 132.804 m3/sec and the HSS Nakayasu method Q100 = 102.444. Furthermore, an analysis of tides on the coast of Cilacap Regency with the results of MWL (Mean Water Level) = 110 cm, LWL (Low Water Level) = 20 cm, HWL (High Water Level) = 220 cm.

1. Introduction

Global warming followed by climate change has become a new disaster in the world. One of the impact of climate change is the rise of sea water to land surface [1]. Cilacap Regency is one of the regions in Central Java, which is located in the southern coastal region bordering the Indonesian Ocean. The topography of Cilacap Regency has caused frequent flooding due to tidal flooding [2]. One of the areas that experienced flooding in Cilacap Regency was Kawunganten District. Most areas in Kawunganten Regency are dominated by lowlands, which results in frequent flooding. Floods that occur due to sea level rise. Tidal floods occur in areas directly adjacent to mangrove forests and are still affected by tides with intensity of 2-3 times a month. The floods occurred because of the high discharge and abundant water supply from the upstream of the Cibereum River and the Kawunganten River which empties into the Parit River with the river estuary located in Ujungmanik village. Parit River has a length of approximately 7 km from the downstream of Segoro Anakan, which is one of the rivers that are prone to flooding in Cilacap Regency. In early 2017, floods in the Parit river occurred several times which resulted in a flood retaining embankment bursting so that the water overflowed to the residents' housing which caused considerable losses to the people living around the Parit river.

Content from this work may be used under the terms of the Creative Commons Attribution 3.0 licence. Any further distribution of this work must maintain attribution to the author(s) and the title of the work, journal citation and DOI.
Published under licence by IOP Publishing Ltd

The problem of tidal flooding in Kawunganten District can be overcome by constructing a moveable weir in the estuary of the Parit River. A moveable weir is a weir construction with a gate that can be moved to contain and drain the air, so as to minimize the increase in flooding [3]. The availability of the moveable weir is expected to reduce flooding in the Kawunganten District and surrounding areas. Tidal flood control planning in a watershed can be done well if the flood discharge plan is known [4]. Therefore, a hydrological analysis is needed for the study of the planned flood discharge in the watershed area and an analysis of the capacity of the Parit river [5]. Where the river hydraulics analysis is intended to analyze the profile of flood water levels in the Parit river with various times from the planned flood discharge.

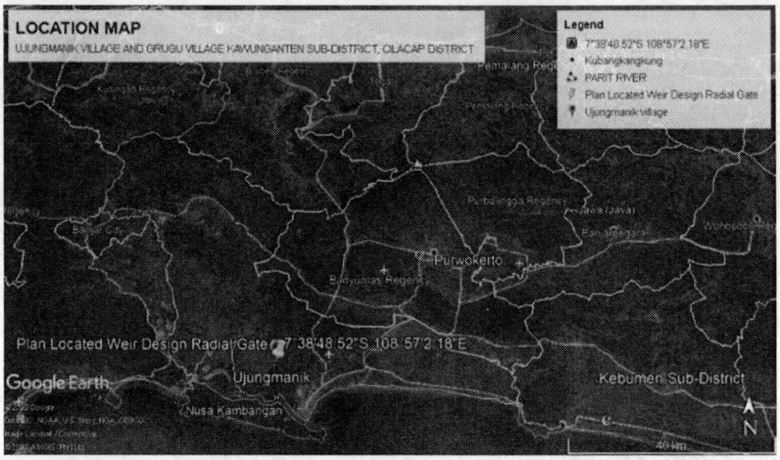

Figure 1. Map of Ujungmanik village (*Google Earth, 2019*)

This research was conducted at the estuary of the Grugu river and the Parit river sub-watershed Cikonde and Sapuragel watershed in Ujungmanik Village, Kawunganten District, Cilacap Regency, Central Java, with a coordinate position of 070 38 '48.9912 "S and 1080 57' 2.0124" E. To go to the study site reached by land. It takes about 1 hour as far as ± 26 Km from Cilacap City to get to the project site by 2 or 4 wheeled vehicles. Whereas from Kawunganten District it can be reached for 15 minutes as far as ± 7.3 km or from the Kubangkangkung highway entering as far as + 3, 0 km to the south up to Ujungmanik Village.

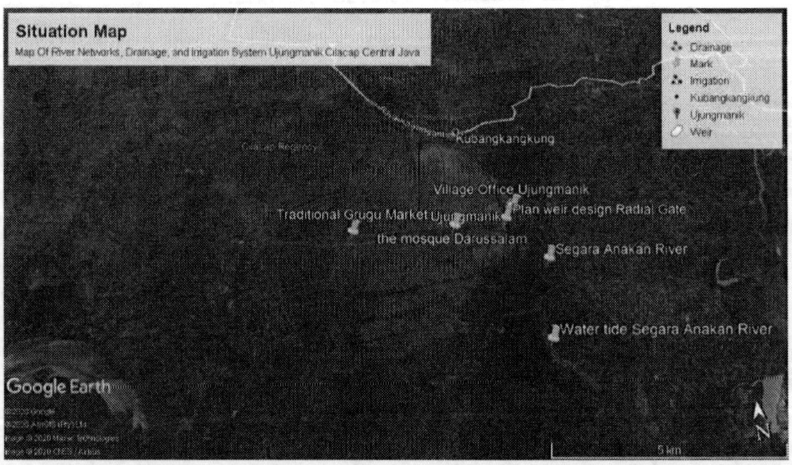

Figure 2. Inventory map of the drainage and irrigation river network (Google Earth, 2019)

2. Literature review

2.1. Hydrological analysis
Hydrological data is a collection of information or facts about hydrological phenomena, such as the amount: rainfall, temperature, evaporation, duration of solar radiation, wind speed, river discharge, river water level, flow velocity, river sediment concentration will always change with respect to time [6].

2.2. Watershead (DAS)
The watershed is determined based on the topography of the area, where the watershed is an area bounded by ridges between two rivers to the river being observed [7]. On the topographic map, it can be determined how to make imaginary lines connecting the points that have the highest contour elevation on the left and right of the river being observed. To determine the area of a watershed a planimeter can be used [8].

2.3. Tide water resources management
Tides are periodic sea level rise and fall caused by the attraction of celestial bodies, especially the moon and the sun against the mass of water on earth in a certain time. Management of water resources is an effort to deal with securing coastal areas due to global climate change [9].

3. Research method
This study uses a quantitative research methodology with detailed, systematic, structured calculations. The data in this study include primary data and secondary data. Primary data obtained by surveys and observations that include bathymetry surveys, tidal measurement surveys, river flow velocity measurements and sedimentation surveys. Secondary data was obtained from the BBWS agency (Balai Besar Sungai Serayu Region) that include rainfall data, evapotranspiration, temperature, wind speed, air humidity, topographic maps, geological maps. Based on these data, a hydrological analysis is then performed which covers: the determination of the Watershed (DAS) and the determination of the rain gauge station using the Thiessen Polygon Method. Then calculating the maximum rainfall, analyzing the return period of the T rainfall year plan and calculating the flood discharge plan based on the amount rainfall plan, calculation of flood discharge and rainfall intensity.

The steps to be taken in the hydrological analysis of the design of Parit river movable Weir are:
1. Hydrological analysis of the calculation of the rainfall plan and flood discharge plan for the tides.
2. Hydraulic analysis simulation using the HEC-RAS program to obtain flood water levels and tides.

Based on the background description, the formulation of the problems that will be discussed in this study are as follows:
1. Hydrological analysis uses maximum daily rainfall data for 10 years.
2. Time periode at Q5, Q10, Q50, Q100 years.
3. Simulation of hydraulic analysis using the HEC-RAS program to obtain flood water levels.

4. Results and discussion

4.1. Watershed (DAS) calculation
The Watershed (DAS) is an area where all water flows into the destination river. Calculation of the area of influence of rain stations on watersheds (DAS) consists of three stations, namely Cilacap Station, Majenang Station, Ujungbarang Station.

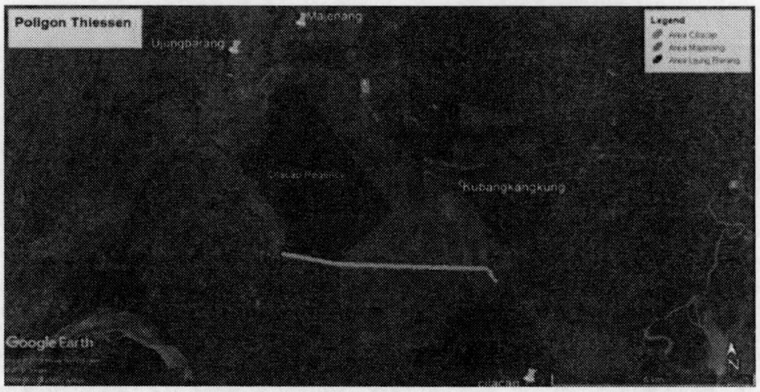

Figure 3. Polygon thiessen

The following table are the results of calculations using the Thiessen Polygon Method on the Topographic map of the Parit River rain station which has an influence on the Sapuregel and DAS Cikonde Sub Watershed.

Table 1. Extent of rain station influence on watershed

No	Rainfall Station	Percentage (%)	DAS Area (km^2)
A1	Sta Cilacap	25.54%	7
A2	Sta Majenang	21.89%	6
A3	Sta Ujungbarang	52.55%	14.4
	Total	100%	27.4

4.2. Hydrology

4.2.1. Calculation of rainfall plan.
For the calculation of rainfall plans use 3 stations, namely: Cilacap station, Majenang Station, and Ujung Barang Station. After proven by the Thiessen Polygon Method on the topographic map of the Parit River, the station estuary which influences the Cikonde Sub-basin and Sapuregel Sub-basin.

The Parit River is the estuary of the Cibereum River and the Kawunganten River Cikonde and Sapuregel Sub-watersheds, which are affected by tides in Segoro Anakan, the length of the Parit River is approximately 7 km from the downstream of Segoro Anakan.

Table 2. Maximum rainfall

Year	Cilacap Sta (mm)	Majenang Sta (mm)	Ujungbarang Sta (mm)
2009	166	124	90
2010	135	172	113
2011	135	44	96
2012	124	49	90
2013	97	48	87
2014	194	78	104
2015	132	91	145
2016	181	99	110
2017	180	118	123
2018	173	148	128

Calculation of rainfall plans using probability distribution namely Gumbel Probability Distribution, Log Normal Probability Distribution and Log Person Type III Probability Distribution.

Table 3. Calculation results of rainfall plans

No.	Periode Repeated (Year)	Distribution System		
		Gumbel I (mm)	Log Normal (mm)	Log Pearson Tipe III (mm)
1	2	1,141,657	1,123,572	117,959
2	5	1,397,087	1,307,634	140,090
3	10	1,566,204	1,440,330	149,801
4	20	1,728,425	1,575,166	157,227
5	25	1,779,884	1,620,111	158,756
6	50	1,938,403	1,759,227	162,821
7	100	2,095,753	1,885,502	165,463

4.2.2. Calculation of flood discharge plan

Plan flood discharge is the maximum flood discharge from a river or channel whose amount is based on a certain return period [10]. Flood discharge plan, used as a basis in planning a hydraulic building with the aim that the planned building is able to receive the amount of flooding that is likely to occur in the planned return period. In analyzing the flood discharge the plan will use 3 methods namely Haspers Method, Weduwen Method, and Nakayau Synthetic Unit Hydrograph Method.

Table 4. Recapitulation of flood discharge plans

		Hasper	Weduwen	Nakayasu
Design Flood Discharge (m3/dt)	**Q 2 Th**	200,317	83,727	61,603
	Q 5 Th	237,900	105,917	80,413
	Q 10 Th	254,392	115,975	88,892
	Q 20 Th	267,002	123,908	95,458
	Q 25 Th	269,599	125,606	96,818
	Q 50 Th	279,502	129,999	100,444
	Q 100 Th	280,989	132,804	102,817

IOP Conf. Series: Materials Science and Engineering **930** (2020) 012078 doi:10.1088/1757-899X/930/1/012078

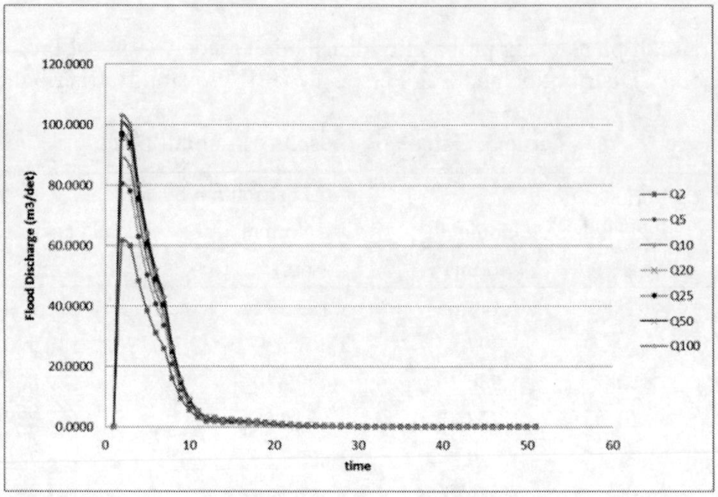

Figure 4. Flood hydrograph syintetic unit nakayasu (HSS nakayasu)

The selection of flood discharges used in the hydraulic analysis planning of the Radial Gate Dam Weir is a flood discharge with a 100-year return period that occurred in the field using the HSS Nakayasu Q100 method = 102.4444.

4.2.3. Tidal of sea water analysis

Calculation of tidal phenomena using tide data belonging to PT (Pelindo III) Tanjung Intan Indonesia III Port Cilacap, Central Java. The mixed tidal type tends to a single daily (Mix Tide Prevailing Semidiurnal), with an average height of sea level ranging from 110 cm, the lowest tide of 20 cm and the highest tide of 220 cm.

4.3. Analysis of HEC-RAS water flooding weir

Depend on the trials using the Hec-Ras software to calculate flood simulations of the tidal plan with the Synthetic Unit Hydrograph (HSS) method of Nakayasu Q100 = 102,817 m3 / s. Tidal flood simulations using the HEC-RAS program obtained Flood Water level (MAB) +4.35 m, Normal Water Level (MAN) elevation +2.20, Low Water Level (MAR) +0.20. Then obtained the water level in the downstream of the weir + 4.55 m.

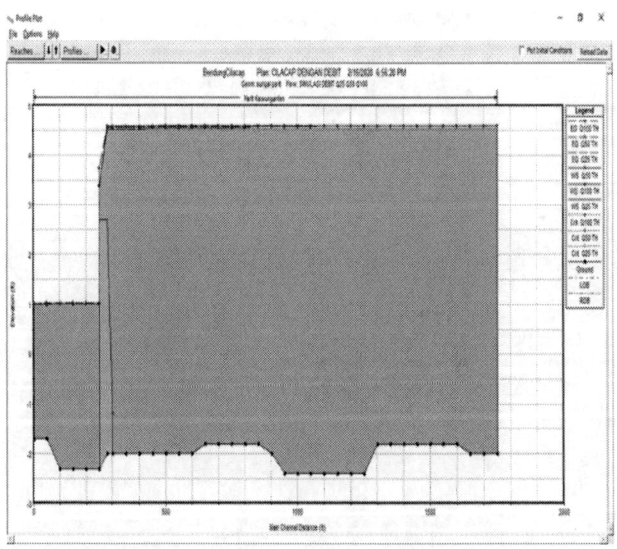

Figure 5. Long profile of HEC-RAS analysis result

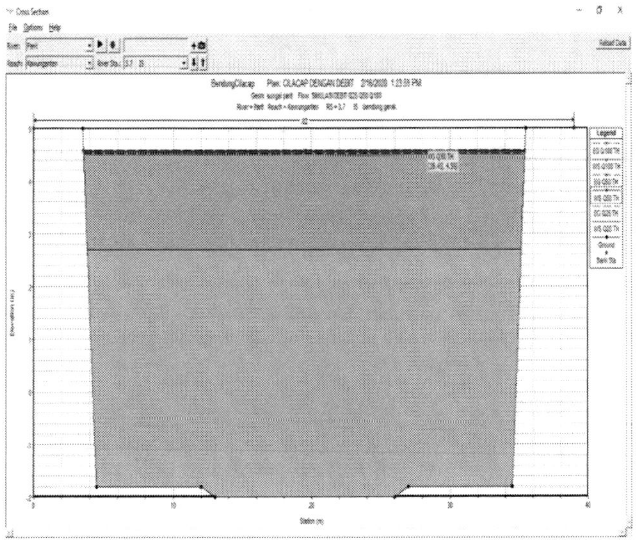

Figure 6. Cross profile of HEC-RAS analysis result

5. Conclusions and recommendations

5.1. Conclusions

Based on the results of the analysis, it can be concluded as follows:

1. The results of the hydrological analysis of the planned rainfall as the basis for tidal flood treatment are used in the Repeat Period with the Gumbel Method T100 = 2,095,753 mm, the Normal Log Method T100 = 1,885,502 mm, the Pearson III Log Method T100 = 165,463 mm. Calculation of rainfall plan is done using the Pearson Log Method III with a value of T100 = 165.463 mm. Calculation of flood discharge plan as a basis for handling flood tides using a 100 year return period. The method used is the haspers method, the Wesuwen method and the Nakayasu HSS method. The results of the calculation of the flood discharge plan with the haspers method Q100 = 280.989 m3/sec, while the weduwen method Q100 = 132.804 m3/sec and the HSS Nakayasu method Q100 = 102.444. From the three calculation results, the chosen flood discharge plan using the Nakayasu HSS method is the most appropriate for the conditions at the study site.
2. The results of tidal analysis on the coast of Cilacap Regency are included in the type of tidal mixture leaning to a single daily (Mix Tide Prevalling Semidiurnal), the results of MWL (Mean Water Level) = 110 cm, LWL (Low Water Level) = 20 cm, HWL (High Water Level) = 220 cm.
3. Tidal flood simulation using the Hec-Ras program is obtained by the surface water flood (MAB) +4.35 m elvation and the normal water surface elevation (MAN) +2.20

5.2. Recommendations

The recommendations that can be submitted are as follows:

1. Need to build a dike with a suitable height to deal with flooding in the river.
2. Routine maintenance needs to be done on the river such as cleaning grass and dredging the bottom of the channel because it can affect the storage capacity.

Acknowledgements

Our gratitude goes to the Directorate of Research and Public Service, Ministry of Research and Technology of Indonesia that has funded this research, Civil Engineering Department of Sultan Agung Islamic University, as well as all those stakeholders who participated and supported.

IOP Conf. Series: Materials Science and Engineering **930** (2020) 012078 doi:10.1088/1757-899X/930/1/012078

References

[1] Adi H P and Wahyudi S I 2018 Tidal flood handling through community participation in drainage management system (a case study of the first water board in Indonesia) *Int. J. Integr. Eng.* **10** Civil & Environmental Engineering 19–23 doi : https://doi.org/10.30880/ijie.2018.10.02.004

[2] Wahyudi S I and Adi H P 2017 Revitalizing and preparing drainage operation and maintenance to anticipate climate change in Semarang heritage city *J. Environ. Sci. Eng. B* **6** 1 17–26

[3] Frahmana B, Santosa B, Sipil J T, Gunadarma U and Gate D 2015 Planning a motion weir with a drum gate type of floodgate for flood control in the river basin of walahan weir, Karawang Regency in *Prosiding Psikologi, Ekonomi, Sastra Arsitektur dan Teknik Sipil* **6** 6–12

[4] Bahri Z and Yuniarti E 2016 Automatic counter weight sluice gate model in tidal rice fields to cope with sinking rice plants at hight tide *in Seminar Nasional Science dan Teknologi November* 1–8

[5] Wahyudi S I and Adi H P 2018 Evaluating environment, erosion and sedimentation aspects in coastal area to determine priority handling (a case study in Jepara regency, Northern Central Java, Indonesia) *IOP Conf. Ser. Earth Environ. Sci.* **140** doi :10.1088/1755-1315/140/1/012042

[6] Wahyudi H P, Rochim A A and Marot D 2014 Aspects of hydrology, tidal and water storage capacity for simulating dike model of channel and retention basin *Int. J. Civ. Environ. Eng.* **14** 05 6–10

[7] Aziza M I, Wardoyo W and Anwar N 2017 Storage simulation of sembayat moveable weir as a longstrorage for meeting the needs raw water and irrigation in lamongan regency and the Northern Region of Gresik Regency *J. Apl. Tek. Sipil* **15** 2 73–89

[8] Wahyudi S I, Adi H P, Lekerkerk J, Ven M V D, Vermeer D and Adnan M S 2019 Assessment of polder system drainage experimentation performance related to tidal floods in Mulyorejo, Pekalongan, Indonesia *Int. J. Integr. Eng.* **9** 73–82

[9] Salazar F 2013 Analysis of the discharge capacity of radial-gated spillways using CFD and ANN — Oliana Dam case study *J. Hydraul. Res.* **2** Juni

[10] Tjahyad D 2004 The hidraulic behavior of the rubber weir is filled with water *Media Komun. Tek. Sipil* **12** 3 100–12

Analysis of water surface profile and flood discharge of Cijangkelok river

A W P Sjarief[1*] and U Lasminto[1]

[1]Department of Civil Engineering, Institut Teknologi Sepuluh Nopember, Surabaya, Indonesia

[*]Corresponding author's e-mail: anjani.sjarief@pu.go.id

Abstract. The downstream of Cisanggarung River Basin is often affected by floods with an increasing degree of intensity every year. For instance, the 2017 severe floods caused economic losses and fatalities, especially along the border of Cirebon and Brebes Regency. The flood was attributed to high rainfall intensity and the meeting point between Cisanggarung River and the Cijangkelok River. Therefore, a study on the flood in this area is highly vital to reduce the affected area in the future. This study aims to analyze the maximum flood discharge capacity in the Cijangkelok River and analyze cross-sectional river capacity using numerical model simulations. Based on the hydrological calculations and analysis, it is known that the planned flood discharge in the Cijangkelok River in the 5-year return period is 256,25 m^3/s, the 10-year return period is 310,35 m^3/s, the 25-year return period is 392,91 m^3/s, and the 50-year return period is 465,93 m^3/s. The analysis of the water surface profile of each flood discharge return period shows that overflow occurs in the cross-section 4090, 8254, 9166, 9753, and 11970. It is also revealed that the existing cross-section of the Cijangkelok River is unable to accommodate hydrological discharge. Field visit results show that the river has siltation or sedimentation, the river embankments are irregular, and land-use changes around the Cijangkelok River that affects the river.

1. Introduction

Based on the Minister of Public Works Regulation No. 04/PRT/M/2015 concerning Criteria and Determination of River Areas, the Cisanggarung watershed area covers the area of 1,000.24 km2. Geographically, this area includes the Cisanggarung River Basin Area which is located at coordinates 105 ° 20' - 108 ° 47' East Longitude and 6 ° 45' - 7 ° 12' South Latitude. Based on its geographical position, the river is located in the eastern part of West Java, on a regional road link that connects Cirebon City with the East Priangan region and an alternative road in the middle lane that connects Bandung-Majalengka with West Java.

Flooding is a severe problem, especially in the area along the Cimanuk-Cisanggarung River Basin. The floods in the Cimanuk Cisanggarung River Basin are primarily due to river morphological damage and high sedimentation rates entering several rivers (Balai Besar Cimanuk-Cisanggarung River Region, 2015). Floods in the Cimanuk Cisanggarung River Basin are almost evenly distributed in all regions; the most flood-prone areas include Indramayu, Majalengka, Cirebon, and Brebes Regencies.

The downstream of Cisanggarung River Basin is often affected by floods with a higher degree of intensity every year. The severe 2017 flood, for example, caused economic losses and fatalities, especially along Cirebon and Brebes Regencies. The eroded upper Cisanggarung watershed and its

Content from this work may be used under the terms of the Creative Commons Attribution 3.0 licence. Any further distribution of this work must maintain attribution to the author(s) and the title of the work, journal citation and DOI.
Published under licence by IOP Publishing Ltd

tributaries and the reduced capacity of the river due to river bed aggradation impeded several places from flowing flood discharge and thus leading to overflowed river capacity. Therefore, it is necessary to study the flood to reduce the affected area in the future, especially in the Cijangkelok River as a tributary of the Cisanggarung watershed that was severely affected by the 2017 flood. It was noted that at that time, the Cijangkelok River highly contributed to the flow of flood discharge in the Cisanggarung River, making the study of this area as an essential thing to do.

Given the above situation, this study aims to analyze the maximum flood discharge capacity in the Cijangkelok River, analyze the discharge capacity at the river's cross-section, and evaluate the flood model on several return periods of flood. Hydrological data processing was carried out to examine the average discharge and explain the flood discharge redesign. The data were used as input to the simulation of the hydraulics model. The hydraulic modeling application is expected to help analyze the river cross-section and study the profile condition before the water in the event of a flood.

2. Literature review

2.1. Catchment area

According to Kodoatie (2002), Watershed Area is a naturally formed water management area, from which the rainfall flows into the rivers and generated tributaries. The Watershed Area is also known as Catchment Area. This area is vital since the utilization of watershed natural resources without taking the ability and environmental sustainability into account will harm the ecosystems and water systems. Therefore, in planning the watershed management, we need to use the most appropriate technological advancement based on additional rules of watershed management (Kodoatie, 2002). According to Susanto (1994), Watersheds are physically and biologically integrated area under the same watershed management, which influences the ecosystem. Consequently, the problems that occur in the upstream, such as organic solution (chemical pollutants), will affect the downstream. At present, Watershed Management becomes very important for maintaining the potential of water resources. The destructed upstream area will lead to the damage in the downstream areas, such as floods, landslides, or increasing groundwater levels.

2.2. Flood

Some flood control systems can indeed reduce the damage caused by small to moderate floods but they incur unaffordable cost during massive waves—that may take place around the high river embankment. Excessive runoff on the dam at the time of flooding out of the predicted flooding may collapse the dam, and thus creating a tremendous flow velocity through the destruction of the dam, causing severe flooding. The causal relationship in the flood problem indicates the complexity of the problem. The current tendency of land use changes, increasing population requiring additional infrastructure, and economic interests drive the decreasing land carrying capacity that leads to the escalating flood frequency and intensity.

The Flood Control Study in the Cisanggarung River Basin (BBWS Cimanuk-Cisanggarung, March 2016) was conducted to determine the efforts of flood mitigation and flood control planning in the Cisanggarung River Basin. This study resulted in a master plan for flood control in the Cisanggarung watershed. Similarly, the study on Preparation of Potential Flood and Drought Map of Cimanuk-Cisanggarung river area (BBWS Cimanuk-Cisanggarung, June 2016) was carried out to provide thematic geospatial data and information about the flood prone and drought prone areas as an integrated preparation for flood and drought mitigation in regencies and cities passed by the Cimanuk Cisanggarung River Basin. These studies result in disaster-prone map for each Regency/City and a multi-map digital map with a scale of 1: 50,000.

2.3. Unit hydrograph
The unit hydrograph is a direct runoff hydrograph produced by an active rain (net rainfall), which occurs evenly throughout the watershed and with fixed integration for a specified unit of time. The unit hydrograph indicates a direct runoff hydrograph at the downstream end of the river flow, which piled up with active rain of one millimetre. It was considered equivalent to the watershed's surface based on the specified provisions (Triatmodjo, 2008). In a purer form of resolution, the watershed unit is a direct runoff caused by an active rain unit, divided equally in time and space (Soemarto, 1999).

2.4. Hydraulics analysis
An analysis of river hydraulics intended aims to study the profile of floodwater levels in rivers with various times of return from the discharge plan. The hydraulic study will examine the extent of the current flood control influence. Based on the analysis support, we can conduct topographic measurements along the river to measure the extent of the problem, and the longitudinal, and transverse cross-section.

The Saint Venant equation was used to model flows in open channels, such as rivers and model of flood tracking (Pratiwi et al., 2013). Flood alignment was calculated using the equation of continuity and steady-flow momentum (Rinaldi et al., 2018). These measurements are described in the following mathematical equation:

$$\frac{dA}{dt} + \frac{\partial Q}{\partial x} - q_\ell = 0 \tag{1}$$

$$\frac{\partial Q}{\partial t} + \frac{\partial(\alpha \frac{Q^2}{A})}{\partial x} + gA\left(\frac{\partial h}{\partial x} + \frac{n^2 gQ|Q|}{AR^{\frac{4}{3}}}\right) = 0 \tag{2}$$

where :

A	= the total cross-sectional area of flow
Q	= discharge
q_ℓ	= lateral discharge per unit length
R	= hydraulic radius
g	= gravity velocity
x	= distance
t	= time
n	= Manning coefficient
α	= distribution momentum coefficient

3. Research Methods

3.1. *Study area*
This research focuses on the Cijangkelok River, a tributary of the Cisanggarung Main River. Incoming Discharge from the Cijangkelok River influences the amount of Discharge in the Cisanggarung River that leads to frequent flooding in the downstream of the Cisanggarung River and the meeting area between Cisanggarung River and Cijangkelok River (figure 1).

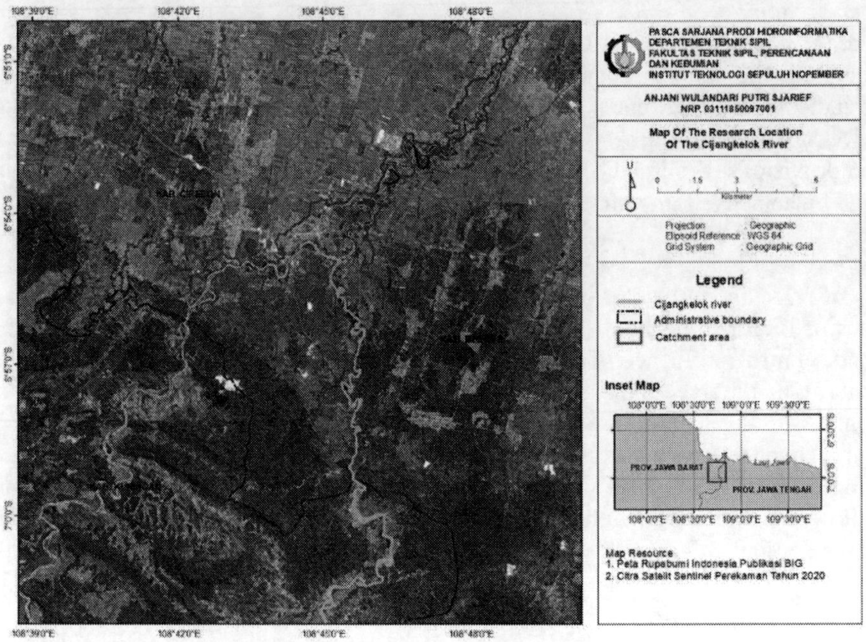

Figure 1. Research locations in the downstream of Cisanggarung sub-watershed, Cijangkelok river

3.2. *Data collection and data processing*

The study used hydrological data, hydraulic data, and topographic data. The steps and stages in the research methods are presented in a flow chart (figure 2) with the following description:

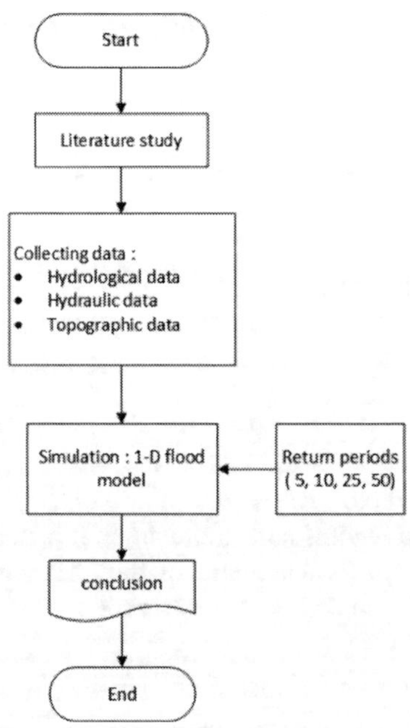

Figure 2. Study flowchart

3.2.1. Hydrology data. The collected rainfall data consists of daily rainfall data taken from each rain station considered on watershed of Cisanggarung and Cijangkelok watershed sub. Flood discharge design was done using rainfall data from several rain stations in the Cisanggarung and Sub Cijangkelok watersheds. The study used rainfall data from 6 (six) rain stations, namely Garawangi, Ciniru, Luragung, Ciawigebang, Cibeureum, and Cibendung Rain Stations. Given the limited availability of data in this study, the researcher used the daily rainfall data recorded for nine years from 2010-2018.

3.2.2. Hydraulic data. River geometry data were used as an input for analyzing water level profiles using spatial data from the Digital Elevation Model (DEM). Data obtained from DEM were in the form of cross-section data, hydraulic structure, and depth.

3.2.3. Return periods scenarios. The calibrated model was simulated with 5, 10, 25, and 50-years return period of discharge inputs.

3.2.4. Conclusion and analysis. This stage provides an explanation of the results of the river cross-sectional water profile simulation. It was conducted to analyze the existing channel capacity to determine the ability of the channel to receive flood discharge. It also aimed to identify whether it was necessary or not to do channel dimensions planning or other alternative solutions to cope with the existing flood discharge. At this hydraulic stage, the analysis was conducted using the MIKE 11 assist program. Input data used include cross-section geometry data, unit hydrographs, and manning coefficients.

4. Results and discussion

4.1 Hydrological analysis

A hydrological analysis aimed to determine the hydrological characteristics of the Cijangkelok River drainage area. In particular, it was done to determine the amount of flood discharge plans in a water building plan. The planned flood discharge in this study was conducted using rainfall data, since the rainfall data are one of the few data that can estimate the magnitude of the proposed flood discharge.

Based on the analysis, the Cijangkelok River sub-Basin has an area of 165.57 km2, with a river length of 36.61 km. A summary of the rainfall plan was used to analyze the flood discharge plan from the rainfall data and the watershed area. Due to their topographic conditions and the number of eligible stations for the Thiessen method that can handle the calculation, the rainfall analysis obtained 6 hours of rain that occurred in the Cijangkelok River sub-Basin.

In general, discharge of flood plans were used to conduct water disposal that leads to zero puddle occurrence. This analysis classifies the soil types according to the current status of land-use in the Cijangkelok River sub-Basin. It was revealed that the Cijangkelok River sub-Basin is classified as A and D soil types. Based on these results, Curve Number and Impervious data were used to analyze flood discharge plans. figure 3 and table 1 present the result of flood discharge analysis with the return periods of 5-years, 10-years, 25-years, and 50-years.

Table 1. Return period flood discharge

Peak Discharge (m^3/s)			
Q_5	Q_{10}	Q_{25}	Q_{50}
256,25	310,35	392,91	465,93

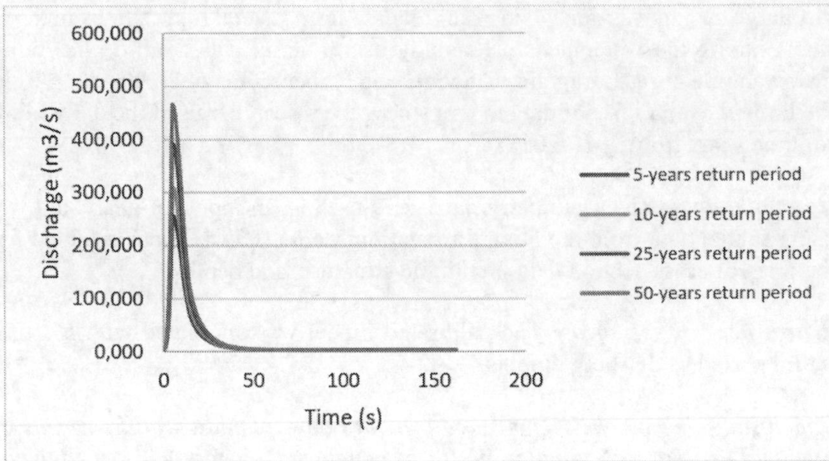

Figure 3. Return period *of* flood discharge *source: Analysis result, 2020*

Figure 3 shows that the peak discharge at the first 24 hours with peak discharge in the return period of 5-years is 256.25 m3/s, the return period of 10-years is 310.35 m3/s, the return period of 25-years is 392.91 m3/s, and the 50-years return period is 465.93 m3/s. On the second day, the graph started to fall, indicating the receding flow of water up to the flow base in the Cijangkelok River.

4.2 Modeling hydraulic

This hydraulic analysis consists of a summary of the existing cross-section of the planned channel—Mike 11 helps analyze the Existing cross-section, with discharge and cross-section data as input data. The analysis using MIKE 11 aims to determine the current condition of the Cijangkelok River. It shows the flow of water discharged through the river and the water level at the cross-section. MIKE 11 will display the model of the channel according to the provided input data. Figure 5 shows geometric data on the Cijangkelok River.

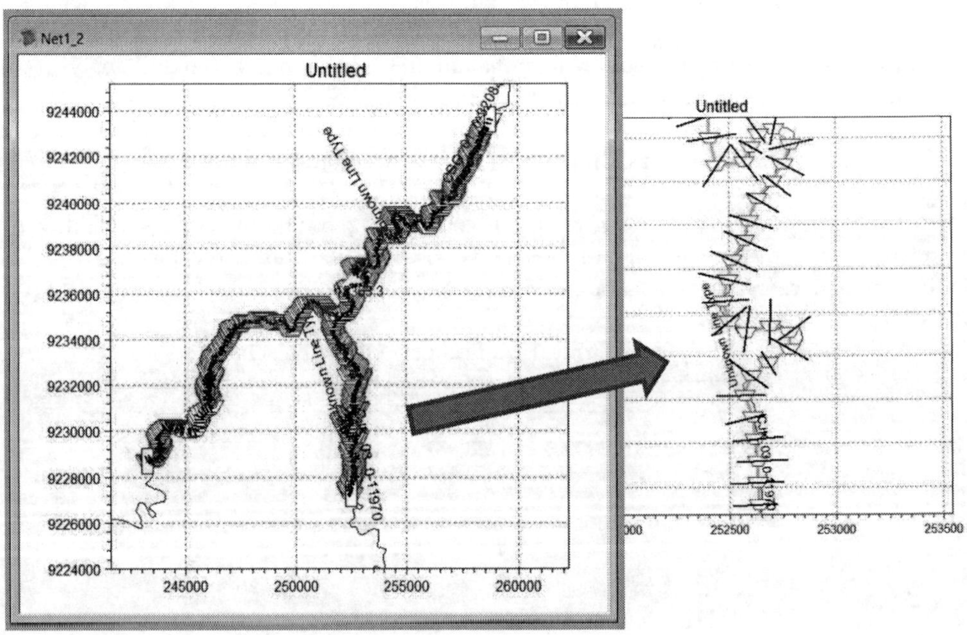

Figure 4. Cijangkelok's geometry

Figure 6 shows an example of a Cijangkelok River's cross-section with irregular shape and siltation due to sedimentation in the river.

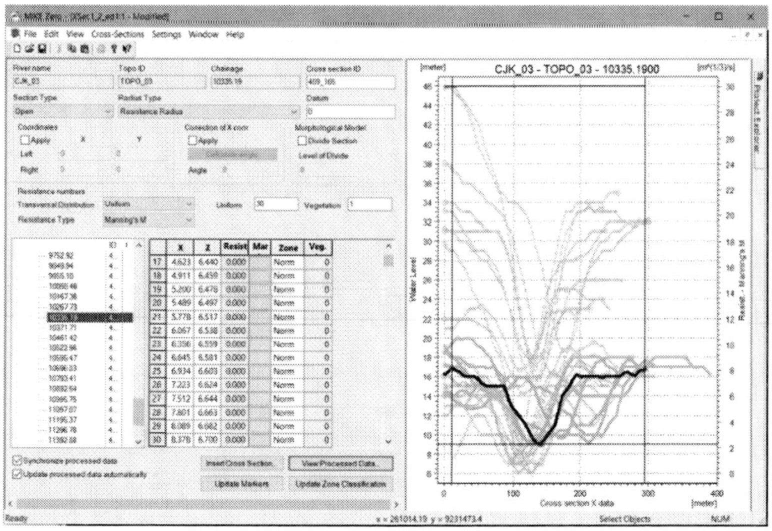

Figure 5. Cijangkelok River cross-section with a transverse look.

The unsteady flow feature helps analyze the existing cross-section, and the input data was conducted using a flood discharge plan based on the return period, including a rough coefficient data and cross-profile data.

4.3 Result

Figure 7 shows the profile of surface simulation by return period of 5, 10, 25, and 50-years of Cijangkelok River. These analyses indicate the overflow occurring in cross-sections 4090, 8254, and 11970 by a 5-year return period. Moreover, there was a significant increase in the return period of 10, 25, and 50-years. It shows the water level of overflow simulation and occurrence in the same cross-sections with a return period of a 5-years, but different in the amount of Discharge.

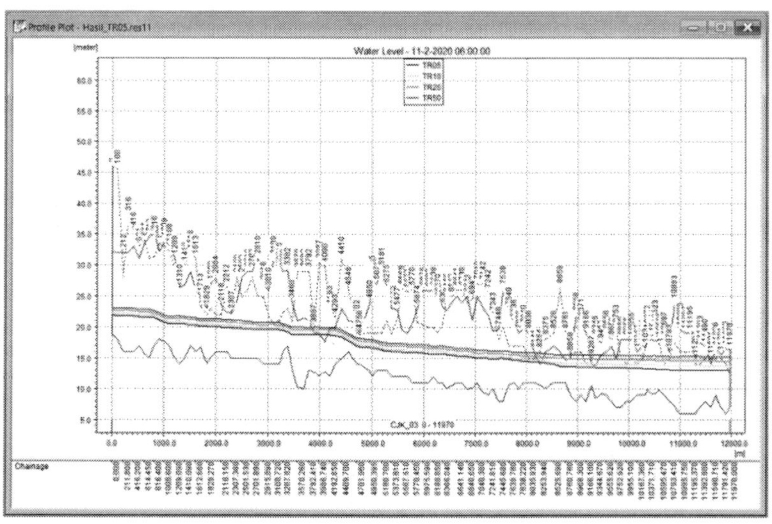

Figure 6. Long-section profile by return period 5, 10, 25, and 50-years

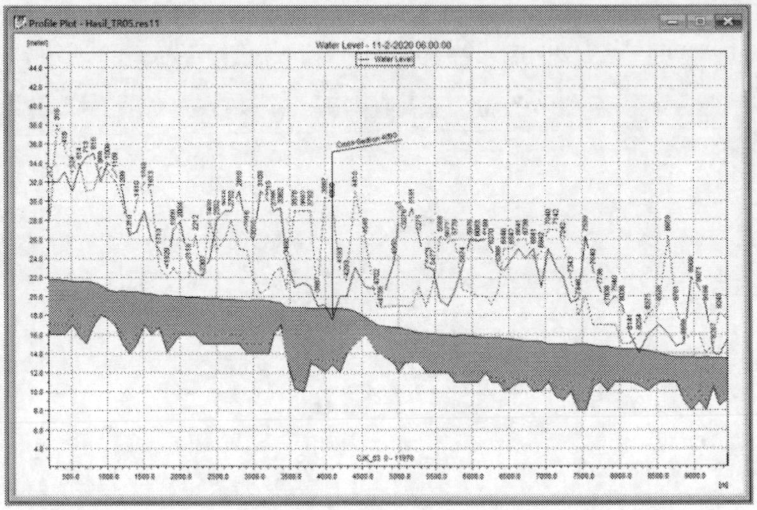

Figure 7. Cross-Section: 4090

From the picture above (figure 8), simulation results show that the existing channel cannot drain the existing Discharge. It is indicated that the maximum water level profile has increased the left and right embankments. This condition is the same as the real condition on the ground where the Cijangkelok River often floods during heavy rains (figure 9).

Figure 8. Cijangkelok river flooding (16 Februari 2020, 19.40 PM)

Once we know the Cijangkelok River's cross-section capacity, we can propose an alternative flood management to solve the problem, namely by way of channel normalization, pump procurement, or pond pool planning. If we decide to use channel normalization as an alternative treatment, we need to re-analyze the river cross-section suitable for flooding in the Cijangkelok River by enlarging or redesigning the cross-section. River normalization can also start by giving embankments to the river to prevent the overflow from entering the land. The data used for embankment planning were obtained from the evaluation on the Cijangkelok River's cross-section capacity. The cross-sectional capacity evaluation results indicates the parts of the river that flood—based on the analysis of the river's cross-section by MIKE 11 application and using data of the return period of flood discharge. We can also use the dike planning. However, in case of flood after normalization, we can use a pump if the river cannot accommodate the existing flood discharge.

5. Conclusion

Based on the above analysis and discussion, we can draw the following points as conclusions, namely:

1. Peak discharge of Cijangkelok river in the return period of 5-years is 256.25 m3/s, the return period of 10-years is 310.35 m3/s, the return period of 25-years is 392.91 m3/s, and the 50-year return period is 465.93 m3/s.
2. The analysis on the existing capacity to accommodate the Cijangkelok River indicates that the location cannot accommodate the design of flood discharge. The overflow at downstream is caused by the siltation on the riverbed and the damaged embankment.
3. Based on the analysis on each flood discharge return period, it is revealed that the existing cross-section of the Cijangkelok River is unable to accommodate the flood discharge. Such condition is attributed to several factors, such as river siltation or sedimentation, irregular river embankments, and land-use changes around the Cijangkelok River.
4. The current handling solution is to normalize the river given the fact that only several cross-sections that overflowed.

References

[1] Asdak C 1995 *Hidrologi dan pengelolaan daerah aliran sungai* Gadjah Mada Univ. Press (Yogyakarta)
[2] BBWS Cimanuk Cisanggarung 2015 *Review pola pengelolaan SDA WS Cimanuk Cisanggarung* (Cirebon)
[3] BBWS Cimanuk Cisanggarung 2016 *Studi pengendalian banjir di DAS Cisanggarung* (Cirebon)
[4] BBWS Cimanuk Cisanggarung 2016 *Studi penyusunan peta potensi banjir dan kekeringan WS Cimanuk Cisanggarung* (Cirebon)
[5] Istiarto 2014 *HEC-RAS Dasar simple geometry river* (Universitas Gadjah Mada: Yogyakarta)
[6] Kodoatie, Robert J, and Roestam S 2010 *Tata ruang air* Andi
[7] Kodoatie, Robert J, and Sugiyanto 2002 *Cetakan 1 : Banjir – beberapa penyebab dan metode pengendaliannya dalam perspektif lingkungan* Pustaka Pelajar (Yogyakarta)
[8] MIKE by DHI 2012 *Mike zero: Project oriented water modelling step-by-step training guide DHI Water Environ.* (Denmark)
[9] Mike by DHI 2017 *Mike 11 a modelling system for rivers and channels user guide DHI Water and Environ.* (Denmark)
[10] Spiegel M R 1981 *3rd edition: Theory and problem of statistics* Kin Keong Printing CO. PTE. LTD. (Singapore)
[11] Idham M, Akbar R, Mohammad F, Arie M and Arno A K 2018 *The use of rapid assessment for flood hazard map development in the upper Citarum river basin* MATEC Web Conf. **229**
[12] Nugroho 2004 *Peranan teknologi dalam pengelolaan sumberdaya air dan mitigasi bencana di Indonesia* Proc. Seminar Nasional Hari Air Sedunia 2004
[13] Linsley R K and Franzini J B 1991 *Teknik sumber daya air* Erlangga (Jakarta)
[14] Br S H 1981 *Hidrologi terapan* (Yogyakarta)
[15] Br S H 1993 *Analisis hidrologi* PT Gramedia Pustaka Utama (Jakarta)
[16] Br S H 2000 *Hidrologi : Teori, masalah dan penyelesaian* Nafiri Offset (Yogyakarta)
[17] Sudjarwadi 1987 *Teknik sumber daya air* Pusat Antar Univ. (PAU) Ilmu Teknik (Universitas Gadjah Mada: Yogyakarta)
[18] Suripin 2004 *Sistem drainase perkotaan yang berkelanjutan* (Yogyakarta)

The flood prediction model using Artificial Neural Network (ANN) and weather Application Programming Interface (API) as an alternative effort to flood mitigation in the Jenelata Sub-watershed

O M Gessang[1*], U Lasminto[1]

[1] Department of Civil Engineering, Faculty of Civil, Planning and Geo Engineering, Sepuluh Nopember Institute of Technology, Surabaya, East Java, Indonesia.

*Corresponding author's e-mail: oscar.gessang@gmail.com

Abstract. Jenelata is a sub of the Jeneberang watershed in South Sulawesi which has rainfall intensity from 2,800 mm to 4,000 mm per year on the upstream area, some of the rainfall occurs in a short period with high intensity, resulting significant rise to river water level. It potentially to cause floods downstream. Topography of the upstream area is mountainous with an average slope of ± 0.024 along the river, with flow length of 38,314 km it has a large velocity flow. Purpose of this research is to providing alternative for flood mitigation using rainfall predictions and runoff calculations, hypoteticaly it can reduce the impact of possible flood. This research was conducted using Artificial Neural Network (ANN) as rainfall predictor and input variable for runoff calculation using SCS method. ANN input variable will use weather prediction data from the global weather API, while SCS method will calculate maximum runoff in the catchment for the next 24 hours. Results of the rainfall prediction model get deviation of 28.81 mm and accuracy of 58% compared to the observations. Meanwhile, runoff model discharge acquires deviation of 1181.7 m^3 and water level ± 0.19 m at the designated location of water level gauge.

1. Introduction

Mitigation is a series of efforts to reduce disaster risks which can be done through physical development, awareness and capacity building to face the threats of disaster (*Undang-Undang No. 24 Tahun 2007, 2007*). Indonesian government regulations stated that mitigation variables for disaster events must be prepared by early warning and mitigation of disaster. The sequence of steps is to observe the possible causes of disaster, analyse and observe the data, then decisions can be made based on results and analysis of the disaster which can be disseminated to take actions (*Peraturan Pemerintah No. 21 Tahun 2008, 2008*). This task is very difficult to do on a daily basis where natural conditions changed dynamically with very sensitive variables such as weather. Rapid development on technology in various fields of applied science has many new possibilities in terms of its application with science methods. One of these is the application of weather prediction using Artificial Neural Network (ANN) (Loucks D. P, 2017). The application of ANN with data-driven system can significantly help to overcome several types of recurring problems. Automation is also increasingly needed these days, as it is important to be able to produce fast, reliable and efficient outputs. However, application of these technologies and systems requires availability of sufficient amount of data to be able to accessed.

Content from this work may be used under the terms of the Creative Commons Attribution 3.0 licence. Any further distribution of this work must maintain attribution to the author(s) and the title of the work, journal citation and DOI.
Published under licence by IOP Publishing Ltd

Flood is one of the disasters that causes the most damage (materially and non-materially) in Indonesia [1]. According to the Indonesian Disaster Risk reports, the data show that the main cause of flood events is rainfall [2]. National Water Resources Information System encourages provision of data and information needed to mitigate flood disasters, using hydrological, hydrometeorological and hydrogeological information as information material in decision making [3]. Apart from that, the systems and technologies owned by each agency and ministry in charge are still very limited. BMKG as the Coordinator of the Information System for Hydrology, Hydrometeorology, Hydrogeology (SIH3) [4] as well as agencies / organizations with authority in climate and weather issues seems already to have, use and apply weather prediction technology and algorithms on national scale which sometimes will not be available at a specific value on times or locations. The Ministry of Public Works and Public Housing (PUPR) also has many hydrological stations, but (also) many still use daily observations, for example, the location of the research at the upstream part of the Jenelata, Gowa Regency, South Sulawesi Province. The location is quite remote and does not have (working) telemetric climatological recorder. One alternative that can be used is the use of global weather predictions that are available as a service and part of the "Big Data". This method can be used and developed as an application for flood mitigation integrated with ANN and Machine Learning (ML).

1.1. Research objectives
The purpose of this research is to provide alternatives for flood mitigation using predictions of rainfall and runoff calculation, which hopefully can reduce the impact of possible flood events, to which the research was conducted using Artificial Neural Network (ANN) as a rainfall predictor and input for runoff calculation. ANN input variable will use weather prediction data from the global weather API. SCS method will be used to calculate maximum runoff in the catchment for the next 24 hours.

1.2. Limitations
The research is carried out through spatial analysis, hydrology and hydraulics as basic variables in the calculation of rainfall - runoff. Limitations on the model are the output will have specific values, such as variables and coefficients on the studied area.

2. Literature review
Literature used as a reference of this research are previous research and theoretical base for Climatology and Weather Prediction, ANN, ML, Deep Learning (DL), Rain - Runoff, Information systems, Decision Support System (DSS) and legislations related to objects of the research

2.1. Weather
Weather is a dynamic and complex state of the atmosphere at certain times and places. Weather can generally be expressed with rain, air temperature, cloud cover, evaporation, humidity, and wind speed from day to day [5]. The weather forecasting method has become a lot better in the past few years. Weather forecasting for 5 days is now more reliable and comparable to 2-day forecasting done 20 years ago. This is now possible by the use of technology in collecting and processing of weather data. Complex modelling of atmospheric conditions can be done using computers that are also increasingly sophisticated. Despite technological advances, accuracy in weather forecasts is still not absolute, with relatively large errors, because the weather is a complex system [6]. To be able to predict the weather, data of atmospheric conditions at the location must be obtained, such as temperature, air pressure and other atmospheric characteristics. From several weather prediction methods, an ANN model is built to predict rainfall according to the composition of atmospheric characteristics that occurs at the research site

2.2. Artificial Neural Network (ANN)
Artificial Neural Network is a set of an algorithm works on a (virtual) machine designed to work like the neural of the human brain, that it can solve certain kind of problems with specific functions. Information and data stored are obtained through the learning process [7]. The ANN is built of neurons, connecting each other by paths called synaptic, thus, forming a network of algorithms with different

paths depending on the value and weight of each synaptic. Each neuron will process input data to specific information as targets.

The use of ANN in predicting weather has been already done by many researchers such as [8], [9], [10], [11], [12]. They are looking for relationships between weather, rain and methods to get (more) accurate predictions. With a combination of sensitive variables, a weather - rainfall prediction is more non-linear in nature and (very) complex to be numerically calculated. Rainfall modelling is one of the most difficult to predict accurately, so that detailed knowledge of atmospheric condition is very important. However, it will still be difficult to predict and observe accurately. The model must be made using simple parameterization to effectively account for processes that can arise as a result of the limitations of the forecast model and other errors in observation [11]. The approach using the ANN method implemented simple, quick, and flexible problem solving. It is also easily adaptable to new environments [12]. ANN has many advantages in terms of time, cost, and ease in production. It can also be integrated with automation systems such as automatic data acquisition systems so that it can be implemented into a real-time forecasting system [8]. The advantages of ANN lie on its ability to process data flexibly in a relatively short time of reproduction and to calculate non-linear relationships between input and output variables. This makes ANN worth to be considered as a tool to estimate long-term runoff in real-time [9].

2.3. Rainfall - runoff

Precipitation or rainfall occurs from the product of evaporation and transpiration processes. It will undergo a process of infiltration, percolation, where some part of the rainfall will become overland flow and the other part will flow (underground) and get infiltrated. The amount or composition of surface runoff and infiltrated flow is affected by topography characteristic of the area. Surface runoff can be divided into 2 (two) groups: sheet runoff and surface runoff. The flow of water that enters the water flow system becomes a stream flow [13].

The value of surface runoff is affected by rainfall, vegetation, land cover and presence of water storage infrastructure / buildings. Calculation method using Curve Number (CN) is based Soil Conservation Service (SCS) where rain runoff (Pe) is a function of cumulative rain (P), land use, soil type and humidity through Maximum Retention (S), and Initial Abstraction (Ia) [14]. Soil ability to store maximum water depends on the characteristics of the soil in the watershed symbolized by the Curve Number (CN). The value is varying from 100 on waterlogged surfaces to 30 on permeable surfaces with high infiltration values.

$$Pe = \frac{(P - Ia)^2}{P - Ia + S} \tag{1}$$

$$CN = \frac{\sum \text{Ai. CNi}}{\sum \text{Ai}} \tag{2}$$

$$S = \frac{25400 - 254.CN}{CN} \tag{3}$$

$$Ia = 0.2.S \tag{4}$$

The SCS method calculates moisture condition of a soil (and its initial water content condition). This value will be calculated based on the amount of rain 5 days in advance. Indonesia has moderate climate usually compatible with CN on Antecedent Moisture Condition (AMC) II type. Other types of AMC can be calculated from CN AMC II using equation (5) or (6) [15].

$$CN_{amc1} = \frac{4.2.CN_{amc2}}{(10 - 0.58\,.CN_{amc2})} \tag{5}$$

$$CN_{amc3} = \frac{23.CN_{amc2}}{(10 - 0.13\,.CN_{amc2})} \tag{6}$$

Rating Curve equation on the Water Level Gauging Station.

$$Q = 829.3(h - 0.614)^2 \tag{7}$$

3. Research methods

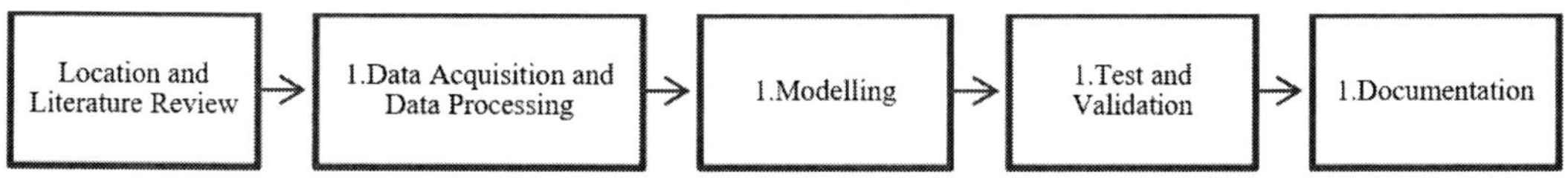

Figure 1 Research Method Flow

3.1. Location, data acquisition and processing

Jenelata as a part of the Jeneberang watershed in South Sulawesi is located at -5.27S, 119.57E to -5.34S, 118.83E. This sub-watershed has significant value of rainfall (2800 - 4000 mm/year) at the upper reaches [16]. Jenelata river is dendritic on flow pattern with mountainous area at the upstream area. The watershed produces an average slope of 0.0238. Length of the main river is 38.31 km, with difference on elevation of ± 912.64 m from the highest point to the outlet. It has large potential for erosion upstream and sedimentation downstream. Downstream Jenelata meets the Jeneberang river which flows into the Makassar Strait. Jenelata is influenced by wet tropical climate, the west monsoon and east monsoon with normal temperatures of 24°C to 33°C. The Jeneberang watershed development guide document states that the watershed is included as an area that is prone to disasters, including soil erosion, flooding and forest fires [17].

The rainy season occurs in November to April and the dry season starts in May until October [17]. The maximum rainfall recorded at the upstream area is 900 mm (28 days) and 600 mm (20 day) in the downstream area. The Lower Jeneberang River Flood Control Project report [16] states that the average rainfall intensity in one hour is 35% of the total daily rainfall at each rain post, and 90% of daily rainfall has a mean rainfall period of 4 to 6 hours.

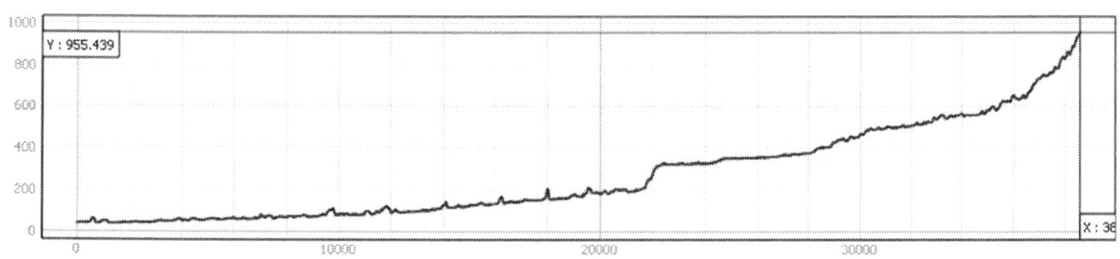

Figure 2 River Basin Elevation

Climatological data used in the ANN model are divided into 2 groups, namely historical climatology data in the modelling phase (training - validation) and API climatological prediction data in the forecasting phase.

IOP Conf. Series: Materials Science and Engineering **930** (2020) 012080 doi:10.1088/1757-899X/930/1/012080

Rainfall data are used in the modelling phase, the amount of which is accordance with the amount of climatological data time and location (closest). The total number of data is 1848 data: 1663 for training data and 185 for testing the model. The number of data is fairly small and, from the analysis and observations, problem comes from the data recording station (or devices). A lot of data of the rainfall station at the research location are incomplete, that it makes rain distribution area processing cannot be done, and the only complete data recording (found and) available at the research area is at Limbunga.

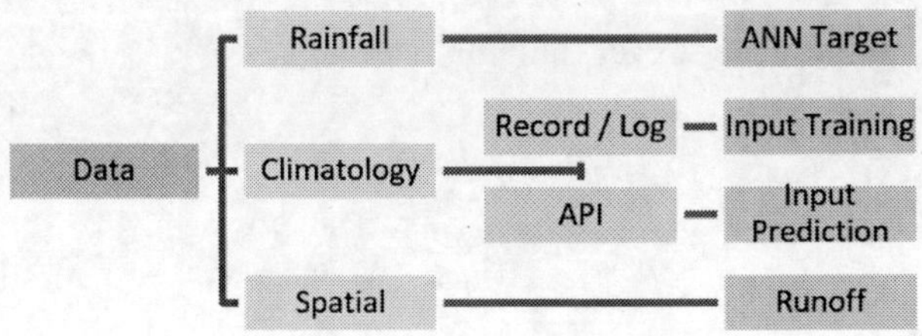

Figure 3 Data type and processing

Other variables used in order to calculate runoff are obtained from spatial data processing, rain prediction results generated by ANN data source models in the form of Digital Elevation Model (DEM), and land cover data obtained from the Pompengan - Jeneberang river basin management authorities (BBWSPJ).

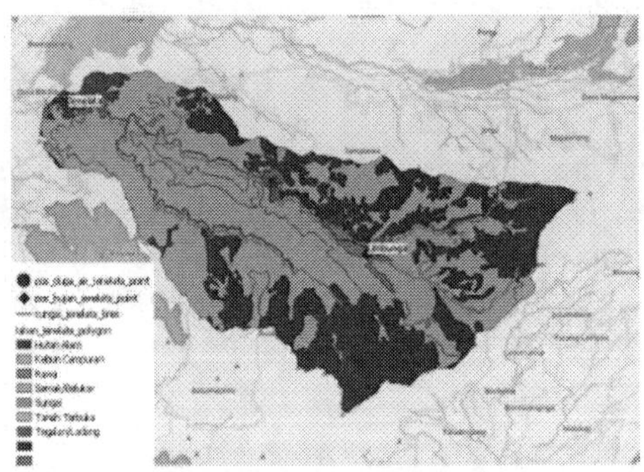

Figure 4 Land Use Map

Table 1 Land use table

Land Use	Area
Natural Forest	83,663,673.30
Mix Fields	69,931,400.62
Swamps	3,862,647.12
Shrubs	67,889,703.90
River	1,802,729.45
Grass (open field)	2,036,172.56
Fields	2,000,926.64

3.2. Modelling

ANN model is built on Python. Data inputs are using climatology historical data within the time frame as rainfall data. The total data are then split into 2 groups: 80% of training data and 20% of validation and test data. Trained ANN model is a result of variable for weight on each input node (neuron) stored in a file (with .h5 extension). This file can be used to make predictions.

The training process is carried out multiple times to get good accuracy predictions with the smallest deviation using training and validation data. Performance of the model is calculated using test data and measured with the value of Root Mean Square Error (RMSE).

After the model training result is considered having good performance and accuracy, rainfall prediction is carried out using input from extracted global weather variable data, that these data are acquired from API request on darksky.net (Apple Inc's). The acquired data are processed as input on ANN model (.h5). The output of this model's process is the prediction value for tomorrow rainfall.

Figure 5 ANN Modelling Flow

Rainfall predictions from the ANN are stored on the database, then they are used as input on the runoff calculation using SCS method. Distribution of hourly rainfall intensity is calculated using Modified Mononobe's method with 4 to 6 hours of rainfall distribution time [16] to calculate estimated hourly discharge. Runoff calculation variables such as CN are extracted from spatial data using soil type and land use. Other variables like AMC's are calculated using previous 5-day rainfall. This step of calculation produces the Maximum Retention (S), Initial Abstraction (Ia) of the location, rainfall depth (Q mm) and volume (Q m3).

3.3. Testing and validation

Validation process is carried out on each of the models. In the ANN model, validation process is conducted on testing phase because on the training phase it already has a validating process to measure the performance on the model's training. The predicted rainfall is validated (or tested) by comparing prediction values to recorded data at given period of time. Meanwhile, at the runoff calculation using SCS, validation process is carried out using rainfall of a specific date / time given as input on the calculation, and compares the result with the value of recorded water level in Jenelata outlet.

4. Result and discussion

4.1. ANN modelling

The ANN model is built on Python, an architecture of the network (figure 6) consisting of 1 input layer for training data (x1) and target data (x2), 4 hidden layers, and 1 output layer (rain prediction).

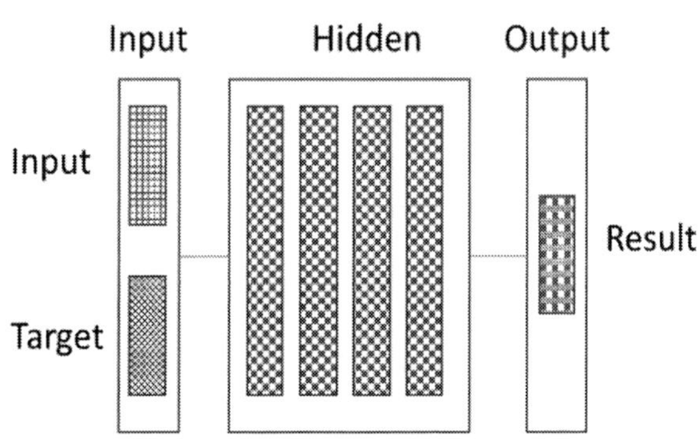

Figure 6 ANN Architecture

Table 2 ANN Architecture

Layer	Node	Activation	Type
1	6	RELU	Input
2	48	RELU	Hidden
3	72	RELU	Hidden
4	36	RELU	Hidden
5	6	RELU	Hidden
6	1	RELU	Output

Data pre-processing phase is carried out to re-format the data to meet the Python ANN modelling requirement. Then, data are divided into 2 groups: the training data (80%) and the validation - test data (20%).

Table 3 Data training format

No	Historical Climatology Data						Recorded Rainfall Data
	Month	Temp (C)	Humid (%)	Pressure (mmhg)	Wind (km/h)	Clouds (%)	
1	10	35.2	81.71	1076.1	5	40	0
2	10	35.2	81.71	1076.1	4	40	0
...	...	...	...	...	...	...	...
1663	10	26	60.58	1076.4	5	20	0

Result of training stage on an ANN modelling.

```
Train on 1663 samples, validate on 185 samples
Epoch 1/5000
1663/1663 [==========] - 0s 227us/step - loss: 247.4028 - accuracy: 0.3067 - val_loss: 25.3907 -
val_accuracy: 0.5946
Epoch 2/5000
1663/1663 [==========] - 0s 100us/step - loss: 144.4335 - accuracy: 0.3957 - val_loss: 25.2943 -
val_accuracy: 0.5946
...
Epoch 5000/5000
1478/1478 [==========] - 0s 114us/step - loss: 2.25 - accuracy: 0.7835 - val_loss: 5.3380 - val_accuracy:
0.5216
```

Figure 7 ANN training Results

Validation data for the ANN are 20% of the total training data, while in network testing used 185 data, which then cross-validated with the recorded field data.

Table 4 ANN Testing

No	Month	Temp (C)	Humid (%)	Press (mmhg)	Wind (km/h)	Clouds (%)	Recorded	Predicted Rainfall
1	12.00	29.71	96.25	1017.00	4.00	76.00	30.00	21.206234
2	12.00	32.00	87.67	1017.00	5.00	56.00	29.00	17.563517
...	...	...	...	...	...	...	...	...
185	7.00	31.00	74.75	1012.00	3.00	40.00	1.00	6.478799

Parameter of the training model is a learning rate = 0.0002, loss = 'mean squared error', optimizer = 'nadam', metrics = 'accuracy', dan epoch = 5000, resulting Loss: 2.25, Accuracy: 64.72.

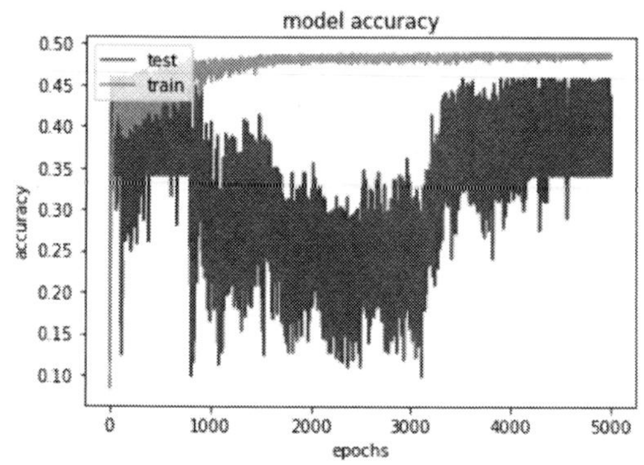

Figure 8 ANN training accuracy curve

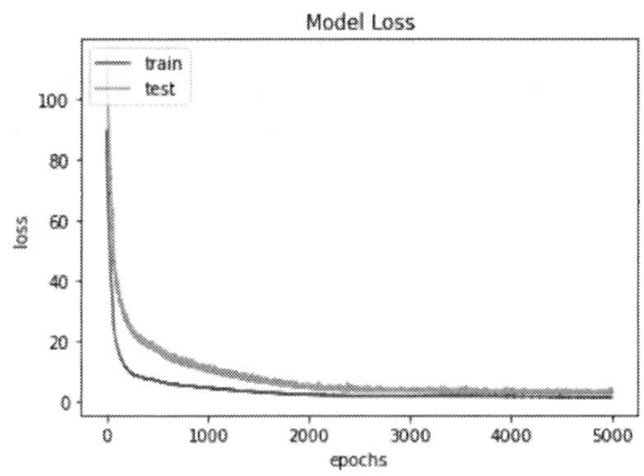

Figure 9 ANN training loss curve

Model testing is similar to validation stage, and the difference is that testing stage is done by entering the value of a new weather variable as input which has never been seen by the model on the training phase or validating phase. Prediction results are then compared with the field recorded data. This test is conducted using 193-day rainfall data, the result of the predictions are (seem to be) less desirable with research expectation, but it is predicted due to quality and lack of data availability in the training model process that will lead the model to fail in predicting the value of rainfall with large intensity or extreme values.

In this case, for some climatological data that should produce extreme rainfall (large values), the model only produces predictions with large deviation values. Meanwhile, for the prediction of average rainfall (<100 mm), the model can make a pretty good prediction with RMSE of 1.45, deviation of 28.81, and the correlation coefficient between rainfall of 0.6. Analysis for this problem is the lack of data used on training phase, especially on high intensity rainfall.

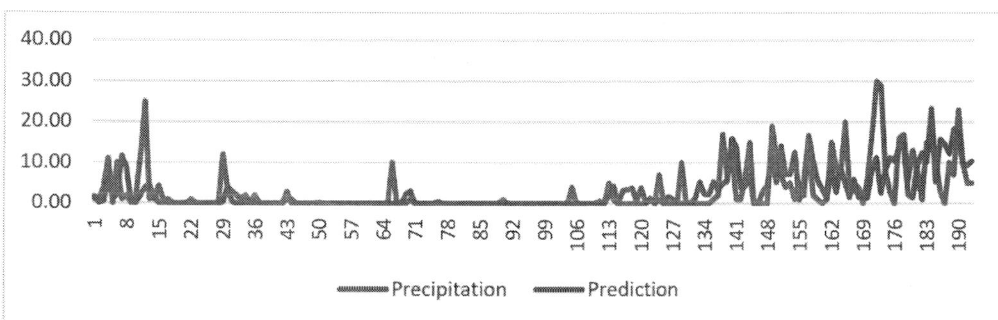

Figure 10 ANN training loss curve

4.2. *Runoff modelling*

Runoff calculation is using spatial variables in the proses. To be able perform runoff calculation, catchment area variables are required by calculating spatial data using processing software *Quantum GIS* (QGIS).

Table 5 Catchment variables

Area (Km2)	Area (m2)	River (Km)	Upstream (masl)	Downstream (masl)	Slope (%)
231.19	231187254.42	38.31	955.44	42.80	2.38%

Calculated CN of the area is 62.70, runoff calculation uses SCS method with AMC from 5-day rainfall value, while for the rain duration assumed to be 4 hours long distributed using the Mononobe method.

Testing phase of the model is conducted using rainfall data at certain period of time, calculated with model, and compared with water level measurement in Jenelata water level gauge downstream.

Calculated value (table 6) is calculated using rainfall data from January 22 2019 as an input, where precipitation value recorded is 364 mm and maximum water level in the water level station gauge record is 4.4 m. Results of maximum volume on daily rainfall calculated using SCS rational method eq. (1), (2), (3), (4) are 53,119,118.21 m^3 and 14,755,.31 m^3/s.

Table 6 Daily maximum

Location	Precipitations	CN	S (mm)	Ia (mm)	Q(mm)	V(m3)	Q(m3/s)
Jenelata	364.00	62.70	151.10	30.22	229.76	53,119,118.21	14,755.31

Distributed rainfall using modified Mononobe's method (table 6) gets total runoff volume of 46,270,478.52 m^3, and maximum discharge is at the 3rd hour for 38,539,168.52 m3, ΔQ = 6,844,668.44 m3 ~ 87%

Table 7 Hourly distribution

Hour	Prediction (mm)	Total (mm)	CN	S (mm)	Ia (mm)	Q(mm)	V(m3)	Q(m3/s)
1	33.28	33.28	79.74	64.52	12.90	4.89	1,130,982.00	314.16
2	59.60	92.89	79.74	64.52	12.90	19.61	4,532,861.60	1,259.13
3	**229.31**	**322.19**	**79.74**	**64.52**	**12.90**	**166.70**	**38,539,168.52**	**10,705.32**
4	41.81	364.00	79.74	64.52	12.90	8.94	2,067,466.40	574.30

Hypothetically, the output of the calculation can be compared to the value of *Principal Hydrological Gauging Stations for Flood Warning and Evacuation of Jeneberang river* [18], with such a large number of discharges (table 7), the authorities should have issued an early warning within 24 hours before it happened.

Principal Hydrological Gauging Stations for Flood Warning and Evacuation

Name of Gauging Station	Hydrological Data to be Gauged	Critical Discharge to Initiate Each of Steps for Flood Warning and Evacuation		
		Step 1	Step 2	Step 3
Bili-Bili Dam	Bili-Bili Dam Inflow Discharge	642.3m^3/s	1,000.0m^3/s	1,200.0m^3/s
Patarikan bridge	Discharge of Jenelata River	400.0m^3/s	900.0m^3/s	1,100.0m^3/s
Kampili Weir	Discharge of Jeneberang River	1,150.0m^3/s	1,800.0m^3/s	2,300.0m^3/s
Maccini Sombala	Water Level of below Crown Level	1.5m (EL. 4.7m)	1.0m (EL. 5.2m)	0.6m (El. 5.6m)

Figure 11 The estimated discharge will provide a flood prediction based on Principal Hydrological Gauging Stations for Flood Warning and Evacuation of Jeneberang river.

To validate the hypothesis and calculation, the runoff model is validated with water level on the observation point data during the rainy period of January 22, 2019 with 364 mm of rainfall intensity.

Table 8 Evaluation for the calculation and field observation shows that water level within the time frame has deviation of 0,12 m and total discharge of 1.370,82 m^3

Location	Observation		Model		ΔWL (m)
	Discharge (m^3)	WL(m)	Discharge (m^3)	WL(m)	
Jenelata	12,076.14	4,43	10,705.32	4,20	0,1208

5. Conclusion

From the research, there are several conclusions which can be drawn:

1. Result on the rain prediction test found 28.81% of prediction error with RMSE of 1.45, and a correlation coefficient between rainfall is 0.6. This result is considerably reasonable as the model is trained with (relatively) small amount of data.
2. Significant value of deviation on extreme rainfall prediction is likely to happen due to the lack of available data used in the learning process.
3. Differences in runoff calculations using the daily SCS method and model occur with a standard deviation of 0.13. This is likely to occur because the model calculates daily rainfall distribution in 4 to 6 hours and AMC.
4. Using (decent amount of) good quality of data will make models produce better precision on the rainfall prediction.
5. Application of the ANN model can be used as an effective and efficient alternative in the decision-making process along with the numerical model.
6. Hopefully, future research can be carried out with (more) decent amount of quality data and (maybe) using other methods to make (more) applicable models that can be used in disaster mitigation as a system.

References

[1] Aldrian E K M 2011 *Adaptasi dan Mitigasi Perubahan Iklim di Indonesia* (Jakarta: Pusat Perubahan Iklim dan Kualitas Udara, Deputi Bidang Klimatologi, Badan Meteorologi dan Geofisika)

[2] Asdak C 2014 *Hidrologi dan Pengelolaan Daerah Aliran Sungai, cetakan ke-6* (Yogyakarta: Gadjah Mada University Press)

[3] Badan Nasional Penanggulangan Bencana 2016 *Risiko Bencana Indonesia (RBI)* (Jakarta: Badan Nasional Penanggulangan Bencana)

[4] BBWS Pompengan Jeneberang 2015 *Pola Pengelolaan Wilayah Sungai Jeneberang* (Makassar: BBWS Pompengan - Jeneberang)

[5] Beven K. 2011 Rainfall-runoff modelling: the primer. In B. K. J *Rainfall-runoff modelling: the primer* (Oxford: John Wiley & Sons)

[6] Campolo M A S 2003 Artificial neural network approach to flood forecasting in the River Arno *Hydrolog Sci J* **48** p 381-98

[7] Chow V T. 1988 *Applied Hydrology* (New York: Mc.Graw Hill Book Company)

[8] Haykin S. 2009 *Neural Networks and Learning Machines 3rd Edition* (Ontario: Pearson Education Inc.)

[9] Iseh A J, W. 2013 Weather forecasting models, methods and applications *Int. J. Eng. Res.* **2** 12 p 1945-56

[10] JICA 1980 *Lower Jeneberang River Flood Control Project* (Jakarta: Japan International Coorporation Agency)

[11] JICA 2005 *The Study on Capacity Development for Jeneberang River Basin Management* (Jakarta: Japan International Cooperation Agency)

[12] Krasnopolsky V M 2013 *The Application of Neural Networks in the Earth System Sciences, Neural Networks Emulations for Complex Multidimensional Mappings* (Dordrecht: Springer)

[13] Loucks D P V B 2017 *Water resource systems planning and management: An introduction to methods, models, and applications* (Delft: Springer International Publishing AG)

[14] Peraturan Pemerintah No 21 Tahun 2008 2008 *Penyelenggaraan Penanggulangan Bencana* (Jakarta: Pemerintah Republik Indonesia)

[15] Peraturan Presiden No.88 Tahun 2012 2012 *Kebijakan Pengelolaan Sistem Informasi Hidrologi, Hidrometeorologi, dan Hidrogeologi pada Tingkat Nasional* (Jakarta: Pemerintah Republik Indonesia)

IOP Conf. Series: Materials Science and Engineering **930** (2020) 012080 doi:10.1088/1757-899X/930/1/012080

[16] Sharma A, K. B. (2014) Adaptive Predicting of Weather Forecasting with ANN using Back Propagation Technique *Int. J. Latest Trends Eng.* **4** 297-300
[17] Toth E. 2009 Classification of hydro-meteorological conditions and multiple artificial neural networks for streamflow forecasting *Hydrol. Earth Syst. Sci.* **13** p 1555–66
[18] Undang Undang No 17 Tahun 2019 2019 *Sumber Daya Air* (Jakarta: Pemerintah Republik Indonesia)
[19] Undang-Undang No. 24 Tahun 2007 2007 *Penanggulangan Bencana* (Jakarta: Republik Indonesia)
[20] US Army Corps of Engineers 2000 *Hydrologic Modeling System HEC-HMS, Technical Reference Manual* (Davis CA: Institute for Water Resources, Hydrologic Engineering Center)

The analysis of water balanced in Bendo Reservoir using Dynamic System

N F Margini[1*], N Anwar[2], and W Wardoyo[2]

[1] Doctoral Student in Department of Civil Engineering, Institute Technology of Sepuluh Nopember, Indonesia
[2] Lecturer in Department of Civil Engineering, Institute Technology of Sepuluh Nopember, Indonesia.

*Corresponding author's e-mail: nastasia@ce.its.ac.id

Abstract. This research was conducted to analyze balancing the storage capacity of the Bendo Reservoir with the quantity of water needed. Bendo Reservoir located in Nginden Sawo Village, Ponorogo Regency is one of the multipurpose reservoirs in Indonesia designed for irrigation, water supply, and simultaneously hydroelectric power. The water available is observed to be decreasing along with climate change, thereby, requiring an appropriate reservoir operational system. Several methods have been designed to achieve this purpose through the use of vast and complex variables but this study used a Dynamic Systems approach for analysis. Hhowever, involved the use of two scenarios in the simulation model to obtain the most optimum results. A period of 5 years was simulated and the result showed water shortage started in the 54th month for scenario 2, 49th month for scenario 1, and 46th month for the base model. Therefore, means scenario 2 has the most optimum conditions and recommended as the appropriate solution for Bendo Reservoir water distribution.

1. Introduction

Nowadays water resources are becoming very limited due to rapid changes in population, and urbanization. [1]. Then, optimization of water resources has been developed through mathematical models to maximize the benefits of clean water. Because land and clean water are key factors for sustainable agricultural development in a country. Significant hydrological change is indicated by the magnitude and duration of rainfall that has changed from the previous year. As we are concerned, the amount of rain will significantly affect the inflow discharge and the amount of the existing storage volume [2]. Changes in the amount of water availability will result in changes in reservoir management and operational systems.

The reservoir management system is usually planned to have the ability of releasing water according to demand in the downstream. This, therefore, means its purpose is to balance the quantity of water available with the demanded quantity but this is mostly affected by an increase in population and rapid economic growth, thereby, causing an imbalance. Meanwhile, inappropriate planning of water management systems threatens agricultural sustainability due to the need for water by several users.

A number of areas in Ponorogo Regency, East Java Province have various water management problems, including flooding and drought [3]. This reservoir will later contain and accommodate Nginden River water precisely in Bendo Hamlet, Nginden Sawo Village, Ponorogo Regency, as shown in figure 1. The Bendo Reservoir project began in 2013 and is planned for completion in 2020. This

Content from this work may be used under the terms of the Creative Commons Attribution 3.0 licence. Any further distribution of this work must maintain attribution to the author(s) and the title of the work, journal citation and DOI.
Published under licence by IOP Publishing Ltd

reservoir is a multipurpose reservoir operated as a supplier of water supply, hydropower, and irrigation. Because it is multipurpose, the availability of water in reservoirs has an important role. Whereas the current hydrological change dramatically affects the amount of water available in the reservoir.

A reservoir is ideally expected to function optimally and due to the fact that Bendo reservoir is designed to meet irrigation, domestic, and non-domestic raw water needs, there is the need for a system to regulate the volume of water to be expended to ensure optimal utilization of water discharge in the reservoir. This involves having a mechanism to analyze each intake and available discharge and for the purpose of this study, a dynamic system approach was used to solve the water distribution problem in the Bendo Reservoir.

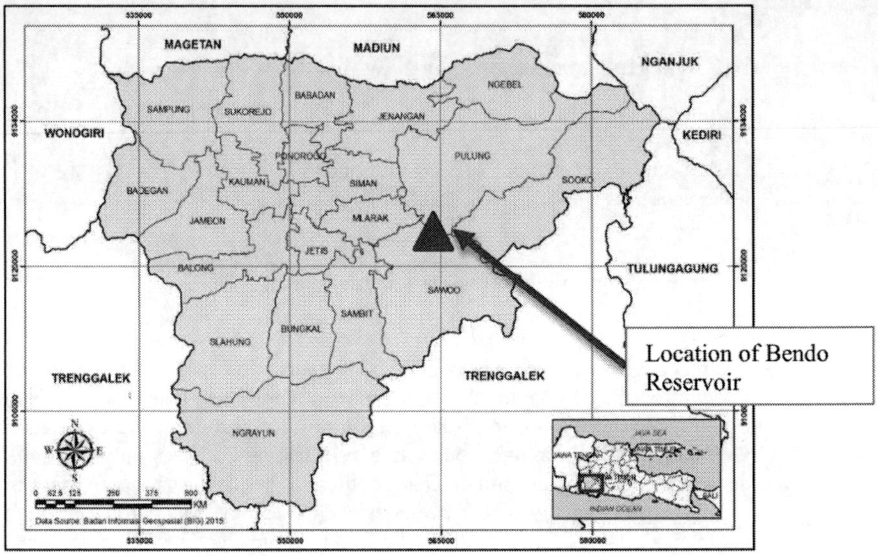

Figure 1. Location of Bendo Reservoir.

2. Literatur Review

Most of the study on reservoir optimization focus on creating operational strategies that can help overcome the difficulties of water variability. The difficulties of various stakeholders in dealing with water variability were also raised by [4] Haimes and Hall in 1977 and have extensively studied the optimization of reservoir operations since 1977.

The development of research on reservoir operations from the sciencedirect.com for keyword: operational optimization reservoir is increasing rapidly. In 2010 there were around 350 articles and then sharply increased to 1200 articles in 2019. From the many studies that have been developed various methods to optimize the amount of water availability and reservoir operation as in writings of [5] Yeh (1985). In this article, Yeh review about Linear Programming (LP) and Dynamic Programming (DP). Then, [6] writing about Nonlinear Programming and Simulation. In 2014, [7] which uses a dynamic system in the analysis process. The difference with this research is the influence of climate hydrological change and the function of the multipurpose reservoir.

The water stored in reservoirs serves multiple objectives such as irrigation which is usually influenced by climate change as well as the increase in population and food needs due to population growth. This research, therefore, used a dynamic system approach to offer a new method to optimize available water.

S. Ahmad & Simonovic in 2000 study about dynamic systems have systems and tools to represent complex systems in analyzing dynamic behaviour [7]. The most important feature of a dynamic system that facilitates the experiment of relationships in a system that is changed to produce different decisions. The dynamic system used in this study has four criteria, namely blocks, stock, flow, connectors, and converters. Stock (level) is used to represent something that has accumulated, for example, is the storage capacity in the reservoir. Then, flows (exchange rates) represent activities that fill and reduce stock.

Next, connectors (arrows) are used to establish relationships between variables in the Dynamic System that show causal/causal relationships [7]. Variable definitions, mathematical formulations, and units are provided in table 1.

The modeling, planning, and management of water resources systems were carried out years ago. The dynamic system approach has also resulted in a system of management, planning, policy, and sustainability of water resources [7], [8], and includes in the management of environmental water resources [9]. Some topics in water resources systems for flood management, hydrological systems and decision support systems for water operations [9]–[12]. [13] implemented the rice management in Central Taiwan using VENSIM Model. Whereas [10] present a new approach to optimizing irrigation management by developing a dynamics system model using VENSIM software. The VENSIM system is considered suitable for modeling and simulation involving significant components [14].

3. Methodology

This research was conducted through several stages including

a. Define the purpose of the model
System dynamics models are developed to understand the relationships between components in a system in order to complex problems. They require a fundamental problem to be solved after which a modeler is used to determine the purpose of the model and gather relevant data both in the form of statistical data and knowledge from experts in the field.
This study aimed to balance the quantity of water available with those required in a Bendo Reservoir operation. This involved dividing the inflow discharge data into outflows in the form of irrigation and clean water requirements.

b. Define the identify variables
The dynamic system is a simulation which focuses on the interaction between variable and feedback loop within the structure [15]. The constituent variables in this system have been identified as having a causal relationship as arranged in a table to facilitate the identification process . In the table, write the name of a variable, a definition in the field of water resources manajement using International Unit System, and an explanation of its position in the system, as shown in table 1.

Table 1. Dynamic system model variable.

No.	Variable	Definition	Unit	Information
1	Storage Capacity	The amount of water that has the power and potential to be used in human life	m^3	Level
2	Discharge Inflow	River minimum input discharge which can be utilized for people needs	m^3/ detik	Auxialary
3	Runoff Coefficient (C)	Surface flow rate to the ability of the soil to absorb water	-	Constanta
4	Rainfall Intensity (I)	The amount of rain expressed in each unit of time	mm/jam	Auxialary
5	Average daily rainfall max (R24)	Maximum rainfall averages 24 hours	mm	Constanta
6	Catchment Area (A)	Wide catchment area for certain rivers	Km^2	Constanta
7	Irrigation	the amount of water needed to irrigation needs	m3/ s	Auxialary
8	Irrigation area	Unity of land that gets water from an irrigation network	Ha	Constanta
9	Discharge Intake (Dr)	The amount of water volume tapped at the intake to irrigation needs	m^3/ s/Ha	Auxiiary
10	Irrigation Efficiency	Comparison of the amount of water supplied minus water loss with the amount given	-	Constanta
11	NFR	The amount of water level tapped to meet the needs of irrigation in the fields	mm/ day	Auxiiary
12	WLR	The amount of water level for preparing paddy fields	mm/ day	Constanta
13	(P)	The amount of volume of water that can seep into the ground	mm/ day	Constanta
14	Paddy consumptive water needs	Amount of water volume needed by paddy plants	mm/ day	Auxialary
15	Eto	Value of environmental needs, a collection of vegetation to conduct evapotranspiration	mm/ day	Constanta
16	Paddy crop coefficient	The coefficient of rice plants depends on the type of variety and its growth stage	-	Constanta
17	Water Supply	The amount of water needed for domestic and urban households	m3	Auxialary
18	Domestic total needs	The amount of total consumptive water for households in an area	l/day	Auxialary
19	per capita domestic water demand	The large amount of water consumption for each individual	l/people/day	Constanta
20	Total population	The number of inhabitants who occupy an area and its growth	people	Auxialary
21	Percentage of Population Increase	The rate of population growth in an area	%	Auxialary
22	Non-domestic total needs	The amount of total consumptive water for public buildings in an area	l/day	Constanta
23	Number of public buildings	the number of public buildings in an area	buildings	Auxialary
24	Percentage of public building increase	the rate of increase in the number of public buildings in an area	%	Auxialary
25	Non-domestic water needs	The large amount of water consumption for each individual in a building	l/people/buildings	Constanta

c. Diagram the underlying mechanisms and feedback loops diagram

The main components affecting the balance of water available in the Bendo Reservoir include large inflow discharges, outflow for clean water, and for irrigation which are interrelated and influenced by sub-components. For example, the inflow discharge which is considered absolute is influenced by the drainage coefficient of rainfall intensity, catchment area, and maximum daily average rainfall. The quantity of inflow discharge can, therefore, be increased by increasing the water supply using another river nearby.

The need for irrigation and clean water also has its constituent components. For example, the clean water required is influenced by the increase in population and number of public buildings in the region while irrigation water needed is affected by the area to be irrigated, irrigation efficiency, DR (discharge ratio at intake), NFR (Net Field Requirements), WLR (Water Level Requirements), Percolation, Potential Evapotranspiration (Eto), and the coefficient of the plant. There is, therefore, the need to model the water quantity for these variables in line with the region's development and growth. All the interrelated components were used to make Causal Loop Diagrams (CLD) and stock-flow diagrams (SFD) as shown in figures 2 and 3.

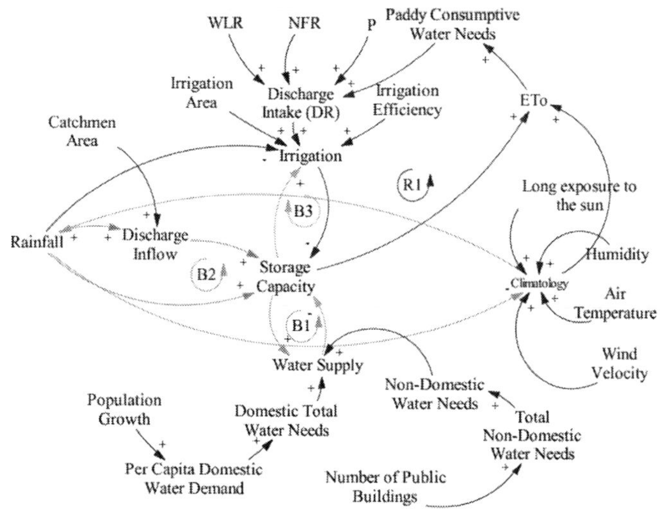

Figure 2. Causal loop diagram of the Bendo Reservoir reservoir water operating system model.

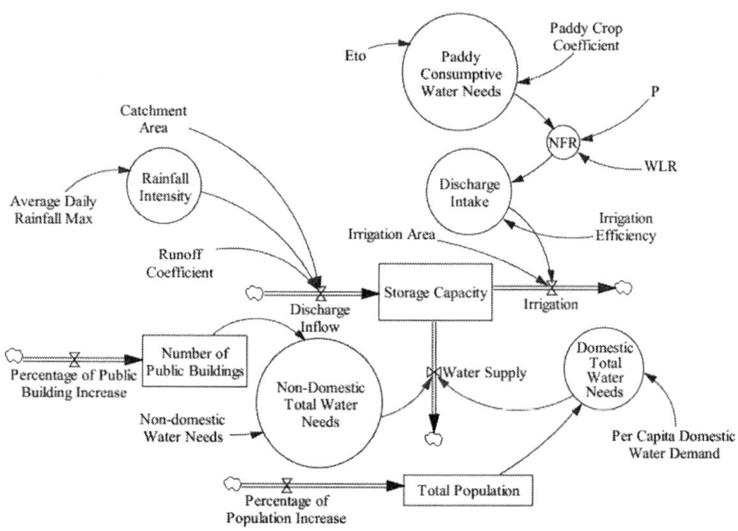

Figure 3. Stock flow diagram of storage capacity in Bendo Reservoir.

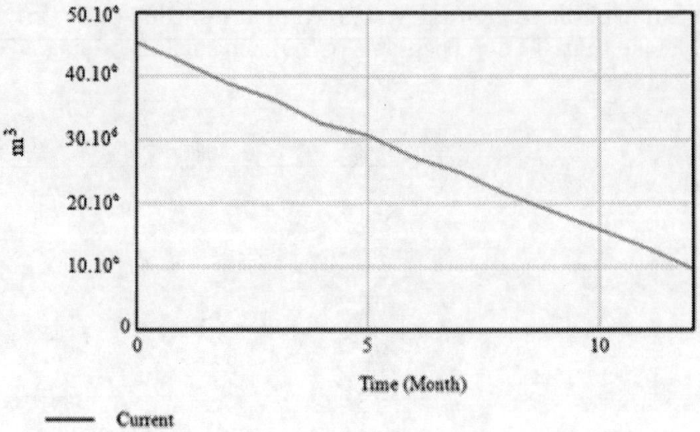

Figure 4. Results of model storage capacity.

Figure 4 shows the initial water in the reservoir amounted to 45,415,000 m3 and the value was observed to have decreased after been used as the source for irrigation and clean water for one year. This is due to the fact that the inflow discharge and initial reservoir quantity is smaller than the quantity needed as well as the irrigation plants using ordinary seeds. These results were later validated using available data.

d. Testing the model
The conceptual model was validated to determine the correctness and accuracy of the theories and assumptions used as well as their compatibility with the purpose of the model. This validation process was intended to ascertain the ability of the model to effectively describe the real system conditions using two methods which are statistical approach involving the Mean Comparison Test and Ampltitudinal Variation Comparison Test (% error variance) [15] formulated as follows:

$$E1 = \frac{\left[\overline{S} - \overline{A}\right]}{\overline{A}} \tag{1}$$

Where,
$\overline{S}$ = average of the data model
$\overline{A}$ = average of existing data

$$E2 = \frac{|Ss - Sa|}{Sa} \tag{2}$$

Where,
Ss = standard deviation of model
Sa = standard deviation of existing data

Each test is explained as follows:

- Mean comparison test
  The model is considered valid if E1 < 5%. Based on the average comparison test in the water distribution system in this reservoir, as in table 2 below shows that the value of E1 on the validation of water availability is 2.911%. The value of E1 on validation of total population is 3.905%. Then the model is considered valid because the E1 value produced is less than 5%.

- Amplitude Variation Comparison Test (% Error Variance)
  The water availability was recorded to have a validation of 2.911% as shown in table 2 and graphically illustrated in figure 5 while the population number variable had 0.504%, therefore, the model is considered valid because the E2 value is less than 30%. The model and actual data are also shown in the figure to have a small difference and share the same trend.

Table 2. Result of the models comparison test.

	Mean Comparison test	**Amplitude Variation Comparison Test**
Water availability	2.911 %	0.098 %
Total Population	3.905 %	0.504%

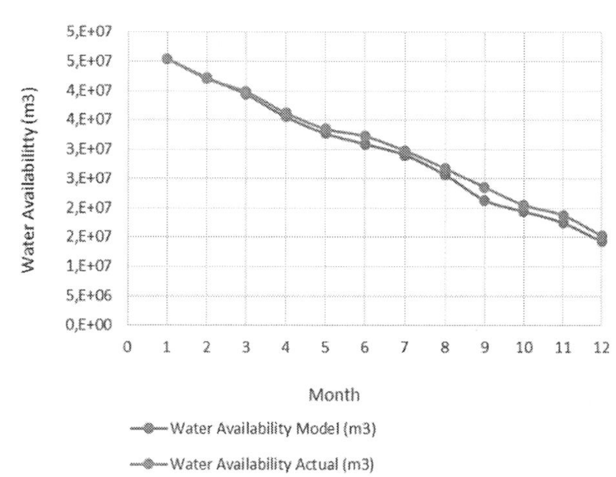

Figure 5. Comparison graph for the model and actual data.

e. Testing the model's response to different policies
Several scenarios were formulated after the model has passed the validation test to provide more optimum conditions and possibilities for the future. The process involved indicating how the structure of the system from the base model can be changed using new parameters through the concept of structural scenario. Moreover, the model parameters were not also changed to determine the impact of other variables in the system through parameter scenarios. The scenarios drafted include:
   - Mean comparison test
   - Scenario 2 adds the possibility of supervision from the Sawoo River close to the reservoir site and increases the percentage of population growth and public buildings in the Ponorogo Regency area as the study location.

The simulation was conducted using these scenarios for 5 years or sixty (60) months because changes in hydrological climate conditions generally occur once every five years.

f. Conclusion
A conclusion was made after the two proposed scenarios have been completely analyzed.

IOP Conf. Series: Materials Science and Engineering **930** (2020) 012081 doi:10.1088/1757-899X/930/1/012081

4. Results and discussion

- Scenario 1

  Scenario 1 adds a superior paddy crop coefficient with 0.9 compared with the estimated 1.1 for average common varieties [13] which are intended to reduce the amount of irrigation water needed. The SFD for the first scenario is presented in figure 6 while the result is in figure 7:

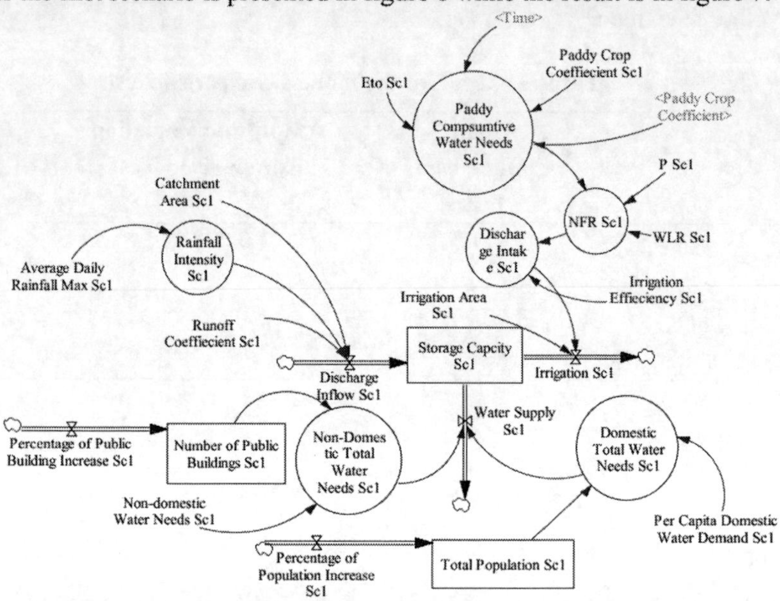

Figure 6. SFD of scenario 1.

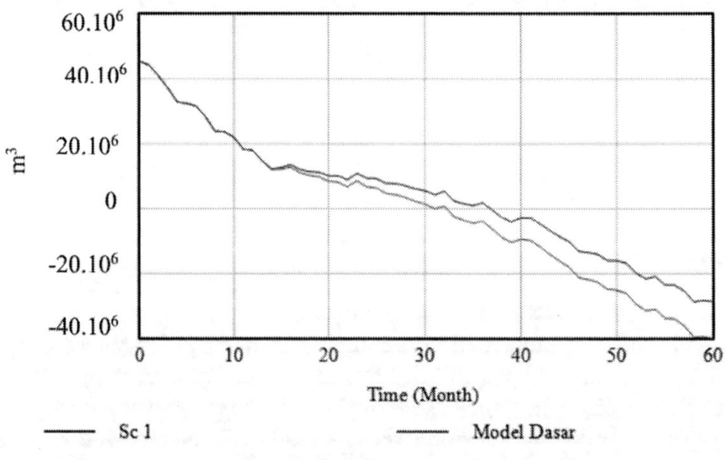

Figure 7. Simulation results of water availability in scenario 1.

- Scenario 2

  This second scenario was developed by adding the Sawo's irrigation area up to 1,000 Ha as an alternative variable, Sawoo River water supply averaging 4.4 m^3/second, and increasing the percentage of population growth and public buildings [16]. The SFD for scenario 2 is shown in figure 8 while the result is presented in figure 9.

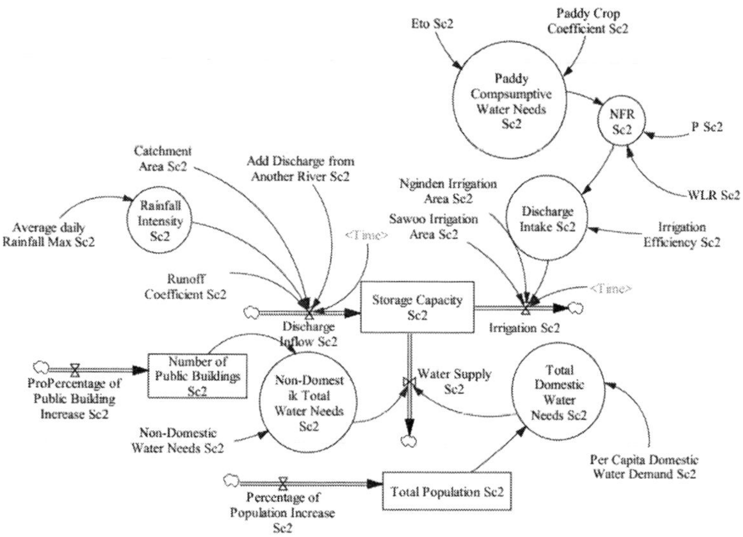

Figure 8. SFD fir scenario 2.

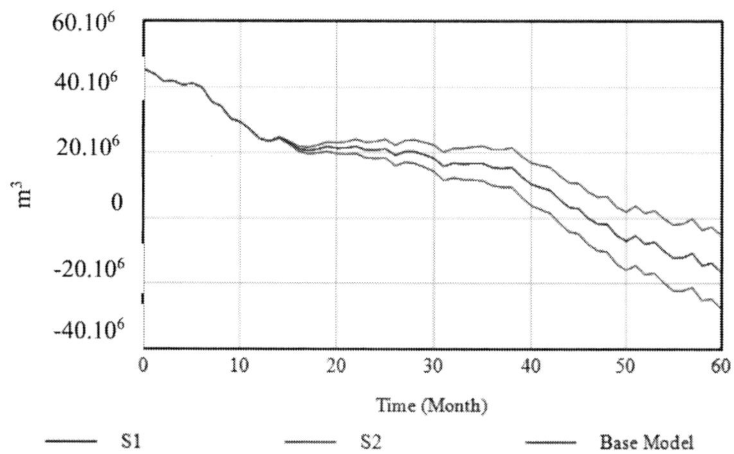

Figure 9. Simulation results for the water availability in scenario 2.

The results showed the quantity of water available for the base model, Scenario 1, and Scenario 2 experiences some differences starting from the 13th to the 60th month. Moreover, Scenario 2 was observed to be more positioned at saving reservoir water usage compared to the others. Meanwhile, water availability deficit was reported to have occurred in the base model, Scenario 1, and Scenario 2 in the 42nd, 46th, and 54th month respectively. The option selected to add more water supply to the Bendo Reservoir is clearly illustrated to have the ability of supplying to 1000 ha of irrigated rice fields.

5. Conclusion

A Dynamic System was developed and implemented in this research to produce water resource solutions due to uncertain inflow by allocating the resources for different uses and in several areas. Bendo Reservoir was used as a case study due to the government's effort towards implementing an initiative to overcome the water crisis in Indonesia. The results provided are better in minimizing the quantity of water shortages and ensure the reservoir satisfies water outflows for the next five (5) years.

Several further studies are required to improve methodology and analysis because this research only considers the uncertainty associated with the reservoir flow and demand even though it can be extended to others such as the operational dates of different water-saving scenarios. Meanwhile, the uncertainty and variability in reservoir flow for case studies are based on limited historical data and may not illustrate long-term trends such as drought and hydrological change, therefore, further research is required.

Acknowledgements
The first author also gratefully acknowledge the funding (BPPDN) with ID number 190006058904 from The Government of Indonesia.

References

[1] Osama S, Elkholy M and Kansoh R M 2017 Optimization of the cropping pattern in Egypt *Alexandria Eng. J.* **56** 4 557–66 doi: 10.1016/j.aej.2017.04.015
[2] Allawi M F, Jaafar O, Mohamad Hamzah F, Koting S B, Mohd N S B and El-Shafie A 2019 Forecasting hydrological parameters for reservoir system utilizing artificial intelligent models and exploring their influence on operation performance *Knowledge-Based Syst.* **163** 907–26 doi:10.1016/j.knosys.2018.10.013
[3] Fajar R, Busisi A, Sangkawati S and Budieny H 2015 Perencanaan Bendungan Bendo Ponorogo *J. Karya Tek. Sipil* **4** 4 428–38
[4] Haimes Y Y and Hall W H 1977 Sensitivity, responsivity, stability and irreversibility as multiple objectives in civil systems *Adv. Water Resour.* **1** 2 71–81 doi: 10.1016/0309-1708(77)90025 2
[5] William W G Y 1985 Reservoir management and operations models *Water Resour. Res.* **21** 12 1797–818
[6] Wu Y and Chen J 2012 An operation-based scheme for a multiyear and multipurpose reservoir to enhance macroscale hydrologic models *J. Hydrometeorol.* **13** 1 270–83 doi: 10.1175/JHM-D 10-05028.1
[7] Ahmad S and Simonovic S P 2000 System dynamics modeling in science *Syst. Dyn. Model. Reserv. Oper. Flood Manag.* **14** July 190–8
[8] Winz I, Brierley G and Trowsdale S 2009 The use of system dynamics simulation in water resources management *Water Resour. Manag.*, **23** 7 1301–23 doi: 10.1007/s11269-008-9328 7
[9] Guo W and Wang H 2010 Optimal operation of Three Gorges reservoir based on Ant Colony algorithm *Proc. - 2010 Int. Conf. Intell. Comput. Cogn. Informatics, ICICCI 2010* 1 360–3 doi: 10.1109/ICICCI.2010.101
[10] Elmahdi A, Malano H, Etchells T and Khan S 2020 System dynamics optimisation approach to irrigation demand management *MODSIM 2005 - Int. Congr. Model. Simul. Adv. Appl. Manag. Decis. Making, Proc.* February 2015 196–202 doi: 10.13140/2.1.3408.4648
[11] Wang J, Huang W, Ma G and Wang Y 2016 Determining the optimal year-end water level of a multi-year regulating storage reservoir: A case study *Water Sci. Technol. Water Supply* **16** 2 284–94 doi: 10.2166/ws.2015.138
[12] Wang H and Liu J 2013 Reservoir operation incorporating hedging rules and operational inflow forecasts *Water Resour. Manag.* **27** 5 1427–38 doi: 10.1007/s11269-012-0246-3
[13] Wu R S, Liu J S, Chang S Y and Hussain F 2017 Modeling of mixed crop field water demand and a smart irrigation system *Water (Switzerland)* **9** 11 1–21 doi: 10.3390/w9110885
[14] Wilcox B P 2010 Ecohydrology Bearing - Invited Commentary Transformation ecosystem change and ecohydrology: ushering in a new era for watershed management *Ecohydrology* **130** February 126–30 doi: 10.1002/eco
[15] Barlas Y 1996 Formal aspects of model validity and validation in system dynamics *Syst. Dyn. Rev.* **12** 3 183–210 doi: 10.1002/(sici)1099-1727(199623)12:3<183::aid-sdr103>3.3.co;2-w
[16] Anwar N, Kusumawati S and Margini N F Goal Programming Approach for Irrigation Interconnection Schemes of Singolatri , Kedawang and Lider Reservoirs Including Porolinggo Springs at Banyuwangi, East Java , Indonesia p 1–9

Numerical modeling of sub-surface drainage systems in Citraland, West Surabaya

S Dwiuntoroadi[1*], U Lasminto[2]

[1] Graduate Student of Hydroinformatics, Departement of Civil Engineering, Institut Teknologi Sepuluh Nopember
[2] Lecturer in Departement of Civil Engineering, Institut Teknologi Sepuluh Nopember

*Corresponding author's e-mail: sih.dwiuntoroadi@pu.go.id

Abstract. Flooding occurs when the existing drainage system is unable to accommodate the water flow, that is due to a decreased drainage system capacity, increased flow of water, or a combination of both. On January 31, 2019, two hours of heavy rain caused Citraland housing to be flooded up to one meter high. This study aims to overcome the existing flooding problems through the assessment of sub-surface drainage systems for hilly areas in the Citraland residential area, Surabaya. Annual rainfall data from the rain stations around the residential area will be processed according to the proportion of the area to get a rain analysis plan that will be used in modeling. Plan rainfall data, area data and types of land cover, and catchment area data are analyzed hydrologically to get the design discharge. The hydraulic data of the dimensions of the existing drainage system will be analyzed for the capacity in flowing of Flow Plan. The Drainage System is then modeled using the Storm Water Management Model (SWMM) application. Through using topographic data and land cover, four alternative solutions have been modeled which include the addition of water reservoirs in various housing locations to overcome existing flooding. The results of this study are the second scenario through the addition of one water reservoirs which can handle flooding most effectively because it reduces 34.61% of the inundation volume that occurs. In addition, the fourth scenario through the addition of three water reservoirs can reduce 57.82% of the inundation volume that occurs.
Keywords: flood, drainage, SWMM, water reservoirs

1. Introduction

Flooding or occurrence of inundation in an area in a residential or urban area still occurs in many cities in Indonesia, one of which occurs in the city of Surabaya. During the rainy season, the western part of Surabaya city has puddle points that disturb residents' activities. On January 31, 2019, heavy rain for two hours caused the Citraland area to be flooded up to 1 meter.

The purpose of this research is to conduct a flood simulation using a numerical model through the assessment of a sub-surface drainage system in the residential area of Citraland Surabaya. It is hoped that this research can help various parties including the government of Surabaya in an effort to reduce and even overcome the flood inundation that occurs.

Content from this work may be used under the terms of the Creative Commons Attribution 3.0 licence. Any further distribution of this work must maintain attribution to the author(s) and the title of the work, journal citation and DOI.
Published under licence by IOP Publishing Ltd

2. Literature Study

Flooding or inundation occurs when the system that accommodates the inundation is not able to accommodate the flowing flow. This is the result of three possibilities that occur: system capacity decreases due to a reduction in the channel cross-sectional area, increased flow of water flow due to increased rainfall or get additional streams, or a combination of both. The system mentioned here means drainage network system of an area. Meanwhile, the drainage system in general can be defined as a series of water structures that functions to reduce and/or remove excess of water (flooding) from an area or land, so that land can function optimally, so the drainage system is an engineered infrastructure in an area to cope with inundation flood (Suripin, 2004).

According to Mulyanto (2012), drainage is a science of drying the soil. Drainage is derived from the word 'to drain' which is a term used to express systems related to handling the problem of excess water, both above and below the land settlement. The definition of drainage is not limited to the technique of excessive water disposal but is more broadly related to its relation to aspects of life that are within the urban area. All matters concerning excess water in the city area can certainly cause quite complex problems. The increasingly complex problems of urban drainage in the planning and construction depend on the ability of each planner. Therefore, in the work, it is required cooperation with several experts in other related fields.

SWMM is a rainfall-runoff model used for simulating the quantity and quality of surface runoff from urban areas. Surface runoff is generated from rain catchment areas. The surface runoff load is then channeled through pipelines, open drains, storage systems, and pumps. The EPA SWMM.5.1 program can model and analyze the quantity and problems of urban runoff. SWMM shows the quantity produced in each sub-watershed, water discharge, and flow depth. By using EPA SWMM, the conditions occuring in the field can be modeled by including the parameters recorded under field conditions. The collection of technical data including channel length, channel dimension, and channel slope is done by direct observation in the field. (SWMM User Manual, 2015)

3. Research Methods

3.1. Study Area

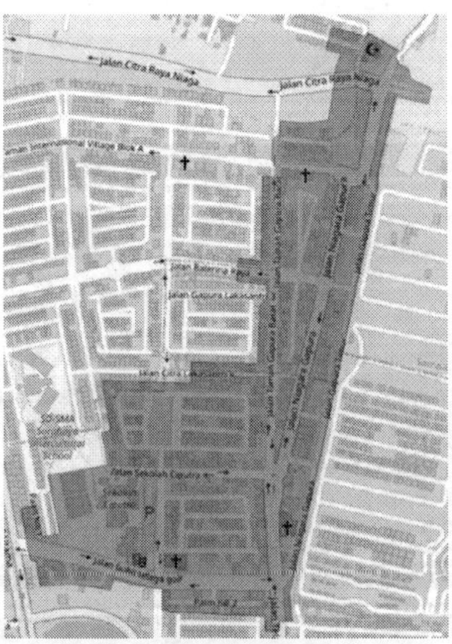

Figure 1. Study Area Location (QGIS)

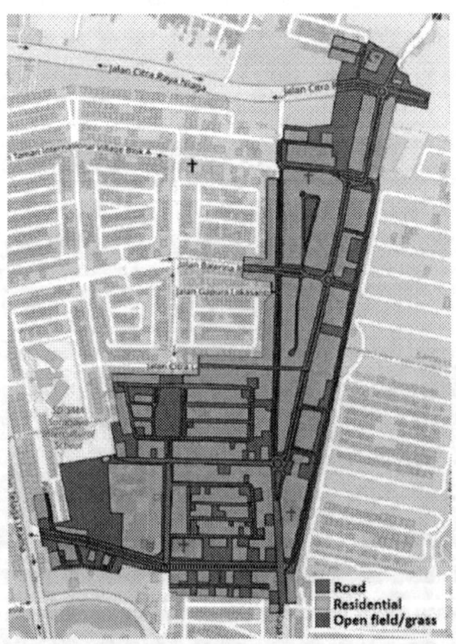

Figure 2. Land Use Digitizing (QGIS)

IOP Conf. Series: Materials Science and Engineering **930** (2020) 012082 doi:10.1088/1757-899X/930/1/012082

The study area is located in the Citraland residential area, West Surabaya with an area of 0.483 km^2 (figure 1).

3.2. Spatial Analysis
Spatial analysis is conducted to get contour information for research location. Digital Elevation Model (DEM) from DEMNAS which has scale 1:50.000 is used for this process. Satellite image and observation are used to digitize the land use (Figure 2) which later will be used for determine Impervious percentage and CN number.

3.3. Hydrology Analysis
Hydrological analysis will produce a rain plan that will be included in the model as input, with a return period of 2 years, 5 years, and 10 years.

3.4. Hydraulic Analysis
The SWMM program is used to model existing sub-surface drainage systems. From the model, it will be analyzed using a rain plan which is calculated through a hydrological analysis to determine the flood points to be overcome.

3.5. Flood Management Scenarios
The scenario to be examined considers the location of the water reservoir based on land availability, flood location, contour and slope of the land, as well as the amount of flood volume that can be reduced.

3.6. Study Flowchart

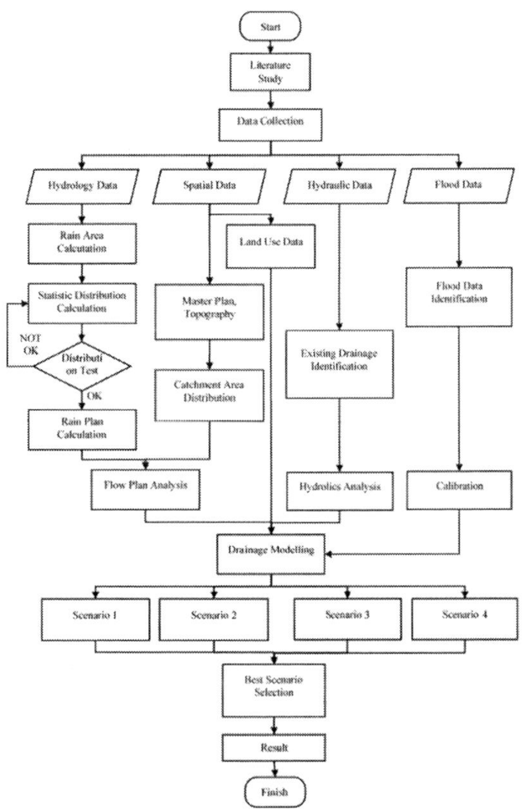

Figure 3. Study Flowchart

4. Results

4.1. Rain Station Influence
There are several rain stations around the study site, including Kandangan, Banyu Urip, Gunungsari, and Kebon Agung rain stations. By using data on the location of the Surabaya city rain station and the location of the study, a rain station that affected the research area was determined using the Thiessen polygon method. Through the use the QGIS application, it is determined that the study area is 100% affected only by one rain station, namely the Kandangan rain Station.

4.2. Rainfall Plan
Of the 10-years of rainfall data at the Kandangan rain station, test is carried out to determine distribution types use, which is Normal, Log-Normal, Gumbel, or Log-Pearson III. And from the test, the one suitable for use in Kandangan rain station is the Pearson III Log distribution and the result is presented on table 1.

Table 1. Calculation Result of Rainfall Plan using Log Peason III

Return Period	Log $\bar{x}$	k	Sd	Log Q	R (mm)
2	1.97	-0.116	0.084	1.9604	91.29
5	1.97	0.79	0.084	2.0365	108.78
10	1.97	1.333	0.084	2.0821	120.82

4.3. Chi-Square and Smirnov-Kolmogorof Test
There are 2 ways to test the suitability of the distribution with existing data, namely using the Chi-Square and Smirnov-Kolmogorof tests. These tests are used to determine the relationship between the depth of rain and the value of probability. From the Chi-Square test the theoretical Chi value > the calculated Chi value = 3.841 > 2 so that the hypotheses tested can be accepted. Meanwhile, from the Smirnov-Kolmogorof test, the value of Do > Dmax = 0.41> 0.1304 is obtained, so the hypothesis being tested can also be accepted.

4.4. SWMM Modelling
Modeling with SWMM is done by entering the required data on components from SWMM such as Rain Gage, Junction Node, Conduit, and Sub-catchment. The result is like in Figure 4.

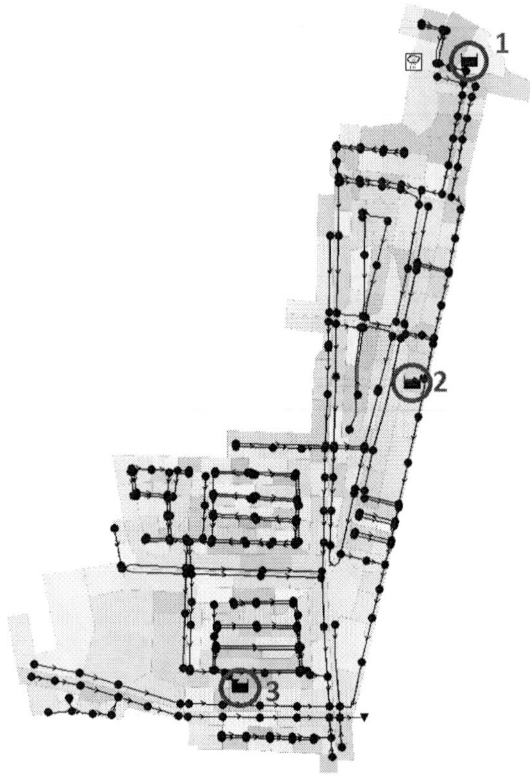

Figure 4. Drainage Model modelled in SWMM, numbers indicate the locations of reservoirs.

4.5. Existing Drainage Analysis
With all the hydraulic data entered along with the rainfall data from the previous hydrology analysis. The simulation was carried out, with the result that there were 10 nodes that flooded with a total flood volume of 5256 m^3 (2-year return period), 8613 m^3 (5-year return period), and 10870 m^3 (10-year return period).

4.6. Scenario 1 Analysis
Scenario 1 includes the addition of a water reservoir at location number 1 (red circle, figure 4) of 750 m^3 over an area of 250 m^2. As shown in Figure 5, the scenario resulted in the decrease of the flooding by shifting back the flood occurring by at least one hour. Scenario 1 results in a reduction in total flood volume of 9.28% (2-year return period), 7.29% (5-year return period) and 6.20% (10-year return period).

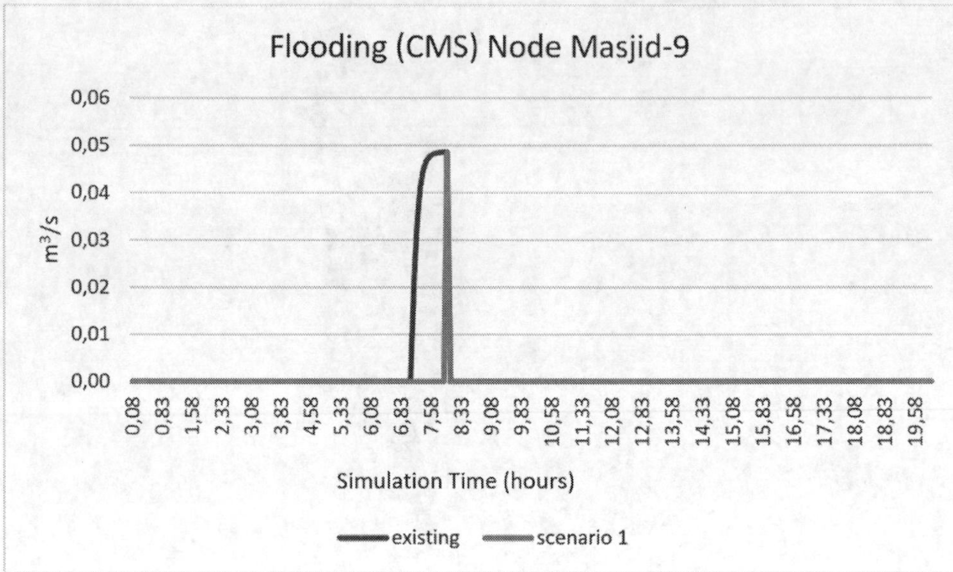

Figure 5. Graph showing the decrease of flooding occurring on scenario 1 in one of the nodes

4.7. Scenario 2 Analysis

Scenario 2 includes the addition of a water reservoir at location number 2 (red circle, figure 4) of 3000 m^3 over an area of 1000 m^2. Figure 6 shows a different result from scenario 1, which is the flooding reduced by decreasing the flow of water coming out of the node, instead of shifting back the time it occurred. It also means that the flood remains in the same time period, but with less water and lower depth. Scenario 2 results in a reduction in total flood volume of 34.61% (2-year return period), 25.83% (5-year return period) and 23% (10-year return period).

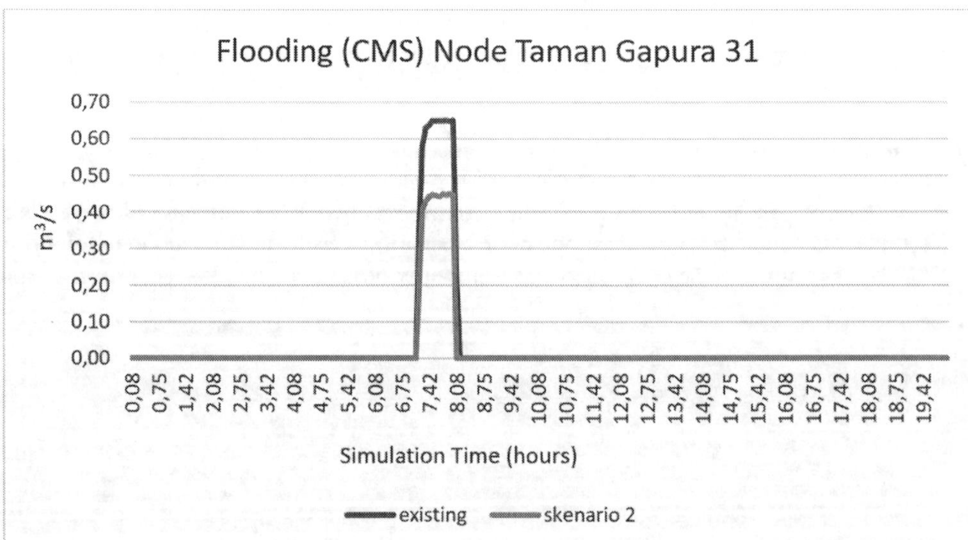

Figure 6. Graph showing the decrease of flooding occurring on scenario 2 in one of the nodes

4.8. Scenario 3 Analysis

Scenario 3 includes the addition of a water reservoir at location number 3 (red circle, figure 4) of 3000 m^3 over an area of 1000 m^2. Similar to scenario 2, Figure 7 shows the decrease of flooding occurring by decreasing the water coming out of the nodes. This also means that the flood remains in the same time period, but with less water and lower depth. Scenario 3 results in a reduction in total flood volume of 16% (2-year return period), 8.29% (5-year return period) and 6.39% (10-year return period).

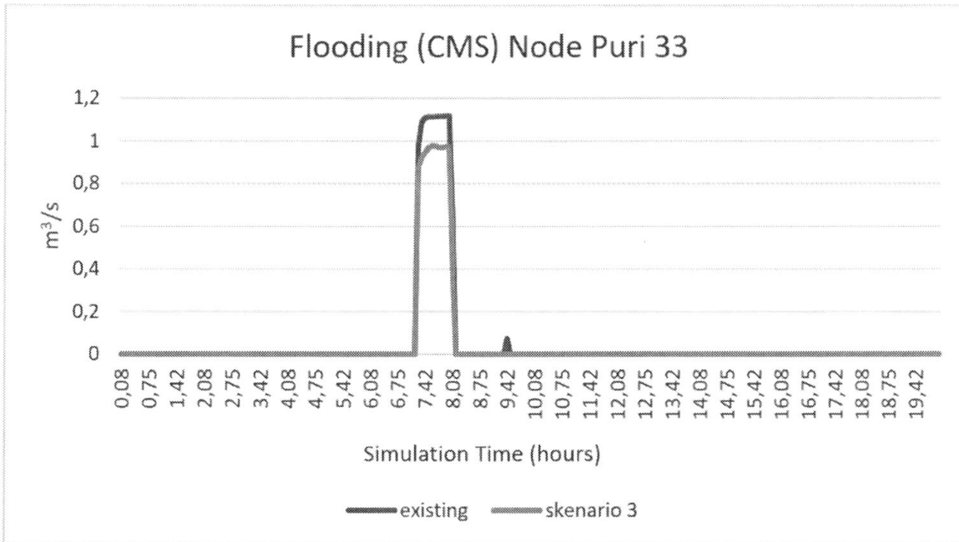

Figure 7. Graph showing the decrease of flooding occurring on scenario 3 in one of the nodes

4.9. Scenario 4 Analysis

Scenario 4 includes the addition of three water reservoirs at all locations, number 1 to 3 (red circle, figure 4) of total 6750 m^3 over an area of 2250 m^2. Scenario 4 results in a reduction in total flood volume of 57.82% (2-year return period), 40.59% (5-year return period) and 35.04% (10-year return period).

4.10.Result

The result of modelling and simulation of the scenarios is shown in the following table:

Table 2. Comparative tables of total flood volume reduced by m^3 and percentage

	Return Period					
	2-years		5-years		10-years	
	m^3	%	m^3	%	m^3	%
Scenario 1	488	9.28%	628	7.29%	674	6.20%
Scenario 2	1819	34.61%	2225	25.83%	2500	23.00%
Scenario 3	841	16.00%	714	8.29%	695	6.39%
Scenario 4	3039	57.82%	3496	40.59%	3809	35.04%

IOP Conf. Series: Materials Science and Engineering **930** (2020) 012082 doi:10.1088/1757-899X/930/1/012082

5. Conclusion

Based on the results of the analysis and discussion above, several conclusions can be drawn:

1. By using the rainfall calculation for the return period of 2, 5, and 10 years and SWMM modeling, it was found that flooding occurred at several node points was caused by the channel which is not large enough to drain the planned rainfall.
2. The second scenario through the addition of one water reservoir can handle flooding most effectively because it reduces 34.61% of the inundation volume that occurs. Besides, the fourth scenario through the addition of three water reservoirs can reduce 57.82% of the inundation volume that occurs.

References

[1] Asdak Chay 2010 *Hidrologi dan Pengelolaan Daerah Aliran Sungai: Edisi Revisi Kelima* (Yogyakarta: Gajah Mada University Press Yogyakarta).
[2] Bidang Penelitian, Pusat Penelitian Promosi dan Kerjasama, Badan Informasi Geospasial 2018 *Laporan Akhir Penelitian Kebencanaan Hidrologis* (Jakarta).
[3] Steven Valerian B 2017 Simulasi Penataan Subsistem Untuk Mengurangi Genangan Pada Sisitem Drainase Kawasan Perumahan Pepelegi Indah Waru Sidoarjo *Jurnal ITS* (Surabaya).
[4] David B dan John W Davies 2011 *Urban Drainage Third Edition* (London: Spon Press).
[5] Halim H 2001 *Drainase Terapan* (Yogyakarta: Penerbit UII Press).
[6] Gilang I 2017 Simulasi Skenario Kolam Retensi untuk Menurunkan Debit Puncak Banjir Kali Ngotok *Jurnal ITS* (Surabaya).
[7] William J, Lewis E Rossman dan W Robert C James 2010 *User's Guide to SWMM5 13th Edition* (Ontario: CHI Publisher).
[8] Kodoatie, Robert J dan Roestam Sjarief 2005 *Pengelolaan Sumber Daya Air Terpadu* (Yogyakarta: Penerbit Andi).
[9] Kodoatie, Robert J dan Sugiyanto 2002 *Banjir* (Yogyakarta: Pustaka Pelajar).
[10] Agus M 2001 *Hidrolika Terapan* (Yogyakarta: Penerbit Andi).
[11] Mulyanto 2010 *Penataan Drainase Perkotaan* (Semarang: Graha Ilmu).
[12] Norman, Panglima Raizal Mahendra dan Edijatno 2017 Evaluasi Kapasitas Saluran Drainase UNESA dengan Adanya Pengembangan Kawasan Surabaya Barat *Jurnal Teknik Hidroteknik* **2** (Surabaya).
[13] Suripin 2004 *Sistem Drainase Perkotaan yang Berkelanjutan* (Yogyakarta: ANDI OFFSET).
[14] Rangga Adi S 2017 Alternative study of type and location of flood control infrastructure in the drainage system, Avfour Kelor channel, Tuban regency *Jurnal ITS* (Surabaya).
[15] Bambang T 2014 *Hidrologi Terapan* (Yogyakarta: Beta Offset).

AUTHOR INDEX